Dynamic Modeling and Econometrics in Economics and Finance

Volume 32

In recent years there has been a rapidly growing interest in the study of dynamic nonlinear phenomena in economic and financial theory, while at the same time econometricians and statisticians have been developing methods for modeling such phenomena. Despite the common focus of theorists and econometricians, both lines of research have had their own publication outlets. The new book series is designed to further the understanding of dynamic phenomena in economics and finance by bridging the gap between dynamic theory and empirics and to provide cross-fertilization between the two strands. The series will place particular focus on monographs, surveys, edited volumes, conference proceedings and handbooks on:

- Nonlinear dynamic phenomena in economics and finance, including equilibrium, disequilibrium, optimizing and adaptive evolutionary points of view; nonlinear and complex dynamics in microeconomics, finance, macroeconomics and applied fields of economics.
- Econometric and statistical methods for analysis of nonlinear processes in economics and finance, including computational methods, numerical tools and software to study nonlinear dependence, asymmetries, persistence of fluctuations, multiple equilibria, chaotic and bifurcation phenomena.
- Applications linking theory and empirical analysis in areas such as macrodynamics, microdynamics, asset pricing, financial analysis and portfolio analysis, international economics, resource dynamics and environment, industrial organization and dynamics of technical change, labor economics, demographics, population dynamics, and game theory.

The target audience of this series includes researchers at universities and research and policy institutions, students at graduate institutions, and practitioners in economics, finance and international economics in private or government institutions.

All titles in this series are peer-reviewed. The series is indexed in Scopus.

Paulo B. Brito

Economic Dynamics and Distributions

Differential Equations, Optimal Control and Applications

Springer

Paulo B. Brito
Lisbon School of Economics
and Management
University of Lisbon
Lisbon, Portugal

ISSN 1566-0419 ISSN 2363-8370 (electronic)
Dynamic Modeling and Econometrics in Economics and Finance
ISBN 978-3-031-94716-2 ISBN 978-3-031-94717-9 (eBook)
https://doi.org/10.1007/978-3-031-94717-9

This Springer imprint is published by the registered company Springer Nature Switzerland AG
The registered company address is: Gewerbestrasse 11, 6330 Cham, Switzerland

Ao Pedro, Vasco, Sofia, Francisco e Artur.
Aos meus pais.

Preface

This book aims to provide the common mathematical foundations of several economic dynamic and distribution phenomena by adopting a pragmatic and heuristic approach. On perspective, several back-thoughts and concerns have influenced its writing.

A first concern relates to the paradoxes arising when confronting economic time series and events, with the relevant economic theory. The first puzzle is that, although every economy embodies a complex interaction of an immense number of "moving parts" (agents, markets, institutions, countries), the shape of most aggregate economic time series—in particularly historical ones—is relatively monotonic and smooth. The second puzzle is related to the fact that, given a sufficiently extended time span, all variables seem to be endogenously related in the long run. However, what lies behind most applied economics is a description of the world in terms of the mechanics of cause and consequence, and potentially plagued by the problem of indeterminacy: several theories can be consistent with a particular "evidence". The third puzzle concerns the never ending debate—with a bearing on economic policy—on the relative importance of the two main interacting drivers in an economy: the granular effect of incentives on the decisions of private agents and economic policy authorities, and the almost mechanical effect of the aggregate state of the economy constraining economic decision making. In an environment in which there is a proliferation of data, information, and debates, those puzzles justify the need to having a solid foundation in economic theory, providing low dimensional models incorporating a wide spectrum of qualitative features, and integrating consistently both dynamic and distributional components.

A second concern relates to the fact that the variational principle provides a common mathematical foundation to several specialized areas in economics (as macroeconomics, microeconomics, growth economics, environmental economics, spatial economics, and statistics). In other words, several forms of calculus underlie most economic concepts and results. In order to bring together the common mathematical structures of different economic sub-fields, the double continuum of time and the domains of other types of heterogeneity distribution is the one that allows for a simpler joint consideration of both micro and macro dimensions, intertemporal and distributional components, and deterministic and stochastic environments. It combines both simplicity and low dimensionality of the models, and allows for invoking an immense pool of results from several fields in mathematics.

A third concern relates to the fact that I was trained as an economist, and the potential readership of this book are also economists with an interest in economic theory. Accepting the risk of displeasing readers with a solid mathematical competence, I follow a heuristic and somewhat informal approach, and restrain to presenting low dimension and/or linear models. I have sacrificed mathematical rigour and generality to the endeavour of providing useful qualitative insights that can be suggested by the different mathematical models. This choice is not only the one that makes sense for the goal of covering a wide range of mathematical structures, with existing or potential application in economic theory and statistics, but also the one that addresses my next concern.

A last concern is pedagogical. As a lecturer on macroeconomics, growth theory, financial economics and mathematical economics—in the University of Lisbon (ISEG) and also episodically in the University of Porto—I come to the conclusion that it is better to fully solve and characterize some simple models, not evading the "hard" technical details, rather than having in the syllabus a huge number of recent research papers which, by their on goal to contribute to moving the research frontier, can only be lightly skimmed. Thus my preference goes to study models in which explicit (or closed form) solutions can be obtained, and providing the tools to enable a geometrical interpretation of their solutions. My hope is to develop a "clinical eye" that, for instance, allows for interpreting the simulation results of large quantitative macromodels (which sometimes have only a very small number of really driving mechanisms).

I expect the main audience of this book to include advanced master or introductory PhD students in economics. But I hope fellow researchers in economics, finance, and other areas using differential equations and optimal control theories may find it useful. In particular, readers interested in the dynamics of heterogeneity and distributions will find useful material.

The book is divided into four parts, assembling nineteen chapters. Each chapter is dedicated to a particular type of differential equation or to optimization problems constrained by different types of differential equations. Some applications related economic models are presented as examples or exercises. Part I presents several different types of low dimensional ordinary differential equations, in particular, linear, non-linear regular, piecewise smooth and constant, and singular differential equations. The last two types of ordinary differential equations are not absent in similar economic textbooks, and can only be found in somewhat specialized applied mathematics textbooks. I include them because they are useful in modeling switching and feedback effects (for instance generated by policy rules) in both intertemporal micro and macroeconomic dynamic models. Part II deals with functionals and functional calculus applied to "static" optimization problems, calculus of variations, and optimal control of ordinary differential equations. As I try to show, a background knowledge in functional calculus is required to both understanding the common elements in several separate concepts and theories in economics dealing with dynamics, distributions and uncertainty. Part III presents an introduction to first-order and parabolic partial differential equations, and their related optimal control problems. I try to show that the usefulness of partial differential equations goes beyond their usual application in finance, and should be in the toolbox of economists for dealing with heterogeneity and aggregation in a comprehensive and powerful way. Part IV is also dedicated to an introductory presentation of stochastic processes, stochastic differential equations and stochastic optimal control. It provides an introduction to the relationship between forward and backward equations, and between the dynamic programming and the Pontriyagin's approaches to stochastic optimal control via the envelope theorem. Clearly, the last two parts try to pinpoint some small drops in a literature that is wide as an ocean. Overall, an effort has been made to highlight the connections between both the solution methods, the characterization of the solutions among different types of differential equations, and their associated optimal control problems.

Too many years were needed to conceive and complete this book. I acknowledge and thank the several cohorts of students for enduring and contributing to the severals rounds of revisions of my lecture notes that were the stepping stones leading to this book. I also thankfully acknowledge the following colleagues that, by several different types of interaction, influenced my work: Bernardino Adão, Costas Azariadis, Emmanuele Augeraud-Veron, Robert Becker, Jess Benhabib, Gaetano Bloise, Stefano Bosi, Raouf Boucekkine, Alan Champneys, Luís F. Costa, Hippolyte d'Albis, Rui Dilão, Jean-Pierre Drugeon, Fátima Fabião, Giorgio Fabbri, Gustav Feichtinger, Ana Fernandes, Rodolphe Santos Ferreira, Pedro Gil, Fausto Gozzi, Huw Dixon, Mike Jeffrey, Cuong Le Van, Teresa Lloyd-Braga, Giancarlo Marini, Leonor Modesto, Alfredo M. Pereira, Alessandro Piergallini, Xavier Raurich, Çağri Sağlam, Carlos D. Santos, Manuel S. Santos, Thomas Seegmuller, Gerhard Sorger, Harutaka Takahashi, Mario Tirelli, Stephen Turnovsky, Vladimir Veliov, Alain Venditti, and Bertrand Wigniolle.

I thankfully acknowledge FCT, I.P., the Portuguese national funding agency for science, research and technology, for its financial support over the years.[1] I also thank GREQAM (Groupement de Recherche en Economie Quantitative d'Aix Marseille), and, in particular Alain Venditti and Thomas Seegmuler, for their regular invitation to participate in the high quality workshops and conferences they organize, where some of the models presented in this book have been presented.

Possibly this book would not have been raised to the daylight without the impetus, intellectual stimulus, benevolence, companionship, and love from Leonor, to whom I am deeply thankful.

Lisbon, Portugal
April 2025

Paulo B. Brito

---

[1] Under the projects UID/ECO/00124/2013, UID/ECO/00436/2013, PEst-OE/EGE/UI0436/2014, PTDC/EGE-ECO/28814/2017, UIDB/05069/2020.

Competing Interests The author has no competing interests to declare that are relevant to the content of this manuscript.

Contents

Part IV Stochastic Differential Equations

Acronyms

AD	Arrow-Debreu Contracts or Economy
BGP	Balanced Growth Path
BK	Backward Kolmogorov Equation
BSDE	Backward Stochastic Differential Equation
BPDE	Backward Partial Differential Equation
CV	Calculus of Variations
CODE	Constrained Ordinary Differential Equations
CRRA	Constant Relative Risk Aversion
DAE	Differential Algebraic Equation
DGE	Dynamic General Equilibrium
DS	Dixit-Stiglitz Aggregator
DSGE	Dynamic Stochastic General Equilibrium
EIS	Elasticity of Intertemporal Substitution
EP	Edgeworth Pareto Substitutability, Independence, or Complementarity
EVS	Elasticity of Variety Substitution
EZ	Epstein-Zin Utility Functional
FBSDE	Forward Backward Stochastic Differential Equation
FSDE	Forward Stochastic Differential Equation
FPDE	Forward Partial Differential Equation
FPK	Fokker-Planck-Kolmogorov Equation
HJB	Hamilton-Jacobi-Bellman Equation
MHDS	Maximized Hamilton Dynamic System
OCODE	Optimal Control of a Ordinary Differential Equation
OCPDE	Optimal Control of a Partial Differential Equation
OCSDE	Optimal Control of a Stochastic Differential Equation
ODE	Ordinary Differential Equation
PDE	Partial Differential Equation
PMP	Pontryagin Maximum Principle
PWSC	Piecewise Smooth or Continuous Ordinary Differential Equation
RCK	Ramsey-Cass-Koopmans Model
SDE	Stochastic Differential Equation

SDF	Stochastic Discount Factor
SDPP	Stochastic Dynamic Programming Principle
SPMP	Stochastic Pontriyagin's Maximum Principle
vNM	von Neumann-Morgenstern Utility Functional

Symbols

t	Independent variable, time $t \in \mathbf{T}$
x	One-dimensional independent variable other than time $x \in \mathbf{X}$
φ	Parameter vector
$\mathbf{x}$	Multi-dimensional independent variable other than time $\mathbf{x} \in \mathbf{X}$
T	Domain of the indepdendent variable time, $\mathbf{T} \subseteq \mathbb{R}_+$
X	One-dimensional space of other independent variables, $\mathbf{X} \subseteq \mathbb{R}$
$\mathbf{X}$	Multi-dimensional space of other independent variables, $\mathbf{X} \subseteq \mathbb{R}^n$
y	Dependent variable, or realizations of stochastic processes
$\mathbf{y}$	Dependent vector variable, or realizations of stochastic vector processes
$\mathbf{Y}$	Domain of a scalar dependent variable, $\mathbf{Y} \subset \mathbb{R}$
$\mathbf{Y}$	Domain of the dependent vector variables, $\mathbf{Y} \subseteq \mathbb{R}^m$
Φ	Domain of the parameter vector
$\mathscr{Y}$	Set of functions $y : \mathbf{X} \to \mathbf{Y}$
$\mathscr{Y}$	Set of vector functions $\mathbf{y} : \mathbf{X} \to \mathbf{Y}$
$(y(x))_{x \in \mathbf{X}}$	Distribution of $y(x)$ across the domain $\mathbf{X}$
$(y(t))_{t \in \mathbf{T}}$	Trajectory of $y(t)$ over the time domain $\mathbf{T}$
$f(y)$	One-dimensional function of the one-dimensional dependent variable
$f(\mathbf{y})$	One-dimensional function of the vector dependent variable
$\mathbf{f}(y)$	Vector function of the one-dimensional dependent variable
$\mathbf{f}(\mathbf{y})$	Vector function of the vector dependent variable
$\mathbf{f}(\mathbf{y}, \varphi)$	Parameterized vector function of the vector dependent variable
a, b, c, d	Real scalar constant coefficients
$a(x), b(x), c(x), d(x)$	Real scalar variable coefficients
$a(t), b(t), c(t), d(t)$	Real scalar time-dependent coefficients

$\mathbf{A}, \mathbf{B}$	Real matrix with constant coefficients
$\mathbf{A}(x), \mathbf{B}(x)$	Real matrix with variable coefficients
$\mathbf{A}(t), \mathbf{B}(t)$	Real matrix with time-dependent coefficients
$\mathbf{A}^\top$	Matrix **A** transposed
$\mathbf{A}^{-1}$	Non-singular matrix inverse
$\mathbf{A}^{+}$	Singular matrix inverse
$\mathbf{P}$	Eigenvector matrix of **A**
Λ	Jordan form of matrix **A**
$\mathrm{tr}(\mathbf{A})$	Trace of matrix **A**
$\det(\mathbf{A})$	Determinat of matrix **A**
$\Delta(\mathbf{A})$	Determinat of matrix **A**
$\sigma(\mathbf{A})$	Spectrum of matrix **A**
$\mathbf{e}^{\mathbf{A}t}$	Exponential matrix function
$C^1(\mathrm{X})$	Set of continuous and first-order continuously differentiable functions with domain X
$C^2(\mathrm{X})$	Set of continuous and first- and second-order continuously differentiable functions with domain X
$C^1(\mathrm{X};\,\mathrm{Y})$	Set of continuous and first-order continuously differentiable functions mapping set X to Y
$C^2(\mathrm{X};\,\mathrm{Y})$	Set of continuous and continuously differentiable until (at least the second order) functions mapping set X to Y
$C^{1,2}(\mathrm{X},\,\mathrm{Y})$	Set of functions with domain X × Y, that are continuous and first-order continuously differentiable in set X, and continuous and second-order continuously differentiable in set Y
$y'(x)$	Derivative of $y(x)$ when x does not represent time: $y'(x) = \frac{dy(x)}{dx}$
$\dot{y}$	Derivative of $y(t)$ when t represents time: $\dot{y} = \frac{dy(t)}{dt}$
$\partial_{x_i} f(\mathbf{x})$	Partial derivative of a function of a vector of variables as regards variable x_i: $\partial_{x_i} f(\mathbf{x}) = \frac{\partial f(\mathbf{x})}{\partial x_i}$
$\partial_{x_j} f_i(\mathbf{x})$	Partial derivative of an element $f_i(\mathbf{x})$ of a vector function of a vector of variables as regards variable x_j: $\partial_{x_j} f_i(\mathbf{x}) = \frac{\partial f_i(\mathbf{x})}{\partial x_j}$
$\mathbf{D}_{x_j}\mathbf{f}(\mathbf{x})$	Gradient of a vector function as regards variable x_j: $\mathbf{D}_{x_j}\mathbf{f}(\mathbf{x}) = (\frac{\partial f_1(\mathbf{x})}{x_j}, \ldots \frac{\partial f_i(\mathbf{x})}{x_j}, \ldots \frac{\partial f_m(\mathbf{x})}{x_j})^\top$
$\mathbf{D}_{\mathbf{x}}\mathbf{f}(\mathbf{x})$	Jacobian of a vector function as regards all variables: $\mathbf{D}_{\mathbf{x}}\mathbf{f}(\mathbf{x}) = \begin{pmatrix} \frac{\partial f_1(\mathbf{x})}{\partial x_1} & \ldots & \frac{\partial f_1(\mathbf{x})}{\partial x_j} & \ldots & \frac{\partial f_1(\mathbf{x})}{\partial x_n} \\ \ldots & \ldots & \ldots & \ldots & \ldots \\ \frac{\partial f_m(\mathbf{x})}{\partial x_1} & \ldots & \frac{\partial f_m(\mathbf{x})}{\partial x_j} & \ldots & \frac{\partial f_m(\mathbf{x})}{\partial x_n} \end{pmatrix}$
$\mathcal{W}^s$	Stable manifold
$\mathcal{W}^u$	Unstable manifold
$\mathcal{W}^c$	Center manifold
$\mathcal{W}_i^{s,loc}$	Local stable manifold in the neighborhood of steady state $\bar{y}_i$

$\mathcal{W}_i^{u,loc}$	Local stable manifold in the neighborhood of steady state $\bar{y}_i$
$\mathcal{W}_i^{c,loc}$	Local center manifold
Σ	Switching boundary in PWSC ODE
$\partial\mathbf{f}(\mathbf{y})$	Generalized Jacobian
$\mathcal{L}_{\mathbf{f}_j}(H(\mathbf{y}))$	Lie derivative of vector field $\mathbf{f}_j$ onto the normal of Σ
Σ_t	Crossing subset of Σ
Σ_s	Sliding subset of Σ
Σ_s^s	Stable sliding subset of Σ
Σ_s^u	Unstable sliding subset of Σ
$f_s(\mathbf{y})$	Sliding vector field
$\mathcal{S}$	Singular set $\mathcal{S} \in \mathbf{Y} \times \Phi$
Γ	Steady state $\Gamma \in \mathbf{Y} \times \Phi$
$\mathcal{S}_0^p$	Fast-slow singularity set
$\mathcal{S}_1^p$	Fast-slow singular points
Γ_E	Projection of $\mathcal{S}_0^p$ onto set $\mathbf{Y}$
$\mathcal{S}_0^i$	Impasse singularity set
Γ_I	Projection of $\mathcal{S}_0^i$ onto set Y
$\mathcal{L}_{\mathbf{f}^d}(\delta(\mathbf{y}))$	Lie derivative of the desingularized vector field
P_I	Impasse steady states $P_I \subset \Gamma_I$
$\mathcal{S}_1^i$	Impasse-singular set
Γ_I^k	Kernel-singular impasse set
Γ_I^i	Image-singular impasse set
$\Gamma_{I,1}^a$	Attractor impasse set
$\Gamma_{I,1}^r$	Repeller impasse set
$\mathrm{F}[y]$	Functional of function y
$\mathrm{F}[\mathbf{y}]$	Functional of vector function $\mathbf{y}$
$\mathbf{D}_{\eta(\cdot)}\mathbf{F}[y]$	Gâteaux differential of functional $\mathrm{F}[y]$ along direction $\eta(\cdot)$
$\mathbf{D}_{\delta(x_0)}\mathbf{F}[y]$	Gâteaux differential of functional $\mathrm{F}[y]$ for a pointwise change in $y(x_0)$
$\frac{\delta\mathrm{F}[y]}{\delta y(x)}$	Gâteaux derivative for a change in $y(x)$
$\frac{\delta\mathrm{F}[y]}{\delta y(x_0)}$	Gâteaux derivative for a pointwise change in $y(x_0)$
$\mathbf{D}_{\eta(\cdot)^2}\mathbf{F}[y]$	Second order "own" Gâteaux differential of functional $\mathrm{F}[y]$
$\mathbf{D}_{\eta_1(\cdot),\eta_2(\cdot)}\mathbf{F}[y]$	Second order "crossed" Gâteaux differential of functional $\mathrm{F}[y]$
$\frac{\delta^2\mathrm{F}[y]}{\delta(x)^2}$	Second order Gâteaux derivative for a change in $y(x)$
$\mathbf{D}_{\eta(\cdot)}\mathbf{F}[y]$	Gâteaux differential of vector functions
$\frac{\delta\mathrm{F}[\mathbf{y}]}{\delta\mathbf{y}(x)}$	Gâteaux gradient for a pointwise change in $\mathbf{y}(x)$
$\mathcal{F}[y(x)](s)$	Fourier transform of function $y(x)$ with parameter s
$\mathcal{F}^{-1}[Y(s)](x)$	Inverse Fourier transform of function $Y(s)$ with variable x
$\mathcal{L}[y(x)](s)$	Laplace transform of function $y(x)$ with parameter s
$\mathcal{L}^{-1}[Y(s)](x)$	Laplace Fourier transform of function $Y(s)$ with variable x
$(\Omega, \mathcal{F}, \mathbb{P}, \mathbb{F})$	Filtered probability space

ω	Elementary event, $\omega \in \Omega$
$y(t, \omega)$	Realization conditional on the event ω at time t
$Y(t)$	One-dimensional stochastic process at time t, where $Y(t) = (y(t, \omega))_{\omega \in \Omega}$
$(Y(t))_{t \in T}$	Stochastic process
$(Y(s))_{s \in [0,t]}$	History of the process stochastic process until time t, for $s \leq t$
$\mathbf{Y}(t)$	Multi-dimensional stochastic process at time t
$(\mathbf{Y}(t))_{t \in T}$	Multi-dimensional stochastic process
$(W(t))_{t \in T}$	Wiener process
$\mathbb{P}[Y(t)]$	Unconditional probability at time t of the process $(Y(t))_{t \in \mathbf{T}}$
$\mathbb{E}[Y(t)]$	Unconditional expected value at time t of the process $(Y(t))_{t \in \mathbf{T}}$
$\mathbb{E}[Y(t)^n]$	Unconditional n-th moment at time t of the process $(Y(t))_{t \in \mathbf{T}}$
$\mathbb{V}[Y(t)]$	Unconditional variance at time t of the process $(Y(t))_{t \in \mathbf{T}}$
$\mathbb{E}_t[Y(s)]$	Conditional expected value at time t of the process $(Y(s))_{s \geq t}$
$\mathbb{P}_t[Y(s)]$	Conditional probability at time t of the process $(Y(s))_{s \geq t}$
$\mathbb{E}_t[Y(s)^n]$	Conditional n-th moment at time t of the process $(Y(s))_{s \geq t}$
$\mathbb{V}_t[Y(s)]$	Conditional variance at time t of the process $(Y(s))_{s \geq t}$

List of Figures

List of Tables

Chapter 1
Introduction and Overview

1.1 Introduction

This is a book on **differential equations** (mostly) applied to economics. It is also a book on **optimal control problems** whose optimum conditions are also differential equations.

Differential equations are a class of functional equations, containing one or more (unknown) functions of one or more independent variables, and their derivatives of first or higher orders. When the independent variable is time, a specific differential equation represents the **dynamics**of the underlying function. When the independent variable is not time we can generically say that a differential equation represents a **profile or a distribution**. When there are two or more independent variables, and one of them is time, a differential equation represents a **profile or distribution dynamics**. If the independent variable belongs to a probability space, the differential equation represents a mechanism for generating sample paths, and together with the underlying probability distribution, it provides a **stochastic dynamic distribution** of sample paths. Therefore, we restrain the book to presenting the mathematical tools for setting up, solving, and characterizing theories and models in continuous time, and/or in continuous profiles or distributions.[1]

The main functions we are interested in represent individual or aggregate economic variables that are associated to quantities (as consumption, production, investment, income, asset positions, population, labor effort, employment), or prices or returns (as interest rates and wages). The main independent variable, at least in macroeconomics and economic growth theories, is time. However, we can also (or instead) have other types of independent variables as space, the previously referred quantity variables (consumption, income, asset position, labor effort, etc), characteristics of agents (age, skills), dimension (measured by the level of one or more stock

[1] The domain of the independent variables is the set of real numbers or vectors. This distinguishes differential equations from difference equations that are functional equations that contain functions in which at least one of the independent variable is the space of integers.

P. B. Brito, *Economic Dynamics and Distributions*, Dynamic Modeling and Econometrics in Economics and Finance 32, https://doi.org/10.1007/978-3-031-94717-9_1

variables), or the states of nature (in the probabilistic sense). In models addressing the dynamics of heterogeneity, as in spatial dynamics, heterogeneous agent models, or distributional models, we can have at least two independent variables (as time and space or skills or income). In stochastic dynamic models, there is an inherent heterogeneity introduced by the multitude of states of nature.

The book tries to highlight both the **common and the contrasting aspects** of modeling dynamics, distributions, and distributive dynamics. Setting up a theory from a locally differentiable **variational mechanism** is the common core feature that underlies the use of differential equations. This means that a local interaction of the dependent variable is specified by its gradient relative to a local change of the independent variable or variables.[2] However, different types of differential equations, and the associated optimal control problems, may require the use of different types of calculus, in particular when dealing with functionals and random variables.[3]

The use of differential equations requires the existence of an **order relationship** in the domains of the independent variables. This means that independent variables are more than just an indexing device, as they require quantitatively specified differences along particular axes, such as between present-future, nearby-faraway, high-low skill, poor-rich, small-large or similar. There is thus a fundamental difference between *varieties*, which relate to qualities, that are not endowed with an order structure, and *variations*, which relate to quantitative differences.[4] However, defining the fundamental variation mechanism by local gradients does not exclude the possible existence of global, non-local, or aggregate background influences. In time-dependent differential equations, the existence of non-linear, non-differentiable or even non-continuous dynamic mechanisms, introduces a distinction and an interaction between local and global dynamics. For non-time dependent variables, the existence of aggregate or normalizing constraints, for instance, also introduces a distinction and an interaction between local and global effects.

Continuous-time environments[5] allow for capturing, in a simpler way, aggregate effects and long-run trends, and thus became the dominant setting in growth economics.[6] In macroeconomics, they were applied the dominant models until the late 1980s.[7] Since the 1990s dynamic general equilibrium (DGE), and dynamic stochastic general equilibrium models (DSGE), became the dominant paradigm in macroe-

---

[2] Modeling with difference equations, which became *de rigueur* in macroeconomics, also involves variational mechanisms. However, in this case, it is not represented by gradients but by local variations.

[3] For the history and the fundamental role of calculus in the history of Western science and mathematics, see Strogatz (2019).

[4] We can deal with varieties by increasing the dimension of the state space, that is, the number of the dependent functions.

[5] See Merton (1982), Isohätälä et al. (2016), and Brunnermeier and Sannikov (2016) for arguments on the continuous versus discrete time modeling in economics.

[6] See Barro and Sala-i-Martin (2004), and Acemoglu (2009).

[7] The state of the art at that time can be seen in v.g. Burmeister and Dobell (1970), and Gandolfo (1997) on the related equations.

conomics, but they have been specified in discrete time.[8] Although discrete time simplifies the quantitative calibration, simulation, and estimation of models, it has the cost of rendering more difficult or impossible the derivation of explicit solutions or characterizing them. Recently, the research in macroeconomics seems to be rediscovering the usefulness of continuous-time modeling for solving DGE and DSGE models with heterogeneous agents.[9] In other areas of economic theory, modeling in continua can be found in spatial economics,[10] in finance,[11] in microeconomics, particularly in contract theory,[12] search and matching in labor markets,[13] and networks. Another important related area is mean-field games.[14]

To materialize the pedagogical concerns with which this book has been written, two options were taken. First, it is a "**methods book**" in the sense that its emphasis is on the common formal structures of models and not on their specific aspects. Economic models are presented as examples or exercises featuring differential equations and optimization problems, whose mathematical solutions have been previously derived. Second, it is an **introductory book** in following sense: it uses a heuristic rather than a rigorous approach, by presenting the analytic, functional or geometrical fundamentals, and it focuses on linear and linearized low dimensional equations. Linear models, even in the case of multi-dimensional and stochastic differential equation, usually have closed form or explicit solutions, which allows for a complete understanding of their properties. More complicated non-linear structures, which are usually the case in economic models, can only be completely understood when simpler linear models have been explored completely beforehand.[15] At last, it tries to provide the **tools for modeling simple economies with heterogeneous agents** that are becoming topical in economics, by invoking their relationship with simple aggregative models. However, given the number and the simplicity of models that can be studied with the previous constraints in mind implies this is **not a survey book**. This explains the fact that most examples that are studied next refer to seminal original contributions to several strands of the economics' literature that tend to be low-dimensional.

Modeling with differential equations has other advantages, in addition to pedagogical simplification and clarity. First, it allows for consistently and jointly dealing, both theoretically and analytically, with the dynamic and distributional components of models. Second, it simplifies obtaining necessary conditions for optimal control

---

[8] See Azariadis (1993), Ljungqvist and Sargent (2018), and Miao (2020). Alogoskoufis (2019) is an exception

[9] See Brunnermeier and Sannikov (2016) and Gabaix et al. (2016), Brunnermeier et al. (2020), and the references therein. In macrofinance see Stokey (2009).

[10] See Fujita and Thisse (2002).

[11] See Merton (1990), Dana and Jeanblanc (2003), Cvitanić and Zapatero (2004), or Dumas and Luciano (2017)).

[12] See Cvitanić and Zhang (2013).

[13] See Rogerson et al. (2005).

[14] See Bensoussan et al. (2013).

[15] This has also the benefit of allowing, for instance, to understand the (usually very few) core driving forces of high-dimensional, but low-complexity, quantitative macroeconomic models.

problems. And, at last, it allows for tapping into the immense pool of insights, theories, methods, and results from mathematics, and a wide span of applied sciences, and understanding the specificities of economic modeling.

1.2 Overview

This section presents a synthesis of deterministic differential equations, in Sect. 1.2.1, of stochastic differential equations, in Sect. 1.2.2, and of optimization of functionals, including optimal control, in Sect. 1.2.3, that are addressed in this book.

1.2.1 Deterministic Differential Equations

Let $\mathbf{y}(\mathbf{z})$ be a function mapping $\mathbf{y} : \mathrm{Z} \to \mathrm{Y}$, where the image set is a subset of the $n-$dimensional Euclidean space, $\mathrm{Y} \subseteq \mathbb{R}^n$, and $\mathbf{z} = (z_1, \ldots, z_m)$ is the vector of independent variables with domain $\mathrm{Z} \subseteq \mathbb{R}^m$. A **differential equation**is a functional equation of the form

$$\overline{\mathbf{F}}(\mathrm{D}^p_{\mathbf{z}}(\mathbf{y}(\mathbf{z})), \ldots, \mathrm{D}_{\mathbf{z}}(\mathbf{y}(\mathbf{z})), \mathbf{y}(\mathbf{z}), \mathbf{z}) = \mathbf{0}, \tag{1.1}$$

where vector function $\overline{\mathbf{F}}(.)$ is known, $\mathrm{D}_{\mathbf{z}}(\mathbf{y}(\mathbf{z}))$ is an appropriately defined first-order gradient $\mathrm{D}_{\mathbf{z}}(\mathbf{y}(\mathbf{z})) \equiv \big(\mathrm{D}_{z_1}(\mathbf{y}(\mathbf{z})), \ldots, \mathrm{D}_{z_m}(\mathbf{y}(\mathbf{z}))\big)$, where $\mathrm{D}_{z_i}(\mathbf{y}(\mathbf{z})) \equiv \left(\frac{\partial y_1}{\partial z_i}, \ldots \frac{\partial y_n}{\partial z_i}\right)^\top$ for $i \in \{1, \ldots, m\}$, and $\mathrm{D}^p_{\mathbf{z}}(\mathbf{y}(\mathbf{z}))$ is the multi-dimensional matrix of pth-order derivatives, with $p > 1$.

Most applications also involve a vector of parameters $\boldsymbol{\varphi} \in \Phi \subseteq \mathbb{R}^q$. Then the differential equations of interest extends equation (1.1) to

$$\overline{\mathbf{F}}(\mathrm{D}^p_{\mathbf{z}}(\mathbf{y}(\mathbf{z})), \ldots, \mathrm{D}_{\mathbf{z}}(\mathbf{y}(\mathbf{z})), \mathbf{y}(\mathbf{z}), \mathbf{z}) = \mathbf{0}. \tag{1.2}$$

A **differential equation problem** is defined by a differential equation (1.2) supplemented with **side conditions**. There are several types of side conditions involving specific values of function $\mathbf{y}$, at one or more boundaries of set Z, or functionals depending on function $\mathbf{y}$.

Solving a differential equation means finding a function $\boldsymbol{\phi}(\mathbf{z})$, mapping $\phi : \mathrm{Z} \to \mathrm{Y}$, that satisfies Eq. (1.2). We can verify that $\boldsymbol{\phi}(\mathbf{z})$ is a solution if it satisfies

$$\overline{\mathbf{F}}(\mathrm{D}^p_{\mathbf{z}}(\boldsymbol{\phi}(\mathbf{z})), \ldots, \mathrm{D}_{\mathbf{z}}(\boldsymbol{\phi}(\mathbf{z})), \boldsymbol{\phi}(\mathbf{z}), \mathbf{z}, \boldsymbol{\varphi}) = \mathbf{0}.$$

If there is a function $\phi(\cdot)$ that verifies Eq. (1.2), we call it **general solution**. Thus $\mathbf{y} = \boldsymbol{\phi}(\mathbf{z})$. For some differential equations, we can obtain $\boldsymbol{\phi}(\mathbf{z})$ explicitly, or in closed-form. Qualitative theories of differential equations allow us to characterize its solutions even if we cannot find explicit solutions.

We call **particular solution** to the general solution of a differential equation problem that is constrained to satisfy specified side conditions. In some cases, although the general solution of a differential equation exists and is unique, a particular solution may not exist, or can exist but be multiple. A problem is said to be **well posed** if a unique general solution exists and satisfies the additional side constrains.

Deterministic differential equations are classified (among other aspects) by the type of the set of independent variables, Z, by the dimension of the state space, m, by the number of the independent variables, n, by the maximum order of differentiation, p, by the dimension of the parameter space, q, and by the functional form of function $\overline{\mathbf{F}}(.)$. Next, we describe the cases dealt in this book.

1.2.1.1 Ordinary Differential Equations

Ordinary differential equation (ODE) are differential equations in which the independent variable is of dimension one and its domain is a subset or the whole set of real numbers: that is $m = 1$, and $z \in \mathrm{Z} \subseteq \mathbb{R}$. Therefore, Eq. (1.1) becomes $\overline{\mathbf{F}}(\mathrm{D}^p(\mathbf{y}(z)), \ldots, \mathrm{D}(\mathbf{y}(z)), \mathbf{y}(z), z) = 0$. The value of p specifies the order of the equation. If $p = 1$ the equation is called **first-order equation**, if $p = 2$, it is called **second-order equation**, and so forth. However, all equations with $p \geq 2$ can be transformed into first-order equations by defining higher-order derivatives as new variables.

The general representation of an ODE in implicit form is, therefore,

$$\overline{\mathbf{F}}(\mathrm{D}_z(\mathbf{y}(z)), \mathbf{y}(z), z) = 0 \tag{1.3}$$

where, again, $\overline{\mathbf{F}}$ is known, and $\mathrm{D}_z(\mathbf{y}(z)) = \left(\frac{dy_1(z)}{dz}, \ldots, \frac{dy_n(z)}{dz}\right)$ is the gradient of $\mathbf{y}$ as regards the independent variable z.

If ODE (1.3) can be written in the form

$$\mathbf{A}(\mathbf{y}, z)\, \mathrm{D}_z(\mathbf{y}(z)) + \mathbf{G}(\mathbf{y}, z) = 0 \tag{1.4}$$

where $\mathbf{A}(.)$ is a $n \times n$ matrix and $\mathbf{G}(.)$ is a $n \times 1$ vector, we call it **quasi-linear ODE**. We can distinguish quasi-linear ODE according to the determinant of $\mathbf{A}(\cdot)$ as a function of $\mathbf{y}$: if $\det(\mathbf{A}(\mathbf{y}, z)) \neq 0$, for every $(\mathbf{y}, z) \in \mathrm{Y} \times \mathrm{Z}$, we say the ODE is **non-singular**; or, if there is a pair of values for $(\mathbf{y}, z) \in \mathrm{Y} \times \mathrm{Z}$ such that $\det(\mathbf{A}(\mathbf{y}, z)) = 0$ then we say the ODE is **singular**. Furthermore, we can also distinguish quasi-linear ODE according to the differentiability and continuity properties of matrices $\mathbf{A}$ and $\mathbf{G}$ as functions of $\mathbf{y}$: if any one of those two matrices is continuous but non-locally non-differentiable then we say the **ODE is piecewise differentiable**; if any one of those two matrices is locally non-continuous then we say the **ODE is piecewise continuous**.

If an ODE is non-singular and if matrices **A** and **G** as functions of **y** are continuous and differentiable we call it a **regular ODE**. In this case, we represent ODE (1.4) in the **explicit form**

$$\mathrm{D}_z(\mathbf{y}(z)) = \mathbf{F}(\mathbf{y}, z). \tag{1.5}$$

Intuitively, if **F** is linear, Eq. (1.5) can be considered a representation of a variation over an Euclidean space, while Eq. (1.3) as a representation of a variation over a generalized surface (which can be singular, locally non-differentiable, locally discontinuous or not). Other classifications of ODE depend on the specific functional forms of the matrices involved.

We use two different notation conventions depending on the nature of the independent variable. If the independent variable is time, then we write $z = t$, and follow the convention of using the Newton notation for derivatives $\mathrm{D}_t(\mathbf{y}) \equiv \dot{\mathbf{y}} = \left(\dot{y}_1 \ldots \dot{y}_n\right)^\top$. If the independent variable is other than time, then we write $z = x$ and use the Lagrange notation for derivatives $\mathrm{D}_x(\mathbf{y}) \equiv \mathbf{y}'(x) = \left(y_1'(x) \ldots y_n'(x)\right)^\top$.

Part I of the book is dedicated to solving, and characterizing the solutions, to low-dimensional ODE. It contains six chapters.

Chapter 2 deals with **scalar linear** ODE over functions $y \in \mathbb{R}$. We consider two cases: the case in which $y = y(x) \in \mathrm{Y}$ is a function of any non-time independent variable $x \in \mathrm{X} \subset \mathbb{R}$, and having image in $\mathrm{Y} \subseteq \mathbb{R}$,

$$y'(x) = a(x)\, y(x) + b(x), \;\; x \in \mathrm{X} \subseteq \mathbb{R}, \tag{1.6}$$

and the case in which $y = y(t) \in \mathrm{Y}$ is a function of time $t \in \mathrm{T} \subseteq \mathbb{R}_+$, and having image in $\mathrm{Y} \subseteq \mathbb{R}$,

$$\dot{y} = a(t)\, y(t) + b(t), \;\; t \in \mathrm{T} \subseteq \mathbb{R}_+. \tag{1.7}$$

While Eq. (1.6) appears, for instance, as a primitive for utility and probability density functions, Eq. (1.7) represents both forward or anticipative (backward) evolution in time for several economic variables such as stocks, flows or prices. Those equations have explicit solutions, which make them the keystone for all types of differential equations. Although the formal structure of both ODE is the same, the side conditions are usually different, which implies that the type of characterization is also different. In particular, in characterizing the solution to time-dependent ODE, we are interested in the asymptotic states and their (global) stability properties.

Chapter 3 studies **scalar non-linear** regular time-dependent ODE,

$$\dot{y} = f(y(t), t), \;\; t \in \mathrm{T} \subseteq \mathbb{R}_+, \tag{1.8}$$

where it is assumed that function $f(y, \cdot)$ is non-linear but regular (i.e., it is non-singular, continuous and differentiable at least until order three). In some cases, these ODE have explicit solutions. We restrain to the time-dependent case because that is the case dealt in the very rich qualitative theory of ODE. Independently from the existence of known explicit solutions, and differently from linear ODE, local and global dynamics properties can differ, essentially because there can be a non-unique finite number of steady and orbits connecting them.

Chapter 4 deals with **planar linear ODE** over functions $\mathbf{y} \in \mathbb{R}^2$ of one independent variable. When the independent variable is not time $\mathbf{y} = \mathbf{y}(x)$, mapping $\mathbf{y} : \mathrm{X} \rightarrow \mathbf{Y} \subseteq \mathbb{R}^2$, is governed by

$$\mathbf{y}'(x) = \mathbf{A}\,\mathbf{y}(x) + \mathbf{B},\ x \in \mathrm{X} \subseteq \mathbb{R}, \tag{1.9}$$

and when the independent variable is time $\mathbf{y} = \mathbf{y}(t)$, mapping $\mathbf{y} : \mathrm{T} \rightarrow \mathbf{Y} \subseteq \mathbb{R}^2$, is governed by

$$\dot{\mathbf{y}} = \mathbf{A}\,\mathbf{y}(t) + \mathbf{B},\ t \in \mathrm{T} \subseteq \mathbb{R}_+, \tag{1.10}$$

where $\mathbf{A} \in \mathbb{R}^{2\times 2}$ and $\mathbf{B} \in \mathbb{R}^2$ are constant coefficient matrices. Planar linear ODE also have closed form solutions and have much richer behavior than scalar ODE. They are particular useful because maximized Hamiltonian systems associated to optimality conditions for (scalar) optimal control problems are planar ODE, for any type of the independent variable. In several cases, planar ODE solutions follow saddle paths combining both forward and backward dynamics along the ordered set of the independent variable.

Chapter 5 extends time-dependent scalar ODE (1.8) to **planar non-linear and regular ODE** of type

$$\dot{\mathbf{y}} = \mathbf{f}(\mathbf{y}),\ \mathbf{y} : \mathrm{T} \rightarrow \mathbf{Y} \subseteq \mathbb{R}^2. \tag{1.11}$$

This allows for enlarging the number of possible global types of dynamics, that do not occur in scalar ODE, like homoclinic and heteroclinic orbits and limit cycles. It also allows for characterizing the dynamics of optimal control problems of non-linear ODE and non-linear dynamic general equilibrium (DGE) models. Most of the qualitative theory of ODE is dedicated to the time-dependent case.

Chapter 6 contains a brief introduction to two types of time-dependent **non-regular piecewise ODE**, such as

$$\dot{\mathbf{y}} = \begin{cases} \mathbf{f}_1(\mathbf{y}), & \text{if } H(\mathbf{y}) \leq 0, \\ \mathbf{f}_2(\mathbf{y}), & \text{if } H(\mathbf{y}) > 0, \end{cases} \tag{1.12}$$

where $\Sigma = \{\mathbf{y} \in \mathrm{Y} : H(\mathbf{y}) = 0\}$ is called the switching surface. Let $\tilde{\mathbf{y}} \in \Sigma$, then two cases are possible: if $\mathbf{f}_1(\tilde{\mathbf{y}}) = \mathbf{f}_2(\tilde{\mathbf{y}})$ and $\mathrm{D}_{\mathbf{y}}\mathbf{f}_1(\tilde{\mathbf{y}}) \neq \mathrm{D}_{\mathbf{y}}\mathbf{f}_2(\tilde{\mathbf{y}})$, we say we have a **piecewise-smooth ODE**, and if $\mathbf{f}_1(\tilde{\mathbf{y}}) \neq \mathbf{f}_2(\tilde{\mathbf{y}})$ then we say we have a **piecewise-continuous or Filippov ODE**. We deal with both the scalar and the planar case. Although these ODE are usually not in the toolbox of economists, they are useful for addressing the existence of several types of switching, some of which resulting from bounds in optimal control problems. Furthermore, they can have advantages as regards non-linear regular ODE, as is the case of ODE (1.11), both because the introduction of non-linearities can be more easily related to economic fundamentals, and also because they allow for obtaining similar types of global dynamics as regular

ODE by concatenating the different explicit solutions from different linear ODE segments (which can be analytically solvable).

A brief introduction to **singular ODE**, for time-dependent equations, can be found in Chap. 7. There are two main types of singular ODE, **fast-slow** ODE and **constrained or impasse** ODE. For scalar equations, over $y = y(t) \in \mathrm{Y} \subseteq \mathbb{R}$, they take the generic form

$$a(y)\,\dot{y} = g(y). \tag{1.13}$$

For fast-slow ODE we have $a(y) = \epsilon$, where ϵ belongs to a small interval containing zero. For constrained or impasse ODE we have instead $a(y) = \delta(y)$, where $\delta(y)$ is a function in which there is a point (or at least a finite number of points), $y^i \in \mathrm{Y}$, called impasse point(s), such that $\delta(y^i) = 0$. In the first case, we have a singularity of a global nature if $\epsilon = 0$, and in the second case we have a local singularity. In both cases, the vector field $f(y) = \frac{g(y)}{a(y)}$ is not Lipschitzian. Planar equations, over $\mathbf{y} = \mathbf{y}(t) \in \mathbf{Y} \subseteq \mathbb{R}^2$, are written as

$$\mathbf{A}(\mathbf{y})\,\dot{\mathbf{y}} = \mathbf{G}(\mathbf{y}). \tag{1.14}$$

In a fast-slow ODE, matrix $\mathbf{A}$ takes the form $\mathbf{A} = \begin{pmatrix} 1 & 0 \\ 0 & \epsilon \end{pmatrix}$. This type of singularity can generate discontinuous solution paths. In addition, by varying the parameter ϵ, which acts like a bifurcation parameter by changing dimension of the stable and unstable manifolds, we can have bifurcations different from regular ODE, that are generically called singularity induced bifurcations. For constrained or impasse ODE, there is at least one point $\mathbf{y}^i \in \mathbf{Y}$ such that $\det(\mathbf{A}(\mathbf{y}^i)) = 0$. To the set of points $\Gamma_I = \{\mathbf{y} \in \mathbf{Y} : \det(\mathbf{A}(\mathbf{y}^i)) = 0\}$, we call impasse or singular surface. Trajectories colliding with this surface either cannot be continued over time, or, if they traverse Γ_I, they evolve within stable or unstable manifolds that can have (or not) a different dimension from those in which they evolved before crossing Γ_I.

These two types of singular ODE are not in the usual toolbox of economists, and their type of solutions can have properties that are different from both regular non-linear ODE and piecewise ODE. We include them in this book because there are two outstanding aspects related to them: first, they can appear very easily in models in which there are strong feedback mechanisms- as models in which there are externalities or there are policy rules-, and, second, their solutions can display a change in the dimension of stable or unstable manifolds along transition paths. In the macroeconomics literature, we can associate them to the existence of temporary indeterminacy. They can also be present in the solution paths of optimal control problems when the benchmark curvature conditions are not satisfied.

1.2.1.2 Partial Differential Equations

A **partial differential equation** (PDE) is a functional equation containing one or more functions of at least two independent variables together with its derivatives. In economic applications, one of the independent variable is time. This is the case of equation (1.1) when $m > 1$, $\mathrm{Z} = \mathrm{T} \times \mathrm{X} \subseteq \mathbb{R}^m$, $p = 2$ and $\overline{\mathbf{F}}(.)$ is monotonic and regular in all the derivative arguments. We restrain to the case in which $m = 2$.

Two main types of equations involving function $\mathbf{y} = \mathbf{y}(t, x)$, where $(t, x) \in \mathrm{T} \times \mathrm{X}$, are studied next: **first-order PDE**

$$\overline{\mathbf{F}}(\mathrm{D}_t(\mathbf{y}(t,x)), \mathrm{D}_x(\mathbf{y}(t,x)), \mathbf{y}(t,x), t, x) = 0, \ \ (t,x) \in \mathrm{T} \times \mathrm{X}, \tag{1.15}$$

containing only first-order derivatives, and **parabolic PDE**

$$\overline{\mathbf{F}}(\mathrm{D}_t(\mathbf{y}(t,x)), \mathrm{D}_{xx}(\mathbf{y}(t,x)), \mathrm{D}_x(\mathbf{y}(t,x)), \mathbf{y}(t,x), t, x) = 0, \ \ (t,x) \in \mathrm{T} \times \mathrm{X}, \tag{1.16}$$

containing second derivative for the non-time independent variable.

Part III includes two chapters on PDE. Chapter 13 presents a brief introduction to scalar **quasi-linear first-order** PDE, over function $y : \mathrm{T} \times \mathrm{X} \to \mathrm{Y} \subseteq \mathbb{R}$,

$$a(x)\, \partial_t\, y(t,x) + b(x)\, \partial_x y(t,x) + c(y(t,x), x) = 0 \tag{1.17}$$

where we use the notation $\partial_t y(t,x) \equiv \frac{\partial y(t,x)}{\partial t}$ and $\partial_x y(t,x) \equiv \frac{\partial y(t,x)}{\partial x}$. This type of PDE models, for instance, transport dynamics or age-dependent distributions over time. When there is a conservative or normalization requirement, they can represent a movement of a frequency density function over time, in the direction stipulated by the sign of $a(x)/b(x)$. We show they offer an appropriate formal representation of aggregate social mobility, measured by income or wealth.

Chapter 14 presents a brief introduction to scalar **quasi-linear parabolic** PDE, over function $y : \mathrm{T} \times \mathrm{X} \to \mathrm{Y} \subseteq \mathbb{R}$,

$$\partial_t y(t,x) = a(x)\, \partial_{xx} y(t,x) + b(x)\, \partial_x y(t,x) + c(y(t,x)) \tag{1.18}$$

where $\partial_{xx} y(t,x) \equiv \frac{\partial^2 y(t,x)}{\partial x^2}$. The sign of $a(\cdot)$ introduces a distinction between two types of parabolic PDE: if $a(x) > 0$ for all $x \in \mathrm{X}$ then the PDE is called **forward parabolic PDE**, and if $a(x) < 0$ for all $x \in \mathrm{X}$ it is called a **backward PDE**. Forward PDE are complemented with an initial condition, at $t = 0$, for instance $y(0,x) \in \mathrm{Y}$, and backward PDE are complemented with a terminal condition, for instance $y(T,x) = h(x)$ if $\mathrm{T} = [0, T]$ for a finite T. The size of the literature and cases involving parabolic PDE dwarfs the already very voluminous literature on ODE. The particular usefulness of parabolic PDE is related to their relationship to the statistics of stochastic differential equations via the Kolmogorov equations. Our goal in this chapter is to present just the simplest equations, allowing for obtaining explicit solutions, and showing they provide a very effective tool to address the dynamics of inequality. In particular, we show how to model the income dynamics in the presence of aggregate and pure social mobility, and its relationship with the Fokker-Planck-Kolmogorov equation.

1.2.2 Stochastic Differential Equations

Let $m = 2$, $\mathrm{Z} \subseteq \mathbb{R} \times \Omega$ where $(\Omega, \mathsf{F}, \mathbb{P})$ is a probability space, $p = 1$ and $\bar{\mathbf{F}}(.)$ is monotonic and regular in all the derivative arguments. A **stochastic differential equation** (SDE) is a functional equation over time-dependent random variables with realizations $\mathbf{y} : \mathrm{T} \times \Omega \to \mathbf{Y} \subseteq \mathbb{R}^n$, generating a stochastic process $(\mathbf{Y}(t))_{t \in \mathrm{T}}$. The image set $\mathbf{Y}$ is the set of all possible realizations, and is sometimes called state space.

Part III presents a brief introduction to SDE and their application in optimal control problems. Chapter 17 contains a brief introduction to stochastic processes, and, in particular, to Wiener processes. If $\mathbf{Y}(t)$ is a function of a Wiener process, then it is not differentiable in the classic sense, and we need a generalized theory of differentiation (or integration) to obtain a valid mathematical specification for a SDE. The most widespread theory of differentiation is the **Itô's or stochastic calculus**. It provides the elements for setting up and solving **Itô stochastic differential equations**, and, in particular to the **scalar diffusion equations** over variable $Y(t) = (y(t, \omega))_{\omega \in \Omega}$,

$$dY(t) = \mu(t, Y(t))\, dt + \sigma(t, Y(t))\, dW(t), \quad t \in \mathrm{T} \tag{1.19}$$

where every realization is denoted by $y(t, \omega) \in \mathrm{Y} \subseteq \mathbb{R}$, and where $(W(t))_{t \in \mathrm{T}}$ is a standard Wiener process, i.e., a stochastic process whose variations follows a normal distribution with zero mean and variance dt, $dW(t) \sim N(0, dt)$. As for the other differential equations, forward and backward forms of SDE are distinguished by the fact that while the first requires an initial side condition, the latter requires a terminal side condition.[16]

There are two important relationships between SDE and PDE. Let

$$P(t, y_t, s, y_s) = \mathbb{P}\big[Y(t) = y_t | Y(s) = y_s]$$

be the probability that the realization of the process at time t is $Y(t) = y_t \in \mathrm{Y}$, when at time $s < t$ the realization was $Y(s) = y_s \in \mathrm{Y}$. Writing $p(t, y) = P(t, y, 0, y_0)$, the probability of realization y at each time t, conditional on y_0 at time $t = 0$, follows the Kolmogorov forward or Fokker-Planck equation

$$\partial_t p(t, y) = \frac{1}{2}\, \partial_{yy}\Big(\sigma(t, y) p(t, y)\Big) - \partial_y\Big(\mu(t, y)\, p(t, y)\Big), \quad (t, y) \in \mathbb{R}_+ \times \mathrm{Y}, \tag{1.20}$$

[16] There is a potential difficulty here due to the fact that Itô's processes are non-anticipating.

that is a parabolic PDE, can be obtained by solving this equation together with the initial constraint a Dirac-delta generalized function centered at y_0, $p(y, 0) = \delta(y - y_0)$. From the solution to this equation, we can find the conditional (on the initial realization of y_0) moments of order i of the solution to the SDE (1.19)

$$\mathsf{m}_i(t) = \mathbb{E}[Y(t)^i] = \int_{\mathrm{Y}} y^i\, p(t, x)\, dx.$$

Furthermore, the probability of each realization y at each time t, conditional on $h(y_T)$ at time $T > t$, denoted by $q(t, y) = P(T, y_T, t, y)$, follows the Kolmogorov backward equation

$$\partial_t q(t, y) = -\sigma(t, y)\frac{1}{2}\,\partial_{yy} q(t, y) - \mu(t, y)\partial_y q(t, y),\ (t, y) \in [0, T] \times \mathrm{Y}, \tag{1.21}$$

which is a backward parabolic PDE, that is solved with the terminal constraint $q(T, y) = h(Y_T)$. This last PDE is frequently used in mathematical finance, and can also appear as an adjoint equation in distributional optimal control problems.

Chap. 18 is dedicated to solving **scalar linear SDE**

$$dY(t) = (\mu_0 + \mu_1 Y(t))\, dt + (\sigma_0 + \sigma_1 Y(t))\, dW(t),\ t \in \mathrm{T}, \tag{1.22}$$

where μ_0, μ_1, σ_0, and σ_1 are constants. If this equation is constrained to satisfy an initial side condition, $Y(0) = y_0$, then we have a **forward SDE**(FSDE). We explicitly solve these SDE and obtain their time-dependent statistic moments. We illustrate both its sample paths and their stochastic dynamic properties. In particular, we find its possible asymptotic states and relate them to the properties of their deterministic (ODE) skeleton $\dot{y} = \mu_0 + \mu_1\, y$. If the SDE (1.22) is constrained to satisfy a terminal condition $Y(T) = Y_T$, that can be deterministic or stochastic, then we have a **backward SDE** (BSDE). The existence of solutions to BSDE is not guaranteed because the non-reversibility of time of SDE creates a mismatch between the forward nature of information, generated by the Wiener process, and the backward progression of the solution. We briefly discuss one method to address this issue and present the solution to some simple linear equations.

1.2.3 Functionals and Optimization of Functionals

Consider the set of functions $\mathcal{Y}$, mapping $y : \mathrm{Z} \to \mathrm{Y}$, where $\mathrm{Y} \subseteq \mathbb{R}$ and Z can be a number space, a probability space, or a product of number and/or probability spaces. A **functional** F is a mapping between a function space and a number space, $\mathsf{F} : \mathcal{Y} \to \mathbb{R}$. The following functionals are frequently applied in economics: the linear functional

$$\mathsf{F}[y] = \int_{\mathrm{X}} f\big(y(x), x\big) dx, \tag{1.23}$$

or a functional that is a transformation of a linear functional

$$\mathsf{F}[y] = g\Big(\int_{\mathrm{X}} f\big(y(x), x\big)dx\Big). \tag{1.24}$$

Utility functionals over a continuum of varieties, over time, or over contingent consumption realizations, and production functions over a continuum of inputs, are all examples of those functionals.

We can extend the previous definitions to functionals over functions of more than one independent variable. For linear functionals, we can have for functionals over two-dimensional functions $y = y(x_1, x_2)$,

$$\mathsf{F}[y] = \int_{\mathrm{X}_1 \times \mathrm{X}_2} f\big(y(x_1, x_2), x_1, x_2\big)\, dx_2\, dx_1, \tag{1.25}$$

or intertemporal utility functional over varieties, or over intertemporal consumption or income of heterogeneous agents,

$$\mathsf{F}[y] = \int_{\mathrm{T} \times \mathrm{X}} f\big(y(t, x), t, x\big)\, dx\, dt, \tag{1.26}$$

or expected utility functionals over stochastic consumption or output processes

$$\mathsf{F}[y] = \int_{\mathrm{T} \times \Omega} f\big(y(t, \omega), t, \omega\big) d\omega\, dt. \tag{1.27}$$

Among several types of economic theories and applications, we are interested in maximizing one of the previous functionals without or with side constraints. However, the classic definition of derivative cannot be applied in those cases because $\mathsf{F}[y]$ is an infinite-dimensional object. In order to perform variational analysis with functionals we need a generalized calculus over infinite-dimensional spaces.

Part II introduces some useful facts from functional calculus, or calculus over functionals, as a necessary prelude to addressing optimal control problems for systems governed by ODE, PDE, and SDE.

In Chap. 8, we introduce the main concepts of functional calculus and provide several applications in utility, production, and other "static" infinite-dimensional economic models. Consider a functional of function y, $\mathsf{F}[y]$, and let function $y(x)$ be perturbed along direction $\eta(x)$ such that $y(x)$ is perturbed to $y(x) + \epsilon\, \eta(x) \in \mathrm{Y}$, where $\epsilon > 0$ is a constant. The variation of the functional is $\mathsf{F}[y + \epsilon\, \eta] - \mathsf{F}[y]$. A generalized direction derivative[17]

$$\mathrm{D}_{\eta(\cdot)}\mathsf{F}[y] = \lim_{\epsilon \to 0} \frac{\mathsf{F}[y + \epsilon \eta] - \mathsf{F}[y]}{\epsilon},$$

[17] On the concepts of directional derivative, see Shapiro (1990).

is called **Gâteaux differential**. For the linear functional (1.23), the Gâteaux differential is

$$\mathrm{D}_{\eta(\cdot)}\mathsf{F}[y] = \int_{\mathrm{X}} \frac{\partial\big(y(x), x\big)}{\partial y(x)}\,\eta(x)dx.$$

A perturbation of the functional by a Dirac-delta generalized function, relative to a point $x \in \mathrm{X}$, also called a spike perturbation,

$$\mathrm{D}_{\delta(x)}\mathsf{F}[y] = \int_{\mathrm{X}} \frac{\partial\big(y(z), z\big)}{\partial y(z)}\,\delta(z-x)\,dz = \frac{\partial\big(y(x), x\big)}{\partial y(x)}$$

is called **Gâteaux derivative**. We provide extensions of Gâteux differentials and derivatives for functions of several independent variables, and for higher order variations and derivatives. We also use functional calculus to solve several (static) infinite-dimensional problems in microeconomics and statistics. In particular, we have the problem of finding a distribution $\big(y^*(x)\big)_{x\in\mathrm{X}}$ such that a functional is maximized,

$$\max_y \mathsf{F}[y] = \int_{\mathrm{X}} f\big(y(x)\big)dx, \tag{1.28}$$

for a known value function $f(\cdot)$. The first-order necessary condition, $\mathrm{D}_\eta\mathsf{F}[y] = 0$, for an arbitrary $\eta(\cdot)$, yields the functional equation $f'\big(y^*(x)\big) = 0$ for each $x \in \mathrm{X}$. The optimum, if it exists, is a function $y^*(x)$. If there is a functional constraint that has to be satisfied, we have instead the problem

$$\max_y \left\{ \int_{\mathrm{X}} f\big(y(x)\big)dx : \int_{\mathrm{X}} g\big(y(x)\big)dx = 0 \right\}, \tag{1.29}$$

where functions $f(\cdot)$ and $g(\cdot)$ are known. The first-order conditions are, in this case,

$$\begin{cases} f'\big(y(x)\big) + \lambda\, g'\big(y(x)\big) = 0, & \text{for each } x \in \mathrm{X}, \\ \int_X g\big(y(x)\big)dx = 0, \end{cases}$$

and the optimum, if it exists, is a function $y^*(x)$, and a number λ^* (a Lagrange multiplier). At last, if there is a point-wise constraint, as in the problem

$$\max_y \left\{ \int_{\mathrm{X}} f\big(y(x)\big)dx : g\big(y(x)\big) = \bar{g} \right\}, \tag{1.30}$$

where functions $f(\cdot)$ and $g(\cdot)$ are known functions and $\bar{g}$ is a number, the first-order conditions are now

$$\begin{cases} f'\big(y(x)\big) + \lambda(x)\, g'\big(y(x)\big) = 0, & \text{for each } x \in \mathrm{X} \\ g\big(y(x)\big) = \bar{g}, & \text{for each } x \in \mathrm{X}, \end{cases}$$

and optimum is a pair of functions, $y^*(x)$ and the Lagrange multiplier $\lambda^*(x)$.

1.2.4 Calculus of Variations and Optimal Control of ODE

1.2.4.1 Calculus of Variations

Calculus of variations problems, briefly introduced in Chap. 9, can be considered extensions of the functional optimization problem (1.28) in which the derivative of function $y(\cdot)$ is also an argument of the value function $F(\cdot)$

$$\max_{y} \int_{\mathrm{X}} F\big(y'(x), y(x), x\big) dx, \tag{1.31}$$

sometimes supplemented with point-wise conditions referring to boundary values of $y(\cdot)$, x, or both. The first-order condition is now a second-order differential equation, the Euler-Lagrange equation

$$\frac{\partial F(y'(x), y(x), x)}{\partial y(x)} + \frac{d}{dx}\Big(\frac{\partial F(y'(x), y(x), x)}{\partial y'(x)}\Big) = 0, \text{ for each } x \in \mathrm{X},$$

that has to be solved jointly with other side conditions.

An important observation is that while the "static" problems (1.28), (1.29), and (1.30) can be specified for independent variables whose domains do not have an order relationship (varieties), a calculus of variations problem can only be specified for independent variables having an order relationship. This means that the space X in problem (1.31) cannot be a space of varieties.[18]

When time is the independent variable, we write the intertemporal calculus of variations problem as

$$\max_{y(.)} \int_{\mathrm{T}} F(\dot{y}(t), y(t), t)\, dt, \tag{1.32}$$

where the derivative $\dot{y}$ is also an argument of the value function $F(\cdot)$, and where $\mathrm{T} = [0, T]$, and T can be finite or infinite. The Euler-Lagrange equation is, in this case,

$$\frac{\partial F(\dot{y}, y, t)}{\partial y(t)} + \frac{d}{dt}\Big(\frac{\partial F(\dot{y}, y, t)}{\partial \dot{y}}\Big) = 0, \text{ for each } t \in \mathrm{T},$$

and it has to be solved together with initial and, sometimes, terminal conditions (it is a mixed initial-terminal value ODE problem). This problem usually differs from problem (1.31) on the type of point-wise conditions: while in time-dependent

[18] However, if the state variable y depends on both time and a variety index, we can specify an intertemporal problem for a bundle of varieties. In this case, we have a value functional similar to (1.26).

problem, there are fixed or constrained values for the dependent variable, y, in non-time dependent problems sometimes conditions on the boundaries of set X can also be chosen optimally (for instance in optimal taxation problems).

1.2.4.2 Optimal Control of ODE

In Part II, we also study **optimal control problems** of systems governed by scalar ODE. When the state and the control variables, $y(x)$ and $u(x)$, are non-time-dependent variables, a generic problem is

$$\max_{u(\cdot)} \mathsf{V}[u, y] = \int_{\mathrm{X}} f(y(x), u(x), x)\, dx \tag{1.33a}$$

subject to

$$y'(x) = g(y(x), u(x), x), \quad \text{for each } x \in \mathrm{X} \tag{1.33b}$$

$$\text{side conditions.} \tag{1.33c}$$

This type of problem has been applied, for instance, in optimal spatial allocation models, and in optimal taxation models when the government and the tax-payer have asymmetric information. In the first type of problems, the ODE constraint (1.33b) specifies spatial allocation costs, and in the second, it specifies an incentive compatibility constraint.

When the state and the control variables, $y(t)$ and $u(t)$, are time-dependent we have,

$$\max_{u(\cdot)} \mathsf{V}[u, y] = \int_{\mathrm{T}} f(y(t), u(t), t)\, dt \tag{1.34a}$$

subject to

$$\dot{y} = g(y(t), u(t), t), \quad \text{for each } t \in \mathrm{T} \tag{1.34b}$$

$$y(0) = y_0, \tag{1.34c}$$

and sometimes supplemented by a terminal condition. This problem is common in macroeconomic and growth optimal allocation problems, for the whole economy, or as a model for representative agents (households or firms) in DGE models. In particular, the discounted autonomous infinite horizon problem is the most common on those literatures,

$$\max_{u(\cdot)} \mathsf{V}[u, y] = \int_0^{\infty} f(y(t), u(t))\, e^{-\rho t}\, dt \tag{1.35a}$$

subject to

$$\dot{y} = g(y(t), u(t)), \quad \text{for each } t \in \mathrm{T} = \mathbb{R}_+ \tag{1.35b}$$

$$y(0) = y_0, \tag{1.35c}$$

where $\rho > 0$ is the rate of time preference.

There are two main methods for obtaining necessary (and sometimes necessary and sufficient) conditions (in addition to transforming the problem into a calculus of variations problem).

First, the **Pontriyagin's maximum principle** (PMP), which is briefly addressed in Chap. 10, introduces a co-state variable $\lambda(\cdot)$ and the Hamiltonian function

$$H(y, u, \cdot) = f(y, u, \cdot) + \lambda\, g(y, u, \cdot),$$

and states that the triple (y^*, u^*, λ) satisfies, at the optimum, the following conditions: u^* maximizes the Hamiltonian function $u^* = \arg\max H(y, u, \cdot)$, λ is driven by the adjoint ODE,

$$\lambda'(x) + \frac{\partial}{\partial y} H(y(x), u^*(x), x) = 0, \text{ for each } x \in \mathrm{X},$$

plus a terminal condition, and the constraints (1.33b)–(1.33b). If the dependent variable is time, we have instead the adjoint ODE

$$\dot{\lambda} + \frac{\partial}{\partial y} H(y(t), u^*(t), t) = 0, \text{ for each } t \in \mathrm{T}.$$

For time-dependent discounted infinite-horizon problems, in which functions $f(.)$ and $g(.)$ are sufficiently differentiable, we can invoke the implicit function theorem to obtain the optimal control as a function of the state and co-state variables $u^* = U(y, q)$, where $q(t) = \lambda(t)\, e^{\rho t}$. If this is the case, the solution can be obtained by solving, or qualitatively characterized by linearizing and solving, the modified Hamiltonian dynamic system (MHDS),

$$\dot{y} = g(y, U(q, y)), \tag{1.36a}$$

$$\dot{q} = \rho q - \frac{\partial}{\partial y} H(y, U(q, y)), \tag{1.36b}$$

jointly with the initial and transversality conditions. The MHDS is usually a nonlinear (regular) planar ODE in (y, q).

Second, the **dynamic programming** principle establishes that necessary conditions for an optimum satisfy an implicit functional equation that is called the **Hamilton-Jacobi-Bellman** (HJB) equation. For the time-dependent problem (1.35a), the HJB equation is the implicit PDE equation involving the value function $V(t, y)$

$$-\,\partial_t V(t, y) = \max_{u(.)} \left\{ f(y, u, t) + \partial_y V(t, y)\, g(y, u, t) \right\}.$$

For the infinite-horizon autonomous problem (1.35), the HJB is an implicit ODE involving the time-independent value function $V(y)$

$$\rho V(y) = \max_{u(.)} \left\{ f(y, u) + V'(y)g(y, u) \right\}.$$

Maximizing for u, and assuming that all functions are sufficiently smooth, we obtain the policy function $u = U(y, V'(y))$, that substituting in the HJB yields the maximized HJB equation which is an implicit ODE

$$\rho\, V(y) = f\left(y, U(y, V'(y))\right) + V'(y)g\left(y, U(y, V'(y))\right).$$

There is a relationship between the necessary conditions according to the two approaches via the envelope theorem: if we take the derivative of the maximized HJB equation relative to y, apply the envelope theorem $(f_u(\cdot)V'(y) + g_u(\cdot)V''(y) = 0)$, and define $q = V'(y)$, we obtain the MHDS (1.36). This equivalence requires strong assumptions regarding the constitutive functions of the optimal control problem. But it allows for using the rich qualitative theory of differential equations to characterize the dynamics of the solution to the problem when we cannot obtain explicit solutions.

Chapter 12 deals with several extensions of the optimal control problem: singular problems, two-stage optimal control problems, problems with pointwise constraints on the state and control variables, optimal control of piecewise differentiable and piecewise continuous ODE, and two-state optimal control problems.

1.2.5 Optimal Control of PDE

In Part III, we provide an introduction to optimal control of PDE problems. We address simple problems by assuming one-dimensional state and control variables, $y(t, x)$, and $u(t, x)$, with domains in $\mathrm{T} \times \mathrm{X} \subseteq \mathbb{R}_+ \times \mathbb{R}$, and images in subsets of the Euclidean space $\mathbb{R}$.

Chapter 14 introduces **optimal control problem of first-order PDE**, having, for instance, the form

$$\max_{u(\cdot)} \mathsf{V}[u, y] = \int_{\mathrm{T}\times\mathrm{X}} f(y(t, x), u(t, x), t, x)\, dx\, dt \tag{1.37a}$$

subject to

$$\partial_t y(t, x) + a(t, x)\partial_x y(t, x) = g(y(t, x), u(t, x)), \tag{1.37b}$$

$$y(0, x) = y_0(x). \tag{1.37c}$$

This problem models the optimal choice of a distribution that is transported over time.[19] The constraint is similar to PDE (1.18) in which some of the coefficients are functions of the controi variable. We derive first-order conditions, from a general-

[19] Observe that the variables in this problem have domains over two sets endowed with order relationships, differently from the case of functional (1.26) that can be maximized for a function of a set of varieties and time.

ization of the PMP. Denoting the co-state variable by $\lambda(\cdot)$, the Hamiltonian function is

$$H(y, u, \lambda, t, x) = f(y, u, t, x) + \lambda\, g(y, u).$$

If the constitutive functions are smooth enough, the necessary first-order conditions can be represented by a MHDS of first-order forward-backward PDE

$$\begin{aligned}
\partial_t y(t,x) &= \partial_x y(t,x) + g(y(t,x), U(y(t,x), \lambda(t,x))),\\
\partial_t \lambda(t,x) &= -\partial_x \lambda(t,x) + \frac{\partial}{\partial y} H(y(t,x), U(y(t,x), \lambda(t,x)), \lambda(t,x), t, x).
\end{aligned}$$

The solution, if it exists, displays an optimal distribution evolving over time that can converge to a bounded asymptotic distribution or not.

Chapter 16 introduces **optimal control problem of parabolic PDE**, having, for instance, the form

$$\max_{u(\cdot)} \mathsf{V}[u, y] = \int_{\mathrm{T}\times\mathrm{X}} f(y(t,x), u(t,x), t, x)\, dx\, dt \tag{1.38a}$$

subject to

$$\partial_t y(t,x) = \partial_{xx} y(t,x) + g(y(t,x), u(t,x)), \tag{1.38b}$$

$$y(0,x) = y_0(x), \tag{1.38c}$$

where the constraint (1.38b) is analogous to PDE (1.18) in which the coefficients can be functions of the control variable. The first-order necessary conditions, from a generalized PMP, can also be represented by a generalized MHDS of forward-backward parabolic PDE

$$\begin{aligned}
\partial_t y(t,x) &= \partial_{xx} y(t,x) + g(y(t,x), U(y(t,x), \lambda(t,x))),\\
\partial_t \lambda(t,x) &= -\partial_{xx} \lambda(t,x) + \frac{\partial}{\partial y} H(y(t,x), U(y(t,x), \lambda(t,x)), \lambda(t,x), t, x).
\end{aligned}$$

A particular interesting case is an optimal control problem of a Fokker-Planck-Kolmogorov equation

$$\max_{u(\cdot)} \mathsf{V}[u, y] = \int_{\mathrm{T}\times\mathrm{X}} f(u(t,x))\, y(t,x)\, e^{-\rho t}\, dx\, dt$$

subject to

$$\partial_t y(t,x) = -\partial_x \big(\mu(u(t,x), x)\, y(t,x)\big) + \frac{1}{2}\partial_{xx}\big(\sigma(u(t,x), x) y(t,x)\big),$$

$$y(0,x) = y_0(x),$$

which, for instance, allows us to model optimal social mobility when the social planer is inequality averse.

1.2.6 Optimal Control of SDE

Chapter 19 contains a brief and pragmatic introduction to **stochastic optimal control**, or optimal control of a stochastic differential equation. The simplest and more well-known problem is the expected discounted infinite horizon problem

$$\max_{U(\cdot)} \mathsf{V}[U, Y] = \mathbb{E}_0 \left[\int_0^{\infty} f(Y(t), U(t))\, e^{-\rho t}\, dt \right] \tag{1.39a}$$

subject to

$$dY(t) = \mu(Y(t), U(t))\, dt + \sigma(Y(t), U(t))\, dW(t), \text{ for each } t \in \mathrm{T}, \tag{1.39b}$$

$$y(0, x) = y_0(x). \tag{1.39c}$$

Comparing the constraint (1.39b) with the diffusion equation (1.19), we observe that the drift and the diffusion functions, $\mu(\cdot)$ and $\sigma(\cdot)$, depend on the control variable $u(\cdot)$. To find the necessary conditions for an optimum, the most well-known approach is the **stochastic dynamic programming principle** (SDPP). It states that, for any realization of the optimal state and control variables at each time t, $Y^*(t) = y$ and $U^*(t) = u$, the value function satisfies the HJB equation

$$\rho\, v(y) = \max_{u(.)} \left\{ f(y, u) + v'(y)\, \mu(y, u) + \frac{1}{2} \sigma(y, u)^2\, v''(y) \right\},$$

which is an implicit second-order ODE. Although much less known in the economic literature, we present a version of the Pontriyagin's maximum principle (SPMP). We also show that a generalization of the envelope theorem allows for finding a stochastic version of the MHDS that is consistent with the SPMP.

Part I
Ordinary Differential Equations

Chapter 2
Scalar Linear ODE

Linear scalar ordinary differential equations (ODE), and related problems, their solutions, and characterization of their behavior, are the subjects covered in this chapter.

Definition 2.1 (*Scalar linear ODE*) A scalar linear ODE is a functional equation of a function $y(x)$, mapping $y : \mathrm{X} \subseteq \mathbb{R} \to \mathrm{Y} \subseteq \mathbb{R}$, defined by

$$y' = a(x)\, y + b(x), \tag{2.1}$$

where $y'(x) \equiv \frac{dy(x)}{dx}$, and $a(\cdot)$ and $b(\cdot)$ are known. If $a(x)$ and $b(x)$ are known real-valued functions with domain X, then the ODE is said to be **non-autonomous**. If a and b are known real numbers, then the ODE is said to be **autonomous**.

Solving an ODE means finding (or proving that there exists) a function $f(x)$, mapping $f : \mathbb{R} \to \mathbb{R}$, and satisfying Eq. (2.1), for at least a subset $\mathrm{X} \subseteq \mathbb{R}$. Given an interval $\mathrm{X} \subseteq \mathbb{R}$, function $f(\cdot)$ traces out an ordered subset $\mathrm{Y} \subset \mathbb{R}$ such that $\mathrm{Y} = f(\mathrm{X})$. We say an ODE has a closed form (or explicit) solution if it can be expressed as a specific function of a finite number of elementary functions. In the case of the linear ODE (2.1), a closed form solution always exists for the whole domain X. From now on, we distinguish the elements of Y traced out by the solution of an ODE, $(y(x))_{x \in \mathrm{X}}$, from a specific element of Y that is associated to a specific value for x_0, $y(x_0)$, and from a specific element of set Y, say y_0. This distinction is important for understanding the difference between a general solution (or a family of solutions), dependent on $y(x_0)$, and particular solutions to an ODE, dependent on y_0, and between forward and backward ODE.

An **ODE problem** is defined by an ODE together with a **side condition**. The side condition provides additional information on the value of the dependent variable—for example $y_0 \in \mathrm{Y}$—for a particular value (or values) of the independent variable—for example $x_0 \in \mathrm{X}$—or for a particular value of a functional over function $y(x)$. Therefore, while the solution of an ODE is a family of functions, the solution to an

P. B. Brito, *Economic Dynamics and Distributions*, Dynamic Modeling and Econometrics in Economics and Finance 32, https://doi.org/10.1007/978-3-031-94717-9_2

ODE problem, if it exists, is a specific unique function $y : X \to Y$. This is the reason justifying the solution to an ODE to be called **general solution**, and the solution to an ODE problem, to be called **particular solution**. For linear ODE, although general solutions always exist—and can be explicitly specified—particular solutions either are unique or do not exist.

The variational nature of the modeling strategy leading to an ODE provides an intuition on the nature of its solution. This property also reveals itself in the two alternative forms associated to Eq. (2.1), resulting from the fundamental theorem of calculus: we have a **forward** ODE if

$$y(x) = y(x_0) + \int_{x_0}^{x} (a(s)y(s) + b(s))\, ds, \text{ for } x \geq x_0,$$

or we have a **backward** ODE if

$$y(x) = y(x_1) - \int_{x}^{x_1} (a(s)y(s) + b(s))\, ds, \text{ for } x \leq x_1,$$

where $y(x_0)$ and $y(x_1)$ are arbitrary elements of Y, that we associate to the arbitrary elements of X, such as x_0 and x_1, respectively. If side conditions allow for the existence of unique solutions, then we can find specific (and not arbitrary) values for functions $y(x_0)$ or $y(x_1)$ within set Y.

In economic applications, we are interested in characterizing the qualitative properties of the (explicit or implicit) solutions to ODE. This means, in general, describing the properties of the curve $(x, y(x))_{x \in X}$ traced out in the space $X \times Y$. As the set X is endowed with an order relationship, this generally entails a description of the **distribution** of a variable over its range Y. For the particular case in which the independent variable is time, the solution to an ODE provides **orbits or trajectories**, $(t, y(t))_{t \in T}$. Characterizing the **dynamics**, or the behavior of a solution of an ODE over time, has been historically the main concern of the qualitative theory of ODE.

In the rest of this chapter, Sects 2.1 and 2.2 present solutions and applications to ODE and ODE problems, and Sect. 2.3 contains a first presentation of the dynamics generated by scalar ODE.

2.1 Solutions to Scalar Linear ODE

In this section, analytical solutions to autonomous and non-autonomous equations are obtained, in Sects 2.1.1 and 2.1.2, respectively.

2.1.1 Autonomous Equations

A **scalar linear autonomous** ODE is a functional equation of the form

$$y' = a\,y + b, \text{ for } x \in \mathrm{X} \subseteq \mathbb{R}, \tag{2.2}$$

where a and b are real numbers. If $b = 0$ the equation is said to be **homogeneous** and if $b \neq 0$ the equation is said to be **non-homogeneous**.[1]

Proposition 2.1 (General solution to the autonomous homogeneous ODE) *Consider the ODE* (2.2) *where a is a real number and* $b = 0$. *Its general solution is,*

$$y(x) = \begin{cases} y(x_0)\, e^{a\,(x-x_0)}, & \text{if } a \neq 0, \quad (2.3a)\\ y(x_0), & \text{if } a = 0, \quad (2.3b)\end{cases}$$

for each $x \in \mathrm{X}$, *where* $(x_0, y(x_0))$ *is an arbitrary pair belonging to* $\mathrm{X} \times \mathrm{Y} \subseteq \mathbb{R}^2$.

Proof Although there are several methods for solving ODE (2.2) with $b = 0$,[2] next we use three alternatives: separation of variables, *ansatz*, and recursive integration methods.

First, the separation of variables method[3] involves integrating separately variables y and x. As $y'(x) \equiv dy(x)/dx$, Eq. (2.2) with $b = 0$ is equivalent to $\frac{dy}{y} = a\,dx$. Interpreting it as a forward equation, integrating by transformation yields

$$\int_{y(x_0)}^{y(x)} \frac{dy}{y} = \int_{x_0}^{x} a\,ds, \text{ for } x \geq x_0.$$

Performing the integration yields $\log\left(\frac{y(x)}{y(x_0)}\right) = a\,(x - x_0)$. At last, solving for $y(x)$ yields function (2.3a) if $a \neq 0$, or function (2.3b) if $a = 0$.

Second, the method of undetermined coefficients[4] involves conjecturing an *ansatz* depending on some parameters that can be explicitly determined if the conjecture is correct. Let us conjecture that the solution is of the form $\phi(x) = \alpha\, e^{\lambda x}$, where α and λ are arbitrary real numbers. If the conjecture is correct, we should be able to determine the coefficients λ and α by substituting function $\phi(x)$ in the ODE. Therefore, it should satisfy

---

[1] The reason for this designation is obvious: while $f(y) = a\,y$ is a homogeneous function, $f(y) = a\,y + b$ is non-homogenous. Recall that function $f(x)$ is homogeneous of degree n if, multiplying the independent variable by an arbitrary real number λ, it is equivalent to multiplying the function by λ^n: $f(\lambda\,x) = \lambda^n\,f(x)$. If $f(\lambda\,x) \neq \lambda^n\,f(x)$ the function $f(\cdot)$ is a non-homogeneous function.

[2] For instance: separation of variables, using Laplace transforms, using Fourier transforms, transforming into an integral equation, using the concept of generating function, just to name a few. See (Zwillinger, 1998).

[3] Used at the first time by Leibniz in the late 1680s.

[4] First used by Euler in the 1740s.

$$\phi'(x) - a\,\phi(x) = 0 \iff \alpha\, e^{\lambda x}\, c(\lambda) = 0, \; x \in \mathrm{X}$$

where $c(\lambda) \equiv \lambda - a$ is called **characteristic**. For an arbitrary α, the equivalence only holds for the roots of the characteristic equation $c(\lambda) = 0$, that is for $\lambda = a$. However, $\phi(x) = \alpha\, e^{a x}$, still depends on the undetermined coefficient α. Taking an arbitrary pair $(x_0, \phi(x_0))$ we obtain α by solving $\phi(x_0) = \alpha\, e^{a x_0}$. Again we obtain functions (2.3a) if $a \neq 0$, or function (2.3b) if $a = 0$.

A third method involves repeated integration.[5] The ODE in forward form is, from the fundamental theorem of calculus, $y(x) = y(x_0) + \int_{x_0}^{x} a\, y(s) ds$. Substituting $y(s)$ inside the integral, and factoring, yields

$$\begin{aligned} y(x) &= y(x_0) + \int_{x_0}^{x} a \left(y(x_0) + \int_{x_0}^{s} a\, y(s')\, ds' \right) ds \\ &= y(x_0) + a\, y(x_0)\,(x - x_0) + \int_{x_0}^{x}\int_{x_0}^{s} a^2\, y(s')\, ds'\, ds. \end{aligned}$$

Repeating the same operation yields

$$\begin{aligned} y(x) &= y(x_0)\left(1 + a\,(x - x_0)\right) + \int_{x_0}^{x}\int_{x_0}^{s} a^2 \left(y(x_0) + \int_{x_0}^{s'} a\, y(s'')\, ds'' \right) ds'\, ds \\ &= y(x_0)\left(1 + a\,(x - x_0) + \frac{a^2\,(x - x_0)^2}{2}\right) + \int_{x_0}^{x}\int_{x_0}^{s}\int_{x_0}^{s'} a^3\, y(s'')\, ds''\, ds'\, ds. \end{aligned}$$

If we continue substituting $y(\cdot)$, yields

$$y(x) = y(x_0) \sum_{n=0}^{\infty} \frac{a^n\,(x - x_0)^n}{n!},$$

which is the series representation of the exponential in solution (2.3a).[6] □

It should be noticed that solution (2.3) represents a **family** of solutions because $y(x_0)$ represents an arbitrary element of the space Y.

Proposition 2.2 (General solution to the autonomous non-homogeneous ODE) *Consider the* ***non-homogeneneous autonomous*** *ODE in Eq.* (2.2) *where* $(a, b) \in \mathbb{R}^2$. *The general solution to that equation is the family of functions*

$$y(x) = \begin{cases} y(x_0)\, e^{a\,(x - x_0)} - \dfrac{b}{a}\left(1 - e^{a\,(x - x_0)}\right), & \text{if } a \neq 0, \quad (2.4a) \\ y(x_0) + b\,(x - x_0), & \text{if } a = 0, \quad (2.4b) \end{cases}$$

[5] This method is related to the series solution to ODE and is attributed to Euler (see Simmons 2017, Chap. A 21). We will use this method again for planar ordinary and stochastic differential equations.

[6] Recall that exponential function is an analytic function, i.e., it has an infinite series representation. In this case, $e^{\lambda x} = \sum_{n=0}^{\infty} \frac{(\lambda x)^n}{n!}$.

for each $x \in \mathrm{X}$, *where* $(x_0, y(x_0))$ *is an arbitrary element of* $\mathrm{X} \times \mathrm{Y} \subseteq \mathbb{R}^2$.

Proof There are several methods to obtain an explicit solution. The transformation of variables method simply transforms the non-homogeneous ODE into a homogeneous ODE. Let $a \neq 0$. If we define function $z(x) = y(x) + \frac{b}{a}$, differentiate it and substitute both $y(x)$ and its derivative from the non-homogeneous ODE (2.2), yields the equivalent homogenous ODE $z'(x) = a\, z(x)$. Then, from Eq. (2.3a), its solution is $z(x) = z(x_0)\, e^{a\,(x-x_0)}$. Transforming back to y, we obtain function (2.4a) because $y(x) = z(x_0)\, e^{a\,(x-x_0)} - \frac{b}{a} = \left(y(x_0) + \frac{b}{a}\right) e^{a\,(x-x_0)} - \frac{b}{a}$. If $a = 0$ we obtain function (2.4b), applying the l'Hôpital's rule

$$y(x) = y(x_0) - \lim_{a \to 0} \frac{b\left(1 - e^{a\,(x-x_0)}\right)}{a} = y(x_0) - \lim_{a \to 0} \left(- b\,(x - x_0) e^{a\,(x-x_0)} \right).$$

Alternatively, we can use the separation of variables method, setting $dy(x) = b\, dx$, that implies $\int_{y(x_0)}^{y(x)} dy = \int_{x_0}^{x} b\, dx$, and integrating to obtain (2.4b) again. □

Exercise 2.1 Use the method of separation of variables to find the solution (2.4) to ODE (2.2)

2.1.2 Non-Autonomous Equations

A **non-autonomous homogeneous** ODE is a particular case of equation (2.1) where $b(x) = 0$ for all $x \in \mathrm{X}$,

$$y'(x) = a(x)\, y(x), \quad \text{for } x \in \mathrm{X}, \tag{2.5}$$

where the coefficient function $a(\cdot)$ is an arbitrary bounded function $a : \mathrm{X} \to \mathbb{R}$, that is not necessarily continuous or differentiable.

Proposition 2.3 (General solution to the non-autonomous homogeneous ODE) *The general solution to ODE* (2.5) *is function*

$$y(x) = y(x_0)\, e^{\int_{x_0}^{x} a(s)\, ds}, \quad \textit{for each } x \in \mathrm{X} \tag{2.6}$$

where $(x_0, y(x_0))$ *is an arbitrary element of* $\mathrm{X} \times \mathrm{Y}$.

Proof Using again the separation of variables method, and integrating equation (2.5), we obtain $\log\left(\frac{y(x)}{y(x_0)}\right) = \int_{x_0}^{x} a(s)\, ds$. Solving for $y(x)$ yields general solution (2.6). □

Proposition 2.4 (General solution to the non-autonomous non-homogeneous ODE) *The general solution to the scalar non-autonomous non-homogenous ODE* (2.1) *is*

$$y(x) = y(x_0)\, e^{\int_{x_0}^{x} a(s)\, ds} + \int_{x_0}^{x} e^{\int_s^x a(z)\, dz}\, b(s) ds, \; for\; x \in \mathrm{X}, \tag{2.7}$$

where $(x_0, y(x_0))$ *is an arbitrary element of* $\mathrm{X} \times \mathrm{Y}$.

Proof We apply the variation of constants method.[7] First, we consider the solution to the homogeneous equation, where $b(x) = 0$ for all $x \in \mathrm{X}$. From Eq. (2.6) its solution is $y_h\big(x, y(x_0)\big) = y(x_0)\, e^{\int_{x_0}^{x} a(s)ds}$, for a fixed $x_0 \leq x$. However, we expect the solution to Eq. (2.1), for an arbitrary $x > x_0$, to be of the form

$$y(x) = y_h(x, y_0(x)) = y_0(x)\, e^{\int_{x_0}^{x} a(s)ds}. \tag{2.8}$$

Taking derivatives with respect to x and using the Leibniz's rule[8] we obtain

$$y'(x) = y_0'(x)\, e^{\int_{x_0}^{x} a(s)ds} + y_0(x)\, a(x)\, e^{\int_{x_0}^{x} a(s)ds} = y_0'(x)\, e^{\int_{x_0}^{x} a(s)ds} + a(x)\, y(x)$$

which should be equal to Eq. (2.1). Equating the right-hand sides of both equations, we obtain the ODE $y_0'(x) = b(x)e^{-\int_{x_0}^{x} a(s)ds}$. As function $y_0(.)$ is continuous, then, from the fundamental theorem of calculus, $\int_{x_0}^{x} y_0'(s)\, ds = y_0(x) - y_0(x_0)$, and $y_0(x) = y_0(x_0) + \int_{x_0}^{x} b(s)e^{-\int_{x_0}^{s} a(z)dz} ds$. Substituting in Eq. (2.8), and because $y_0(x_0) = y(x_0)$, yields function (2.7). □

Example 2.1 (*Local changes modeled by ODE*) In economics, there are several functions that are based on local changes modeled by ODE. The three ODE-based functions are: variations, $\nu(x) \equiv y'(x)$, growth rates, $\gamma(x) \equiv \frac{y'(x)}{y(x)}$, decay rates $\delta(x) \equiv -\frac{y'(x)}{y(x)}$, and elasticities, $\epsilon(x) \equiv \frac{y'(x)\, x}{y(x)}$. We say function $y(\cdot)$ is a **constant variation function** if it satisfies $\nu = y'(x)$, it is a **constant growth (decay) function** if it satisfies $\gamma = \frac{y'(x)}{y(x)}$ ($\delta = -\frac{y'(x)}{y(x)}$), and it is a **constant elasticity function** if it satisfies $\epsilon \equiv \frac{y'(x)\, x}{y(x)}$, where ν, γ and ϵ are real numbers. Therefore, general constant variation, growth, and elasticity functions can be obtained by solving scalar linear ODE. In particular, a generic constant variation function is

$$y(x) = y(x_0) + \nu\, (x - x_0), \;\; \text{for } x > x_0,$$

a generic constant growth function (if $\gamma > 0$) or decay function (if $\delta > 0$) are,

$$y(x) = y(x_0)\, e^{\gamma\, (x - x_0)}, \text{ or } y(x) = y(x_0)\, e^{-\delta(x - x_0)}, \text{ for } x > x_0,$$

---

[7] Due to Lagrange (1811).

[8] Let $h(x) = \int_{a(x)}^{b(x)} f(x, s)ds$. The Leibniz rule states that

$$\frac{dh(x)}{dx} = b'(x) f(x, b(x)) - a'(x) f(x, a(x)) + \int_{a(x)}^{b(x)} \frac{\partial}{\partial x} f(x, s)ds.$$

from Proposition 2.1, and a generic constant elasticity function is

$$y(x) = y(x_0) \left(\frac{x}{x_0}\right)^{\epsilon} \text{ if } \text{sign}(x) = \text{sign}(x_0)$$

from Proposition 2.3. More generally, we can specify models in which variations, growth/decay rates or elasticities are endogenous to function $y(\cdot)$, such as $\nu(x) = V(x, y(x))$, $\gamma(x) = G(x, y(x))$, or $\epsilon(x) = E(x, y(x))$. However, in these cases, the associated ODE are non-linear in general.

2.2 ODE Problems

The general solutions to scalar linear ODE, provided by Propositions 2.1 to 2.4, depended on the arbitrary pair $(x_0, y(x_0))$. Therefore, they represent families of functions with the domain X and range Y. An **ODE problem** is defined by an ODE together with a side condition. If the side condition is a functional equation we have

$$\begin{cases} y' = a(x)\, y + b(x), \text{ for each } x \in \mathrm{X}, & (2.9a) \\ \int_{\mathrm{X}} C(y(x), x)\, dx = \text{constant}, & (2.9b) \end{cases}$$

where $C(\cdot)$ is a known function. If set X is unbounded, we require $c(x) = C(y(x), x)$ to be a $L^1(\mathrm{X})$ function.[9] The solution to an ODE problem is called a **particular solution** to the ODE (2.1). It exists and is unique if the side condition allows us to obtain explicit values for the pair $(x_0, y_0) \in \mathrm{X} \times \mathrm{Y}$, which requires compatibility between the ODE and the side condition. If this is not the case, although there is an explicit general solution to the ODE, a particular solution to the associated ODE problem does not exist.

The functional (2.9b) is sufficiently general to encompass several particular cases. For instance, if we fix a value for $y_0 \in \mathrm{Y}$, and require $y(x_0) = y_0$, for a fixed x_0, this is equivalent to assuming the side condition

$$\int_{\mathrm{X}} \delta(x - x_0)\, y(x)\, dx = y_0 \tag{2.10}$$

where $\delta(\cdot)$ is the Dirac-delta generalized function, and $y(x)$ is the solution to an ODE. In the particular case of a non-autonomous and homogeneous ODE (2.5) with side condition $\int_{\mathrm{X}} \delta(x - x_0)\, y(x_0)\, e^{a\,(x-x_0)}\, dx = y_0$, the particular solution is $y(x) = y_0\, e^{\int_{x_0}^{x} a(s) ds}$, where (x_0, y_0) is a fixed pair within the set $\mathrm{X} \times \mathrm{Y}$. The solution traces out a unique curve within set Y.

Several examples of ODE problems relating to spatial allocations, statistic distributions, and utility functions are presented next.

[9] We require that $C(\cdot)$ satisfies the condition $\int_{\mathrm{X}} |C(y(x), x)| dx < \infty$.

Example 2.2 (*Spatial distribution of production*) Consider a firm that seeks to produce and distribute its product across a continuum of locations. The geographical span of its market is a (Salop, 1979) circle of length $L > 0$, with locations indexed by $x \in \mathrm{X} = [-\frac{L}{2}, \frac{L}{2}]$. The production and sales in location x are denoted by $y(x)$ and $u(x)$, respectively. It is assumed that moving the product from its production site, $x = 0$, to the consumer locations entails a cost, which implies that the redistribution rate decays proportionally. The rate of decay, $D(x, y(x))$ has a constant component, $\delta > 0$—a transport cost—and an endogenous component that decreases with the ratio between local sales and production, $u(x)/y(x)$. Therefore,

$$D(x, y(x)) \equiv -\frac{y'(x)}{y(x)} = \delta - \frac{u(x)}{y(x)}, \text{ for } x \in \mathrm{X}.$$

The total output of the firm, to be distributed across all locations, is assumed to be a positive constant denoted by Y. The spatial distribution of the firm's output, $y : \mathrm{X} \to \mathrm{Y} = \mathbb{R}_+$, is the solution to the following ODE problem:

$$\begin{cases} y'(x) = u - \delta\, y(x), \; x \in \mathrm{X} & (2.11a) \\ \int_{-L/2}^{L/2} y(x)\, dx = Y. & (2.11b) \end{cases}$$

where we let sales to be constant across space, $u(x) = u > 0$. Equation (2.11a) has the general solution $y(x) = \frac{u}{\delta} + \left(y(0) - \frac{u}{\delta}\right) e^{-\delta x}$, where $y(0) \in \mathrm{Y}$ is arbitrary. Substituting in the side condition (2.11b) yields $y(0) = \frac{u}{\delta} + \frac{\delta Y - u L}{e^{\delta L/2} - e^{-\delta L/2}}$. Therefore, the solution to the sales distribution problem is

$$y(x) = \frac{u}{\delta} + \left(\frac{\delta Y - u L}{e^{\delta L/2} - e^{-\delta L/2}}\right) e^{-\delta x}, \text{ for } x \in \mathrm{X}.$$

This model formalizes a distribution technology for a firm with equal sales at every market location. However, this model is misspecified for two reasons: first, it may not be in the firm's interest to sell the same amount across all locations, because there are spatial frictions, and, second, the solution does not satisfy a requirement for distributions in the (Salop, 1979) topology, $y(-L/2) = y(L/2)$, because we have

$$y\left(-\frac{L}{2}\right) = \frac{u}{\delta} + \frac{\delta Y - u L}{1 - e^{-\delta L}} \neq y\left(\frac{L}{2}\right) = \frac{u}{\delta} + \frac{\delta Y - u L}{e^{\delta L} - 1}.$$

We will address this problem again in Example 4.1.

Example 2.3 (*Statistic distributions*) ODE provide a method for generating families of continuous probability density (PDF) or cumulative density functions (CDF).[10] The starting contribution to this approach is Pearson (1894), for PDFs, and Burr (1942) for CDFs.

[10] See Kotz and Vicari (2005) and Lee et al. (2013).

Let $p(x)$ be a mapping $\mathrm{X} \subseteq \mathbb{R} \to [0, 1]$. Function $p(\cdot)$ is a general Pearson PDF[11] if it is specified by the x-dependent growth rate

$$\gamma(x) = \frac{a_0 + a_1\, x}{b_0 + b_1\, x + b_2\, x^2}, \quad \text{for } x \in \mathrm{X},$$

and furthermore it satisfies the normalization condition $\int_X p(x)\, dx = 1$. Recalling that $\gamma(x) \equiv \frac{p'(x)}{p(x)}$, we present two cases.

First, let the support be $X = \mathbb{R}_+$, and $a_1 = b_1 = b_2 = 0$ and $a_0 = -\lambda$ and $b_0 = 1$. Then $\gamma(x) = -\lambda \in (0, 1)$, and the PDF follows a **exponential distribution** with parameter λ,

$$p(x) = \lambda\, e^{-\lambda x}, \quad \text{for } x \in [0, \infty) \text{ if } 0 < \lambda < 1. \tag{2.12}$$

To prove this, we write the ODE problem for the PDF,

$$\begin{cases} p' = -\lambda\, p \text{ for } x \in \mathrm{X} = [0, \infty), & (2.13a) \\ \int_0^\infty p(x)\, dx = 1. & (2.13b) \end{cases}$$

The general solution to ODE (2.13a), for an arbitrary $p(0)$, is $p(x) = p(0)\, e^{-\lambda x}$. Substituting in the side condition, we conclude that the particular solution exists if and only if $0 < \lambda < 1$—the integral only exists if $\lambda > 0$, and, $p(0) = \lambda \in (0, 1)$ if $\lambda < 1$-, —and it is the exponential distribution (2.12). Figure 2.1a represents the solution for a particular value of λ.

Second, let the support be $X = \mathbb{R}_+$, and $a_0 = b_1 = b_2 = 0$ and $a_1 = -1$ and $b_0 = 1$. Then $\gamma(x) = -x$, and the PDF follows a **standard normal distribution**

$$p(x) = \frac{e^{-\frac{x^2}{2}}}{\sqrt{2\,\pi}}, \quad \text{for } x \in \mathrm{X} = \mathbb{R}, \tag{2.14}$$

represented in Fig. 2.1b. To prove this, we write the ODE problem for the PDF

$$\begin{cases} p' = -x\, p(x), \ \text{ for } x \in \mathrm{X} = (-\infty, \infty), & (2.15a) \\ \int_{-\infty}^\infty p(x)\, dx = 1. & (2.15b) \end{cases}$$

The general solution to Eq. (2.15a) is $p(x) = p(x_0)\, e^{-\int_{x_0}^x s\, ds} = p(x_0)\, e^{-\frac{x^2}{2} + \frac{x_0^2}{2}}$ for an arbitrary point $(x_0, p(x_0))$. Substituting in the constraint (2.15b), yields $y(x_0)\, e^{\frac{x_0^2}{2}} \int_{-\infty}^\infty e^{-\frac{x^2}{2}}\, dx = p(x_0)\, e^{\frac{x_0^2}{2}} \sqrt{2\pi} = 1$,[12] allows us to find $p(x_0)\, e^{\frac{x_0^2}{2}} = (2\pi)^{-\frac{1}{2}} \in (0, 1)$.

Exercise 2.2 Prove that the normal PDF, $p(\cdot) \sim N(\mu, \sigma^2)$ is generated by growth rate $\gamma(x) = \frac{\mu - x}{\sigma^2}$ defined over the support $\mathrm{X} = \mathbb{R}$.

---

[11] See Johnson et al. (1994, pp. 15–25) and Kleiber and Kotz (2003, Sect. 2.3).

[12] Because the Gaussian integral satisfies $\int_{-\infty}^\infty e^{-\frac{x^2}{2}}\, dx = \sqrt{2\pi}$, as can be seen in Appendix Sect. 15.7.2.

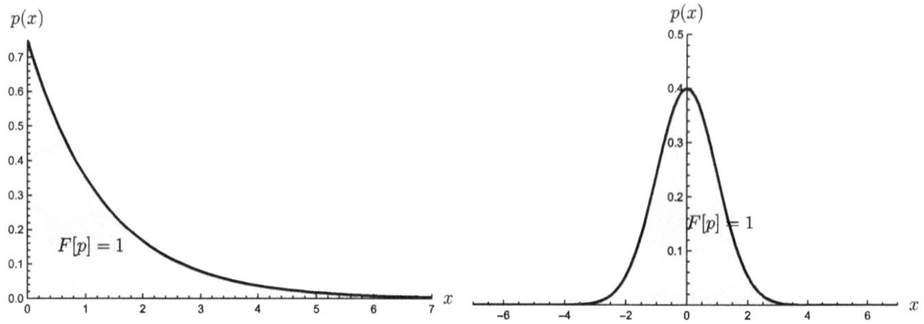

(a) Exponential density for $\lambda = 0.75$ (b) Standard normal density

Fig. 2.1 Examples of Pearson probability distribution functions

Exercise 2.3 Prove that the Pareto PDF, with scale parameter $x_0 > 0$ and shape parameter $\epsilon > 0$, $p(\cdot) \sim P(x_0, \epsilon)$, is generated by the elasticity function $\epsilon(x) = -(1+\epsilon)$ with support $\mathrm{X} = [x_0, \infty)$.

Exercise 2.4 Prove that the log-normal PDF $p(\cdot) \sim LN(\mu, \sigma^2)$, where $\sigma > 0$, is generated by the elasticity function $\epsilon(x) = \frac{\mu - \sigma^2 - \log x}{\sigma^2}$ with support $\mathrm{X} = \mathbb{R}_+$.

Example 2.4 (*Utility functions*) Let $x \in \mathrm{X} \subseteq \mathbb{R}_+$ be the quantity of an item, and $u(\cdot)$ be an utility function, a cardinal measure of the value of that item. If $u'(x) > 0$ for a specific level of consumption x, we say that it is a good. If $u'(x) > 0$ for every $x \in \mathrm{X}$, we say there is no satiation, meaning that it is a good for any arbitrary level of consumption. When the consumption level is a random variable, X, $u(x)$ refers to a particular realization $X = x$, and according to the seminal paper (Pratt, 1964) the curvature of $u(\cdot)$, called a Bernoulli utility function, can be taken as a measure of the underlying attitude towards risk. In particular, that paper introduces two types of measures of (both local and global) risk aversion, that involve (second-order) ODE: the coefficient of risk aversion $a(x) = -\frac{u''(x)}{u'(x)}$, and the coefficient of proportional risk aversion $r(x) = x\, a(x)$.

In the economics literature, a Bernoulli utility function is said to be of the constant absolute risk aversion (CARA) type if it is represented by the exponential function

$$u(x) = -\frac{e^{-cx}}{c}, \quad \text{for } x \geq 0 \tag{2.16}$$

where $a(x) = c > 0$ is the coefficient of absolute risk aversion, and it is said to be of the constant relative risk aversion (CRRA) type if it is represented by a generalized logarithm

$$u(x) = \log_{c^*}(x) = \frac{x^{1-c^*} - 1}{1 - c^*}, \quad \text{for } x \geq 0, \tag{2.17}$$

where $r(x) = c^* > 0$ is the coefficient of relative risk aversion. Both utility functions are solutions to ODE problems. Utility function (2.16) is the unique solution to the problem

$$\begin{cases} u''(x) = -c\,u'(x), & \text{for } x \in X = (0, \infty), \\ u'(0) = -c\,u(0) = 1. \end{cases} \quad \begin{matrix} (2.18a) \\ (2.18b) \end{matrix}$$

To show this, we write the marginal utility as $v(x) = u'(x)$, and write the risk aversion function as $a(x) = c = -\frac{v'(x)}{v(x)}$. Therefore, the marginal utility is specified as a constant decay function. Solving this ODE, for an arbitrary value $v(0) = u'(0)$, associated to $x = 0$, we obtain $u'(x) = u'(0)\,e^{-cx}$. Integrating, and invoking the fundamental theorem of calculus, we obtain the general solution $u(x) = u(0) + \int_0^x u'(s)\,ds = u(0) + \frac{u'(0)}{c}\left(1 - e^{-cx}\right)$. The particular form (2.16) is obtained by substituting the side conditions (2.18b) in the associated general solution.

Using a similar approach, we can also show that the utility function (2.17) is the unique solution to the problem

$$\begin{cases} u''(x)\,x = -c^*\,u'(x), & \text{for } x \in X = (0, \infty) \\ u'(1) - 1 = u(0) = 0 \end{cases} \quad \begin{matrix} (2.19a) \\ (2.19b) \end{matrix}$$

Define again $v(x) = u'(x)$ and observe that the constant risk aversion utility function is a constant (negative) elasticity function, $r(x) = c^* = -\frac{v'(x)\,x}{v(x)}$. Therefore, it has the general solution $u'(x) = u'(x_0)\left(\frac{x}{x_0}\right)^{-c^*}$, for an arbitrary value $u'(x_0)$, such that $x_0 > 0$. Integrating, we find the general constant relative aversion utility function $u(x) = u(0) + \int_0^x u'(s)\,ds = u(0) + \frac{u'(x_0)}{1-c^*}\,x_0^{c^*}\left(x^{1-c^*} - x_0^{1-c^*}\right)$. The particular form (2.17) is obtained by substituting the side conditions (2.19b).

Exercise 2.5 If the absolute risk aversion coefficient is specified by function $a(x) = \frac{\alpha\,(1-\gamma)}{\alpha\,x+\beta\,(1-\gamma)}$ a family of hyperbolic absolute risk aversion utility functions[13] can be obtained. Find the general utility function and the specific side conditions that yield $u(x) = \frac{1-\gamma}{\gamma}\left(\beta + \frac{\alpha}{1-\gamma}\,x\right)^{\gamma}$.

2.3 Scalar Linear ODE Dynamics

In this section, we address dynamics of autonomous linear ODE. In Sect. 2.3.1, we represent geometrically their solutions, in Sect. 2.3.2, we briefly present the main definitions and methods of the geometric or qualitative theory of ODE. The types of problems of time-dependent ODE, and several applications to economics, are gathered in Sect. 2.3.3.

[13] See Cass and Stiglitz (1970).

2.3.1 ODE Solutions and Their Geometry

The linear scalar **autonomous** ODE, when time t is the independent exogenous variable, is defined by

$$\dot{y} = a\,y + b, \text{ for } t \in \mathrm{T} = [0, \infty) \tag{2.20}$$

where $\dot{y} = \frac{dy(t)}{dt}$,[14] and the parameters, (a, b), belong to a known subset of the set of real tuples $\Phi \subseteq \mathbb{R}^2$. From now on it is assumed that initial value of the independent variable is equal to zero, that is $t_0 = 0$. The explicit solutions can take the following forms:

Proposition 2.5 *Consider the autonomous ODE* (2.20)*, with parameters* $(a, b) \in \Phi$*. Depending on the parameters, the explicit solutions to the ODE can take the following forms,*

$$y(t) = \begin{cases} y(0)\,e^{a\,t} - \dfrac{b}{a}\left(1 - e^{a\,t}\right), \; \textit{if } a \neq 0, \textit{ and } b \neq 0, & (2.21a) \\ y(0)\,e^{a\,t}, \; \textit{if } a \neq 0, \; \textit{and } b = 0, & (2.21b) \\ y(0) + b\,t, \; \textit{if } a = 0, \; \textit{and } b \neq 0, & (2.21c) \\ y(0), \; \textit{if } a = b = 0, & (2.21d) \end{cases}$$

for each $t \in \mathrm{T}$*, and* $y(t \mid t = 0) = y(0)$ *is an arbitrary element of* Y.

Proof See Propositions 2.1 and 2.2. □

The solution to a particular scalar linear autonomous ODE is a function $y(t) = y(t; y(0); a, b)$, mapping $\mathrm{T} \times \mathrm{Y} \times \Phi \rightarrow \mathrm{Y}$, thus depending on time, on an arbitrary element $y(0)$ of set Y, and on the values of parameters. Figures 2.2 and 2.3 illustrate the behavior over time for some representative cases, associated to different values of the parameters, and for several initial arbitrary values $y(0)$. For every case, each subfigure presents the time evolution of solutions in a $(t, y(t))$-diagram, called **orbits**, and in a $(y, \dot{y})$ diagram. The projection of the solution onto the y-axis is called a **phase diagram**, and it is a representation of the direction of change of $y(t) \in \mathrm{Y}$ when time increases. This relationship between y and the corresponding sign of $\dot{y}$ can be observed clearly in the right-hand subfigures.

Figure 2.2 illustrates trajectories (also called paths or orbits) and phase diagrams for **homogeneous** ODE, $\dot{y} = a\,y$, for different signs of a: for $a < 0$ in Fig. 2.2a, and for $a > 0$, in Fig. 2.2b. We observe, in the left-hand subfigures, that when $a < 0$, for any initial value $y(0)$, all trajectories converge asymptotically, or are permanently equal, to $y = 0$, and when $a > 0$ all the trajectories diverge to $+\infty$ if $y(0) > 0$ and to $-\infty$ if $y(0) < 0$. This behavior is consistent with the right-hand subfigures, where line $\dot{y} = a\,y$ is negatively slopped in Fig. 2.2a, and is positively slopped in Fig. 2.2b. In both cases, if $y(0) = 0$, the trajectories are stationary, that is $y(t) = y(0) = 0$ for

[14] It became standard to use the Newtonian notation for the derivative for time-dependent ODE.

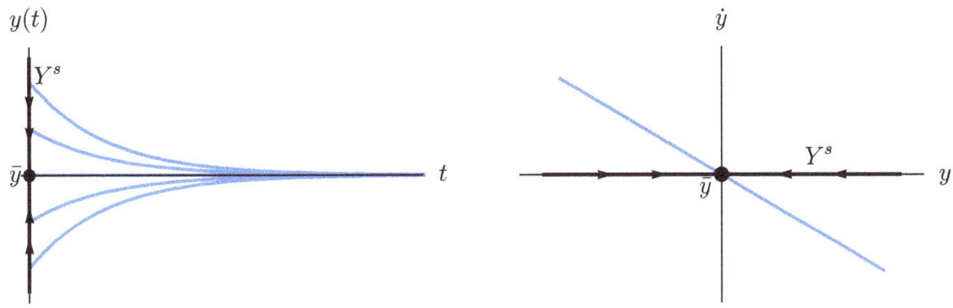

(a) Case: $a < 0$

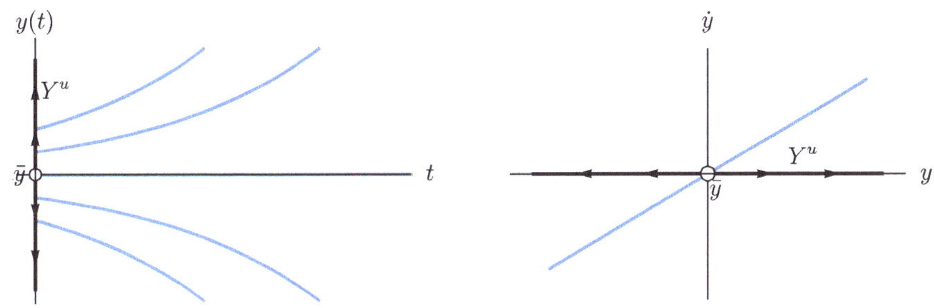

(b) Case: $a > 0$

Fig. 2.2 Trajectories for several values of $y(0)$ (left), and representations (right) for the ODE $\dot{y} = a\,y$. The phase diagrams are represented by the thick arrowed segments

every $t \in [0, \infty)$. In the associated right-hand subfigures the line $\dot{y} = a\,y$ crosses the horizontal axis at this same point, $y = 0$, indicating that this is a steady state.

Figure 2.3 illustrates the trajectories and phase diagrams for **non-homogeneous** ODE, $\dot{y} = a\,y + b$, for different signs of a, and b. Figure 2.3a illustrates a case in which $a < 0$ and $b > 0$, and Fig. 2.3b depicts a case in which $a > 0$ and $b < 0$. Comparing to the analogous homogenous cases (in Fig. 2.2a and b) we observe that the trajectories are qualitatively similar but quantitatively different, as far as the asymptotic behavior is concerned: if $a < 0$, independently of $y(0)$, all the trajectories converge asymptotically to a point $\bar{y} = -\frac{b}{a}$, and if $a > 0$, all the trajectories again diverge to $+\infty$ if $y(0) > -\frac{b}{a}$ and to $-\infty$ if $y(0) < -\frac{b}{a}$. Furthermore, in both cases, if $y(0) = -\frac{b}{a}$, the trajectories are stationary, that is $y(t) = -\frac{b}{a}$ for every $t \in [0, \infty)$. Comparing the right-hand side subfigures of Figs. 2.3 and 2.2, we observe that the first displays a parallel vertical shift, which leads only to a change in the steady state level and not to the dynamics of the solutions, for the cases in which a has the same sign.

If $a = b = 0$, then the ODE degenerates to $\dot{y} = 0$, and the solution, $y(t) = y(0)$, is **always** time-invariant, i.e., it is independent from the exogenous variable t. Intu-

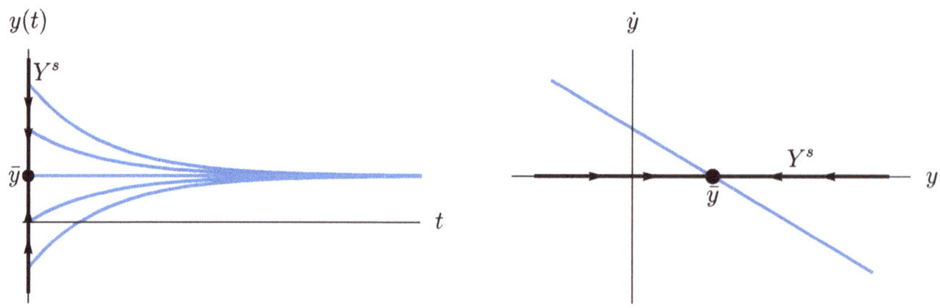

(a) Case: $a > 0$ and $b < 0$

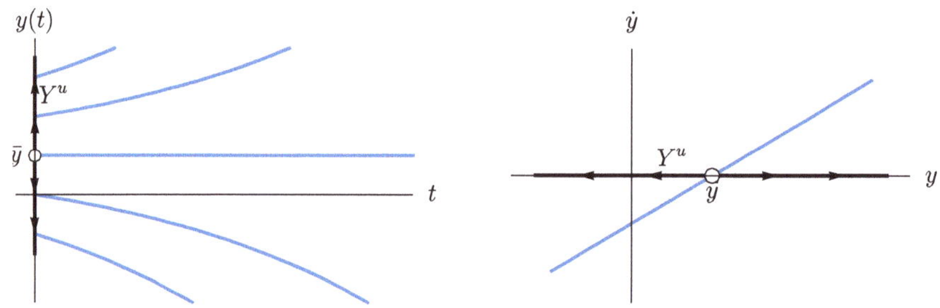

(b) Case: $a < 0$ and $b > 0$

Fig. 2.3 Trajectories for several values of $y(0)$ (left), and phase diagram (right) for the ODE $\dot{y} = a\,y + b$

itively, we can say that there are no dynamics, and that this corresponds to a boundary case between convergence and divergence.

Exercise 2.6 Illustrate the solutions for the cases in which $a = 0$ and $b \neq 0$.

2.3.2 *Qualitative Theory and Dynamics*

Characterizing the dynamics generated by the ODE refers to describing the behavior of the flows $\big(y(t)\big)_{t\in \mathrm{T}}$—also called trajectories, orbits or paths—within set Y that are generated by an ordered sequence of times $t \in \mathrm{T}$. As we just saw, the qualitative behavior of those paths is determined by the parameters, and is also influenced by $y(0)$. If we write the ODE as $\dot{y} = f(y, \varphi)$, where $\varphi \in \Phi \subseteq \mathbb{R}^p$ is a parameter set, the dynamic properties of the solution depend on the properties of $f(y, \cdot)$. Next, we introduce some concepts and results from the **qualitative theory of ODE**.

Definition 2.2 A **steady state** (or equilibrium point) is an element of Y satisfying $f(y, \cdot) = 0$. The set of steady states is

$$y^{ss} = \{y \in \mathrm{Y} : f(y, \cdot) = 0\}.$$

We can distinguish **stationary**, or time-independent, solutions to an ODE from **non-stationary**, or time-dependent solutions. If $f(y)$ is such that all elements of Y are steady states, that is if $y^{ss} = \mathrm{Y}$, then stationarity of solutions is a generic feature of an ODE. If only for specific values of $y(0)$ there is a stationary orbit, then most solutions to an ODE are non-stationary, and a further characterization of the solution to an ODE is needed.

Let y^{ss} be non-empty, and let $\bar{y} \in y^{ss}$ be a steady state. We can characterize a generic solution of an ODE by qualifying $\bar{y}$ according to its **stability properties**.

Definition 2.3 A steady state $\bar{y} \in y^{ss}$ is **asymptotically stable** if, for any $y(0) \in \mathrm{Y}$, the flow or trajectory $(y(t))_{t \in \mathrm{T}}$ generated by the ODE $\dot{y} = f(y, \cdot)$ has the property that a forward trajectory converges to the steady state $\bar{y}$,

$$\lim_{t \to \infty} y(t) = y(\infty) = \bar{y}.$$

A steady state is **unstable** if, for any $y(0)$ in a neighborhood of $\bar{y}$, a forward trajectory does not converge to the steady state $\bar{y}$. However, a backward trajectory can converge to the steady state, $\lim_{t \to -\infty} y(t) = y(-\infty) = \bar{y}$.

The stability properties of steady states lead to the introduction of a partition of Y into **invariant subsets**.

Definition 2.4 We call **attractor set** (or basin of attraction) to the invariant subset of Y, denoted by $\mathcal{W}^s$, such that *for each* $y(0) \in \mathcal{W}^s / y^{ss}$, the solution path converges asymptotically to a steady state in $y^{ss} \in \mathrm{Y}$. We call **repelling set** to the invariant subset of Y, denoted by $\mathcal{W}^u$, such that *for each* $y(0) \in \mathcal{W}^u$, and $y(0) \notin y^{ss}$, the solution path is asymptotically unbounded. We call **center set** to the set of points such that the solution is neither stable nor unstable, and denote it by $\mathcal{W}^c$.

The qualitative theory of ODE includes two important geometrical representations: the phase, and the bifurcation diagrams:

Definition 2.5 A **phase diagram** is a graphical representation of the set Y specifying the steady states, and the attracting and/or repelling invariant sets. The nature of the invariant sets is indicated by the local forward time evolution of the flow of y within set Y.

As $f(y, \varphi)$ is a function of parameters $\varphi \in \Phi$, the stationarity of solutions, the steady state set specification, and the stability properties of the steady states, depend on the values of $\varphi \in \Phi$. The number and stability properties of the steady states can change for different values of $\varphi \in \Phi$. If the number of steady states and their stability properties do not change for different values of φ, we say the dynamics is **generic**. If they change we are in the presence of bifurcations.

Definition 2.6 A **bifurcation point** is pair $(y, \varphi) \in y^{ss} \times \Phi$ such that a small change in φ induces a qualitative change in the number and the stability properties of steady states.

Definition 2.7 A **bifurcation diagram** is a graphical representation of the steady state set $y^{ss}(\varphi)$, as a function of $\varphi \in \Phi$, together with an identification of their stability properties.

Regarding the linear autonomous ODE (2.20), we readily observe in Eq. (2.21) that solutions are stationary only if $a = b = 0$ (see Eq. (2.21c)). In all other cases, the solutions are, in general, non-stationary. Consider first the **homogeneous equation** $\dot{y} = a\, y$, and its geometry in Fig. 2.2. The existence and number of **steady states** depend on the parameter a: if $a = 0$ there is an **infinite number** of steady states, consisting in all elements of Y, that is $y^{ss} = \mathrm{Y}$, and if $a \neq 0$ there is a **single** steady state, $\bar{y} = 0$, and therefore $y^{ss} = \{0\}$ if $0 \in \mathrm{Y}$, or no steady state, i.e., $y^{ss} = \emptyset$, if $0 \notin \mathrm{Y}$. If $a \neq 0$ and $0 \in \mathrm{Y}$ then a steady state exists and its **stability properties** depend on the sign of a: (1) if $a < 0$ then $\lim_{t\to\infty} y(t) = \bar{y} = 0$, for any $y(0) \neq 0$, and the steady state $\bar{y} = 0$ is asymptotically stable, and the solution paths and the **phase diagram** are depicted in Fig. 2.2a, or (2) if $a > 0$ then

$$\lim_{t\to\infty} y(t) = \begin{cases} \pm\infty\ , & \text{if } y(0) \neq 0, \\ 0, & \text{if } y(0) = 0, \end{cases}$$

and the equilibrium point $\bar{y} = 0$ is unstable, as shown in Fig. 2.2b.

Therefore, if $a \neq 0$, and $\bar{y} = 0 \in \mathrm{Y}$, there are only two kinds of possible **invariant sets**. If $a < 0$ the basin of attraction for $\bar{y}$ is the whole set Y, that is $\mathcal{W}^s = \mathrm{Y}$, and the repeller set $\mathcal{W}^u$ is empty, as in the right Fig. 2.2a. In this case, we say $\bar{y}$ is **globally asymptotically stable**. If $a > 0$ then $\bar{y}$ is repelling and unstable, and, because $\mathcal{W}^u = \mathrm{Y}$ and $\mathcal{W}^s$ is empty, instability has a global nature, as can be observed in the right Fig. 2.2b. If $a = 0$ we have a center set that is coincident with the image space, thus $\mathcal{W}^c = \mathrm{Y}$.

For the **non-homogeneous** ODE $\dot{y} = a\, y + b$, the existence and number of steady states depends on a and b: (1) if $a \neq 0$ we have one **unique** steady state, $y^{ss} = \{\bar{y}\}$ where $\bar{y} \equiv -\frac{a}{b}$ if $\bar{y} \in \mathrm{Y}$; (2) if $a = b = 0$ there is an **infinite number** of steady states, $y^{ss} = \mathrm{Y}$; or (3) if $a = 0$ and $b \neq 0$ steady states do not exist, and therefore y^{ss} is empty. The qualitative dynamics is similar to the homogeneous equation, except for the last case in which $a = 0$ and $b \neq 0$. For $a \neq 0$, and assuming that $-\frac{b}{a} \in \mathrm{Y}$, the non-homogeneous ODE only introduces quantitative differences when comparing to the related homogeneous ODE: first, the steady state is shifted from $\bar{y} = 0$, if $b = 0$, to $\bar{y} = -\frac{b}{a}$, if $b \neq 0$, and, second, the stability behavior is qualitatively the same but instead relative to the equilibrium point $\bar{y} = -b/a$, with asymptotically stability if $a < 0$ and instability if $a > 0$. The left graphs in Fig. 2.3a and 2.3b show several solutions paths for the cases in which $a < 0$ and $a > 0$, and the right graph in Fig. 2.3a shows the global asymptotically stable invariant attractor set $\mathcal{W}^s$ for the first case, and right graph in Fig. 2.3b shows the unstable repeller invariant set $\mathcal{W}^u$ for the second case. For $a = 0$, while in the homogenous ODE case (i.e., if $b = 0$), the solution is stationary and there is an infinite number of steady states (all the elements of Y), in the non-homogeneous ODE case (i.e., if $b \neq 0$), **there are no steady states** and the solution to the ODE is always non-stationary.

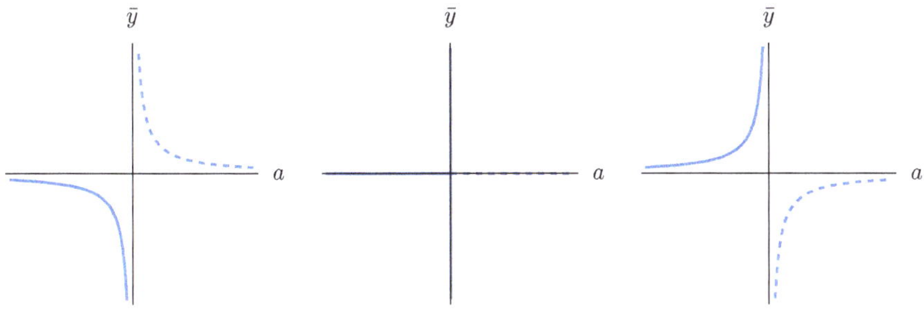

Fig. 2.4 Bifurcation diagrams for ODE $\dot{y} = a\,y + b$, with $b < 0$, $b = 0$ and $b > 0$, where solid (dashed) lines represent asymptotically stable (unstable) steady states

Table 2.1 Bifurcation analysis for the linear scalar ODE (2.20)

	$a < 0$	$a = 0$	$a > 0$
$b = 0$	One asymptotically stable steady state	Infinite number of steady states	One unstable steady state
$b \neq 0$	One asymptotically stable steady state	Zero steady states	One unstable steady state

Summing up: first, in both Figs. 2.2a and 2.3a, there is one unique asymptotically stable steady state (represented by a solid circle), and in both Figs. 2.2b and 2.3b, there is one unique unstable steady state (represented by an empty circle); second, the y-axes are geometrical representations of attractor sets, in both Figs. 2.2a and 2.3a, and of repelling sets in both Fig. 2.2b and b, and, at last, in right graphs in Figs. 2.2 and 2.3, the **phase diagrams** are the geometrical representation of the solutions in which time is represented implicitly, and coincide with their projections onto the $y-$axes, with an arrow indicating the direction of the local variation of the solution $y(t)$ when time moves forward.

Figure 2.4 presents **bifurcation diagrams** for the cases in which $b < 0$ (left graph), $b = 0$ (center graph), and $b > 0$ (right graph). If we take the case in which b is fixed and negative (left panel) we observe that the steady state depends on a in the way we already discussed: the steady state is unique and asymptotically stable if $a < 0$ (represented by a solid curve), a steady state does not exist if $a = 0$, and the steady state is unique but unstable if $a > 0$ (represented by a dashed curve). The **bifurcation Table** 2.1 summarizes the types of possible dynamics for the scalar linear autonomous ODE for all values of the parameters $\Phi = (a, b)$:

In economic applications, **comparative dynamics** refers to the characterization of the changes in the solution of an ODE problem when a parameter (or an exogenous variable) varies such that a bifurcation point (in this case $a = 0$) is not crossed. The aim of those exercises is to study the quantitative response to shocks in the absence of qualitative changes in the dynamics, that is, when the phase diagram—the number of steady states and their stability properties—does not change qualitatively.

2.3.3 Dynamic Scalar ODE Problems

As we saw, an ODE problem is defined by an ODE together with a side condition. In economic dynamic models, the most frequent side conditions take the form of an initial condition, when the point $(0, y(0)) = (0, y_0) \in \mathrm{T} \times \mathrm{Y}$ is fixed, or of a terminal condition, when the point $(T, y(T)) = (T, y_T) \in \mathrm{T} \times \mathrm{Y}$ is fixed, or both initial and terminal conditions. When there are terminal conditions, we distinguish between finite-horizon problems, when $\mathrm{T} = [0, T]$, with a finite T, and infinite-horizon problems, when $\mathrm{T} = [0, \infty)$ and $\lim_{t\to\infty} y(t) = y_\infty$.[15]

An **initial-value problem**, for the scalar linear autonomous linear ODE (2.20), is defined by

$$\begin{cases} \dot{y} = a\, y + b, & (2.22a) \\ y(0) = y_0 \in \mathrm{Y}, & (2.22b) \end{cases}$$

where y_0 is fixed. The particular solution to this problem is

$$y(t) = \begin{cases} \bar{y} + \big(y_0 - \bar{y}\big)\, e^{a\,t}, \text{ if } a \neq 0, & (2.23a) \\ y_0 + b\,t, \text{ if } a = 0, & (2.23b) \end{cases}$$

for each $t \in \mathrm{T}$, where $\bar{y} = -\frac{b}{a}$ is the unique steady state $y^{ss} = \{\bar{y}\}$. Associated to the same ODE (2.20) we call scalar **forward ODE** to the integral equation

$$y(t) = y_0 + \int_0^t \big(a\, y(s) + b\big) ds,$$

that has the same solution (2.23). Observe that in both cases the solution is unique, in the sense that it displays a single trajectory and not a family of trajectories as the solution to the ODE (2.20). From Definition 2.3, the steady state of equation (2.22a) is asymptotically stable if $a < 0$, and it is unstable if $a > 0$. The qualitative properties of the solution to an initial-value problem, and to the equivalent forward ODE, depend on a: if $a < 0$ then $\lim_{t\to\infty} y(t) = \bar{y}$, for any y_0; if $a > 0$ then $\lim_{t\to\infty} y(t) = \pm\infty$ if $y_0 \neq \bar{y}$; or if $a = 0$ the solution is stationary if $b = 0$ or $\lim_{t\to\infty} y(t) = \pm\infty$ if $b \neq 0$.

[15] Observe that those two types of constraints are consistent with a Dirac-delta constraint as in Eq. (2.10).

A **terminal-value problem**, for the same ODE, is defined by

$$\begin{cases} \dot{y} = a\,y + b, & (2.24a)\\ y(T) = y_T \in \mathrm{Y}, & (2.24b) \end{cases}$$

where $y(T) = y(t|t = T)$ and y_T is fixed. The particular solution to this problem is instead

$$y(t) = \begin{cases} \bar{y} + \big(y_T - \bar{y}\big)\, e^{-a\,(T-t)}, \text{ if } a \neq 0, & (2.25a)\\ y_T + b\,(t - T), \text{ if} a = 0, & (2.25b) \end{cases}$$

for $t \in \mathrm{T} = [0, T]$. In this case, the initial value $y(0) = \bar{y} + \big(y_T - \bar{y}\big)\, e^{-a\,T}$ is endogenous, and not exogenously fixed as in the initial-value problem's solution. Associated to the same ODE (2.20), we call scalar **backward ODE** to the integral equation

$$y(t) = y_T - \int_t^T \big(a\,y(s) + b\big)ds,$$

that has the same solution as in Eq. (2.25). In macro-economic and growth models, a frequent **infinite-horizon problem** is, assuming that $\mu > 0$,

$$\begin{cases} \dot{y} = a\,y + b, \text{ for} t \in \mathrm{T} = [0, \infty), & (2.26a)\\ \lim_{t\to\infty} e^{-\mu\,t}\, y(t) = 0. & (2.26b) \end{cases}$$

In this literature, if a solution exists and is unique it is said to be **determinate**, and if more than one solution exists then it is said to be **indeterminate**.

Proposition 2.6 *Consider problem* (2.26a)–(2.26b) *and assume that* $\mu > 0$. *Then the solution always exists and can be of one of the two types: (1) if* $a < \mu$ *then the solution*

$$y(t) = \bar{y} + (y(0) - \bar{y})\, e^{a\,t}, \textit{ for } t \in [0, \infty), \tag{2.27}$$

is indeterminate, because $y(0) \in \mathrm{Y}$ *is arbitrary; or (2) if* $a \geq \mu$ *then the solution*

$$y(t) = \bar{y} \textit{ for all } t \in [0, \infty) \tag{2.28}$$

is determinate, and stationary (time-independent).

Proof The general solution to Eq. (2.26a) is $y(t) = \bar{y} + \big(y(0) - \bar{y}\big)\, e^{a\,t}$ for $t \in \mathrm{T} = \mathbb{R}_+$. Then $\lim_{t\to\infty} e^{-\mu\,t}\, y(t) = \lim_{t\to\infty} \Big(\bar{y}\, e^{-\mu\,t} + \big(y(0) - \bar{y}\big)\, e^{-(\mu-a)\,t}\Big) = 0 + \lim_{t\to\infty} \big(y(0) - \bar{y}\big)\, e^{-(\mu-a)\,t}$. If $\mu - a < 0$ the side condition (2.26b) is satisfied, for any value of $y(0)$, which implies that the solution is in Eq. (2.27). But, if $\mu - a \geq 0$ the side condition is satisfied if and only if $y(0) = \bar{y}$, implying the solution is in Eq. (2.28). In the first case, the solution to the problem is indeterminate, and in the second, it is determinate. □

In applications, the asymptotic side condition (2.26b) formalizes several types of constraints: ruling-out speculative bubbles, solvability or sustainability conditions, or transversality conditions in infinite-horizon optimal control problems.

2.3.3.1 Dynamics and Classification of Dynamic General Equilibria

Dynamic general equilibrium (DGE) theory introduces a specific nomenclature for variables concerning the information on its level at time $t = 0$: a variable is said to be **pre-determined** if it is observed (there is perfect information), or a variable is said to be **non-pre-determined** if its level is anticipated (there are only beliefs about its level). Regarding its asymptotic level (at $t \rightarrow \infty$): a variable is said to be **asymptotically stationary** if it is converges to a steady state, or it is said to be **asymptotically non-stationary** if it is asymptotically unbounded. In growth theory, if a variable is asymptotically unbounded it is said to evolve along or converge to a to a balanced growth path (BGP) if its time derivative is constant or asymptotically stable. At last, regarding its uniqueness properties, a variable is said to **determinate** if the particular solution to the equivalent forward or backward ODE is unique, or it is said to be **indeterminate** if the particular solution is non-unique.

In models displaying non-stationary solutions, the literature frequently distinguishes **transitional dynamics**, when there is convergence to a steady state or to a BGP, from **long-run dynamics** when a variable is asymptotically unbounded.

Therefore, the specific characterization of economic equilibria in macro models depends on the existence and the stability properties of steady states, belonging to set Y. Some examples are frequent in the literature. First, if a variable is pre-determined the trajectory described by the solution is always determinate, however, it can be stationary (if $a < 0$) or non-stationary (if $a > 0$). The first case is frequent in models with adaptive expectations in which there is only transitional dynamics. For instance, if $\dot{p} = \lambda(\bar{p} - p)$, if $\lambda > 0$, where p is the log of price. The second case is frequent in endogenous growth models in which there is no transitional dynamics and there is only long-run dynamics. For instance this is the case if $\dot{k} = A\,k$, where k is the capital stock. Second, if a variable is non-predetermined the trajectory can be determinate, if $y(0)$ is determined uniquely, or can be indeterminate if $y(0)$ can take any value within set Y. For scalar models, the solutions are usually stationary, and there is neither transitional nor long-run dynamics, if the terminal condition is of the type $\lim_{t\rightarrow\infty} y(t)e^{-\mu t} = 0$ for $\mu > 0$. Table 2.2 summarizes the type of equilibria, we could expect in a dynamic general equilibrium model (DGE).

Table 2.2 Classification of equilibrium paths in DGE models

y	$a < 0$	$a = 0$	$a > 0$
Pre-determined	Determined Stable	Determined Stationary	Determined Non-stationary
Non-pre-determined	Indeterminate	Ambiguous	Determined

2.3.3.2 Comparative Dynamics

In this section, we provide some detail on comparative dynamic exercises—that is, finding the multipliers—for permanent changes in the parameters a and b for ODE problems with determinate solutions. Consider first the problem for a pre-determined variable y that is governed by ODE (2.20), where $a < 0$ and $b \neq 0$. This ODE has one unique steady state $\bar{y} = -\frac{b}{a}$. The multipliers are different for initial-value and terminal-value problems.

First, for an initial-value problem, we take derivatives to Eq. (2.23) to obtain

$$\frac{\partial y(t)}{\partial a} = \frac{\partial \bar{y}}{\partial a}\left(1 - e^{a\,t}\right) + (y(t) - \bar{y})\,t, \text{ for each } t \in \mathrm{T},$$

and

$$\frac{\partial y(t)}{\partial b} = \frac{\partial \bar{y}}{\partial b}\left(1 - e^{a\,t}\right), \text{ for each } t \in \mathrm{T},$$

where $\frac{\partial \bar{y}}{\partial a} = -\frac{\bar{y}}{a}$ and $\frac{\partial \bar{y}}{\partial b} = \frac{\bar{y}}{b}$. We observe that in both cases there is a change in the steady state, a common time-varying transition effect, given by the difference $\left(1 - e^{a\,t}\right)$, and in the case of a the multiplicative nature of the shock introduces an additional time-varying effect $(y(t) - \bar{y})\,t$. This effect stems from the fact that the transition dynamics of the pre-shock solution may be still operating before the shock takes place. If the comparative dynamics multipliers are computed relative to the steady state then this effect disappears, and we have

$$\frac{\partial y(t)}{\partial a} = \frac{\partial \bar{y}}{\partial a}\left(1 - e^{a\,t}\right), \text{ for each } t \in \mathrm{T}.$$

If $\mathrm{T} = [0, \infty)$, and $a < 0$ there is no impact response, as $\frac{\partial y(0)}{\partial a} = \frac{\partial y(0)}{\partial b} = 0$, and the long run multipliers are equal to the derivatives of the steady state function, $\frac{\partial y(\infty)}{\partial a} = \frac{\partial \bar{y}}{\partial a}$ and $\frac{\partial y(\infty)}{\partial b} = \frac{\partial \bar{y}}{\partial b}$. For $t \in (0, \infty)$, the multipliers are time-dependent, that is transitional.

Second, consider now a scalar ODE model in which the state variable is non-predetermined, but there is determinacy. The solution to the infinite-horizon terminal value problem is given in Eq. (2.28). In this case, the comparative dynamics expressions are

$$\frac{\partial y(t)}{\partial a} = \frac{\partial \bar{y}}{\partial a}, \text{ for cach } t \in (0, \infty)$$

and

$$\frac{\partial y(t)}{\partial b} = \frac{\partial \bar{y}}{\partial b} \text{ for each } t \in (0, \infty),$$

and there is not transition dynamics: the solution anticipates perfectly the terminal value of y.

2.3.3.3 Applications

Next we present some applications to demography, market adjustment, macroeconomics, and growth economics.

Example 2.5 (*Population dynamics: the Malthusian model*) The simplest model of population dynamics[16] assumes that the size of a population follows a constant growth rate process. Denoting the size of the population, at time t, by $N(t)$, the rate of change, $\eta = \beta - \mu$, is equal to the difference between the time- and age-independent fertility rate, $\beta \geq 0$, and the mortality rate, $\mu \geq 0$. In the absence of emigration or immigration, the population size follows the ODE

$$\dot{N} = \eta N, \quad \text{for } t \in [0, \infty). \tag{2.29}$$

This ODE is usually interpreted as a forward equation, where the initial population, $N(0) = N_0$, is fixed. The population size at time t is

$$N(t) = N_0 \, e^{\mu t}, \text{ for } t \in (0, \infty).$$

Therefore: the population becomes asymptotically unbounded if $\mu > 0$, and stationary for all $t \geq 0$ if $\mu = 0$, with $N(t) = N_0$ for all t, or converges to extinction, that is $\lim_{t \to \infty} N(t) = 0$, if $\mu < 0$.

A backward interpretation provides the path that the population size should follow in order to reach a target N_T at time $T > 0$:

$$N(t) = N_T \, e^{-\mu (T-t)}, \text{ for } t \in [0, T].$$

We can discuss the realism of that population target by comparing $N(0)$ with an observed population size N_0.

Example 2.6 (*Market adjustment dynamics*) The formalization of the "principle of exchange" by Jevons (1907)[17] was possibly the first application of ODE to economics. It is concerned with the dynamics of adjustment of prices and quantities towards attaining an equilibrium, in a market for any good. This topic remained the main early application of ODE in economics in the early twenty century, as can be seen in the space dedicated to it in Samuelson (1947), in the existence of contributions until the 1970s, as Beckmann and Ryder (1969), and in the numerous versions of the model in Gandolfo (1997). Two types of adjustments were considered. A Walrasian adjustment takes place when the disequilibrium between quantities demanded and supplied leads to a change in the price biding, formalized by the ODE

$$\dot{p} = f\Big(Q_d(p) - Q_s(p)\Big) \tag{2.30}$$

---

[16] See Banks (1994), or Hritonenko and Yatsenko (2013, Sect. 6.1.1), and Bacaër (2011) for historical notes.

[17] See Grattan-Guinness (2009).

where $Q_d(\cdot)$ and $Q_s(\cdot)$ are the quantities demand and supplied at price p. In this model, $f'(\cdot) > 0$ and $f(0) = 0$, and $Q_d'(p) < 0$ and $Q_s'(p) > 0$. A Marshallian adjustment takes place when the disequilibrium between the bid and ask prices in resolved by changes in quantities, formalized by the ODE

$$\dot{q} = g\Big(P_d(q) - P_s(q)\Big)$$

where $P_d(\cdot)$ and $P_s(p)$ are the inverse demand and supply functions. In this model, $g'(\cdot) > 0$ and $g(0) = 0$, and $P_d'(p) < 0$ and $P_s'(p) > 0$. Consider a simple version of the Walrasian adjustment model, in which all the functions are linear. Specifically, assume that $Q_d = d_0 - d\, p$, $Q_s = s_0 + s\, p$, and $f(\Delta Q) = \phi\, \Delta Q$, where all parameters are positive, and $d_0 > s_0$. Then the price adjustment Eq. (2.30) is the linear forward ODE $\dot{p} = \lambda\, (\bar{p} - p)$, where $\lambda = \phi\, (d + s) > 0$. The equilibrium price, at which transactions take place, is $\bar{p} = \frac{d_0 - s_0}{d+s}$, and the instantaneous price level announcements, over the *tâtonement* process, is $p(t) = \bar{p} + (p_0 - \bar{p})\, e^{-\lambda t}$ for $t \in [0, \infty)$. The steady state equilibrium price and quantities $(\bar{p}, \bar{q})$, where $\bar{q} = Q_d(\bar{p}) = Q_s(\bar{p})$, are reached only asymptotically. Computing the half-life, that is the time taken to halve the initial disequilibrium $p_0 - \bar{p}$, we find $\tau = \frac{\log 2}{\lambda}$, meaning that the adjustment is faster if ϕ, d and s are greater.

Example 2.7 (*Linear capital accumulation dynamics*) In equilibrium, capital accumulation dynamics is driven by the equality between gross real investment and savings at every point in time, $i(t) = s(t)$. Assuming a Keynesian constant savings function, $s(t) = s\, y(t)$, and a constant returns to scale production technology,[18] $y(t) = A\, k(t)$, where $A > 0$ is the marginal (and average) productivity of capital, then savings is $s(t) = s\, Ak(t)$, where $s \in (0, 1)$ is the constant savings rate, and $k(t)$ is the capital stock, at time t. The gross investment is equal to net investment plus depreciation, $i(t) = \dot{k}(t) + \delta\, k(t)$, where δ is the depreciation rate of capital. Therefore, the capital accumulation equation is also governed by a constant growth process, γ, leading to the capital accumulation ODE

$$\dot{k} = \gamma\, k(t), \;\; \text{for } (t, k) \in \mathbb{R}^2_+,$$

where $\gamma \equiv s\, A - \delta$. In general, this equation is specified as a forward ODE, in which capital is a pre-determined variable, with a known initial level $k(0) = k_0$. Therefore, the stock of capital evolves over time according to the function

$$k(t) = k_0\, e^{\gamma t}, \;\; \text{for } t \in \mathrm{T} = [0, \infty).$$

There is economic growth if $s\, A > \delta$, and the economy will evolve along a BGP over time, meaning that there is no transitional dynamics.

[18] The first linear model of capital accumulation is attributed to Frankel (1962).

Example 2.8 (*Budget constraint dynamics*) A fundamental differential equation in economics is the budget constraint equation.[19] Let $a(t) \in \mathbb{R}$ be the net asset position of an economic agent at time t,[20] which is a stock variable that is reported in its balance sheet. If $a > 0$ we say the agent is a net creditor and if $a < 0$ it is a net debtor. Assume that the asset has an instantaneous return $r(t)$ and that the agent has a flow of non-financial income denoted by $y(t)$, and a flow of expenditures, denoted by $e(t)$. One of the iron "laws" of economics is that the change in the asset position, or investment, is equal to savings, $s(t)$, and savings are equal to total income (financial and non-financial) net of expenditure. A general budget constraint is a non-autonomous ODE

$$\dot{a}(t) = s(t) = r(t)\, a(t) + y(t) - e(t), \text{ for every } t \in \mathrm{T}, \tag{2.31}$$

where we call $y(t) - e(t)$ the primary surplus. If the interest rate and the primary surplus are time-independent, we have the autonomous budget constrain

$$\dot{a}(t) = s(t) = r\, a(t) + y - e, \text{ for every } t \in \mathrm{T}. \tag{2.32}$$

This equation allows for finding an answer to the question: given an initial asset position $a(0) = a_0$ what will be net asset positions over time ? We can answer by setting a as a pre-determined variable, and Eq. (2.32) as a forward ODE, and solve the initial-value problem

$$\begin{cases} \dot{a}(t) = r\, a(t) + y - e, \text{ for } t \in \mathrm{T} & (2.33a) \\ a(0) = a_0, \text{ for } t = 0. & (2.33b) \end{cases}$$

The solution is

$$a(t) = \bar{a} + (a_0 - \bar{a})e^{r\,t}, \quad t \in [0, \infty),$$

where $\bar{a} = -\frac{y-e}{r}$ is the steady state net asset position. If $r > 0$ we can observe that the flow $\big(a(t)\big)_{t\in[0,\infty)}$ is determinate and unstable: if $a_0 > \bar{a}$, then $\lim_{t\to\infty} a(t) = +\infty$ and the agent will become a very large (indeed unboundedly large) creditor; if $a_0 < \bar{a}$, then $\lim_{t\to\infty} a(t) = -\infty$ and the agent will become a very large (indeed unboundedly large) debtor; or if $a_0 = \bar{a}$ its asset position will be stationary $a(t) = \bar{a}$ for any t.

---

[19] As far I am aware, the first time a budget constraint has been modeled as an ODE has been in the constraint to the Ramsey (1928) optimal capital accumulation model. In the macroeconomic literature of the late 1960s and 1970s, static ISLM models were extended to dynamics by incorporating the stock-flow dynamics represented by the government budget constraint (see Blinder and Solow (1973) model, following Christ (1968), and the controversy on the long run effects of fiscal policy Infante and Stein (1976) and Blinder and Solow (1976)). Turnovsky (1977) presented the state of the art of this literature before the paradigm of macroeconomics turned to rational expectations models after the Lucas (1976) critique.

[20] It can be a household, a government, or a whole economy. In the first case, a represents the net asset position, in the second, it is usually the net government debt, and in the third, it is the net asset position of a country as regards the rest of the world.

If a represents a ratio of debt over another indexing variable (as population, prices, GDP, etc) then r represents the interest rate net of rate of growth of the indexing variable. If the indexing rate of growth is greater than the interest rate, then it can be the case that $r < 0$. In this case, the dynamics change radically: the process for $\left(a(t)\right)_{t\in[0,\infty)}$ becomes asymptotically stable and converges to $\bar{a}$, for any initial asset position a_0.

Example 2.9 (*Debt sustainability*) We could ask a different question, as in the "debt sustainability" literature: is the present level of the asset position a_0 sustainable? The question posed like this is close to meaningless, before we specify a mathematical measure for "sustainability". One frequently used definition of sustainability is: the currently observed asset position is sustainable if the present value of the asset position is asymptotically equal to zero

$$\begin{cases} \dot{a}(t) = r\,a(t) + y - e, \ \text{for}\, t \in \mathrm{T} = [0, \infty) & (2.34a) \\ \lim_{t\to\infty} a(t)e^{-\rho t} = 0, \ \rho > 0, & (2.34b) \end{cases}$$

where $e^{-\rho t}$ is a discount factor, and $\rho > 0$ is the rate of discount. Using our previous example, the answer regarding the sustainability of the debt depends on the relationship between r and ρ: if $r < \rho$ then the solution is

$$a(t) = \bar{a} + (a(0) - \bar{a})e^{rt}, \ t \in [0, \infty),$$

and any initial asset position, $a(0)$ is sustainable; however, if $r \geq \rho$ then the solution is

$$a(t) = \bar{a}, \ \text{for all}\ t \in [0, \infty),$$

and we should have $a(0) = \bar{a}$. If the agent is an initial debtor, having $a_0 < 0$, then this level of debt is sustainable only if it satisfies, for every point in time, $y - e = -r\,a_0 > 0$, i.e., its income is permanently higher than its expenditure. Differently from Example 2.8, the net wealth is a non-pre-determined variable but can be determinate if $r \geq \rho$ or indeterminate if $r < \rho$.

Exercise 2.7 (Rational expectations) The simplest rational expectation model in Sargent and Wallace (1973) provides one of the first examples of perfect foresight dynamics using ODE. Let p denote the logarithm of the price index and assume that it is governed by the ODE

$$\dot{p} = \beta\,(p - m), \ p : \mathrm{T} \to \mathbb{R}, \tag{2.35}$$

where the adjustment coefficient, β, is positive, and m is the log of the nominal stock of money (money supply controlled by the central bank), that is assumed to be constant. Prove that introducing the initial price $p(0) = p_0$, and interpreting the ODE (2.35) as a forward ODE leads to the meaningless result, from the economic point of view, that an increase in money supply will lead to price deflation. Prove that introducing an absence of speculative bubbles constraint $\lim_{t\to\infty} e^{-\beta t}\,p(t) = 0$, leads to the economic meaningful conclusion that price should be permanently at its steady state level: $p(t) = \bar{p} = m$. In the first case, the price is seen as a predetermined variable, and in the second case as a non-predetermined variable.

2.3.4 Non-Autonomous Time-Dependent Equations

This section addresses time-dependent (regular) scalar linear ODE, with domain in $\mathrm{T} = [0, T) \subseteq \mathbb{R}_+$. The non-autonomous and non-homogeneous ODE, involving function $y : \mathrm{T} \to \mathrm{Y} \subseteq \mathbb{R}$, is[21]

$$\dot{y} = a(t)\, y + b(t), \ \text{ for } t \in \mathrm{T}. \tag{2.36}$$

The time-dependent scalar linear **non-autonomous** ODE is defined in (2.36). The general solution to the **homogeneous** equation is, from Proposition 2.3,

$$y(t) = y(0)\, e^{\int_0^t a(s)\, ds}, \ \text{ if } b(t) = 0 \text{ for all } t \in \mathrm{T} \tag{2.37}$$

and the general solution to the **non-homogeneous** equation is, from Proposition 2.4,

$$y(t) = y(0)\, e^{\int_0^t a(s)\, ds} + \int_0^t e^{\int_s^t a(z)\, dz}\, b(s) ds, \ \text{ if } b(t) \neq 0, \ \text{ for at least one } t \in \mathrm{T}, \tag{2.38}$$

where $y(0) \in \mathrm{Y}$ is arbitrary in both cases. As the number of potential equations is very large, we restrict our attention to particular cases, and related examples, of forward and backward ODE, and to ODE with discontinuous coefficients, that model anticipated and temporary changes in exogenous variables.

2.3.4.1 Forward ODE

The **initial value problem**, for a non-autonomous forward ODE, is

$$\begin{cases} \dot{y} = a(t)\, y + b(t), \ \text{for}\, t \in \mathrm{T} & (2.39\text{a}) \\ y(0) = y_0 \in \mathrm{Y} \text{ for}\, t = 0, & (2.39\text{b}) \end{cases}$$

where y_0 is fixed. From Proposition 2.4, the analytical solution is

$$y(t) = y_0\, e^{\int_0^t a(s)\, ds} + \int_0^t e^{\int_s^t a(z)\, dz}\, b(s) ds, \ \ t \in \mathrm{T}. \tag{2.40}$$

---

[21] If we redefine the independent variable as $t = \tau$ we can transform the non-autonomous scalar linear ODE into a planar non-linear equation: $\dot{y_1} = \lambda(y_2) y_1 + \beta(y_2)$ $\dot{y}_2 = 1$ where $y_2(t) = t$. As we saw in the last section, the solutions to non-autonomous equations can be quite different from the solutions of analogous autonomous equations.

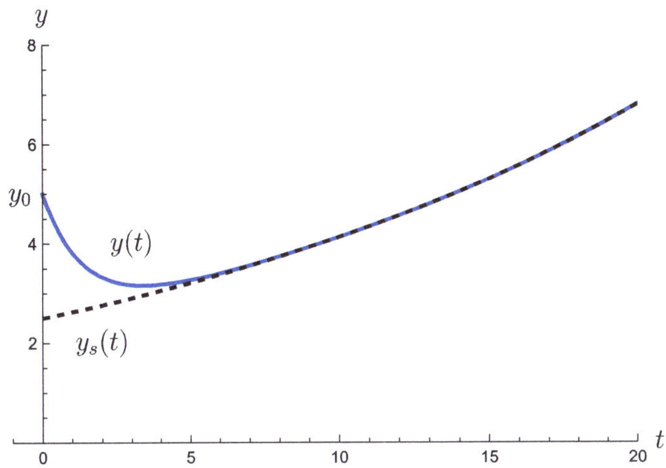

Fig. 2.5 Trajectories of y and y_s from Eq. (2.41), for $\lambda < 0$, $\mu > 0$, and $y_0 > \frac{\beta}{\mu-\lambda}$

If $b(.) = 0$ we say we have time dependence of multiplicative type, and if $a(t) = a$, a constant, and $b(t)$ is a function, we say we have time-dependence of additive type. The following particular cases are frequent in economics.[22] Their solutions display convergence to a long-run trajectory, represented by a function $y_s(t)$, that can or cannot be stationary.

First, a particular **additive** time-dependent model is

$$\dot{y} = \lambda\, y + b\, z(t), \quad \text{for } z(t) = e^{\mu\, t}, \tag{2.41}$$

where $\lambda < 0$, and z grows exponentially at rate of growth $\mu > 0$. The solution to a forward equation with $y(0) = y_0$ fixed, is

$$y(t) = \begin{cases} e^{\lambda\, t}\left(y_0 - \dfrac{\beta}{\mu - \lambda}\right) + \dfrac{\beta}{\mu - \lambda}\, e^{\mu\, t} & \text{if } y_0 \neq \frac{\beta}{\gamma-\lambda}, \\ y_0\, e^{\mu\, t} & \text{if } y_0 = \frac{\beta}{\mu-\lambda}. \end{cases} \tag{2.42}$$

Figure 2.5 illustrates the solution with $y_0 \neq \frac{\beta}{\mu-\lambda}$ converging to an unbounded function growing at the same same rate of growth as $z(t)$, $\lim_{t\to\infty} y(t) = y_s(t) \equiv \frac{\beta}{\mu-\lambda} e^{\mu\, t} = \frac{\beta}{\mu-\lambda}\, z(t)$.

Second, time-dependent exogenous growth processes provide a general case of ODE in which there is an exogenous **multiplicative** time-dependent variable

$$\dot{y} = \gamma(t)\, y, \quad \text{where } \gamma(t) = \bar{\gamma} + (\gamma_0 - \bar{\gamma})\, e^{-\zeta\, t} \tag{2.43}$$

[22] For other possible functions and their meaning in economics, see Gandolfo (1997).

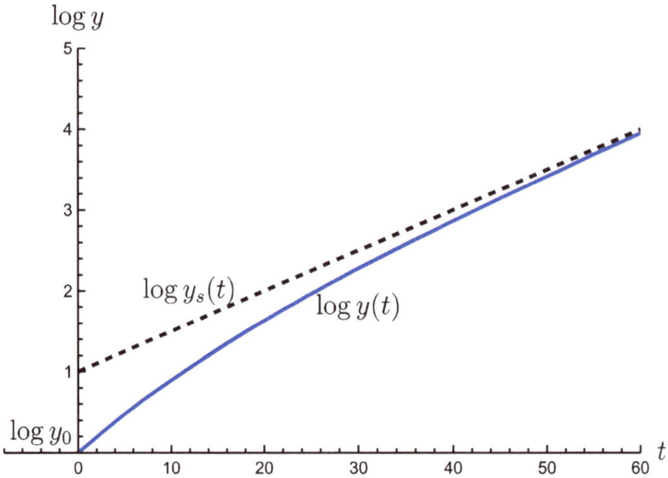

Fig. 2.6 Trajectories of solutions to (2.43) in a log-scale, for $\bar{\gamma} > 0$

where the growth rate, γ, enters multiplicatively. This ODE features a mean reverting process for the rate of growth satisfying, $\dot{\gamma} = \zeta\,(\gamma - \bar{\gamma})$, converging asymptotically to a constant growth rate $\bar{\gamma}$, that is $\lim_{t\to\infty}\gamma(t) = \bar{\gamma}$.

The solution to the initial-value problem, specified by Eq. (2.43) and $y(0) = y_0$, is

$$y(t) = \begin{cases} y_0 \exp\left\{\bar{\gamma}\,t - \dfrac{\gamma_0 - \bar{\gamma}}{\zeta}\left(e^{-\zeta\,t} - 1\right)\right\} & \text{if } \gamma_0 \neq \bar{\gamma} \\ y_0\, e^{\bar{\gamma}\,t} & \text{if } \gamma_0 = \bar{\gamma}. \end{cases} \tag{2.44}$$

If $\gamma_0 \neq \bar{\gamma} > 0$, then $\lim_{t\to\infty} y(t) = y_s(t) \equiv y_0\, e^{\frac{\gamma_0-\bar{\gamma}}{\zeta}+\bar{\gamma}\,t}$, as illustrated in Fig. 2.6, in the logarithmic scale. The solution converges to an exponential function with growth rate $\bar{\gamma}$. If $\bar{\gamma} = 0$, the solution converges to a constant, $\lim_{t\to\infty} y(t) = \bar{y}_s \equiv y_0\, e^{\frac{\gamma_0}{\zeta}}$.

One important conclusion from the above examples is that solutions to scalar non-autonomous ODE (2.36) can be asymptotically unbounded, or can converge to a constant,[23] depending on the specification of functions $a(\cdot)$ and $b(\cdot)$.

Example 2.10 (*Gompertz population model*) The Gompertz (1825) population growth model[24] extends the Malthusian model, in Example 2.5, by addressing its unrealistic property that a population can reach an infinite size asymptotically. Differently from the Malthusian model it assumes that the rate of growth of the population is not constant but instead decays at a constant rate. The resulting model is a particular case of the forward ODE (2.43) in which $\bar{\gamma} = 0$. Then the particular solution is

$$N(t) = N_0 \exp\left\{-\frac{\gamma_0}{\zeta}\left(e^{-\zeta\,t} - 1\right)\right\},$$

[23] That is not a steady state in the sense of Definition 2.2.

[24] See Banks (1994, Sect. 3.1).

where N_0 and γ_0 denote the initial size of the population and of its growth rate, respectively, and ζ is the rate of decay of the growth rate of the population. In this model, the population size will not be asymptotically unbounded as in the Malthusian model.

Example 2.11 *(Debt dynamics with time-varying interest rate and primary deficit)* Denote the nominal government debt by $b(\cdot)$, and observe that if $b(t) > 0$ $(b(t) < 0)$ the government is a net debtor (creditor), and if $\dot{b} > 0$ $(\dot{b}(t) < 0)$ the government has a budget deficit (surplus), at time $t \geq 0$. If we let the nominal interest rate and the primary deficit, denoted by $r(t)$ and $d(t)$ respectively, be time-varying, we can extend the government budget constraint in equations (2.33a) and (2.33b) in order to obtain a general model for the government debt dynamics

$$\begin{cases} \dot{b} = r(t)\, b + d(t), \text{ for } t \in \mathrm{T} & (2.45a)\\ b(0) = b_0 \in \mathbb{R}, \text{ for } t = 0. & (2.45b) \end{cases}$$

The solution is a specific case of function (2.40),

$$b(t) = b_0\, e^{R(t;0)\,t} + \int_0^t e^{R(t;s)(t-s)}\, d(s)ds, \;\; t \in \mathrm{T}, \tag{2.46}$$

where $R(t;\tau) = \frac{1}{t-\tau}\int_\tau^t r(s)\, ds$ is the average interest rate between dates $\tau < t$ and t. Given the initial nominal debt level b_0, and a forecast on the future paths of the interest rate and the primary deficit, $\big(r(t), d(t)\big)_{t\in[0,\infty)}$, function (2.46) provides a forecast, at time $t = 0$, of the future path of the government debt $(b(t))_{t\in[0,\infty)}$. We observe that debt increases with the initial debt level and the cumulative primary deficits, both capitalized by the average interest rates.

2.3.4.2 Backward ODE

A frequent **terminal value problem**, or non-autonomous backward ODE, in economics can be formally specified by

$$\begin{cases} \dot{y} = a(t)\, y + b(t) \text{ for } t \in t \in \mathrm{T} = [0, \infty), & (2.47a)\\ \lim_{t\to\infty} e^{-\int_0^t a(s)ds}\, y(t) = 0 \text{ for } t = 0, & (2.47b) \end{cases}$$

Proposition 2.7 *The (formal) solution to the terminal-value problem* (2.47a)–(2.47b) *is*

$$y(t) = -\int_t^\infty e^{-\int_t^s a(z)dz} b(s)ds, \;\; \textit{for each } t \in \mathrm{T} = [0, \infty). \tag{2.48}$$

Proof The general solution to Eq. (2.47a), from Proposition 2.4, is equivalent to $e^{\int_0^t a(s)\, ds}\, y(t) = y(0) + \int_0^t e^{-\int_0^s a(z)\, dz}\, b(s)ds$, where $y(0) \in \mathrm{Y}$ is arbitrary. Taking the limit $t \to \infty$, using the terminal constraint (2.47b), and solving for $y(0)$, yields

$y(0) = -\int_0^\infty e^{-\int_0^s a(z)\,dz}\, b(s)ds$. Then $e^{\int_0^t a(s)\,ds}\, y(t) = -\int_t^\infty e^{-\int_0^s a(z)\,dz}\, b(s)d$ and multiplying both sides by $e^{-\int_0^t a(s)\,ds}$ yields function (2.48). □

Example 2.12 (*Debt sustainability with time-varying interest rate and primary deficit*) Extending the debt sustainability model of Example 2.9 to the time-varying interest and primary deficit case, we have the terminal-value problem, or forward ODE,

$$\begin{cases} \dot{b} = r(t)\, b + d(t), \text{ for } t \in \mathrm{T} & (2.49a) \\ \lim_{t\to\infty} e^{-\int_0^t r(s)\,ds}\, b(t) = 0, & (2.49b) \end{cases}$$

where the side condition translates mathematically the sustainability requirement. Solving the problem, we find that the sustainable debt level path to be

$$b(t) = -\int_t^\infty e^{-\int_t^s r(z)\,dz}\, d(s)ds, \text{ for each } t \in [0, \infty), \tag{2.50}$$

given the beliefs, at time $t = 0$, on the future paths of the interest rate, and the primary deficit, $\big(r(t), d(t)\big)_{t\in[0,\infty)}$. The sustainable debt level has to be symmetric to the present value of the future primary deficits. That is, if the present value of future primary deficits is negative (positive) the government needs to have a positive (negative) initial asset position—i.e., it needs to have a sovereign wealth or investment fund, at time $t = 0$. Again, while in problem (2.45a)–(2.45b) debt is taken as pre-determined variable, in problem (2.49a)–(2.49b) it is specified as a non-predetermined variable.

Exercise 2.8 (Rational expectations with varying money supply) Take the data of Exercise 2.7 but assume instead that the money supply can vary over time. Prove that, when price is specified as a non-predetermined variable, the price at time t anticipates the beliefs concerning the future money supply: $p(t) = \beta \int_t^\infty e^{\beta\,(t-s)} m(s)\, ds$.

Exercise 2.9 (Asset price and the non-Ponzi-game condition) Let the asset price, $S(t)$, be governed by ODE $\dot{S} = -\mu(t) + r(t)\, S$, where $r(\cdot)$ is the rate of return and $\mu(\cdot)$ is a dividend. Assuming that the non-Ponzi-game condition, $\lim_{t\to\infty} e^{-\int_0^t r(s)ds}$ $S(t) = 0$, is satisfied, find the asset price at $t = 0$.

2.3.4.3 Anticipated and Temporary Variations

A common application of non-autonomous ODE in macroeconomics is related to studying the effects of anticipated changes in exogenous variables. Again, the economic phenomena studied by initial or terminal value problems are different. We can distinguish two types of classes of time-dependent changes: first, concerning the expected **timing** of the shocks, we can distinguish between non-anticipated and anticipated shocks, and, second, concerning the **duration** of the shocks, we can distinguish between temporary and permanent shocks.

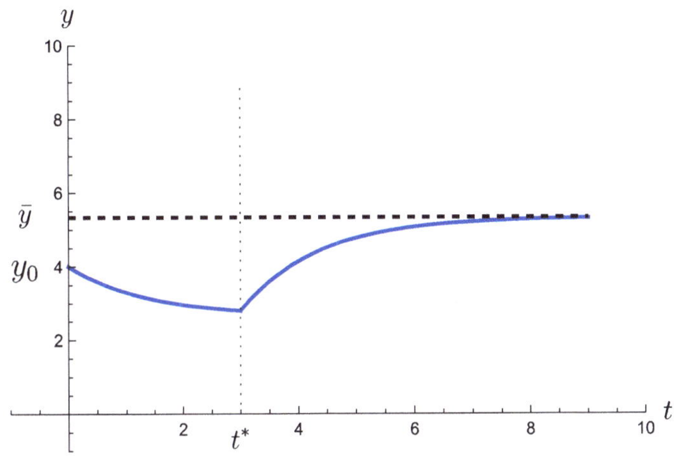

Fig. 2.7 Trajectories for problem (2.51a)–(2.51b) for $a = -0.75$, $b_0 = 2$ and $b_1 = 4$, and $y_0 = 4$, $t^* = 3$

Example 2.13 (*Anticipated variations for a pre-determined variable*) Let y be a pre-determined variable, for which $y(0)$ is observed, and assume there will be an additive shock at a prescribed future date, $t^* > 0$, in an exogenous variable that affects the dynamics of y. This is a special case of the time-additive process modeled by Eq. (2.41):

$$\begin{cases} \dot{y} = a\,y + b(t), \text{ where } b(t) = \begin{cases} b_0 & \text{if } 0 \le t < t^* \\ b_1 & \text{if } t^* \le t < \infty, \end{cases} & \text{(2.51a)} \\ y(0) = y_0 \text{ given}, & \text{(2.51b)} \end{cases}$$

where it is assumed that $a \neq 0$. The solution to this problem is

$$y(t) = \begin{cases} y_0 e^{at} + \dfrac{b_0}{a}\left(e^{at} - 1\right), & \text{if } 0 \le t < t^* \\ y_0 e^{at} + \dfrac{b_0}{a} e^{at} + \left(\dfrac{b_1 - b_0}{a}\right) e^{a(t-t^*)} - \dfrac{b_1}{a}, & \text{if } t^* \le t < \infty. \end{cases}$$

Figure 2.7 illustrates the solution when $b_1 > b_0$. Observe that the solution, at any point in time, capitalizes the past changes of the variable $b(t)$. This means that it only responds to the shock $b_1 - b_0$ **after** it is observed at time $t = t^*$ ($t = 3$ in the example in the figure).

Example 2.14 (*Anticipated shock for a non-predetermined variable*) Differently from Example 2.13 we now assume that y is a non-pre-determined variable that is governed by the same differential equation (2.51a), where the coefficient $a > 0$, that is constrained by the terminal condition

$$\lim_{t \to \infty} y(t)e^{-at} = 0, \tag{2.52}$$

instead of the initial condition (2.51b). In this case, the variable y is non-predetermined and not predetermined as in Example 2.13. The solution to that backward ODE is now

$$y(t) = \begin{cases} -\frac{b_0}{a} - \left(\frac{b_1 - b_0}{a}\right) e^{a(t-t^*)}, & \text{if } 0 \leq t < t^* \\ -\frac{b_1}{a}, & \text{if } t^* \leq t < \infty. \end{cases}$$

Observe that, in this case, the solution for the period $[0, t^*)$ depends on the shock $b_1 - b_0$ **before** it has actually taken place, and at time t^* the solution is not discontinuous, meaning that the effect of the shock has already been anticipated in the previous $[0, t^*)$-period. From this point onwards the solution is permanently stationary at the steady state level $\bar{y} = -b_1/a$.

Example 2.15 (*Anticipated shock for a non-predetermined variable with fixed initial level*) Let us assume that the variable y is a non-predetermined variable, as in Example 2.14, but that we know it has a fixed value y_0 at time $t = 0$. In this case, we say we have a **mixed initial-terminal value problem**, being specified jointly by the ODE (2.51a), by the initial condition (2.51b), and by the terminal condition (2.52).

This problem has no solution in the space of continuous functions of time, $C(\mathbb{R}_+)$. However, if we allow for solutions in the space of piecewise continuous functions of time, $PC(\mathbb{R}_+)$, we have the following solution

$$y(t) = \begin{cases} y_0, & \text{if } t = 0 \\ -\frac{b_0}{a} - \left(\frac{b_1 - b_0}{a}\right) e^{a(t-t^*)}, & \text{if } 0 < t < t^* \\ -\frac{b_1}{a}, & \text{if } t^* \leq t < \infty. \end{cases}$$

We observe that the solution is right-continuous, satisfying

$$y(0) = y_0 \neq y_{0^+} = \lim_{t \downarrow 0} y(t) = -\frac{b_0}{a} - \left(\frac{b_1 - b_0}{a}\right) e^{-at^*}$$

which means there is a discontinuous jump of size $y_{0^+} - y_0$ at time $t = 0$. See Fig. 2.8.

Comparing with the solution to a related initial-value problem, for example in Eq. (2.40), we see that this solution has an anticipating nature: for $0 < t < t^*$ the solution depends on the anticipated levels of the variable $y(t)$ **after** its change, b_1, and the time of the shock, for $t \geq t^*$, it is not influenced by its the value before the change, y_0. The previous two cases illustrate two fundamental types of dynamics in dynamic general equilibrium (DGE) models that usually couple both forward and backward types of dynamics.

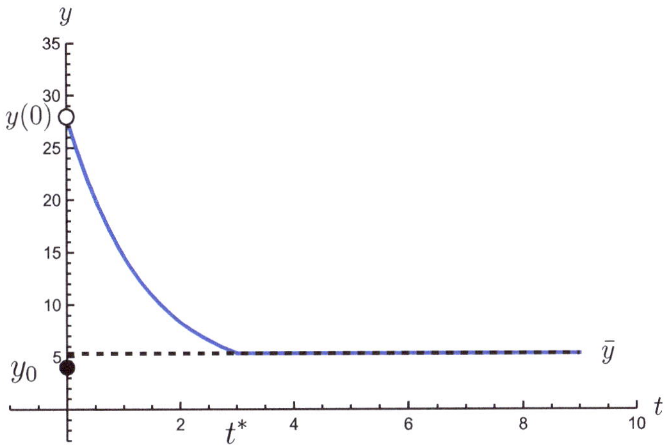

Fig. 2.8 Trajectories for the mixed problem (2.51a)–(2.51b)–(2.52). At time $t = 0$ there is a "jump" from the initial value y_0 to y_{0^+}

2.4 References

There is a huge mathematics literature on scalar linear ODE. Hale and Koçak (1991, Chap. 1) is a great modern textbook. See, for instance, Hubbard (1994) on the history and meaning of differential equations. More detailed presentations of non-autonomous equations can be found in Hale and Koçak (1991, Chap. 2) and Hubbard and West (1991). Gandolfo (1997) is an useful textbook with numerous applications to economics.

Chapter 3
Scalar Non-Linear ODE: The Regular Case

In this chapter, we address regular time-dependent non-linear scalar ordinary differential equations (ODE). Although some of the results are valid for ODE having any independent variable, in a domain $\mathrm{X} \subseteq \mathbb{R}$, in particular for equations having closed form solutions, we focus on time as the independent variable, because most of the contributions come from dynamic systems theory.

Consider the mapping $y : \mathrm{T} \subset \mathbb{R}_+ \rightarrow \mathrm{Y} \subseteq \mathbb{R}$, satisfying the (autonomous) ODE $F\big(\dot{y}, y\big) = 0$. We say the ODE is **regular** if $F\big(\dot{y}, y, \cdot\big)$ is continuous and continuously differentiable in $(\dot{y}, y)$, and non-singular, that is $F_{\dot{y}}(\dot{y}, y) \neq 0$ for every $y \in \mathrm{Y}$.[1] In this chapter we consider specifically ODE of type

$$\dot{y} = f\big(y, \boldsymbol{\varphi}\big), \; f : \mathrm{Y} \times \Phi \rightarrow \mathrm{Y}, \tag{3.1}$$

where $\boldsymbol{\varphi} \in \Phi \subseteq \mathbb{R}^p$ is a set of real-valued p parameters. In this case, the solution is a mapping $y : \mathrm{T} \times \Phi \rightarrow \mathrm{Y} \subseteq \mathbb{R}$.

Most dynamic economic models contain explicitly specified non-linear (and regular) ODE, or non-explicitly specified non-linear ODE, in which only growth and curvature properties of function $f(\cdot)$ are postulated. In most cases, even in the case of explicitly specified ODE, analytic solutions—also called explicit or closed-form solutions—do not exist or are not known. In order to solve those ODE, some questions immediately arise: Does a solution exist? Does it have an explicit analytical solution ? Even if we cannot find an explicit solution, can we characterize its dynamic properties?

The existence problem is beyond the object of this book. However, it has been proved that if $f(y, \cdot)$ is continuous and sufficiently smooth (i.e., it is continuously differentiable up to a higher order) then one unique solution, i.e., a function $y : \mathrm{T} \rightarrow$

[1] Chapters 6 and 7 contain introductions to ODE in which continuity, differentiability or non-singularity assumptions may not hold.

P. B. Brito, *Economic Dynamics and Distributions*, Dynamic Modeling and Econometrics in Economics and Finance 32, https://doi.org/10.1007/978-3-031-94717-9_3

Y, exists and is unique for an interval or for the whole domain T.[2] Furthermore, the degree of differentiability of the solution $y(t)$ is the same as that of $f(y)$.

In Chap. 2, we showed that linear ODE have explicit solutions and they are defined for the whole set T. In the rest of this chapter, we will see that, for some non-linear regular ODE, called **normal forms**, analytic solutions can also be obtained. The knowledge of analytical solutions matters because they allow for characterizing the global dynamic properties of solutions, meaning that we can completely characterize the mapping $T \to Y$, for every point in Y, generated by the solution of non-linear ODE. For ODE that are not in normal form, resorting to a qualitative (or numerical) approximation is the only method to uncover or characterize their solutions.

Analytic approximation is at the core of the **qualitative (or geometrical) theory of ODE**, whether explicit solutions are known or not. For any ODE, we can characterize its solution by topological equivalence to an ODE for which we already know the explicit solution, and, in particular to a linear ODE. Next, we briefly present the main ideas.

The findings of Chap. 2, concerning linear scalar ODE, can be summarized as follows: first, we can find explicit solutions to linear scalar ODE; second, when a steady state exists and it is unique it is either (asymptotically) stable or unstable; third, there is a degenerate case in which there is an infinite number of steady states; and fourth, the whole image space Y is an attractor set when the steady state is stable, it is a repeller set if the steady state is unstable, or it a set of steady states when we have the degenerate case. Therefore, either $\mathcal{W}^s = Y$, or $\mathcal{W}^u = Y$ or $\mathcal{W}^c = Y$.[3] This means that we have a characterization of the **global dynamic** properties associated to a linear ODE, in the sense that we can find the dynamic properties of the mapping $y : T \to Y$, and they are the same for every point of set Y.

Non-linear ODE, as we will show in this chapter, display some differences; first, explicit solutions may or may not be known[4]; second, there can be zero, one, or any **finite number** of steady states; third, **local and global dynamics** can differ. That is, we can have one or more steady states that can be stable, one or more unstable steady states (both called hyperbolic), or we can have steady states that are neither stable nor unstable (called non-hyperbolic). The existence of multiple steady states introduces partitions of the state space Y, in which the mapping $y : T \to Y$ has different dynamic properties depending on location of a point belonging to set Y we are considering. We call local stable manifold of steady state $\bar{y}_i$ to the set of initial points such that the solution to the ODE converges to that steady state in forward time, $\mathcal{W}^{s,loc}_{\bar{y}_i} = \{y(0) \in Y : \lim_{t\to\infty} y(t) = \bar{y}_i\}$ and local unstable manifold of steady state $\bar{y}_i$ to the set of initial points such that the solution of the ODE does not converge to it in forward time but converges in backward time, $\mathcal{W}^{u,loc}_{\bar{y}_i} = \{y(0) \in Y : \lim_{t\to-\infty} y(t) = \bar{y}_i\}$. The (global) stable manifold, $\mathcal{W}^s$, is the union of all or subsets of the local stable manifolds and the (global) unstable manifold, $\mathcal{W}^u$, is the union

[2] See e.g., Hirsch and Smale (1974, Chap. 8).

[3] See the Definition 2.4.

[4] There are complete lists of differential equations with known explicit solutions in Zwillinger (1998), Zaitsev and Polyanin (2003), or Canada et al. (2004).

of all or subsets of local unstable manifolds. The center manifold, $\mathcal{W}^c$, is the union all steady states that are neither stable nor unstable, which in the case of non-linear regular ODE correspond to isolated points of Y, differently from the linear case.

In the rest of this chapter, in Sect. 3.1, we solve and characterize the most relevant normal forms, and, in Sect. 3.2, we summarize the main concepts and results from the qualitative theory of ODE.

3.1 Normal Forms

A **normal form** is the simplest ODE whose exact solution is usually known and represents a whole family of ODE by topological equivalence.[5] The normal forms for regular scalar ODE that we study next are particular cases of the polynomial ODE

$$\dot{y} = f(y, \boldsymbol{\varphi}) \equiv a_0 + a_1\, y + a_2\, y^2 + a_3\, y^3, \tag{3.2}$$

where $\boldsymbol{\varphi} \equiv (a_0, a_1, a_2, a_3) \in \mathbb{R}^4$, and $y : \mathrm{T} \to \mathrm{Y}$, for $\mathrm{Y} \subseteq \mathbb{R}$, and $\mathrm{T} \subseteq \mathbb{R}_+$. Linear scalar ODE, studied in Chap. 2, are particular cases in which $a_2 = a_3 = 0$. In this chapter, we address cases in which at least one of coefficients a_2, or a_3 is different from zero. Hale and Koçak (1991, Chap. 2) show that the most relevant cases are[6] ODE depending on a single parameter (a), as the Ricatti's equation $\dot{y} = a + y^2$ the quadratic Bernoulli equation, $\dot{y} = ay + y^2$, the cubic Bernoulli equation, $\dot{y} = ay - y^3$, and as the Abel equation $\dot{y} = a + y - y^3$, and ODE depending on two parameters (a and b) as Abel's equation, $\dot{y} = a + by - y^3$.

To study those ODE, we start by using an **analytical approach** that consists in finding their explicit solutions (when known). Next, by using the **qualitative (or geometric) approach**, we characterize solutions. The qualitative approach, that can provide an analysis of the solution to an ODE without solving it explicitly, includes the following components: finding the set of steady state $y^{ss} = \{y \in \mathrm{Y} : f(y, \varphi) = 0\}$; characterizing their local dynamics, by computing $f_y(y, \cdot)$,[7] for every $y \in y^{ss}$, by linearizing the ODE in the neighborhood of every steady state; finding bifurcation points; obtaining a partition of the image set Y between stable, unstable and center manifolds; drawing phase diagrams for the qualitatively different cases; and drawing the bifurcation diagram. We adopt the convention of identifying an ODE with its associated bifurcation diagram.

---

[5] In heuristic terms, we say functions $f(y)$ and $g(x)$ are topologically equivalent if there exists a diffeomorphism, i.e., a smooth map h with a smooth inverse h^{-1}, such that if $y = h(x)$ then $h(g(x)) = f(h(x))$. This property may hold globally of locally, as in the Grobman-Hartmann theorem.

[6] See also Golubitsky and Schaeffer (1985, Chap. 4).

[7] In the rest of this chapter, we use the notation for derivatives, $f_y(y, \cdot) \equiv \frac{\partial f(y,\cdot)}{\partial y}$, $f_{yy}(y, \cdot) \equiv \frac{\partial^2 f(y,\cdot)}{\partial y^2}$.

3.1.1 The Ricatti's Equation: Saddle-Node or Fold Bifurcation

The **Ricatti** ODE is defined by

$$\dot{y} = f(y, a) \equiv a + y^2, \text{ for } a \in \mathbb{R}. \tag{3.3}$$

This equation is a particular case of equation (3.2) in which $a_0 = a$, $a_2 = 1$, and $a_1 = a_3 = 0$. It has explicit solutions depending on the sign of parameter a[8] :

$$y(t) = \begin{cases} \dfrac{y(0)}{1 - y(0)\,t} & \text{if } a = 0, & (3.4a) \\ \sqrt{a}\left(\tan\left(\sqrt{a}(t - y(0)^{-1})\right)\right) & \text{if } a > 0, & (3.4b) \\ -\sqrt{-a}\left(\tanh\left(\sqrt{-a}(t - y(0)^{-1})\right)\right) & \text{if } a < 0, & (3.4c) \end{cases}$$

for each $t \in$ T, where $y(0)$ is an arbitrary element of set Y. The time path displayed by the solution to the ODE depends on parameter a: (1) if $a = 0$ then there is a unique steady state $y^{ss} = \{0\}$, the solution tends to the steady state if $y(0) < 0$, but if $y(0) > 0$ the solution becomes infinite in finite time $t^s = \dfrac{1}{y(0)}$.[9] That is, $\lim_{t \to t^s} y(t) = \pm\infty$, if $t > t^s$, or $\lim_{t \to \infty} y(t) = 0$ independently of the level of $y(0)$ if $t < t^s$; (2) if $a > 0$ then there are no steady states, and the solution becomes infinite at a periodic sequence of times $t \in \{t^s, \pi + t^s, 2\pi + t^s, \ldots, n\pi + t^s, \ldots\}$, and $\lim_{t \to n\pi + t^s} y(t) = \pm\infty$ for $n \in \mathbb{N}$; or (3) if $a < 0$ then there are two steady states, $y^{ss} = \{-\sqrt{-a}, \sqrt{a}\}$, and

$$\lim_{t \to \infty} y(t) = \begin{cases} -\sqrt{-a}, & \text{if } {}^{1}\!/_{y(0)} < \sqrt{-a} \text{ or } -\sqrt{-a} < {}^{1}\!/_{y(0)} < \sqrt{-a}, \\ +\infty, & \text{if } {}^{1}\!/_{y(0)} > \sqrt{-a}. \end{cases}$$

From a qualitative or geometric perspective, the ODE (3.3) has the following properties. First, by solving equation $a + y^2 = 0$ for y we find again that: (a) if $a > 0$ there are no steady states, set y^{ss} is empty; (b) if $a = 0$ there is one steady state $\bar{y} = 0$, that is $y^{ss} = \{0\}$; and (c) if $a < 0$ there are two steady states $y^{ss} \in \{\bar{y}_1, \bar{y}_2\}$ where $\bar{y}_1 = -\sqrt{-a} < 0$, and $\bar{y}_2 = \sqrt{-a} > 0$. This last case cannot exist in linear ODE. Second, as $f_y(y, a) = 2\,y$, when there are steady states, we have: if $a = 0$, as $f_y(0, 0) = 0$, then steady state $\bar{y} = 0$ is neither stable nor unstable, or, if $a < 0$, the steady state $\bar{y}_1$ is asymptotically stable, because $f_y(-\sqrt{-a}, a) = -2\sqrt{-a} < 0$, and the steady state $\bar{y}_2$ is unstable, because $f_y(\sqrt{-a}, a) = 2\sqrt{-a} > 0$.

---

[8] See Appendix Sect. 3.5.1

[9] This is different to what occurs in the linear case, v.g., $\dot{y} = y$, whose solution, $y(t) = y(0)\,e^t$, if $y(0) \neq 0$, only tends to an infinite level in infinite time.

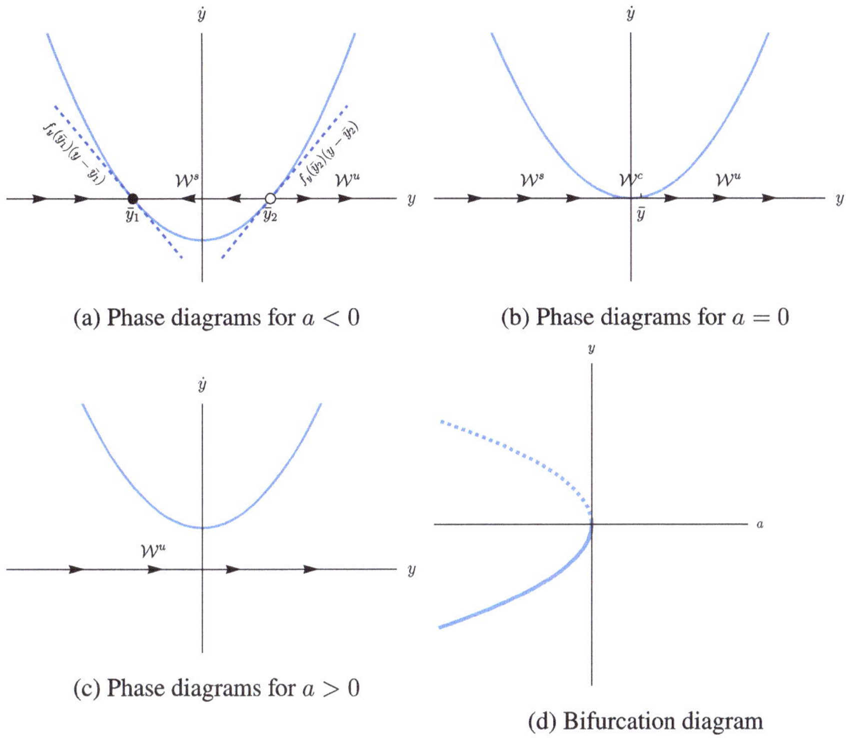

Fig. 3.1 Phase diagrams and bifurcation diagram for Eq. (3.3)

We can now provide a global partition of image set Y (see Fig. 3.1), depending on the sign of a. If (see Fig. 3.1a) $a < 0$ we can partition the state space as $\mathrm{Y} = \mathcal{W}^s \cup \mathcal{W}^u$, where the stable manifold is coincident to the local stable manifold (or the basin of attraction) associated to the steady state $\bar{y}_1$, $\mathcal{W}^s = \mathcal{W}^{s,loc}_{\bar{y}_1} = \{y \in \mathrm{Y} : y < \sqrt{-a}\}$ and the unstable manifold is contained in the unstable manifold associated to steady state $\bar{y}_2$, $\mathcal{W}^u = \{y \in \mathrm{Y} : y > \sqrt{-a}\} \subset \mathcal{W}^{u,loc}_{\bar{y}_2}$. This means that if $y(0) \in \mathcal{W}^s$ then $\lim_{t\to\infty} y(t) = \bar{y}_1$ and if $y(0) \in \mathcal{W}^u$ then $\lim_{t\to\infty} y(t) = \infty$. If $a > 0$ (see Fig. 3.1c) as, for any initial point $y(0)$ the solution is asymptotically unbounded, then $\mathcal{W}^s$ and $\mathcal{W}^c$ are empty and $\mathcal{W}^u = \mathrm{Y}$. At last, if $a = 0$ (see Fig. 3.1b) the steady state $\bar{y} = 0$ is at the boundary between the stable and the unstable manifolds associated to steady state $\bar{y} = 0$, $\mathcal{W}^s = \mathcal{W}^{s,loc}_{\bar{y}} = \{y \in \mathrm{Y} : y < 0\}$ and $\mathcal{W}^u = \mathcal{W}^{u,loc}_{\bar{y}} = \{y \in \mathrm{Y} : y > 0\}$ and the center manifold is non-empty $\mathcal{W}^c = \{0\}$.

There are several elements that distinguish this ODE from a linear ODE.[10] First, the set Y can be partitioned into subsets with different stability properties; second,

[10] Compare to the related phase diagrams for an analogous linear ODE (in the right diagrams of Figs. 2.2 and 2.3).

the local manifolds can be different from the global manifolds (in this case $\mathcal{W}^{u,loc}_{\bar{y}_2} \neq \mathcal{W}^u$); third, center manifolds, differently from the linear case, are associated to a degenerate steady state having the property $f_y(y^{ss}, \cdot) = 0$, and not to the whole set Y, as in ODE $\dot{y} = 0$.[11]

However, Fig. 3.1a also provides an illustration of the difference between local and global dynamics analyses. At every steady state, the vector field of a non-linear ODE can be locally approximated by the vector field of a linear ODE, that is tangent to the former at that steady state. That is, the non-linear Ricatti ODE (3.3) can be approximated by a linear $\dot{y} = f_y(\bar{y}_i)\,(y - \bar{y}_i)$ at every steady state $\bar{y}_i$, if a steady state exists. For the Ricatti equation, this is the case only if $a < 0$, and this is illustrated in Fig. 3.1a by ODE $\dot{y} = f_y(\bar{y}_1)\,(y - \bar{y}_1)$, that is negatively slopped and is tangent to steady state $\bar{y}_1$, and by ODE $\dot{y} = f_y(\bar{y}_2)\,(y - \bar{y}_2)$, that is positively slopped and is tangent to the steady state $\bar{y}_2$.

We call **bifurcation point** to an element of the subset of points in the state space and the space of the parameters such there is a change in the phase diagram, $\{(y, a) \in Y \times \Phi : f(y, a) = f_y(y, a) = 0\}$. Different ODE have different types of bifurcation points. In the case of the Ricatti ODE (3.3), there is a bifurcation point at $(y, a) = (0, 0) \in Y \times \Phi$, that is called **saddle-node bifurcation**. It is characterized by the fact that at the bifurcation point, a small variation in the bifurcating parameter, a in this case, changes the number of steady states from zero to two steady states with different stability properties. We determine the bifurcation point by solving the following system of equations for (y, a):

$$\begin{cases} f(y, a) = 0 \\ f_y(y, a) = 0 \end{cases} \Leftrightarrow \begin{cases} a + y^2 = 0 \\ 2\,y = 0. \end{cases}$$

Figure 3.1d illustrates a bifurcation diagram for ODE (3.3), in a (a, y)–diagram. It is built by plotting a curve satisfying the steady state condition $a + y^2 = 0$, $\bar{y}(a) = \pm\sqrt{-a}$, and decorating by a solid curve its segment referring to stable steady states, satisfying $f_y(\bar{a}) < 0$, and by a dashed curve its segment referring to unstable steady states, satisfying $f_y(\bar{y}(a)) > 0$. Observe that the curve does not cross the positive quadrant for a. The saddle-node bifurcation point corresponds to the origin $(0, 0)$. This point separates the values for a such that there are two steady states (for $a < 0$) from the points of a such that there is no steady state (for $a > 0$). At last, note that it satisfies $f_{yy}(0, 0) = 2 \neq 0$.

Exercise 3.1 Show that the ODE $\dot{y} = a - y^2$ also displays a saddle-node bifurcation. Draw the phase and the bifurcation diagrams.

[11] The interpretation of Fig. 3.1b should bear in mind the fact that the solution to the ODE for $a = 0$, in Eq. (3.4a), is unbounded when $t = y(0)^{-1}$, which means that it depicts the solution path for $t < y(0)^{-1}$.

3.1.2 *Quadratic Bernoulli Equation: Transcritical Bifurcation*

The **quadratic Bernoulli** ODE is defined by

$$\dot{y} = ay + y^2, \text{ for } a \in \mathbb{R}. \tag{3.5}$$

This equation is a particular case of equation (3.2) with $a_1 = a$, $a_2 = 1$, and $a_0 = a_3 = 0$. It has the explicit solution[12]

$$y(t) = \begin{cases} \dfrac{y(0)}{1 - y(0)\,t}, & \text{if } a = 0, \quad (3.6a) \\ \dfrac{a}{(1 + a\,y(0)^{-1})\,e^{-at} - 1}, & \text{if } a \neq 0, \quad (3.6b) \end{cases}$$

for each $t \in \mathrm{T}$, where $y(0)$ is an arbitrary element of Y.

The steady state set is $y^{ss} = \{y \in \mathbb{R} : a\,y + y^2 = 0\}$. Both the number of steady states and their stability properties depend on the parameter a, but in a different way than the Ricatti ODE. First, if $a = 0$ there is only one steady state $y^{ss} = \{0\}$, if $a \neq 0$ there are two steady states $y^{ss} = \{\bar{y}_1, \bar{y}_2\} = \{0, -a\}$, meaning that the two steady states, $\bar{y}_1$ and $\bar{y}_2$, coalesce into $\bar{y}_1 = \bar{y}_2 = 0$ if $a = 0$. Second, stability properties of steady states depend on the sign of a: as $f_y(\bar{y}_1, a) = a$ and $f_y(\bar{y}_2, a) = -a$ then if $a < 0$ ($a > 0$) steady state $\bar{y}_1$ is asymptotically stable (unstable) and steady state $\bar{y}_2$ is unstable (asymptotically stable). If $a = 0$ the unique steady state, $\bar{y} = 0$ is neither stable nor unstable, because $f_y(0, 0) = 0$, because it is attracting for $y(0) < 0$ and is repelling for $y(0) > 0$.

The phase diagrams in Fig. 3.2 show how the dynamics change with different values of a: first, if $a < 0$ Fig. 3.2a shows that $\lim_{t\to\infty} y(t) = 0$ if $y(0) < -a$ and $\lim_{t\to\infty} y(t) = +\infty$ if $y(0) > -a$; second, if $a = 0$, Fig. 3.2b shows that $\lim_{t\to\infty} y(t) = 0$ if $y(0) < 0$ and $\lim_{t\to\infty} y(t) = +\infty$ if $y(0) > 0$; and Fig. 3.2c shows that $\lim_{t\to\infty} y(t) = -a$ if $y(0) < 0$ and $\lim_{t\to\infty} y(t) = +\infty$ if $y(0) > 0$. We observe again the partition of the phase space Y into different submanifolds, related to the difference between local and global dynamics, that differentiate this non-linear ODE from linear ODE. However, there is a difference as regards the Ricatti ODE: if $a < 0$ the stable manifold is $\mathcal{W}^s = \{y \in Y : y < -a\}$ and the unstable manifold is $\mathcal{W}^u = \{y \in Y : y > -a\}$; if $a > 0$ the stable manifold is $\mathcal{W}^s = \{y \in Y : y < 0\}$ and the unstable manifold is $\mathcal{W}^u = \{y \in Y : y > 0\}$, and if $a = 0$ there is a non-empty center manifold $\mathcal{W}^c = \{0\}$.

[12] See Appendix Sect. 3.5.2 for the explicit solution to the general Bernoulli ODE.

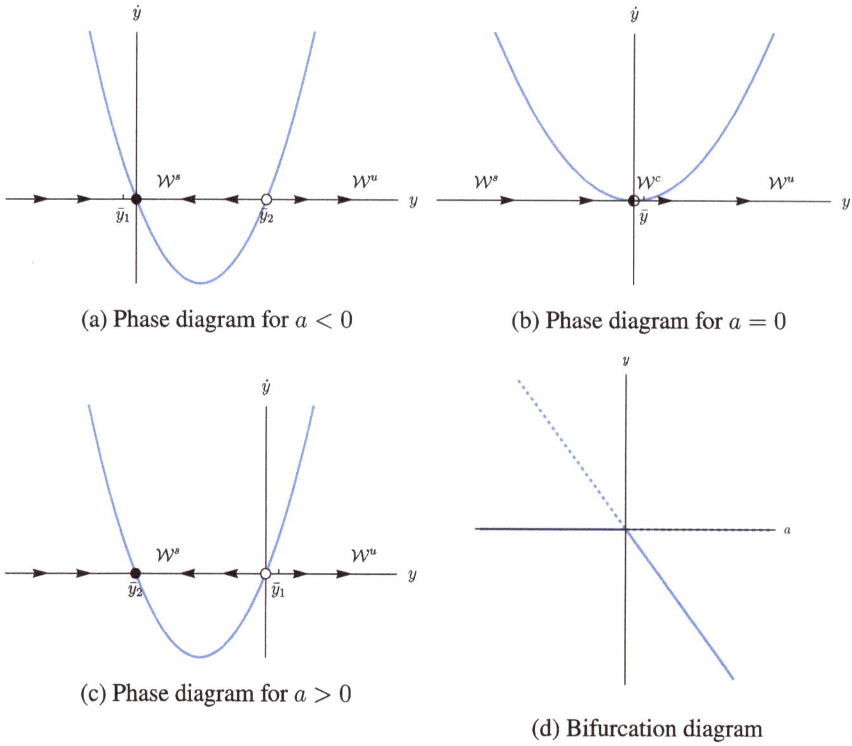

Fig. 3.2 Phase diagrams and bifurcation diagram for Eq. (3.5)

The bifurcation diagram Fig. 3.2d suggests calling it a **transcritical bifurcation**: the bifurcation point is $(y, a) = (0, 0)$,[13] unfolds in a way in which there are always two steady states, but their local stability properties are switched when a passes through the bifurcation value $a = 0$—the stable steady state becomes unstable and the unstable steady state becomes stable. Also observe that $f_{yy}(0, 0) = 2$, similarly to the Ricatti equation, meaning that the bifurcation point is a minimum for curve $f(y, a)$.

Exercise 3.2 Show that $\dot{y} = a\,y - y^2$ also displays a transcritical bifurcation. Draw the phase and the bifurcation diagrams.

---

[13] We determine the bifurcation point by solving the non-linear system for (y, a):

$$\begin{cases} f(y, a) = 0, \\ f_y(y, a) = 0, \end{cases} \Leftrightarrow \begin{cases} a\,y + y^2 = 0, \\ a + 2\,y = 0, \end{cases} \Leftrightarrow \begin{cases} 2\,a\,y + 2y^2 = 0, \\ a\,y + 2\,y^2 = 0, \end{cases} \Leftrightarrow \begin{cases} a\,y = 0, \\ a + 2\,y = 0. \end{cases}$$

3.1.3 Bernoulli's Cubic Equation: Subcritical Pitchfork

The **cubic Bernoulli** ODE is defined by

$$\dot{y} = a\, y - y^3. \tag{3.7}$$

This equation is a particular case of equation (3.2) with $a_1 = a$, $a_3 = -1$, and $a_0 = a_2 = 0$. As it is a Bernoulli equation then it also has explicit solutions:

$$y(t) = \begin{cases} \left(y(0)^{-2} + 2\,t \right)^{-\frac{1}{2}}, & \text{if } a = 0, \quad (3.8a) \\ \pm\sqrt{|a|}\left[1 - \left(1 - \frac{a}{y(0)^2} \right) e^{-2\,a\,t} \right]^{-\frac{1}{2}}, & \text{if } a \neq 0, \quad (3.8b) \end{cases}$$

for each $t \in \mathrm{T}$, where $y(0)$ is an arbitrary element of Y.

The steady state set is $y^{ss} = \{y \in \mathbb{R} : f(y, a) = a\, y - y^3 = 0\}$. Therefore: (1) if $a \leq 0$ there is **one** steady state $y^{ss} = \{\bar{y}\} = \{0\}$, and (2) if $a > 0$ there are **three** steady states $y^{ss} = \{\bar{y}_1, \bar{y}_2, \bar{y}_3\} = \{-\sqrt{a}, 0, \sqrt{a}\}$.

The local dynamics at every steady state are determined by $f_y(y, a) = a - 3\, y^2$ for $y \in y^{ss}$. Therefore: if $a < 0$, the steady state $\bar{y} = 0$ is asymptotically stable (as $f_y(0, a) = a < 0$); if $a = 0$, the steady state $\bar{y} = 0$ is neither stable nor unstable (as $f_y(0, 0) = 0$), and if $a > 0$, steady state $\bar{y}_2 = 0$ is unstable (as $f_y(0, a) = a > 0$), and the other two steady states $\bar{y}_1 = -\sqrt{a}$ and $\bar{y}_3 = \sqrt{a}$ are asymptotically stable (as $f_y(-\sqrt{a}, a) = f_y(\sqrt{a}, a) = -2a < 0$).

Figure 3.3 illustrates the phase diagrams for the three previous cases. If $a \leq 0$, see Fig. 3.3a and b as for any $y(0) \in \mathrm{Y}$ we have $\lim_{t\to\infty} y(t) = 0$ then the stable manifold is coincident with the whole state space, $\mathcal{W}^s = \mathrm{Y}$. If $a > 0$, see Fig. 3.3c, we have $\lim_{t\to\infty} y(t) = -\sqrt{a}$ if $y(0) < 0$ and $\lim_{t\to\infty} y(t) = \sqrt{a}$ if $y(0) > 0$. In this case, the stable manifold is $\mathcal{W}^s = \mathrm{Y}/\{0\}$. However, in this case $\mathcal{W}^s = \mathcal{W}^{s,loc}_{\bar{y}_1} \cup \mathcal{W}^{s,loc}_{\bar{y}_3}$, the stable manifold is the union of two local stable manifolds, associated to steady state $\bar{y}_1$, $\mathcal{W}^{s,loc}_{\bar{y}_1} = \{y \in \mathrm{Y} : y < 0\}$, and to steady state $\bar{y}_3$, $\mathcal{W}^{s,loc}_{\bar{y}_3} = \{y \in \mathrm{Y} : y > 0\}$.

Figure 3.3d illustrates the bifurcation diagram for the **subcritical pitchfork bifurcation**: it has a bifurcation point at $(y, a) = (0, 0)$, at this point a small variation in parameter a changes the number of steady states from one to three, and, because $f_{yy}(y, a) = -6\, y$, function $f(y, a)$ has an inflection point at the bifurcation point, $f_{yy}(0, 0) = 0$, differently from the previous cases. However, $f_{yyy}(y, a) - 3$ for any $(y, a) \in \mathrm{Y} \times \Phi$.[14]

---

[14] To find the bifurcation points, $\{(y, a) \in \mathrm{Y} \times \Phi : f(y, a) = 0,\ f_y(y, a) = 0\}$, we solve the system of equations for (y, a):

$$\begin{cases} a\, y - y^3 = 0 \\ a - 3\, y^2 = 0 \end{cases} \iff \begin{cases} 3\, a\, y - 3\, y^3 = 0 \\ a\, y - 3\, y^3 = 0 \end{cases} \iff \begin{cases} 2\, a\, y = 0 \\ a - 3\, y^2 = 0. \end{cases}$$

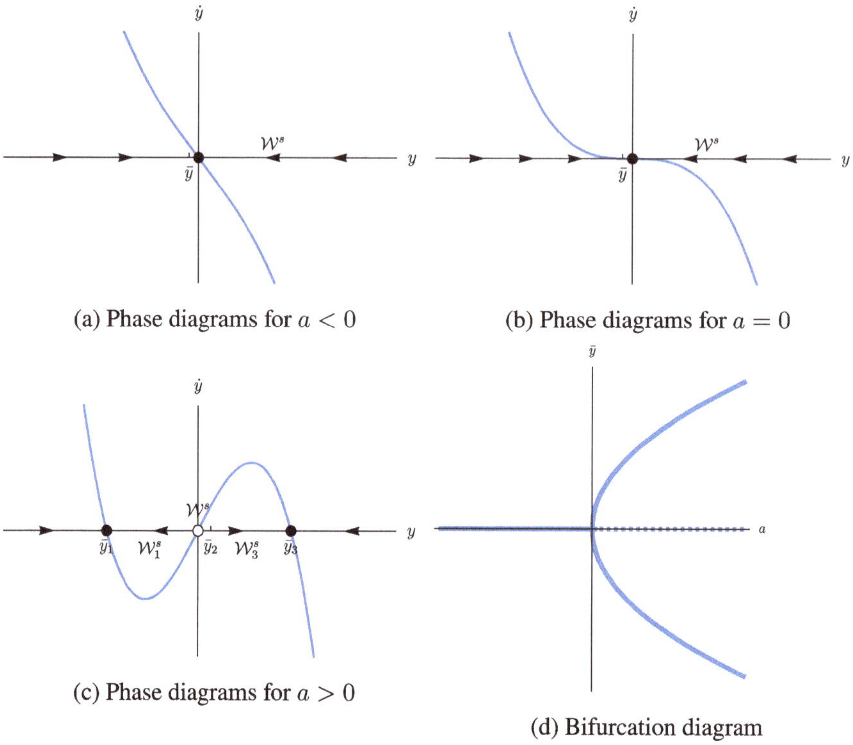

Fig. 3.3 Phase diagrams and bifurcation diagram for Eq. (3.7)

Exercise 3.3 The equation $\dot{y} = ay + y^3$ has also a bifurcation point called **super-critical pitchfork**. Find its solution and draw the phase and bifurcation diagrams.

3.1.4 Abel's Equation: Hysteresis

The ODE

$$\dot{y} = a + y - y^3 \tag{3.9}$$

is called an **Abel equation of the first kind**. Although closed form solutions have been found recently[15] they are too cumbersome to report. If $a = 0$ the Abel's equation reduces to the particular cubic Bernoulli's equation (3.7).

[15] For known closed form solutions for ODE see, Zaitsev and Polyanin (2003) or Zwillinger (1998).

This ODE can have one, two or three steady states. The number of steady states is equal to the number of real roots of the polynomial equation $f(y, a) \equiv a + y - y^3 = 0$, that depends on the sign of the discriminant[16] $\Delta(a) \equiv \left(\frac{a}{2}\right)^2 - \left(\frac{1}{3}\right)^3$: if $a \in \left(-\frac{2}{3\sqrt{3}}, \frac{2}{3\sqrt{3}}\right)$ then $\Delta(a) < 0$ and there are three steady states, $y^{ss} = \{\bar{y}_1, \bar{y}_2, \bar{y}_3\}$, satisfying $\bar{y}_1 < -\frac{1}{\sqrt{3}}$, $\bar{y}_2 \in \left(-\frac{1}{\sqrt{3}}, \frac{1}{\sqrt{3}}\right)$ and $\bar{y}_3 > \frac{1}{\sqrt{3}}$; if $a \in \left\{-\frac{2}{3\sqrt{3}}, \frac{2}{3\sqrt{3}}\right\}$ then $\Delta(a) = 0$ and there are two steady states, $y^{ss} = \{\bar{y}_1, \bar{y}_2\}$, satisfying $\bar{y}_1 = -\frac{2}{\sqrt{3}}$ and $\bar{y}_2 = \frac{1}{\sqrt{3}}$ if $a = -\frac{2}{3\sqrt{3}}$, or $\bar{y}_1 = -\frac{1}{\sqrt{3}}$ and $\bar{y}_2 = \frac{2}{\sqrt{3}}$ if $a = \frac{2}{3\sqrt{3}}$; or, if $a \notin \left[-\frac{2}{3\sqrt{3}}, \frac{2}{3\sqrt{3}}\right]$ then $\Delta(a) > 0$ and there is only one steady state, $y^{ss} = \{\bar{y}\}$, satisfying $\bar{y} < -\frac{2}{\sqrt{3}}$ if $a < -\frac{2}{3\sqrt{3}}$ or $\bar{y} > \frac{2}{\sqrt{3}}$ if $a > \frac{2}{3\sqrt{3}}$.

Taking the derivative of the right-hand-side of equation (3.9), yields $f_y(y, a) = 1 - \left(\sqrt{3}\, y\right)^2$, for any $a \in \mathbb{R}$. Then, $f_y(y, a) > 0$ if $y \in \left(-\frac{1}{\sqrt{3}}, \frac{1}{\sqrt{3}}\right)$, $f_y(y, a) = 0$ if $y = \pm\frac{1}{\sqrt{3}}$; and $f_y(y, a) < 0$, if $y < -\frac{1}{\sqrt{3}}$ or $y > \frac{1}{\sqrt{3}}$. This is a generic determination of the slope of the vector field for any value of a, depending on the level for y.

In order to characterize fully the local dynamics and the features of the phase diagrams, it is convenient to perform a bifurcation analysis. The set of bifurcation points is $H = \{(y, a) \in \mathrm{Y} \times \Phi : f(y, a) = 0,\ f_y(y, a) = 0\}$. Differently from the previous cases, we have *two* bifurcation points, called **hysteresis points**[17]

$$H = \left\{\left(-\frac{1}{\sqrt{3}}, \frac{2}{3\sqrt{3}}\right), \left(\frac{1}{\sqrt{3}}, -\frac{2}{3\sqrt{3}}\right)\right\}.$$

We readily observe that those points satisfy the conditions $\Delta(a) = f_y(y, a) = 0$.

From the previous findings, we can illustrate a complete unfolding of the hysteresis points in Fig. 3.4. It shows that: if $a < -\frac{2}{3\sqrt{3}}$, there is only one asymptotically stable steady state $\bar{y} < -\frac{1}{\sqrt{3}}$, see Fig. 3.4a; if $a = -\frac{2}{3\sqrt{3}}$, there are two steady states $\bar{y}_1 = -\frac{1}{\sqrt{3}}$, that is asymptotically stable, and $\bar{y}_2 = \frac{1}{\sqrt{3}}$, that is neither stable nor unstable, see Fig. 3.4b; if $-\frac{2}{3\sqrt{3}} < a < \frac{2}{3\sqrt{3}}$, there are three steady states where $\bar{y}_1 < -\frac{1}{\sqrt{3}}$ and $\bar{y}_3 > \frac{1}{\sqrt{3}}$ are asymptotically stable and $\bar{y}_2 \in \left(-\frac{1}{\sqrt{3}}, \frac{1}{\sqrt{3}}\right)$ is unstable, see Fig. 3.4c; if $a = \frac{2}{3\sqrt{3}}$, there are two steady states $\bar{y}_1 = -\frac{1}{\sqrt{3}}$ that is neither stable nor unstable and $\bar{y}_2 = \frac{1}{\sqrt{3}}$ that is asymptotically stable, see Fig. 3.4d; and if $a > \frac{2}{3\sqrt{3}}$, there is only one asymptotically stable steady state $\bar{y} > \frac{1}{\sqrt{3}}$, see Fig. 3.4e.

---

[16] See the Appendix Sect. 3.5.3.

[17] We solve the system

$$\begin{cases} a + y - y^3 = 0, \\ 1 - 3y^2 = 0, \end{cases} \Leftrightarrow \begin{cases} 3(a + y) - 3y^3 = 0, \\ y - 3y^3 = 0, \end{cases} \Leftrightarrow \begin{cases} 3a + 2y = 0 \\ y = \pm(\sqrt{3})^{-1}. \end{cases}$$

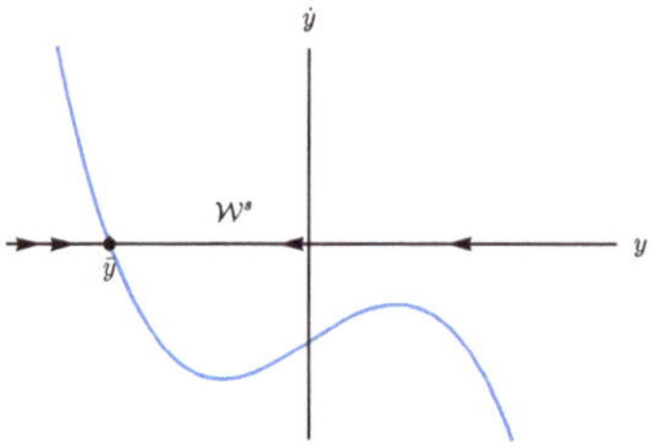

(a) Phase diagrams for $a < -\frac{2}{3\sqrt{3}}$

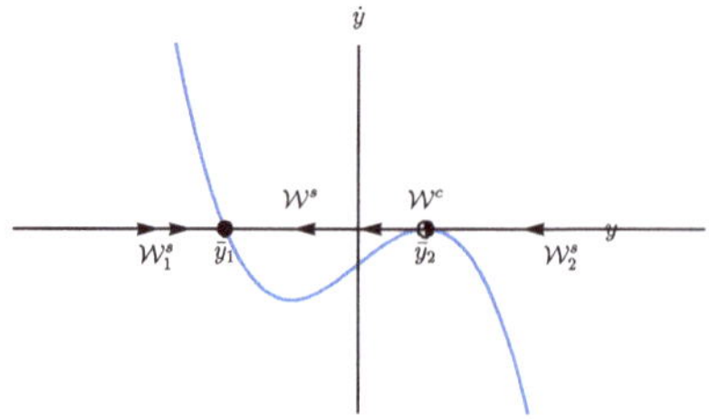

(b) Phase diagrams for $a = -\frac{2}{3\sqrt{3}}$

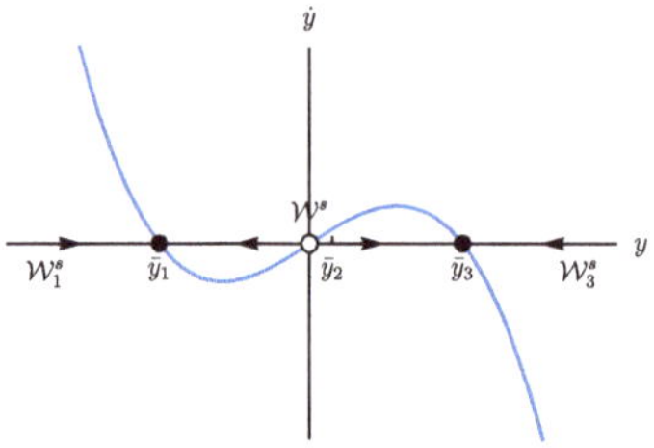

(c) Phase diagrams for $-\frac{2}{3\sqrt{3}} < a < \frac{2}{3\sqrt{3}}$

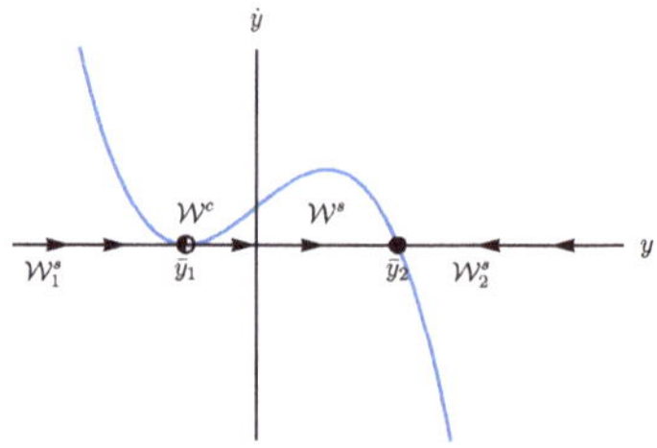

(d) Phase diagrams for $a = \frac{2}{3\sqrt{3}}$

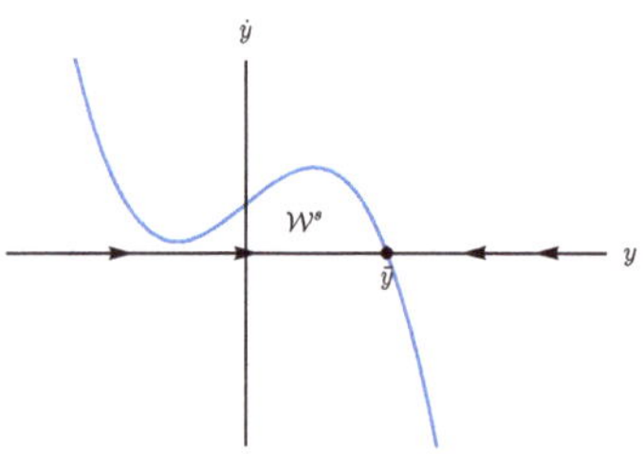

(e) Phase diagrams for $a > \frac{2}{3\sqrt{3}}$

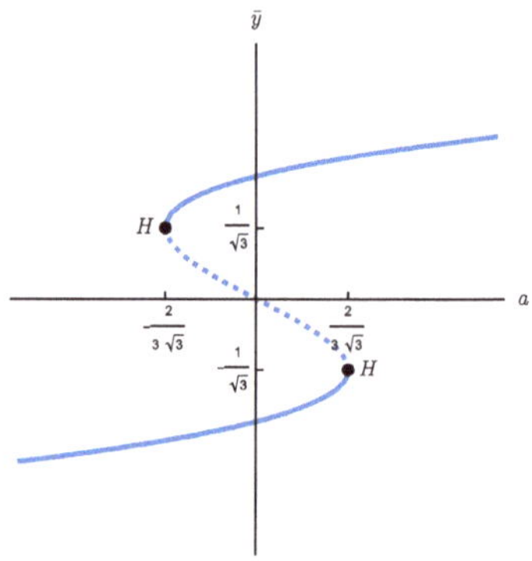

(f) Bifurcation diagram, showing the hysteresis points H.

Fig. 3.4 Phase diagrams and bifurcation diagram for Eq. (3.9)

The phase diagrams for the hysteresis bifurcation, in Fig. 3.4, can be compared to the phase diagrams for the pitchfork bifurcation, in Fig. 3.3. The phase diagrams not related to the critical values of the parameter a look similar—compare Fig. 3.3c with Fig. 3.4c. However, at the bifurcations, three steady states coalesce into one at the pitchfork bifurcation point, and two steady states coalesce into one at the hysteresis bifurcation points—compare phase diagrams in Fig. 3.4b and d with Fig. 3.3b . Furthermore, in both cases, the stable manifold, $\mathcal{W}^s$, coincides with the set

Y/y^{ss} when the parameter a is non-critical, or, when there are three steady, it is the union of the local stable manifolds of two asymptotically stable steady states, $\mathcal{W}^s = \mathcal{W}^{s,loc}_{\bar{y}_1} \cup \mathcal{W}^{s,loc}_{\bar{y}_3}$. Finally, at bifurcation points, there is a non-empty center manifold.

The reason for the difference in this bifurcation is related to the type of non-linearity. In the case of the ODE (3.9) we have $f_{yy}(y, a) = -6\,y$ and $f_{yyy}(y, a) = -6$, and at the bifurcation points we have a local maximum and a local minimum for $f(y, a)$ as $f_{yy}\left(-\frac{1}{\sqrt{3}}, \frac{2}{3\sqrt{3}}\right) = \frac{6}{\sqrt{3}} > 0$ and $f_{yy}\left(\frac{1}{\sqrt{3}}, -\frac{2}{3\sqrt{3}}\right) = -\frac{6}{\sqrt{3}} < 0$.

The bifurcation diagram in Fig. 3.4f displays the **hysteresis curve**. If we start with a small a, and increase it, we have a unique negative asymptotically stable steady state, until we reach an hysteresis point associated to negative a. At this point, a new non-generic positive steady state emerges and is split into two, with different stability properties, for a sligh increase in a. This scenario, with two asymptotically stable and one unstable steady states, is maintained for higher values of a until a new hysteresis point is reached. At this point, two negative steady states, one unstable and one asymptotically stable, disappear, and for slightly larger values of a only one positive asymptotically stable steady state remains. Thus, there are two "jumps" in the branches of the hysteresis curve at those two specific values of a.

Exercise 3.4 The equations $\dot{y} = a - y - y^3$ or $\dot{y} = a - y + y^3$ also display **hysteresis** bifurcations. Study the local dynamics and draw the phase and bifurcation diagram.

3.1.5 Cubic Equation: Cusp Bifurcation

The previous ODE had only one parameter that were bifurcating parameter. In this subsection, we consider another Abel ODE of the first kind having two parameters, $(a, b) \in \mathbb{R}^2$,

$$\dot{y} = f(y, a, b) \equiv a + b\,y - y^3. \tag{3.10}$$

This ODE can have one, two, or three steady states. Solving the cubic polynomial equation $f(y, a) = a + b\,y - y^3 = 0$[18] we find again that the number of real roots of this equation depends on the sign of the discriminant

$$\Delta(a, b) \equiv \left(\frac{a}{2}\right)^2 - \left(\frac{b}{3}\right)^3. \tag{3.11}$$

If $\Delta(a, b) < 0$, there are three real roots, if $\Delta = 0$, there are two real roots (one is multiple), and if $\Delta > 0$, there is one real root and a pair of complex conjugate roots. Therefore, we have the following cases: if $\Delta < 0$, there are three steady states, if

[18] See again appendix section 3.5.3.

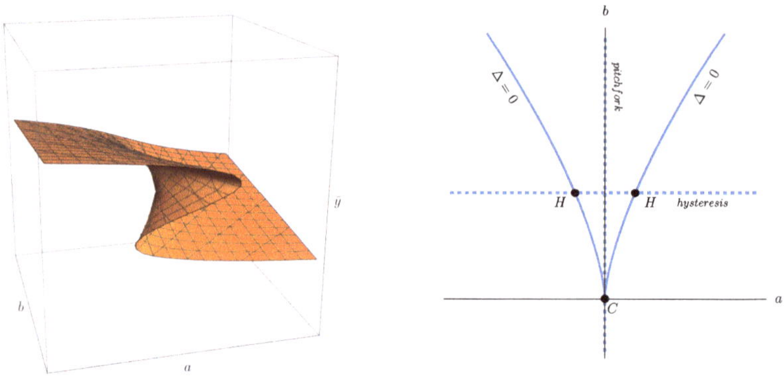

(a) Bifurcation diagram for the cusp

(b) Bifurcation diagram for co-dimension 2 cusp bifurcation

Fig. 3.5 Bifurcation diagram and projection onto Φ for ODE (3.10). Pitchfork and hysteresis bifurcations cases, for ODE (3.7) and (3.9), are shown in (b)

$\Delta = 0$ there are two steady states, and if $\Delta > 0$, there is one steady state. [19] Plotting the manifold $f(y, a, b) = 0$ in the three-dimensional space with axes (a, b, y), see Fig. 3.5a, we obtain a surface called **cusp**.[20] In that curve we have a geometrical illustration of the previous analysis regarding the number of steady states.

We can project the cusp surface into the two-dimensional space of parameters, in Fig. 3.5b, to obtain a **projected bifurcation diagram**. This figure suggests the existence of two types of bifurcations. First, we can determine a locus of hysteresis points by solving the system

$$f(y, a, b) = 0, \ \ f_y(y, a, b) = 0. \tag{3.12}$$

yielding $H(a, b) = \{(a, b) \in \mathbb{R}^2 : \Delta(a, b) = 0\}$.[21] As $\Delta(a, b) = 0$ defines a one-dimensional manifold in the two-dimensional parameter space, we say that $H(a, b)$ correspond to **co-dimension one bifurcation** set. The previous two types of bifurcations, that are unfolded by only one parameter, are special cases of this case: first, the hysteresis bifurcation, in Fig. 3.4f, is a special case in the projected bifurcation diagram cor-

[19] Again, the steady states can be determined explicitly but the expressions are again too cumbersome.

[20] This model played an important role in the catastrophe theory modeling in the 1980s, starting with Thom (1975) and Zeeman (1977).

[21] To prove this we solve the system

$$\begin{cases} a + b\,y - y^3 = 0, \\ b - 3\,y^2 = 0, \end{cases} \Leftrightarrow \begin{cases} 3\,a + 2\,b\,y = 0, \\ b - 3\,y^2 = 0, \end{cases} \Leftrightarrow \begin{cases} 3\,a + 2\,b\,y = 0 \\ 2\,b^2 + 9\,a\,y = 0, \end{cases} \Leftrightarrow \begin{cases} 27\,a^2 + 18\,a\,b\,y = 0 \\ 4\,b^3 + 18\,a\,b\,y = 0. \end{cases}$$

The solutions to the system must satisfy $18\,a\,b\,y = -27\,a^2 = -4\,b^3 \Leftrightarrow \left(\frac{a}{2}\right)^2 - \left(\frac{b}{3}\right)^3 = \Delta(a, b) = 0$.

responding to a horizontal line $b = 1$, and the sub-critical pitchfork bifurcation, in Fig. 3.3d, is a special case corresponding to a vertical line $a = 0$.

Points in $H(a, b)$ refer to just one level of degeneracy and trace out a curve in the two parameter space Φ. The cusp point corresponds to a **co-dimension two bifurcation**, a higher level of degeneracy, that corresponds to a point (or zero-dimensional manifold) in space Φ. Those points can be found by solving, for (y, a, b), the system

$$f(y, a, b) = 0, \;\; f_y(y, a, b) = 0, \;\; f_b(y, a, b) = 0, \tag{3.13}$$

and call them **co-dimension two bifurcation points**. We find that there is an unique point $(y, a, b) = (0, 0, 0)$, that is called a **cusp point**, see point C in Fig. 3.5b. This point corresponds to kind of degenerate pitchfork bifurcation point.

Exercise 3.5 Characterize the dynamics for equation $\dot{y} = 1 + by - y^3$, where $b \in \mathbb{R}$.

Exercise 3.6 Unfold the bifurcation diagram Fig. 3.5 for the ODE (3.10), that is, draw the representative phase diagrams.

3.2 Characterizing Dynamics

In this section, we consider again the regular non-linear scalar ODE (3.1) in which function $f(y, \cdot)$ is continuous, and continuously differentiable at least until the third order, $f(y, \cdot) \in C^3(\mathbb{R})$, and where $\boldsymbol{\varphi} \in \Phi \subseteq R^p$, is the $p-$dimensional vector of parameters. We showed that the dynamics is determined by the properties of the vector field $f(y, \boldsymbol{\varphi})$. Section 3.2.1 contains a brief introduction to the qualitative theory of ODE, highlighting the difference between local and global dynamics. In Sect. 3.2.2, we clarify the approach taken in comparative dynamics analyses.

Assuming that $\boldsymbol{\varphi} = (a, b) \in \mathbb{R}^2$, the ODE we have studied were generically characterized by the partial derivatives of the vector field $f(\cdot)$ over $\mathrm{Y} \times \Phi$: (1) the linear ODE assumed that $f_y(y, a) \neq 0$, and $f_{yy}(y, a) = f_{yyy}(y, a) = 0$; (2) the Ricatti equation assumed that $f_{yy}(y, a) \neq 0$, $f_a(y, a) \neq 0$, and $f_{yyy}(y, a) = f_y(y, a) = 0$; (3) the quadratic Bernoulli equation assumed that $f_{yy}(y, a) \neq 0$, $f_{ya}(y, a) \neq 0$ and $f_{yyy}(y, a) = 0$; (4) the cubic Bernoulli equation assumed that $f_{yyy}(y, a) \neq 0$ and $f_{ya}(y, a) \neq 0$; (5) the Abel first kind equation with one parameter assumed that $f_{yyy}(y, a) \neq 0$, and $f_a(y, a) \neq 0$; and (6) the cusp equation assumed that $f_{yyy}(y, a, b) \neq 0$ and $f_a(y, a, b) \neq 0$. The other derivatives, which were not specified for each case, can have any sign.

3.2.1 Qualitative Theory of ODE

The **qualitative theory of ODE** is concerned with the time-dependent behavior of solutions. It provides an useful approach to describe the stability properties of the solution to an ODE, both when explicit solutions are known or are not known. As we saw, their main components consist in finding the steady states, characterizing their local dynamics, finding bifurcation points, and unfolding the bifurcations points as a tool for describing the global dynamics. Next, we summarize and extend the contents of Sect. 3.1 to any scalar non-linear ODE.

The local dynamics analysis in the neighborhood of steady states is the first step in the application of the qualitative theory. The **steady state set** is defined by the

$$\Gamma = \{(y, \boldsymbol{\varphi}) \in Y \times \Phi : \; f(y, \boldsymbol{\varphi}) = 0\}. \tag{3.14}$$

We already found cases in which Γ is empty, has one element, or can have more than one element. Assume that for a particular value for the parameter $\boldsymbol{\varphi}_0 \in \Phi$, at least one steady state exists, denoted by $\bar{y}_0 = \bar{y}(\boldsymbol{\varphi}_0)$:

Definition 3.1 (*Hyperbolic steady state*) Given a specific parameter value $\boldsymbol{\varphi} = \boldsymbol{\varphi}_0$, the steady state $\bar{y}_0 = \bar{y}(\boldsymbol{\varphi}_0)$ is called **hyperbolic** if $f_y(\bar{y}_0, \boldsymbol{\varphi}_0) \neq 0$. An hyperbolic steady state is **locally asymptotically stable** if $f_y(\bar{y}_0, \boldsymbol{\varphi}_0) < 0$, and it is **locally unstable** if $f_y(\bar{y}_0, \boldsymbol{\varphi}_0) > 0$.

Definition 3.2 (*Non-hyperbolic steady state*) Given a specific parameter level $\boldsymbol{\varphi} = \boldsymbol{\varphi}_0$, the steady state $\bar{y}_0 = \bar{y}(\boldsymbol{\varphi}_0)$ is called **non-hyperbolic** if $f_y(\bar{y}_0, \boldsymbol{\varphi}_0) = 0$.

This initial classification of steady states can be done independently from the number of steady states, provided that they exist. However, to obtain a full characterization of the dynamics, we need a global dynamics characterization of the steady states, including classification of non-hyperbolic steady states corresponding to particular bifurcation points.

If $f(y, \cdot) \in C^4(\mathbb{R})$, we can perform a Taylor expansion in a neighborhood of a steady state $\bar{y}_0 \in y^{ss}$,[22] in which $f(\bar{y}_0, \boldsymbol{\varphi}_0) = 0$,

$$f(y, \boldsymbol{\varphi}_0) = f_y(\bar{y}_0, \boldsymbol{\varphi}_0)\,\epsilon + f_{yy}(\bar{y}_0, \boldsymbol{\varphi}_0)\,\frac{\epsilon^2}{2!} + f_{yyy}(\bar{y}_0, \boldsymbol{\varphi}_0)\,\frac{\epsilon^3}{3!} + O\big(\epsilon^4\big). \tag{3.15}$$

where $\epsilon = |y - \bar{y}_0|$ is close to zero.

If, for a given parameter value $\boldsymbol{\varphi}_0$, the steady state $\bar{y}_0 = \bar{y}(\boldsymbol{\varphi}_0)$ is **hyperbolic**, a non-linear ODE can be locally first-order approximated as[23]

---

[22] We can use the Landau little-o or Bachmann big-O notations. A function $f(x)$ is said to be of order $g(x)$, denoted by $f(x) = O(g(x))$ if $\lim_{x\to 0} \frac{f(x)}{g(x)} = c$ where c is a constant. A function $f(x)$ is of weak order $g(x)$, denoted $f(x) = o(g(x))$ if $\lim_{x\to 0} \frac{f(x)}{g(x)} = 0$.

[23] This is an application of the Grobman-Hartmann Theorem 5.1 for a scalar ODE.

$$\dot{y} = f_y(\bar{y}_0, \boldsymbol{\varphi}_0)\,(y - \bar{y}_0) + O((y - \bar{y}_0)^2).$$

That hyperbolic steady state is **locally asymptotically stable** (LAS) if $f_y(\bar{y}, \boldsymbol{\varphi}_0) < 0$ and it is **unstable** if $f_y(\bar{y}, \boldsymbol{\varphi}_0) > 0$.

If the steady state is unique, that is if $y^{ss} = \{\bar{y}\}$, or there are no parameter values $\boldsymbol{\varphi} \in \Phi$ such that $f_y(\bar{y}, \boldsymbol{\varphi}) = 0$, then the ODE is topologically equivalent to the linear ODE

$$\dot{y} = \lambda\,(y - \bar{y}), \ \text{ for } \lambda = f_y(\bar{y}, \boldsymbol{\varphi}).$$

Then, for any arbitrary value for parameter $\boldsymbol{\varphi} \in \Phi$, the phase diagram will have the same properties as the approximated linear ODE. In this case, we say that the steady state is **globally asymptotically stable** (GAS) if $\lambda < 0$, or **globally unstable** if $\lambda > 0$. In the first case, we generally have $\mathcal{W}^s = \mathcal{W}^{s,loc}_{\bar{y}} = \mathrm{Y}$, and in the second case, we have $\mathcal{W}^u = \mathcal{W}^{u,loc}_{\bar{y}} = \mathrm{Y}$. Therefore, when the steady state is unique, linearization captures the main qualitative aspects of an ODE.

Example 3.1 (*Approximating a non-linear ODE*) Consider the following ODE, whose form is frequently found in economics,

$$\dot{y} = f(y) \equiv y^{\alpha} - a, \ \ y : \mathrm{T} \rightarrow \mathbb{R}_+ \tag{3.16}$$

where $a > 0$ and $\alpha \neq 0$ are two constants, and $f_y(y) = \alpha y^{\alpha-1}$. Then there is one unique steady state $\bar{y} = a^{\frac{1}{\alpha}}$. As $\lambda \equiv f_y(\bar{y}) = \alpha\, a^{\frac{\alpha-1}{\alpha}} \neq 0$ the steady state is hyperbolic. That hyperbolic steady state is GAS if $\alpha < 0$ and is globally unstable if $\alpha > 0$. In the neighborhood of the steady state, the ODE (3.16) is topologically equivalent to the linear ODE $\dot{y} = \lambda(y - \bar{y})$. Then, the general solution to ODE (3.16) can be locally approximated by the solution to this linear ODE, $y(t) = \bar{y} + (y(0) - \bar{y})e^{\lambda t}$ for $t \in [0, \infty)$, for any $y(0) \in \mathbb{R}_+$. In particular, if we fix $y(0) = y_0$ the particular solution can be approximated by $y(t) = \bar{y} + (y_0 - y)e^{\lambda t}$.

If the steady states are not unique, linearization can still provide a local characterizing of every the steady state, although it should be complemented with a bifurcation analysis if we would like to obtain a global characterization of the dynamics. Let $\boldsymbol{\varphi}_0 \in \Phi$ and the steady state set be $y^{ss}(\boldsymbol{\varphi}_0) = \{\bar{y}_{i,0}\}_{i=1}^{n_{ss}}$, where n_{ss} is the number of steady states. If $f_y(\bar{y}_{i,0}, \boldsymbol{\varphi}_0) < 0$ we say that the steady state $\bar{y}_{i,0}$ is hyperbolic and **locally asymptotically stable** (LAS), and if $f_y(\bar{y}_{i,0}, \boldsymbol{\varphi}_0) > 0$ the steady state is hyperbolic and **locally unstable**. In the first case we have an associated local stable manifold $\mathcal{W}^{s,loc}_i$, and in the second case we have a local unstable manifold $\mathcal{W}^{u,loc}_i$, that are subsets of the image set Y.

Assume there is parameter value $\boldsymbol{\varphi}_0 \in \Phi$ such that there is a steady state $\bar{y}_0$ that is non-hyperbolic, that is satisfying $f_y(\bar{y}_0, \boldsymbol{\varphi}_0) = 0$. In the neighborhood of the non-hyperbolic steady state $\bar{y}^0$, the higher-order derivatives $f_{yy}(\bar{y}_0, \boldsymbol{\varphi}_0)$, or $f_{yyy}(\bar{y}_0, \boldsymbol{\varphi}_0)$, in the Taylor expansion from Eq. (3.15) allow us to characterize the solution to the non-linear ODE by equivalence to the normal forms presented in Sect. 3.1. For regular scalar ODE, we have found the following types of bifurcations[24]:

Definition 3.3 (*Fold bifurcation*) Let $f(y, \varphi) \in C^{2,1}(\mathrm{Y} \times \Phi)$. A point $(y, \varphi) = (\bar{y}, \varphi_0) \in \Gamma$, satisfying $f(\bar{y}, \varphi_0) = f_y(\bar{y}, \varphi_0) = 0$, $f_{yy}(\bar{y}, \varphi_0) \neq 0$, and $f_\varphi(\bar{y}, \varphi_0) \neq 0$, is called a fold bifurcation point.

If ODE (3.1) has a fold bifurcation point, then it is topologically equivalent to the Ricatti's ODE, $\dot{y} = \varphi \pm y^2$.

Definition 3.4 (*Transcritical bifurcation*) Let $f(y, \varphi) \in C^{2,1}(\mathrm{Y} \times \Phi)$. A point $(y, \varphi) = (\bar{y}, \varphi_0) \in \Gamma$, satisfying $f(\bar{y}, \varphi_0) = f_y(\bar{y}, \varphi_0) = 0$, $f_{yy}(\bar{y}, \varphi_0) \neq 0$, and $f_{y\varphi}(\bar{y}, \varphi_0) \neq 0$, is called a transcritical bifurcation point.

If ODE (3.1) has a transcritical bifurcation point, then it is topologically equivalent, to the quadratic Bernoulli ODE $\dot{y} = \varphi y \pm y^2$.

Definition 3.5 (*Pitchfork bifurcation*) Let $f(y, \varphi) \in C^{3,1}(\mathrm{Y} \times \Phi)$. A point $(y, \varphi) = (\bar{y}, \varphi_0) \in \Gamma$, satisfying $f(\bar{y}, \varphi_0) = f_y(\bar{y}, \varphi_0) = f_{yy}(\bar{y}, \varphi_0) = 0$, $f_{yyy}(\bar{y}, \varphi_0) \neq 0$, and $f_{y\varphi}(\bar{y}, \varphi_0) \neq 0$, is called a pitchfork bifurcation point.

If ODE (3.1) has a pitchfork bifurcation point, then it is topologically equivalent to the cubic Bernoulli ODE $\dot{y} = \varphi y \pm y^3$

Definition 3.6 (*Hysteresis bifurcation*) Let $f(y, \varphi) \in C^{3,1}(\mathrm{Y} \times \Phi)$. A point $(y, \varphi) = (\bar{y}, \varphi_0) \in \Gamma$, satisfying $f(\bar{y}, \varphi_0) = f_{yy}(\bar{y}, \varphi_0) = 0$, $f_y(\bar{y}, \varphi_0) \neq 0$, $f_{yyy}(\bar{y}, \varphi_0) \neq 0$, and $f_\varphi(\bar{y}, \varphi_0) \neq 0$, is called an hysteresis bifurcation point.

If ODE (3.1) has a hysteresis bifurcation point, then it is topologically equivalent to the Abel ODE $\dot{y} = \varphi \pm y \pm y^3$.

All the previous bifurcations are co-dimension one bifurcations. A co-dimension two bifurcation is

Definition 3.7 (*Cusp bifurcation*) Let $f(y, \boldsymbol{\varphi}) \in C^{3,1}(\mathrm{Y} \times \Phi)$, where $\boldsymbol{\varphi} = (\varphi_1, \varphi_2) \in \Phi \subseteq \mathbb{R}^2$. A point $(y, \boldsymbol{\varphi}) = (\bar{y}, \boldsymbol{\varphi}_0) \in \Gamma$, satisfying $f(\bar{y}, \boldsymbol{\varphi}_0) = f_{yy}(\bar{y}, \boldsymbol{\varphi}_0) = 0$, $f_y(\bar{y}, \boldsymbol{\varphi}_0) \neq 0$, $f_{yyy}(\bar{y}, \boldsymbol{\varphi}_0) \neq 0$, $f_{\varphi_1}(\bar{y}, \boldsymbol{\varphi}_0) \neq 0$ and $f_{y\varphi_2}(\bar{y}, \boldsymbol{\varphi}_0) \neq 0$ is called a cusp bifurcation point.

If ODE (3.1) has a cusp bifurcation point, then it is topologically equivalent, and the dynamics is qualitatively equivalent, to the Abel ODE $\dot{y} = \varphi_1 \pm \varphi_2\, y \pm y^3$. As we saw in Sect. 3.1 if, for particular values for the parameters, there are n_{ss} steady states and $n_{ss} = n_s + n_u + n_c$, where n_s is the number of LAS steady states, n_u the number of LU steady states, and n_c the number of non-hyperbolic steady states, then the stable manifold is defined by the union of all local stable manifolds, $\mathcal{W}^s = \cup_{i=1}^{n_s} \mathcal{W}_i^{s,loc}$, and center manifold $\mathcal{W}^c$ is the union of all the non-hyperbolic steady states. Therefore, the following partition is introduced in the state space $\mathrm{Y} = \mathcal{W}^s \cup \mathcal{W}^c \cup \mathrm{Y}/(\mathcal{W}^s \cup \mathcal{W}^u)$. In an informal sense, a global dynamics analysis of equation (3.1) entails finding all the possible partitions of the state space Y for all $\boldsymbol{\varphi} \in \Phi$, or, in other words, performing a global bifurcation analysis

[24] See Kuznetsov (2023, Chap. 3).

3.2.2 Comparative Dynamics in Economics

In Sect. 2.3.3.2, we clarified, for linear ODE, the difference between comparative dynamics and bifurcation analyses in economics. For a non-linear ODE, the purpose of a **comparative dynamics** exercise is to compute qualitatively of quantitatively the variation of the solution to an ODE **in the neighborhood of an hyperbolic steady state** when there is a variation of a parameter away from a bifurcation value.

Consider the ODE (3.1) with only one parameter $\varphi \in \Phi$, and assume that for $\varphi = \varphi_0 \in \Phi$ there is an hyperbolic steady state $\bar{y}(\varphi_0) \in y^{ss}$. Let us introduce a variation in the parameter value such that $d\varphi = \varphi_1 - \varphi_0$, and assume there is a variation in the steady state $d\bar{y} = \bar{y}'(\varphi)\, d\varphi$, where $d\bar{y} = \bar{y}_1 - \bar{y}_0$ and $\bar{y}'(\varphi) \equiv \frac{d\bar{y}(\varphi)}{d\phi}$.

For a non-anticipated, constant and permanent change $d\varphi$, the **variational ODE** is

$$\dot{y} = \lambda\, dy(t) + b\, d\varphi, \text{ for } \lambda \equiv f_y(\bar{y}_0, \varphi_0), \text{ and } b \equiv f_\varphi(\bar{y}_0, \varphi_0),$$

where $dy(t) = y(t) - \bar{y}_0$. The related ODE problem depends on the side condition associated to the economic type of variable y.

If variable y is a pre-determined variable, we introduce the side condition $dy(0) = y(0) - \bar{y}_0 = \bar{y}_0 - \bar{y}_0 = 0$, and, obtain the solution to this linear (variational) ODE

$$dy(t) = d\bar{y}\left(1 - e^{\lambda t}\right), \text{ for } t \in [0, \infty)\ d\bar{y} = \bar{y}'(\varphi_0)\, d\varphi$$

where $\bar{y}'(\varphi_0) = -\frac{b}{\lambda} = -\frac{f_y(\bar{y}_0,\varphi_0)}{f_\varphi(\bar{y}_0,\varphi_0)}$ or, equivalently, we can obtain the short-run multipliers

$$\frac{\partial y(t)}{\partial \varphi} = \bar{y}'(\varphi_0)\left(1 - e^{\lambda t}\right), \text{ for } t \in [0, \infty).$$

If $\lambda < 0$, there is transitional dynamics in the sense that the short-run multipliers converge to the long-run multiplier $\lim_{t\to\infty} \frac{\partial y(t)}{\partial \varphi} = \frac{d\bar{y}}{d\varphi} = \bar{y}'(\varphi_0)$. If $\lambda > 0$, the multipliers are unbounded (as is the case in economic growth models).

If variable y is a non-pre-determined variable, the side condition usually requires that the perturbed solution should be asymptotically bounded. As in most cases $\lambda > 0$ then the solution to the comparative dynamics exercise is

$$dy(t) = d\bar{y} \iff \frac{\partial y(t)}{\partial \varphi} = \bar{y}'(\varphi_0), \text{ for each } t \in (0, \infty),$$

and the solution "jumps" at time $t = 0^+$ to the new steady state, meaning that there is no transitional dynamics.

In both cases, a comparative dynamics analysis entails introducing a small perturbation on a phase diagram, without changing its type. This distinguishes a comparative dynamics exercise from a bifurcation analysis, because the last one seeks to discover changes in the phase diagram's type.

3.3 Applications to Economics

Example 3.2 (*Capital accumulation with Inada production function*) Solow (1956) is a benchmark contribution to the theory of economic growth, or more precisely, to the theory of capital accumulation. It states that, in equilibrium, there are two drivers to net capital accumulation: technology and savings. The technology is neoclassical, meaning that output per worker, y, is produced by physical capital per worker, k, by means of an increasing (capital is productive) and concave (there are diminishing marginal returns) production function

$$y = f(k),\ \ f(0) = 0,\ \ \text{and} f'(k) > 0,\ \ \text{and}\ \ f''(k) < 0,\ \ \text{for}\ k > 0.$$

The output can be consumed or can be saved. In Solow (1956)'s model, it is assumed that savings per capita is, somewhat mechanically,[25] proportional to income: $s\,y$ where $0 < s < 1$. In a closed economy, the goods market equilibrium condition states that gross investment per capital—equal to net investment plus depreciation $\dot{k} + \delta\,k$ – should be equal to savings. Assuming a constant population, the capital accumulation equation is formalized by the non-linear non-explicit scalar ODE,

$$\dot{k} = s\,f(k) - \delta\,k,\ \ k \in \mathbb{R}_+. \tag{3.17}$$

If we assume that the production function is globally concave, and that the marginal productivity satisfies the Inada (see e.g., Inada 1964) conditions, $\lim_{k\to 0} f'(k) = \infty$ and $\lim_{k\to\infty} f'(k) = 0$, then the ODE (3.17) has two steady states, thus $y^{ss} = \{0, \bar{k}\}$, where $\bar{k} \equiv \bar{k}(s, \delta) = \{k : s\,f(k) = \delta\,k\}$ is finite. In the neighborhood of the steady $\bar{k}$ the non-linear (and non-explicit) Solow equation can be approximated by the linear ODE

$$\dot{k} = \lambda\,(k - \bar{k}),\ \ \text{where}\ \lambda \equiv s\,f'(\bar{k}) - \delta.$$

As $s\,f'(\bar{k}) < s\,\frac{f(\bar{k})}{\bar{k}} = \delta$, from the Inada conditions, then $\lambda < 0$ for any $k > 0$, and the steady state $\bar{k}$ is asymptotically stable. The other steady state $k = 0$, corresponding to the non-existence or the colapse of the economy, is unstable and is never reached if $k(0) > 0$. Performing a comparative dynamics exercise, for an increase in the savings rate, we find

$$\frac{dk(t)}{ds} = \bar{k}'(s)\left(1 - e^{\lambda\,t}\right),\ \ \text{for}\ t \in [0, \infty),$$

where the long-run multiplier is $\bar{k}'(s) = \frac{dk}{ds} = -\frac{\delta\,\bar{k}}{\lambda\,s} > 0$. Equivalently, the long-run elasticity of capital is directly proportional to the elasticity of the marginal propensity to save $\epsilon_k = -\frac{\delta}{\lambda}\,\epsilon_s$.

Example 3.3 (*Cobb-Douglas production function*) A benchmark Inada production function is the Cobb-Douglas function

[25] Labeled ad-hoc savings function by the post-1990 literature.

$$y = f(k) = A\,k^{\alpha},\ \ A > 0,\ 0 < \alpha < 1, \tag{3.18}$$

where A is a total factor productivity, and $\alpha \in (0, 1)$ is the share of capital in national income, y. Figure 3.6a illustrates this production function for fixed values of A and α. In this case, the capital accumulation equation is a Bernoulli ODE

$$\dot{k} = s\,A\,k^{\alpha} - \delta\,k,\ \ k \in \mathbb{R}_{+}. \tag{3.19}$$

It is a particular case of ODE (3.21), which has the closed form (global) solution (3.22). Therefore, the general solution to ODE (3.19) is

$$k(t) = \left(\bar{k}^{1-\alpha} + \left(k(0)^{1-\alpha} - \bar{k}^{1-\alpha}\right) e^{\lambda\,t}\right)^{\frac{1}{1-\alpha}},\ \text{ for each } t \in [0, \infty),$$

where the steady state capital intensity is $\bar{k} = \left(\frac{s\,A}{\delta}\right)^{\frac{1}{1-\alpha}}$, and the coefficient providing the speed of adjustment is $\lambda \equiv -\delta\,(1-\alpha)$. This theory highlights two forces for stability: the depreciation of capital and the concavity of the production function. Figure 3.6b illustrates the phase diagram. We observe that the steady state $\bar{k}$ is globally asymptotically stable for $k > 0$. The variational ODE in the neighborhood of steady state is $\dot{k} = \lambda\,(k - \bar{k})$, has the solution $k(t) = \bar{k} + (k_0 - \bar{k})\,e^{\lambda\,t}$. In Fig. 3.6c, we illustrate the fact that the global (exact) and the linearly approximated solutions are quantitatively very close.

Example 3.4 (*Capital accumulation with concave-convex production functions*) The Solow model displays convergence to a steady state. However, it does not capture well the process of structural transformation that takes place in the process of economic development. In particular, there seems to be some form of stability for poor and rich economies, with low rates of growth, while the process of industrialization is characterized by high rates of growth. The transition period, from poor to rich, appears to be characterized by some sort of instability.

This process can be captured by assuming instead a concave-convex production function, displaying locally increasing marginal returns for intermediate levels of the capital intensity, as in Skiba (1978). For instance, assume the following production function

$$y = f(k, B) = k^{\alpha}\left(A + \frac{k^{3-\alpha}}{2\,(k^3 + B^3)}\right),\ \text{for } A > 0,\ 0 < \alpha < 1,\ B > 0, \tag{3.20}$$

where B can be taken as a parameter representing generic inefficiencies in transforming the capital stock into output. This production function is illustrated in Fig. 3.7a. The capital accumulation equation, assuming a Solow economy, in which the savings function is ad-hoc, is now

$$\dot{k} = s\,f(k, B) - \delta\,k,\ \ k{:}\mathrm{T} \to \mathbb{R}_{+}$$

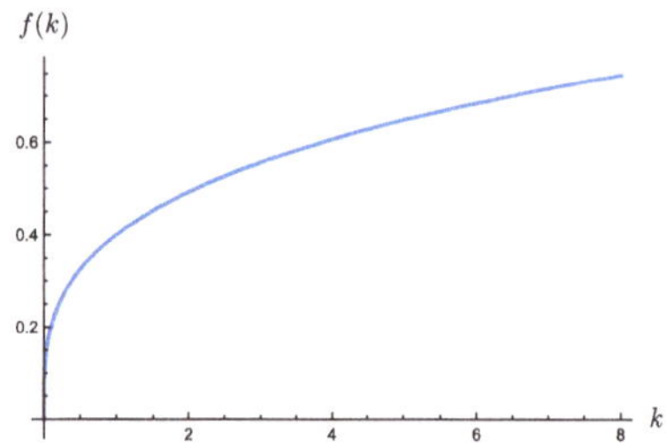

(a) Production function

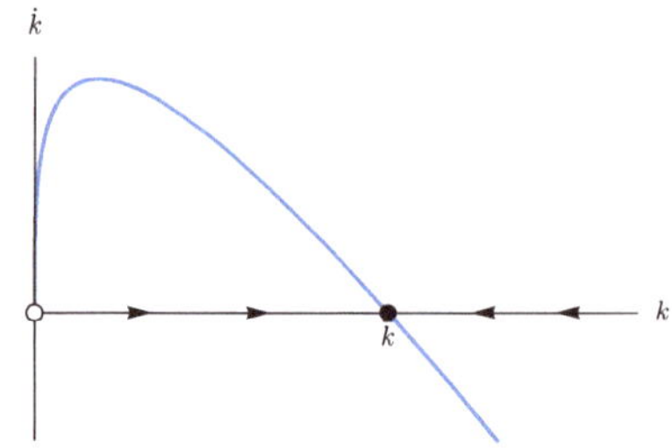

(b) Phase diagram

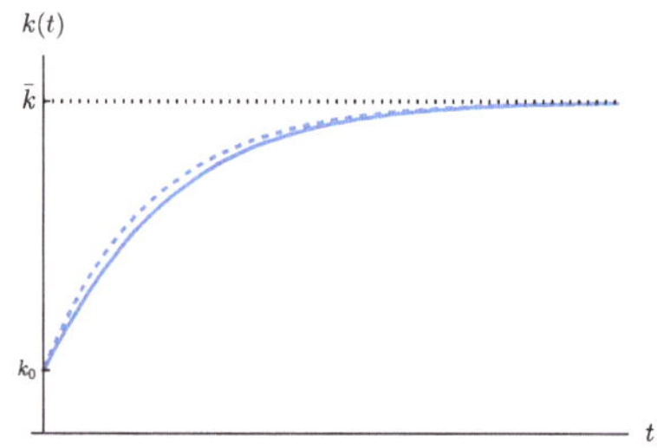

(c) Exact solution (solid) and linear approximation

Fig. 3.6 Solow (1956) model: Cobb-Douglas production function and the accumulation of capital, for $A = 0.4$, $\alpha = 0.3$, $s = 0.3$ and $\delta = 0.05$

where $k(0) = k_0 > 0$ is given. Figure 3.7 contains three different generic phase diagrams for different levels of B: first, if there are high inefficiencies in the production process and B is large (see Fig. 3.7b) then there is only one asymptotically stable steady state for a low but positive level of k—and therefore of output y; second, for an intermediate level of inefficiencies and a medium level of B (see Fig. 3.7c) then there are three steady states, two asymptotically stable steady states corresponding to low or high levels of k, separated by an unstable steady state for a medium level of k; and, at last, if there are low inefficiencies in production and B is small (see Fig. 3.7d) there will be again one asymptotically stable steady state for a high level of k (and y). This seems to capture several apparent facts: (a) there is a poverty trap which seems hard to escape from; (b) several technological and institutional changes are required to successfully complete the process of development rates, without the economy falling back to poverty; and (c) mature economies also seem to be in a relatively stable state.

Plotting the bifurcation diagram (see Fig. 3.7e), we find we are in the presence of hysteresis (similar to Fig. 3.4f): the asymptotic state of the economy depends on its starting point, or on its history. Choosing B as the bifurcation parameter, and denoting $F(k, B) = s\, f(k, B) - \delta\, k$, the hysteresis points $H = \{(k_1^c, B_1^c), (k_2^c, B_2^c)\}$ are found by solving the two non-linear equation system

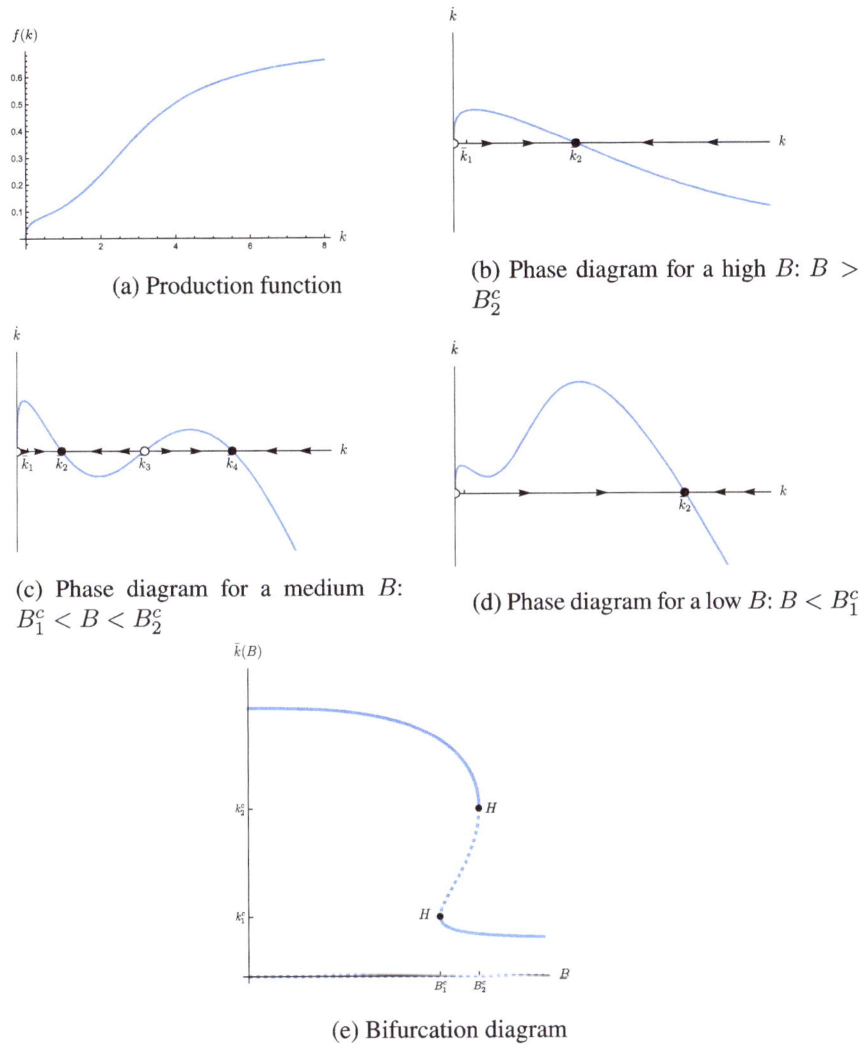

(a) Production function

(b) Phase diagram for a high B: $B > B_2^c$

(c) Phase diagram for a medium B: $B_1^c < B < B_2^c$

(d) Phase diagram for a low B: $B < B_1^c$

(e) Bifurcation diagram

Fig. 3.7 Concave-convex production function and the accumulation of capital. Assumptions: $A = 0.4$, $\alpha = 0.3$, $s = 0.3$ and $\delta = 0.05$

$$\begin{cases} F(k, B) = 0 & \iff s\, f(k, B) = \delta\, k \iff \dfrac{f(k, B)}{k} = \dfrac{\delta}{s}, \\ F_k(k, B) = 0 & \iff s\, f_k(k, B) = \delta \iff f_k(k, B) = \dfrac{\delta}{s}. \end{cases}$$

Both bifurcation points satisfy $f(k, B) = f_k(k, B)\, k$ meaning that the production function is locally concave. Although we cannot determine explicitly the bifurcation points, we can find that a steady state $\bar{k}$ is locally asymptotically stable if

$F_k(\bar{k}, B) < 0$ and the production function is locally concave, and that the steady state $\bar{k}$ is locally unstable if $F_k(\bar{k}, B) > 0$ and the production function is locally convex. This is because

$$F(k, B) = 0, \text{ and } F_k(k, B) \lesseqgtr 0 \iff f(k, B) \gtreqless f_k(k, B)\, k.$$

As we see in Fig. 3.7c a locally unstable steady state only occurs for intermediate values of B and the production function is locally convex, i.e., displaying locally increasing marginal returns. This steady state, say k^c, is indeed a critical point that separates the basins of attraction to the lower steady state from the higher steady state. This means that if $k(0) < \bar{k}_2$, such that $k_1^c < \bar{k}_2 < k_2^c$, the economy will converge to the lower steady state, and if $k(0) > \bar{k}_2$ it will converge to the higher steady state. A small variation in the other parameters, s, δ, or A will change k^c. Therefore, we can only determine the change in the development phase of the economy from changes in those parameters (and in B) by means of a bifurcation analysis, not by a comparative dynamics exercise.

Exercise 3.7 (*Capital accumulation with concave-convex production functions*) Consider Example 3.4 and let production function be instead $f(k) = \frac{1}{2}\big(1 + \text{erf}(\log(k) - \mu\big)$, where $\text{erf}(x) = \frac{1}{2\pi}\int_0^x e^{-s^2} ds$. Study the local and global dynamic properties of the capital accumulation model.

Exercise 3.8 (*Population dynamics: the logistic model*) The Malthusian model of population growth in Example 2.5 was found to be at odds with data on population since early XVIII century. Verhulst (1845), and latter researchers, found that the non-linear ODE, called logistic equation

$$\dot{n} = \gamma(n)\, n, \text{ where } \gamma(n) = \eta\left(1 - \frac{n}{k}\right),$$

where η and k (the carrying capacity) are parameters, provided a better description of the population data.[26] Study the local and global dynamic properties of the logistic population model.

Exercise 3.9 (*Logistic function*) The logistic probability function $p : \mathbb{R} \to (0, 1)$ is determined from a variable growth model

$$\frac{p'(x)}{p(x)} = \gamma(p(x)) \equiv \beta_1 (1 - p(x))\, p(x), \text{ for each } x \in \mathbb{R},$$

Assuming that $p(0) = \frac{1}{1+e^{-\beta_0}}$, for $x = 0$, solve the ODE problem to find the logistic function, assuming that β_0 and β_1 are arbitrary constants.

[26] See Banks (1994) for this and other population models represented by differential equations.

3.4 References

Hale and Koçak (1991), Part I , III) contains a very good introduction to non-linear dynamics. Kuznetsov (2023) is a more advanced reference. Brock and Malliaris (1989) is an important reference with applications in economics, and Grass et al. (2008, Chap. 2) has a compact presentation of all the important results, together with several examples in economics and related fields.

3.5 Appendix: Solutions to Some ODE

3.5.1 Solution to Ricatti's Equation

Consider ODE (3.3). First, let $a = 0$. Separating variables, we have $\frac{dy}{y^2} = dt$, and integrating, $\int_{y(0)}^{y(t)} \frac{dy}{y^2} = \int_0^t ds$, yields $\frac{1}{y(0)} - \frac{1}{y(t)} = t$, and solution (3.4a). Second, let $a \neq 0$. At this point, it is convenient to note that $\frac{d\tan^{-1}(x)}{dx} = \frac{1}{1+x^2}$ and $\frac{d\tanh^{-1}(x)}{dx} = \frac{1}{1-x^2}$, where $\tan(x) = \frac{\sin(x)}{\cos(x)} = \frac{e^{ix}-e^{-ix}}{i(e^{ix}+e^{-ix})}$ and $\tanh(x) = \frac{e^x-e^{-x}}{e^x+e^{-x}}$. Therefore, if $a > 0$ integrating yields

$$\int \frac{dy}{a+y^2} = dt \Leftrightarrow \frac{1}{\sqrt{a}} \int \frac{1}{1+x^2} dx = t + k \Leftrightarrow \frac{1}{\sqrt{a}} \tan^{-1}(x) = t + k$$

where $x = y/\sqrt{a}$. Solving the last equation for x and mapping back to y yields solution (3.4b). If $a < 0$ we integrate equation (3.3) by using a similar transformation, $x = y/\sqrt{-a}$, we obtain

$$\int \frac{dy}{a + y^2} = dt \Leftrightarrow -\frac{1}{\sqrt{-a}} \int \frac{1}{1-x^2} dx = t + k \Leftrightarrow -\frac{1}{\sqrt{-a}} \tanh^{-1}(x) = t + k,$$

which, by transforming back to y, yields solution (3.4c).

3.5.2 Bernoulli ODE

A Bernoulli ODE is a non-linear ODE specified by

$$\dot{y} = ay + by^{\eta}, \quad \text{for } y : T \to \mathbb{R} \tag{3.21}$$

where $\eta \neq 1$, $a \neq 0$ and $b \neq 0$ are real numbers. We can obtain the explicit solution by transforming it to a simple linear ODE. A first transformation, $z(t) = y(t)^{1-\eta}$, yields an equivalent non-homogenous linear ODE, $\dot{z} = (1-\eta)y^{-\eta}\dot{y} = (1-\eta)(ay^{1-\eta}+b) = (1-\eta)(az+b)$. A second transformation, $w(t) = z(t) + \frac{b}{a}$, yields the equivalent homogeneous ODE $\dot{w} = a(1-\eta)w$, that has the solution $w(t) = w(0)\, e^{a(1-\eta)t}$. Inverting the second transformation yields $z(t) = -\frac{b}{a} + \left(z(0) + \frac{b}{a}\right) e^{a(1-\eta)t}$. We finally obtain the solution to the Bernoulli equation (3.21) by inverting the first transformation, yielding

$$y(t) = \left[-\frac{b}{a} + \left(y(0)^{1-\eta} + \frac{b}{a} \right) e^{a(1-\eta)t} \right]^{\frac{1}{1-\eta}}. \tag{3.22}$$

If $a = 0$ the solution is the particular case $y(t) = \left(y(0)^{1-\eta} + b\,(1-\eta)\,t \right)^{\frac{1}{1-\eta}}$. If $\eta = 2$ and $b = 1$ we have the quadratic Bernoulli equation (3.5), with solutions (3.6), and if $\eta = 3$ and $b = -1$ we have the cubic Bernoulli equation (3.7) with solutions (3.8).

3.5.3 *Solution to the Cubic Polynomial Equation*

In this section, we present the solutions for the cubic polynomial equation

$$x^3 + \alpha_2\, x^2 + \alpha_1\, x + \alpha_0 = 0.$$

By performing a Tschirnhausen transformation (see King 1996), we can transform into a simpler, irreducible, equation. Defining $y = x + \frac{\alpha_2}{3}$ and $a_1 \equiv \frac{\alpha_2^2}{3} - \alpha_1$, and $a_0 \equiv \alpha_0 - \frac{\alpha_1\,\alpha_2}{3} + \frac{2\,\alpha_2^3}{27}$, we obtain the following (monic) cubic polynomial equation

$$y^3 - a_1 y + a_0 = 0. \tag{3.23}$$

If $a_0 = 0$ we have $y\,(y^2 - a_1\, y) = 0$. The solutions are $y_1 = 0$ and the roots of the quadratic equation, $y^2 - a_1\, y = 0$, $y_2 = \sqrt{a_1}$, and $y_3 = -\sqrt{a_1}$. If $a_0 \neq 0$, we prove that the solutions for the monic cubic equation are

$$y_j = w^{j-1}\theta^{\frac{1}{3}} + \frac{b}{3}\left(\omega^{j-1}\theta^{\frac{1}{3}}\right)^{-1}, \quad j = 1, 2, 3 \tag{3.24}$$

where θ is in Eq. (3.26), and $\{\omega^j\}_{j=0,1,2}$ is in Eq. (3.27). It can be shown that the solutions to the monic equation (3.23), denoted by y_1, y_2, and y_3, depend on the discriminant

$$\Delta \equiv \left(\frac{a_0}{2}\right)^2 - \left(\frac{a_1}{3}\right)^3. \tag{3.25}$$

In particular: (1) if $\Delta < 0$, there are three real roots; (2) if $\Delta = 0$, there are two real roots (with two multiple roots, i.e., equal); or (3) if $\Delta > 0$, there is one real root and two complex conjugate roots. To find the solutions, write $y = u + v$. Then we get the equivalent representation

$$u^3 + v^3 + 3\left(u\,v - \frac{a_1}{3}\right)(u+v) + a_0 = 0.$$

As $y = u + v \neq 0$, u and v solve simultaneously

$$\begin{cases} u^3 + v^3 + a_0 = 0 \\ u\,v - \dfrac{a_1}{3} = 0 \end{cases} \Leftrightarrow \begin{cases} u^3\,u^3 + u^3\,v^3 + u^3\,a_0 = 0 \\ u^3\,v^3 = \left(\frac{a_1}{3}\right)^3 \end{cases} \Leftrightarrow \begin{cases} u^6 + a_0 u^3 + \left(\frac{a_1}{3}\right)^3 = 0 \\ u\,v = \frac{a_1}{3}. \end{cases}$$

The first equation is a quadratic polynomial in u^3 which has roots $u^3 = -\frac{a_0}{2} \pm \sqrt{\Delta}$, where Δ is the discriminant of equation (3.23). We can take any solution to the previous equation and set

$$\theta \equiv -\frac{a_0}{2} + \sqrt{\Delta}. \tag{3.26}$$

Three cases are possible: (1) if $\Delta > 0$ then $\theta > -\frac{a_0}{2}$ is a real number, (2) if $\Delta = 0$ then $\theta = -\frac{a_0}{2}$; or (3) if $\Delta < 0$ then θ is a complex number $\theta = -\frac{a_0}{2} + \sqrt{|\Delta|}\,i$ where $i = \sqrt{-1}$.

At this stage, it is useful to observe that the solutions to equation $x^3 = 1$ are $x_1 = 1 = \omega^0$, $x_2 = \omega$, and $x_3 = \omega^2$. where

$$\omega^0 = 1,\ \omega = e^{\frac{2\pi i}{3}} = -\frac{1}{2}(1 - \sqrt{3}i),\ \text{and}\omega^2 = e^{\frac{4\pi i}{3}} = -\frac{1}{2}(1 + \sqrt{3}i). \tag{3.27}$$

Therefore, $u^3 = \theta$ has also three solutions $u_1 = \omega^0\,\theta^{\frac{1}{3}}$, $u_2 = \omega\theta^{\frac{1}{3}}$, and $u_3 = \omega^2\theta^{\frac{1}{3}}$, and because $v = \frac{a_1}{3u}$, we finally obtain the solutions (3.24) to the polynomial equation (3.23).

3.5.4 *Steady State for the Abel Equation*

In Abel's equation (3.9), the steady state set y^{ss} is determined from the solutions to the equation $y^3 - y - a = 0$. Applying the previous results, we have $\theta = \frac{a}{2} + \sqrt{\Delta}$, where $\Delta = \left(\frac{a}{2}\right)^2 - \left(\frac{1}{3}\right)^3$. Therefore, $\Delta = 0$ if and only if $a = \pm 2\,3^{-\frac{3}{2}}$, which implies $\theta = \pm 3^{-\frac{3}{2}}$, and there are two sets of two real roots: (1) if $a = -2\,3^{-\frac{3}{2}}$, the two real solutions are $-3^{-\frac{1}{2}}$ and $2\,3^{-\frac{1}{2}}$; or (2) if $a = 2\,3^{-\frac{3}{2}}$, they are $3^{-\frac{1}{2}}$ and $-2\,3^{-\frac{1}{2}}$.

Chapter 4
Planar Linear ODE

Planar linear ordinary differential equations (ODE) extend scalar linear ODE, dealt in Chap. 2, by adding a new dependent variable, and coupling two linear ODE. In this chapter, we restrain the analysis to autonomous linear ODE, of the following form,

$$y_1'(x) = a_{11}\, y_1(x) + a_{12}\, y_2(x) + b_1, \tag{4.1}$$

$$y_2'(x) = a_{21}\, y_1(x) + a_{22}\, y_2(x) + b_2, \tag{4.2}$$

where the coefficients are real numbers. Planar linear ODE also have closed form solutions. However, the greater dimension of the dependent variable not only originates new types of solutions but also adds new qualitative features to them.

In this chapter, it is convenient to write equation (4.1) in matrix form:

Definition 4.1 A **planar linear autonomous ODE** is a functional equation, involving the vector function $\mathbf{y} : \mathrm{X} \subseteq \mathbb{R} \to \mathbf{Y} \subseteq \mathbb{R}^2$ and its gradient $\mathbf{y}'$, of the form

$$\mathbf{y}' = \mathbf{A}\,\mathbf{y} + \mathbf{B}, \;\; \text{for each } x \in \mathrm{X}, \tag{4.3}$$

where the coefficient matrix $\mathbf{A} \in \mathbb{R}^{2\times 2}$ and vector $\mathbf{B} \in \mathbb{R}^2$ are fixed. We say the ODE is **homogeneous** if $\mathbf{B} = \mathbf{0}$, and that it is **non-homogeneous** if $\mathbf{B} \neq \mathbf{0}$.

Extending the previous vector and matrices, we have

$$\mathbf{y}(x) \equiv \begin{pmatrix} y_1(x) \\ y_2(x) \end{pmatrix}, \; \mathbf{y}' \equiv \begin{pmatrix} y_1'(x) \\ y_2'(x) \end{pmatrix}, \; \mathbf{A} \equiv \begin{pmatrix} a_{11} & a_{12} \\ a_{21} & a_{22} \end{pmatrix}, \; \mathbf{B} \equiv \begin{pmatrix} b_1 \\ b_2 \end{pmatrix}. \tag{4.4}$$

If time is the independent variable, we have the mapping $\mathbf{y} = \mathbf{y}(t)$, where $t \in \mathrm{T} \subseteq \mathbb{R}_+$, and specify instead the ODE as

P. B. Brito, *Economic Dynamics and Distributions*, Dynamic Modeling and Econometrics in Economics and Finance 32, https://doi.org/10.1007/978-3-031-94717-9_4

$$\begin{aligned} \dot{y}_1 &= a_{11}\, y_1(t) + a_{12}\, y_2(t) + b_1, \\ \dot{y}_2 &= a_{21}\, y_1(t) + a_{22}\, y_2(t) + b_2, \end{aligned} \tag{4.5}$$

or, in matrix notation,

$$\dot{\mathbf{y}} = \mathbf{A}\,\mathbf{y} + \mathbf{B}, \text{ for each } t \in \mathrm{T}. \tag{4.6}$$

Next, Sect. 4.1 addresses the solution to planar linear ODE and problems for a generic independent variable, Sect. 4.2 presents an introduction to dynamics of time-dependent linear planar autonomous ODE, and Sect. 4.3 contains some applications to economics.

4.1 Generic Planar ODE

We start with the simplest homogeneous equations in Sect. 4.1.1, continue with non-homogeneous equations, in Sect. 4.1.2, and conclude this section with the solution to ODE problems, in Sect. 4.1.3.

4.1.1 Homogeneous ODE

Any planar linear homogeneous ODE, containing $\mathbf{y} = \mathbf{y}(x)$, has an explicit general solution:

Proposition 4.1 (General solution to the planar linear homogeneous ODE) *Consider the mapping* $\mathbf{y} : \mathrm{X} \subseteq \mathbb{R} \to \mathbf{Y} \subseteq \mathbb{R}^2$ *and let* $\mathbf{A} \in \mathbb{R}^{2\times 2}$ *be any real matrix. The general solution to the planar linear homogeneous autonomous ODE* $\mathbf{y}' = \mathbf{A}\,\mathbf{y}$ *is the vector function*

$$\mathbf{y}(x) = \mathbf{e}^{\mathbf{A}\,(x-x_0)}\,\mathbf{y}(x_0), \text{ for each } x \in \mathrm{X}, \tag{4.7}$$

where $\mathbf{e}^{\mathbf{A}\,(x-x_0)}$ *is the* (2×2)*-matrix exponential function,*

$$\mathbf{e}^{\mathbf{A}\,(x-x_0)} \equiv \sum_{n=0}^{+\infty} \frac{1}{n!}\mathbf{A}^n (x-x_0)^n = \mathbf{I} + \mathbf{A}\,(x-x_0) + \frac{1}{2}\mathbf{A}^2\,(x-x_0)^2 + \ldots, \tag{4.8}$$

and $(x_0, \mathbf{y}(x_0))$ *is an arbitrary element of* $\mathrm{X} \times \mathbf{Y}$.

Proof Let us conjecture that the solution to the homogeneous ODE is a vector function $\mathbf{y}(x) = \mathbf{f}(x)$, for $\mathbf{f}(x) = \mathbf{e}^{\mathbf{A}\,x}\,\mathbf{f}_0$, where $\mathbf{f}_0$ is a vector of constants. From Lemma 4.6 (iii), in Appendix 4.6, we have $\mathbf{f}'(x) = \mathbf{A}\,\mathbf{e}^{\mathbf{A}\,x}\,\mathbf{f}_0 = \mathbf{A}\,\mathbf{f}(x)$. Therefore, our conjecture is right for every $x \in \mathrm{X}$, and therefore, it is also right for an arbitrary x_0 if $\mathbf{y}(x_0) = \mathbf{e}^{\mathbf{A}\,x_0}\,\mathbf{f}_0$. Solving for $\mathbf{f}_0$, yields $\mathbf{f}_0 = \mathbf{y}(x_0)\,\mathbf{e}^{-\mathbf{A}\,x_0}$, and substituting in our conjectured function, yields equation (4.7). □

The fundamental component of the solution to a planar linear ODE is the exponential matrix function $\mathbf{e}^{\mathbf{A}x}$ associated to coefficient matrix $\mathbf{A}$. The existence of a taxonomy of matrix exponential functions allows us to establish a taxonomy of the solutions to the homogeneous ODE.

We need to recall some facts from matrix algebra.[1] Let λ be a complex number. The **characteristic polynomial** of matrix $\mathbf{A}$ is the square polynomial function of λ,

$$c(\lambda) = \det\left(\mathbf{A} - \lambda\,\mathbf{I}\right) = \lambda^2 - \text{tr}(\mathbf{A})\,\lambda + \det(\mathbf{A}),$$

where $\text{tr}(\mathbf{A}) = a_{11} + a_{22}$ is the trace, and $\det(\mathbf{A}) = a_{11}\,a_{22} - a_{12}\,a_{21}$ is the determinant of matrix $\mathbf{A}$. We call **characteristic equation** to the polynomial equation $c(\lambda) = 0$, and **spectrum of matrix A** to the subset of the set of two-dimensional complex numbers $\sigma(\mathbf{A}) = \left\{\lambda \in \mathbb{C}^2 : \det\left(\mathbf{A} - \lambda\,\mathbf{I}\right) = 0\right\}$ that are solutions to the characteristic equation. The elements of the spectrum $\sigma(\mathbf{A})$ are called the **eigenvalues** of $\mathbf{A}$, and are

$$\lambda_- = \frac{\text{tr}(\mathbf{A})}{2} - \sqrt{\Delta(\mathbf{A})}, \text{ and } \lambda_+ = \frac{\text{tr}(\mathbf{A})}{2} + \sqrt{\Delta(\mathbf{A})} \tag{4.9}$$

where $\Delta(\mathbf{A}) \equiv \left(\frac{\text{tr}(\mathbf{A})}{2}\right)^2 - \det(\mathbf{A}) \in \mathbb{R}$ is the **discriminant** of matrix $\mathbf{A}$.

The sign of the discriminant allows us to distinguish between three cases: if $\Delta(\mathbf{A}) > 0$ then both eigenvalues are **real and distinct**, satisfying $\lambda_+ > \lambda_-$; if $\Delta(\mathbf{A}) = 0$ then both eigenvalues are **real and equal**—or multiple—satisfying $\lambda_+ = \lambda_- = \lambda = \frac{\text{tr}(\mathbf{A})}{2}$; and, if $\Delta(\mathbf{A}) < 0$ then the eigenvalues are **complex conjugate**, satisfying $\lambda_\pm = \alpha \pm \beta\,i$, where $\alpha = \frac{\text{tr}(\mathbf{A})}{2}$, $\beta = \sqrt{|\Delta(\mathbf{A})|}$, and $i \equiv \sqrt{-1}$ is the imaginary number. Figure 4.1 illustrates the previous three cases in a $(\text{tr}(\mathbf{A}), \det(\mathbf{A}))$ cartesian graph.

The following matrices are called **Jordan canonical forms**

$$\mathbf{\Lambda}_+ = \begin{pmatrix} \lambda_- & 0 \\ 0 & \lambda_+ \end{pmatrix}, \; \mathbf{\Lambda}_0 = \begin{pmatrix} \lambda & 1 \\ 0 & \lambda \end{pmatrix}, \; \mathbf{\Lambda}_- = \begin{pmatrix} \alpha & \beta \\ -\beta & \alpha \end{pmatrix}. \tag{4.10}$$

We call **canonical matrix** to any matrix that is equal or has the same spectrum as a Jordan canonical form.

All the other matrices are called **non canonical** matrices. Any **canonical matrix** $\mathbf{A}$ that is not in the Jordan canonical form can be transformed into a Jordan canonical form by a similarity transformation. Matrix $\mathbf{A}$ is **similar** to a Jordan canonical form $\mathbf{\Lambda}$, if there is a (non-singular) matrix $\mathbf{P} \in \mathbb{R}^{2\times 2}$ such that the following relationship between matrix $\mathbf{A}$ and a Jordan canonical form $\mathbf{\Lambda}$ is satisfied

$$\mathbf{A} = \mathbf{P}\,\mathbf{\Lambda}\,\mathbf{P}^{-1} \Leftrightarrow \mathbf{\Lambda} = \mathbf{P}^{-1}\,\mathbf{A}\,\mathbf{P}. \tag{4.11}$$

[1] See Appendix 4.5.1.

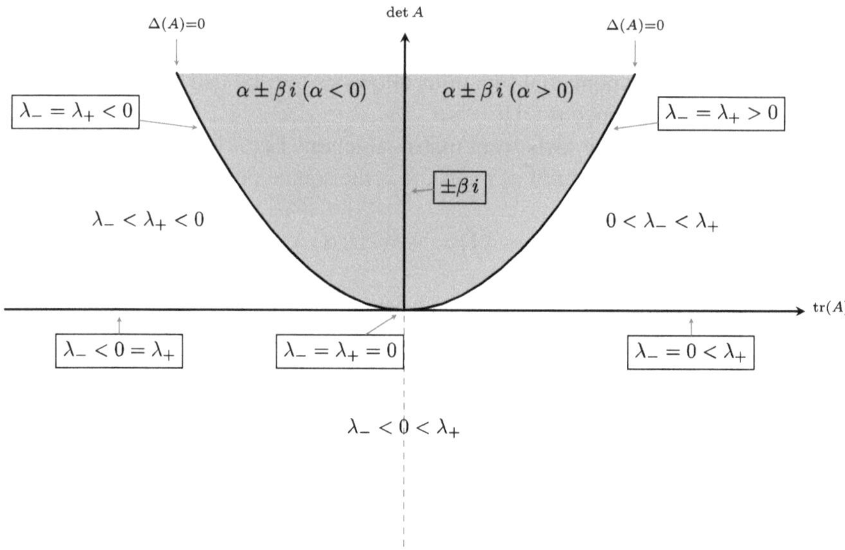

Fig. 4.1 Eigenvalues of **A** in the (tr(**A**), det (**A**)) space. The gray area corresponds to the existence of complex conjugate eigenvalues

If matrix **P** is the **eigenvector matrix** associated to matrix **A**, then matrix **A** can be related to its similar Jordan canonical form by the following Lemma[2]:

Lemma 4.1 (Jordan canonical form of a matrix **A**) *Consider a canonical matrix* **A**. *Then:*

1. *if* $\Delta(\mathbf{A}) > 0$ *then matrix* **A** *is similar to the Jordan canonical form* $\mathbf{\Lambda}_+$, *and* $\sigma(\mathbf{A}) = \sigma(\mathbf{\Lambda}_+) = \{\lambda_-, \lambda_+\}$, *where* $\lambda_- = \frac{\mathrm{tr}(\mathbf{A})}{2} - \sqrt{\Delta(\mathbf{A})}$ *and* $\lambda_+ = \frac{\mathrm{tr}(\mathbf{A})}{2} + \sqrt{\Delta(\mathbf{A})}\}$;
2. *if* $\Delta(\mathbf{A}) = 0$ *then matrix* **A** *is similar to the Jordan canonical form* $\mathbf{\Lambda}_0$, *and* $\sigma(\mathbf{A}) = \sigma(\mathbf{\Lambda}_0) = \{\lambda, \lambda\}$, *where* $\lambda = \frac{\mathrm{tr}(\mathbf{A})}{2}$;
3. *if* $\Delta(\mathbf{A}) < 0$ *then matrix* **A** *is similar to the Jordan canonical form* $\mathbf{\Lambda}_-$, *and* $\sigma(\mathbf{A}) = \sigma(\mathbf{\Lambda}_-) = \{\alpha - \beta i, \alpha + \beta i\}$, *such that* $\alpha = \frac{\mathrm{tr}(\mathbf{A})}{2}$, *and* $\beta = \sqrt{\mid \Delta(\mathbf{A}) \mid}$ *are the real and the complex part of the eigenvalues.*

Proof Proof see the Appendix 4.5.1. □

A **non-canonical square matrix** is a matrix with the form

$$\mathbf{A}_h = \begin{pmatrix} \alpha & \beta \\ \beta & \alpha \end{pmatrix}, \tag{4.12}$$

where α and β are real numbers. In particular, if $\beta = 0$ then matrix $\mathbf{A}_d = \alpha\, \mathbf{I}_{2\times 2}$ is called **diagonal matrix**. Non-canonical matrices are distinct from canonical matrices because their diagonals and anti-diagonals have equal elements, that is $a_{11} = a_{22}$

[2] Some authors call this the eigen decomposition theorem.

and $a_{12} = a_{21}$. As $\mathrm{tr}(\mathbf{A}) = 2\,\alpha$ and $\det(\mathbf{A}) = \alpha^2 - \beta^2$, then $\Delta(\mathbf{A}) = \beta^2 \geq 0$, and the eigenvalues $\mathbf{A}_h$ are always real: $\sigma(\mathbf{A}_h) = \{\lambda_{h,-}, \lambda_{h,+}\}$, where $\lambda_{h,-} = \alpha - |\beta|$ and $\lambda_{h,+} = \alpha + |\beta|$. There is multiplicity for the case of the diagonal matrix $\mathbf{A}_d$ in which $\sigma(\mathbf{A}_d) = \{\alpha\}$.

The transformation similar to (4.11), $\mathbf{A}_h = \mathbf{P}_h\,\mathbf{\Lambda}_h\,\mathbf{P}_h^{-1}$, also applies with

$$\mathbf{\Lambda}_h = \begin{pmatrix} \alpha - |\beta| & 0 \\ 0 & \alpha + |\beta| \end{pmatrix}, \text{ and } \mathbf{P}_h = (\mathbf{P}_h^-, \mathbf{P}_h^+) = \begin{pmatrix} -\beta & \beta \\ |\beta| & |\beta| \end{pmatrix}. \tag{4.13}$$

The **exponential matrix functions** for canonical matrices, both in the Jordan canonical form or not, are provided by the next Lemma:

Lemma 4.2 (Matrix exponential functions for Jordan canonical forms) *Consider the Jordan canonical forms in Eq. (4.10). Then the exponential matrix functions* $\mathbf{e}^{\mathbf{\Lambda}_i x}$*, associated to* $\mathbf{\Lambda}_i$*, for* $i \in \{+, 0, -\}$*, are*[3]

$$\mathbf{e}^{\mathbf{\Lambda_+ x}} = \begin{pmatrix} e^{\lambda_- x} & 0 \\ 0 & e^{\lambda_+ x} \end{pmatrix}, \ \mathbf{e}^{\mathbf{\Lambda_0 x}} = e^{\lambda x}\begin{pmatrix} 1 & x \\ 0 & 1 \end{pmatrix}, \ \mathbf{e}^{\mathbf{\Lambda_- x}} = e^{\alpha x}\begin{pmatrix} \cos\beta x & \sin\beta x \\ -\sin\beta x & \cos\beta x \end{pmatrix} \tag{4.14}$$

If matrix $\mathbf{A}$ *is similar to a Jordan form* $\mathbf{\Lambda}_i$*, for* $i \in \{+, 0, -\}$*, then its matrix exponential function is*

$$\mathbf{e}^{\mathbf{A}\,x} = \mathbf{P}\,\mathbf{e}^{\mathbf{\Lambda}_i\,x}\,\mathbf{P}^{-1}, \ \textit{for } i \in \{+, 0, -\}. \tag{4.15}$$

Proof See Appendix 4.6. □

Corollary 4.1 (Matrix exponential functions for non-canonical matrices) *Let the coefficient matrix be a non-canonical matrix as in Eq. (4.12). Then its associated exponential matrix is*[4]

$$\mathbf{e}^{\mathbf{A}_h x} = e^{\alpha\,x}\begin{pmatrix} \cosh(\beta\,x) & \sinh(\beta\,x) \\ \sinh(\beta\,x) & \cosh(\beta\,x) \end{pmatrix}. \tag{4.16}$$

Table 4.1 provides a taxonomy of exponential matrices that appear in the general solution to the homogeneous planar linear ODE, Eq. (4.7).

---

[3] From Euler's formula as $e^{\alpha+\beta i} = e^{\alpha\,x}(\cos\beta x + i\,\sin\beta x)$, and $e^{\alpha-\beta i} = e^{\alpha\,x}(\cos\beta x - i\,\sin\beta x)$. It can be convenient to consider, instead, the canonical form of the exponential matrix as a $\mathbb{C}^{2\times 2}$—rather than a $\mathbb{R}^{2\times 2}$ matrix—$\mathbf{\Lambda}_- = \begin{pmatrix} e^{(\alpha+\beta i)x} & 0 \\ 0 & e^{(\alpha-\beta i)x} \end{pmatrix} = e^{\alpha\,x}\begin{pmatrix} \cos\beta x + i\,\sin\beta x & 0 \\ 0 & \cos\beta x - i\,\sin\beta x \end{pmatrix}$. In this case, the eigenvector matrix $\mathbf{P}$ will also be a $\mathbb{C}^{2\times 2}$, not a $\mathbb{R}^{2\times 2}$, matrix, as in the main text. See Appendix 4.5.1.

[4] Proof: As $\mathbf{A}_h = \mathbf{P}_h\,\mathbf{\Lambda}_h\,\mathbf{P}_h^{-1}$ and as $\mathbf{\Lambda}_h$ is a diagonal matrix—see Eq. (4.13)—then $\mathbf{e}^{\mathbf{\Lambda}_h\,x} = \begin{pmatrix} e^{\lambda_{h,-}\,x} & 0 \\ 0 & e^{\lambda_{h,+}\,x} \end{pmatrix}$, and $\mathbf{e}^{\mathbf{A}_h x} = \mathbf{P}_h\,\mathbf{e}^{\mathbf{\Lambda}_h\,x}\,\mathbf{P}_h^{-1} = \frac{e^{\alpha\,x}}{2}\begin{pmatrix} e^{-|\beta|\,x} + e^{|\beta|\,x} & -\frac{\beta}{|\beta|}\left(e^{-|\beta|\,x} - e^{|\beta|\,x}\right) \\ -\frac{|\beta|}{\beta}\left(e^{-|\beta|\,x} - e^{|\beta|\,x}\right) & e^{-|\beta|\,x} + e^{|\beta|\,x} \end{pmatrix}$, that simplifies to Eq. (4.16) if we recall $\cosh(\beta\,x) = \frac{1}{2}\left(e^{\beta x} + e^{-\beta x}\right)$ and $\sinh(\beta\,x) = \frac{1}{2}\left(e^{\beta x} - e^{-\beta x}\right)$. For the diagonal matrix, we have $\mathbf{e}^{\mathbf{A_d x}} = e^{\alpha x}\left(\begin{smallmatrix} 1 & 0 \\ 0 & 1 \end{smallmatrix}\right)$.

Table 4.1 Coefficient matrices and their exponential functions

	Matrix algebra $\mathbf{A}$ similar to Jordan normal forms			Other matrices	
	$\Delta(\mathbf{A}) > 0$	$\Delta(\mathbf{A}) = 0$	$\Delta(\mathbf{A}) < 0$		
Jordan forms	$\mathbf{\Lambda}_+ = \begin{pmatrix} \lambda_- & 0 \\ 0 & \lambda_+ \end{pmatrix}$	$\mathbf{\Lambda}_0 = \begin{pmatrix} \lambda & 1 \\ 0 & \lambda \end{pmatrix}$	$\mathbf{\Lambda}_- = \begin{pmatrix} \alpha & \beta \\ 0 & -\beta\ \alpha \end{pmatrix}$		
$\mathbf{A}$	$\mathbf{P}\,\mathbf{\Lambda}_+\,\mathbf{P}^{-1}$	$\mathbf{P}\,\mathbf{\Lambda}_0\,\mathbf{P}^{-1}$	$\mathbf{P}\,\mathbf{\Lambda}_-\,\mathbf{P}^{-1}$	$\mathbf{A}_d = \begin{pmatrix} \lambda & 0 \\ 0 & \lambda \end{pmatrix}$	$\mathbf{A}_h = \begin{pmatrix} \alpha & \beta \\ \beta & \alpha \end{pmatrix}$
$\mathbf{e}^{\mathbf{\Lambda}\,x}$	$\mathbf{e}^{\mathbf{\Lambda}_+ x} = \begin{pmatrix} e^{\lambda_- x} & 0 \\ 0 & e^{\lambda_+ x} \end{pmatrix}$	$\mathbf{e}^{\mathbf{\Lambda}_0 x} = e^{\lambda x} \begin{pmatrix} 1 & x \\ 0 & 1 \end{pmatrix}$	$\mathbf{e}^{\mathbf{\Lambda}_- x} = e^{\alpha x} \begin{pmatrix} \cos\beta x & \sin\beta x \\ -\sin\beta x & \cos\beta x \end{pmatrix}$		
$\mathbf{e}^{\mathbf{A}\,x}$	$\mathbf{P}\,\mathbf{e}^{\mathbf{\Lambda}_+ x}\,\mathbf{P}^{-1}$	$\mathbf{P}\,\mathbf{e}^{\mathbf{\Lambda}_0 x}\,\mathbf{P}^{-1}$	$\mathbf{P}\,\mathbf{e}^{\mathbf{\Lambda}_- x}\,\mathbf{P}^{-1}$	$e^{\lambda x} \begin{pmatrix} 1 & 0 \\ 0 & 1 \end{pmatrix}$	$e^{\alpha x} \begin{pmatrix} \cosh(\beta x) & \sinh(\beta x) \\ \cosh(\beta x) & \sinh(\beta x) \end{pmatrix}$

Superposition principle. It is useful to consider the relationship between solutions to ODE with canonical matrices and their relation to the similar matrices in the Jordan canonical form. Let us start with the case in which $\mathbf{A} = \mathbf{\Lambda}_+$, and the planar homogeneous ODE $\mathbf{w}' = \mathbf{\Lambda}_+ \, \mathbf{w}$, where $\mathbf{w} = (w_-, w_+)^\top$. The general solution to this ODE is $\mathbf{w}(x) = \mathbf{e}^{\mathbf{\Lambda}_+} \, \mathbf{w}(0)$. As $\mathbf{\Lambda}_+$ is a diagonal matrix, this planar ODE assembles two uncoupled homogeneous scalar linear ODE,

$$\begin{cases} w_-' = \lambda_- \, w_-, & (4.17a) \\ w_+' = \lambda_+ \, w_+, & (4.17b) \end{cases}$$

implying that its solution also involves two decoupled exponential functions

$$\begin{cases} w_-(x) = w_-(x_0) \, e^{\lambda_-(x-x_0)}, & (4.18a) \\ w_+(x) = w_+(x_0) \, e^{\lambda_+(x-x_0)}. & (4.18b) \end{cases}$$

Now, let $\mathbf{A}$ satisfy $\Delta(\mathbf{A}) > 0$ but let it be different from its similar Jordan form, that is $\mathbf{A} \neq \mathbf{\Lambda}_+$, and consider the differential equation $\mathbf{y}' = \mathbf{A} \, \mathbf{y}$. As $\mathbf{A}$ is similar to $\mathbf{\Lambda}_+$, we introduce the transformation $\mathbf{y}(x) = \mathbf{P} \, \mathbf{w}(x)$, where matrix $\mathbf{P}$ is the (invertible) eigenvector matrix of $\mathbf{A}$. Then, as $\mathbf{w}' = \mathbf{P}^{-1} \, \mathbf{y}' = \mathbf{P}^{-1} \, \mathbf{A} \, \mathbf{y} = \mathbf{P}^{-1} \, \mathbf{A} \, \mathbf{P} \, \mathbf{w} = \Lambda_+ \, \mathbf{w}$, we obtain the previous decoupled ODE (4.17). The inverse transformation, $\mathbf{y} = \mathbf{P} \, \mathbf{w}$, yields the general solution

$$\mathbf{y}(x) = \mathbf{P}^- \, w_-(x_0) \, e^{\lambda_-(x-x_0)} + \mathbf{P}^+ \, w_+(x_0) \, e^{\lambda_+(x-x_0)}, \quad \text{for each } x \in \mathrm{X}, \tag{4.19}$$

where $\mathbf{P}^-$ ($\mathbf{P}^+$) is the eigenvector associated to the eigenvalue λ_- (λ_+), and $\mathbf{w}(x_0) = \big(w_-(x_0), \, w_+(x_0)\big)^\top = \mathbf{P}^{-1} \mathbf{y}(x_0)$. The fact that the solution to the planar ODE can be expressed as the weighted sum of the solutions of two decoupled scalar ODE is called the **superposition principle**.

This principle applies to ODE having canonical coefficient matrices $\mathbf{A}$ with positive discriminants, that are similar to the $\mathbf{\Lambda}_+$ Jordan canonical form. It also applies to ODE with non-canonical coefficient matrices, even for the case in which $\beta = 0$. In this case,—see Eq. (4.13)—we have $\lambda_- = \lambda_{h,-} = \alpha - |\beta|$ and $\lambda_+ = \lambda_{h,+} = \alpha + |\beta|$, and the eigenvectors $\mathbf{P}^- = \mathbf{P}_h^-$ and $\mathbf{P}^+ = \mathbf{P}_h^+$.

For ODE having canonical coefficient matrices with non-positive discriminants, a slightly different type of superposition applies, as can be seen in the next Exercises:

Exercise 4.1 Let the coefficient matrix $\mathbf{A}$ have $\Delta(\mathbf{A}) = 0$. Prove that the solution of the homogeneous ODE is

$$\mathbf{y}(x) = \left(\Big(w_-(x_0) + w_+(x_0) \, (x - x_0) \Big) \, \mathbf{P}^s + w_+(x_0) \, \mathbf{P}^g \right) e^{\lambda(x-x_0)}, \quad \text{for each } x \in \mathrm{X} \tag{4.20}$$

where, $\mathbf{P}^s$ ($\mathbf{P}^g$) is the simple (generalized) eigenvector of $\mathbf{A}$,[5] and $\mathbf{w}(x_0)$ is the solution to the ODE $\mathbf{w}' = \mathbf{\Lambda}_0\, \mathbf{w}$ evaluated at $x = x_0$.

Exercise 4.2 Let the coefficient matrix $\mathbf{A}$ have $\Delta(\mathbf{A}) < 0$. Prove that the solution can be expressed as

$$\mathbf{y}(x) = e^{\alpha(x-x_0)} \Big(w_-(x_0) \Big(\cos\beta(x - x_0)\, \mathbf{P}^r - \sin\beta(x - x_0)\, \mathbf{P}^i \Big) +$$

$$+\, w_+(x_0) \Big(\sin\beta(x - x_0)\, \mathbf{P}^r + \cos\beta(x - x_0)\, \mathbf{P}^i \Big) \Big), \text{ for each } x \in \mathrm{X}, \quad (4.21)$$

where, $\mathbf{P} = (\mathbf{P}^r, \mathbf{P}^i)$ is the eigenvector matrix of $\mathbf{A}$,[6] and $\mathbf{w}(x_0)$ is the solution to the ODE $\mathbf{w}' = \mathbf{\Lambda}_-\, \mathbf{w}$ evaluated at $x = x_0$.

Exercise 4.3 Coefficient matrices with zero eigenvalues or complex eigenvalues with zero real parts are important special cases. From equations (4.9) and Lemma 4.3, we can conclude that if $\det(\mathbf{A}) = 0$, there is one or there are two zero eigenvalues, and if $\mathrm{tr}(\mathbf{A}) = 0$ and $\det(\mathbf{A}) > 0$, there are two complex conjugate eigenvalues with zero real parts. The first cases exist for canonical matrices similar to Jordan forms $\mathbf{\Lambda}_+$, $\mathbf{\Lambda}_0$, and for diagonal matrices $\mathbf{A}_d$, and the third case exists for canonical matrices similar to Jordan forms $\mathbf{\Lambda}_-$, and for hyperbolic matrices $\mathbf{A}_h$. Show that for all those cases the general solution to the planar linear homogeneous ODE in matricial form is (4.7), with exponential matricial functions in which $e^{\lambda_- x} = 1$, $e^{\lambda_+ x} = 1$ or $e^{\lambda x} = 1$ for the first cases, and $e^{\alpha x} = 1$ for the second cases (see Table 4.1), and $\cosh(0) = 1$ and $\sinh(0) = 0$ for non-canonical matrices.

Existence of x-independent solutions. Function (4.7) specifies a family of solutions to the homogeneous ODE, $\mathbf{y}' = \mathbf{A}\,\mathbf{y}$, that are functions of the independent variable $x \in \mathrm{X}$, depending on the coefficient matrix $\mathbf{A}$, and on an arbitrary vector $\mathbf{y}(x_0) \in \mathbf{Y}$. Table 4.1 shows that solutions to linear ODE involve exponential, linear, and trigonometric functions of $x \in \mathrm{X}$. However, as $\mathbf{e}^{\mathbf{A}\,(x-x_0)} > \mathbf{0}$ for every finite $x \in \mathrm{X}$, if $\mathbf{y}(x_0) = \mathbf{0}$ then the solution (4.7) is independent of x, that is, it satisfies $\mathbf{y}(x) = \mathbf{0}$ for every $x \in \mathrm{X}$. This property holds even in the case in which matrix $\mathbf{A}$ is singular, that is, even if $\det(\mathbf{A}) = 0$. This property may not hold for non-homogeneous planar ODE.

4.1.2 Non-Homogeneous ODE

Non-homogeneous linear planar ODE (4.3) also have closed form solutions. In addition to coefficient matrix $\mathbf{A}$, the form of the coefficient matrix $\mathbf{B} \neq \mathbf{0}$ can introduce

[5] See the Appendix 4.6 for its computation.

[6] See the Appendix 4.6.

significant changes in the form of the solution to an ODE, in particular when matrix $\mathbf{A}$ is singular.[7]

Proposition 4.2 (General solution to the planar linear non-homogeneous ODE) *Consider the ODE (4.3) where* $\mathbf{A} \in \mathbb{R}^{2\times 2}$, *and* $\mathbf{B} \in \mathbb{R}^2$ *are arbitrary. Then the general solution to the ODE is a vector function* $\mathbf{y} : \mathrm{X} \subseteq \mathbb{R} \rightarrow \mathbf{Y} \subseteq \mathbb{R}^2$, *taking one of the following forms:*

1. *if* $\det(\mathbf{A}) \neq 0$, *the solution is*

$$\mathbf{y}(x) = -\mathbf{A}^{-1}\,\mathbf{B} + \mathbf{e}^{\mathbf{A}\,(x-x_0)}\left(\mathbf{y}(x_0) + \mathbf{A}^{-1}\,\mathbf{B}\right),\ \textit{for each } x \in \mathrm{X}, \tag{4.22}$$

where $\mathbf{A}^{-1}$ *is the elementary inverse of matrix* $\mathbf{A}$, *and* $\mathbf{e}^{\mathbf{A}\,x}$ *is a matrix exponential function of* $\mathbf{A}$, *and* $(x_0, \mathbf{y}(x_0))$ *is an arbitrary element of* $\mathrm{X} \times \mathbf{Y}$;

2. *if* $\det(\mathbf{A}) = 0$ *the solution is*

$$\mathbf{y}(x) = -\mathbf{A}^{+}\,\mathbf{B} + \mathbf{e}^{\mathbf{A}\,(x-x_0)}\left(\mathbf{y}(x_0) + \mathbf{A}^{+}\,\mathbf{B}\right) + \left(\mathbf{I} - \mathbf{A}\,\mathbf{A}^{+}\right)\mathbf{B}\,(x - x_0),\ \textit{for each } x \in \mathrm{X}, \tag{4.23}$$

where $\mathbf{A}^{+}$ *is the Moore-Penrose generalized inverse of matrix* $\mathbf{A}$, *and* $(x_0, \mathbf{y}(x_0))$ *is an arbitrary element of* $\mathrm{X} \times \mathbf{Y}$.

Proof We transform the generic ODE (4.3) into a simpler ODE, with known solution, and do the inverse transformation, separating the cases in which matrix $\mathbf{A}$ is non-singular or singular. First, assume that matrix $\mathbf{A}$ is non-singular, that is $\det(\mathbf{A}) \neq 0$. Then matrix $\mathbf{A}$ has an unique inverse, $\mathbf{A}^{-1}$, satisfying $\mathbf{A}\,\mathbf{A}^{-1} = \mathbf{A}^{-1}\,\mathbf{A} = \mathbf{I}$. Two cases are possible: (1) if matrix $\mathbf{A}$ is a Jordan form, $\mathbf{A} = \mathbf{\Lambda}$, or is an hyperbolic matrix, $\mathbf{A} = \mathbf{A}_h$, we define $\mathbf{w}(x) = \mathbf{y}(x) + \mathbf{A}^{-1}\,\mathbf{B}$. Then $\mathbf{w}' = \mathbf{y}' = \mathbf{A}\,\mathbf{y} + \mathbf{B} = \mathbf{A}\,(\mathbf{w} - \mathbf{A}^{-1}\,\mathbf{B}) + \mathbf{B} = \mathbf{A}\,\mathbf{w}$. The solution to this ODE, from Proposition 4.1, is $\mathbf{w}(x) = e^{\mathbf{A}\,(x-x_0)}\mathbf{w}(x_0)$. Transforming back to $\mathbf{y}$, as $\mathbf{y}(x) = \mathbf{w}(x) - \mathbf{A}^{-1}\,\mathbf{B}$, yields the solution function (4.22); (2) if matrix $\mathbf{A}$ is a canonical matrix, but not a Jordan form, we define the vector function $\mathbf{w}(x) = \mathbf{P}^{-1}\left(\mathbf{y}(x) + \mathbf{A}^{-1}\,\mathbf{B}\right)$, where $\mathbf{P}$ is the eigenvector matrix of $\mathbf{A}$. Taking derivatives, we have $\mathbf{w}' = \mathbf{P}^{-1}\,\mathbf{y}' = \mathbf{P}^{-1}\left(\mathbf{A}\,\mathbf{y} + \mathbf{B}\right)$. As $\mathbf{P}^{-1}\,\mathbf{A}\,\mathbf{P} = \mathbf{\Lambda}$, where $\mathbf{\Lambda}$ is the Jordan canonical similar to $\mathbf{A}$, and $\mathbf{A}\,\mathbf{A}^{-1} = \mathbf{I}$, then we obtain a homogeneous equation $\mathbf{w}' = \mathbf{\Lambda}\,\mathbf{w}$, that has the general solution $\mathbf{w}(x) = e^{\mathbf{\Lambda}\,(x-x_0)}\,\mathbf{w}(x_0)$. Transforming back to $\mathbf{y}$, as $\mathbf{y}(x) = \mathbf{P}\,\mathbf{w}(x) - \mathbf{A}^{-1}\,\mathbf{B}$, yields again the solution function (4.22). Second, assume that matrix $\mathbf{A}$ is singular, that is $\det(\mathbf{A}) = 0$. Then matrix $\mathbf{A}$ does not have an unique inverse. Denoting $\mathbf{A}^{+}$ its Moore-Penrose (or pseudo-) inverse, we have $\mathbf{A}\,\mathbf{A}^{+} = \mathbf{A}^{+}\mathbf{A} \neq \mathbf{I}$, and we can use the fact that $\mathbf{A}^{+} = \mathbf{P}\,\mathbf{\Lambda}^{+}\,\mathbf{P}^{-1}$ where $\mathbf{\Lambda}^{+}$ is the Moore-Penrose inverse of the similar Jordan form.[8] Two cases are possible: (1) If matrix $\mathbf{A}$ is a Jordan form, $\mathbf{\Lambda}$, or is a diagonal matrix $\mathbf{A}_d$, it is a diagonal matrix with at least one zero in the principal diagonal. Defining $\mathbf{w}(x) = \mathbf{y}(x) + \mathbf{A}^{+}\,\mathbf{B}$, and taking derivatives we get $\mathbf{w}' = \mathbf{y}' = \mathbf{A}\,\mathbf{y} + \mathbf{B} = \mathbf{A}\,(\mathbf{w} - \mathbf{A}^{+}\,\mathbf{B}) + \mathbf{B} = \mathbf{A}\,\mathbf{w} + (\mathbf{I} - \mathbf{A}\,\mathbf{A}^{+})\,\mathbf{B}$ which is a non-homogeneous ODE. However, as $(\mathbf{I} - \mathbf{A}\,\mathbf{A}^{+})\,\mathbf{B}$ is a vector of constants, the solution to this ODE is

[7] Compare with the cases in Exercise 4.3.

[8] See Magnus and Neudecker (1988, p. 36).

$\mathbf{w}(x) = e^{\mathbf{A}\,(x-x_0)}\mathbf{w}(x_0) + (\mathbf{I} - \mathbf{A}\,\mathbf{A}^+)\,\mathbf{B}\,(x - x_0)$. Transforming back to $\mathbf{y}$, as $\mathbf{y}(x) = \mathbf{w}(x) - \mathbf{A}^+\,\mathbf{B}$, yields the solution function (4.23); (2) if matrix $\mathbf{A}$ is a canonical matrix, but not a Jordan form, or if it is a non-canonical matrix $\mathbf{A}_h$, we define the vector function $\mathbf{w}(x) = \mathbf{P}^{-1}\left(\mathbf{y}(x) + \mathbf{A}^+\,\mathbf{B}\right)$, where $\mathbf{P}$ is the eigenvector matrix associated to $\mathbf{A}$. Taking derivatives we have $\mathbf{w}' = \mathbf{P}^{-1}\,\mathbf{y}' = \mathbf{P}^{-1}\left(\mathbf{A}\,\mathbf{y} + \mathbf{B}\right)$. As $\mathbf{P}^{-1}\,\mathbf{A}\,\mathbf{P} = \mathbf{\Lambda}$, where $\mathbf{\Lambda}$ is the Jordan canonical form similar to $\mathbf{A}$, $\mathbf{P}^{-1}\mathbf{A}\,\mathbf{A}^+ = \mathbf{\Lambda}^+\,\mathbf{P}^{-1}$ and $\mathbf{P}\,\mathbf{P}^{-1} = \mathbf{P}^{-1}\,\mathbf{P} = \mathbf{I}$, then we obtain an ODE $\mathbf{w}' = \mathbf{\Lambda}\,\mathbf{w} + (\mathbf{I} - \mathbf{\Lambda}\,\mathbf{\Lambda}^+)\,\mathbf{P}^{-1}\,\mathbf{B}$, that has the general solution $\mathbf{w}(x) = e^{\mathbf{\Lambda}\,(x-x_0)}\mathbf{w}(x_0) + (\mathbf{I} - \mathbf{\Lambda}\,\mathbf{\Lambda}^+)\,\mathbf{P}^{-1}\,\mathbf{B}\,(x - x_0)$. Transforming back to $\mathbf{y}$, as $\mathbf{y}(x) = \mathbf{P}\,\mathbf{w}(x) - \mathbf{A}^+\,\mathbf{B}$, and observing that $\mathbf{P}\,\mathbf{\Lambda}\,\mathbf{\Lambda}^+\,\mathbf{P}^{-1} = \mathbf{A}\,\mathbf{A}^+$ yields again the solution function (4.23). □

This proposition refers to two cases, depending on the determinant of matrix $\mathbf{A}$. If $\det(\mathbf{A}) \neq 0$ then $\mathbf{A}$ is non-singular, and the general solution is (4.22). In this case, matrix $\mathbf{A}$ has the unique inverse $\mathbf{A}^{-1} = \frac{\text{adj}(\mathbf{A})^\top}{\det(\mathbf{A})}$ where $\text{adj}(\mathbf{A})$ is its adjoint.[9] An important property of this matrix inverse is that it satisfies $\mathbf{A}\,\mathbf{A}^{-1} = \mathbf{A}^{-1}\,\mathbf{A} = \mathbf{I}$. In solution (4.22) the matrix exponential is one of the cases in Table 4.1, depending on the spectrum of matrix $\mathbf{A}$.

If $\det(\mathbf{A}) = 0$ then $\mathbf{A}$ is singular. The eigenvalues of a singular square matrix $\mathbf{A}$ are always real and at least one of them is equal to zero. Therefore, matrix $\mathbf{A}$ can only take some specific forms: (1) it is equal or similar to the Jordan form $\mathbf{\Lambda}_+$ in which one of the elements is equal to zero; (2) or is a diagonal matrix, of type $\mathbf{A}_d$, but with all elements equal to zero, (3) or it is equal or similar to a Jordan form $\mathbf{\Lambda}_0$ with $\lambda = 0$. The general solutions for the first two cases are particular cases of equation (4.23). Their specific solutions are provided in Exercise 4.4. The solution to the third case is the only specific non-homogeneous ODE that is not covered in Proposition (4.2), but is provided in Exercise 4.5.

Exercise 4.4 In this exercise, we present the specific cases of ODE with singular coefficient matrices $\mathbf{A}$, consistent with solution (4.23). First, we have matrices similar to degenerate $\mathbf{\Lambda}_+$ Jordan forms, or degenerate non-canonical matrices (see Corollary 4.1 and Exercise 4.6), $\mathbf{A} = \mathbf{P}\,\mathbf{\Lambda}\,\mathbf{P}^{-1}$, where $\mathbf{\Lambda} \in \left\{ \begin{pmatrix} \lambda_- & 0 \\ 0 & 0 \end{pmatrix}, \begin{pmatrix} 0 & 0 \\ 0 & \lambda_+ \end{pmatrix} \right\}$. The Moore-Penrose inverse of the coefficient matrix is $\mathbf{A}^+ = \mathbf{P}\,\mathbf{\Lambda}^+\,\mathbf{P}^{-1}$ where $\mathbf{\Lambda}^+ \in \left\{ \begin{pmatrix} \lambda_-^{-1} & 0 \\ 0 & 0 \end{pmatrix}, \begin{pmatrix} 0 & 0 \\ 0 & \lambda_+^{-1} \end{pmatrix} \right\}$ and the matrix exponential functions $\mathbf{e}^{\mathbf{A}\,x} = \mathbf{P}\,\mathbf{e}^{\mathbf{\Lambda}\,x}\,\mathbf{P}^{-1} \in \left\{ \begin{pmatrix} e^{\lambda_- x} & 0 \\ 0 & 1 \end{pmatrix}, \begin{pmatrix} 1 & 0 \\ 0 & e^{\lambda_+ x} \end{pmatrix} \right\}$. Defining the transformed coefficient matrix $\tilde{\mathbf{B}} = \mathbf{P}^{-1}\,\mathbf{B} = (\tilde{b}_1, \tilde{b}_2)^\top$, and the transformed arbitrary vector function $\mathbf{w}(x) = \mathbf{P}^{-1}\,\mathbf{y}(x) = (w_-(x), w_+(x))^\top$, we have the specific general solutions for the two previous cases

$$\mathbf{y}(x) = \mathbf{P}\begin{pmatrix} -\dfrac{\tilde{b}_1}{\lambda_-} + e^{\lambda_-(x-x_0)}\left(w_-(x_0) + \frac{\tilde{b}_1}{\lambda_-}\right) \\ w_+(x_0) + \tilde{b}_2\,(x - x_0) \end{pmatrix} \tag{4.24}$$

or

[9] For the square matrix $\mathbf{A}$ in Eq. (4.4) we have $\text{adj}(\mathbf{A}) = \begin{pmatrix} a_{22} & -a_{21} \\ -a_{12} & a_{11} \end{pmatrix}$.

$$\mathbf{y}(x) = \mathbf{P} \begin{pmatrix} w_-(x_0) + \tilde{b}_1\,(x - x_0) \\ -\dfrac{\tilde{b}_2}{\lambda_+} + e^{\lambda_+ (x - x_0)} \left(w_+(x_0) + \frac{\tilde{b}_2}{\lambda_+} \right) \end{pmatrix}. \tag{4.25}$$

If matrix $\mathbf{A}$ is a degenerate Jordan form, we can use the previous formulas by substituting $\mathbf{P} = \mathbf{P}^{-1} = \mathbf{I}$. Second, for a degenerate diagonal matrix $\mathbf{A}_d = \mathbf{0}$, we can also apply the general solution (4.23) observing that $\mathbf{A}^+ = \mathbf{P} = \mathbf{P}^{-1} = \mathbf{e}^{\mathbf{A}\,x} = \mathbf{I}$, which yields the solution

$$\mathbf{y}(x) = \mathbf{y}(x_0) + \mathbf{B}\,(x - x_0) \tag{4.26}$$

Prove the previous results.

Exercise 4.5 Non-homogeneous ODE in which the coefficient matrix is similar to a degenerate Jordan form of type $\mathbf{\Lambda}_0$, satisfying $\Delta(\mathbf{A}) = 0$ and $\mathrm{tr}(\mathbf{A}) = 0$, take the generic form,

$$\begin{cases} y_1' = a\,y_2 + b_1, \quad \text{where } a \neq 0, & (4.27\text{a}) \\ y_2' = b_2. & (4.27\text{b}) \end{cases}$$

The general solution to this ODE (show this) is

$$\begin{cases} y_1(x) = y_1(x_0) + \big(a\,y_2(x_0) + b_1\big)\,(x - x_0) + \dfrac{a\,b_2}{2}\,(x - x_0)^2 & (4.28\text{a}) \\ y_2(x) = y_2(x_0) + b_2\,(x - x_0). & (4.28\text{b}) \end{cases}$$

Differently from the solutions to the other types of degenerate cases, presented in Eq. (4.23), this solution depends on $(x - x_0)^2$.

Exercise 4.6 For non-canonical matrices $\mathbf{A} = \mathbf{A}_h$, we have $\det(\mathbf{A}) = 0$ if $\beta = \alpha$ or $\beta = -\alpha$. For the first case, i.e., if $\beta = \alpha$, we have

$$\mathbf{y}(t) = \frac{1}{2}\Bigg(-\frac{b_1 + b_2}{2\,\alpha} + \left(y_1(x_0) + y_2(x_0) + \frac{b_1 + b_2}{2\,\alpha} \right) e^{2\,\alpha\,(x - x_0)} \Bigg) \begin{pmatrix} 1 \\ -1 \end{pmatrix}$$
$$- \frac{1}{2}\Big(y_1(x_0) - y_2(x_0) - (b_1 - b_2)\,(x - x_0) \Big) \begin{pmatrix} 1 \\ 1 \end{pmatrix}. \tag{4.29}$$

For the second case, i.e., if $\beta = -\alpha$, we have

$$\mathbf{y}(t) = \frac{1}{2}\Bigg(-\frac{b_1 - b_2}{2\,\alpha} + \left(y_1(x_0) - y_2(x_0) + \frac{b_1 - b_2}{2\,\alpha} \right) e^{2\,\alpha\,(x - x_0)} \Bigg) \begin{pmatrix} 1 \\ 1 \end{pmatrix}$$
$$+ \frac{1}{2}\Big(y_1(x_0) + y_2(x_0) + (b_1 + b_2)\,(x - x_0) \Big) \begin{pmatrix} 1 \\ -1 \end{pmatrix}. \tag{4.30}$$

Prove this.

Two important properties of homogeneous ODE, regarding decomposability and/or existence of x-independent solutions, may not be satisfied by related non-homogeneous ODE, in particular when matrix $\mathbf{A}$ is non-singular.

Decomposing solutions. Non-homogeneous ODE, with canonical $\mathbf{A}$ matrices, also allow for a decomposition of solutions analogous to those presented for homogeneous ODE in Eqs. (4.19), (4.20), and (4.21), if matrix $\mathbf{A}$ is non-singular. In this case, we introduce the following transformations: $\mathbf{y}(x) \mapsto \mathbf{y}(x) - \bar{\mathbf{y}}$ and $\mathbf{w}(x) = \mathbf{P}^{-1}\,\mathbf{y}(x) \mapsto \mathbf{w}(x) = \mathbf{P}^{-1}\,(\mathbf{y}(x) - \bar{\mathbf{y}})$. Therefore, a non-zero vector $\mathbf{B}$ causes a parallel shift in the solution of the ODE and does not change its functional dependence on the independent variable, x. If matrix $\mathbf{A}$ is singular this may not be the case, as we see next.

Existence of x-independent solutions. For the non-homogeneous ODE (4.3), function (4.22) presents the explicit general solution for the case in which matrix $\mathbf{A}$ is non-singular, satisfying $\det(\mathbf{A}) \neq 0$. Then matrix $\mathbf{A}$ has a unique inverse $\mathbf{A}^{-1}$, and we can define $\bar{\mathbf{y}} = -\mathbf{A}^{-1}\,\mathbf{B}$. Function (4.22) specifies a family of solutions that are functions of the independent variable $x \in X$, parameterized by the coefficient matrices, that depend on an arbitrary vector $\mathbf{y}(x_0) \in \mathbf{Y}$. An important fact about that solution is the following: if $\mathbf{y}(x_0) = \bar{\mathbf{y}}$ then the particular solution to the non-homogeneous ODE (4.3) is **independent from** x, that is $\mathbf{y}(x) = \bar{\mathbf{y}}$ for every $x \in X$.

For non-homogeneous ODE in which matrix $\mathbf{A}$ is singular this property does not hold in general. If $\det(\mathbf{A}) = 0$ then matrix $\mathbf{A}$ does not have a classical inverse. In this case, there is a pseudo inverse, denoted by $\mathbf{A}^{+}$. If the coefficient matrix is of the types studied in Exercise 4.4, we have two possibilities: first, if $\left(\mathbf{I} - \mathbf{A}\,\mathbf{A}^{+}\right)\mathbf{B} = \mathbf{0}$ and $\mathbf{y}(x_0) = -\mathbf{A}^{+}\,\mathbf{B}$ then the solution is independent from x, specifically $\mathbf{y}(x) = -\mathbf{A}^{+}\,\mathbf{B}$, for every $x \in X$; alternatively, if $\left(\mathbf{I} - \mathbf{A}\,\mathbf{A}^{+}\right)\mathbf{B} \neq \mathbf{0}$ then the solution **cannot be independent** from x, even if $\mathbf{y}(x_0) = -\mathbf{A}^{+}\,\mathbf{B}$. This second case cannot occur in homogeneous ODE. At last, if the coefficient matrix is of the type studied in Exercise 4.5 such that $b_2 \neq 0$, then this is also a case in which there is no solution independent from x.

4.1.3 Planar ODE Problems

For finding a particular solution to an ODE, we have to introduce a number of side conditions that is equal to the dimension of the image space, $\mathbf{Y}$. If the ODE problem has a solution, we can obtain $\mathbf{y}(x_0)$ as a function of the data provided by the side conditions. For scalar ODE problems we saw in Sect. 2.2 that we need to add one side condition. Solving a problem involving a planar ODE requires adding **two side conditions**. When the independent variable is time (see next Sect. 4.2) side conditions refer to initial or terminal—or both—values for the dependent variables. For other types of independent variables, as we saw in Chap. 2, some side conditions may take the form of functional constraints. As there are too many different types of side-conditions, we choose a specific example to illustrate how to solve a spatial distribution planar ODE problem.

Example 4.1 (*A firm's allocation of production across locations*) In Example 2.2 addressing the location of production, $y(x)$, and sales, $u(x)$, over locations $x \in \mathrm{X} = [-\frac{L}{2}, \frac{L}{2}]$, we saw that a space independent distribution of sales was unrealistic. Specifying a spatial optimization problem in Example 10.1 allow us to endogenize sales, and obtain optimality conditions the are represented by the following ODE problem

$$\begin{cases} y'(x) = u(x) - \delta\, y(x), & \text{for every } x \in \mathrm{X} \\ u'(x) = u(x) - p, & \text{for every } x \in \mathrm{X} \\ y(-\frac{L}{2}) = y(\frac{L}{2}), & \text{for } x \in \{-\frac{L}{2}, \frac{L}{2}\}, \\ \int_{-L/2}^{L/2} y(x)\, dx = Y, \ \text{for all } x \in \mathrm{X}, & \end{cases}$$

where δ, Y, and p are positive numbers. The first two parameters—already defined in Example 2.2—represent a constant transportation cost and the total production, and the last represents the selling price. In matrix form, the first two conditions take the following form

$$\begin{pmatrix} y'(x) \\ u'(x) \end{pmatrix} = \begin{pmatrix} -\delta & 1 \\ 0 & \delta \end{pmatrix} \begin{pmatrix} y(x) \\ u(x) \end{pmatrix} + \begin{pmatrix} 0 \\ -p \end{pmatrix}.$$

The coefficient matrix has $\mathrm{tr} = 0$, $\det = -\delta^2 < 0$ and $\Delta = \delta^2 > 0$. Therefore, the eigenvalues are real and have opposite signs: $\lambda_- = -\delta < 0 < \lambda_+ = \delta$. The Jordan canonical form of the coefficient matrix, and the eigenvector matrix are $\mathbf{\Lambda} = \begin{pmatrix} -\delta & 0 \\ 0 & \delta \end{pmatrix}$ and $\mathbf{P} = \begin{pmatrix} 1 & 1 \\ 0 & 2\delta \end{pmatrix}$, respectively. As the coefficient matrix is invertible, there is an unique invariant solution $\begin{pmatrix} \bar{y} \\ \bar{u} \end{pmatrix} = -\begin{pmatrix} -\delta & 1 \\ 0 & \delta \end{pmatrix}^{-1} \begin{pmatrix} 0 \\ -p \end{pmatrix} = \frac{p}{\delta^2} \begin{pmatrix} 1 \\ \delta \end{pmatrix}$.

Using the superposition principle, in Eq. (4.19), the general solution to the ODE is

$$\begin{pmatrix} y(x) \\ u(x) \end{pmatrix} = \frac{p}{\delta^2} \begin{pmatrix} 1 \\ \delta \end{pmatrix} + w_- \begin{pmatrix} 1 \\ 0 \end{pmatrix} e^{-\delta\,(x - x_0)} + w_+ \begin{pmatrix} 1 \\ 2\,\delta \end{pmatrix} e^{\delta\,(x - x_0)},$$

where w_- and w_+ are arbitrary constants. To find the particular solution to the firm's problem, we substitute this general solution in the two side-conditions, yielding a system in (w_-, w_+),

$$\begin{cases} \int_{-L/2}^{L/2} y(x)\, dx = Y, \\ y(-\frac{L}{2}) = y(\frac{L}{2}), \end{cases} \iff \begin{cases} \dfrac{p\,L}{\delta^2} + \dfrac{w_-\, e^{\delta x_0} + w_+\, e^{-\delta x_0}}{\delta} \left(e^{\delta L/2} - e^{-\delta L/2} \right) = Y, \\ w_-\, e^{\delta x_0} = w_+\, e^{-\delta x_0}, \end{cases}$$

that has the solution $w_-\, e^{\delta x_0} = w_+\, e^{-\delta x_0} = \frac{Y\,\delta^2 - p\,L}{2\,\delta} \left(e^{\delta\, L/2} - e^{-\delta\, L/2} \right)^{-1}$. Substituting in the general solution, we obtain the optimal distributions of output, and sales location

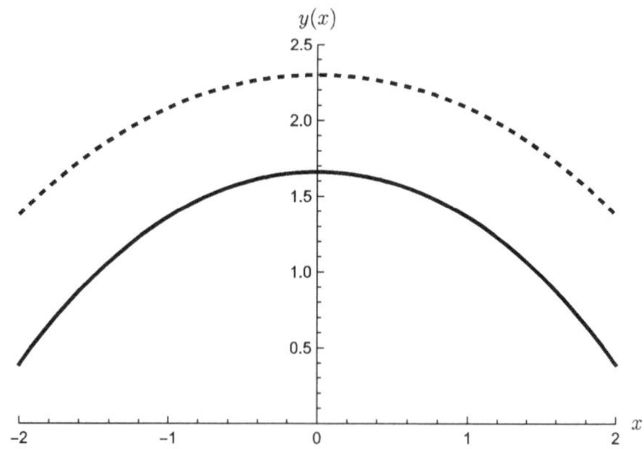

Fig. 4.2 Illustration of equation (4.31a) for $p = 1$, $\delta = 0.5$, $L = 4$, and $Y = 5$ (solid curve) and $Y = 8$ (dashed curve).

$$y(x) = \frac{p}{\delta^2} + \frac{Y\delta^2 - p\,L}{2\,\delta}\left(\frac{e^{-\delta\,x} + e^{\delta\,x}}{e^{\delta\,L/2} - e^{-\delta\,L/2}}\right), \text{ for } x \in \mathbb{X}, \tag{4.31a}$$

$$u(x) = \frac{p}{\delta} + \left(Y\delta^2 - p\,L\right)\left(\frac{e^{\delta\,x}}{e^{\delta\,L/2} - e^{-\delta\,L/2}}\right), \text{ for } x \in \mathbb{X}. \tag{4.31b}$$

The solution to the output distribution, in Eq. (4.31a), is illustrated in Fig. 4.2, for particular values of the parameters. Observe that we solve the inconsistency with the Salop (1979) topology that was present in the related Example 2.2. That figure also makes clear that production should be located in $x = 0$.

4.2 Planar Autonomous ODE Dynamics

Linear planar ODE, when compared to linear scalar ODE studied in Sect. 2.3, display some new types of dynamics: first, in addition to global asymptotic stability and instability, as in the scalar case, conditional stability or saddle-path dynamics is a new type of dynamics for the planar case; second, in addition to monotonic solution paths, as in the scalar case, several types of non-monotonic solution paths exist; at last, in addition to co-dimension one bifurcations, co-dimension two bifurcations can also exist. Saddle-path dynamics is a very common type of dynamics in both macroeconomic and growth models, and is the type of solution dynamics for most optimal control problems (as will be seen in Chap. 10).

This section has two aims: first, to present the qualitative theory on the dynamics generated by planar linear ODE, tracing out the behavior over time of the solution trajectories $\left(\mathbf{y}(t)\right)_{\mathbb{T}}$; and, second, presenting the main types of particular solutions with relevance to economics.

The general autonomous **linear planar ordinary differential equation**[10] is a linear functional equation defined over vector function $\mathbf{y} : \mathrm{T} \to \mathbf{Y} \subseteq \mathbb{R}^2$, where the independent variable is time with domain $\mathrm{T} \subseteq \mathbb{R}_+$, usually $\mathrm{T} = [0, \infty)$,

$$\dot{\mathbf{y}} = \mathbf{A}\mathbf{y} + \mathbf{B}, \;\; \mathbf{y} : \mathrm{T} \subseteq \mathbb{R}_+ \to \mathbf{Y} \subseteq \mathbb{R}^2, \tag{4.32}$$

where $\mathbf{A} \in \mathbb{R}^{2\times 2}$ and $\mathbf{B} \in \mathbb{R}^2$ were defined in Eq. (4.4). In **expanded form** we have

$$\begin{aligned} \dot{y}_1 &= a_{11}y_1 + a_{12}y_2 + b_1 \\ \dot{y}_2 &= a_{21}y_1 + a_{22}y_2 + b_2. \end{aligned} \tag{4.33}$$

We already proved, in Proposition 4.2, that linear planar ODE have explicit general solutions whose form depends on matrix $\mathbf{A}$, and, in particular on its determinant:

1. if $\det(\mathbf{A}) \neq 0$, the general solution is

$$\mathbf{y}(t) = -\mathbf{A}^{-1}\,\mathbf{B} + \mathbf{e}^{\mathbf{A}\,(t-t_0)}\left(\mathbf{y}(t_0) + \mathbf{A}^{-1}\,\mathbf{B}\right), \;\; \text{for each } t \in \mathrm{T}, \tag{4.34}$$

   where $\mathbf{A}^{-1}$ is the elementary inverse of matrix $\mathbf{A}$, and $\mathbf{e}^{\mathbf{A}\,t}$ is a matrix exponential function of $\mathbf{A}$, as in Lemma 4.2 for a canonical matrix or in Lemma 4.1 for a non-canonical matrix, and $(t_0, \mathbf{y}(t_0))$ is an arbitrary element of $\mathrm{T} \times \mathbf{Y}$;
2. if $\det(\mathbf{A}) = 0$, the general solution is

$$\mathbf{y}(t) = -\mathbf{A}^{+}\,\mathbf{B} + \mathbf{e}^{\mathbf{A}\,(t-t_0)}\left(\mathbf{y}(t_0) + \mathbf{A}^{+}\,\mathbf{B}\right) + \left(\mathbf{I} - \mathbf{A}\,\mathbf{A}^{+}\right)\mathbf{B}\,(t - t_0), \;\; \text{for each } t \in \mathrm{T}, \tag{4.35}$$

   where $\mathbf{A}^{+}$ is the Moore-Penrose generalized inverse of matrix $\mathbf{A}$.

Table 4.1 presents the possible forms of the exponential matrix $\mathbf{e}^{\mathbf{A}t}$.

Example 4.2 (*Second-order linear ODE*) A scalar second-order linear ODE is

$$\ddot{y} - a_1\dot{y} - a_0 y - b = 0. \tag{4.36}$$

It can be solved by transforming it to a planar linear ODE. Defining $y_1 = y$ and $y_2 = \dot{y} = \dot{y}_1$, then Eq. (4.36) is equivalent to the planar ODE $\dot{\mathbf{y}} = \mathbf{A}\,\mathbf{y} + \mathbf{B}$ where $\mathbf{A} = \left(\begin{smallmatrix} 0 & 1 \\ a_0 & a_1 \end{smallmatrix}\right)$, and $\mathbf{B} = \left(\begin{smallmatrix} 0 \\ b \end{smallmatrix}\right)$. We readily find that $\mathrm{tr}(\mathbf{A}) = a_1$ and $\det(\mathbf{A}) = -a_0$, and the eigenvalues are $\lambda_{\mp} = \frac{a_1}{2} \mp \sqrt{\left(\frac{a_1}{2}\right)^2 + a_0}$. If $a_0 \neq 0$ the general solution is of the form (4.34), and, in particular, if $a_0 > -\left(\frac{a_1}{2}\right)^2$ the solution is

$$y(t) = -\frac{b}{a_0} + \frac{1}{\lambda_+ - \lambda_-}\left(\left(\lambda_+ e^{\lambda_- t} - \lambda_- e^{\lambda_+ t}\right)\left(y(0) + \frac{b}{a_0}\right) + \left(e^{\lambda_+ t} - e^{\lambda_- t}\right)\dot{y}(0)\right),$$

[10] This chapter only covers the autonomous case. For the dynamics associated to the planar non-autonomous equations, $\dot{\mathbf{y}} = \mathbf{A}(t)\,\mathbf{y} + \mathbf{B}(t)$, see Hubbard and West (1991).

if $a_0 = -\left(\frac{a_1}{2}\right)^2$, the solution is

$$y(t) = b\left(\frac{2}{a_1}\right)^2 + e^{\frac{a_1}{2}t}\left(\frac{2-a_1 t}{2}\left(y(0) - b\left(\frac{2}{a_1}\right)^2\right) + \dot{y}(0)\right),$$

if $a_0 < -\left(\frac{a_1}{2}\right)^2$, and writing $\beta = \sqrt{\left|\left(\frac{a_1}{2}\right)^2 + a_0\right|}$, the solution is

$$y(t) = -\frac{b}{a_0} + e^{\frac{a_1}{2}t}\left(\left(\cos\beta t - \frac{a_1}{2\beta}\sin\beta t\right)\left(y(0) + \frac{b}{a_0}\right) + \frac{\sin\beta t}{\beta}\,\dot{y}(0)\right).$$

If $a_0 = 0$ the solution is of form (4.35), and, in particular

$$y(t) = y(0) - \frac{1}{a_1}\left(\dot{y}(0) + \frac{b}{a_1}\right)\left(1 - e^{a_1 t}\right) - \frac{b}{a_1}t.$$

Exercise 4.7 (*Linear second-order ODE*) Solve the previous Example (hint, Appendix 4.5.1 details the relevant algebraic facts).

Example 4.3 (*Non-autonomous scalar equations*) In Chap. 2, we studied the non-autonomous linear scalar equation (2.41), $\dot{y} = \lambda\, y + b\, z(t)$, for $z(t) = e^{\gamma t}$, where $\lambda < 0$, $b > 0$ and $\gamma > 0$, and $z(t)$ was an additive time-dependent shock. We can express this model as a planar forward ODE

$$\begin{aligned}\dot{y} &= \lambda\, y + b\, z,\\ \dot{z} &= \gamma\, z.\end{aligned} \tag{4.37}$$

The steady state, $(\bar{y}, \bar{z}) = (0, 0)$, is a saddle-point. The general solution is

$$\begin{aligned}y(t) &= \left(y(0) - \frac{b}{\gamma-\lambda}\right)e^{\lambda t} + \frac{b}{\gamma-\lambda}\,z(0)\,e^{\gamma t},\\ z(t) &= z(0)\,e^{\gamma t}.\end{aligned} \tag{4.38}$$

4.2.1 Qualitative Theory

In this section, we extend the elements and the approach of the qualitative theory of ODE already presented in Sect. 2.3.2 and 3.2, for scalar linear and regular non-linear ODE, to planar linear ODE.

4.2.1.1 Steady States

The $\Omega_{\mathbf{Y}}$-limit ($\mathrm{A}_{\mathbf{Y}}$-limit) set is the subset of $\mathbf{Y}$ to which trajectories converge asymptotically forward (backward) in time. Differently from non-linear (regular) ODE (see Chap. 5), the limit sets for planar linear ODE only contain steady states. Steady states are time-independent solutions of an ODE. The first step in dynamics' analyses addresses the existence and uniqueness of steady states.[11]

Definition 4.2 (*Steady state set*) The steady state set ODE (4.32) is the set of elements of $\mathbf{Y}$ such that $\dot{\mathbf{y}} = \mathbf{0}$. Formally,

$$\mathbf{y}^{ss} = \left\{\mathbf{y} \in \mathbf{Y} : \ \mathbf{A}\,\mathbf{y} + \mathbf{B} = \mathbf{0}\right\}. \tag{4.39}$$

We can have one, an infinite number, or no steady states. That is, the set $\mathbf{y}^{ss}$ can have one element, an infinite number of elements, or be empty.

Proposition 4.3 (Characterization of set $\mathbf{y}^{ss}$) *Let the set of steady states be as in definition 4.2. Then:*

(1) If $\det(\mathbf{A}) \neq 0$ *then there is one unique steady state,* $\bar{\mathbf{y}} = -\mathbf{A}^{-1}\,\mathbf{B}$, *and, therefore,* $\mathbf{y}^{ss} = \{\bar{\mathbf{y}}\}$.

(2) If $\det(\mathbf{A}) = 0$ *and the following conditions are satisfied, then there is an infinite number of steady states belonging to one of the following one-dimensional (linear) manifolds*

(a) if $\mathrm{tr}(\mathbf{A}) < 0$ *and* $P_1^- b_2 = P_2^- b_1$, *then the steady state set is*

$$\mathbf{y}^{ss} = \{\mathbf{y} \in \mathbf{Y} : \ P_1^+(\lambda_- y_2 + b_2) = P_2^+(\lambda_- y_1 + b_1)\};$$

(b) if $\mathrm{tr}(\mathbf{A}) > 0$ *and* $P_1^+ b_2 = P_2^+ b_1$, *then the steady state set is*

$$\mathbf{y}^{ss} = \{\mathbf{y} \in \mathbf{Y} : \ P_1^-(\lambda_+ y_2 + b_2) = P_2^-(\lambda_+ y_1 + b_1)\};$$

(c) if $\mathrm{tr}(\mathbf{A}) = 0$ *and* $P_2^s b_1 = P_1^s b_2$, *then the steady state set is*[12]

$$\mathbf{y}^{ss} = \{\mathbf{y} \in \mathbf{Y} : \ P_1^s y_2 - P_2^s y_1 = P_1^g b_2 - P_2^g b_1\}.$$

(3) If $\det(\mathbf{A}) = \mathrm{tr}(\mathbf{A}) = 0$ *and* $\mathbf{A} = \mathbf{0}$, *and* $P_2^+ b_1 - P_1^+ b_2 = P_1^- b_2 - P_2^- b_1 = 0$ *then* $\mathbf{y}^{ss} = \mathbf{Y}$ *(all elements of the two-dimensional space* $\mathbf{Y}$ *are steady states).*

(4) If none of the previous conditions is satisfied, then $\mathbf{y}^{ss}$ *is empty, and steady states do not exist.*

---

[11] In the mathematical literature, they are also called equilibrium points. In economic applications, and in this book, we reserve the word equilibrium to market clearing.

[12] See Appendix Sect. 4.6 for the determination of the eigenvector matrices in the different cases.

Proof A steady state is a point $\mathbf{y}$ such that $\mathbf{Ay} + \mathbf{B} = \mathbf{0}$. If $\det(\mathbf{A}) \neq 0$ then there is a unique inverse matrix $\mathbf{A}^{-1}$ and therefore a unique fixed point exists $\bar{\mathbf{y}} = -\mathbf{A}^{-1}\mathbf{B}$. If matrix $\mathbf{A}$ is singular, that is $\det(\mathbf{A}) = 0$, then a classic inverse of matrix $\mathbf{A}$ does not exist. In this case, observe that $\mathbf{Ay} + \mathbf{B} = \mathbf{0}$ is equivalent to $\mathbf{P\Lambda P}^{-1}\mathbf{y} + \mathbf{B} = \mathbf{0}$, then as $\mathbf{P}$ is non-singular, we write $\mathbf{\Lambda P}^{-1}\mathbf{y} + \mathbf{P}^{-1}\mathbf{B} = \mathbf{0}$. When $\det(\mathbf{A}) = 0$ there are only real eigenvalues and the expansion of this equation can take only one of two forms, depending on the discriminant of $\mathbf{A}$. First, if $\Delta(\mathbf{A}) > 0$ we can expand $\mathbf{\Lambda}_+\mathbf{P}^{-1}\mathbf{y} + \mathbf{P}^{-1}\mathbf{B}$ to obtain

$$
\begin{aligned}
P_2^+ (\lambda_- y_1 + b_1) &= P_1^+ (\lambda_- y_2 + b_2), \\
P_2^- (\lambda_+ y_1 + b_1) &= P_1^- (\lambda_+ y_2 + b_2).
\end{aligned}
$$

Then: (1) if $\lambda_+ = 0 > \lambda_-$ and $P_2^- b_1 = P_1^- b_2$, the existence of steady states requires $P_1^+(\lambda_- y_2 + b_2) = P_2^+(\lambda_- y_1 + b_1)$, or if $\lambda_+ > 0 = \lambda_-$ and $P_1^+ b_2 = P_2^+ b_1$, the existence of steady state requires $P_2^-(\lambda_+ y_1 + b_1) = P_1^-(\lambda_+ y_2 + b_2)$, or (3) if $\lambda_+ = \lambda_- = \lambda = 0$, and $\mathbf{A} = \mathbf{0}$, there are steady states only if $P_2^+ b_2 - P_1^+ b_1 = P_1^- b_2 - P_2^- b_1 = 0$. Second, if $\Delta(\mathbf{A}) = 0$ and the Jacobian matrix is $\mathbf{\Lambda}_0$, we can expand $\mathbf{\Lambda}_0\mathbf{P}^{-1}\mathbf{y} + \mathbf{P}^{-1}\mathbf{B}$ to obtain

$$
\begin{aligned}
P_2^g (\lambda_- y_1 + b_1) + P_1^s y_2 &= P_1^g (\lambda_- y_2 + b_2) + P_2^s y_1, \\
P_2^s (\lambda\, y_1 + b_1) &= P_1^s (\lambda_+ y_2 + b_2).
\end{aligned}
$$

Then, if $\lambda = 0$, steady states exist if and only if $P_2^s b_1 = P_1^s b_2$ and $P_2^g b_1 + P_1^s y_2 = P_1^g b_2 + P_2^s y_1$. In all other cases, steady states do not exist. □

If the ODE (4.32) is **homogenous**, i.e., if $\mathbf{B} = \mathbf{0}$, then the above conditions $P_1^- b_2 = P_2^- b_1$, $P_1^+ b_2 = P_2^+ b_1$, and $P_1^- b_2 = P_2^s b_1$ are always satisfied, which implies that steady states always exist, but are multiple if $\det(\mathbf{A}) = 0$.

If the ODE (4.32) is **non-homogenous**, i.e., if $\mathbf{B} \neq \mathbf{0}$, and if none of the conditions (1) to (3) in Proposition 4.3 are satisfied, then we have case (4) and a **steady state does not exist** and $\mathbf{y}^{ss} = \emptyset$. This is the case if $\det(\mathbf{A}) = 0$, $\mathbf{B} \neq \mathbf{0}$, and other conditions are not true: in 2(a) if $P_1^- b_2 \neq P_2^- b_1$, or in case 2(b) if $P_1^+ b_2 \neq P_2^+ b_1$, or in case 2(c) if $P_1^- b_2 \neq P_2^s b_1$, or in case 3 if $P_1^- b_2 \neq P_2^- b_1$, or $P_1^+ b_2 \neq P_2^+ b_1$.

Table 4.2 sumarizes our previous results:

Table 4.2 Existence and number of steady states for the planar linear ODE

	$\det(\mathbf{A}) \neq 0$	$\det(\mathbf{A}) = 0$	
		$\mathrm{tr}(\mathbf{A}) \neq 0$	$\mathrm{tr}(\mathbf{A}) = 0$
$\mathbf{B} = \mathbf{0}$	Unique	Infinite (dim 1)	Infinite (dim 1 or 2)
$\mathbf{B} \neq \mathbf{0}$		Infinite (dim 1)	Infinite (dim 1 or 2)
		Zero	Zero

4.2.1.2 Stability of Steady States

If a steady state does not exist the solution to the ODE is always time-dependent, independently from $\mathbf{y}(0)$. If a steady state exists, and $\mathbf{y}(0) \in \mathbf{y}^{ss}$ then the solution $\mathbf{y}(t) \in \mathbf{y}^{ss}$ is time-independent. In this case, we say it is **stationary**.

If a steady state exists and $\mathbf{y}(0) \notin \mathbf{y}^{ss}$ then the general solution to the ODE is time-dependent and generates a family of solution paths $\left(\mathbf{y}(t)\right)_{t\in[0,\infty)}$. In this case, there are two properties of the solution paths over time that we will focus next: their stability and monotonicity properties.

Stability analysis refers to the characterization properties of the solution trajectories of a planar ODE regarding their convergence properties, that slightly differ from the scalar case.

Definition 4.3 (*Stability definitions*) Consider the planar linear ODE (4.32), and assume it has at least one steady state $\bar{\mathbf{y}} \in \mathbf{y}^{ss}$. Similarly to a linear scalar ODE, the (general) solution to a linear planar: is **asymptotically stable** if, for any arbitrary $\mathbf{y}(0) \neq \bar{\mathbf{y}} \in \mathbf{Y}$, the solution converges asymptotically to a steady state $\bar{\mathbf{y}}$: i.e., $\lim_{t\to\infty} \mathbf{y}(t) = \bar{\mathbf{y}}$, or is **unstable** if, for $\mathbf{y}(0) \neq \bar{\mathbf{y}} \in \mathbf{Y}$, the solution is asymptotically unbounded, i.e., $\lim_{t\to\infty} \mathbf{y}(t) = \pm\infty$. Differently from a linear planar ODE a solution to a linear planar ODE can be: **conditionally stable** if there there is a subset $\mathbf{Y}^s \subset \mathbf{Y}$ such that if $\mathbf{y}(0) \in \mathbf{Y}^s$, then the solution converges asymptotically to $\bar{\mathbf{y}}$, but if $\mathbf{y}(0) \notin \mathbf{Y}^s$ the solution is unstable; or, it can be **stable**, but not asymptotically stable, if for $\mathbf{y}(0)$ in a neighborhood of $\bar{\mathbf{y}}$, the solution stays close to $\bar{\mathbf{y}}$, for every $t \in (0, \infty)$ but does not converges asymptotically to $\bar{\mathbf{y}}$, i.e., there is a constant δ such that $||\mathbf{y}(t) - \bar{\mathbf{y}}|| < \delta$ for all $t > 0$ if $||\mathbf{y}(0) - \bar{\mathbf{y}}|| < \epsilon$ for a given constant ϵ.

As ODE (4.32) is linear then the stability properties of the solutions are **global**, in the sense they are the same for any $\mathbf{y}(0) \in \mathbf{Y}$.

The eigenvalues of $\mathbf{A}$ not only determine the number of steady states of ODE (4.32) but also their stability properties:

Proposition 4.4 *Let the steady state set $\mathbf{y}^{ss}$ have one unique element or be a linear manifold of dimension one, and let $\mathbf{y}(0) \notin \mathbf{y}^{ss}$. Then the stability properties of the solution to Eq. (4.32) can be the following:*

(1) if all the eigenvalues of matrix $\mathbf{A}$ have negative real parts then there is one unique steady state and all solutions of the ODE are asymptotically stable;
(2) if all eigenvalues of matrix $\mathbf{A}$ have positive real parts then there is one unique steady state and all solutions are unstable;
(3) if matrix $\mathbf{A}$ has one negative and one positive eigenvalue then there is one unique steady state and there is conditional stability;
(4) if the eigenvalues of matrix $\mathbf{A}$ are complex with zero real parts then there is an unique steady state and the solutions are stable but not asymptotically stable;
(5) if matrix $\mathbf{A}$ has one zero eigenvalue then the steady state set is a one-dimensional manifold (a center manifold) and the other eigenvalue is negative (i.e., in case $\lambda_- < 0$ and $\lambda_+ = 0$), then all solutions are asymptotically stable and converge

to that manifold; however, if the other eigenvalue is positive (i.e., in case $\lambda_+ > 0$ and $\lambda_- = 0$) then all solutions are unstable.

Proof The solution to the ODE (4.32) can take one of the following three forms: First, if $\Delta(\mathbf{A}) > 0$, the general solution is (see Eq. (4.19))

$$\mathbf{y}(t) = \bar{\mathbf{y}} + w_-(0)\, e^{\lambda_- t}\, \mathbf{P}^- + w_+(0)\, e^{\lambda_+ t}\, \mathbf{P}^+,$$

where $\mathbf{w} = \mathbf{P}^{-1}\mathbf{y}$ and $\mathbf{y}(0) \neq \bar{\mathbf{y}}$. Then: (1) if $\lambda_- < \lambda_+ < 0$, as $\lim_{t\to\infty} e^{\lambda_+ t} = \lim_{t\to\infty} e^{\lambda_- t} = 0$, then $\lim_{t\to\infty} \mathbf{y}(t) = \bar{\mathbf{y}}$ for any $\mathbf{y}(0) \neq \bar{\mathbf{y}}$, and the solution is asymptotically stable; (2) if $0 < \lambda_- < \lambda_+$, as $\lim_{t\to\infty} e^{\lambda_+ t} = \lim_{t\to\infty} e^{\lambda_- t} = +\infty$, then $\lim_{t\to\infty} \mathbf{y}(t) = \pm\infty$ for any $\mathbf{y}(0) \neq \bar{\mathbf{y}}$, and the solution is unstable; (3) if $\lambda_- < 0 < \lambda_+$, as $\lim_{t\to\infty} e^{\lambda_+ t} = +\infty$ and $\lim_{t\to\infty} e^{\lambda_- t} = 0$, then $\lim_{t\to\infty} \mathbf{y}(t) = \pm\infty$ if $\mathbf{w}_+ \neq 0$ or $\lim_{t\to\infty} \mathbf{y}(t) = \bar{\mathbf{y}}$ if $\mathbf{w}_+ = 0$ and the solution is conditionally stable; (4) if $\lambda_- < 0 = \lambda_+$, as $e^{\lambda_+ t} = 1$ and $\lim_{t\to\infty} e^{\lambda_- t} = 0$, then the solution will converge to the linear manifold in 2(b) of Proposition 4.3; (5) if $\lambda_- = 0 < \lambda_+$, as $e^{\lambda_- t} = 1$ and $\lim_{t\to\infty} e^{\lambda_+ t} = \infty$, then the solution is unstable. Second, if $\Delta(\mathbf{A}) = 0$, the general solution is (see Eq. (4.20))

$$\mathbf{y}(t) = \bar{\mathbf{y}} + e^{\lambda t}\left(\mathbf{P}^s\,(w_1(0) + w_2(0)\,t) + w_2(0)\,\mathbf{P}^g\right),$$

where $\mathbf{w}(0) = (w_1(0), w_2(0))^\top = \mathbf{P}^{-1}\mathbf{y}(0)$, where $\mathbf{P} = (\mathbf{P}^s, \mathbf{P}^g)$. Then: (1) if $\lambda < 0$, as $\lim_{t\to\infty} e^{\lambda t} = 0$, then $\lim_{t\to\infty} \mathbf{y}(t) = \bar{\mathbf{y}}$ and the solution is asymptotically stable; (2) if $\lambda > 0$, as $\lim_{t\to\infty} e^{\lambda t} = \infty$, then $\lim_{t\to\infty} \mathbf{y}(t) = \pm\infty$ the solution is unstable. At last, if $\Delta(\mathbf{A}) < 0$, the general solution is

$$\begin{aligned}\mathbf{y}(t) &= \bar{\mathbf{y}} + e^{\alpha t}\left((w_1(0)\cos\beta t + w_2(0)\sin\beta t)\mathbf{P}^r + (w_2(0)\cos\beta t - w_1(0)\sin\beta t)\mathbf{P}^i\right) = \\ &= \bar{\mathbf{y}} + e^{\alpha t}\left(w_1(0)(\cos\beta t\mathbf{P}^r - \sin\beta t\mathbf{P}^i) + w_2(0)(\sin\beta t\mathbf{P}^r + \cos\beta t\mathbf{P}^i)\right).\end{aligned}$$

Then: (1) the solution is asymptotically stable if $\alpha < 0$; (2) it is unstable if $\alpha > 0$; or (3) it is stable but not asymptotically stable if $\alpha = 0$. In the last case, the solution is periodic.[13] □

The dynamic behavior of the solution to Eq. (4.32) is therefore essentially related to the algebraic properties of matrix $\mathbf{A}$:

Theorem 4.1 (Stability properties for planar linear ODE) *Consider the planar ODE (4.32). Assume that $\mathbf{y}^{ss}$ is not empty. The asymptotic properties of the solution as a function of the trace and determinant of* **A** *display:*

[13] See Appendix Sect. 4.7.

Table 4.3 Stability of steady states and the characteristics of matrix **A**

	det (**A**) < 0	det (**A**) = 0	det (**A**) > 0
tr(**A**) < 0	Conditionally stable	Asymptotically stable	Asymptotically stable
		Stationary solutions	Stable
tr(**A**) > 0		Unstable	Unstable

(1) *asymptotic stability if and only if* $\text{tr}(\mathbf{A}) < 0$ *and* $\det(\mathbf{A}) \geq 0$;
(2) *saddle path (or conditional) stability if and only if* $\det(\mathbf{A}) < 0$;
(3) *instability if and only if* $\text{tr}(\mathbf{A}) > 0$ *and* $\det(\mathbf{A}) \geq 0$;
(4) *stability but not asymptotic stability if* $\text{tr}(\mathbf{A}) = 0$ *and* $\det(\mathbf{A}) \geq 0$.

While the existence of steady states depends on both coefficient matrices **A** and **B**, the time-dependence of the general solutions only depends on matrix **A**. Table 4.3 tabulates the findings in Theorem 4.1.

4.2.1.3 Manifolds, Eigenspaces, and Stability Analysis

Assuming an initial arbitrary value $\mathbf{y}(0) \notin \mathbf{y}^{ss}$, we just concluded that the solution has three types of forward time properties: it converges asymptotically to a steady state, it diverges over time (from a steady state or not) or it has a limited amplitude, neither converging nor diverging by much. As in linear scalar ODE we can partition set **Y** into two invariant subsets: the **attracting set** or **stable manifold** $\mathcal{W}^s = \{\mathbf{y} \in \mathbf{Y} : \lim_{t\to\infty} \mathbf{y}(t; \mathbf{y}) = \bar{\mathbf{y}}\}$ or the **repelling set** or **unstable manifold** $\mathcal{W}^u = \{\mathbf{y} \in \mathbf{Y} : \lim_{t\to\infty} \mathbf{y}(t; \mathbf{y}) = \pm\infty\}$. However, differently from the linear scalar ODE, there can be a **center manifold**, denoted by $\mathcal{W}^c$, which is a subset of points which are neither asymptotically stable nor unstable, and the state space **Y** can be globally partitioned into two different submanifolds in the case of the existence of conditional stability $\mathbf{Y} = \mathcal{W}^s \cup \mathcal{W}^u$. As for linear scalar ODE, the **stable**, **unstable**, and **center** manifolds are **global** manifolds because there is a unique steady state or a unique center manifold.[14]

An important aspect of linear ODE with dimensions larger than one is the close relationship between the previous manifolds and eigenspaces.

We call **eigenspaces** to the subsets of space **Y** that are spanned by the associated eigenvectors, $\mathbf{P}^-$ and $\mathbf{P}^+$ if $\Delta(\mathbf{A}) > 0$, or $\mathbf{P}^s$ and $\mathbf{P}^g$ if $\Delta(\mathbf{A}) = 0$, or $\mathbf{P}^r$ and $\mathbf{P}^i$ if $\Delta(\mathbf{A}) < 0$. In particular, we can determine those eigenspaces by cancelling the effect of the unrelated exponential functions and setting the transformed variables arbitrary

[14] In the case of non-linear ODE, that can have more than one, but finite in number, of steady states, we distinguished between **local manifolds** and **global manifolds** when they refer to a particular steady state or to the whole space. See Chap. 3.

values as $w_-(0) = 0$ or $w_+(0) = 0$.[15] If there are two real distinct eigenvalues, the eigenspace associated to λ_- is

$$\mathcal{E}^- = \{\mathbf{y} \in \mathbf{Y} : P_1^-(y_2 - \bar{y}_2) = P_2^-(y_1 - \bar{y}_1)\},$$

and the eigenspace associated to λ_+ is

$$\mathcal{E}^+ = \{\mathbf{y} \in \mathbf{Y} : P_1^+(y_2 - \bar{y}_2) = P_2^+(y_1 - \bar{y}_1)\}.$$

We have to distinguish the previous eigenspaces from the stable, unstable, and center eigenspaces, that are denoted by $\mathcal{E}^s$, $\mathcal{E}^u$, and $\mathcal{E}^c$, respectively.

In the case of linear ODE, the **stable manifold** is spanned by the eigenspaces which are associated to the eigenvalues with negative real parts, $\mathcal{W}^s \equiv \bigoplus_{j\in\pm}\{\mathcal{E}^j : \text{Re}(\lambda_j) < 0\}$, the **unstable manifold** is spanned by the eigenspaces which are associated to the eigenvalues with positive real parts, $\mathcal{W}^u \equiv \bigoplus_{j\in\pm}\{\mathcal{E}^j : \text{Re}(\lambda_j) > 0\}$, and the **center manifold** is spanned by the eigenspaces which are associated to the complex eigenvalues with zero real parts, $\mathcal{W}^c \equiv \bigoplus_{j\in\pm}\{\mathcal{E}^j : \text{Re}(\lambda_j) = 0\}$.

Therefore, for the planar linear ODE (4.32), we have the following cases:

1. if all eigenvalues have negative real parts, then $\mathcal{W}^s = \mathcal{E}^- \oplus \mathcal{E}^+ = \mathbf{Y}$, and $\mathcal{W}^u$ and $\mathcal{W}^c$ are empty, which means that $\mathcal{W}^s$ is spanned by $\mathcal{E}^-$ and $\mathcal{E}^+$ (i.e, the elements in $\mathcal{W}^s$ are a weighted sum of elements of $\mathcal{E}^-$ and $\mathcal{E}^+$). Then $\mathbf{Y}$ is an **attracting set**;
2. if all eigenvalues have positive real parts, then $\mathcal{W}^u = \mathcal{E}^- \oplus \mathcal{E}^+ = \mathbf{Y}$, and $\mathcal{W}^s$ and $\mathcal{W}^c$ are empty. Then $\mathbf{Y}$ is a **repeller set**;
3. if there is a saddle point, then the eigenspace $\mathcal{E}^-$ is called a **stable eigensapce**, and we write $\mathcal{E}^s = \mathcal{E}^-$. In this case, the stable manifold is equal to the stable eigenspace $\mathcal{W}^s = \mathcal{E}^s$, and $\mathcal{W}^u = \mathbf{Y}/\mathcal{W}^s$, and set $\mathcal{W}^c$ is empty. Then $\mathcal{E}^s$ is an **attracting set** and $\mathcal{W}^u$ is the **repelling set**;
4. if there is at least one eigenvalue with zero real part, then $\mathcal{W}^c$ is non-empty. Three cases are possible (see the proof of Proposition 4.3):
   a. first, if $\lambda_- < 0 = \lambda_+$ then $\mathcal{W}^c = \mathcal{E}^c = \{\mathbf{y} \in \mathbf{Y} : P_1^-(\lambda_- y_2 - b_2) = P_2^-(\lambda_- y_1 - b_1)\}$, $\mathcal{W}^s = \mathbf{Y}/\mathcal{W}^c$ and $\mathcal{W}^u$ is empty;
   b. second, if $\lambda_- = 0 < \lambda_+$ then $\mathcal{W}^c = \{\mathbf{y} \in \mathbf{Y} : P_2^+(\lambda_+ y_1 - b_1) = P_1^+(\lambda_+ y_2 - b_2)$, and $\mathcal{W}^u = \mathbf{Y}/\mathcal{W}^c$ and $\mathcal{W}^s$ is empty,
   c. third, $\mathcal{W}^c = \mathbf{Y}$ and $\mathcal{W}^s$ and $\mathcal{W}^u$ are both empty if there are two eigenvalues with zero real parts. If there are two real zero eigenvalues $\lambda_- = \lambda_+ = 0$ then $\mathcal{W}^c = \mathbf{Y}$ and all the elements of the set are steady states (there are no time-dependent solutions), and the associated model is static. However, if the eigenvalues are imaginary conjugate, $\lambda_\pm = \pm\beta i$ then $\mathcal{W}^c = \mathbf{Y}$ and contains one steady state, for $\mathbf{y}(0) = \mathbf{0}$, and an infinite number of periodic trajectories, for $\mathbf{y}(0) \neq \mathbf{0}$. To

[15] We can determine the eigenspace $\mathcal{E}^-$ if we set $w_+(0) = 0$, yielding $w_-(0)\, e^{\lambda_- t}\, P_1^- = y_1(t) - \bar{y}_1$ and $w_-(0)\, e^{\lambda_- t}\, P_2^- = y_2(t) - \bar{y}_2$. Thus $w_-(0)\, e^{\lambda_- t} = \frac{y_1(t)-\bar{y}_1}{P_1^-} = \frac{y_2(t)-\bar{y}_2}{P_2^-}$, for every $t \in \mathrm{T}$. We proceed in an analogous way for $\mathcal{E}^+$.

prove this we define a function (called Dulac function) $H(\mathbf{y}) = \frac{1}{2}(y_1^2 + y_2^2)$ such that the solution to the ODE satisfies $H(\mathbf{y}(t)) = H(\mathbf{y}(0))$ for each $t \in \mathrm{T}$.[16] Therefore, the center manifold

$$\mathcal{W}^c = \{\mathbf{y} \in \mathbf{Y} : H(\mathbf{y}) = H_0\},$$

for an arbitrary $H_0 = \frac{y_1(0)^2 + y_2(0)^2}{2}$, contains an infinite number of periodic trajectories.

Three important observations, that will be useful when addressing non-linear dynamics in Chap. 5, can be made at this point. First, eigenspaces provide information on the specific time behavior of the solutions both for linear and non-linear ODE, Second, for linear ODE, manifolds can coincide with eigenspaces, which is not the case for non-linear ODE. This is the case for saddles, where $\mathcal{W}^s = \mathcal{E}^s = \mathcal{E}^-$, and for saddle-node ODE, where $\mathcal{W}^c = \mathcal{E}^-$ or $\mathcal{W}^c = \mathcal{E}^+$, as will be seen in the next subsection. Third, if there are eigenvalues with zero real parts, the associated center manifolds contain an infinite number of steady states, or of periodic trajectories, which is generally not the case in non-linear regular ODE.

4.2.2 *Phase Diagrams*

A **phase diagram** for a planar autonomous ODE is a geometrical representation of the dynamics in the two-dimensional space $\mathbf{Y} \subseteq \mathbb{R}^2$. It contains the following elements (see Figs. 4.3 and 4.4). First, it contains **isoclines (or nullclines)** that are the geometrical loci in the space $\mathbf{Y}$ at which one variable, y_1 or y_2, is constant. There are two isoclines, the first associated to y_1, and the second associated with y_2, $\mathcal{I}_1 = \{\mathbf{y} \in \mathbf{Y} : \dot{y}_1 = 0\}$, and $\mathcal{I}_2 = \{\mathbf{y} \in \mathbf{Y} : \dot{y}_2 = 0\}$. By looking at Eq. (4.33) it should be evident that every isocline is a line in the space $\mathbf{Y}$. The steady states are the loci where isoclines intersect or are coincident. Second, isoclines divide the set $\mathbf{Y}$ into **quadrants**. If there is a unique steady state, there are four quadrants, in which each quadrant represents a particular joint change in time for both variables (increasing if $\dot{y}_j > 0$ and decreasing if $\dot{y}_j < 0$):$\{\mathbf{y} \in \mathbf{Y} : \dot{y}_1 > 0,\ \dot{y}_2 > 0\}$, $\{\mathbf{y} \in \mathbf{Y} : \dot{y}_1 < 0,\ \dot{y}_2 > 0\}$, $\{\mathbf{y} \in \mathbf{Y} : \dot{y}_1 > 0,\ \dot{y}_2 < 0\}$, and $\{\mathbf{y} \in \mathbf{Y} : \dot{y}_1 < 0,\ \dot{y}_2 < 0\}$. If there is an infinite number of steady states, or there is no steady state, the number of quadrants reduces to two or three, respectively. Finding those quadrants allows us to represent the direction of the forward time evolution of both variables in a representative grid of points in $\mathbf{Y}$, that is called the **vector field**, that display the local direction of evolution of the solution $\mathbf{y}(t)$. Third, **eigenspaces** $\mathcal{E}^-$ and $\mathcal{E}^+$, for the case in which $\Delta(\mathbf{A}) > 0$, are also usually drawn. They are lines in the space $\mathbf{Y}$ passing through the steady

[16] This is the case if and only if $\dot{H}(\mathbf{y}(t)) = 0$ for any t. But $\dot{H}(\mathbf{y}(t)) = \nabla H(y)\,\dot{\mathbf{y}} = H_{y_1}\dot{y}_1 + H_{y_2}\dot{y}_2 = y_1\beta y_2 + y_2(-\beta y_1) = 0$.

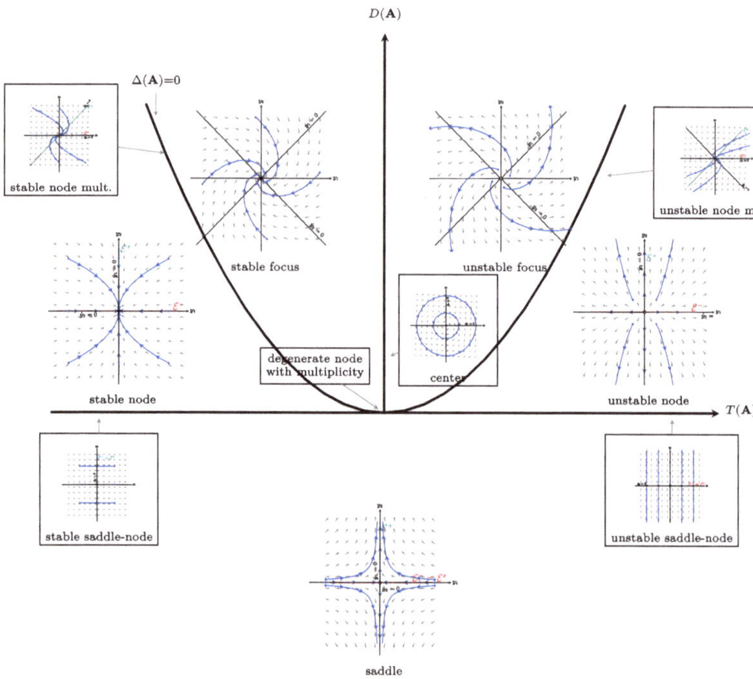

Fig. 4.3 Bifurcation diagram: phase portraits in the $(\mathrm{tr}(\mathbf{A}), \det(\mathbf{A}))$-plane for the transformed ODE, $\dot{\mathbf{w}} = \mathbf{\Lambda}\,\mathbf{w}$, where $\mathbf{\Lambda}$ is a Jordan canonical form, and $\mathbf{B} = \mathbf{0}$. The red (green) lines correspond to the eigenspace $\mathcal{E}^-$ ($\mathcal{E}^+$) for the cases without multiplicity, or to the simple (generalized) eigenspaces for the cases with multiplicity. For the node it is assumed that $\beta > 0$

state (or representative steady states) with slopes corresponding to the eigenvectors $\mathbf{P}^-$ and $\mathbf{P}^+$. Fourth, related to the previous eigenspaces we can also draw the **stable**, **unstable**, and **center manifolds**, $\mathcal{W}^s$, $\mathcal{W}^u$, and $\mathcal{W}^c$, which can be one- or two-dimensional subsets of $\mathbf{Y}$, that introduce another partition to set $\mathbf{Y}$ associated to loci of points that converge or not to a steady state (assuming a forward ODE). At last, some representative trajectories—or particular trajectories for ODE problems—called **integral curves**—are also usually drawn, starting from particular points $\mathbf{y}(0)$ located in each quadrant. They are parametric curves representing the solution to the ODE within space $\mathbf{Y}$, in which the forward time evolution is implicitly represented by arrows. They have two noteworthy features: when an integral path crosses one isocline the associated component of $\mathbf{y}(t)$ changes slope (from increasing to decreasing or the opposite), and when $\Delta(\mathbf{A}) \geq 0$ there is a leading asymptotic direction that is tangent to one of the eigenspaces, which can result in a hump-shaped trajectory over time.

Let $\mathbf{y}^{ss}$ be non-empty. In an autonomous planar linear ODE (4.32), the following **types of phase diagrams** (see Table 4.4) can exist:

- **node** if $\det(\mathbf{A}) \geq 0$ and $\Delta(\mathbf{A}) \geq 0$. In particular, we can have: a **stable node** if $\Delta(\mathbf{A}) > 0$ and $\mathrm{tr}(\mathbf{A}) < 0$; an **unstable node** if $\Delta(\mathbf{A}) > 0$ and $\mathrm{tr}(\mathbf{A}) > 0$; a

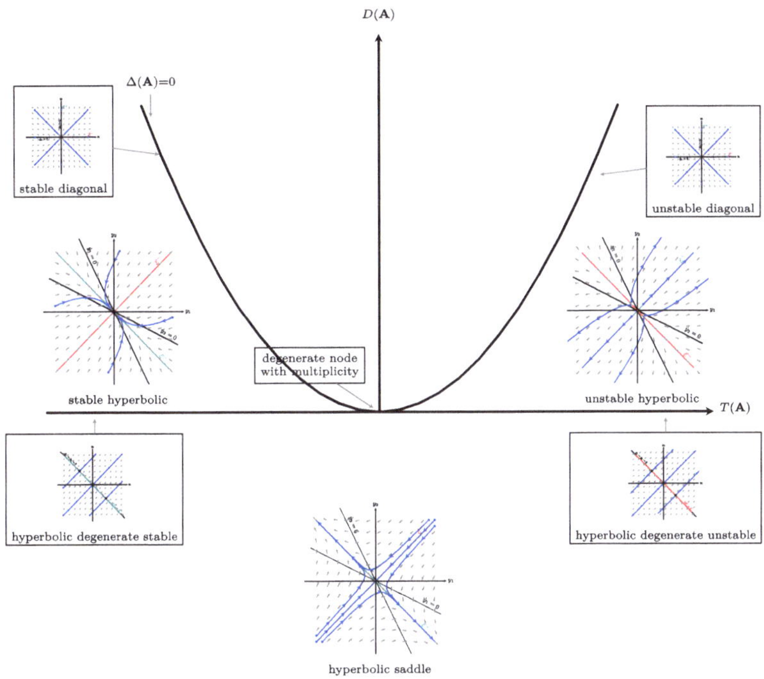

Fig. 4.4 Bifurcation diagram: phase portraits in the $(\mathrm{tr}(\mathbf{A}), \det(\mathbf{A}))$-plane for the ODE, $\dot{\mathbf{y}} = \mathbf{A}_h\,\mathbf{y}$, where $\mathbf{A}_h$ is an hyperbolic matrix for $\beta > 0$, and $\mathbf{B} = \mathbf{0}$. If $-\alpha < |\beta| < \alpha$, the phase diagram is a saddle, if $|\beta| > |\alpha|$, it is a stable (unstable) node if $\alpha < 0$ ($\alpha > 0$), if $|\beta| = |\alpha|$, it is a degenerate stable (unstable) node if $\alpha < 0$ ($\alpha > 0$), and if $\beta = 0$, it is a stable or unstable diagonal case

stable node with multiplicity if $\Delta(\mathbf{A}) = 0$ and $\mathrm{tr}(\mathbf{A}) < 0$; an **unstable node with multiplicity** if $\Delta(\mathbf{A}) = 0$ and $\mathrm{tr}(\mathbf{A}) > 0$; a **stable saddle-node** if $\det(\mathbf{A}) = 0$ and $\mathrm{tr}(\mathbf{A}) < 0$; an **unstable saddle-node** if $\det(\mathbf{A}) = 0$ and $\mathrm{tr}(\mathbf{A}) > 0$; or a **degenerate node** if $\det(\mathbf{A}) = \mathrm{tr}(\mathbf{A}) = 0$;
- a **saddle** if $\det(\mathbf{A}) < 0$;
- **foci** if $\Delta(\mathbf{A}) < 0$. In particular, we can have: a **stable focus** if in addition $\mathrm{tr}(\mathbf{A}) < 0$; an **unstable focus** if in addition $\mathrm{tr}(\mathbf{A}) > 0$; or a **center** if in addition $\mathrm{tr}(\mathbf{A}) = 0$.

4.2.3 Bifurcation Analysis

Bifurcation analysis, as we saw in the two previous chapters, studies the qualitative changes in the dynamics of the solution to the ODE (4.40) for variations of parameters within the set Φ. In other words, it studies which types of phase diagrams can occur. This is done by finding bifurcations, that is by identifying parameters values associated to non-generic dynamics or phase diagrams, and, in particular, to those

Table 4.4 Phase diagrams for linear planar ODE, when a steady state exists

	$\det(\mathbf{A}) < 0$	$\det(\mathbf{A}) = 0$	$\det(\mathbf{A}) > 0$		
			$\Delta(\mathbf{A}) < 0$	$\Delta(\mathbf{A}) = 0$	$\Delta(\mathbf{A}) > 0$
$\text{tr}(\mathbf{A}) < 0$	Saddle	Stable saddle-node	Stable focus	Stable node with multiplicity	Stable node
$\text{tr}(\mathbf{A}) = 0$	Saddle	Degenerate node	Center		
$\text{tr}(\mathbf{A}) > 0$	Saddle	Unstable saddle-node	Unstable focus	Unstable node with multiplicity	Unstable node

associated to changes in the dimension of the stable manifold. From our previous results, this is tantamount to finding the parameter values that are associated to qualitative changes in the eigenvalues of matrix **A**, that is, to changes in the signs of the trace and the determinant of matrix **A**.[17]

Figures 4.3 and 4.4 present two **bifurcation diagrams** (also called Poicaré diagrams), containing all the possible phase diagrams, in the $(\text{tr}(\mathbf{A}), \det(\mathbf{A}))$-space, for homogeneous ODE with Jordan coefficient matrices—that is for equations $\dot{\mathbf{w}} = \mathbf{\Lambda}\,\mathbf{w}$—and for equation $\dot{\mathbf{y}} = \mathbf{A}_h\,\mathbf{y}$ having a non-canonical coefficient matrix, respectively. For ODE (4.32), in which matrix **A** is canonical but not in a Jordan form, and vector **B** is non-zero, the phase diagrams would display two geometrical transformations that are analogous to $\mathbf{y} - \bar{\mathbf{y}} = \mathbf{P}\,\mathbf{w}$: first, the isoclines would rotate, by action of the eigenvector matrix **P**, and, second, the steady state will be shifted away from the origin. Therefore, a phase diagram for ODE (4.32) would be qualitatively equivalent as the one for the related similarity transformed system $\dot{\mathbf{w}} = \mathbf{\Lambda}\,\mathbf{w}$, that is represented in Fig. 4.3. Several taxonomies for ODE dynamics, and the associated phase diagrams, can be established.

First, according to the **dimension of the manifolds**, we can have the following cases that are related to the signs of the real parts of the associated eigenvalues: (1) two-dimensional stable manifolds and empty stable and center manifolds for stable nodes, including nodes with multiplicity, and stable foci; (2) one-dimensional stable and unstable manifolds and empty center manifolds in the case of saddles; (3) two-dimensional unstable manifolds and empty stable and center manifolds in the case of unstable nodes, including nodes with multiplicity, and foci; (4) one-dimensional stable and center manifolds and empty unstable manifolds in the case of stable saddle-node; (5) one-dimensional unstable and center manifolds and empty stable manifolds in the case of unstable saddle nodes; or (6) two-dimensional center manifolds and empty stable and unstable manifolds in the case of centers and degenerate nodes.

Second, according to the **time-profile** of solution trajectories, we can have several types of paths: (1) **stationary**, or constant over time, solution paths, that exist for

[17] Observe that in the scalar ODE, we only needed a parameter to characterize the stability properties of the ODE. The trace and the determinant are the extension of the coefficient a in Sect. 2.3.2.

degenerate nodes for any initial value of $\mathbf{y}(0)$, or for the other phase diagrams if $\mathbf{y}(0) \in \mathbf{y}^{ss}$; (2) **asymptotically stationary** solution paths if there is convergence to a steady state over time, that exist for stable nodes (including with multiplicity and saddle nodes), for any initial value $\mathbf{y}(0)$, or for saddles if $\mathbf{y}(0) \in \mathcal{E}^-$; (3) **non-stationary** solution paths satisfying $\lim_{t\to\infty} \mathbf{y}(t) = \pm\infty$, that exist for unstable nodes (including with multiplicity and saddle nodes) when $\mathbf{y}(0) \notin \mathbf{y}^{ss}$, or for saddles if $\mathbf{y}(0) \notin \mathcal{E}^-$; (4) **monotonic** solution paths if $\text{sign}(\dot{\mathbf{w}}(t))$ is the same for all $t \in \text{T}$, that can be monotonically increasing if $\dot{\mathbf{w}}(t) > \mathbf{0}$ for all t, or monotonically decreasing if $\dot{\mathbf{w}}(t) < \mathbf{0}$ for all t. This case occurs always for ODE with non-canonical $\mathbf{A}$ matrices or for all the types of nodes associated to canonical matrices. A stationary solution can be seen as a particular type of monotonic solution; (5) **oscillatory** solution paths if $\mathbf{w}(t) = \mathbf{w}(t + p(t))$, for $t \in \text{T}$ and time-dependent period $p(t) \in \text{T}$, in which a particular level of the solution is repeated in increasing intervals if $p'(t) > 0$, for unstable foci, or in decreasing intervals if $p'(t) < 0$, for stable foci, and there is a sequence of points, increasing or decreasing over time $\tau \in \{t_0, t_1, \ldots, t_s, \ldots\}$ such that $\dot{\mathbf{w}}(\tau) = 0$; (6) **periodic** solutions paths with period p if the solution satisfies $\mathbf{w}(t) = \mathbf{w}(t + p)$ for $t \in \text{T}$ and p is a constant, which is the type of solution for centers, for which $\text{tr}(\mathbf{A}) = 2\alpha = 0$. Observe that in this case, if we transform the system into polar coordinates[18] we have $r(t) = r_0$ constant and $\theta(t) = \theta_0 - \beta t$; at last, (7) we can have **hump-shaped** paths if the solution path of a planar equation is such that only one variable satisfies $\dot{y}_i(t) = 0$ for a finite $t \in \text{T}$ and the other variable y_{-i} is monotonic. This can be the case for nodes with multiplicity, and for nodes associated to canonical matrices $\mathbf{A}$ that are similar but different from Jordan canonical forms. Geometrically, a necessary condition for the existence of these trajectories is that the eigenspaces $\mathcal{E}^-$ or $\mathcal{E}^+$ are not co-incident with the isoclines $\mathcal{I}_1$ or $\mathcal{I}_2$.

For any matrix $\mathbf{A}$, we can extend the bifurcation analysis we already performed for linear scalar ODE, in Sect. 2.3.2, to the planar case. We distinguished phase diagrams regarding their **genericity**: a phase diagram is **generic** (**non-generic**) if a small change in any parameter does not change (changes) the type of the phase diagram. In both Figs. 4.3 and 4.4, the **non-generic cases** cases are drawn inside a box: they include saddle nodes, when $\det(\mathbf{A}) = 0$, centers, when $\Delta(\mathbf{A}) < 0$ and $\text{tr}(\mathbf{A}) = 0$, or nodes with multiplicity, when $\Delta(\mathbf{A}) = 0$. There is a fundamental distinction between the first two cases and the third: while the first two cases are **bifurcations** because the dimensions of the stable and unstable manifolds change by one (in the case of saddle nodes) or two (in the case of centers), the third case is **not a bifurcation** because (in the case of nodes with multiplicity) the dimension of the stable and unstable manifolds does not change—the time profile of trajectories change from monotonic (or hump shaped) to oscillatory.

It is convenient to write ODE (4.32) as

$$\dot{\mathbf{y}} = F(\mathbf{y}, \varphi) = \mathbf{A}(\varphi)\,\mathbf{y} + \mathbf{B}(\varphi) \tag{4.40}$$

[18] See Sect. 4.7 in the Appendix.

where φ is a parameter vector $\varphi \in \Phi \subset \mathbb{R}^p$. This implies that the solution to the ODE is a mapping $\mathbf{y} : \mathrm{T} \times \Phi \rightarrow \mathbf{Y} \subseteq \mathbb{R}^2$. Our classification of the phase diagrams in Table 4.4 allows us to classify bifurcations according to the number of parameters that should change to unfold the non-generic cases associated to eigenvalues with zero real parts.

Let us denote the trace, determinant, and the discriminant as functions of parameters by

$$T(\varphi) = \mathrm{tr}(\mathbf{A}(\varphi)),\ D(\varphi) = \det(\mathbf{A}(\varphi)),\ \text{and}\ \Delta(\varphi) = \left(\frac{T(\varphi)}{2}\right)^2 - D(\varphi).$$

The **co-dimension** of a bifurcation refers to the number of parameters that define a bifurcation point.

In planar ODE, there are only bifurcations of co-dimension one and two. A **bifurcation of co-dimension one** exists if there is one parameter, φ, called **bifurcating parameter**, such that when it takes a specific value, say $\varphi = \varphi^\star$, we have

$$D(\varphi^*) = \det(\mathbf{A}(\varphi^*)) = 0, \tag{4.41}$$

and $D(\varphi) \neq 0$ for $\varphi \neq \varphi^*$. If $\varphi = \varphi^*$ then matrix $\mathbf{A}$ has a zero eigenvalue and the phase diagram is a saddle-node, and if $\varphi \neq \varphi^*$ then the phase diagram can be a saddle, a stable or an unstable node.

A **bifurcation of co-dimension two** exists if there are two bifurcation parameters, say φ_1 and φ_2, such that, if $(\varphi_1, \varphi_2) = (\varphi_1^*, \varphi_2^*)$, we have

$$\begin{cases} T(\varphi_1^*, \varphi_2^*) = \mathrm{tr}(\mathbf{A}(\varphi_1^*, \varphi_2^*) = 0, \\ D(\varphi_1^*, \varphi_2^*) = \det(\mathbf{A}(\varphi_1^*, \varphi_2^*)) \geq 0, \end{cases} \tag{4.42}$$

and $T(\varphi_1, \varphi_2) \neq 0$ for $(\varphi_1 \varphi_2) \neq (\varphi_1^*, \varphi_2^*)$. If $(\varphi_1 \varphi_2) = (\varphi_1^*, \varphi_2^*)$ then matrix $\mathbf{A}$ has two eigenvalues with zero real parts, and the phase diagram can be a degenerate node or a center.

In both cases, we have **global bifurcations** because the qualitative properties of the dynamics is the same for the whole set $\mathbf{Y}$ and not only for a neighborhood of a steady state, as can be the case in non-linear models. This means that the stable, unstable, and center manifolds, $\mathcal{W}^s$, $\mathcal{W}^u$ and $\mathcal{W}^c$, also have a global nature.

There are two types of representations of **bifurcation diagrams** in the economics' literature:

1. as partitions of the Φ space, when there are at least two parameters. In this space we represent the lines $\{\varphi \in \Phi : D(\varphi) = 0\}$ and $\{\varphi \in \Phi : T(\varphi) = 0\}$, and $\{\varphi \in \Phi : \Delta(\varphi) = 0\}$,
2. as an implicit plot of $T(\varphi)$ and $D(\varphi)$ in a trace-determinant cartesian diagram as in 4.3, when parameters vary within their domains. Geometrically bifurcations

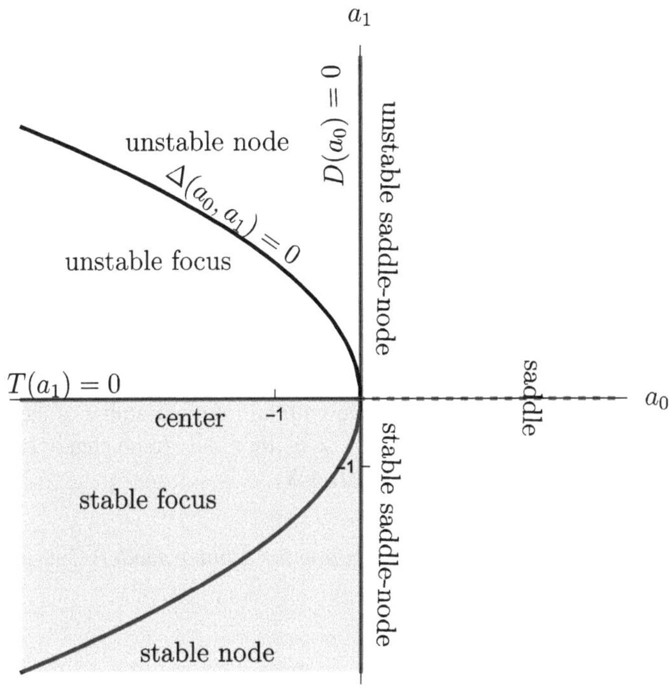

Fig. 4.5 Bifurcation diagram for ODE (4.36) in the space (a_0, a_1)

exist if those lines cross the horizontal axis or the positive half of the vertical axis.[19]

Example 4.4 (*Linear second-order ODE*) Consider again the scalar second-order linear ODE in Example 4.2, in which $\varphi = (a_0, a_1, b) \in \mathbb{R}^3$. We already determined the trace, determinant, and the eigenvalues of matrix $\mathbf{A}$, as a function of the parameters, a_0, a_1 and b: $T(a_1) = \text{tr}(\mathbf{A}) = a_1$, $D(a_0) = \det(\mathbf{A}) = -a_0$, $\Delta(a_0, a_1) = \left(\frac{a_1}{2}\right)^2 + a_0$, and $\lambda_{\mp} = \lambda_{\mp}(a_0, a_1) = \frac{a_1}{2} \pm \sqrt{\left(\frac{a_1}{2}\right)^2 + a_0}$. The complete unfolding of the bifurcation diagram is shown in Fig. 4.5. The phase diagram is a saddle if $a_0 > 0$, the stable manifold has dimension two if $a_0 < 0$ and $a_1 < 0$, and the unstable manifold has dimension two if $a_0 < 0$ and $a_1 > 0$. The bifurcating parameters are a_0, and a_1 and we can have bifurcations of co-dimension one for $a_0 = 0$, or bifurcations of co-dimension two for $a_0 \leq 0$ and $a_1 = 0$. In particular, if $b = 0$, there is a stable saddle-node if $a_0 = 0$ and $a_1 < 0$, there is an unstable saddle node if $a_0 = 0$ and $a_1 > 0$, a center if $a_0 < 0$ and $a_1 = 0$, and a degenerate node if $a_0 = a_1 = 0$.

[19] This approach follows Grandmont (1988) for planar difference equations. See also Grandmont (2008).

Example 4.5 (*Co-dimension one bifurcation for a linear model*) Consider the following planar ODE, where $\varphi = (\mu, b_1, b_2)$ is a vector of parameters with arbitrary real values,

$$\dot{y}_1 = \mu\, y_1 + y_2 - b_1,$$
$$\dot{y}_2 = y_2 - b_2.$$

The coefficient matrix $\mathbf{A} = \left(\begin{smallmatrix} \mu & 1 \\ 0 & 1 \end{smallmatrix}\right)$, has trace $T(\mu) = 1 + \mu$ and determinant $D(\mu) = \mu$. Then, it has eigenvalues are $\lambda_-(\mu) = \mu$ and $\lambda_+(\mu) = 1$. The bifurcating parameter is μ, and the other parameters although not influencing the qualitative dynamics at a steady state, determine its existence. The phase diagram is a saddle if $\mu < 0$, or it is an unstable node if $\mu > 0$. There is a co-dimension one bifurcation for $\mu = \mu^* = 0$. In this case: if $b_1 = b_2 = b$, then there is an infinite number of steady states and the phase diagram is an unstable saddle node with an infinite number of steady states, as $\mathbf{y}^{ss}(\mu^*) = \{\mathbf{y} \in \mathbb{R}^2 : y_2 = b\}$; but if $b_1 \neq b_2$ then the are no steady state and all solutions are non-stationary for every $\mathbf{y}(0) \in \mathbf{Y}$.

Example 4.6 (*Co-dimension two bifurcation for a linear model*) Consider the following planar ODE,

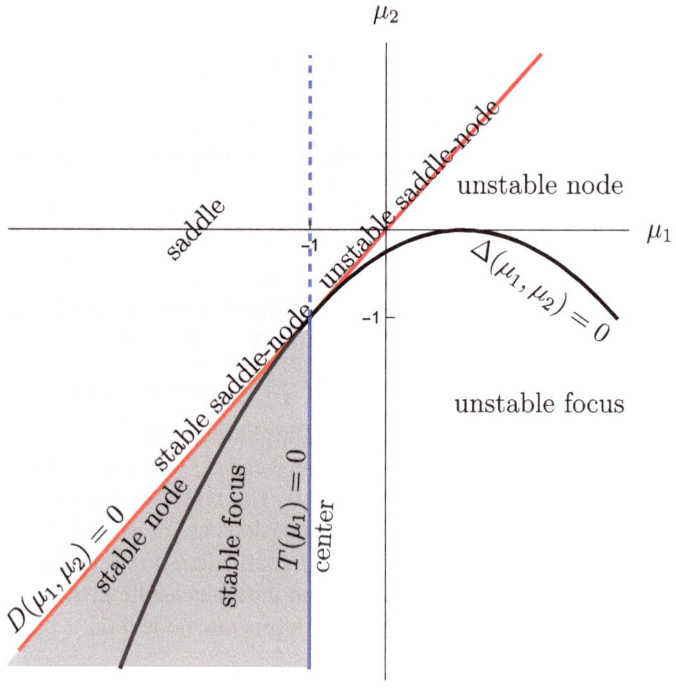

Fig. 4.6 Bifurcation diagram for ODE (4.43) in the space (μ_1, μ_2). The subset of the parameter space associated to stable manifolds of dimension two (one) are colored in dark (light) gray

$$\begin{aligned}\dot{y}_1 &= \mu_1\, y_1 + y_2 - b\\ \dot{y}_2 &= \mu_2\, y_1 + y_2 - b.\end{aligned} \tag{4.43}$$

where parameter vector is $\varphi = (\mu_1, \mu_2, b) \in \mathbb{R}^3$. The coefficient matrix is now $\mathbf{A} = \begin{pmatrix} \mu_1 & 1 \\ \mu_2 & 1 \end{pmatrix}$, the trace is $T(\mu_1) = 1 + \mu_1$ and $D(\mu_1, \mu_2) = \mu_1 - \mu_2$, and the eigenvalues of $\mathbf{A}$ are $\lambda_{\pm} = \frac{1+\mu_1}{2} \pm \sqrt{\mu_2 + \left(\frac{1-\mu_1}{2}\right)^2}$. The complete unfolding of the bifurcation diagram is shown in Fig. 4.6. In particular, the phase diagram is a saddle if $\mu_2 > \mu_1$, the stable manifold has dimension two if $\mu_2 < \mu_1 < -1$, and the unstable manifold has dimension two if $\mu_1 > -1$ and $\mu_2 < \mu_1$. The bifurcating parameters are μ_1, and μ_2 and we can have bifurcations of co-dimension one for $(\mu_1, \mu_2) = \big(-1, (-\infty, -1)\big)$, or a bifurcation of co-dimension two for $(\mu_1, \mu_2) = (-1, -1)$. In particular, there is a stable saddle-node if $\mu_2 = \mu_1 < -1$, there is an unstable saddle node if $\mu_2 = \mu_1 > -1$, there is a center if $\mu_2 < \mu_1 = -1$, and a degenerate node if $\mu_1 = \mu_2 = -1$.

Exercise 4.8 (*Non-autonomous scalar equations*) Draw the phase diagrams and the bifurcation diagram for Example 4.3.

4.2.4 Problems for Planar ODE

All the previous general solutions for linear planar ODE depend on an arbitrary vector $\mathbf{y}(0) \in \mathbf{Y}$. This means that they always exist but are not unique, in the sense that their solution does not trace an unique trajectory $\big(\mathbf{y}(t)\big)_{t\in\mathrm{T}}$ within $\mathbf{Y}$ but a family of solutions.

In economic dynamic models, planar ODE problems are specified by introducing two side conditions. The following problems can be found in the literature: first, **initial-value problems** if the side condition is $\mathbf{y}(0) = \mathbf{y}_0$ where $\mathbf{y}_0 = (y_{1,0}, y_{2,0})$ is a fixed element of $\mathbf{Y}$ at $t = 0$, and we are interested in solving a forward ODE; second, **terminal-value problems** if the side condition is $\mathbf{y}(T) = \mathbf{y}_T$, where $\mathbf{y}(T) = (y_1(T), y_2(t) = (y_{1,T}, y_{2,T}$, is a fixed element of $\mathbf{Y}$ at a terminal time $t = T \in (0, \infty)$, and we are interested in solving a backward ODE; or a **mixed initial-terminal value problems**, where we know the initial value for one variable, say $y_1(0) = y_{1,0}$, and a terminal or asymptotic value for the other variable, say $y_2(T) = y_{2,T}$, where both $y_{1,0}$ and $y_{2,T}$ are fixed elements of the set $\mathbf{Y}$. However, other type of restrictions for the terminal value of y_2 are possible. In this case, the ODE has a forward-backward feature, which does not exist in scalar ODE problems.

As for scalar ODE, the existence of solutions to linear ODE problems is not always the case, and, even when existing, they may not be unique. Next we show that: first, initial value problems have unique solutions; second, terminal-value problems have solutions—not necessarily determinate—if matrix $\mathbf{A}$ is non-singular, or if it is singular and $\mathbf{B} = \mathbf{0}$, but may not have solutions if $\mathbf{A}$ is singular and $\mathbf{B} \neq \mathbf{0}$; and, third, that mixed problems may have solutions—determinate or indeterminate—or not, even in the case in which $\mathbf{A}$ is non-singular.

Proposition 4.5 (Solutions to initial-value problems) *Consider ODE problem specified by Eq. (4.32) and initial condition* $\mathbf{y}(0) = \mathbf{y}_0 \in \mathbf{Y}$, *where* $\mathbf{y}_0$ *is fixed for* $t = 0$. *Then the solution for the initial-value problem is unique, and takes one of the following two forms:*

1. *if* $\det(\mathbf{A}) \neq 0$ *it is* $\mathbf{y}(t) = \bar{\mathbf{y}} + \mathbf{P}\mathbf{e}^{\mathbf{\Lambda t}}\mathbf{P}^{-1}(\mathbf{y}_0 - \bar{\mathbf{y}})$, *for each* $t \in \mathrm{T}$, *where* $\bar{\mathbf{y}} = -\mathbf{A}^{-1}\mathbf{B}$; *or*
2. *if* $\det(\mathbf{A}) = 0$, *it is* $\mathbf{y}(t) = -\mathbf{A}^+\mathbf{B} + \mathbf{P}\mathbf{e}^{\mathbf{\Lambda t}}\mathbf{P}^{-1}(\mathbf{y}_0 + \mathbf{A}^+\mathbf{B}) + (\mathbf{I} - \mathbf{A}\mathbf{A}^+)\mathbf{B}\,t$, *for each* $t \in \mathrm{T}$, *where* $\mathbf{A}^+$ *is the Moore-Penrose inverse of* $\mathbf{A}$.

Proof In the general solutions for both non-singular and singular $\mathbf{A}$ matrices, in Eqs. (4.34) and (4.35), if we solve $\mathbf{y}(t) = \mathbf{y}_0$ for $t = 0$, and note that $\mathbf{e}^{\mathbf{\Lambda t}}|_{t=0} = \mathbf{I}$, we find $\mathbf{y}(0) = \mathbf{y}_0$. Substituting back in general solution (4.34) or (4.35) we find unique particular solutions. □

Proposition 4.6 (Solutions to terminal-value problems with infinite horizon) *Consider ODE problem specified by Eq. (4.32), and the terminal condition* $\lim_{t\to\infty} \mathbf{y}(t) = \bar{\mathbf{y}}$. *If* $\det(\mathbf{A}) \neq 0$, *then*

(1) if the phase diagram is a stable node or a stable focus then the solution is $\mathbf{y}(t) = \bar{\mathbf{y}} + \mathbf{P}\mathbf{e}^{\mathbf{\Lambda}t}\,\mathbf{P}^{-1}(\mathbf{y}(0) - \bar{\mathbf{y}})$, *for each* $t \in \mathrm{T}$ *and for any* $\mathbf{y}(0) \in \mathbf{Y}$, *and therefore, it exists but is indeterminate;*
(2) if the phase diagram is an unstable node or an unstable focus then the solution is $\mathbf{y}(t) = \bar{\mathbf{y}}$, *for each* $t \in \mathrm{T}$, *and, therefore, it exists and is determinate;*
(3) if $\bar{\mathbf{y}}$ *is a saddle-point then the solution is* $\mathbf{y}(t) = \bar{\mathbf{y}} + w_-(0)\,\mathbf{P}^-\,e^{\lambda_- t}$, *for each* $t \in \mathrm{T}$, *and for arbitrary* $w_-(0)$, *and, therefore, it exists but is indeterminate.*

Proof If $\det(\mathbf{A}) \neq 0$ then there is an unique steady state $\bar{\mathbf{y}}$. Furthermore: (1) for stable nodes or foci, as all the eigenvalues of $\mathbf{A}$ have negative real parts, then $\lim_{t\to\infty} \mathbf{e}^{\mathbf{\Lambda t}} = \mathbf{I}_{2\times 2}$. This implies that $\lim_{t\to\infty} \mathbf{y}(t) = \bar{\mathbf{y}}$ independently of the value of $\mathbf{y}(0)$; (2) for unstable nodes and foci, as all the eigenvalues of $\mathbf{A}$ have positive real parts then all the exponential functions $e^{\lambda_+ t}$, $e^{\lambda_- t}$, $e^{\lambda t}$ or $e^{\alpha t}$ are asymptotically unbounded. Then $\lim_{t\to\infty} \mathbf{P}\mathbf{e}^{\mathbf{\Lambda t}}\mathbf{w}(0) = \mathbf{0}$ if and only if $\mathbf{w}(0) = \mathbf{P}^{-1}\,(\mathbf{y}(0) - \bar{\mathbf{y}}) = \mathbf{0}$. Then as $\mathbf{y}(0) = \bar{\mathbf{y}}$ is uniquely determined, the solution, $\mathbf{y}(t) = \bar{\mathbf{y}}$ is unique. (3) For a saddle, as the Jacobian form of $\mathbf{A}$ is $\mathbf{\Lambda}_+$, the solution takes the form $\mathbf{y}(t) = \bar{\mathbf{y}} + w_-(0)\,\mathbf{P}^-\,e^{\lambda_- t} + w_+(0)\,\mathbf{P}^+\,e^{\lambda_+ t}$, that implies $\lim_{t\to\infty} e^{\lambda_+ t} = +\infty$ and $\lim_{t\to\infty} e^{\lambda_- t} = 0$. Therefore $\lim_{t\to\infty} \mathbf{y}(t) = \bar{\mathbf{y}}$ if and only if $w_+(0) = 0$, and the solution is $\mathbf{y}(t) = \bar{\mathbf{y}} + w_1(0)\,\mathbf{P}^-\,e^{\lambda_- t}$. □

Exercise 4.9 (*Terminal-value problems for singular* $\mathbf{A}$ *matrices*) Consider the same problem as in Proposition 4.6 but assume instead that $\det(\mathbf{A}) = 0$. Prove the following: (1) if $\mathbf{B} = 0$ then a solution to the terminal-value problem always exists, and it is determinate if $\mathrm{tr}(\mathbf{A}) > 0$, or it is indeterminate $\mathrm{tr}(\mathbf{A}) < 0$; and (2) if $\mathbf{B} \neq 0$ then a solution to the terminal value problem **may not exist**.

One of the most common problems in economics is the mixed initial-terminal value problem:

Proposition 4.7 (Solutions to mixed-value problems) *Consider ODE problem specified by Eq. (4.32), the initial condition* $y_1(0) = y_{1,0}$, *where* $y_{1,0}$ *is fixed, and the terminal condition* $\lim_{t\to\infty} \mathbf{y}(t) = \bar{\mathbf{y}}$. *If* $\det(\mathbf{A}) \neq 0$, *then*

1. *if the phase diagram is a stable node then the solution is indeterminate, as the solution depends on an arbitrary* $y_2(0)$;
2. *if the phase diagram is a saddle then the solution is determinate, at it takes the form*

$$\mathbf{y}(t) = \bar{\mathbf{y}} + \frac{(y_{1,0} - \bar{y}_1)}{P_1^-}\mathbf{P}^-\, e^{\lambda_- t}; \tag{4.44}$$

3. *if the phase diagram is an unstable node then a unique solution exists if and only if* $y_{1,0} = \bar{y}_1$ *and* $y_2(0) = \bar{y}_2$. *If* $\mathbf{y}(0) \neq \bar{\mathbf{y}}$ *then there is no solution that is continuous in time.*

Proof Let matrix $\mathbf{A}$ be similar to the Jordan form $\mathbf{\Lambda}_+$. Taking the general solution (4.34) expressed in the transformed variables as in Eq. (4.19), we have $\mathbf{y}(t) = \bar{\mathbf{y}} + \mathbf{P}e^{\mathbf{\Lambda}t}\mathbf{w}(0)$, where $\mathbf{w}(0) = \mathbf{P}^{-1}\big(\mathbf{y}(0) - \bar{\mathbf{y}}\big)$ if and only if $\begin{pmatrix} w_-(0) \\ w_+(0) \end{pmatrix} = \mathbf{P}^{-1}\begin{pmatrix} y_1(0)-\bar{y}_1 \\ y_2(0)-\bar{y}_2 \end{pmatrix}$. Therefore: (1) if the phase diagram is a stable node—with $\mathbf{A}$ having two negative eigenvalues—then $\lim_{t\to\infty}\mathbf{y}(t) = \bar{\mathbf{y}}$ is satisfied for any vector $\mathbf{w}(0)$. As we have a fixed value for $y_1(0)$, then the vector $\mathbf{w}(0)$ depend on an arbitrary value for $y_2(0)$, meaning that the solution to the mixed-value problem exists but is indeterminate; (2) if the phase diagram is a saddle—with the eigenvalues of $\mathbf{A}$ satisfying $\lambda_- < 0 < \lambda_+$—we can only have convergence to the steady state if $w_+(0) = 0$ then $\mathbf{w}(0) = \mathbf{P}^{-1}\big(\mathbf{y}(0) - \bar{\mathbf{y}}\big)$ if and only if $\begin{pmatrix} P_1^- & 0 \\ P_2^- & 1 \end{pmatrix}\begin{pmatrix} w_-(0) \\ \bar{y}_2 - y_2(0) \end{pmatrix} = \begin{pmatrix} y_1(0)-\bar{y}_1 \\ 0 \end{pmatrix}$. Then $\begin{pmatrix} w_-(0) \\ \bar{y}_2 - y_2(0) \end{pmatrix} = \begin{pmatrix} 1 \\ -P_2^- \end{pmatrix}\frac{(y_{1,0}-y_1)}{P_1^-}$, yielding the unique solution (4.44); (3) if the phase diagram is an unstable node—with $\mathbf{A}$ having two positive eigenvalues—then matrix $\mathbf{e}^{\mathbf{A}t}$ is asymptotically unbounded, and the solution can only reach the steady state $\bar{\mathbf{y}}$ if it is stationary, which can only be the case if $\mathbf{w}(0) = \mathbf{0}$, if and only if $\mathbf{y}(0) = \bar{y}$. This is not possible, in general, for a fixed $y_1(0) = y_{1,0} \neq \bar{y}_1$, if $\mathbf{y}(t) \in C^0(\mathbb{R})$, i.e., if it belongs to the space of continuous functions of time. However, in the space of piecewise continuous functions, a unique solution $y_1(0) = y_{1,0}$, $\mathbf{y}(t) = \bar{\mathbf{y}}$ for $t \in (0, \infty)$ exists. □

If the steady state is a saddle point, the solution (4.44), allows for the determination of the initial value of $y_2(0)$

$$y_2(0) = \bar{y}_2 + \frac{P_2^-}{P_1^-}(y_{1,0} - \bar{y}_1),$$

where $\frac{P_2^-}{P_1^-}$ is the slope of the eigenspace $\mathcal{E}^-$ that is co-incident with the stable manifold $\mathcal{W}^s$.

Some other mixed-problems are of interest in macroeconomic and growth theory applications:

Example 4.7 (*Mixed-value problem with conditions on both variables*) Some variants of the mixed-value problems in Proposition 4.7, for the ODE (4.32), appear in the economic literature:

1. problems in which the side conditions are $y_1(0) = y_{1,0}$ and $\lim_{t\to\infty} y_2(t) = \bar{y}_2$. We can prove that the solutions are the same as in that proposition when we assume that $\det(\mathbf{A}) \neq 0$;
2. problems in which the side conditions are $\mathbf{y}(0) = \mathbf{y}_0$ and $\lim_{t\to\infty} \mathbf{y}(t) = \bar{\mathbf{y}}$. In this case, there are no solutions in the space of continuous functions of time. However, we can find a solution in the space of piecewise continuous functions of time, if we allow one or both functions $y_1(t)$ or $y_2(t)$ to "jump" at time $t = 0$. For instance, if the steady state is a saddle point, and the variable y_1 is pre-determined, the solution is

$$y_1(t) = \bar{y}_1 + (y_{1,0} - \bar{y}_1)\, e^{\lambda_- t}, \text{ for each } t \in [0, \infty),$$

$$y_2(t) = \begin{cases} y_{2,0}, & \text{for } t = 0, \\ \bar{y}_2 + \dfrac{P_2^-}{P_1^-}(y_{1,0} - \bar{y}_1)\, e^{\lambda_- t}, & \text{for each } t \in (0, \infty). \end{cases}$$

   Thus, there is an initial "jump" of variable y_2, equal to $y_{2,0^+} - y_{2,0}$, where $y_{2,0^+} = \bar{y}_2 + \frac{P_2^-}{P_1^-}(y_{1,0} - \bar{y}_1)$. This is a common solution to "overshooting models" (see Example 4.10);
3. problems in which the ODE is an unstable node, $\mathbf{B} = 0$, and the side conditions are $y_1(0) = y_{1,0}$ and $\lim_{t\to\infty} e^{-\rho t} \dfrac{y_1(t)}{y_2(t)^\sigma} = 0$, where ρ and σ are two positive constants. This is the case of the endogenous growth model in Example 4.11. We show that this problem has a unique solution although it is not stationary over time.

Example 4.8 (*Non-autonomous scalar equation problems*) In example 4.3, we obtained the general solution (4.38) to a scalar non-autonomous problem by transforming it to a planar ODE. We can specify two alternative problems, associated to two different time-dependent ODE:

1. a **forward** ODE, specified as an initial-value problem by the ODE (4.37), and side conditions $y(0) = y_0$ and $z(0) = 1$ for $t = 0$. The particular solution,

$$y(t) = \left(y_0 - \frac{\beta}{\gamma - \lambda}\right) e^{\lambda t} + \frac{\beta}{\gamma - \lambda}\, e^{\gamma t},$$
$$z(t) = e^{\gamma t},$$

   exists as is unique, as proved in Proposition 4.6 (1). Then it is clear that although the phase diagram is a saddle, we are interested in a non-stationary solution because $\lim_{t\to\infty} \begin{pmatrix} y(t) \\ z(t) \end{pmatrix} = \lim_{t\to\infty} \begin{pmatrix} e^{\gamma t} \frac{\beta}{\gamma-\lambda} \\ 1 \end{pmatrix} = \pm\infty$. The solution converges to a line in the (z, y)-space with slope $y = \frac{\beta}{\gamma-\lambda}\, z$.

2. or a **backward** ODE specified as a mixed-value problem by the ODE (4.37), and the side conditions $y(0) = y_0$ for $t = 0$ and $\lim_{t\to\infty} e^{-\gamma t} z(t) = 0$. In this case, the particular solution

$$y(t) = \left(y_0 - \frac{\beta}{\gamma - \lambda}\right) e^{\lambda t},$$
$$z(t) = 0,$$

asymptotically converges to the steady state $\lim_{t\to\infty}(y(t), z(t))^\top = \mathbf{0}$, within the stable eigenspace $\mathcal{E}^-$.

4.3 Applications to Macroeconomics

Pre-RE macroeconomic models. ISLM models were the benchmark models in macroeconomics between early 1950s and middle 1970s (and still are the core of some undergraduate macroeconomic courses). They were static (non-dynamic) models. Their extension to dynamics ISLM models was accomplished by assuming sluggish adjustment by some variables. A central aspect of those models, which make them the target of the Lucas critique, is that they assumed **static expectations**: agents are not aware or have no beliefs concerning the future state of the economy, and, in particular, on the future consequences of economic policy. As behavioral functions refer directly to aggregate relationships and are not obtained from consistent microeconomic foundations, these models have been called **ad-hoc macro models**. Rational expectation macro models overcame the first shortcoming, and dynamic general equilibrium (DGE) macroeconomic models, by explicitly incorporating the microeconomic foundations of agents' behavior, overcame the second shortcoming.[20]

According to our previous definitions, ad-hoc dynamic macro models had the structure of initial-value problems in which the dynamic system was a stable node or stable focus (see Turnovsky, 1977, Takayama, 1994, Tu, 1994, or Gandolfo, 1997). The next example is a typical pre-RE macro model.

Example 4.9 (*A benchmark dynamic ISLM model*) The economy has three markets: a good's market, a money market, and a risk-free asset market. In the good's market, the aggregate demand has a private and a public component, where the private demand is assumed to be a linear function of income and the real interest rate, $D(y, r) = d_0\, y - d_1\, r$, where $0 < d_0 < 1$ and $d_1 \geq 0$, and g is the (exogenous) government demand for goods, and the aggregate supply is y. The demand for money is also a linear function of income and the interest rate, $L(y, r) = l_0\, y - l_1\, r$, where $l_0 \geq 0$

[20] However, the increased complexity of modeling whole economies with rational and interacting agents were solved by introducing strong assumptions, as perfect information and homogeneity of agents. The recent heterogeneous agent macroeconomic models extend DGE models by assuming imperfect information, agent heterogeneity, among other features. Some of those aspects will be dealt in the next chapters.

and $l_1 > 0$, and the supply of money, m, is exogenous. Invoking the Walras law, we only need to model the equilibrium adjustments in the good's and the money markets. It is assumed that both markets do not clear instantaneously, and there are two temporary disequilibrium adjustment functions, $\dot{y} = \gamma_1(D(y,r) - y)$, and $\dot{r} = \gamma_2(L(y,r) - m)$, where γ_1 and γ_2 are the speeds of adjustment in the good's and in the monetary markets, respectively. The short-run dynamics of this economy is represented by a planar linear ODE, in which both variables are pre-determined:

$$\begin{aligned}
\dot{y} &= \gamma_1 \left(D(y,r) + g - y\right) = \gamma_1 \left(-(1-d_0)\, y - d_1\, r + g\right), \text{ for } t \in \mathbb{R}_+,\\
\dot{r} &= \gamma_2 \left(L(y,r) - m\right) = \gamma_2 \left(l_0\, y - l_1\, r - m\right), \text{ for } t \in \mathbb{R}_+,\\
y(0) &= y_0 \text{ fixed},\\
r(0) &= r_0 \text{ fixed}.
\end{aligned}$$

In matrix notation, we denote $\mathbf{y} = (y, r)$, and the coefficient matrices are

$$\mathbf{A} = \begin{pmatrix} -\gamma_1\,(1-d_0) & -\gamma_1\, d_1 \\ \gamma_2\, l_0 & -\gamma_2\, l_1 \end{pmatrix}, \text{ and } \mathbf{B} = \begin{pmatrix} \gamma_1\, g \\ -\gamma_2\, m \end{pmatrix}.$$

As $D(\mathbf{A}) = \gamma_1\, \gamma_2\, \delta > 0$, where $\delta \equiv l_1\,(1 - d0) + l_0\, d_1 > 0$, then there is a unique steady state $(\bar{y}, \bar{r}) = \left(\frac{l_1\, g + d_1\, m}{\delta}, \frac{l_0\, g - d_1\, m}{\delta}\right)$, and, because $T(\mathbf{A}) = -(\gamma_1\,(1-d_0) + \gamma_2\, l_1) < 0$, it is a stable node or a stable focus. As the discriminant is $\Delta(\mathbf{A}) = \gamma_1\,\gamma_2\,(\ell^* - \ell)$, were $\ell \equiv l_0\, d_1$, and $\ell^* \equiv \frac{1}{\gamma_1\,\gamma_2}\left(\frac{\gamma_1\,(1-d_0)}{2} - \frac{\gamma_2\, l_1}{2}\right)^2 > 0$, then the existence of oscillatory trajectories, associated to the existence of complex eigenvalues, depends on the size of the "crossed effects" of the interest rate on goods' demand (d_1), and on the transactions' demand for money (l_0): if $0 \le \ell < \ell^*$ then the steady state is a stable node, and if $\ell > \ell^*$ it is a stable focus. The explicit solution in the first case is

$$\begin{pmatrix} y(t) \\ r(t) \end{pmatrix} = \begin{pmatrix} \bar{y} \\ \bar{r} \end{pmatrix} + \begin{pmatrix} \gamma_2\, l_1 + \lambda_- \\ \gamma_2\, l_0 \end{pmatrix} e^{\lambda_- t}\, w_-(0) + \begin{pmatrix} \gamma_2\, l_1 + \lambda_+ \\ \gamma_2\, l_0 \end{pmatrix} e^{\lambda_- t}\, w_+(0)$$

where

$$\begin{pmatrix} w_-(0) \\ w_+(0) \end{pmatrix} = \frac{1}{\gamma_2 l_0\,(\lambda_- - \lambda_+)} \begin{pmatrix} \gamma_2 l_0\,(y_0 - \bar{y}) - (\gamma_2\, l_1 + \lambda_+)\,(r_0 - \bar{r}) \\ -\gamma_2 l_0\,(y_0 - \bar{y}) + (\gamma_2\, l_1 + \lambda_-)\,(r_0 - \bar{r}) \end{pmatrix}.$$

Figure 4.7a illustrates the phase diagram for the adjustment starting from two alternative points at time $t = 0$: point A (B) in which there is an excess demand (supply) for goods and an excess supply (demand) for money. We observe that in both cases there is a leading direction of the solution provided by the eigenspace associated to the (negative) eigenvalue that is smaller in absolute value, $\mathcal{E}^+ = \{(y, r) : \gamma_2\, l_0\,(y - \bar{y}) = (\gamma_2\, l_1 + \lambda_+)\,(r - \bar{r})\}$. As this eigenspace is positively sloped, then in both cases the

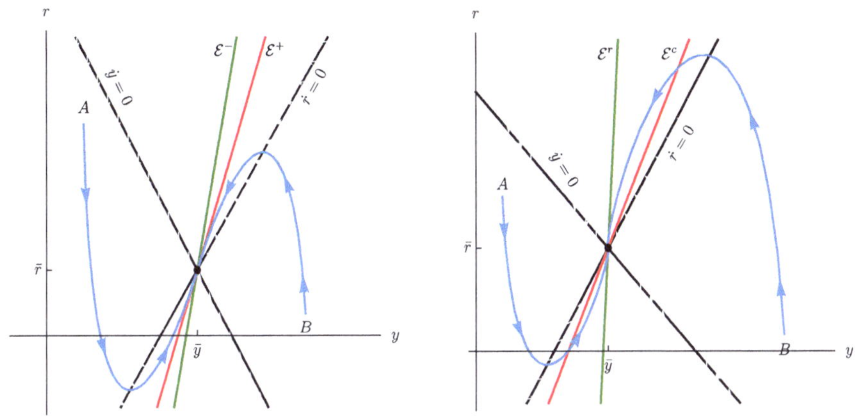

(a) Phase diagram: for $0 \leq \ell < \ell^*$ (b) Phase diagram: for $\ell > \ell^*$

Fig. 4.7 The dynamic ISLM model for the stable node (a) and the stable focus case (b)

adjustment of r is hump-shaped, because the solution path crosses the isocline $\dot{r} = 0$ at some finite t. In both cases, this occurs when there is a change from excess supply to excess demand (in case A) for money, or the opposite (in case B).

Figure 4.7b illustrates the phase diagram when the steady state is a stable focus. In this case, the solution is

$$\begin{pmatrix} y(t) \\ r(t) \end{pmatrix} = \begin{pmatrix} \bar{y} \\ \bar{r} \end{pmatrix} + \mathbf{P}\, e^{\alpha t} \begin{pmatrix} \cos\beta t & \sin\beta t \\ -\sin\beta t & \cos\beta t \end{pmatrix} \mathbf{P}^{-1} \begin{pmatrix} y_0 - \bar{y} \\ r_0 - \bar{r} \end{pmatrix}$$

where $\alpha = \frac{T(\mathbf{A})}{2}$ $\beta = \sqrt{|\Delta(\mathbf{A})|}$ and the eigenvector matrix is

$$\mathbf{P} = \begin{pmatrix} -\gamma_1\, d_1\, (\gamma_1\,(1-d_0) + \alpha + \beta) & -\gamma_1\, d_1\, (\gamma_1\,(1-d_0) + \alpha - \beta) \\ (\gamma_1\,(1-d_0) + \alpha)^2 + \beta^2 & (\gamma_1\,(1-d_0) + \alpha)^2 + \beta^2. \end{pmatrix}$$

The slope of the adjustment in the neighborhood of the steady state, that is given by the eigenspace $\mathcal{E}^r$, is the most significant difference between the two phase diagrams. It indicates that the speed of the adjustment of y close to the steady state is bigger in the second case.

Exercise 4.10 Prove the results in Example 4.9. Consider the simplifying assumption $d_1 = l_0 = 0$. In this case, prove that the steady state is a stable node. Furthermore, show that, in this case, the steady state levels satisfy $\overline{y} = \overline{y}(g)$ and $\overline{r} = \overline{r}(m)$. This means that the fiscal policy is efficient for controlling y and the monetary policy is only efficient for controlling r. This was the main contentious issue in the debate between monetarists and Keynesians, that lasted for three decades. Find the short run multipliers for a permanent increase in g.

Post RE ad-hoc macroeconomic models. In the early 70s, it became clear, particularly as a result of the events in the currency markets after the colapse of the Bretton Woods system, that the introduction of agent anticipations in macromodels were essential. The simplest way to introduce beliefs is by assuming that there is only one right belief at the aggregate level, and that it is consistent with the prediction of the relevant existing macroeconomic theory. Thus the designation of rational expectations (RE) equilibrium. The Dornbusch (1976) model became a benchmark for RE ad-hoc macroeconomic models.[21]

In that strand of models, which can also be labeled as ad-hoc, the dynamics is generated by the existence of slow adjustments for pre-determined variables and of perfect foresight belief dynamics for non-predetermined variables. Mathematically, they are initial-terminal value ODE problems in which the dynamic system is usually a saddle.

Example 4.10 (*The exchange rate overshooting model*) It is a macroeconomic model for a small open economy having a flexible exchange rate, and in which there are free international movements of financial assets and goods. It addresses the joint dynamics of the nominal exchange rate and domestic prices, driven by perfect foresight in the asset market and a sluggish adjustment in the domestic goods market. A simplified version of the model is the following. The (log) nominal exchange rate (national currency per unit of foreign currency), e, is determined by the Fisher open equation, in which its expected change is equal to the difference between the domestic and the foreign nominal interest rates, i and i^*, respectively: $\dot{e} = E_t\left[\frac{de(t)}{dt}\right] = i - i^*$. The domestic nominal interest rate clears instantaneously the domestic money market, $m - p = -i$, where the log real supply of money is $m - p = \ln M/P$—where M is the money stock and P is the price level—and the demand for real cash balances is Keynesian, $L(i) = i$ to simplify. In the product market, the real aggregate demand is a function of the log income, of the nominal interest rate, and the log real exchange rate (assuming that the log international price satisfies $p^* = 0$), $d(y, e, i, p) = \mu\,(e - p) - \sigma\, i + \delta\, y$, where $\mu > 0$, $\sigma > 0$ and $0 < \delta < 1$. The log real aggregate supply, y, is exogenous. The adjustment of the goods market is sluggish, but, differently from the previous model, this economy has flexible prices, adjusting according to the ODE $\dot{p} = \gamma\big(d(y, e, i, p) - y\big)$, implying that if there is excess demand (supply) then prices increase (decrease). From those assumptions, the following planar linear ODE for the joint adjustment of prices and the exchange rate is obtained,

$$\dot{p} = \gamma\big(- (\mu + \sigma)\, p + \mu\, e + \sigma m - (1 - \delta)\, y\big), \tag{4.45a}$$

$$\dot{e} = p - m - i^*. \tag{4.45b}$$

As p is a pre-determined variable and e is non-predetermined variable, the model is complemented by the initial condition, $p(0) = p_0$ fixed, and by a so-called "ruling-out speculative bubble condition" requiring the convergence of the nominal exchange rate to its steady state level, $\lim_{t\to\infty} e(t) = \bar{e}$. Therefore, we have a mixed ODE

[21] See Turnovsky (1995) for more RE models.

problem, where $\mathbf{y} = (p, e)$, the coefficient matrix and vector are

$$\mathbf{A} = \begin{pmatrix} -\gamma\,(\delta+\sigma) & \gamma\,\delta \\ 1 & 0 \end{pmatrix}, \text{ and } \mathbf{B} = \begin{pmatrix} \gamma\,(\sigma\,m - (1-\gamma)\,y) \\ -m - i^* \end{pmatrix},$$

and (m, y, i^*) are exogenous variables. As $\det(\mathbf{A}) = -\gamma\,\delta < 0$, then there is a unique steady state $(\bar{p}, \bar{e}) = \left(m + i^*, \bar{p} + \frac{\sigma\,i^* + (1-\gamma)\,y}{\delta}\right)$. This steady state is a saddle point, because the eigenvalues of $\mathbf{A}$ satisfy $\lambda_- < 0 < \lambda_+$, where $\lambda_\mp = -\frac{\pi\,(\delta+\sigma)}{2} \mp \sqrt{\left(\frac{\pi\,(\delta+\sigma)}{2}\right)^2 + \pi\,\delta}$. The solution to the model is

$$p(t) = \bar{p} + p_0\,e^{-\lambda_- t},\ t \in \mathbb{R}_+, \tag{4.46a}$$

$$e(t) = \bar{e} + \frac{p_0}{\lambda_-}\,e^{-\lambda_- t}, t \in \mathbb{R}_+. \tag{4.46b}$$

Figure 4.8a shows the phase diagram. If the initial price level is given at point $p_0 > \bar{p}$, the nominal exchange rate can only satisfy the terminal condition, if its initial level $e(0^+)$ is placed at the stable eigenspace $\mathcal{E}^-$. Thus, the expectation at time $t = 0$ for the nominal exchange rate is a function of the long-run exchange rate $e(0) = \bar{e} + \frac{p_0}{\lambda_-}$. As that eigenspace is the only locus of points converging to the steady state, then $\lambda_-(e(t) - \bar{e}) = p(t) - \bar{p}$ for every $t \in (0, \infty)$, meaning that the model predicts that the exchange rate and inflation should be negatively correlation over time.

We can also perform a comparative dynamics analysis for the effect over time of a monetary shock. Let the economy be at a steady state, as in point $(\bar{p}_0, \bar{e}_0)$

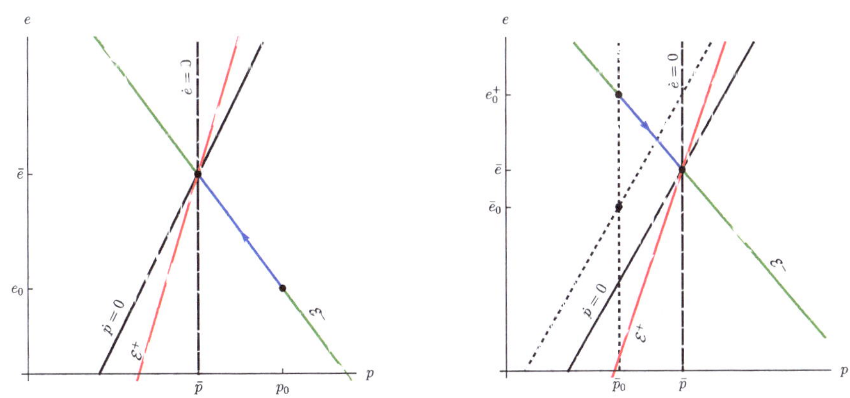

(a) Phase diagram: trajectory for $p_0 > \bar{p}$.

(b) Comparative dynamics for a shock in m.

Fig. 4.8 The Dornbusch (1976) overshooting in panel (**b**): starting from a steady state $(\bar{p}_0, \bar{e}_0)$, at the moment of the shock the economy "jumps" to $(\bar{p}_0, e_0^+)$, and adjusts over time to the new steady state $(\bar{p}, \bar{e})$

in Fig. 4.8b, and introduce a non-anticipated and permanent increase in the stock of money, Δm. That initial point, that was a steady state before the shock, is not a steady state after the shock, which implies that the economy will move away from it. As the price is a pre-determined variable, which implies that the exchange rate is non-predetermined, then, according to this model, the economy will move instantaneously to point $(\bar{p}_0, e(0^+))$, where $e(0^+) = \bar{e}_0 + \left(\frac{\partial \bar{e}}{\partial m} - \frac{1}{\lambda_-}\frac{\partial \bar{p}}{\partial m}\right)\Delta m$, where $\frac{\partial \bar{p}}{\partial m} = \frac{\partial \bar{e}}{\partial m} = 1$ are the long-run multipliers. Over time, the change in the two variables are $\Delta p(t) = \frac{\partial p(t)}{\partial m}\Delta m$ and $\Delta e(t) = \frac{\partial e(t)}{\partial m}\Delta m$, where the dynamic multipliers are

$$\frac{\partial p(t)}{\partial m} = \frac{\partial \bar{p}}{\partial m}\left(1 - e^{\lambda_- t}\right), \tag{4.47a}$$

$$\frac{\partial e(t)}{\partial m} = \frac{\partial \bar{e}}{\partial m} - \frac{1}{\lambda_-}\frac{\partial \bar{p}}{\partial m}e^{\lambda_- t}. \tag{4.47b}$$

Asymptotically there is convergence to point $(\bar{p}, \bar{e}) = (\bar{p}_0 + \Delta p(\infty), \bar{e}_0 + \Delta e(\infty)) = (\bar{p}_0 + \Delta m, \bar{e}_0 + \Delta m)$. The literature called "overshooting" to the adjustment illustrated in Fig. 4.8b : as prices are sticky at time $t = 0$, a monetary shock forces the nominal exchange rate to over-devalue by "jumping" to a level higher than in the long run. Formally, this evolution is necessary because the exchange rate and prices should be negatively related—there is a inflationary process together with a nominal appreciation—for the existence of convergence to the new steady state, as required by the absence of speculative bubbles assumption.

Exercise 4.11 Prove that (4.46) is the solution to the version of the Dornbusch (1976) in Example 4.10. Draw the phase diagram in Fig. 4.8a. Obtain the dynamic multipliers (4.47a)–(4.47b) and draw the perturbed phase diagram in Fig. 4.8b.

DGE modeling. Ramsey-Cass-Koopmans (RCK) models started a strand of the so-called non-ad-hoc modeling in macroeconomics. The seminal paper was Ramsey (1928), and Cass (1965) and Koopmans (1965) rediscovered and systematized it in the 1960s using advances in the optimal control theory.[22] The inconsistency of the microeconomic foundations of the ad-hoc macro models has been solved by assuming that the economy is populated by homogeneous agents, with perfect information, and the institutional setting enables it to operate efficiently.[23] Although the difference between normative and positive economics is sometimes not clear in particular applications, the RCK model is at the origin of what is called today representative-agent DGE (dynamic general equilibrium) modeling.

Invoking our previous definitions, these models feature a mixed initial-terminal value ODE problem that is obtained from the first-order conditions of optimal control problems. When the equilibrium is Pareto optimal, the optimality conditions (or the

[22] See Chap. 10. For early textbook versions of continuous time Ramsey model, see Blanchard and Fischer (1989), Brock and Malliaris (1989), or Turnovsky (1995).

[23] See Blanchard (2000) and Sargent (2008) on the evolution of macroeconomic modeling.

equilibrium in a DGE interpretation) is represented by a non-linear planar ODE, together with an initial and a terminal condition[24]

$$\dot{k} = F(k) - c, \tag{4.48a}$$

$$\dot{c} = \frac{c}{\sigma}\,(r(k) - \rho), \tag{4.48b}$$

$$k(0) = k_0 \text{ given}, \tag{4.48c}$$

$$\lim_{t\to\infty} k(t)\,c(t)^{-\sigma}\,e^{-\rho t} = 0, \tag{4.48d}$$

where k is the stock of capital (a pre-determined variable) and c is consumption (a non-predetermined variable). The first ODE (4.48a) represents the budget constraint, equating capital accumulation to savings as in the (3.17) capital accumulation equation—but without the Keynesian savings function. The ODE (4.48b) is an Euler equation, or an intertemporal arbitrage condition between present and future consumption. It equates the change in marginal utility (that is, the value) of present consumption and the change in the value of time to the net increase in production capacity allowing for future increases in consumption, $\frac{du'(c(t))}{u'(c(t))} + \rho\,dt = F'(k)\,dt$. There are two side conditions, (4.48c) and (4.48d), stating that: the initial level of capital is fixed, and the capital accumulation process should be bounded in present value terms. The RCK model is a non-linear regular ODE. As it is proved in Example 5.4, the phase diagram in RCK models is a saddle, which implies that the solution to the mixed initial-value problem is unique, as was proved in Proposition 4.7. There is a direct relationship between uniqueness of the saddle path, for a given initial level of the stock of capital, and the existence and uniqueness of an optimum (or DGE) path.

Example 4.11 (*The AK model*) A linear version of the RCK model is the benchmark endogenous growth model, the so-called AK model. Endogenous growth models, with one stock variable, are usually initial-terminal value problems in which there are no positively valued steady states. However, detrending the model yields an unstable saddle-node. The simplest, well known AK model[25] is a planar linear ODE in the stock of capital and consumption, $(K, C) \in \mathbb{R}^2_+$, together with an initial condition on capital (a pre-determined variable) and a transversality condition,[26]

$$\dot{K} = A\,K - C, \tag{4.49a}$$

$$\dot{C} = C\,(A - \rho), \tag{4.49b}$$

$$K(0) = k_0, \tag{4.49c}$$

$$\lim_{t\to\infty} \frac{K(t)}{C(t)}\,e^{-\rho t} = 0, \tag{4.49d}$$

[24] Its solution is studied in next Chap. 5.

[25] See Rebelo (1991), or Acemoglu (2009, Chap. 11).

[26] This problem corresponds to the first-order conditions for an optimal control problem as will be shown in Chap. 10.

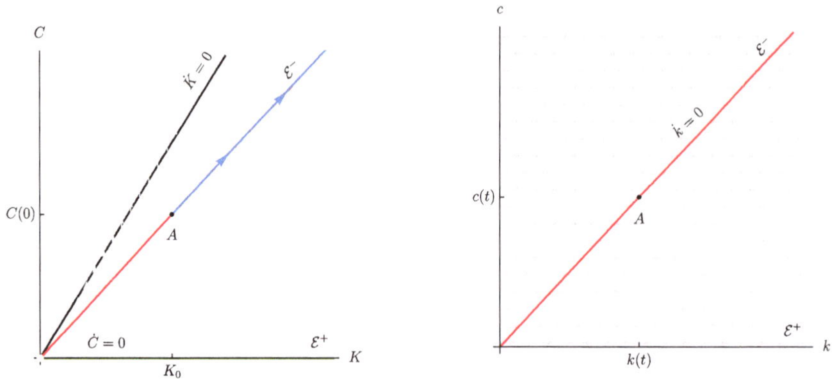

(a) Phase diagram: AK model in levels.

(b) Phase diagram: AK model in detrended variables.

Fig. 4.9 The Rebelo (1991) AK model in level and detrended variables

where A and ρ denote productivity and the rate of time preference. It is usually assumed that $A > \rho > 0$. Defining the growth rate by $\gamma = A - \rho > 0$, the general solution to Eqs. (4.49a) and (4.49b) is $K(t) = w_-(0)\, e^{\gamma t} + w_+(0)\, e^{\gamma t}$ and $C(t) = w_-(0)\, \rho\, e^{\gamma t}$. Substituting in the side condition in Eqs. (4.49c)–(4.49d) and solving for $\mathbf{w}(0)$ yields $w_+(0) = 0$, and $w_-(0) = k_0$, yields the eigenspaces $\mathcal{E}^- = \{(C, K) : C = \rho K\}$ and $\mathcal{E}^+ = \{(C, K) : C = 0\}$, and the solution to the AK model

$$K(t) = k_0\, e^{\gamma t}, \text{ and } C(t) = \rho\, k_0\, e^{\gamma t} \text{ for each } t \in [0, \infty). \tag{4.50}$$

The solution is illustrated in the phase diagram in Fig. 4.9a . We observe that starting from point A, the solution evolves along the eigenspace $\mathcal{E}^-$ passing through the steady state $\bar{K}, \bar{C} = (0, 0)$, which means that the ratio $\frac{C(t)}{K(t)} = \rho$ is constant over time. This is the reason why, in growth theory, it is labeled a balanced growth path (BGP). An alternative way of solving the model is frequently used when more than one type of capital are present (or there is imperfect substitutability between different types of capital)[27]: assuming that the level variables, K and C, can be seen as having multiplicative trends and no transition components, $K(t) = k(t)\, e^{\gamma_k t}$, $C(t) = c(t)\, e^{\gamma_c t}$, and substituting in Eqs. (4.49a) and (4.49b), such that the resulting ODE is autonomous, we have to set $\gamma_k = \gamma_c = \gamma$, we obtain the initial-terminal value problem

[27] In this sense, the K in the AK model should be conceived as total capital, including physical, human, natural, and social capital, under the assumption that they are perfect Edgeworth-Pareto substitutes.

$$\dot{k} = \rho k - c, \tag{4.51a}$$

$$\dot{c} = 0, \tag{4.51b}$$

$$k(0) = k_0, \tag{4.51c}$$

$$\lim_{t\to\infty} \frac{k(t)}{c(t)} e^{-\rho t} = 0. \tag{4.51d}$$

Solving the ODE problem we obtain a stationary solution $(k(t), c(t)) = \left(k_0, \rho k_0\right)$ for each $t \in (0, \infty)$, represented in phase diagram Fig. 4.9b. This solution indicates that there is no transition dynamics and shows again that the solution follows a balance growth path as in Eq. (4.50). That is, the economy grows at a constant positive growth rate and becomes unbounded only in infinite time. This model is said to be a model of endogenous growth because the growth rate of the economy is derived from the solution to an ODE problem, and is not given in advance.

Neo-Keynesian (NK) DGE models keep most of the structure of RCK models with the exception of the representative-agent assumption. An early example is the Romer (1986b) model,[28] that is an extension of the RCK model in which there are externalities in production. In Example 5.6, we study an extension of this model in which there are both externalities in production (positive or negative) and consumption (positive or negative). If K and C represent the aggregate capital stock and consumption, the HK-DGE can be represented by the mixed initial-terminal ODE problem for the household capital stock and consumption, k, and c, respectively

$$\dot{k} = F(k, K) - c,$$

$$\dot{c} = \frac{c}{\sigma}\left(r(k, K) - \rho - \beta\frac{\dot{C}}{C}\right),$$

$$k(0) = k_0 \text{ given},$$

$$\lim_{t\to\infty} k(t)\, c(t)^{-\sigma} e^{-\rho t} = 0.$$

At the equilibrium, a micro-macro consistency condition, $K = k$ and $C = c$, should be satisfied. The phase diagram for this model can be a saddle, as in the RCK model, but can also be a stable node or focus. In the last case, as was proved in Proposition 4.7, the solution to the mixed initial-value problem is indeterminate (not unique). If a HK-DGE path is not unique then the equilibrium is said to be indeterminate, meaning that self-fulfilling prophecies can be sustained at the macroeconomic equilibrium, and these are consequence of the existence of imperfections in the markets (externalities, incompleteness of contracts, policy rules, etc). Finding conditions allowing for indeterminacy became one of the main contributions of HK models.

[28] Other earlier continuous-time NK models with production externalities are Benhabib and Farmer (1996) and Weder (2001)

4.4 References

A classic reference on linear ODE is Hirsch and Smale (1974). For textbooks on ordinary differential equations that cover linear planar equations, see Hale and Koçak (1991, Chap. 8) and Perko (1996, Chap. 1). A reference on matrix algebra and its application to ODE is Barnett (1990).

For references on economic applications of ODE, and in particular linear ODE, see Tu (1994, Chap. 5), Simon and Blume (1994, Chap. 24,25), Gandolfo (1997, Chap. 18), or de la Fuente (2000, Chap. 10).

4.5 Matrix Algebra and Exponential Matrices

4.5.1 A Review of Matrix Algebra

Consider the square matrix $\mathbf{A} \in \mathbb{R}^{2\times 2}$[29]

$$\mathbf{A} = \begin{pmatrix} a_{11} & a_{12} \\ a_{21} & a_{22} \end{pmatrix}.$$

The **trace**, the **determinant**, and the **discriminant** of matrix $\mathbf{A}$ are $\mathrm{tr}(\mathbf{A}) \equiv a_{11} + a_{22}$, $\det(\mathbf{A}) \equiv a_{11}a_{22} - a_{12}a_{21}$, and $\Delta(\mathbf{A}) \equiv \left(\frac{\mathrm{tr}(\mathbf{A})}{2}\right)^2 - \det(\mathbf{A})$, respectively.

The kernel (or null space) of matrix $\mathbf{A}$ is a vector $\mathbf{v}$ defined as $\mathrm{kern}(\mathbf{A}) = \{\mathbf{v} : \mathbf{A}\mathbf{v} = \mathbf{0}\}$. The dimension of the kernel provides a measure of the linear independence between the rows of $\mathbf{A}$. Another way to assess it is by solving the characteristic equation.

The **characteristic polynomial of matrix A** is

$$\det(\mathbf{A} - \lambda \mathbf{I}_2) = \lambda^2 - \mathrm{tr}(\mathbf{A})\lambda + \det(\mathbf{A}) \tag{4.52}$$

where $\lambda \in \mathbb{C}$ is an eigenvalue. The **spectrum** of $\mathbf{A}$ is the **set of eigenvalues**, that are the solutions to the characteristic equation, $\sigma(\mathbf{A}) \equiv \{\lambda \in \mathbb{C} : \det(\mathbf{A} - \lambda\mathbf{I}_2) = 0\}$. The elements of the spectrum are called **eigenvalues** of $\mathbf{A}$. For square matrix $\mathbf{A}$ the eigenvalues are

$$\lambda_+ = \frac{\mathrm{tr}(\mathbf{A})}{2} + \Delta(\mathbf{A})^{\frac{1}{2}},\ \lambda_- = \frac{\mathrm{tr}(\mathbf{A})}{2} - \Delta(\mathbf{A})^{\frac{1}{2}}. \tag{4.53}$$

A useful result on the relationship between the eigenvalues and the trace and the determinant of $\mathbf{A}$ is:

Lemma 4.3 *Let λ_- and λ_+ be the eigenvalues of a 2×2 matrix $\mathbf{A}$. Then they satisfy:*

---

[29] A standard reference on matrix algebra Strang (1988).

$$\lambda_- + \lambda_+ = \text{tr}(\mathbf{A}), \textit{and } \lambda_-\lambda_+ = \det(\mathbf{A}). \tag{4.54}$$

Proof $c(\lambda) = \det(\mathbf{A} - \lambda\mathbf{I}_2)$ is a square polynomial. Therefore $c(\lambda) = (\lambda - \lambda_-)(\lambda - \lambda_+) = \lambda^2 - \text{tr}(\mathbf{A})\,\lambda + \det(\mathbf{A})$ if and only if $\lambda_- + \lambda_+ = \text{tr}(\mathbf{A})$ and $\lambda_-\lambda_+ = \det(\mathbf{A})$. □

Three cases are possible:

1. if $\Delta(\mathbf{A}) > 0$ then λ_- and λ_+ are real and distinct and $\lambda_- < \lambda_+$,
2. if $\Delta(\mathbf{A}) = 0$ then $\lambda_- = \lambda_+ = \lambda = \text{tr}(\mathbf{A})/2$ are real and multiple,
3. if $\Delta(\mathbf{A}) < 0$ then λ_- and λ_+ are complex conjugate $\lambda_- = \alpha - \beta i$ and $\lambda_+ = \alpha + \beta i$ where $\alpha = \frac{\text{tr}(\mathbf{A})}{2}$, $\beta = \sqrt{|\Delta(\mathbf{A})|}$ and $i = \sqrt{-1}$. We can write the eigenvalues in polar coordinates by defining $r = \sqrt{\alpha^2 + \beta^2}$ and $\tan\theta = \beta/\alpha$, or equivalently $\alpha = r\cos\theta$, and $\beta = r\sin\theta$, then

$$\lambda_- = r\,(\cos\theta - \sin\theta\, i), \;\; \lambda_+ = r\,(\cos\theta + \sin\theta\, i). \tag{4.55}$$

Two matrices $\mathbf{A}$ and $\mathbf{A}'$ are said to be **similar** if their spectrums (or their eigenvalues) are equal. This allows for classifying matrices according to their eigenvalues. It is useful to recall that the Jordan canonical forms for 2×2 matrices are

$$\mathbf{\Lambda}_+ = \begin{pmatrix} \lambda_- & 0 \\ 0 & \lambda_+ \end{pmatrix}, \;\; \mathbf{\Lambda}_0 = \begin{pmatrix} \lambda & 1 \\ 0 & \lambda \end{pmatrix}, \;\; \text{and } \mathbf{\Lambda}_- = \begin{pmatrix} \alpha & \beta \\ -\beta & \alpha \end{pmatrix}. \tag{4.56}$$

The relationship of any square matrix and its similar Jordan canonical form is provided by the following:

Lemma 4.4 (Jordan canonical from of matrix $\mathbf{A}$) *Consider any 2×2 matrix with real entries and its discriminant $\Delta(\mathbf{A})$. Then*

1. *If $\Delta(\mathbf{A}) > 0$ then the Jordan canonical form associated to $\mathbf{A}$ is $\mathbf{\Lambda}_+$.*
2. *If $\Delta(\mathbf{A}) = 0$ then the Jordan canonical form associated to $\mathbf{A}$ is $\mathbf{\Lambda}_0$.*
3. *If $\Delta(\mathbf{A}) < 0$ then the Jordan canonical form associated to $\mathbf{A}$ is $\mathbf{\Lambda}_-$.*

The Jordan canonical form $\mathbf{\Lambda}_-$ can also be represented by a diagonal matrix with complex entries $\mathbf{\Lambda}_- = \begin{pmatrix} \alpha-\beta i & 0 \\ 0 & \alpha+\beta i \end{pmatrix}$. In the space of complex square matrices, we can say that matrix $\mathbf{A}$ is diagonalizable if $\Delta(\mathbf{A}) \neq 0$ then, and it is not diagonalizable if $\Delta(\mathbf{A}) = 0$. Figure 4.1 presents the different cases in a $(\text{tr}(\mathbf{A}), \det(\mathbf{A}))$ diagram.

A useful **taxonomy of the eigenvalues of A** can be established by computing the trace and the determinant of $\mathbf{A}$ (see Fig. 4.1) and observing the following: First, both eigenvalues are **real** numbers if $\Delta(\mathbf{A}) \geq 0$, and they are **complex conjugate** numbers if $\Delta(\mathbf{A}) < 0$. In particular, if $\Delta(\mathbf{A}) > 0$ the eigenvalues are real and **distinct**, and satisfy $\lambda_- < \lambda_+$, and if $\Delta(\mathbf{A}) = 0$ they are equal - or multiple - and satisfy $\lambda = \lambda_- = \lambda_+ = \frac{\text{tr}(\mathbf{A})}{2}$. If $\Delta(\mathbf{A}) < 0$ they are complex conjugate and satisfy $\lambda_\pm = \alpha \pm \beta\, i$, for $i \equiv \sqrt{-1}$, where $\alpha = \frac{\text{tr}(\mathbf{A})}{2}$ and $\beta = \sqrt{|\Delta(\mathbf{A})|}$. Second, the **signs of the real part** of both eigenvalues is the **same** if $\det(\mathbf{A}) > 0$, and they are **different** if $\det(\mathbf{A}) < 0$.

In the first case they are both positive if $\det(\mathbf{A}) > 0$ and $\mathrm{tr}(\mathbf{A}) > 0$, and they are both negative if $\det(\mathbf{A}) > 0$ and $\mathrm{tr}(\mathbf{A}) < 0$.Third, the eigenvalues are **generic**, in the sense that they will not change their type or sign for small changes in the elements of the coefficient matrix $\mathbf{A}$, if $\Delta(\mathbf{A}) \neq 0$, or $\det(\mathbf{A}) \neq 0$, or $\mathrm{tr}(\mathbf{A}) \neq 0$ and $\det(\mathbf{A}) \geq 0$, and they are **not generic** otherwise, that is if $\Delta(\mathbf{A}) = 0$, or $\det(\mathbf{A}) = 0$, or $\mathrm{tr}(\mathbf{A}) = 0$ and $\det(\mathbf{A}) \geq 0$.

We can combine the previous facts into eleven different cases concerning matrix $\mathbf{A}$ in a $(\mathrm{tr}(\mathbf{A}), \det(\mathbf{A}))$ diagram (see Figure 4.1). There are five generic cases (corresponding to two-dimensional subsets), four non-generic cases of co-dimension-one (corresponding to lines), and two non-generic cases of co-dimension-two (the origin):

1. the five generic cases are: (1) if $\mathrm{tr}(\mathbf{A}) > 0$, $\det(\mathbf{A}) > 0$ and $\Delta(\mathbf{A}) > 0$ the two eigenvalues are real, different and positive, $\lambda_+ > \lambda_- > 0$; (2) if $\mathrm{tr}(\mathbf{A}) > 0$, $\det(\mathbf{A}) > 0$ and $\Delta(\mathbf{A}) < 0$ the two eigenvalues are complex conjugate with positive real parts, $\lambda_\pm = \alpha \pm \beta i$ with $\alpha > 0$; (3) if $\mathrm{tr}(\mathbf{A}) < 0$, $\det(\mathbf{A}) > 0$ and $\Delta(\mathbf{A}) > 0$ the two eigenvalues are real, different, and negative, $0 > \lambda_+ > \lambda_-$; (4) if $\mathrm{tr}(\mathbf{A}) < 0$, $\det(\mathbf{A}) > 0$ and $\Delta(\mathbf{A}) < 0$ the two eigenvalues are complex conjugate with negative real parts, $\lambda_\pm = \alpha \pm \beta i$ with $\alpha < 0$; and (5) if $\det(\mathbf{A}) < 0$ the two eigenvalues are real and with opposite signs, $\lambda_+ > 0 > \lambda_-$;
2. the six non-generic cases are: (1) if $\mathrm{tr}(\mathbf{A}) > 0$ and $\Delta(\mathbf{A}) = 0$ the two eigenvalues are real, equal and positive, $\lambda_+ = \lambda_- > 0$; (2) if $\mathrm{tr}(\mathbf{A}) < 0$ and $\Delta(\mathbf{A}) = 0$ the two eigenvalues are real, equal and negative, $\lambda_+ = \lambda_- < 0$; (3) if $\mathrm{tr}(\mathbf{A}) = 0$ and $\det(\mathbf{A}) > 0$ then the two eigenvalues are complex conjugate with zero real part, $\lambda_\pm = \pm\beta i$; (4) if $\mathrm{tr}(\mathbf{A}) > 0$ and $\det(\mathbf{A}) = 0$ the two eigenvalues are real one is positive and the other is equal to zero, $\lambda_+ > 0 = \lambda_-$; (5) if $\mathrm{tr}(\mathbf{A}) < 0$ and $\det(\mathbf{A}) = 0$ the two eigenvalues are real one is negative and the other is equal to zero, $\lambda_+ = 0 < \lambda_-$; and (6) if $\mathrm{tr}(\mathbf{A}) = \det(\mathbf{A}) = 0$ both eigenvalues are real and equal to zero, $\lambda_+ = \lambda_- = 0$.

We emphasize the fact that $\det(\mathbf{A}) < 0$ is a sufficient condition for the existence of two real eigenvalues, and it is a necessary and sufficient condition for $\lambda_- < 0 < \lambda_+$. This is a very important fact which is useful for economic applications.

The following equivalence relationship between matrix $\mathbf{A}$ and its Jordan canonical form $\mathbf{\Lambda}$ holds:

Lemma 4.5 *Let $\mathbf{A}$ be a 2×2 matrix with real entries. Then, there exists a non-singular matrix $\mathbf{P}$ such that*

$$\mathbf{A} = \mathbf{P}\mathbf{\Lambda}\mathbf{P}^{-1}$$

where $\mathbf{\Lambda}$ is the Jordan canonical form of $\mathbf{A}$, and matrix $\mathbf{P}$ is a 2×2 ***eigenvector*** *matrix associated to $\mathbf{A}$.*

There are two types of eigenvectors:

1. if $\Delta(\mathbf{A}) > 0$, the eigenvector matrix is $\mathbf{P} = (\mathbf{P}^-, \mathbf{P}^+)$. It concatenates two simple eigenvectors $\mathbf{P}^-$ and $\mathbf{P}^+$, that are associated to the eigenvalues λ_- and λ_+. They are obtained by solving the homogeneous system

$$(\mathbf{A} - \lambda_j \mathbf{I}_2)\mathbf{P}^j = 0, \text{ for } j \in \{-, +\}$$

where $\mathbf{I}_2$ is the identity matrix of order 2. Observe that $\mathbf{P}^j = \text{kern}(\mathbf{A} - \lambda_j \mathbf{I}_2)$, i.e, it is the null space of matrix $(\mathbf{A} - \lambda_j \mathbf{I}_2)$;

2. if $\Delta(\mathbf{A}) < 0$, the case in which the eigenvalues are complex conjugate, we can use one of the following two approaches for finding the eigenvector matrix:

   a. either we write the Jordan matrix as a complex-valued matrix

$$\Lambda_- = \begin{pmatrix} \alpha - \beta i & 0 \\ 0 & \alpha + \beta i \end{pmatrix}$$

   and compute $\mathbf{P}^j$, for $j = 1, 2$, as a complex-valued vector, by solving $(\mathbf{A} - \lambda_j \mathbf{I}_2)\mathbf{P}^j = 0$ for each $j = 1, 2$;

   b. or we write the Jordan matrix as a real-valued matrix as in Eq. (4.56) and compute $\mathbf{P}$ as a real-valued matrix by setting $\mathbf{P} = (\mathbf{u}, \mathbf{v})$ where $\mathbf{Q} = \mathbf{u} + \mathbf{v}i$ is the solution to the homogeneous system

$$(\mathbf{A} - (\alpha + \beta i)\mathbf{I}_2)\mathbf{Q} = 0,$$

   or, equivalently, $\mathbf{u}$ solves $(\mathbf{A} - \alpha\,\mathbf{I}_2)\mathbf{u} = 0$, and $\mathbf{v}$ solves $(\mathbf{A} - \beta\,\mathbf{I}_2)\mathbf{v} = 0$. In the main text, we write the eigenvector associated to a complex-valued eigenvalue, when $\mathbf{\Lambda}_-$ is real-valued, as $\mathbf{P} = (\mathbf{P}^r, \mathbf{P}^c) = (\mathbf{u}, \mathbf{v})$.

3. if $\Delta(\mathbf{A}) = 0$, we have multiple eigenvalues $\lambda_+ = \lambda_- = \lambda$. In this case, we determine $\mathbf{P} = (\mathbf{P}^s, \mathbf{P}^g)$ where $\mathbf{P}^s$ is a simple eigenvalue and $\mathbf{P}^g$ is a generalized eigenvalue. They are obtained in the following way: first, $\mathbf{P}^s$ solves $(\mathbf{A} - \lambda\mathbf{I})\mathbf{P}^s = 0$, where $\mathbf{I} = \mathbf{I}_2$; second, if $(\mathbf{A} - \lambda\mathbf{I})^2 \neq \mathbf{0}$ we determine $\mathbf{P}^g$ from $(\mathbf{A} - \lambda\mathbf{I})^2\mathbf{P}^g = 0$; however, if $(\mathbf{A} - \lambda\mathbf{I})^2 = \mathbf{0}$ we determine $\mathbf{P}^g$ from $(\mathbf{A} - \lambda\mathbf{I})\mathbf{P}^g = \mathbf{P}^s$.

Conclusion: given a matrix $\mathbf{A}$, we can find matrices $\mathbf{\Lambda}$ and $\mathbf{P}$ such that $\mathbf{A} = \mathbf{P}\,\mathbf{\Lambda}\,\mathbf{P}^{-1}$. As $\mathbf{P}$ is invertible then $\mathbf{\Lambda} = \mathbf{P}^{-1}\mathbf{A}\mathbf{P}$.

Proposition 4.8 *The eigenvector matrices associated to the Jordan canonical forms* $\mathbf{\Lambda}_+$, $\mathbf{\Lambda}_0$, *and* $\mathbf{\Lambda}_-$ *are*

$$\mathbf{P}_+ = \begin{pmatrix} 1 & 0 \\ 0 & 1 \end{pmatrix}, \quad \mathbf{P}_0 = \begin{pmatrix} 1 & 1 \\ 0 & 1 \end{pmatrix}, \quad \mathbf{P}_- = \begin{pmatrix} 1 & 1 \\ -1 & 1 \end{pmatrix}, \tag{4.57}$$

respectively.

Proof For $\mathbf{\Lambda} = \mathbf{\Lambda}_+$, because $(\mathbf{\Lambda}_+ - \lambda_-\mathbf{I})\mathbf{P}^- = 0$ and $(\mathbf{\Lambda}_+ - \lambda_+\mathbf{I})\mathbf{P}^+ = 0$, that is $\begin{pmatrix} 0 & 0 \\ 0 & \lambda_+ - \lambda_- \end{pmatrix}\begin{pmatrix} P_1^- \\ P_2^- \end{pmatrix} = \begin{pmatrix} 0 \\ 0 \end{pmatrix}$, and $\begin{pmatrix} \lambda_- - \lambda_+ & 0 \\ 0 & 0 \end{pmatrix}\begin{pmatrix} P_1^+ \\ P_2^+ \end{pmatrix} = \begin{pmatrix} 0 \\ 0 \end{pmatrix}$ then we get $\mathbf{P} = (\mathbf{P}^-, \mathbf{P}^+) = \mathbf{I}$, because $\lambda_+ \neq \lambda_-$. For $\mathbf{\Lambda} = \mathbf{\Lambda}_0$, there is a multiple eigenvalue equal to λ, and we determine the simple eigenvector from $(\mathbf{\Lambda}_0 - \lambda\mathbf{I})\mathbf{P}^s = 0$. To determine the second eigenvector as $(\mathbf{\Lambda}_0 - \lambda\mathbf{I})^2 = \mathbf{0}$, then we solve $(\mathbf{\Lambda}_0 - \lambda\mathbf{I})\mathbf{P}^g = \mathbf{P}^s$. Therefore, solving $\begin{pmatrix} 0 & 1 \\ 0 & 0 \end{pmatrix}\begin{pmatrix} P_1^s \\ P_2^s \end{pmatrix} = \begin{pmatrix} 0 \\ 0 \end{pmatrix}$, and $\begin{pmatrix} 0 & 1 \\ 0 & 0 \end{pmatrix}\begin{pmatrix} P_1^g \\ P_2^g \end{pmatrix} = \begin{pmatrix} P_1^s \\ P_2^s \end{pmatrix}$, yields $\mathbf{P}^s = (1, 0)$ and $\mathbf{P}^g = (1, 1)$.

For $\mathbf{\Lambda} = \mathbf{\Lambda}_-$ consider eigenvalue $\lambda_+ = \alpha + \beta i$ and assume that there is a complex vector $\mathbf{z} = (u_1 + v_1 i, u_2 + v_2 i)$ that solves $(\mathbf{\Lambda}_- - (\alpha + \beta i)I)\mathbf{z} = 0$. Expanding, yields[30]

$$\begin{cases} \beta\,(u_2 + v_1 + (v_2 - u_1)i) & = 0 \\ \beta\,((v_2 - u_1) - (u_2 + v_1)i) & = 0 \end{cases}$$

then we should have $u_1 = v_2$ and $u_2 = -v_1$. We can arbitrarily set $u_1 = 1$ and $v_1 = 1$, in $\mathbf{P}^1 = (u_1, u_2)^\top$ and $\mathbf{P}^2 = (v_1, v_2)^\top$, to get the third eigenvector matrix. □

If a matrix $\mathbf{A}$ can be transformed into a Jordan canonical form we say matrix $\mathbf{A}$ is **canonical**. **Non-canonical matrices** cannot be transformed into a Jordan canonical form. Square non-canonical matrices take the form of an **hyperbolic matrix**

$$\mathbf{A}_h = \begin{pmatrix} \alpha & \beta \\ \beta & \alpha \end{pmatrix}.$$

If $\alpha = 0$ then $\mathbf{A}_h$ is said to be a diagonal matrix, denoted by $\mathbf{A}_d$.

4.6 Matrix Exponential Functions

The matrix exponential function $\mathbf{e}^{\mathbf{A}x}$ defined in (4.8) has following properties[31]:

Lemma 4.6 (Properties of matrix exponentials $\mathbf{e}^{\mathbf{Ax}}$) *Matrix exponential function $\mathbf{e}^{\mathbf{Ax}}$, defined in Eq. (4.8), has the following properties:*

(i) it satisfies the semigroup property: $\mathbf{e}^{\mathbf{A(x+s)}} = \mathbf{e}^{\mathbf{Ax}}\mathbf{e}^{\mathbf{As}}$;
(ii) the inverse of the matrix exponential is the exponential matrix of the inverse: $\left(\mathbf{e}^{\mathbf{Ax}}\right)^{-1} = \mathbf{e}^{-\mathbf{Ax}}$;
(iii) the derivative of the exponential function commutes: $\frac{d}{dx}\mathbf{e}^{\mathbf{Ax}} = \mathbf{A}\mathbf{e}^{\mathbf{Ax}} = \mathbf{e}^{\mathbf{Ax}}\mathbf{A}$;
(iv) if matrices $\mathbf{A}$ and $\mathbf{B}$ commute, (i.e., if $\mathbf{A}\,\mathbf{B} = \mathbf{B}\,\mathbf{A}$) then $\mathbf{e}^{(\mathbf{A}+\mathbf{B})\mathbf{x}} = \mathbf{e}^{\mathbf{Ax}}\mathbf{e}^{\mathbf{Bx}}$;
(v) let $\mathbf{P}$ be a non-singular,[32] and square matrix. Then $\mathbf{e}^{\mathbf{P}^{-1}\mathbf{APx}} = \mathbf{P}^{-1}\mathbf{e}^{\mathbf{Ax}}\mathbf{P}$.

Corollary 4.2 *If $\mathbf{A}$ is a canonical matrix then its matrix exponential function satisfies*

$$\mathbf{e}^{\mathbf{Ax}} = \mathbf{P}\,\mathbf{e}^{\mathbf{\Lambda x}}\,\mathbf{P}^{-1}, \tag{4.58}$$

where $\mathbf{\Lambda}$ is the Jordan canonical form and $\mathbf{P}$ is the eigenvector matrix associated to $\mathbf{A}$.

---

[30] We use the rules for sums and multiplications of complex numbers: if $x_1 = a_1 + b_1 i$ and $x_2 = a_2 + b_2 i$, then $x_1 + x_2 = (a_1 + a_2) + (b_1 + b_2)i$ and $x_1 x_2 = (a_1 a_2 - b_1 b_2) + (a_1 b_2 + a_2 b_1)i$ because $i^2 = -1$.

[31] See Hirsch and Smale (1974, pp. 82–87).

[32] A non-singular matrix has a determinant different from zero.

Proof From the eigen decomposition theorem we have $\mathbf{P}^{-1}\mathbf{A}\mathbf{P} = \mathbf{\Lambda}$. Therefore, from Lemma 4.6 *(v)* we have $\mathbf{e}^{\mathbf{\Lambda x}} = \mathbf{e}^{\mathbf{P}^{-1}\mathbf{A}\mathbf{P}\mathbf{x}} = \mathbf{P}^{-1}\mathbf{e}^{\mathbf{A x}\mathbf{P}}$. As $\mathbf{P}$ is invertible we have the transformation formula (4.58). □

Proof *(of Lemma* 4.2*)* Consider the definition of matrix exponential, Eq. (4.8) and the Jordan canonical forms (4.56). If $\mathbf{\Lambda} = \mathbf{\Lambda}_+$ we have

$$\mathbf{e}^{\mathbf{\Lambda}_+\mathbf{x}} = \mathbf{I}_2 + \mathbf{\Lambda}_+ x + \frac{1}{2}(\mathbf{\Lambda}_+)^2 x^2 + \ldots = \begin{pmatrix} 1 & 0 \\ 0 & 1 \end{pmatrix} + \begin{pmatrix} \lambda_- x & 0 \\ 0 & \lambda_+ x \end{pmatrix} + \frac{1}{2}\begin{pmatrix} \lambda_-^2 x^2 & 0 \\ 0 & \lambda_+^2 x^2 \end{pmatrix} + \ldots$$

then, performing the matrix additions, yields $\mathbf{e}^{\mathbf{\Lambda}_+\mathbf{x}} = \begin{pmatrix} e^{\lambda_- x} & 0 \\ 0 & e^{\lambda_+ x} \end{pmatrix}$ because $e^y = \sum_{n=0}^{+\infty} \frac{y^n}{n!}$. This result is straightforward to obtain because the Jordan matrix is diagonal. This is not the case for Jordan matrix $\mathbf{\Lambda}_0$. Decomposing $\mathbf{\Lambda}_0$, we find $\mathbf{\Lambda}_0 = \mathbf{\Lambda}_{0,1} + \mathbf{\Lambda}_{0,2} = \left(\begin{smallmatrix} \lambda & 0 \\ 0 & \lambda \end{smallmatrix}\right) + \left(\begin{smallmatrix} 0 & 1 \\ 0 & 0 \end{smallmatrix}\right)$. As the two matrices commute, i.e., $\mathbf{\Lambda}_{0,1}\mathbf{\Lambda}_{0,2} = \mathbf{\Lambda}_{0,2}\mathbf{\Lambda}_{0,1}$, then applying property (iv) of Lemma 4.6 yields $\mathbf{e}^{\mathbf{\Lambda}_0\mathbf{x}} = \mathbf{e}^{(\mathbf{\Lambda}_{0,1}+\mathbf{\Lambda}_{0,2})\mathbf{x}} = \mathbf{e}^{\mathbf{\Lambda}_{0,1}\mathbf{x}}\mathbf{e}^{\mathbf{\Lambda}_{0,2}\mathbf{x}}$. As $\mathbf{e}^{\mathbf{\Lambda}_{0,1}\mathbf{x}} = \begin{pmatrix} e^{\lambda x} & 0 \\ 0 & e^{\lambda x} \end{pmatrix} = e^{\lambda x}\mathbf{I}_2$, and using again formula (4.8), $\mathbf{e}^{\mathbf{\Lambda}_{0,2}\mathbf{x}} = \left(\begin{smallmatrix} 1 & 0 \\ 0 & 1 \end{smallmatrix}\right) + \left(\begin{smallmatrix} 0 & x \\ 0 & 0 \end{smallmatrix}\right) + \frac{x^2}{2}\left(\begin{smallmatrix} 0 & 0 \\ 0 & 0 \end{smallmatrix}\right) + \ldots = \left(\begin{smallmatrix} 1 & x \\ 0 & 1 \end{smallmatrix}\right)$, we obtain the matrix exponential in Eq. (4.14). As $\mathbf{\Lambda}_-$ is non-diagonal, but it is decomposable as the sum of two matrices $\mathbf{\Lambda}_{-,1} = \left(\begin{smallmatrix} \alpha & 0 \\ 0 & \alpha \end{smallmatrix}\right)$ and $\mathbf{\Lambda}_{-,2} = \begin{pmatrix} 0 & \beta \\ -\beta & 0 \end{pmatrix}$ that commute, then applying again property (iv) of Lemma 4.6 $\mathbf{e}^{\mathbf{\Lambda}_-\mathbf{x}} = \mathbf{e}^{\mathbf{\Lambda}_{-,1}\mathbf{x}}\mathbf{e}^{\mathbf{\Lambda}_{-,2}\mathbf{x}}$, where $\mathbf{e}^{\mathbf{\Lambda}_{-,1}\mathbf{x}} = e^{\alpha x}\left(\begin{smallmatrix} 1 & 0 \\ 0 & 1 \end{smallmatrix}\right)$, and from formula (4.8)

$$\mathbf{e}^{\mathbf{\Lambda}_{-,2}\mathbf{x}} = \begin{pmatrix} 1 & 0 \\ 0 & 1 \end{pmatrix} + \begin{pmatrix} 0 & \beta x \\ -\beta x & 0 \end{pmatrix} + \frac{x^2}{2}\begin{pmatrix} \beta^2 x^2 & 0 \\ 0 & -\beta^2 x^2 \end{pmatrix} + \ldots = \begin{pmatrix} \cos\beta x & \sin\beta x \\ -\sin\beta x & \cos\beta x \end{pmatrix},$$

because $\sin y = \sum_{n=0}^{+\infty} \dfrac{y^{2n+1}}{(2n+1)}$ and $\cos y = \sum_{n=0}^{+\infty} \dfrac{y^{2n}}{(2n)}$, we obtain the third matrix in (4.14). □

Proof *(of Lemma* 4.1) We know that $\mathbf{A} = \mathbf{P}\mathbf{\Lambda}\mathbf{P}^{-1}$, where $\mathbf{\Lambda}$ is the Jordan form of $\mathbf{A}$. Then $\mathbf{e}^{\mathbf{A x}} = \mathbf{e}^{\mathbf{P}\mathbf{\Lambda}\mathbf{P}^{-1}\mathbf{x}} = \mathbf{P}\mathbf{e}^{\lambda \mathbf{x}}\mathbf{P}^{-1}$ by property (v) of Lemma 4.6. Matrix $\mathbf{A} = \mathbf{A}_d$ has two equal real eigenvalues equal to λ and, because it is diagonal, it satisfies $\mathbf{A}_d\,\mathbf{P}_d = \mathbf{P}_d\,\mathbf{A}_d$. Therefore $\mathbf{P}_d = \mathbf{I}$ and $\mathbf{e}^{\mathbf{A}_d x} = \mathbf{P}\,e^{\lambda x}\,\mathbf{I}\,\mathbf{P}^{-1} = e^{\lambda x}\,\mathbf{I}$. Matrix $\mathbf{A} = \mathbf{A}_h$ has the real spectrum $\sigma = \{\alpha + \beta, \alpha - \beta\}$ and has eigenvector matrix $\mathbf{P}_h = \left(\begin{smallmatrix} 1 & -1 \\ 1 & 1 \end{smallmatrix}\right)$. Therefore, the exponential matrix is

$$\mathbf{e}^{\mathbf{A}_h x} = \begin{pmatrix} 1 & -1 \\ 1 & 1 \end{pmatrix}\begin{pmatrix} e^{(\alpha+\beta)x} & 0 \\ 0 & e^{(\alpha+\beta)x} \end{pmatrix}\begin{pmatrix} 1 & -1 \\ 1 & 1 \end{pmatrix}^{-1}$$

which, expanding the matrix multiplication, yields matrix (4.16). □

4.7 Solving Foci in Polar Coordinates

Consider the planar ODE (4.32) in which matrix $\mathbf{A}$ has a positive determinant but a negative discriminant. Therefore, there is one unique steady state $\bar{\mathbf{y}}$ and $\mathbf{A}$ is similar to the Jordan canonical form $\mathbf{\Lambda}_-$. Using the transformation $\mathbf{w} = \mathbf{y} - \bar{\mathbf{y}}$ yields the normal form $\dot{\mathbf{w}} = \mathbf{\Lambda}_- \, \mathbf{w}$, that is

$$\dot{w}_1 = \alpha\, y_1 + \beta\, y_2, \tag{4.59a}$$

$$\dot{w}_2 = -\beta\, y_1 + \alpha\, y_2. \tag{4.59b}$$

We say this system is in **cartesian coordinates** $(w_1, w_2) \in \mathbb{R}$. We transform it to **polar coordinates** $(r, \theta) \in \mathbb{R}$, by using the transformation, already mentioned in Eq. (4.55)

$$w_1(t) = r(t)\cos(\theta(t)), \text{ and } w_2(t) = r(t)\sin(\theta(t)), \tag{4.60}$$

where $r^2 = w_1^2 + w_2^2$ measures the distance from a reference point (the radius) and θ, is the angular coordinate such that $\tan(\theta) = \frac{\sin(\theta)}{\cos(\theta)} = \frac{w_1}{w_2}$, that is $\theta = \arctan\left(\frac{w_1}{w_2}\right)$. The inverse transformation is

$$r(t) = \sqrt{w_1(t)^2 + w_2(t)^2}, \text{ and } \theta(t) = \arctan\left(\frac{w_1(t)}{w_2(t)}\right). \tag{4.61}$$

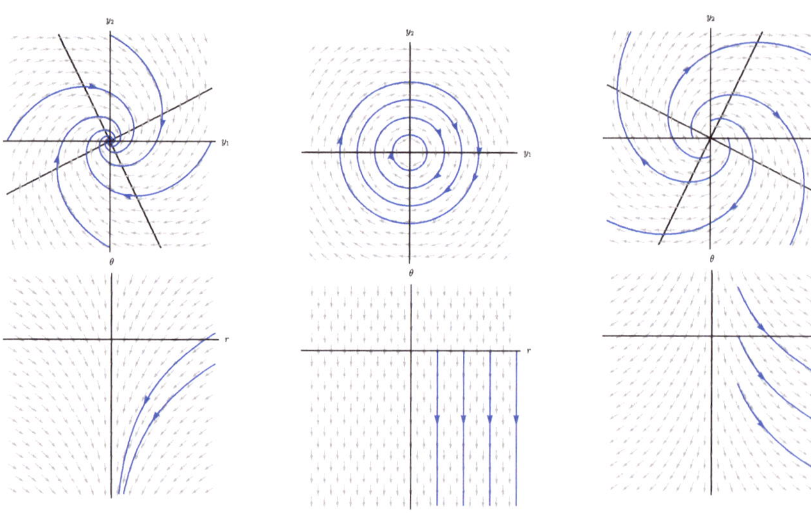

(a) Case $\alpha < 0$: stable focus

(b) Case $\alpha = 0$: center

(c) Case $\alpha > 0$: unstable focus

Fig. 4.10 The foci in cartesian and polar coordinates for $\beta > 0$

Taking time-derivatives for both functions, we find[33]

$$\dot{r} = \alpha\, r, \tag{4.62a}$$

$$\dot{\theta} = -\beta. \tag{4.62b}$$

The solution to this planar ODE is $r(t) = r(0)\, e^{\alpha\, t}$, and $\theta(t) = \theta(0) - \beta\, t$. Using the transformation (4.60), we obtain the general solution to Eqs. (4.59a)–(4.59b) in cartesian coordinates,

$$w_1(t) = e^{\alpha\, t}\Big(w_1(0)\ \cos(\beta\, t) + w_2(0)\ \sin(\beta\, t)\Big), \tag{4.63a}$$

$$w_2(t) = e^{\alpha\, t}\Big(-\ w_1(0)\ \sin(\beta\, t) + w_2(0)\ \cos(\beta\, t)\Big). \tag{4.63b}$$

Using the transformation (4.60),[34] as $\cos\,(\theta(0) - \beta\, t) = \cos\,(\theta(0))\cos\,(\beta\, t)$ $+ \sin\,(\theta(0))\cos\,(\beta\, t)$, and $\sin\,(\theta(0) - \beta\, t) = -\cos\,(\theta(0))\sin\,(\beta\, t) + \sin\,(\theta(0))$ $\cos\,(\beta\, t)$ we find $w_1(0) = r(0)\ \cos\,(\theta(0))$ and $w_2(0) = r(0)\ \sin\,(\theta(0))$, yields solutions (4.63).

Figure 4.10 illustrates the phase diagrams for ODE (4.59), in cartesian coordinates, and for ODE (4.62), in polar coordinates.

33 $\dot{r} = \frac{1}{2}\left(w_1^2 + w_2^2\right)^{\frac{1}{2}-1}\left(2\, w_1\, \dot{w}_1 + 2\, w_2\, \dot{w}_2\right) = \frac{w_1\, \dot{w}_1 + w_2\, \dot{w}_2}{r} = \alpha \frac{w_1^2 + w_2^2}{r} = \alpha\, \frac{r^2}{r} = \alpha\, r,$ and $\dot{\theta} = \frac{w_2\, \dot{w}_1 - w_1\, \dot{w}_2}{w_1^2 + w_2^2} = -\beta\, \frac{w_1^2 + w_2^2}{w_1^2 + w_2^2} = -\beta.$ Recall that the derivative of $\arctan\left(\frac{f(x)}{g(x)}\right)$ is $\frac{d}{dx}\left(\arctan\left(\frac{f(x)}{g(x)}\right)\right) = \frac{f'(x)\, g(x) - f(x)\, g'(x)}{f(x)^2 + g(x)^2}.$

34 Recall the trigonometric equivalences: $\cos\,(x - y) = \cos\,(x)\ \cos\,(y) + \sin\,(x)\ \sin\,(y)$ and $\sin\,(x - y) = -\cos\,(x)\ \sin\,(y) + \sin\,(x)\ \cos\,(y)$.

Chapter 5
Planar Non-Linear ODE: The Regular Case

In this chapter, we address the dynamics of planar non-linear regular ordinary differential equations (ODE), by keeping time as the only independent variable. It extends Chap. 3 by adding another state variable, and extends Chap. 4 by dealing with non-linear ODE. In this chapter, the dependent variable is the mapping $\mathbf{y} : \mathrm{T} \subseteq \mathbb{R}_+ \to \mathbf{Y} \subset \mathbb{R}^2$ generated by the solution to the ODE

$$\dot{\mathbf{y}} = \mathbf{f}\big(\mathbf{y}, \varphi\big), \tag{5.1}$$

where $\mathbf{f}(\mathbf{y}, \cdot) : \mathbf{Y} \to \mathbb{R}^2$ is a continuous, and continuously differentiable, vector function $\mathbf{f} \in C(\mathbb{R})$, and $\varphi \in \mathbb{R}^p$ is a parameter vector of size $p \geq 1$. In an expanded form, we have

$$\dot{y}_1 = f_1(y_1, y_2, \varphi), \tag{5.2a}$$
$$\dot{y}_2 = f_2(y_1, y_2, \varphi). \tag{5.2b}$$

We solved some scalar non-linear regular ODE, in Chap. 3, by using both analytical and qualitative or geometrical approaches. For planar non-linear ODE there are not, in general, analytical solutions, except for some simple cases. In this chapter, we use a qualitative approach to characterize the most well-known types of dynamics for planar ODE, in particular regarding those properties that differentiate them from scalar ODE. First, the Poincaré-Bendixson theory Bendixson (1901) allows us to show that there are three types of **limit sets**, $\Omega_{\mathbf{Y}}$: steady states, periodic trajectories, and cycles composed of steady states and connecting trajectories (see Shilnikov et al. (1998, p. 14)). As we saw, non-linear scalar and planar linear ODE display only the first type of asymptotic limit set.[1] Second, there are three types of **invariant orbits** that do not exist in planar linear ODE: homoclinic, and heteroclinic orbits. As a consequence new types of both local and global bifurcations exist for non-linear

[1] Although for planar linear ODE periodic trajectories can be solutions, but for finite time.

P. B. Brito, *Economic Dynamics and Distributions*, Dynamic Modeling and Econometrics in Economics and Finance 32, https://doi.org/10.1007/978-3-031-94717-9_5

planar ODE models. However, as qualitative theory is also based on linearization, then some results from Chap. 4 can be leveraged both for approximating solutions, and/or for finding bifurcation conditions.[2]

Next, we present and study some important normal forms in Sect. 5.1, introduce the qualitative theory of the dynamics of regular ODE in Sect. 5.2, and apply it to some economic models in Sect. 5.3.

5.1 Normal Forms

A **normal form** is the simplest ODE, whose exact solution is sometimes known, that represents a whole family of ODE by topological equivalence. In this chapter, we consider several normal forms that are particular cases of ODE whose right-hand side is specified by the two-dimensional polynomial vector field

$$\dot{\mathbf{y}} = \mathbf{f}(\mathbf{y}, \varphi) \equiv \mathbf{A}_0 + \mathbf{A}_1 \mathbf{y} + \mathbf{A}_2 \mathbf{y}^{\otimes 2} + \mathbf{A}_3 \mathbf{y}^{\otimes 3}, \tag{5.3}$$

where $\mathbf{y} = (y_1, y_2) : \mathrm{T} \to \mathbf{Y} \subseteq \mathbb{R}^2$, $\mathbf{y}^{\otimes 2} = \left(y_1^2, y_1 y_2, y_2^2\right)^\top$, $\mathbf{y}^{\otimes 3} = \left(y_1^3, y_1^2 y_2, y_1 y_2^2, y_2^3\right)^\top$, $\mathbf{y} : \mathrm{T} \to \mathbf{Y} \subseteq \mathbb{R}^2$, and the coefficient matrices, $\mathbf{A}_0 \in \mathbb{R}^2$, $\mathbf{A}_1 \in \mathbb{R}^{2\times 2}$, $\mathbf{A}_2 \in \mathbb{R}^{2\times 3}$, and $\mathbf{A}_3 \in \mathbb{R}^{2\times 4}$, are functions of the vector of primitive parameters $\varphi \in \Phi \subseteq \mathbb{R}^p$. That is, $\mathbf{A}_i = \mathbf{A}_i(\varphi)$ for $i = 0, \ldots, 3$. The linear planar ODE is a special case of ODE (5.3) in which $\mathbf{A}_2 = \mathbf{0}_{(2\times 3)}$ and $\mathbf{A}_3 = \mathbf{0}_{(2\times 4)}$.

The ODE (4.32) can be written as $\dot{\mathbf{y}} = \mathbf{f}(\mathbf{y}, \varphi) \equiv \mathbf{A}_0 + \mathbf{A}_1 \mathbf{y}$. We found in Chap. 4 that: planar linear ODE have explicit solutions, all solutions correspond to global solutions, and the phase diagrams display global dynamics (global asymptotic stability, global instability, or a global center manifold).

If any of the matrices $\mathbf{A}_2$ or $\mathbf{A}_3$ is different from the zero matrices with the same dimensions, there are in general no explicit solutions, and the potential number of ODE is huge. A large number of non-linear planar ODE—as in several models applied to economics—have qualitative dynamics that are equivalent to linear planar ODE, analogously to what we had considered for the non-linear scalar ODE in Chap. 3. Therefore, we select a small number of ODE displaying generic or non-generic invariant sets, and bifurcations, that cannot exist in linear planar or non-linear scalar ODE. These include, heteroclinic and homoclinic orbits, in Sects. 5.1.1 and 5.1.2, and limit cycles, in Sect. 5.1.3. In Sect. 5.1.4, we briefly present a co-dimension two bifurcation.

[2] Recall, in particular, Fig. 4.3 and Table 4.4.

5.1.1 *Heteroclinic Orbits*

Consider a planar ODE in which there are, at least, two steady states, say $\bar{\mathbf{y}}^1$ and $\bar{\mathbf{y}}^2$. We say there is a **heteroclinic orbit**, and denote it $\text{Het}(\mathbf{y})$, if there are solutions $\mathbf{y}(t)$ that entirely lie in a curve joining $\bar{\mathbf{y}}^1$ and $\bar{\mathbf{y}}^2$. Therefore, if $\mathbf{y}(0) \in \text{Het}(\mathbf{y})$ then $\mathbf{y}(t) \in \text{Het}(\mathbf{y})$, for each $t > 0$, and either $\lim_{t\to\infty} \mathbf{y}(t) = \bar{\mathbf{y}}^1$ and $\lim_{t\to-\infty} \mathbf{y}(t) = \bar{\mathbf{y}}^2$ or $\lim_{t\to\infty} \mathbf{y}(t) = \bar{\mathbf{y}}^2$ and $\lim\limits_{t\to-\infty} \mathbf{y}(t) = \bar{\mathbf{y}}^1$. Heteroclinics can exist both when the stability type of the two steady states are different, or if they are both the same, and are not stable nodes or foci. For instance, a heteroclinic orbit can connect one stable node and one unstable node or a saddle point, or they can connect two saddle points. In the last case, we say there is a **saddle connection**. **Heteroclinic networks** can also exist when there are more than two steady states that are connected by heteroclinic orbits.

Heteroclinic orbits can be **generic**, if they exist for a non-degenerate subset of the parameter set or they can be **non-generic**, if they only exist for a degenerate subset of parameters. In the last case, we say we have a **heteroclinic bifurcation**.

5.1.1.1 Generic Heteroclinic Orbits

Although there are several normal forms generating generic heteroclinic orbits, we focus next on the following case[3]:

$$\dot{y}_1 = a\, y_1\, y_2, \tag{5.4a}$$

$$\dot{y}_2 = 1 + y_1^2 - y_2^2, \tag{5.4b}$$

where it is assumed that $a \neq 0$.[4] The vector field and its Jacobian, evaluated at any point $\mathbf{y} = (y_1, y_2) \in \mathbf{Y} \subseteq \mathbb{R}^2$, are

$$\mathbf{f}(\mathbf{y}) = \begin{pmatrix} a\, y_1\, y_2 \\ 1 + y_1^2 - y_2^2 \end{pmatrix} \text{ and } J(\mathbf{y}) \equiv \mathrm{D}_{\mathbf{y}}\mathbf{f}(\mathbf{y}) = \begin{pmatrix} a\, y_2 & a\, y_1 \\ 2y_1 & -2y_2 \end{pmatrix}.$$

The trace and determinant of the Jacobian, both depending on the parameter a,

$$\mathrm{tr}\,(J(\mathbf{y})) = (a-2)y_2, \text{ and } \det(J(\mathbf{y})) = -2a(y_1^2 + y_2^2),$$

satisfy $\det\big(J(\mathbf{y})\big) \geq 0$ if $a < 0$ and $\det\big(J(\mathbf{y})\big) \leq 0$ if $a > 0$.

The ODE (5.4) has two steady states: $\bar{\mathbf{y}}^1 = (\bar{y}_1^1, \bar{y}_2^1) = (0, -1)$ and $\bar{\mathbf{y}}^2 = (\bar{y}_1^2, \bar{y}_2^2) = (0, 1)$. Although the values of the steady states do not depend on a, this parameter has

[3] This is a particular case of equation (5.3) in which $\mathbf{A}_0 = \left(\begin{smallmatrix} 0 \\ 1 \end{smallmatrix}\right)$, $\mathbf{A}_1 = \mathbf{0}_{(2\times2)}$, $\mathbf{A}_2 = \left(\begin{smallmatrix} 0 & a & 0 \\ 1 & 0 & -1 \end{smallmatrix}\right)$, and $\mathbf{A}_3 = \mathbf{0}_{(2\times4)}$. Furthermore, $\mathrm{D}_a\mathbf{A}_0 = \mathbf{0}_{(2\times1)}$, $\mathrm{D}_a\mathbf{A}_1 = \mathbf{0}_{(2\times2)}$, $\mathrm{D}_a\mathbf{A}_2 = \left(\begin{smallmatrix} 0 & 1 & 0 \\ 0 & 0 & 0 \end{smallmatrix}\right) \neq \mathbf{0}_{(2\times3)}$, and $\mathrm{D}_a\mathbf{A}_3 = \mathbf{0}_{(2\times4)}$.

[4] If $a = 0$, this equation reduces to a scalar Ricatti equation.

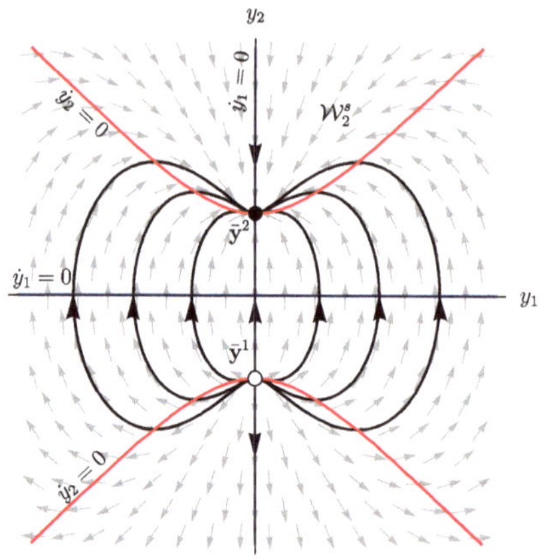

(a) Phase diagram for $a < 0$ (b) Phase diagram for $a > 0$

Fig. 5.1 Phase diagrams for ODE (5.4). In each phase diagram, the isoclines $\dot{y}_1 = 0$ (blue) and $\dot{y}_2 = 0$ (red), the vector field, the stable manifolds associated to steady states $\bar{\mathbf{y}}^1$ and $\bar{\mathbf{y}}^2$, $\mathcal{W}_1^s$ and $\mathcal{W}_2^s$, and some representative integral curves (in black) are represented

a bearing on their local dynamics, and therefore on the characteristics of the phase diagrams.

If $a < 0$, then the steady state $\bar{\mathbf{y}}^1$ is an unstable node—because $\text{tr}\big(J(\bar{\mathbf{y}}^1)\big) = -a > 0$ and $\det\big(J(\bar{\mathbf{y}}^1)\big) > 0$—and steady state $\bar{\mathbf{y}}^2$ is a stable node—because $\text{tr}\big(J(\bar{\mathbf{y}}^2)\big) = a < 0$ and $\det\big(J(\bar{\mathbf{y}}^2)\big) > 0$. Then, for any trajectory starting from any $\mathbf{y}(0) \neq \bar{\mathbf{y}}^2$, there is convergence to steady state $\bar{\mathbf{y}}^2$ (see the phase diagram in Fig. 5.1a). That phase diagram also shows that there is an infinite number of heteroclinic curves, connecting $\bar{\mathbf{y}}^2$ to $\bar{\mathbf{y}}^1$. As $\text{Het}(\mathbf{y}) = \mathbf{Y}/\{\bar{\mathbf{y}}_1\}$ then this set is coincident with the local stable sub-manifold associated to steady state $\bar{\mathbf{y}}^2$, that is $\text{Het}(\mathbf{y}) = \mathcal{W}_2^s$.[5]

If $a > 0$ both steady states, $\bar{\mathbf{y}}^1$ and $\bar{\mathbf{y}}^2$, are saddle points, because $\det\big(J(\bar{\mathbf{y}}^1)\big) = \det\big(J(\bar{\mathbf{y}}^2)\big) < 0$. In this case, there is one unique heteroclinic curve $\text{Het}(\mathbf{y}) = \{(y_1, y_2) : y_1 = 0,\ -1 \le y_2 \le 1\}$, connecting $\bar{\mathbf{y}}^1$ to $\bar{\mathbf{y}}^2$ (see phase diagram Fig. 5.1b). In this case, $\text{Het}(\mathbf{y})$ is the set of points belonging to the intersection of the local unstable manifold of $\bar{\mathbf{y}}^1$ and the local stable manifold of $\bar{\mathbf{y}}^2$: $\text{Het}(\mathbf{y}) = \mathcal{W}_1^u \cap \mathcal{W}_2^s$.

Those heteroclinic orbits, although their number depended on the sign of a, are **generic**, in both cases, because they persist for a wide interval (with positive measure) of values for parameter a. This is not the case in the next example.

---

[5] In this section we use the notation $\mathcal{W}_i^s$ instead of $\mathcal{W}_i^{s,loc}$ for local stable manifolds associated to steady state $\mathbf{y}^i$, and analogously for unstable and center manifolds. For the definition of $\mathcal{W}_2^s$ see Sect. 5.2.2.

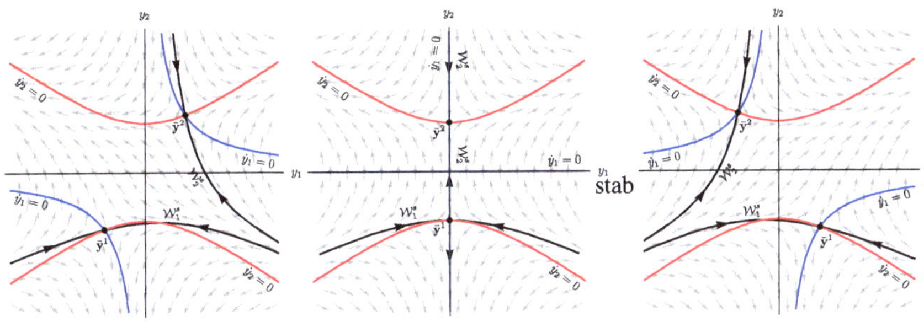

(a) Phase diagrams for $a < 0$ (b) Phase diagrams for $a = 0$ (c) Phase diagrams for $a > 0$

Fig. 5.2 Phase diagrams for ODE (5.5)

5.1.1.2 Non-Generic Case: Heteroclinic Saddle Connection Bifurcation

Let again $\mathbf{y} \in \mathbb{R}^2$ and $a \in \mathbb{R}$,[6] and consider the ODE, that is studied, for instance, in Hale and Koçak (1991, p. 210),

$$\dot{y}_1 = a + 2\, y_1\, y_2, \tag{5.5a}$$

$$\dot{y}_2 = 1 + y_1^2 - y_2^2. \tag{5.5b}$$

The vector field and its Jacobian, for any point $\mathbf{y} = (y_1, y_2) \in \mathbf{Y} \subseteq \mathbb{R}^2$, are

$$\mathbf{f}(\mathbf{y}) = \begin{pmatrix} a + 2\, y_1\, y_2 \\ 1 + y_1^2 - y_2^2 \end{pmatrix} \text{ and } J(\mathbf{y}) \equiv \mathrm{D}_{\mathbf{y}}\mathbf{f}(\mathbf{y}) = \begin{pmatrix} 2\, y_2 & 2\, y_1 \\ 2y_1 & -2y_2 \end{pmatrix}.$$

The trace and determinant of the Jacobian are now $\mathrm{tr}\,(J(\mathbf{y})) = 0$ and $\det\,(J(\mathbf{y})) = -4(y_1^2 + y_2^2) < 0$. As Fig. 5.2. shows, there are always two steady states and they are both saddle points. If $a \neq 0$, as can be observed in phase diagrams Fig. 5.2a and c , there are no trajectories connecting the two steady states. However, if $a = 0$—see phase diagram Fig. 5.2b- , then there is a heteroclinic trajectory connecting the two saddle points (indeed this is the same case as the phase diagram Fig. 5.1b) .

In this case, we say there is a **heteroclinic saddle connection bifurcation**, for $a = 0$. Assuming a related but slightly different normal form of ODE (5.4) we have a heteroclinic bifurcation, instead of generic heteroclinics, meaning that a heteroclinic orbit only exists for a specific value (or some specific values) of paramater a, and an infinitesimal change from its bifurcation value, $a = 0$, eliminates it.

Exercise 5.1 Study the existence of generic and non-generic heteroclinic orbits for the planar ODE $\dot{y}_1 = a + b\, y_1\, y_2$ and $\dot{y}_2 = 1 + y_1^2 - y_2^2$.

[6] This is a particular case of equation (5.3) in which $\mathbf{A}_0 = \left(\begin{smallmatrix} a \\ 1 \end{smallmatrix}\right)$, $\mathbf{A}_1 = \mathbf{0}_{(2\times 2)}$, $\mathbf{A}_2 = \left(\begin{smallmatrix} 0 & 2 & 0 \\ 1 & 0 & -1 \end{smallmatrix}\right)$, and $\mathbf{A}_3 = \mathbf{0}_{(2\times 4)}$. Furthermore, the Jacobians of those matrices, regarding parameter a, are $\mathrm{D}_a\mathbf{A}_0 = \left(\begin{smallmatrix} 1 \\ 0 \end{smallmatrix}\right) \neq \mathbf{0}_{(2\times 1)}$, $\mathrm{D}_a\mathbf{A}_1 = \mathbf{0}_{(2\times 2)}$, $\mathrm{D}_a\mathbf{A}_2 = \mathbf{0}_{(2\times 3)}$, and $\mathrm{D}_a\mathbf{A}_3 = \mathbf{0}_{(2\times 4)}$.

5.1.2 *Homoclinic Orbits*

We say there is a **homoclinic curve** if there is a subset of points $\text{Hom}(\mathbf{y}) \in \mathbf{Y}$, connecting a steady state with itself. This is only possible if the steady state $\bar{\mathbf{y}}$ is a saddle point for which the intersection between its local stable and unstable manifolds is non-empty. Therefore, a homoclinic curve, is a closed curve. We say we have an homoclinic orbit if the initial value $\mathbf{y}(0)$ belongs to a homoclinic curve. As a consequence of the form of a homoclinic curve, homoclinic orbits exist jointly with periodic trajectories. Again, homoclinic curves (and orbits) can be generic or non-generic. Next we illustrate both cases.

5.1.2.1 Generic Homoclinic Orbits

Consider the non-linear planar ODE depending on one parameter, $a \in \mathbb{R}$, of form[7]

$$\dot{y}_1 = y_2, \tag{5.6a}$$

$$\dot{y}_2 = y_1 - a\, y_1^2. \tag{5.6b}$$

The Jacobian, evaluated at any point $\mathbf{y} \in \mathbf{Y}$, $J(\mathbf{y}) \equiv \mathrm{D}_{\mathbf{y}}\mathbf{f}(\mathbf{y}) = \begin{pmatrix} 0 & 1 \\ 1-2ay_1 & 0 \end{pmatrix}$, has trace and determinant equal to $\text{tr}\,(J(\mathbf{y})) = 0$, and $\det\,(J(\mathbf{y})) = 2ay_1 - 1$, respectively. If $a = 0$ this ODE reduces to a linear planar ODE, having a unique steady state, $\bar{\mathbf{y}} = \mathbf{0}$, that is a saddle point (see phase diagram in Fig. 5.3b).

If $a \neq 0$ there are two steady states $\bar{\mathbf{y}}^1 = (0, 0)$ and $\bar{\mathbf{y}}^2 = (1/a, 0)$, where $\bar{\mathbf{y}}^1$ is a saddle point, because $\det\left(J(\bar{\mathbf{y}}^1)\right) = -1 < 0$, and $\bar{\mathbf{y}}^2$ is locally a center, because $\det\left(J(\bar{\mathbf{y}}^2)\right) = 1 > 0$ and $\text{tr}\left(J(\bar{\mathbf{y}}^2)\right) = 0$, for any value of $a \neq 0$. Furthermore, we can prove that there is an invariant subset of $\mathbf{Y}$, such that solutions within it satisfy a potential or first integral curve that is time-invariant.

To prove this fact, we introduce a **Lyapunov function**, a differentiable function of $\mathbf{y}$, denoted by $L(\mathbf{y})$, having the time derivative $\dot{L} = \mathrm{D}_{\mathbf{y}}L \cdot \dot{\mathbf{y}} = L_{y_1}\dot{y}_1 + L_{y_2}\dot{y}_2$. A **first integral** is a set of points $\mathbf{y} \in \mathbf{Y}$, satisfying $\mathbf{y} \neq \bar{\mathbf{y}}$, such that $\dot{L} = 0$. If this is the case, then orbits passing through the integral curve satisfy $L(\mathbf{y}(t)) = \text{constant}$, for any value of $t \in [0, \infty)$.

For ODE (5.6a)-(5.6b), we consider the function[8]

$$L(\mathbf{y}) = -\frac{y_1^2}{2} + \frac{y_2^2}{2} + a\,\frac{y_1^3}{3}.$$

---

[7] This is a particular case of equation (5.3) in which $\mathbf{A}_0 = \mathbf{0}_{(2\times1)}$, $\mathbf{A}_1 = \begin{pmatrix} 0 & 1 \\ 1 & 0 \end{pmatrix}$, $\mathbf{A}_2 = \begin{pmatrix} 0 & 0 & 0 \\ -a & 0 & 0 \end{pmatrix}$, and $\mathbf{A}_3 = \mathbf{0}_{(2\times4)}$. Furthermore, $\mathrm{D}_a\mathbf{A}_0 = \mathbf{0}_{(2\times1)}$, $\mathrm{D}_a\mathbf{A}_1 = \mathbf{0}_{(2\times2)}$, $\mathrm{D}_a\mathbf{A}_2 = \begin{pmatrix} 0 & 0 & 0 \\ -1 & 0 & 0 \end{pmatrix} \neq \mathbf{0}_{(2\times3)}$, and $\mathrm{D}_a\mathbf{A}_3 = \mathbf{0}_{(2\times4)}$.

[8] There is not a constructive way to find the Lyapunov function. If it exists we need to come up with the right conjecture.

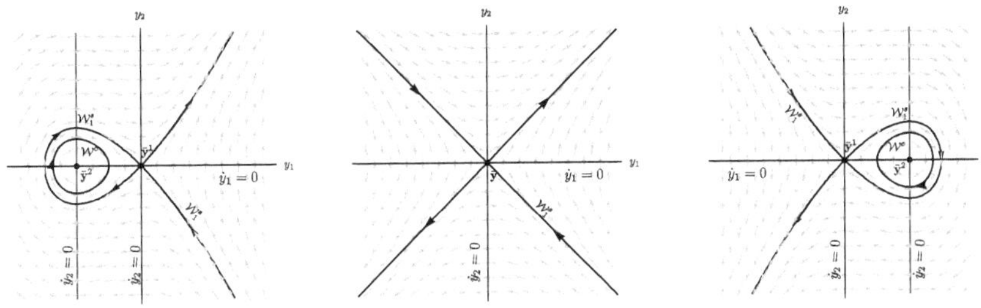

(a) Phase diagram for $a < 0$ (b) Phase diagram for $a = 0$ (c) Phase diagram for $a > 0$

Fig. 5.3 Phase diagrams for ODE (5.6)

Taking the time derivative, and substituting ODE (5.6), we obtain $\dot{L} = (ay_1 - 1)y_1\dot{y}_1 + y_2\dot{y}_2 = (ay_1 - 1)y_1 y_2 + y_2 y_1(1 - ay_1) = 0$, for any $\mathbf{y} \in \mathbb{R}^2$ and $a \neq 0$. We call homoclinic curve (or surface) to the set of points containing homoclinic orbits. In our case, homoclinic orbits converge, both for $t \to \infty$ and $t \to -\infty$, to steady state $\bar{\mathbf{y}}^1$. Therefore, the homoclinic surface is the set of points

$$\mathrm{Hom}(\bar{\mathbf{y}}^1) = \{(y_1, y_2) : L(y_1, y_2) = 0, \ \mathrm{sign}(\bar{y}_1) = \mathrm{sign}(a)\}.$$

In Fig. 5.3a and c, we illustrate the phase diagrams in which there are homoclinic loops (and orbits) for $a < 0$ and $a > 0$, respectively. In both those phase diagrams, the steady state $\bar{\mathbf{y}}^2$ is always a center, they contain an infinite number of closed curves filling up the subspace of $\mathbf{Y}$ between that steady state and the homoclinic loop $\mathrm{Hom}(\bar{\mathbf{y}}^1)$, and this homoclinic loop connects $\bar{\mathbf{y}}^1$ with itself. Then, there is an infinite number of periodic trajectories that are stable but not asymptotically stable. Therefore, curve $\mathrm{Hom}(\bar{\mathbf{y}}^1)$ is the stable manifold associated, which is coincident with the local stable manifold associated to steady state $\bar{\mathbf{y}}^1$, $\mathcal{W}^s = \mathcal{W}^s_1 = \mathrm{Hom}(\bar{\mathbf{y}}^1)$, the subset of $\mathbf{Y}$ inside the curve $\mathrm{Hom}(\bar{\mathbf{y}}^1)$ is a center manifold, $\mathcal{W}^c$, and the remaining subset of $\mathbf{Y}$ is the unstable manifold.

We showed that homoclinic trajectories are generic, i.e., they exist for non-degenerate subset of values of the parameter a, in this case for $a \in \mathbb{R}/\{0\}$. This is not always the case as we show next.[9]

5.1.2.2 Non-Generic Case: Homoclinic or Saddle-Loop Bifurcation

Consider the non-linear ODE, depending on one parameter, $a \in \mathbb{R}$, that is studied, for instance, in Hale and Koçak (1991, p. 210) and Kuznetsov (2005, Chap. 6.2),[10]

---

[9] The Bendixson-Dulac criterium requires $\mathrm{tr}\,(J(\mathbf{y})) = 0$ for for any $\mathbf{y} \in \mathbf{Y}$.

[10] It is a particular case of equation (5.3) in which $\mathbf{A}_0 = \mathbf{0}_{(2\times1)}$, $\mathbf{A}_1 = \left(\begin{smallmatrix} 0 & 1 \\ 1 & a \end{smallmatrix}\right)$, $\mathbf{A}_2 = \left(\begin{smallmatrix} 0 & 0 & 0 \\ -1 & 0 & 0 \end{smallmatrix}\right)$, and $\mathbf{A}_3 = \mathbf{0}_{(2\times4)}$. Furthermore, $\mathrm{D}_a\mathbf{A}_0 = \mathbf{0}_{(2\times1)}$, $\mathrm{D}_a\mathbf{A}_1 = \left(\begin{smallmatrix} 0 & 0 \\ 0 & 1 \end{smallmatrix}\right) \neq \mathbf{0}_{(2\times2)}$, $\mathrm{D}_a\mathbf{A}_2 = \mathbf{0}_{(2\times3)}$, and $\mathrm{D}_a\mathbf{A}_3 = \mathbf{0}_{(2\times4)}$.

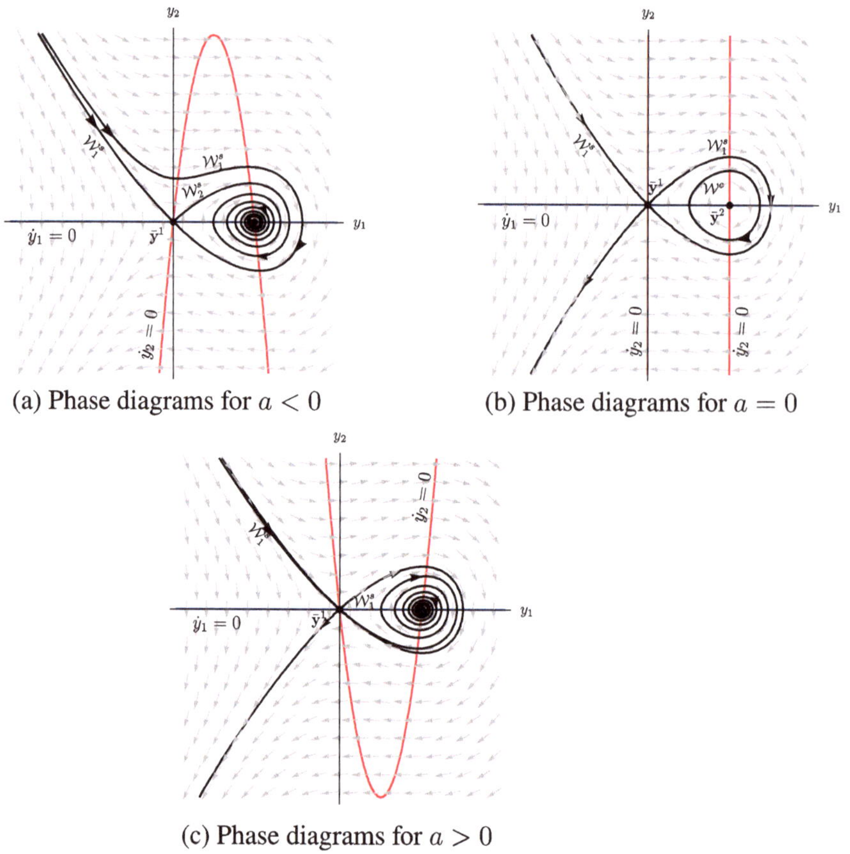

(a) Phase diagrams for $a < 0$

(b) Phase diagrams for $a = 0$

(c) Phase diagrams for $a > 0$

Fig. 5.4 Phase diagrams for Eq. (5.7)

$$\dot{y}_1 = y_2, \tag{5.7a}$$

$$\dot{y}_2 = y_1 + a\, y_2 - y_1^2. \tag{5.7b}$$

The Jacobian, for any $\mathbf{y}$, is $\mathbf{J}(\mathbf{y}) = \mathrm{D}_{\mathbf{y}}\mathbf{f}(\mathbf{y}) = \begin{pmatrix} 0 & 1 \\ 1-2y_1 & a \end{pmatrix}$, has $\mathrm{tr}(\mathbf{J}(\mathbf{y})) = a$, $\det(\mathbf{J}(\mathbf{y})) = 2\,a\,y_1 - 1$, and discriminant $\Delta(\mathbf{J}(\mathbf{y})) = \left(\frac{a}{2}\right)^2 + 1 - 2y_1$.

For any value of a there are two steady states $\bar{\mathbf{y}}^1 = (0, 0)$ and $\bar{\mathbf{y}}^2 = (1, 0)$. While steady state $\bar{\mathbf{y}}^1$ is always a saddle point, the local dynamic properties of steady state $\bar{\mathbf{y}}^2$ depend on the sign of a. As $\lambda_{\pm}(\bar{\mathbf{y}}^2) = \frac{a}{2} \pm \sqrt{\left(\frac{a}{2}\right)^2 - 1}$, then $\bar{\mathbf{y}}^2$ is a stable node if $a < 0$, it is a unstable node if $a > 0$, or it is a centre if $a = 0$.

If $a = 0$ then an homoclinic loop also exists. To prove this, we introduce the function $L(\mathbf{y}) = -\frac{y_1^2}{2} + \frac{y_2^2}{2} + \frac{y_1^3}{3}$. Taking time derivatives, and substituting ODE (5.7a) and (5.7b) yields

$$\dot{L} = (y_1 - 1)y_1\dot{y}_1 + y_2\dot{y}_2 = (y_1 - 1)y_1y_2 + y_2y_1(1 - y_1) + ay_2^2 = ay_2^2,$$

implying that $L(\mathbf{y})$ is a Lyapunov function if and only if $a = 0$. Then we have a homoclinic loop, connecting steady state $\bar{\mathbf{y}}^1$ with itself, $\text{Hom}(\bar{\mathbf{y}}^1) = \{\mathbf{y} \in \mathbf{Y} : L(\mathbf{y}) = 0, \text{ and } a = 0\}$, only for a specific value of parameter a, $a = 0$, and not for subsets of positive measure of the parameter space, as in ODE (5.6).

Figure 5.4 shows the phase diagrams for $a < 0$, $a = 0$ and $a > 0$, in which there is a non-generic homoclinic loop and generic heteroclinic orbits. In phase diagrams Fig. 5.4a and c , for $a < 0$ and $a > 0$, respectively, there is a saddle point and a stable focus, and a heteroclinic orbit connecting the two steady states. In phase diagram Fig. 5.4b , for $a = 0$, there is a saddle point, and an infinite number of centres surrounded by a homoclinic loop.

5.1.3 *Limit Cycles: Andronov-Hopf Bifurcation*

Consider the ODE, depending on the parameter $\lambda \in \mathbb{R}$, as studied, for instance, in Hale and Koçak (1991, p. 212) and Kuznetsov (2005, Chap. 3.4),[11]

$$\dot{y}_1 = y_2 + y_1(a - y_1^2 - y_2^2), \tag{5.8a}$$

$$\dot{y}_2 = -y_1 + y_2(a - y_1^2 - y_2^2). \tag{5.8b}$$

This ODE has a unique steady state $\bar{\mathbf{y}} = \mathbf{0}$, for any value of a. However, there is another invariant curve, a limit cycle, and both are elements of the $\Omega_{\mathbf{Y}}$ set. The Jacobian, of the vector field of (5.8), for any $\mathbf{y} \in \mathbf{Y}$,

$$\mathbf{J}(\mathbf{y}) = \begin{pmatrix} a - 3y_1^2 - y_2^2 & 1 - 2y_1y_2 \\ -1 - 2y_1y_2 & a - y_1^2 - 3y_2^2 \end{pmatrix}, \tag{5.9}$$

has $\text{tr}(\mathbf{J}(\mathbf{y})) = 2\left(a - 2\left(y_1^2 + y_2^2\right)\right)$ and $\Delta(\mathbf{J}(\mathbf{y})) = (y_1^2 + y_2^2 + 1)\,(y_1^2 + y_2^2 - 1)$. Therefore, the eigenvalues, for any pair $(\mathbf{y}, a)$, are $\lambda_\pm(\mathbf{y}, a) = a - 2(y_1^2 + y_2^2) \pm \sqrt{(y_1^2 + y_2^2 + 1)\,(y_1^2 + y_2^2 - 1)}$. The eigenvalues evaluated at the steady state, for any a, are complex conjugate, having the form $\lambda_\pm(\mathbf{0}, a) = a \pm \sqrt{-1} = a \pm i$, where $i = \sqrt{-1}$, with the real part equal to a. Therefore, if $a = 0$, the steady state is locally a center $\lambda_\pm(\mathbf{0}, 0) = \pm i$. This case separates cases in which the steady state is a stable focus (when $a < 0$) from cases in which it is an unstable focus (when $a > 0$). Therefore, at point $(\mathbf{y}, a) = (\mathbf{0}, 0)$, we say there is an **Andronov-Hopf bifurcation** point: $\{(\mathbf{y}, a) : \mathbf{f}(\mathbf{y}, a) = \mathbf{0}, \text{tr}(J(\mathbf{y})) = 0,\ \Delta(J(\mathbf{y})) < 0\}$. This is a co-dimension

---

[11] It is a particular case of ODE (5.3) with $\mathbf{A}_0 = \mathbf{0}_{(2\times1)}$, $\mathbf{A}_1 = \begin{pmatrix} \lambda & 1 \\ -1 & \lambda \end{pmatrix}$, $\mathbf{A}_2 = \mathbf{0}_{(2\times3)}$, and $\mathbf{A}_3 = \begin{pmatrix} -1 & 0 & -1 & 0 \\ 0 & -1 & 0 & -1 \end{pmatrix}$. Furthermore, $\text{D}_\lambda\mathbf{A}_0 = \mathbf{0}_{(2\times1)}$, $\text{D}_\lambda\mathbf{A}_1 = \begin{pmatrix} 1 & 0 \\ 0 & 1 \end{pmatrix} \neq \mathbf{0}_{(2\times2)}$, $\text{D}_\lambda\mathbf{A}_2 = \mathbf{0}_{(2\times3)}$, and $\text{D}_\lambda\mathbf{A}_3 = \mathbf{0}_{(2\times4)}$.

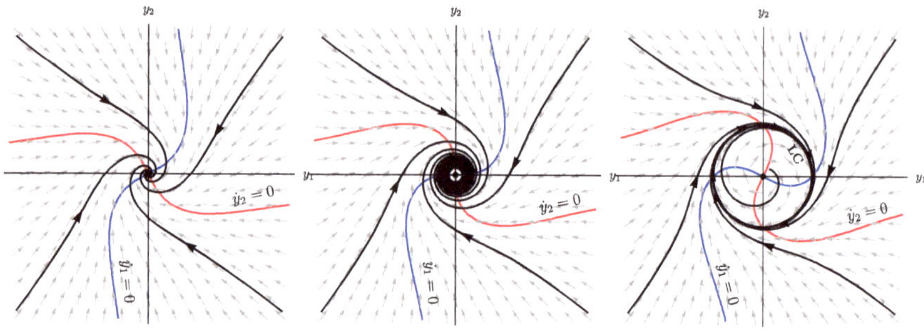

(a) Phase diagrams for $a < 0$ (b) Phase diagrams for $a = 0$ (c) Phase diagrams for $a > 0$

Fig. 5.5 Phase diagrams for ODE (5.8)

one bifurcation that can only exist in non-linear higher-dimensional ODE. Figure 5.5 shows the phase diagrams for the three previous cases: Fig. 5.5a for $a < 0$, Fig. 5.5b for $a = 0$, and Fig. 5.5c for $a > 0$.

In the phase diagram Fig. 5.5c , for $a > 0$, although the steady state is an unstable focus, there is a very slow convergence following a periodic orbit that seems analogous to an isolated center. This is another generic invariant, also having a global nature, that is called a **limit circle**. To prove analytically that it exists we can apply the **Bendixson-Dulac criterium** (see Theorem 5.3): if there are elements $\mathbf{y} \in \mathbf{Y}$ such that $\text{tr}(J(\mathbf{y}))$ changes sign then a limit cycle exists such that it attracts neighbouring trajectories. For ODE (5.8), as $\text{tr}J(\mathbf{y}) = 2\left(a - 2\left(y_1^2 + y_2^2\right)\right)$, then, if $a > 0$, there is a set of points in $\mathbf{Y}$ belonging to the**limit-cycle set**

$$LC = \left\{\mathbf{y} \in \mathbf{Y} : y_1^2 + y_2^2 = \frac{a}{2}\right\},$$

that represents the limit cycle. To prove that there is a closed curve that is a solution to the ODE, we transform the ODE (5.8) from cartesian coordinates, $\mathbf{y}$, into polar coordinates (r, θ)[12] to obtain

$$\dot{r} = r(a - r^2), \text{ and } \dot{\theta} = -1.$$

As $\dot{r} = 0$ for $r^2 = a$, then there is a periodic orbit with radius $\bar{r} = \sqrt{a}$ that is the equivalent to curve LC in polar coordinates. Therefore, the diameter of the limit cycle increases with a, and collapses to a point—a center—when $a \to 0$, consistently with our previous local analysis.

---

[12] See Appendix 4.7. We define $r^2 = y_1^2 + y_2^2$ and $\theta = \arctan \frac{y_2}{y_1}$, and take time derivatives to get $\dot{r} = \frac{y_1\dot{y}_1 + y_2\dot{y}_2}{r}$ and $\dot{\theta} = \frac{y_1\dot{y}_2 - y_2\dot{y}_1}{r^2}$.

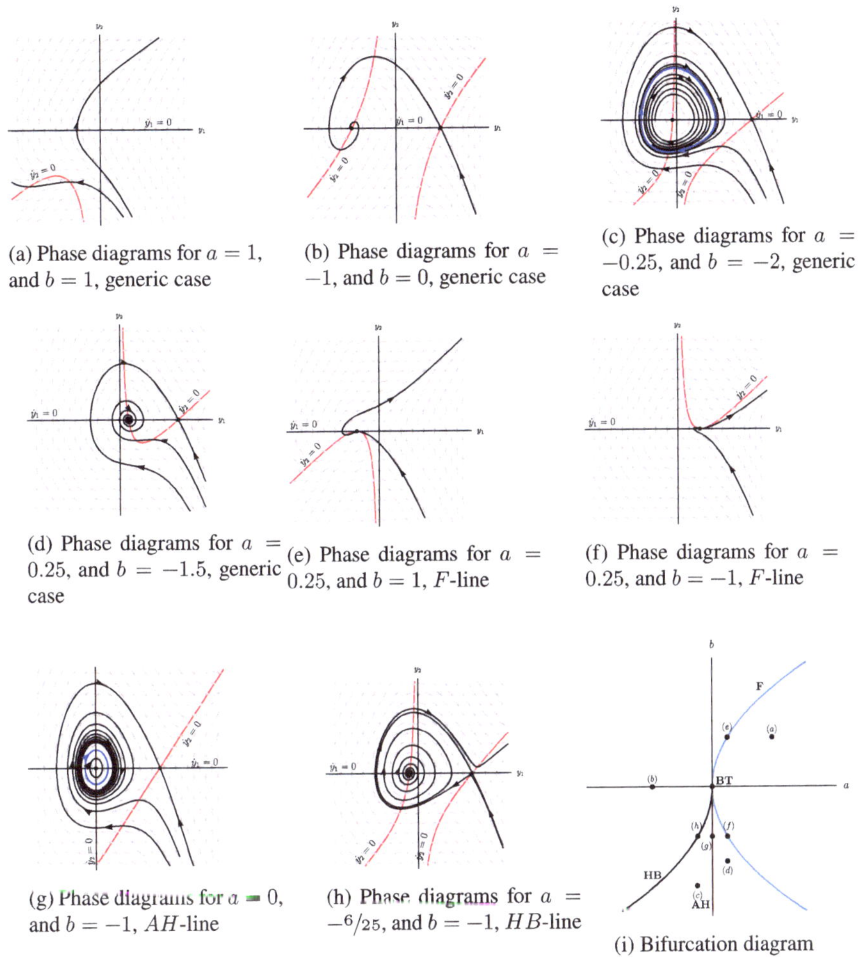

(a) Phase diagrams for $a = 1$, and $b = 1$, generic case

(b) Phase diagrams for $a = -1$, and $b = 0$, generic case

(c) Phase diagrams for $a = -0.25$, and $b = -2$, generic case

(d) Phase diagrams for $a = 0.25$, and $b = -1.5$, generic case

(e) Phase diagrams for $a = 0.25$, and $b = 1$, F-line

(f) Phase diagrams for $a = 0.25$, and $b = -1$, F-line

(g) Phase diagrams for $a = 0$, and $b = -1$, AH-line

(h) Phase diagrams for $a = -6/25$, and $b = -1$, HB-line

(i) Bifurcation diagram

Fig. 5.6 Phase diagrams for ODE (5.10). Lines F, AH, HB refer to saddle-node, Andronov-Hopf, and homoclinic bifurcations, and BT refers to the Bogdanov-Takens bifurcation point. Points (a) to (h), in subfigure (i), refer to the different phase diagrams

5.1.4 The Bogdanov-Takens Co-dimension Two Bifurcation

Until this subsection, we have only considered both generic and non-generic ODE orbits—in this chapter coincident with bifurcations[13]—that were unfolded by changes in one parameter. Now we assume that there are two bifurcating parameters.

---

[13] Recall the distinction between non-generic ODE and bifurcations for the planar linear ODE in Chap. 4.

The normal form of the co-dimension two Bogdanov-Takens bifurcation is[14]

$$\dot{y}_1 = y_2, \tag{5.10a}$$

$$\dot{y}_2 = a + b\, y_1 + y_1^2 - \, y_1\, y_2. \tag{5.10b}$$

where $\varphi \equiv (a, b) \in \mathbb{R}^2$. The Jacobian, evaluated at any $\mathbf{y} \in \mathbf{Y}$ is

$$\mathbf{J}(\mathbf{y}) = \begin{pmatrix} 0 & 1 \\ b + 2\, y_1 - \, y_2 & -y_1 \end{pmatrix}, \tag{5.11}$$

and has $\text{tr}(\mathbf{J}(\mathbf{y})) = -y_1$, $\det(\mathbf{J}(\mathbf{y})) = -b - 2\, y_1 + \, y_2$, and $\Delta(\mathbf{J}(\mathbf{y})) = \frac{y_1^2}{4} + b + 2\, y_1 - \, y_2$. There is a co-dimension two bifurcation at point at point $(\mathbf{y}, a, b) = (\mathbf{0}, 0, 0)$, belonging to the set, $\{(\mathbf{y}, \varphi) : \mathbf{f}(\mathbf{y}, \varphi) = \mathbf{0},\ \text{tr}(\mathbf{J}(\mathbf{y})) = \det(\mathbf{J}(\mathbf{y})) = 0\}$,—see Fig. 5.6 for the bifurcation diagram, and, in particular, point **BT** in Fig. 5.6i. This is the reason this is also called **double-zero bifurcation**. Figure 5.6i also shows other types of bifurcations: local bifurcations of co-dimension one, as fold **F** and Andronov-Hopf **AH**, and a global homoclinic bifurcation **HB**. There are two types of asymptotic invariants, steady states and limit cycles—as $\text{tr}(\mathbf{J}(\mathbf{y}))$ can take any sign for $\mathbf{y} \in \mathbf{Y}$, then periodic trajectories can exist according to the Bendixson-Dulac criterium (see Fig. 5.6c). We will see next under which conditions a limit cycle exists.

It is convenient to inspect the existence and number of steady states. It is easy to see that: if $a > \left(\frac{b}{2}\right)^2 > 0$ there are no steady states (see Fig. 5.6a), if $a = \left(\frac{b}{2}\right)^2 > 0$ there is one unique steady state, $\bar{\mathbf{y}}^f = \left(-\frac{b}{2}, 0\right)$; or if $a\left(\frac{b}{2}\right)^2$ there are two steady states, $\mathbf{y}^{ss} = \{\bar{\mathbf{y}}^1, \bar{\mathbf{y}}^2\} = \{(y_1^1, y_2^1), (y_1^2, y_2^2)\} = \left\{\left(-\frac{b}{2} + \sqrt{\Delta}, 0\right), \left(-\frac{b}{2} - \sqrt{\Delta}, 0\right)\right\}$ for $\Delta \equiv \left(\frac{b}{2}\right)^2 - a > 0$.

As $\text{tr}(\mathbf{J}(\bar{\mathbf{y}}^f)) = \frac{b}{2}$ and $\det(\mathbf{J}(\bar{\mathbf{y}}^f)) = 0$ then steady state $\bar{\mathbf{y}}^f$ is an unstable saddle-node if $b > 0$ (see Fig. 5.6e for $b = 1$), or a stable saddle-node if $b < 0$ (see Fig. 5.6f for $b = -1$). For the multiple steady state cases, we have: as $\text{tr}(\mathbf{J}(\bar{\mathbf{y}}^1)) = \frac{b}{2} - \sqrt{\Delta}$ and $\det(\mathbf{J}(\bar{\mathbf{y}}^1)) = -2\sqrt{\Delta} < 0$ then steady state $\bar{\mathbf{y}}^1$ is always a saddle point, and, as $\text{tr}(\mathbf{J}(\bar{\mathbf{y}}^2)) = \frac{b}{2} + \sqrt{\Delta}$ and $\det(\mathbf{J}(\bar{\mathbf{y}}^2)) = 2\sqrt{\Delta} > 0$, steady state $\bar{\mathbf{y}}^2$ can be a center if $\frac{b}{2} + \sqrt{\Delta} = 0$, or can be asymptotically stable if $\frac{b}{2} + \sqrt{\Delta} < 0$ or unstable if $\frac{b}{2} + \sqrt{\Delta} > 0$. In particular, in Fig. 5.6b, c, and h it is an unstable focus, in Fig. 5.6g, it is a center, and in Fig. 5.6d, it is a stable focus.[15]

Therefore, the fold bifurcation points are

$$(\bar{\mathbf{y}}^f, \varphi^f) = \{(\mathbf{y}, \varphi) : \ \mathbf{f}(\mathbf{y}, \varphi) = \mathbf{0}, \ \text{and} \ \det(\mathbf{J}(\mathbf{y}, \varphi)) = 0\},$$

---

[14] See Kuznetsov (2005, p. 311). This is a particular case of equation (5.3) in which $\mathbf{A}_0 = \left(\begin{smallmatrix} 0 \\ a \end{smallmatrix}\right)$, $\mathbf{A}_1 = \left(\begin{smallmatrix} 0 & 1 \\ b & 0 \end{smallmatrix}\right)$, $\mathbf{A}_2 = \left(\begin{smallmatrix} 0 & 0 & 0 \\ 1 & -1 & 0 \end{smallmatrix}\right)$, and $\mathbf{A}_3 = \mathbf{0}_{(2\times 4)}$. Furthermore, $\text{D}_a\mathbf{A}_0 = \left(\begin{smallmatrix} 0 \\ 1 \end{smallmatrix}\right) \neq \mathbf{0}_{(2\times 1)}$, $\text{D}_a\mathbf{A}_1 = \mathbf{0}_{(2\times 2)}$, $\text{D}_a\mathbf{A}_2 = \text{D}_b\mathbf{A}_2 = \mathbf{0}_{(2\times 3)}$, $\text{D}_a\mathbf{A}_3 = \text{D}_b\mathbf{A}_3 = \mathbf{0}_{(2\times 4)}$, and $\text{D}_b\mathbf{A}_0 = \mathbf{0}_{(2\times 1)}$, and $\text{D}_b\mathbf{A}_1 = \left(\begin{smallmatrix} 0 & 0 \\ 1 & 0 \end{smallmatrix}\right) \neq \mathbf{0}_{(2\times 2)}$.

[15] There are non-generic cases if $\left(\frac{b}{2}\right)^2 - \frac{a}{2} + \left(\frac{b}{2} - 4\right)\sqrt{\Delta} = 0$ for which $\bar{\mathbf{y}}^2$ is a node with multiplicity.

where $\varphi^f \in \mathbf{F} = \left\{(a, b) : a - \left(\frac{b}{2}\right)^2 = 0\right\}$. The fold bifurcation curve **F** is illustrated in phase diagram Fig. 5.6e , unfolding phase diagram Fig. 5.6a (no steady state) into phase diagram Fig. 5.6b (one saddle point and one unstable focus), and in phase diagram Fig. 5.6f unfolding phase diagram Fig. 5.6a (no steady state) into phase diagram Fig. 5.6d (one saddle point and one stable focus).

The Andronov-Hopf bifurcation points $(\bar{\mathbf{y}}^h, \varphi^h) = \{(\bar{\mathbf{y}}, \varphi) : \mathbf{f}(\mathbf{y}, \varphi) = \mathbf{0}, \ \text{tr}(\mathbf{J}(\mathbf{y})) = 0, \text{ and } \det(\mathbf{J}(\mathbf{y}, \varphi)) > 0\}$ correspond to the (multiple) steady state $\bar{\mathbf{y}}^2 = \mathbf{0}$ if $\varphi^h \in \mathbf{AH} = \left\{(a, b) : a = 0, \ b < 0\right\}$. The Andronov-Hopf bifurcation curve **AH** is illustrated in phase diagram Fig. 5.6g , unfolding phase diagram Fig. 5.6d (where there is a stable focus) into phase diagram Fig. 5.6c (where there is an unstable focus).

From a global dynamics perspective, when there are two steady states, several types of generic phase diagrams can occur. First, when steady state $\bar{\mathbf{y}}^1$ is a saddle point and steady state $\bar{\mathbf{y}}^2$ is asymptotically stable (focus or a node) there are no heteroclinic or homoclinic orbits. This is the case of phase diagram Fig. 5.6d in which the stable manifold of steady state $\bar{\mathbf{y}}^1$, $\mathcal{W}_1^s$, is the boundary of the basin of attraction of steady state $\bar{\mathbf{y}}^2$, $\mathcal{W}_2^s$, and the stable manifold in the union of the two stable manidolds, $\mathcal{W}^s = \mathcal{W}_1^s \cup \mathcal{W}_2^s$. Second, if steady state $\bar{\mathbf{y}}^1$ is a saddle point and steady state $\bar{\mathbf{y}}^2$ is unstable (focus or a node) and a generic heteroclinic orbit (saddle-node or saddle-focus connections) connecting the two steady states exist, as shown in Fig. 5.6b. The stable manifold $\mathcal{W}_1^s$ has a non-empty intersection with the unstable manifold $\mathcal{W}_2^u$ and $\mathcal{W}^s = \mathcal{W}_1^s$. Third, if the intersection between the two previous manifolds is empty, then the unstable manifold $\mathcal{W}_2^u$ is also bordered by a limit cycle (as in Fig. 5.6c). In this case, the stable manifold $\mathcal{W}_1^s$ is at the boundary of the basin of attraction of the limit cycle, say $\mathcal{W}_{LC}^s$ and $\mathcal{W}^s = \mathcal{W}_1^s \cup \mathcal{W}_{LC}^s$.

By continuity, there should be a transition, of a global nature and by non-generic phase diagrams, between the former two types of generic phase diagrams. First, between the first two types of phase diagrams, we find phase diagram Fig. 5.6g , in which $\bar{\mathbf{y}}^1$ is a saddle point and $\bar{\mathbf{y}}^2$ is a center and there is an Andronov-Hopf bifurcation: there is an infinite number of centers bounded by a limit cycle—in this case, the stable manifold of the limit cycle is bounded only by $\mathcal{W}_1^s = \mathcal{W}^s$ and inside the limit cycle there is a center manifold, $\mathcal{W}^c$. Second, between the previous second and third generic phase diagrams, we find phase diagram Fig. 5.6h , in which $\bar{\mathbf{y}}^1$ is a saddle point and $\bar{\mathbf{y}}^2$ is an unstable focus, and a limit cycle does not exist. This is only possible if there is a non-generic homoclinic orbit connecting steady state $\bar{\mathbf{y}}^1$ with itself, as in Fig. 5.6h. As this is a non-generic case, we have a homoclinic bifurcation. Although it is not possible to determine it analytically, Kuznetsov (2023, p. 359) provides an approximation to the bifurcation paramaters, $\mathbf{HB} = \left\{(a, b) : a + \frac{6}{25} b^2, \ b < 0\right\}$, that we use in that figure.

Exercise 5.2 Let $\dot{y}_1 = y_2$ and $\dot{y}_2 = a + b\, y_1 + y_1^2 + y_1\, y_2$. Show that it is another normal form for a Bogdanov-Takens bifurcation.

5.2 Qualitative Theory of ODE

Using the concepts and results introduced in Sect. 3.2, for the scalar non-linear ODE, and in Sect. 4.2.1 for planar linear ODE, we briefly introduce the qualitative theory of the dynamics of ODE for the planar non-linear regular ODE (5.2a)–(5.2b). Recall that the qualitative theory of ODE consists in finding a (topological) equivalence between a non-linear (or even incompletely specified) function $\mathbf{f}(.)$, and a normal form ODE that can be linear or non-linear. Whenever there is equivalence to a linear ODE, Sect. 4.2.1 reported a complete taxonomy of the phase diagrams. It is useful to recall several facts concerning linear planar ODE: first, they have zero, one or an infinite number of steady states; second, orbits can converge to a steady state, can be asymptotically unbounded or (in the case of a center) can be periodic; third, bifurcations relate to the change in the dimension of the stable and unstable manifolds, or the number of steady states, or to the number of periodic orbits between zero, one or infinity; and, fourth, the stable, unstable or center manifolds have a global nature. If the ODE is non-linear (and regular), we can expect to obtain the same type of dynamics. However, several different types of phenomena can be present: first, the limit set can also contain limit cycles in addition to steady states; second, there can be a finite number of steady states larger than one, having the same of different local dynamic properties; third, other invariant orbits of a global nature can exist, as homoclinic or heteroclinic orbits; fourth, local bifurcations similar to the linear case can exist, but several new types of local bifurcations can exist as well, related to the change in the number of steady states or the emergence of limit cycles; fifth, several new types of global bifurcations can also be present—as homoclinic or heteroclinic bifurcations; and, at last, non-linear ODE introduce a distinction between local and global stable, unstable, and center manifolds.

Put simply, in non-linear ODE, the local dynamics analysis can provide a complete qualitative characterization of some types of non-linear ODE, or can provide a partial characterization of the dynamics. However, if the vector field $\mathbf{f}(\cdot)$ is topologically equivalent to one of the non-linear normal forms we presented in Sect. 5.1, neglecting global dynamics phenomena not present in linear ODE can lead to a misleading characterization of the dynamics.[16] Next, we provide a short review of the main concepts and results from the qualitative theory of ODE.

5.2.1 *Local Analysis*

Local dynamics characterization of the non-linear regular ODE (5.2) close to a steady state or a periodic orbit draws on the Grobman-Hartmann, the manifold, and the Poincaré-Bendixson theorems. The first two theorems involve using the properties of the solution of an approximated linear ODE to study the local properties close to

[16] As most economic models are non-linear being aware of this fact is crucial for correctly solving, characterizing, or understanding simulations of quantitative macroeconomic models.

a steady-state for a non-linear ODE, and the third introduces a criterium for finding periodic orbits.

Equivalence with linear ODE. Assume there is (at least) one steady state, $\bar{\mathbf{y}} = \bar{\mathbf{y}}(\varphi_0) \in \mathbf{y}^{ss} = \{\mathbf{y} \in \mathbf{Y} \subseteq \mathbb{R}^2 : \mathbf{f}(\mathbf{y}, \varphi_0) = \mathbf{0},\ \varphi = \varphi_0\}$, for a specific parameter value $\varphi_0 \in \Phi$. The Jacobian of $\mathbf{f}(.)$ evaluated at that steady state is matrix

$$\mathbf{J}(\bar{\mathbf{y}}) = \mathrm{D}_{\mathbf{y}}\mathbf{f}(\bar{\mathbf{y}}) = \begin{pmatrix} \frac{\partial f_1(\bar{\mathbf{y}})}{\partial y_1} & \frac{\partial f_1(\bar{\mathbf{y}})}{\partial y_2} \\ \frac{\partial f_2(\bar{\mathbf{y}})}{\partial y_1} & \frac{\partial f_2(\bar{\mathbf{y}})}{\partial y_2} \end{pmatrix}. \tag{5.12}$$

The local characterization of steady states is made by analyzing the spectrum of Jacobian matrix $\mathbf{J}(\bar{\mathbf{y}})$, $\sigma(\mathbf{J}(\bar{\mathbf{y}})) = \{\lambda \in \mathbb{C}^2 : \det(\mathbf{J}(\bar{\mathbf{y}}) - \lambda\,\mathbf{I}) = 0\}$, that is, the eigenvalues

$$\bar{\lambda}_{\pm} = \frac{\mathrm{tr}(\mathbf{J}(\bar{\mathbf{y}}))}{2} \pm \sqrt{\Delta(\mathbf{J}(\bar{\mathbf{y}}))},\ \text{ for } \Delta(\mathbf{J}(\bar{\mathbf{y}})) = \left(\frac{\mathrm{tr}(\mathbf{J}(\bar{\mathbf{y}}))}{2}\right)^2 - \det(\mathbf{J}(\bar{\mathbf{y}})).$$

A steady state is **non-hyperbolic** if the Jacobian has at least one eigenvalue with zero real part. Therefore, $\bar{\mathbf{y}}$ is non-hyperbolic if

$$\det(\mathbf{J}(\bar{\mathbf{y}})) = 0,\ \text{ or } \mathrm{tr}(\mathbf{J}(\bar{\mathbf{y}})) = 0 \text{ and } \Delta(\mathbf{J}(\bar{\mathbf{y}})) < 0. \tag{5.13}$$

A steady state is **hyperbolic** if the Jacobian $\mathbf{J}(\bar{\mathbf{y}})$ has no eigenvalues with zero real parts. Therefore, $\bar{\mathbf{y}}$ is hyperbolic if condition (5.13) does not hold.

Theorem 5.1 (Grobman-Hartmann theorem) *Let $\bar{\mathbf{y}}$ be a hyperbolic steady state. Then there is a neighborhood U of $\bar{\mathbf{y}}$ and a neighborhood U_0 of $\mathbf{y}(0)$ such that the ODE restricted to U is topologically equivalent to the linear variational ODE*

$$\dot{\mathbf{y}} = \mathbf{J}(\bar{\mathbf{y}})(\mathbf{y} - \bar{\mathbf{y}}). \tag{5.14}$$

The original papers are Grobman (1959) and Hartman (1964).

The local (topological) equivalence of ODE (5.2) at a hyperbolic steady state to a linear planar ODE allows for using the facts from Chap. 4, and in particular Fig. 4.3 and Tables 4.3 and 4.4, to characterize its local dynamic properties. Therefore, if $\bar{\mathbf{y}}$ is a hyperbolic steady state, then: it is **locally asymptotically stable** (LAS) if all eigenvalues of $\mathbf{J}(\bar{\mathbf{y}})$ have negative real parts, it is **locally unstable** (LU) if all eigenvalues of $\mathbf{J}(\bar{\mathbf{y}})$ have positive real parts; it is a **local saddle point** (or **locally conditionally stable**) if $\mathbf{J}(\bar{\mathbf{y}})$ has a one positive and one negative eigenvalue; if a steady state is LAS or LU then we can classify it further as a **local node**, if all the eigenvalues of $\mathbf{J}(\bar{\mathbf{y}})$ are real numbers, or as a **local focus**, if they are complex conjugate.

Local manifolds. Let $U \subset \mathbf{Y}$ be a neighborhood of steady state $\bar{\mathbf{y}}^i$, for $i = 1, 2, \ldots$. The **local stable manifold** is the set $\mathcal{W}_i^{s,loc} = \{\mathbf{y} \in U :\ \lim_{t\to\infty} \mathbf{y}(t, \mathbf{y}) = \bar{\mathbf{y}}^i\}$, the **local unstable manifold** is set the $\mathcal{W}_i^{u,loc} = \{\mathbf{y} \in U :\ \lim_{t\to-\infty} \mathbf{y}(t, \mathbf{y}) = \bar{\mathbf{y}}^i\}$, and the **local center manifold** is set the $\mathcal{W}_i^{c,loc} = \{\mathbf{y} \in U :\ ||\mathbf{y}(t, \mathbf{y}) - \bar{\mathbf{y}}^i|| < \delta,\ \text{for each } t \in \mathrm{T}\}$, for a constant $\delta > 0$.

Let n_-, n_+ and n_0 denote the number of eigenvalues of the Jacobian evaluated at steady state $\bar{\mathbf{y}}$ with negative, positive and zero real parts.

Theorem 5.2 (Manifold Theorem) *Suppose there is a steady state $\bar{\mathbf{y}}^i$ and let $\mathbf{J}\bar{\mathbf{y}}^i$ be the Jacobian of the ODE (5.1) evaluated at that steady state. Then there are local stable, unstable and center manifolds, $\mathcal{W}_i^{s,loc}$, $\mathcal{W}_i^{u,loc}$ and $\mathcal{W}_i^{c,loc}$, of dimensions n_-, n_+ and n_0, respectively, such that $n_- + n_+ + n_0 = 2$.*[17] *The local manifolds are tangent to the local eigenspaces $\mathcal{E}^s$, $\mathcal{E}^u$, $\mathcal{E}^c$ of the (topologically) equivalent linearized ODE (5.14).*

The first two eigenspaces $\mathcal{E}^s$ and $\mathcal{E}^u$, are unique, and $\mathcal{E}^c$ need not be unique (see Grass et al. (2008, Chap. 2)). The eigenspaces are spanned by the eigenvectors of the Jacobian matrix $\mathbf{J}(\bar{\mathbf{y}})$ associated to the eigenvalues with negative, positive and zero real parts, respectively.

If the steady state $\bar{\mathbf{y}}$ is unique, and there are no homoclinic trajectories or limit cycles, then the local dynamic properties are also global, meaning that they are valid for the whole state space $\mathbf{Y}$ and not only for a neighborhood U of the steady state. The next example provides an illustration of Theorem 5.2.

Example 5.1 (*Global approximation of a non-linear ODE by a linear ODE*) Consider the non-linear planar ODE that has a non-linearity frequently arising in economic models,

$$\dot{y}_1 = y_1^a - y_2,\ 0 < a < 1, \tag{5.15a}$$

$$\dot{y}_2 = -y_1 + y_2, \tag{5.15b}$$

where $\mathbf{y} : \mathrm{T} \to \mathbf{Y}$, for $\mathrm{T} = \mathbb{R}_+$ and $\mathbf{Y} = \mathbb{R}_+$. It has an unique steady state $\bar{\mathbf{y}} = (\bar{y}_1, \bar{y}_2) = (1, 1)$ has the Jacobian evaluated at the steady state is $\mathbf{J}(\bar{\mathbf{y}}) = \mathrm{D}_{\mathbf{y}}\mathbf{f}(\bar{\mathbf{y}}) = \begin{pmatrix} a-1 & -1 \\ -1 & 1 \end{pmatrix}$. As $\det(\mathbf{J}(\bar{\mathbf{y}})) = a - 2 < 0$, then the steady state is hyperbolic and is locally a saddle point. The eigenvalues of $\mathbf{J}(\bar{\mathbf{y}})$ are $\lambda_\pm = \frac{a}{2} - \sqrt{\left(\frac{a}{2} - 1\right)^2 + 1}$. Therefore, the non-linearity in ODE (5.15) does not add qualitative properties different from those of the linear variational ODE $\dot{\mathbf{y}} = \mathbf{J}(\bar{\mathbf{y}})\,(\mathbf{y} - \bar{\mathbf{y}})$,

$$\dot{y}_1 = (a-1)(y_1 - 1) - (y_2 - 1), \tag{5.16a}$$

$$\dot{y}_2 = -(y_1 - 1) + (y_2 - 1). \tag{5.16b}$$

We already saw that the solution to this equation is $\mathbf{y}(t) = \mathbf{y} + \mathbf{P}e^{\mathbf{J}(\bar{\mathbf{y}})t}\mathbf{w}$, for each $t \in [0, \infty)$, where $\mathbf{P}$ is the eigenvector matrix associated to $\mathbf{J}(\bar{\mathbf{y}})$ and $\mathbf{w} = \mathbf{P}^{-1}(\mathbf{y}(0) - \bar{\mathbf{y}})$. Therefore, the approximate solution is

$$\begin{pmatrix} y_1(t) \\ y_2(t) \end{pmatrix} = \begin{pmatrix} 1 \\ 1 \end{pmatrix} + w_- \begin{pmatrix} 1 - \lambda_1 \\ 1 \end{pmatrix} e^{\lambda_- t} + w_+ \begin{pmatrix} 1 - \lambda_+ \\ 1 \end{pmatrix} e^{\lambda_+ t},$$

[17] These definitions extend to $n > 2$.

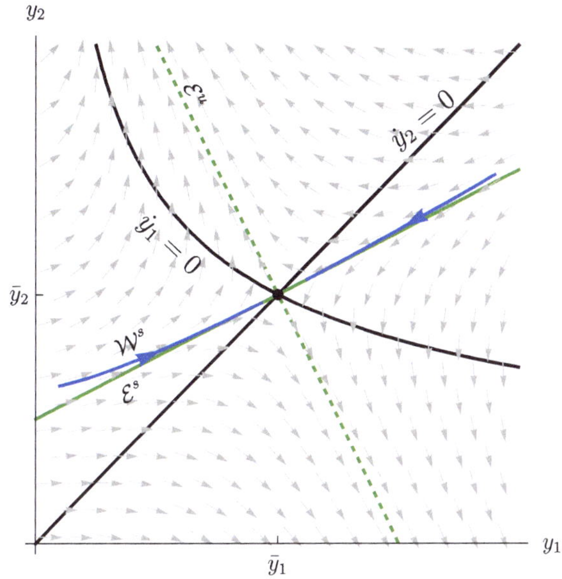

Fig. 5.7 Phase diagrams for ODE (5.15)

where $w_- = y_1(0) - (1 - \lambda_+)\, y_2(0) - \lambda_+$, and $w_+ = -y_1(0) + (1 - \lambda_-)\, y_2(0) + \lambda_-$. The stable and unstable eigenspaces are $\mathcal{E}^s = \{\mathbf{y} : -y_1 + (1 - \lambda_-)\, y_2 + \lambda_- = 0\}$ and $\mathcal{E}^u = \{\mathbf{y} : y_1 - (1 - \lambda_+)\, y_2 - \lambda_+ = 0\}$, respectively. The phase diagram is illustrated In Fig. 5.7. In this figure, we observe that the stable eigenspace is tangent to the stable manifold $\mathcal{W}^s$ at the steady state, and the unstable manifold is the whole set $\mathbf{Y}$ excluding $\mathcal{W}^s$, that is $\mathcal{W}^u = \mathbf{Y}/\mathcal{W}^s$. In such an example, the variational ODE (5.16) captures completely the qualitative properties of ODE (5.15).

Existence of periodic orbits. An infinity of periodic orbits can exist globally (in linear ODE) or locally (in non-linear ODE as we saw in Sect. 5.1.4). For non-linear ODE, the **Poincaré-Bendixson** theorem (see Hale & Koçak, 1991, p. 367) states that if the limit set is bounded, and it is not a steady state, it should be a periodic orbit. Therefore, differently from linear planar ODE, we can have periodic orbits that only emerge asymptotically. This is the case of limit cycles, that were already presented in Sects. 5.1.3 and 5.1.4. In those sections, we saw that limit cycles are trajectories for which the Lyapunov function $L(\mathbf{y}(t))$ is constant for each t, although we do not have a constructive method to finding it.

The Bendixson criterium provides a method for determining the existence of a periodic orbit in a compact subset of $\mathbf{Y}$ (see Hale & Koçak, 1991, p. 373):

Theorem 5.3 (Bendixson-Dulac criterium) *Let D be a compact region of $\mathbf{Y} \subseteq \mathbb{R}^2$. If*

$$div(\mathbf{f}(\mathbf{y})) = \mathrm{tr}\big(\mathrm{D}_{\mathbf{y}}(\mathbf{f}(\mathbf{y}))\big), \text{ has constant sign for all } \mathbf{y} \in D \tag{5.17}$$

*then $\dot{\mathbf{y}} = \mathbf{f}(\mathbf{y})$ does **not** display a constant orbit lying entirely within D.*

While local analysis consists in studying local dynamics in the neighborhood of steady states or periodic orbits, it may only provide an incomplete characterization of the dynamics for non-linear ODE.

5.2.2 Global Analysis

We can conceive global analysis as an approach to obtaining an approximation of the solution to a non-linear ODE (5.1) by drawing on an equivalence to a solution of a non-linear normal form ODE as (5.3). Assuming that the ODE has a steady state $\bar{\mathbf{y}}$, a variational ODE, using a higher-order Taylor approximation in a neighborhood that steady state, takes the form

$$\dot{\mathbf{y}} \approx D_{\mathbf{y}}\mathbf{f}(\bar{\mathbf{y}})\,(\mathbf{y}-\bar{\mathbf{y}}) + D_{\mathbf{y}^{\otimes 2}}\mathbf{f}(\bar{\mathbf{y}})(\mathbf{y}^{\otimes 2}-\bar{\mathbf{y}}^{\otimes 2}) + D_{\mathbf{y}^{\otimes 3}}\mathbf{f}(\bar{\mathbf{y}})(\mathbf{y}^{\otimes 3}-\bar{\mathbf{y}}^{\otimes 3}), \tag{5.18}$$

where $D_{\mathbf{y}}\mathbf{f}(\bar{\mathbf{y}})$ was defined in Eq. (5.12) and

$$D_{\mathbf{y}^{\otimes 2}}\mathbf{f}(\bar{\mathbf{y}}) = \begin{pmatrix} \frac{\partial^2 f_1(\bar{\mathbf{y}})}{\partial y_1^2} & \frac{\partial^2 f_1(\bar{\mathbf{y}})}{\partial y_1 y_2} & \frac{\partial^2 f_1(\bar{\mathbf{y}})}{\partial y_2^2} \\ \frac{\partial^2 f_2(\bar{\mathbf{y}})}{\partial y_1^2} & \frac{\partial^2 f_2(\bar{\mathbf{y}})}{\partial y_1 y_2} & \frac{\partial^2 f_2(\bar{\mathbf{y}})}{\partial y_2^2} \end{pmatrix}, \quad D_{\mathbf{y}^{\otimes 3}}\mathbf{f}(\bar{\mathbf{y}}) = \begin{pmatrix} \frac{\partial^3 f_1(\bar{\mathbf{y}})}{\partial y_1^3} & \frac{\partial^3 f_1(\bar{\mathbf{y}})}{\partial y_1^2 y_2} & \frac{\partial^3 f_1(\bar{\mathbf{y}})}{\partial y_1 y_2^2} & \frac{\partial^3 f_1(\bar{\mathbf{y}})}{\partial y_2^3} \\ \frac{\partial^3 f_2(\bar{\mathbf{y}})}{\partial y_1^3} & \frac{\partial^3 f_2(\bar{\mathbf{y}})}{\partial y_1^2 y_2} & \frac{\partial^3 f_2(\bar{\mathbf{y}})}{\partial y_1 y_2^2} & \frac{\partial^3 f_2(\bar{\mathbf{y}})}{\partial y_2^3} \end{pmatrix}.$$

The next example clarifies the forms of those gradients:

Example 5.2 (*Taylor approximations to steady states in Andronov-Hopf and Bogdanov-Takens ODE*) For the Andronov-Hopf ODE (5.8), we have the Jacobian $D_{\mathbf{y}}\mathbf{f}(\mathbf{y})$ in Eq. (5.9) and the higher-order Hessean matrices, for any $\mathbf{y}$,

$$D_{\mathbf{y}^{\otimes 2}}\mathbf{f}(\mathbf{y}) = \mathbf{0}_{2\times 3}, \ D_{\mathbf{y}^{\otimes 3}}\mathbf{f}(\bar{\mathbf{y}}) = \begin{pmatrix} -6 & 0 & -2 & 0 \\ 0 & -2 & 0 & -6 \end{pmatrix}.$$

For the Bogdanov-Takens ODE (5.10), we have the Jacobian $D_{\mathbf{y}}\mathbf{f}(\mathbf{y})$ in Eq. (5.11) and the higher-order Hessean matrices, for any $\mathbf{y}$

$$D_{\mathbf{y}^{\otimes 2}}\mathbf{f}(\bar{\mathbf{y}}^i) = \begin{pmatrix} 0 & 0 & 0 \\ 1 & -1 & 0 \end{pmatrix}, \ D_{\mathbf{y}^{\otimes 3}}\mathbf{f}(\bar{\mathbf{y}}^i) = \mathbf{0}_{(2\times 4)} \text{ for } i = 1, 2.$$

Global type of dynamics. If higher order derivative matrices $D_{\mathbf{y}^{\otimes 2}}\mathbf{f}(\mathbf{y})$ and/or $D_{\mathbf{y}^{\otimes 3}}\mathbf{f}(\mathbf{y})$ are non-zero matrices, an ODE can display several new types of orbits featuring global dynamic properties that cannot be completely uncovered by linearization. In the previous section, we saw several possible cases: heteroclinic orbits linking two different steady stateshomoclinic orbits linking a steady state with itself, or, limit cycles as a new type of asymptotic invariant.

Local and global manifolds. Non-linear ODE can present a significant difference between local and global manifolds. Several examples were presented in Sect. 5.1.

First, if there are several steady states, such that $\mathbf{y}^{ss} = \{\bar{\mathbf{y}}^1, \ldots \bar{\mathbf{y}}^1, \ldots, \bar{\mathbf{y}}^{nss}\}$, with non-empty local stable manifolds, $\mathcal{W}_i^s = \mathcal{W}_i^{s,loc}$ for $i = 1, \ldots n_{ss}$, then the stable manifold belongs to the union of the local stable manifolds $\mathcal{W}^s \subseteq \cup_{i=1}^{n_{ss}} \mathcal{W}_i^{s,loc}$. Second, if there is one homoclinic loop, the stable manifold can be the intersection of the of local stable and unstable manifold of a steady state $\bar{\mathbf{y}}^i$, that is $\mathcal{W}^s = \mathcal{W}_i^{s,loc} \cap \mathcal{W}_i^{u,loc}$. Third, when there is a limit cycle (as in the ODE (5.8)) the global stable manifold can be coincident with the local stable manifold of the limit cycle. Fourth, if there are closed orbits, it is possible to have center manifolds of dimension two.

Qualitative approach: global bifurcation detection. The number of cases, and the methods to solve them[18], is not only too large to enumerate here but also they are usually based on numerical procedures. However, bifurcation theory provides a qualitative approach to addressing non-linear dynamics. This approach relies on the (topological) equivalence between a given non-linear ODE, and a known normal form, determined at a bifurcation point.

Consider the planar ODE, depending on the vector of parameters $\varphi \in \Phi$,

$$\dot{\mathbf{y}} = \mathbf{f}(\mathbf{y}, \varphi), \ \mathbf{y} \in \mathrm{Y} \subseteq \mathbb{R}^2,$$

and let us assume there is a bifurcation point $(\bar{\mathbf{y}}, \varphi_0)$ such that $\mathbf{f}(\bar{\mathbf{y}}, \varphi_0) = \mathbf{0}$ and some other bifurcation conditions are met. The Taylor expansion equation (5.18) now takes the form

$$\begin{aligned}\dot{\mathbf{y}} = \mathrm{D}_{\mathbf{y}}\mathbf{f}(\bar{\mathbf{y}}, \varphi_0)\,(\mathbf{y} - \bar{\mathbf{y}}) + \mathrm{D}_{\mathbf{y}^{\otimes 2}}\mathbf{f}(\bar{\mathbf{y}}, \varphi_0)(\mathbf{y}^{\otimes 2} - \bar{\mathbf{y}}^{\otimes 2}) + \mathrm{D}_{\mathbf{y}^{\otimes 3}}\mathbf{f}(\bar{\mathbf{y}}, \varphi_0)(\mathbf{y}^{\otimes 3} - \bar{\mathbf{y}}^{\otimes 3}) + \\ + \mathrm{D}_{\varphi}\mathbf{f}(\bar{\mathbf{y}}, \varphi_0)\,(\varphi - \varphi_0) + \mathrm{D}_{\varphi\otimes\mathbf{y}}\mathbf{f}(\bar{\mathbf{y}}, \varphi_0)\,(\varphi \otimes \mathbf{y} - \varphi_0 \otimes \bar{\mathbf{y}}), \quad (5.19)\end{aligned}$$

where the dimension of vector φ_0 is equal to co-dimension of the bifurcation.

In the previous section, we saw that there are two types local bifurcations of co-dimension one[19] associated to a value $\varphi_0 \in \mathbb{R}$ for a single parameter. The first type of bifurcations includes fold, transcritical, pitchfork bifurcations, if there is a bifurcation point $(\bar{\mathbf{y}}, \varphi_0)$, such that, the following conditions are satisfied

$$(\bar{\mathbf{y}}, \varphi_0) = \{(\mathbf{y}, \varphi) : \mathbf{f}(\mathbf{y}, \varphi) = \mathbf{0}, \ \text{and}\det(\mathrm{D}_{\mathbf{y}}\mathbf{f}(\mathbf{y}, \varphi)) = 0\}; \quad (5.20)$$

and the second type of bifurcations includes the Andronov-Hopf bifurcation, for which

$$(\bar{\mathbf{y}}, \varphi_0) = \{(\mathbf{y}, \varphi) : \mathbf{f}(\mathbf{y}, \varphi) = \mathbf{0}, \ \mathrm{tr}(\mathrm{D}_{\mathbf{y}}\mathbf{f}(\mathbf{y}, \varphi)) = 0, \ \text{and } \Delta\mathrm{D}_{\mathbf{y}}\mathbf{f}(\mathbf{y}, \varphi) < 0\}. \quad (5.21)$$

A co-dimension two bifurcation is associated to a vector of a two-dimensional parameter value $\varphi_0 = (\varphi_{1,0}, \varphi_{2,0}) \subset \mathbb{R}^2$, if there is a bifurcation point, $(\bar{\mathbf{y}}, \varphi_0)$, such that

[18] In the next Example 5.4, we present a simple method, reminiscent from the Ważewski (1947) approach, to prove the existence of a heteroclinic orbit.

[19] Compare with their counterparts for scalar ODE in Chap. 3.

$$(\bar{\mathbf{y}}, \varphi_0) = \{(\mathbf{y}, \varphi) : \mathbf{f}(\mathbf{y}, \varphi) = \mathbf{0},\ \mathrm{tr}(\mathrm{D}_{\mathbf{y}}\mathbf{f}(\mathbf{y}, \varphi)) = 0,\ \text{and}\ \det(\mathrm{D}_{\mathbf{y}}\mathbf{f}(\mathbf{y}, \varphi)) = 0\}, \tag{5.22}$$

as is the case for the Bogdanov-Takens bifurcation. Those conditions only provide necessary (and local) conditions for topological equivalence with non-linear normal forms. The next example provides an illustration of other higher-order conditions that are required to fully characterize bifurcations in non-linear regular ODE.

Example 5.3 (*Taylor approximations to bifurcation points in Andronov-Hopf and Bogdanov-takens ODE*) For Andronov-Hopf ODE (5.8), the bifurcation point is $(\bar{\mathbf{y}}, \lambda_0) = (\mathbf{0}, 0)$. The matrix derivatives, evaluated at that bifurcation point, are

$$\mathrm{D}_{\mathbf{y}}\mathbf{f} = \begin{pmatrix} 0 & 1 \\ -1 & 0 \end{pmatrix},\ \mathrm{D}_{\mathbf{y}^{\otimes 2}}\mathbf{f} = \mathbf{0}_{(2\times 3)},\ \mathrm{D}_{\mathbf{y}^{\otimes 3}}\mathbf{f} = \begin{pmatrix} -6 & 0 & -2 & 0 \\ 0 & -2 & 0 & -6 \end{pmatrix},$$

for the state variables, and for the parameters

$$\mathrm{D}_{\lambda}\mathbf{f} = \mathbf{0}_{(2\times 1)},\ \text{and}\ \mathrm{D}_{\mathbf{y}\otimes\lambda}\mathbf{f} = \begin{pmatrix} 1 & 0 \\ 0 & 1 \end{pmatrix}.$$

Observe that the eigenvalues of $\mathrm{D}_{\mathbf{y}}\mathbf{f}$ are complex conjugate with zero real parts, $\lambda_{\pm} = \pm i$. Therefore, an ODE satisfying the local conditions (5.21), and having $\mathrm{D}_{\mathbf{y}^{\otimes 2}}\mathbf{f}$ and $\mathrm{D}_{\varphi}\mathbf{f}$ equal to zero, and the other derivative matrices with the same structure (and in particular being different from zero), are topologically equivalent to the Andronov-Hopf ODE (5.8).

For Bogdanov-Takens ODE (5.10), the bifurcation point is $(\bar{\mathbf{y}}, \varphi_0) = \mathbf{0}_{(4\times 1)}$, where $\varphi = (a, b)$. The matrix derivatives, evaluated at that bifurcation point, are

$$\mathrm{D}_{\mathbf{y}}\mathbf{f} = \begin{pmatrix} 0 & 1 \\ 0 & 0 \end{pmatrix},\ \mathrm{D}_{\mathbf{y}^{\otimes 2}}\mathbf{f} = \begin{pmatrix} 0 & 0 & 0 \\ 1 & -1 & 0 \end{pmatrix},\ \mathrm{D}_{\mathbf{y}^{\otimes 3}}\mathbf{f} = \mathbf{0}_{(2\times 4)},$$

for the state variables, and for the parameters a and b they are[20]

$$\mathrm{D}_{\varphi}\mathbf{f} = \begin{pmatrix} 0 & 0 \\ 1 & 0 \end{pmatrix}\ \text{and}\ \mathrm{D}_{\mathbf{y}\otimes\varphi}\mathbf{f} = \begin{pmatrix} 0 & 0 & 0 & 0 \\ 0 & 0 & 1 & 0 \end{pmatrix}.$$

The eigenvalues of $\mathrm{D}_{\mathbf{y}}\mathbf{f}$ are both equal to zero. An ODE satisfying the local conditions (5.22), and having $\mathrm{D}_{\mathbf{y}^{\otimes 3}}\mathbf{f}$ equal to zero, and the other derivative matrices with the same structure, are topologically equivalent to the Bogdanov-Takens ODE (5.10).

[20] In this case, we have $\mathrm{D}_{\varphi}\mathbf{f} = \begin{pmatrix} \frac{\partial f_1}{\partial a} & \frac{\partial f_1}{\partial b} \\ \frac{\partial f_2}{\partial a} & \frac{\partial f_2}{\partial b} \end{pmatrix}$ and $\mathrm{D}_{\mathbf{y}\otimes\varphi}\mathbf{f} = \begin{pmatrix} \frac{\partial^2 f_1}{\partial y_1 \partial a} & \frac{\partial^2 f_1}{\partial y_1 \partial b} & \frac{\partial^2 f_1}{\partial y_2 \partial a} & \frac{\partial^2 f_1}{\partial y_2 \partial b} \\ \frac{\partial^2 f_2}{\partial y_1 \partial a} & \frac{\partial^2 f_2}{\partial y_1 \partial b} & \frac{\partial^2 f_2}{\partial y_2 \partial a} & \frac{\partial^2 f_2}{\partial y_2 \partial b} \end{pmatrix}$.

5.2.3 Comparative Dynamics in Economics

As already mentioned,[21] **comparative dynamics** exercises consist in introducing a perturbation to a parameter in a dynamic economic model, usually in the neighborhood of an hyperbolic steady state $\bar{y}_0 = \bar{y}(\varphi_0)$ associated to parameter value $\varphi = \varphi_0$. If we consider the effect of changing parameter φ from its initial value, φ_0, this causes the solution to the ODE to move away from $\bar{y}_0$. The dynamics of the perturbation can be approximated by the solution to the variational ODE

$$\dot{\mathbf{y}} = \mathrm{D}_{\mathbf{y}}\mathbf{f}(\bar{\mathbf{y}}_0, \varphi_0)(\mathbf{y} - \bar{\mathbf{y}}_0) + \mathrm{D}_{\varphi}\mathbf{f}(\bar{\mathbf{y}}_0, \varphi_0)(\varphi - \varphi_0). \tag{5.23}$$

Comparing it with the more general Taylor approximation (5.19), we observe that it involves a truncation of the higher order terms. Therefore, if there is only one steady state the truncation has not a qualitative effect on the analysis. However, if there are more than one steady state, and there are local or global bifurcations, the analysis should be conducted carefully by avoiding perturbations that "cross" bifurcation values of the parameter.

The variational ODE (5.23) can be solved as a linear ODE. Assuming a permanent variation in a parameter, from φ_0 to $\varphi = \varphi_0 + \delta_\varphi$, and writing the new steady state as $\bar{\mathbf{y}} = \bar{\mathbf{y}}(\varphi)$, the Jacobian

$$\mathrm{D}_{\varphi}\bar{\mathbf{y}}(\varphi_0) = \lim_{\delta_\varphi \to 0} \frac{\bar{\mathbf{y}}(\varphi_0 + \delta_\varphi) - \bar{\mathbf{y}}(\varphi_0)}{\delta_\varphi} = -\mathrm{D}_{\mathbf{y}}^{-1}\mathbf{f}(\bar{\mathbf{y}}_0, \varphi_0)\mathrm{D}_{\varphi}\mathbf{f}(\bar{\mathbf{y}}_0, \varphi_0)$$

is the vector of **long-run multipliers**. Solving the linearized system allows us to have a general solution to the problem of finding the **short-run** or **transition multipliers**, $d\mathbf{y}(t) \equiv \mathbf{y}(t) - \bar{\mathbf{y}}_0$ for a change in the parameter φ. The particular solution 4.3 depends on the type of problem and on the nature of the side conditions.

5.3 Applications to Economics

Example 5.4 (*The Ramsey-Cass-Koopmans model*) In Sect. 4.3, we briefly presented the Ramsey (1928) model, in the Cass (1965) and Koopmans (1965) version, or RCK model. It is a normative model of capital accumulation (but can also be taken as a positive model if its behavior fits the data) that seeks to maximize the present value of the utility of consumption, constrained by the macroeconomic balance constraint between savings and investment (and possibly other constraints as will be seen in Chap. 10).

The RCK model is a non-linear planar ODE system in the variables $(k, c) \in \mathbb{R}^2_+$, denoting the per-capita physical capital and consumption,

[21] In Sects. 2.3.3.2 and 3.2.2 for linear and non-linear scalar ODE, respectively, and in Sect. 4.3 for linear planar ODE.

$$\dot{k} = f(k) - c - \delta\, k, \tag{5.24a}$$
$$\dot{c} = \eta(c)\, c\, \left(f'(k) - \rho - \delta\right), \tag{5.24b}$$

where $\eta(c) = -\frac{u'(c)}{u''(c)\,c}$ is the elasticity of intertemporal substitution of consumption. Furthermore, it is assumed that k is a pre-determined variable, and c is a non-predetermined variable that should satisfy the initial condition $k(0) = k_0$ (given) and the transversality, condition $\lim_{t\to\infty} u'(c)k(t)e^{-\rho t} = 0$. The ODE (5.24), is a modified Hamiltonian dynamic system (MHDS)[22] in which Eq. (5.24a) represents an economic constraint between savings and capital accumulation, like in the Solow (1956) model, and Eq. (5.24b) represents an arbitrage relationship between consumption and capital accumulation (replacing the Keynesian consumption function that is postulated in the Solow model). In this general version, the two primitive functions, the production and the elasticity of intertemporal substitution functions $f(k)$ and $\eta(c)$, are not completely specified, but are assumed to have the following properties: (1) the elasticity of intertemporal substitution is constant and positive, $\eta(c) = \eta > 0$[23]; and (2) the production function is of the Inada type: it is positive for positive levels of capital, it is monotonically increasing and globally concave. Formally: $f(0) = 0$, $f(k) > 0$ for $k > 0$, $f'(k) > 0$, $\lim_{k\to 0} f'(k) = +\infty$, $\lim_{k\to+\infty} f'(k) = 0$, and $f''(k) < 0$ for all $k \in \mathbb{R}_+$.[24] It is further assumed that the rate of time preference is positive, $\rho > 0$, and the capital depreciation rate is non-negative $\delta \geq 0$.

Figure 5.8 depicts the phase diagram for the case in which $\delta > 0$. As we can see there are three steady states: two steady states satisfying $c = 0$, that is $(k, c) = \{(0, 0), (k', 0)\}$ where $k' = \{k : f(k) = \delta\, k\}$, and a steady state $(\bar{k}, \bar{c}) = \{(k, c) \in \mathbb{R}^2_+ : k = (f')^{-1}(\rho + \delta),\ c = f(k) - \delta\, k\}$. It can be proved that no trajectory converging to steady state $(k', 0)$ satisfies the transversality condition and therefore cannot be a solution to the RCK model. The solution to the RCK model is represented by the stable manifold associated to steady state $(\bar{k}, \bar{c})$. Therefore, phase diagrams illustrating the solutions to optimal control problems should be interpreted with care.

From now on we assume that $\delta = 0$ and study it in more detail. Given the smoothness of the vector field, i.e., of functions $f_1(k, c) \equiv f(k) - c$ and $f_2(k, c) \equiv \eta\, c\, (f'(k) - \rho)$, we know that a solution exists and it is unique. To characterize qualitatively, using the approach summarized in Sect. 5.2, we need to: (1) determine existence and number of steady states; (2) characterize their hyperbolicity and local dynamics by performing, if necessary, a local bifurcation analysis; (3) try to find other invariant trajectories of a global nature; and (4) conduct a comparative dynamics exercise.

---

[22] To be derived in Chap. 10.

[23] Recall from Chap. 2 that this assumption, together with the definition of function η implies that this is equivalent to constraining the utility function to be a power function, see Example 2.4.

[24] Observe that $f(k)$ is locally but not globally Lipschitz, i.e., a small change in k close to zero induces a large change in $f(k)$.

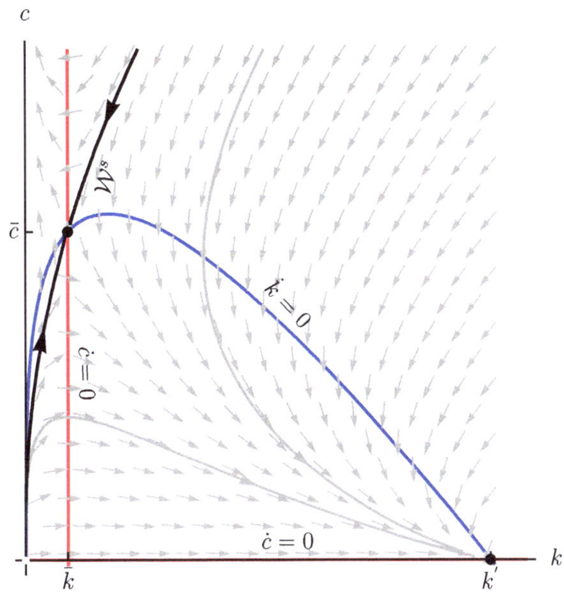

Fig. 5.8 Phase diagram for the RCK model with $\delta > 0$

Steady states and local dynamics. The steady state set is $(\bar{k}^{ss}, \bar{c}^{ss}) = \{(k, c) \in \mathbb{R}^2_+ : \dot{k} = \dot{c} = 0\} = \{(0, 0), (\bar{k}, \bar{c})\}$ where $\bar{k} = g(\rho)$, denoting $g(.) = (f')^{-1}(.)$, and $\bar{c} = f(\bar{k}) = f(g(\rho))$. To prove the existence and uniqueness of a positive steady state level for k we invoke the Inada and global concavity properties of the production function: first, $\dot{c} = 0$ if there is a value k that solves the equation $f'(k) = \rho$; second, because $\rho > 0$ is finite and $f'(k) \in (0, \infty)$ then there is at least one value for k that solves that equation; at last, because the function $f(.)$ is globally strictly concave then $f'(k)$ is monotonically decreasing which implies that the solution is unique.

The Jacobian of system (5.24a)–(5.24b) is

$$\mathrm{D}_{(k,c)}\mathbf{F}(k, c) = \begin{pmatrix} f'(k) & -1 \\ \eta c f''(k) & \eta\left(f'(k) - \rho\right) \end{pmatrix}. \tag{5.25}$$

Therefore: the eigenvalues of Jacobian $\mathrm{D}_{(k,c)}\mathbf{F}(0, 0)$ are $\lambda^0_s = \eta(f'(0) - \rho) = +\infty$, and $\lambda^0_u = f'(0) = +\infty$; and because $\mathrm{tr}(\mathrm{D}_{(k,c)}\mathbf{F}(\bar{k}, \bar{c})) = \rho > 0$ and $\det(\mathrm{D}_{(k,c)}F(\bar{k}, \bar{c})) = \eta\bar{c}f''(\bar{k}) < 0$ the eigenvalues of Jacobian $\mathrm{D}_{(k,c)}\mathbf{F}(\bar{k}, \bar{c})$ are

$$\bar{\lambda}_s = \frac{\rho}{2} - \left(\left(\frac{\rho}{2}\right)^2 - \eta\bar{c}f''(\bar{k})\right)^{\frac{1}{2}} < 0,\ \bar{\lambda}_u = \frac{\rho}{2} + \left(\left(\frac{\rho}{2}\right)^2 - \eta\bar{c}f''(\bar{k})\right)^{\frac{1}{2}} > 0.$$

Therefore, the steady state (0, 0) is unstable,[25] and the steady state $(\bar{k}, \bar{c})$ is a saddle point, and, as a consequence, hyperbolic. The intuition behind this property is transparent when we look at the expression for the determinant: the mechanism generating stability is related to the existence of decreasing marginal returns in production.[26] As the Jacobian (5.25), at every steady state, does not have eigenvalues with zero real parts, then there are no local bifurcation points, meaning that the local stable manifold for steady state $(\bar{k}, \bar{c})$ of ODE (5.24a)–(5.24),

$$\mathcal{W}^{s,loc} = \{(k, c) \in N : \lim_{t\to\infty} (k(t), c(t)) = (\bar{k}, \bar{c})\}$$

for $N = \{(k, c) \in \mathbb{R}^2_+ : ||(k, c) - (\bar{k}, \bar{c})|| < \delta\}$ for a small δ, has a global nature.

From the Grobman-Hartmann theorem, the non-linear ODE (5.24) can be approximated locally in the neighborhood of steady state $(\bar{k}, \bar{c})$ by the (topologically equivalent) linear **variational ODE**, with coefficient matrix $\mathrm{D}_{(k,c)}\mathbf{F}(\bar{k}, \bar{c})$,

$$\begin{pmatrix} \dot{k} \\ \dot{c} \end{pmatrix} = \begin{pmatrix} \rho & -1 \\ \eta\bar{c} f''(\bar{k}) & 0 \end{pmatrix} \begin{pmatrix} k - \bar{k} \\ c - \bar{c} \end{pmatrix}. \tag{5.26}$$

As the eigenvector associated to the "stable" eigenvalue of $\mathrm{D}_{(k,c)}\mathbf{F}(\bar{k}, \bar{c})$ is[27] $\bar{\mathbf{P}}^s = (1, \bar{\lambda}_u)^\top$, then the stable eigenspace of the variational ODE (5.26) is

$$\mathcal{E}^s = \{(k, c) \in N : c = \bar{\lambda}_u k\}$$

gives the locus of points that are tangent to the local stable manifold in the neighborhood N of the steady state $(\bar{k}, \bar{c})$.

Global dynamics. The stable manifold of the RCK model is $\mathcal{W}^s = \{(k, c) \in \mathbb{R}^2_+ : \lim_{t\to\infty}(k(t), c(t)) = (\bar{k}, \bar{c})\}$. Next we prove that for the subset of points $k \in (0, \bar{k})$ the stable manifold is a heteroclinic curve $c = h(k)$, connecting steady states (0, 0) and $(\bar{k}, \bar{c})$, meaning that any solution to the RCK model with $k(0) < \bar{k}$ will grow over time within set $\mathcal{W}^s$.[28]

Although we cannot determine function $h(.)$ explicitly we can prove that it exists. From the local analysis, we know that: as steady state (0, 0) is an unstable node there is an infinite number of paths diverging from it, and as steady state $(\bar{k}, \bar{c})$ is a saddle point then there is one unique path converging to it. To prove that a

---

[25] And singular as we will see in Chap. 7, as a consequence of the fact that $f(k)$ is not locally Lipschitz close to $k = 0$.

[26] Because capital accumulation is equal to savings, and savings finances future increases in consumption by increasing production, the existence of decreasing marginal returns implies that the marginal increase in production will tend to zero, thus eliminating the incentives for future capital accumulation.

[27] We use the useful relationships between eigenvalues and the trace and determinant of a square matrix: $\bar{\lambda}_s + \bar{\lambda}_u = \rho$ and $\bar{\lambda}_s\bar{\lambda}_u = \eta\bar{c} f''(\bar{k}) < 0$.

[28] This is a property of the model that is very rarely recognized in textbook presentations of the model.

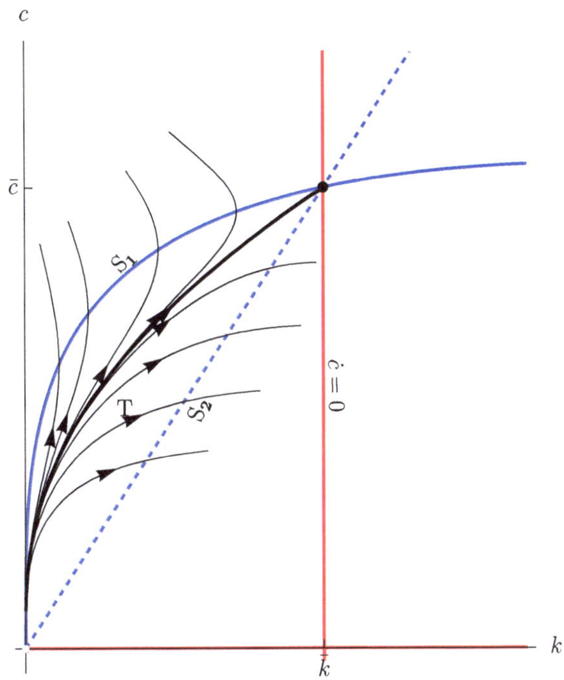

Fig. 5.9 Heteroclinic orbit in the Ramsey model

heteroclinic orbit exists we need to prove that there is an intersection between the local unstable manifold of the first steady state and the local stable manifold of the second steady state. To prove this is the case, we define a "trapping area" $T = \{(k, c) : c \le f(k),\ 0 \le k \le \bar{k}\}$, where the isoclines $\dot{k} = 0$ and $\dot{c} = 0$ define the boundaries $S_1 = \{(k, c) : c = f(k), 0 \le k \le \bar{k}\}$ and $S_2 = \{(k, c) : 0 \le c \le \bar{c},\ k = \bar{k}\}$. We can observe that all the trajectories coming from inside will exit T, with the exception of one trajectory. To prove this observe that: first, the trajectories that cross S_1 will exit T because $\dot{k}|_{S_1} = 0$ and $\dot{c}|_{S_1} = \eta c(f'(k) - \rho) = \eta f(k)(f'(k) - f'(\bar{k})) > 0$, as $f'(k) > f'(\bar{k})$ for $k < \bar{k}$; second, all trajectories that cross S_2 will exit T, as $\dot{k}|_{S_2} = f(\bar{k}) - c = \bar{c} - c > 0$ and $\dot{c}|_{S_2} = 0$; and, by continuity there is only one trajectory that does not cross both boundaries S_1 and S_2, and that trajectory is a heteroclinic trajectory, which is a curve separating the set T into two subsets.

In Fig. 5.9, we illustrate the existence of a heteroclinic orbit for the case in which the production function is Cobb-Douglas, $f(k) = k^{\alpha}$, for $0 < \alpha < 1$, and an isoelastic utility function, $u(c) = \log_{\theta}(c)$. In this case, if $\theta > \alpha$ we can consider instead the trapping area $T = \{(k, c) : \frac{\rho}{\alpha} k \le c \le k^{\alpha}\}$, to prove that the solution to the RCK model is a heteroclinic orbit for $0 < k_0 < \bar{k}$.

Comparative dynamics. Assume that the economy is at the steady state $(\bar{k}, \bar{c})$ and introduce a small change in ρ. As the steady state is a function of ρ, this means that the economy will move away from that steady state. Which path will it follow? From the hyperbolicity property, the economy should asymptotically converge to a new steady state that is also a saddle point, and the new stable manifold will move and pass through that new steady state. As k is a pre-determined variable and does not change discontinuously, c will have to "jump" to the new stable manifold. If the new steady state value for k is bigger than the initial steady state, then there is a "jump" to the new heteroclinic curve. However, as the explicit expression for the stable manifold is not known, we approximate locally the jump to the new stable eigenspace by solving the variational ODE. This is the intuition behind the comparative dynamics exercise in most perfect foresight DGE macro models after Blanchard and Khan (1980) and Buiter (1984), when there is determinacy.

For a variation $d\rho$ in the rate of time preference, the linear variational ODE (5.23) takes the form

$$\begin{pmatrix} \dot{k} \\ \dot{c} \end{pmatrix} = \mathrm{D}_{(k,c)}\mathbf{F}(\bar{k}, \bar{c}) \begin{pmatrix} (k(t) - \bar{k}) \\ (c(t) - \bar{c}) \end{pmatrix} + \mathrm{D}_{\rho}\mathbf{F}(\bar{k}, \bar{c}) d\rho \tag{5.27}$$

where $\mathrm{D}_{\rho}\mathbf{F}(\bar{k}, \bar{c}) = (0, -\eta\bar{c})^{\top}$. Writing $\partial_{\rho}k(t) = \frac{dk(t)}{d\rho} \approx \frac{k(t)-\bar{k}}{d\rho}$, and analogously for c, the general solution to the variational ODE is

$$\begin{pmatrix} \partial_{\rho}k(t) \\ \partial_{\rho}c(t) \end{pmatrix} = \begin{pmatrix} \partial_{\rho}\bar{k} \\ \partial_{\rho}\bar{c} \end{pmatrix} + w_s \begin{pmatrix} 1 \\ \bar{\lambda}_u \end{pmatrix} e^{\bar{\lambda}_s t},$$

where the long run multipliers, $(\partial_{\rho}\bar{k}, \partial_{\rho}\bar{c})^{\top} = -\left(\mathrm{D}_{(k,c)}\mathbf{F}(\bar{k}, \bar{c})\right)^{-1} \mathrm{D}_{\rho}\mathbf{F}(\bar{k}, \bar{c})$, are $\partial_{\rho}\bar{k} = \frac{1}{f''(\bar{k})} < 0$ and $\partial_{\rho}\bar{c} = \rho\partial_{\rho}\bar{k} < 0$, and w_s is an arbitrary constant. As variable k is pre-determined, we set $\partial_{\rho}k(0) = \partial_{\rho}\bar{k} + w_s = 0$ implying $w_s = -\partial_{\rho}\bar{k}$. This finally yields the short-run multipliers

$$\partial_{\rho}k(t) = \frac{1}{f''(\bar{k})}\left(1 - e^{\bar{\lambda}_s t}\right), \text{ and } \partial_{\rho}c(t) = \frac{1}{f''(\bar{k})}\left(\rho - \lambda_u e^{\bar{\lambda}_s t}\right), \text{ for } t \in [0, \infty).$$

In particular, we obtain the impact multipliers $(\partial_{\rho}k(0), \partial_{\rho}c(0)) = (0, \partial_{\rho}\bar{k}\lambda_s)$ that quantifies the positive "jump" of consumption to the new stable eigenspace, and the long-run multipliers $\lim_{t\to\infty}(\partial_{\rho}k(t), \partial_{\rho}c(t)) = (\partial_{\rho}\bar{k}, \partial_{\rho}\bar{c})$. Therefore, on impact consumption increases, which reduces capital accumulation, and therefore reduces consumption over time. The process progressively slowed because the reduction in the per-capita capital stock will increase marginal productivity, which reduces the incentives for further reduction in consumption. This path is qualitatively but not quantitatively exact, because we cannot determine the solution explicitly as in the linear models of Chap. 4.

Example 5.5 (*The* Skiba (1978) *model*) In this example, we present an explicitly specified version of the Skiba (1978) model. The model has the same structure of the RCK model, with the exception of the technology, in which the production function is concave-convex, as in Example 3.4, and is not globally concave.

The MHDS is

$$\dot{k} = f(k, B) - c - \delta k,$$
$$\dot{c} = \frac{c}{\sigma}\left(f_k(k, B) - \rho - \delta\right),$$

where $\rho > 0$ is the rate of time preference, δ is the capital depreciation rate, and σ is the inverse of the elasticity of intertemporal substitution of consumption. The initial and transversality conditions are the same as in the RCK model. Recall, from Example 3.4, that the production function is $f(k, B) = A\,k^{\alpha} + \frac{k^3}{2\,(k^3+B^3)}$.[29]

As in the RCK model, there is one zero steady state $(\bar{k}_1, \bar{c}_1) = (0, 0)$, and there are one, two or three other steady states satisfying the transversality condition, in which the capital stock is determined from the intertemporal arbitrage condition such that $\dot{c} = 0$. Therefore, the other steady states belong to the set $(k^{ss}, c^{ss}) \in \Big\{(c, k) > \mathbf{0} : \; f_k(k, B) = \rho + \delta, \; \text{and} c = f(k, B) - \delta k\Big\}$. In the non-optimizing version of the model, in Example 3.4, the steady state stock of capital was determined instead from the equilibrium between savings and investment, when net investment was equal to zero, and therefore is not optimally determined. The steady state values for positive k, and therefore for c, cannot be determined explicitly. Figure 5.10a illustrates that we can have one, two, or three steady states, $(\bar{k}_2, \bar{c}_2)$, $(\bar{k}_3, \bar{c}_3)$, and $(\bar{k}_4, \bar{c}_4)$, depending in the values of the parameters.[30]

As in "ad-hoc" Skiba model, in Example 3.4, we have an **hysteresis bifurcation**, where the bifurcation points are

$$\{(k_1^*, B_1^*), (k_2^*, B_2^*)\} = \Big\{(k, B) = \det\left(\mathbf{J}(k, B)\right) = \frac{\bar{c}(k)\, f_{kk}(k, B)}{\sigma}, \; f_k = \rho + \delta\Big\},$$

that cannot be determined explicitly. For specific values of the other parameters, the bifurcation diagram in Fig. 5.10b shows that for very large or very small values of B, i.e., for $B > B_1^*$ or $B < B_2^*$, there is only one positive steady state, and for intermediate values of B, i.e., for $B_1^* > B > B_2^*$, there are three steady states. Furthermore, we observe that large values of B correspond to a low value of the steady state capital stock (a "poverty trap"), and very low value of B correspond to a higher level of steady state capital ("economic maturity"). However, comparing the two bifurcation diagrams, in Figs. 3.7 and 5.10b, we observe that the bifurcation curve is decreasing for lower values of B. This is a result of the fact that the savings rate becomes endogenous in the optimizing model.

The phase diagram for intermediate values of B is illustrated in Fig. 5.10c . As in the "ad-hoc" model, in Example 3.4, there are again three steady states. However, the two extreme steady states, $\bar{k}_4$ and $\bar{k}_2$ are now saddle points, locally similar to the RCK

[29] The marginal return to capital is $f_k(k, B) = \alpha\, A\, k^{\alpha-1} + \frac{3\,B^3\,k^2}{2\,(k^3+B^3)^2}$ and $f_{kk}(k, B) = \alpha\,(\alpha - 1)\, A\, k^{\alpha-2} + \frac{3\,B^3\,k\,(B^3-2\,k^3)}{(k^3+B^3)^3}$.

[30] The figure shows another steady state $\bar{k}_5$ that does not satisfy the transversality condition, as in the RCK model.

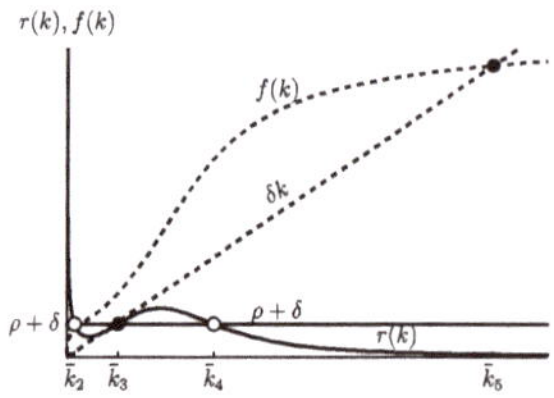

(a) Steady state determination $r(k) \equiv f_k(k, B) = \rho + \delta$

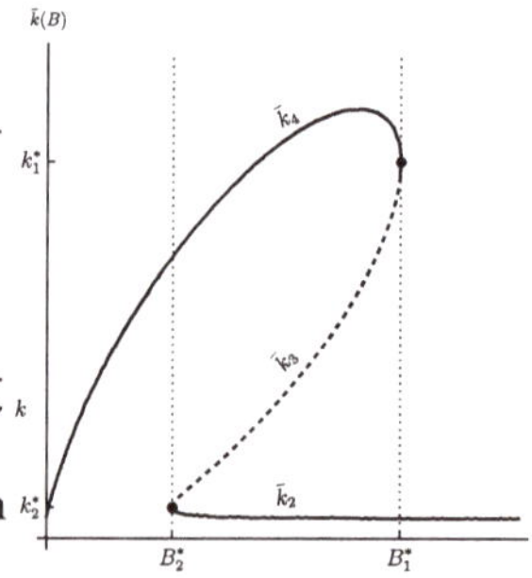

(b) Bifurcation diagram

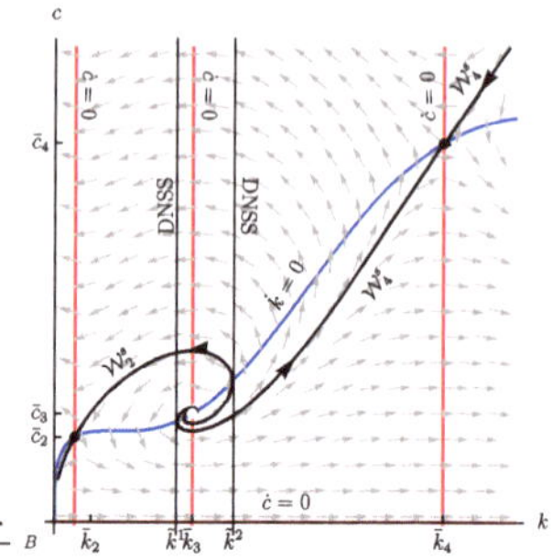

(c) Phase diagram for intermediate values of B

Fig. 5.10 Phase diagram for the Skiba model

model, and the intermediate state, $\bar{k}_3$, is an unstable focus. Furthermore, differently from the "ad-hoc" model, the separation between the convergence to the lower and upper steady states is done along the stable manifolds passing through those steady states. Those two curves are **heteroclinic orbits** connecting the intermediate steady state to one of the two extreme steady states. This implies that given an initial level for the capital stock coincident with the intermediate steady states, there are three possible paths, depending on the initial level of consumption: first, if c is at the steady state level the capital stock will remain stationary, second, if it is lower than in the intermediate steady state level then there will be convergence to the lower steady state, and, third, if it is higher then there will be convergence to the higher steady state. There are two threshold levels of the stock of capital, illustrated by two vertical lines called DNSS-curves[31], that separate different types of dynamics: if $k_0 < \tilde{k}_1$ there will be convergence to the steady state $(\bar{k}_2, \bar{c}_2)$ along the local stable manifold $\mathcal{W}_2^s$, if $k_0 > \tilde{k}_2$ there will be convergence to the steady state $(\bar{k}_4, \bar{c}_4)$ along the stable manifold $\mathcal{W}_4^s$, and if $\tilde{k}_1 \leq k_0 \leq \tilde{k}_2$ there are **multiple optimal trajectories** leading to one of the two extreme steady states.

Example 5.6 (*the* Romer (1986b) *model: bifurcations*) Romer (1986b) is a seminal paper that presents a model in which endogenous growth is generated by productive externalities. We introduce and extend that model by incorporating not only a (stock) production externality but also a (flow) externality on preferences. Preference externalities are introduced by assuming that the utility function is $u = u(c, C)$, where c household's consumption and C represents aggregate consumption. This means that the value of consuming c, at the household level, depends on aggregate consumption. This covers two cases: (i) the case in which private and aggregate consumption are Edgeworth-Pareto (EP) substitutes, $u_{cC}(c, C) = \frac{\partial^2 u}{\partial c \partial C} < 0$, as in *keeping-along with the Joneses* utility function, in which utility depends on the difference between the

[31] Named after Dechert and Nishimura (1983), Sethi (1977) and Skiba (1978). See Grass et al. (2008, Chap. 5).

household private and aggregate or reference consumption; (ii) the case in which private and aggregate consumption are EP complements, $u_{cC}(c, C) = \frac{\partial^2 u}{\partial c \partial C} > 0$, in which aggregate consumption C is a public good. Technological externalities are introduced by assuming that the production function is $y = f(k, K)$, where k is the capital stock at the firm's level and K is the aggregate capital stock. We can assume that both the level of production and the marginal productivity of capital depend on the aggregate capital stock, K. This also covers two cases: (i) the case in which firm level and aggregate capital are EP complements, $f_{kK}(k, K) = \frac{\partial^2 f}{\partial k \partial K} > 0$, in which agglomeration generates external economies to a firm with capital k, as assumed in Romer (1986b); (ii) the case in which firm level and aggregate capital are EP substitutes, $f_{kK}(k, K) = \frac{\partial^2 f}{\partial k \partial K} < 0$, where the aggregation generates congestion. Next, let us assume that the utility function is $u(c, C) = \frac{\left(cC^b\right)^{1-\theta} - 1}{1-\theta}$, where $\theta > 0$ and b can take any real number, the production function is $f(k, K) = k^\alpha K^\beta$, where $0 < \alpha < 1$ and β can be any real number, that the rate of time preference is $\rho > 0$, and that there is no capital depreciation.

Letting households be homogeneous, and having a mass equal to one, and the micro-macro consistency conditions, $K = k$ and $C = c$, be satisfied, then the DGE of this economy is represented by the non-linear planar ODE, together with the transversality and the initial conditions,

$$\dot{k} = A\, k^{\alpha+\beta} - c, \tag{5.28a}$$

$$\dot{c} = \frac{c}{\theta + b\,(\theta - 1)} \left(\alpha A k^{\alpha+\beta-1} - \rho\right), \tag{5.28b}$$

$$\lim_{t \to +\infty} c(t)^{\theta + b\,(\theta-1)}\, k(t)\, e^{-\rho t} = 0, \tag{5.28c}$$

$$k(0) = k_0 \text{ given.} \tag{5.28d}$$

Steady states and local dynamics. Depending on the value of the parameters, we have two cases: (1) if $\alpha + \beta = 1$ then there is a unique steady state is $(\bar{k}, \bar{c}) = \mathbf{0}$; or (2) if $\alpha + \beta \neq 1$ there is an unique positive steady state $(\bar{k}, \bar{c}) = \left(\left(\frac{\alpha A}{\rho}\right)^{\frac{1}{1-(\alpha+\beta)}}, A\bar{k}^{\alpha+\beta}.\right)$.

Let $\alpha + \beta \neq 1$. The Jacobian of ODE (5.28a)–(5.28b), evaluated at the steady state is

$$\mathbf{J}(\bar{k}, \bar{c}) = \begin{pmatrix} \frac{\rho\,(\alpha+\beta)}{\alpha} & -1 \\ \frac{\rho^2\,(\alpha+\beta-1)}{\alpha\left(\theta+b(\theta-1)\right)} & 0 \end{pmatrix},$$

has trace $\text{tr}(\mathbf{J}(\bar{k}, \bar{c})) = \frac{\rho(\alpha+\beta)}{\alpha}$ and determinant $\det(\mathbf{J}(\bar{k}, \bar{c})) = \frac{\rho^2\,(\alpha+\beta-1)}{\alpha\left(\theta+b(\theta-1)\right)}$. Comparing with a RCK model, with an isoelastic utility and Cobb-Douglas production functions, we see that the trace is different from ρ and can have any sign, and the determinant, instead of being negative, can also have any sign. In particular, $\text{tr}(\mathbf{J}) > 0$ if $\beta > -\alpha$, and $\text{tr}(\mathbf{J}) \leq 0$ if $\beta > -\alpha$, and $\det(\mathbf{J}) > 0$ if $\text{sign}(\alpha + \beta - 1) =$

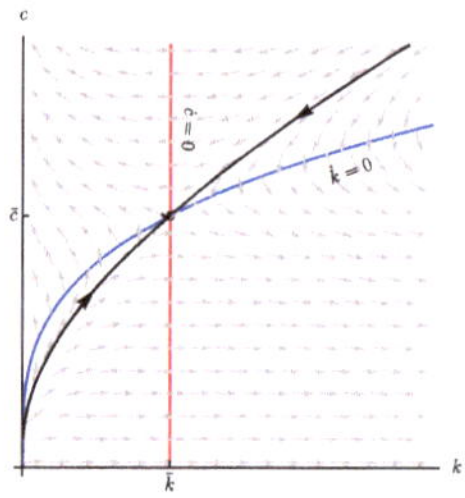

(a) Phase diagram: saddle for $\beta = b = 0$

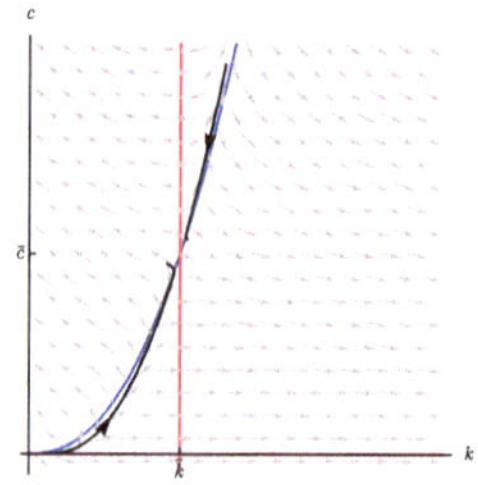

(b) Phase diagram: saddle for $b < 0 < \beta$

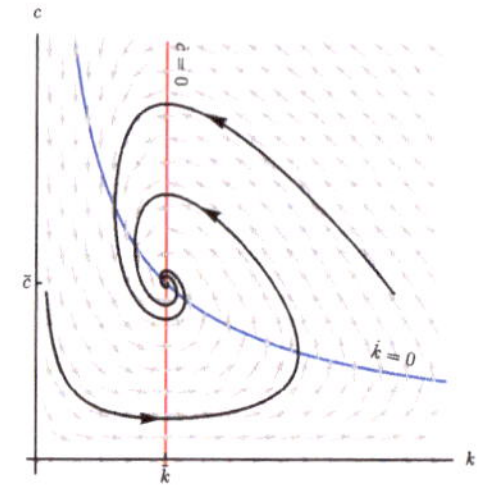

(c) Phase diagram: stable focus for $\beta < 0$ and $b < 0$

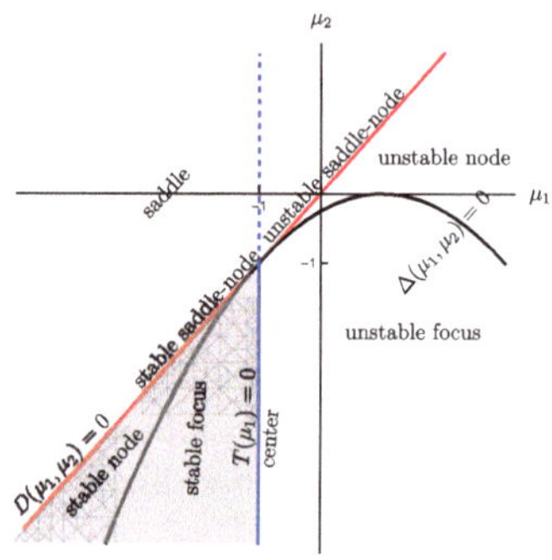

(d) Bifurcation diagram

Fig. 5.11 Bifurcation diagram in (b, β) for and $0 < \alpha < 1$, $\theta > 1$. The red lines separate negative from positive externalities at the consumption level (the vertical line) and at the production level (horizontal line)

$\text{sign}(\theta + b(\theta - 1)$, and $\det(\mathbf{J}) < 0$ if $\text{sign}(\alpha + \beta - 1) \neq \text{sign}(\theta + b(\theta - 1))$, and $\det(\mathbf{J}) = 0$ if $\alpha + \beta = 1$, which is a case in which a BGP exists.

Figure 5.11 presents a bifurcation diagram, and some representative phase diagrams, illustrating several types of macroeconomic equilibria. Equilibrium is **determinate** when the steady state is a saddle point, because there is a unique trajectory converging to the steady state for a given initial level of k. This case exists if $b(\theta - 1) > -\theta$ and $\beta < 1 - \alpha$ or $b(\theta - 1) < -\theta$ and $\beta > 1 - \alpha$. It means that determinacy can exist with several types of combinations of externalities: (1) congestion in production, mild negative externalities in consumption or positive externalities in consumption; (2) positive externalities in production and negative externalities in consumption. Figure 5.11a shows phase diagrams for a determinate equilibrium in which there are no externalities (this is the RCK case), and Fig. 5.11b shows a phase diagram for a determinate equilibrium in which there are positive production externalities and negative preference externalities. Equilibrium is **indeterminate** when the steady state is a stable node or a stable focus, as in Fig. 5.11c . In this case, there is an infinite number of trajectories converging to the steady state, resulting from the fact that expectations are self-fulfilling, and, even in the absence of coordination, the economy converges to the steady state. This case occurs only if $\alpha + \beta < 0$ and

$b(\theta - 1) < -\theta < 0$, that is, when there is congestion in production and negative externalities in consumption.

A **balanced growth path** (BGP), featuring unbounded long-run growth, exists if $\alpha + \beta = 1$, and $\alpha A > \rho$.[32] This case occurs when the production function displays constant returns at the aggregate level, even though it displays decreasing marginal returns at the private level, and it can occur with or without externalities in consumption.

If the steady state is an unstable node or focus, then the equilibrium is **overdeterminate** and is constrained to be coincident with the steady state. Any deviation from it will violate the transversality condition and, therefore, cannot be an equilibrium (in the economic sense). This case would occur, for instance, with high positive externalities in both consumption and production.

Example 5.7 (*The* Benhabib et al. (2001) *model*) The Benhabib et al. (2001) model addresses the potential problems that can emerge if a Taylor monetary rule is adopted. The Taylor rule, $R = R(\pi)$, is a rule involving a change to the nominal interest rate in respond to increases in the inflation rate π. It is a new-Keynesian DGE model in which firms are monopolistic in the market for goods, there is nominal price rigidity (of the Rotemberg (1982) type), and labor is the only production input. Households have benchmark preferences in which money is Edgeworth-Pareto complementary with consumption, supply labor inelastically to firms, pay taxes, receive income from government liabilities, and save by accumulating government liabilities (money and bonds). The government adopts a tax rate policy consistent with a Ricardian monetary-fiscal regime.

The perfect foresight DGE is represented by the non-linear ODE, governing two non-predetermined variables, the inflation rate and the marginal utility of consumption, (π, λ),

$$\dot{\pi} = r\,(\pi - \pi^*) - n(\lambda), \tag{5.29a}$$

$$\dot{\lambda} = \lambda\Big(r + \pi - R(\pi)\Big), \tag{5.29b}$$

where r is the fundamental real interest rate (the rate of time preference) and p^* is the inflation rate at which the Fisher equation is satisfied. ODE (5.29a) is an equilibrium product market condition, where function $n(\lambda) = \psi_0\,\lambda^{\omega} - \psi_1\,\lambda^{\beta}$, with its parameters depending on several primitive behavioral and technological parameters. It is assumed that $\beta < 0$, $\psi_0, \psi_0 < 0$, and ω can take any sign. ODE (5.29b) is a Euler equation where $R(\pi) = R^* \exp\left(\frac{A\,(\pi-\pi^*)}{R^*}\right)$ is the open market intervention rule for anchoring the nominal interest rate R, where A models the degree of response of monetary policy, and $R^* = r + \pi^*$.

The Jacobian of ODE (5.29), evaluated at any point $\mathbf{y} = (\pi, \lambda)$ is $\mathbf{J(y)} = \begin{pmatrix} r & -n'(\lambda) \\ \lambda\,(1-R'(\pi)) & r+\pi-R(\pi) \end{pmatrix}$. If we consider the real interest rate, r, and the policy parame-

[32] if that this observe that the Jacobian evaluated at the steady state $(\bar{k}; \bar{c}) = \mathbf{0}$ has eigenvalues $\lambda_- = \alpha\,A - \rho$ and $\lambda_+ = A > 0$. There is a BGP if that steady state is unstable, which requires $\alpha\,A > \rho$.

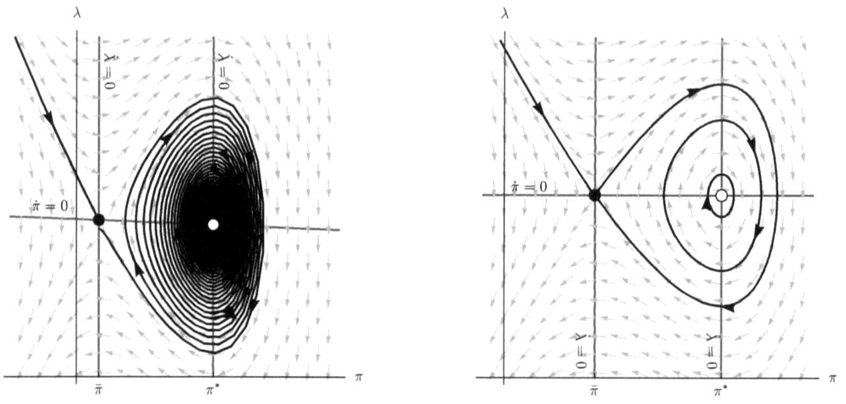

(a) Phase diagram for $r = 0.0045$ and $A = 1.5$

(b) Phase diagram $r = 0$ and $A = 1.5$

Fig. 5.12 Bifurcation diagram for ODE (5.29) for the parameters in Benhabib et al. (2001): $R^* = 0.015$, $\pi^* = 0.0105$, $\omega = 0.5$, $\beta = -1.42857$, $\psi_0 = -0.0564$, $\psi_1 = -0.0826$

ter A, we can prove that there is a Bogdanov-Takens bifurcation point $(\pi, \lambda, r, A) = \left(\pi^*, \tilde{\lambda}, 0, 1\right)$ where $\tilde{\lambda} = \left(\frac{\psi_1}{\psi_0}\right)^{\frac{1}{\omega-\beta}}$.

The Bogdanov-Takens bifurcation can be unfolded by $r > 0$ and/or $A > 1$. In all cases, there will be two steady states (π^*, λ^*) and $(\bar{\pi}, \bar{\lambda})$, such that $\pi^* > \bar{\pi}$ and $\lambda^* < \bar{\lambda}$, in which the first corresponds to an active monetary policy (with a higher nominal interest rate) and the second to a passive monetary policy (with a lower interest rate). The passive policy steady state (π^*, λ^*), is always a saddle point, and the active policy steady state, $(\bar{\pi}, \bar{\lambda})$, can be an unstable node (if $(A-1)(\omega-\beta)\psi_0(\lambda^*)^{\omega} + \left(\frac{r}{2}\right)^2 > 0$), an unstable focus ((if $(A-1)(\omega-\beta)\psi_0(\lambda^*)^{\omega} + \left(\frac{r}{2}\right)^2 < 0$), or a center (if $r = 0$ and $A > 1$).

As we saw in Sect. 5.1.4, the dynamic characterization should be complemented by a global dynamics analysis. Figure 5.12a was built by using the calibrated parameters in Benhabib et al. (2001). This bifurcation is equivalent to case 5.6b in Sect. 5.1.4: performing an active monetary policy for an economy close to the unstable steady state (π^*, λ^*) will eventually lead to a reduction in the inflation rate, in which the economy follows along a heteroclinic orbit (a focus-saddle connection in this case), although there will be several fluctuations along the transition, as illustrated in Fig. 5.12a . This would be the case if $r > 0$. However, if the monetary policy rule is the same (i.e., A is the same) and the economy is again close to (π^*, λ^*) but $r = 0$, the economy will be trapped in a periodic trajectory and there will be no permanent reduction of inflation, in the sense that there is no convergence to $(\bar{\pi}, \bar{\lambda})$. This is illustrated in the phase diagram Fig. 5.12b , that is equivalent to the non-generic phase diagram Fig. 5.6g .

Resorting to a continuity argument, phase diagrams similar to phase diagrams, Fig. 5.6c and h , but not illustrated in Fig. 5.12, can exist for $r > 0$ and a value of $A \in (1, 1.5)$, for the same values of the other (non-bifurcating) parameters. In both cases, trajectories starting close to the high inflation equilibrium may not converge to the low inflation equilibrium, as a consequence of the existence of a limit cycle, even if the monetary policy is active. This is the "peril of the Taylor rule" mentioned by the authors.

Exercise 5.3 (*Two-dimensional Solow model*) Consider the dynamics of a two-sector capital accumulation model

$$\begin{aligned}\dot{k}_1 &= A_1 k_1^{\alpha} k_2^{1-\alpha} - \delta_1 k_1, \text{ for } 0 < \alpha < 1\\ \dot{k}_2 &= A_2 k_1^{1-\beta} k_2^{\beta} - \delta_2 k_2, \text{ for } 0 < \beta < 1\end{aligned}$$

where k_i is the capital stock of sector i, $A_i > 0$ and $\delta_i > 0$ are the total factor productivity and the rate of depreciation for section $i \in \{1, 2\}$. Assume that $k_1 > 0$ and $k_2 > 0$. Under which conditions equilibrium points of equation exist? Provide a bifurcation analysis. If the previous condition(s) don't hold, which type of dynamics is displayed by the solution?

Exercise 5.4 (*The SIR model*) The benchmark model for epidemics is the so-called SIR model. It considers the division of total population, N, between subjected, S, infected, and removed, R, by immunization or death. Therefore $N(t) = S(t) + I(t) + R(t)$ at every moment in time t. The dynamic model of epidemics is

$$\dot{S} = -\beta S I, \tag{5.30}$$

$$\dot{I} = \beta S I - \gamma I, \tag{5.31}$$

$$\dot{R} = \gamma I, \tag{5.32}$$

where all the parameters are positive, and β is the infection rate and γ is the recovery rate: (a) Prove that if the total population is constant, Eq. (5.32) is redundant, i.e., it can be obtained from the other two. (b) Find the steady state(s) of the model (consider only Eqs. (5.30) and (5.31)). Study analytically and geometrically the dynamics of the model. (c) Perform a bifurcation analysis. (d) Assume that, for a given value of the parameter β, e.g., $\beta = \beta_0$, the values of (S, I), at some point in time t, are such that $I(t)$ is close zero and $\beta_0 S(t) < \gamma$. Now introduce a perturbation to β, from β_0 to β_1, such that $\beta_1 > \beta_0$, implying $\beta_1 S(t) > \gamma$. Which type of trajectory for the epidemics one should expect instead?

Exercise 5.5 (*The predator-prey model*) The Lotka-Volterra model (see Kot (2001)) is the benchmark model in mathematical biology. It addresses the dynamics of the interaction between predators and preys. In its simplest formulation it is

$$\begin{aligned}\dot{N}_1 &= N_1(\alpha - \beta N_2),\\ \dot{N}_2 &= -N_2(\gamma - \delta N_1),\end{aligned}$$

where N_1 is the number of prey (for example, rabbits) and N_2 is the number of some predator (for example, foxes), and the rates of growth of preys and predators are $\gamma_1 = \alpha - \beta N_2$ and $\gamma_2 = \delta N_1 - \gamma$. (a) Find the steady states and study the local dynamics. (b) Draw the phase diagram. (c) Prove that function $V = -\delta\, N_1 + \gamma\, \log(N_1) - \beta\, N_2 + \alpha\, \log(N_2)$ is a Lyapunov function, and find and characterize global bifurcations.

Exercise 5.6 (*Equilibrium unemployment dynamics*) Mortensen (1999) extends the Pissarides (1990) model of aggregate unemployment dynamics by considering increased returns to scale in production and constant returns in the labor matching process. Employment, n, and match surplus, p, dynamics are governed by two coupled non-linear ODE

$$\dot{n} = h(p)\,(1-n) - \delta\, n,$$
$$\dot{p} = (r+\delta)\, p + g(p) - f(n),$$

where $h(p)$ is the matching rate per unemployed worker function, $f(n)$ is the production function, and $g(p)$ is the net income procured by the search activity. Mortensen (1999) consider the following functional forms: $h(p) = k\, p^{\kappa}$, $f(n) = a\, n^{\alpha}$, and $g(p) = k\left(\frac{\beta}{1-\eta}\right)\left(\frac{p^{1+\kappa}}{1+\kappa}\right)$. Perform a bifurcation analysis taking κ as the bifurcating parameter. In particular, show under which conditions a limit cycle, or a saddle-loop bifurcation exists. Choose another bifurcating parameter, and unfold a Bogdanov-Takens bifurcation.

5.4 References

A very good introduction to non-linear planar ODE can be found in Hale and Koçak (1991, Part I , III). A classic reference on the field is Guckenheimer and Holmes (1990, Chap. 1, 3, 6). Shilnikov et al. (1998), Chua et al. (2001), and Kuznetsov (2023) contain a very complete presentation of bifurcations for planar and higher dimensional ODE. Brock and Malliaris (1989) is a benchmark book on non-linear economic models. Grass et al. (2008, Chap. 2) offer a compact presentation of all the important results with examples in economics and management science.

Chapter 6
Piecewise Smooth or Continuous ODE

In the previous chapters, we considered regular ODE whose vector fields $\mathbf{f}(\mathbf{y})$ were globally continuous and continuously differentiable.[1] In this chapter, we briefly introduce piecewise smooth or continuous (PWSC) ODE, in which time is the independent variable. PWSC ODE are characterized by having vector fields that are not globally continuous and/or continuously differentiable, although they can be Lipschitzian. This implies that their solutions display some kind of discontinuity or some types of dynamics that do not exist in regular ODE. There are different types of discontinuities in ODE, maps, differential algebraic equations (DAE), or partial differential equations (PDE).

PWSC have been developed in several areas in applied mathematics, control theory, engineering, or biology, in the last 20 years, and were used to formalize several phenomena as friction, impact, switching, sliding or grazing. Existing, or potential, applications in economics include: modeling endogenous dynamic regime switches; solving dynamic optimization problems with side constraints; characterizing solutions in dynamic games; substituting ad-hoc non-linear functions in non-linear growth models by regime switches; or as a way of approximating non-linear dynamics by simpler, almost linear functions. Although "... the subject is huge that we cannot classify everything..."(di Bernardo et al. 2008a) we present the bare minimum to understand these ODE. This material is, surprisingly, not covered in economics textbooks.

Next, we define PWSC ODE, and generically clarify how their solutions differ from those of regular ODE addressed in the previous chapters.

Definition 6.1 (*Piecewise smooth or continuous (PWSC) ODE*) Let set $\mathbf{Y} \subseteq \mathbf{R}^n$ be partitioned into a finite number of open sets $\mathbf{Y}_j$, $j = 1, \ldots, m$, called **branches**, such that $\mathbf{Y} = \cup_{j=1}^{m} \mathbf{Y}_j$, and consider the vector function $\mathbf{y} : \mathrm{T} \rightarrow \mathbf{Y}$. A PWSC is the functional equation $\dot{\mathbf{y}} = \mathbf{f}(\mathbf{y})$ where the known $\mathbf{f}(\cdot)$ vector function has the form

[1] This was also the case for non-autonomous ODE of type $\dot{y} = ay + b(t)$ where $b(t)$ is piecewise continuous, dealt in Chap. 2.

P. B. Brito, *Economic Dynamics and Distributions*, Dynamic Modeling and Econometrics in Economics and Finance 32, https://doi.org/10.1007/978-3-031-94717-9_6

$$\mathbf{f}(\mathbf{y}) = \begin{cases} \mathbf{f}_1(\mathbf{y}) & \text{if } \mathbf{y} \in \mathbf{Y}_1, \\ \cdots & \\ \mathbf{f}_j(\mathbf{y}) & \text{if } \mathbf{y} \in \mathbf{Y}_j, \\ \cdots & \\ \mathbf{f}_m(\mathbf{y}) & \text{if } \mathbf{y} \in \mathbf{Y}_m, \end{cases}$$

and **branch functions** $\mathbf{f}_j(\mathbf{y})$, $j = 1, \ldots, m$ are assumed to be regular, i.e., continuous and continuously differentiable.

Definition 6.2 (*Switching boundary*) A **switching boundary** is the intersection of the closure of two (or more) adjacent sets $\Sigma_{ij} = \overline{\mathbf{Y}}_i \cap \overline{\mathbf{Y}}_j,\ i \neq j = 1, \ldots, m$.

Usually, a switching boundary between two adjacent branchs $\mathbf{Y}_i$ and $\mathbf{Y}_j$, Σ_{ij}, is defined by a $n-1$-dimensional manifold $\mathbf{H}_{ij}(\mathbf{y}) = 0$, called **switching manifold**, where every $\mathbf{H}_{ij}$ is smooth. Formally,

$$\Sigma_{ij} = \{\mathbf{y} : \mathbf{H}_{ij}(\mathbf{y}) = 0\}, \text{ for } i \neq j \in \{1, \ldots, m\}.$$

There are some specificities to the solutions of PWSC ODE. Consider an arbitrary initial point, $\mathbf{y}(0) = \mathbf{y}_0$ belonging to the particular subdomain $\mathbf{Y}_j$, then two types of orbits can occur: (1) if the solution trajectory lies inside the same set, and **does not collide** with any switching boundary Σ_{ij}, then the solution behaves as in a regular ODE, as studied in the previous chapters; (2) but if the solution trajectory **collides** with any particular switching boundary Σ_{ij}, then several (single or multiple) forms of interaction can occur, some of them giving birth to **new dynamic phenomena**, such as crossing, sliding, or grazing.

Then, different types of solutions, **new types of invariant sets**, and new types of local and global bifurcations are possible, that are called generically **discontinuity induced bifurcations**.

From now on, we consider **the two-branch case** in which the state domain is partitioned into two branches, labelled $\mathbf{Y}_1$ and $\mathbf{Y}_2$, and a switching boundary, Σ, such that $\mathbf{Y} = \mathbf{Y}_1 \cup \Sigma \cup \mathbf{Y}_2 \subseteq \mathbb{R}^2$, and

$$\mathbf{Y}_1 = \{\mathbf{y} : H(\mathbf{y}, \varphi) \leq 0\}, \tag{6.1a}$$
$$\mathbf{Y}_2 = \{\mathbf{y} : H(\mathbf{y}, \varphi) \geq 0\}, \tag{6.1b}$$
$$\Sigma = \{\mathbf{y} : H(\mathbf{y}, \varphi) = 0\} \tag{6.1c}$$

where $\varphi \in \Phi \subseteq \mathbb{R}^p$, is a parameter vector and p is the number of parameters. The switching function $H(\mathbf{y}, \cdot)$ is assumed to satisfy the condition $\mathrm{D}_{\mathbf{y}} H(\mathbf{y}, \cdot) \neq \mathbf{0}$ for all $\mathbf{y} \in \Sigma$. The two-branch PWSC ODE is the functional equation with the form

$$\dot{\mathbf{y}} = \mathbf{f}(\mathbf{y}, \varphi) = \begin{cases} \mathbf{f}_1(\mathbf{y}, \varphi), & \text{if } \mathbf{y} \in \mathbf{Y}_1 \text{ or } \mathbf{Y}_1 \cup \Sigma, \\ \mathbf{f}_2(\mathbf{y}, \varphi), & \text{if } \mathbf{y} \in \mathbf{Y}_2 \text{ or } \mathbf{Y}_2 \cup \Sigma. \end{cases} \tag{6.2}$$

In equations (6.1), at least one of the inequalities defining the branches of set $\mathbf{Y}$ can be strict. That is we can have either the partition defined by $H(\mathbf{y}, \varphi) \leq 0$ and $H(\mathbf{y}, \varphi) > 0$, or by $H(\mathbf{y}, \varphi) < 0$ and $H(\mathbf{y}, \varphi) \geq 0$.

In the rest of the chapter, we present an introduction to (linear and non-linear) scalar equations in Sect. 6.1, and to planar equations in Sect. 6.2, and applications to economics in Sect. 6.3.

6.1 Scalar Equations

In this section, we consider the piecewise continuous or piecewise smooth ODE, for the mapping $y : \mathrm{T} \to \mathrm{Y} \subseteq \mathbb{R}$ where $\mathrm{Y} = \mathrm{Y}_1 \cup \mathrm{Y}_2$,

$$\dot{y} = \begin{cases} f_1(y, \varphi), & \text{if } h(y, \varphi) \leq 0, \\ f_2(y, \varphi), & \text{if } h(y, \varphi) > 0, \end{cases} \tag{6.3}$$

where $\varphi \in \Phi \subseteq \mathbb{R}^p$, with $p \geq 1$, is a parameter vector, and the branch functions $f_j : \mathrm{Y}_j \times \phi \to \mathbb{R}$, for $j = 1, 2$, are continuous and smooth functions. The switching boundary is

$$\Sigma = \{y \in \mathrm{Y} : h(y, \varphi) = 0\}$$

where it is assumed that the switching function, $h(\cdot)$, satisfies $h_y(y, \varphi) \neq 0$ for every pair $(y, \varphi) \in \mathrm{Y} \times \Phi$. Then the switching boundary is reduced to a **switching point** $\Sigma = \{\tilde{y}\} = \{y \in \mathrm{Y} : y = \tilde{y}\}$, and the two branches of the state space are simply defined as $\mathrm{Y}_1 = \{y \in \mathrm{Y} : y \leq \tilde{y}\}$, and $\mathrm{Y}_2 = \{y \in \mathrm{Y} : \ y > \tilde{y}\}$.

There are two main types of PWSC ODE:

Definition 6.3 (*Scalar piecewise-smooth (PWS) ODE*) A piecewise-smooth (PWS) ODE is a PWSC ODE such that function $f(y)$ is **continuous but not differentiable** at the boundary Σ, that is [2]

$$f_1(\tilde{y}) = f_2(\tilde{y}) \text{ and } f_{1,y}(\tilde{y}) \neq f_{2,y}(\tilde{y}) \tag{6.4}$$

Definition 6.4 (*Scalar Filippov ODE*) A piecewise-continuous (PWC) or piecewise continuous and piecewise smooth (PWSC) are called Filippov ODE if function $f(y)$ is **not continuous and usually also not differentiable** at the boundary Σ, that is

$$f_1(\tilde{y}) \neq f_2(\tilde{y}) \text{ and } f_{1,y}(\tilde{y}) = f_{2,y}(\tilde{y}) \text{ or } f_{1,y}(\tilde{y}) \neq f_{2,y}(\tilde{y}). \tag{6.5}$$

In Sect. 6.1.1, we show that linear equations can be solved explicitly, and present the characterization of their solutions, and in Sect. 6.1.2 we provide some remarks on non-linear equations.

[2] When there is no room for confusion we use the notation $f_{j,y} \equiv \frac{\partial f_j(y)}{\partial y}$.

6.1.1 *The Linear Scalar Case*

The linear scalar (two-branch) PWSC ODE is specified by

$$\dot{y} = f(y, \varphi) = \begin{cases} \lambda_1 y + m_1, & \text{if } cy + n \leq 0, \\ \lambda_2 y + m_2, & \text{if } cy + n > 0, \end{cases} \tag{6.6}$$

where $\varphi = (m_1, m_2, \lambda_1, \lambda_2, n, c) \in \mathbb{R}^5 \times \mathbb{R}/\{0\}$. This is a special case of equation (6.3) where both the branch functions and the switching function are linear

$$f_j(y) = \lambda_j y + m_j, \text{ for } j = 1, 2, \text{ and } h(y) = c\, y + n, \; c \neq 0. \tag{6.7}$$

As $c \neq 0$ then the solution to $cy + n = 0$ is $\tilde{y} = -n/c$.

Although the ODE is linear in each branch, the solution to equation $\dot{y} = f(y, \varphi)$ behaves in an analogous way as a non-linear ODE, with the difference that we can always solve the PWSC linear ODE (6.3) explicitly.

6.1.1.1 Explicit Solutions

Although we can solve the linear ODE for the two branches, particular solutions can only be obtained for ODE problems, initial-value or terminal-value problems. The general solution to (6.6) while in branch Y_j is

$$y_j(k, t) = \begin{cases} \bar{y}_j + (k_j - \bar{y}_j)e^{\lambda_j t}, & \text{if } \lambda_j \neq 0 \\ k_j + m_j t & \text{if } \lambda_j = 0 \end{cases}$$

where $\bar{y}_j = -m_j/\lambda_j$ and $k_j \in \mathbf{y}$ is an arbitrary element of the state space. We call **hitting time** to the time of contact of a solution, coming from any branch Y_j, with the switching boundary Σ, $\tilde{t} \equiv \{t \in \mathrm{T} : y_j(k, t) = \tilde{y}, \; k \in \mathrm{Y}_j\}$.

Solutions to initial value problems Let $y(0) = y_0 \in \mathrm{Y}_1$ be fixed.[3] The type of continuation over time depends on the existence and type of collision of the solution with Σ, depending on the hitting time $\tilde{t} = \{t : y_1(y_0, t) = \tilde{y}\}$.

First, if $\tilde{t} < 0$, then the solution will remain in the branch Y_1 and it is $y(t) = \bar{y}_1 + (y_0 - \bar{y}_1)e^{\lambda_1 t}$, behaving like the solution to a regular w ODE. Two cases are possible: (a) the solution is bounded and tends asymptotically to the steady state $\bar{y}_1 \in \mathrm{Y}_1$. In this case, we say there is an **asymptotically stable regular steady state** in branch Y_1; (b) or the solution tends to $\pm\infty$ asymptotically. This case occurs in the following situations: if either there is no regular steady state in branch Y_1, or if there is a **unstable regular steady state** and $\bar{y}_1 \neq y_0$, or if there is a **pseudo-steady state** belonging to Σ that is unstable (we will see the next).

[3] If $y(0) = y_0 \in \mathrm{Y}_2$ we can find the solution applying an analogous reasoning.

Second, if $\tilde{t} = \infty$, that is the case if $\lim_{t\to\infty} y_1(y_0, t) = \tilde{y}$, the solution converges **asymptotically**to $\tilde{y}$, called **boundary steady state**. In this case we write $\bar{y}_\Sigma = \tilde{y}$ and call it a **boundary steady state**. A necessary condition is $\lambda_1 < 0$, and the solution to the ODE is $y(t) = \bar{y}_\Sigma + (y_0 - \bar{y}_\Sigma)\, e^{\lambda_1 t}$.

At last, if $\tilde{t} > 0$ and is finite, then the solution **collides in finite time** with Σ in **finite** time. Two types of continuation are possible: (a) either the solution stops at Σ, in which case we say that there is a stable **pseudo-steady state**, denoted by $\bar{\bar{y}}_\Sigma$, and

$$y(y_0, t) = \begin{cases} y_1(y_0, t), & \text{if } 0 \le t < \tilde{t}, \\ y^p, & \text{if } t \ge \tilde{t}; \end{cases}$$

or, (b) the solution continues into the branch Y_2, for $t > \tilde{t}$, yielding the particular solution

$$y(y_0, t) = \begin{cases} y_1(y_0, t) = \bar{y}_1 + (y_0 - \bar{y}_1)\, e^{\lambda_1 t}, & \text{for } 0 \le t < \tilde{t}, \\ \tilde{y}, & \text{for } t = \tilde{t}, \\ y_2(\tilde{y}, t - \tilde{t}) = \bar{y}_2 + (\tilde{y} - \bar{y}_2)\, e^{\lambda_2 (t - \tilde{t})}, & \text{for } t > \tilde{t}, \end{cases}$$

that can be expressed as a composition of the solution in the two branches $y(y_0) = y_1(y_0, t) \circ y_2(\tilde{y}, t - \tilde{t})$. We can conclude that, if $y_0 \neq \tilde{y}$, and the solution path of an initial value problem collides with Σ, with speed $\left.\frac{dy_1(y_0,t)}{dt}\right|_{t=\tilde{t}} = \lambda_1 (y_0 - \bar{y}_1) e^{\lambda_1 \tilde{t}}$ then it has two types of continuations: (a) either it converges to a steady state in branch Y_2, or (b) will be unbounded in Y_2.

Solutions to terminal value problems. In economics, we are sometimes interested in the solutions of boundary-value problems. This is the case when we have a transversality condition. In this case, a similar reasoning can be made. In Sect. 2.3.3, we saw that terminal-value problems, for regular linear scalar ODE could be unique or be indeterminate (continuum of solutions). From the description of the types of solutions to initial-value problems, we can conclude that: (1) solutions can be unique or multiple, depending on the stability properties of the steady state; (2) and can collide and traverse the switching boundary or not.

6.1.1.2 Normal Forms for Linear Scalar PWSC ODE

Next we will see how we can qualitatively relate different types of solutions to two canonical forms of piecewise continuous or smooth (PWSC) ODE: piecewise-smooth (PWS), and piecewise-continuous (PWC) or **Filippov**ODE.

Proposition 6.1 (Canonical forms of non-smooth ODE) *Consider Eq.* (6.3) *where functions $f_j(.)$ and $h(.)$ take the form in Eq.* (6.7)*, and let $c \neq 0$. Then the following two cases are possible:*

1. *If* $(\lambda_1 - \lambda_2)n = c(m_1 - m_2)$ *and* $\lambda_1 \neq \lambda_2$ *then Eq.* (6.3) *is equivalent to the* ***piecewise-smooth****(PWS) equation*

$$\dot{y} = \begin{cases} \lambda_1 y + \beta, & cy \leq 0, \\ \lambda_2 y + \beta, & cy > 0; \end{cases} \tag{6.8}$$

2. *If* $(\lambda_1 - \lambda_2)n \neq c(m_1 - m_2)$ *then Eq.* (6.3) *is equivalent to the piecewise-continuous (PWC) or* ***Filippov*** *equation*

$$\dot{y} = \begin{cases} \lambda_1 y + \beta_1 & \textit{if } cy \leq 0, \\ \lambda_2 y + \beta_2 & \textit{if } cy > 0. \end{cases} \tag{6.9}$$

Proof Consider Eq. (6.6) with $c \neq 0$ and transform $y - \tilde{y} \mapsto y$. If $\beta = (m_1 c - \lambda_1 n)/c = (m_2 c - \lambda_2 n)/c$ then ODE (6.6) satisfies the properties of the PWS ODE (6.4). It satisfies $f_1(\tilde{y}) = f_2(\tilde{y})$, because $\lambda_1 \tilde{y} + \beta = \lambda_2 \tilde{y} + \beta = 0$, and it also satisfies $f_{1,y}(\tilde{y}) \neq f_{2,y}(\tilde{y})$ because $\lambda_1 \neq \lambda_2$. Both conditions imply it is equivalent to ODE (6.8). If $\beta_1 \equiv (m_1 c - \lambda_1 n)/c$, and $\beta_2 \equiv (m_2 c - \lambda_2 n)/c$ then ODE (6.6) satisfies the properties of the PWC ODE (6.5): $f_1(\tilde{y}) \neq f_2(\tilde{y})$, because $\lambda_1 \tilde{y} + \beta_1 \neq \lambda_2 \tilde{y} + \beta_2$, and $f_{1,y}(\tilde{y}) \neq f_{2,y}(\tilde{y})$ because $\lambda_1 \neq \lambda_2$. In this case it is equivalent to ODE (6.8).

In both cases, we are setting the switching point at $\tilde{y} = 0$, and $\Sigma = \{0\}$ separates Y in two subspaces depending on the level of c: if $c > 0$ then $Y_1 = \{y \in Y : y < 0\}$ and $Y_2 = \{y \in Y : y > 0\}$; or, if $c < 0$ then $Y_1 = \{y \in Y : y > 0\}$, and $Y_2 = \{y \in Y : y < 0\}$.

Next we characterize the solutions to PWS and Filippov equations by determining the number of steady states, their stability properties, and perform a bifurcation analysis for paramaters $\varphi = (\lambda_1, \lambda_2, \beta, c)$ of PWS ODE, and for $\varphi = (\lambda_1, \lambda_2, \beta_1, \beta_2 c)$ of PWC ODE

6.1.1.3 Dynamics for PWS ODE

All the possible phase diagrams for ODE (6.8) are shown in Fig. 6.1 for $\beta > 0$, in Fig. 6.2 for $\beta = 0$, and in Fig. 6.3 for $\beta < 0$.

By inspection, we can also observe that steady states are the only invariants, as in linear scalar ODE, but, differently from these equations, we can have multiplicity of steady states. In this sense, PWS equations share some type of dynamics with non-linear regular ODE, in particular, as regards the existence of different types of local dynamics and the non-coincidence of stable and unstable manifolds with the state space. In Figs. 6.1 and 6.3, the steady states, when they exist, are **regular**, i.e., $\bar{y} \in Y_1 \cup Y_2$, and in Fig. 6.2 we have **boundary steady states**, i.e, $\bar{y}_\Sigma = \tilde{y} = 0 \in \Sigma$.

Regular steady states exist only if $\beta \neq 0$, and belong to the interior of Y_1, Y_2, or both. In particular, we can conclude as regards their existence and stability

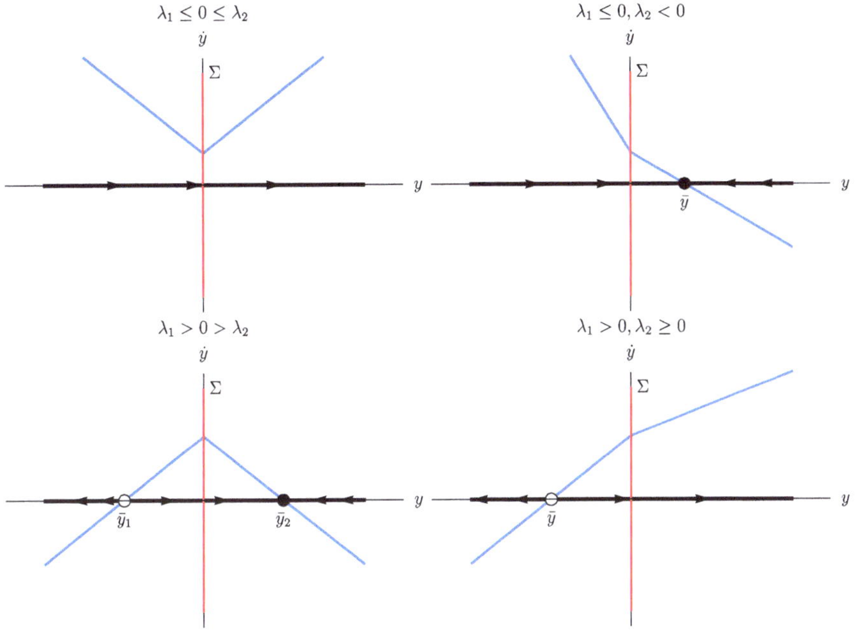

Fig. 6.1 Phase diagrams for PWS ODE where $\beta > 0$

properties that: (1) there are two regular steady states, $\bar{y}_1 < 0 < \bar{y}_2$, with different stability properties, such that, if $\beta > 0$ and $\lambda_1 > 0 > \lambda_2$, $\bar{y}_1$ is unstable and $\bar{y}_2$ is stable, or if $\beta < 0$ and $\lambda_1 < 0 < \lambda_2$, $\bar{y}_1$ is stable and $\bar{y}_2$ is unstable; (2) there is one unique regular steady state, $\bar{y}_1 < 0$ or $\bar{y}_2 > 0$, such that, if $\beta > 0$, $\lambda_1 \le 0$ and $\lambda_2 < 0$, $\bar{y}_2$ is stable, or if $\beta > 0$, $\lambda_1 > 0$ and $\lambda_2 > 0$, $\bar{y}_1$ is unstable, or if $\beta < 0$, $\lambda_1 < 0$ and $\lambda_2 \le 0$, $\bar{y}_1$ is stable, or if $\beta < 0$, $\lambda_1 \ge 0$ and $\lambda_2 > 0$, $\bar{y}_2$ is unstable; (3) there is an infinite number of regular steady states if $\beta = 0$ and $\lambda_1 = 0$ or $\lambda_2 = 0$; or (4) there are no regular steady states if $\beta > 0$ and $\lambda_1 < 0 < \lambda_2$.

We can obtain a generic condition for **local stability**:

Proposition 6.2 (Local asymptotic stability) *If there is at least one steady state, $\bar{y}_j$, it is asymptotically stable if $\lambda_j < 0$.*

Boundary steady states, $\bar{y}_\Sigma = \tilde{y} = 0 \in \Sigma$, exist only if $\beta = 0$ and λ_1 or λ_2 are different from zero. The following cases are possible: first there is a (co-dim 1) **boundary equilibrium bifurcation** if $\beta = 0$, and $\lambda_1 \neq 0$ and $\lambda_2 \neq 0$. We can distinguish two cases: first, there is **persistence** (or border-crossing) if $\beta = 0$ and $\lambda_1 < 0$ and $\lambda_2 < 0$, or $\lambda_1 > 0$ and $\lambda_2 > 0$, in particular there is **stable-stable transition** if $\lambda_1 < 0$ and $\lambda_2 < 0$, or there is **unstable-unstable transition** if $\lambda_1 > 0$ and $\lambda_2 > 0$; or, second, there is a **non-smooth fold** bifurcation if $\beta = 0$ and $\lambda_1 < 0 < \lambda_2$ or $\lambda_1 > 0 > \lambda_2$. Summing up:

Proposition 6.3 (Unfolding of the BEB) *Consider a BEB point, $(y, \beta) = (\tilde{y}, \tilde{\beta}) = (0, 0)$. Then, for a small perturbation of the parameter β, there is:*

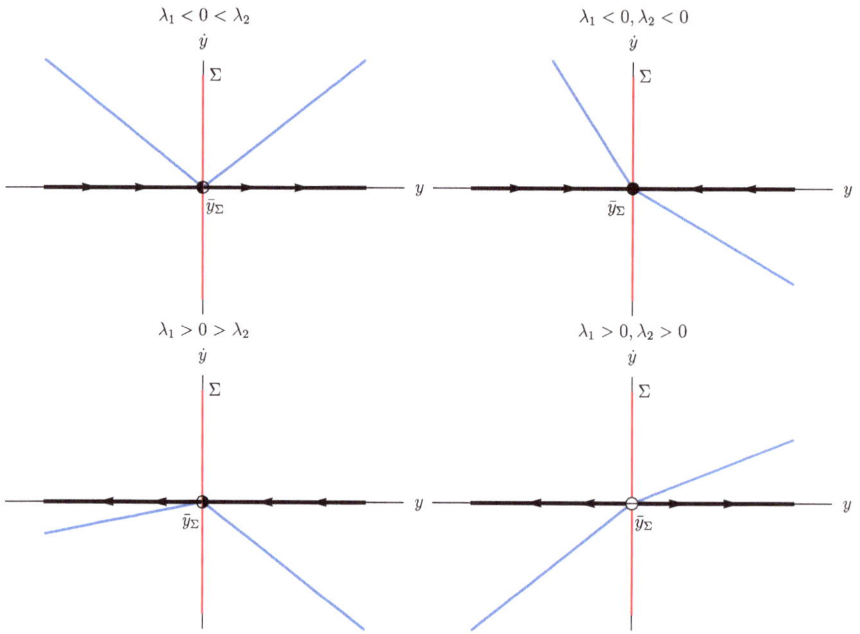

Fig. 6.2 PWS ODE for $\beta = 0$: this is a degenerate case in which there are boundary steady state bifurcations, i.e., $\bar{y} = \tilde{y} = 0$. We can have persistence (second and fourth cases) or a BEB (first and third cases)

1. *persistence if* $\lambda_1 \lambda_2 > 0$;
2. *a non-smooth fold if* $\lambda_1 \lambda_2 < 0$

There is an alternative approach , using differential inclusions, (see Leine and van de Wouw (2008)) that allows for a direct way of finding which type of boundary-equilibrium bifurcation exists. We define a **generalized gradient of the vector field** by

$$\partial \mathbf{f}(y) = \begin{cases} \lambda_1, & \text{if } y < 0, \\ \{\alpha\lambda_1 + (1-\alpha)\lambda_2 : \; 0 \leq \alpha \leq 1\}, & \text{if } y = 0, \\ \lambda_2 & \text{if } y > 0, \end{cases} \tag{6.10}$$

where $\partial \mathbf{f}$ is a weighted average of the coefficients of the two branches with weights α and $1-\alpha$ summing to one. In this approach, we can simply state that there is persistence if $0 \notin \partial \mathbf{f}(0)$, and there is a non-smooth fold if $0 \in \partial \mathbf{f}(0)$. That is, there is a non-smooth fold if there is a weight $\alpha \in (0, 1)$ such that $\alpha\lambda_1 + (1-\alpha)\lambda_2 = 0$.

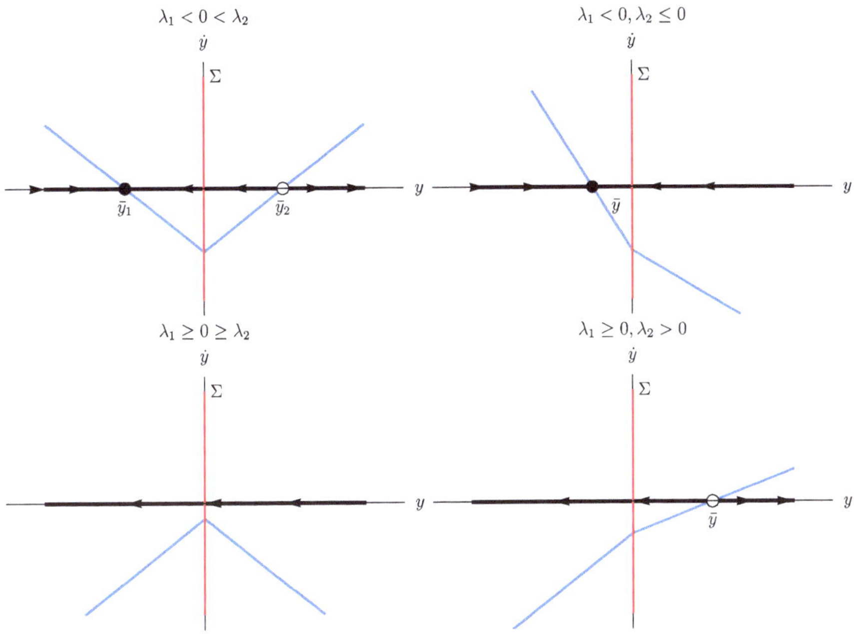

Fig. 6.3 Phase diagrams for PWS ODE where $\beta < 0$: similar to the case in which $\beta > 0$, but with orbits running in the opposite direction

6.1.1.4 Dynamics of Filippov ODE

The Filippov ODE (6.9), assuming that $c \neq 0$, satisfies the following properties for $\tilde{y} = 0$: $f_1(\tilde{y}) \neq f_2(\tilde{y})$ if $\lambda_1 \tilde{y} + \beta_1 \neq \lambda_2 \tilde{y} + \beta_2$, and $f_{1,y}(\tilde{y}) \neq f_{2,y}(\tilde{y})$ if $\lambda_1 \neq \lambda_2$. This ODE generates new dynamics that do not exist in PWS or regular non-linear ODE. In particular, pseudo-steady states can exist.

Definition 6.5 (*Pseudo-steady state*) Consider the Filippov ODE (6.9). A **pseudo-steady state** is a point in the switching boundary, $y^p \in \Sigma$, that is a steady state for the ODE, although it is not a boundary steady state for any of the branches of function $f(y)$.

Therefore, a pseudo-steady state satisfies the following conditions: first, $f_1(y^p) \neq 0$ and $f_2(y^p) \neq 0$; second, there is a convex combination of $f_1(y)$ and $f_2(y)$ allowing for defining the pseudo-steady state set,

$$y^{pss} = \{y \in \Sigma : f_s(y) = (1 - \eta) f_1(y) + \eta f_2(y) = 0, \text{ for } 0 < \eta < 1\}. \quad (6.11)$$

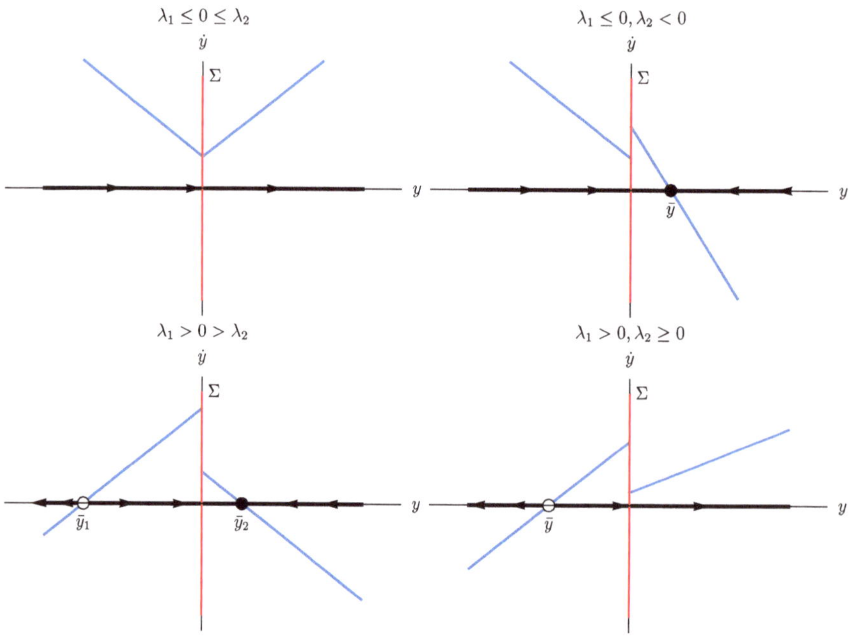

Fig. 6.4 Phase diagrams for Filippov ODE with $\beta_1 > 0$ and $\beta_2 > 0$

Then a pseudo-steady state is a kind of "hidden" steady state in set Σ.[4] As for regular steady states, we can classify it further as stable, if $f_s'(y^p) < 0$, or unstable, $f_s'(y^p) > 0$, for $y^p \in y^{pss}$.

The number of possible phase diagrams in a Filippov ODE is much larger than for the PWS ODE (6.8), as shown in Figs. 6.4, 6.5, 6.6, 6.7, 6.8, 6.9, 6.10, and 6.11. Analyzing those phase diagrams, we can generally find that: first, concerning the number regular steady states: there are no regular steady states, or there is a single steady state, or there are two steady states (as in the PWS system with $\beta \neq 0$) if $\text{sign}(\beta_1) = \text{sign}(\beta_2)$, and $\beta_1 \neq 0$ and $\beta_2 \neq 0$ (see Figs. 6.4 and 6.11); second, when a boundary steady state exists: it can be isolated (as in the PWS case with $\beta = 0$) or it can coexist with another regular equilibrium (differently from the PWS case with $\beta = 0$) if there is one β_j that is equal to zero (see Figs. 6.5, 6.7, 6.8, and 6.10); third, there is a pseudo-steady state (that do not exist for PWS ODE), that can be isolated or co-exist with a regular equilibrium, if $\text{sign}(\beta_1) \neq \text{sign}(\beta_2)$ (see Figs. 6.6 and 6.9).

Assuming that $\beta_1 \neq 0$ and $\beta_2 \neq 0$, we have the following cases: first, only regular steady states exist in the following cases:

1. two regular steady states: $\bar{y}_1 < 0$ and $\bar{y}_2 > 0$:

[4] We can determine the weight to be $\eta = \frac{\beta_1}{\beta_1 - \beta_2}$.

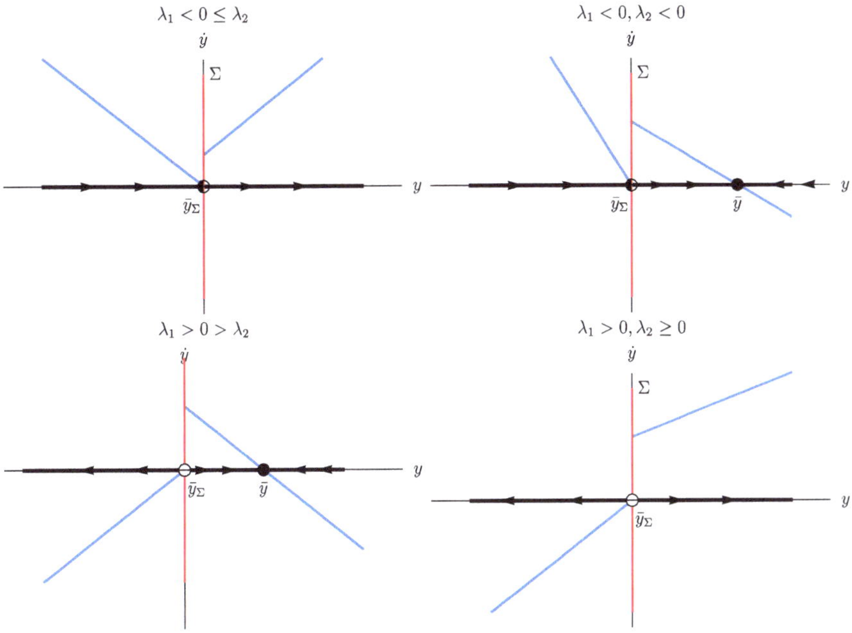

Fig. 6.5 Phase diagrams for Filippov ODE with $\beta_1 = 0$ and $\beta_2 > 0$

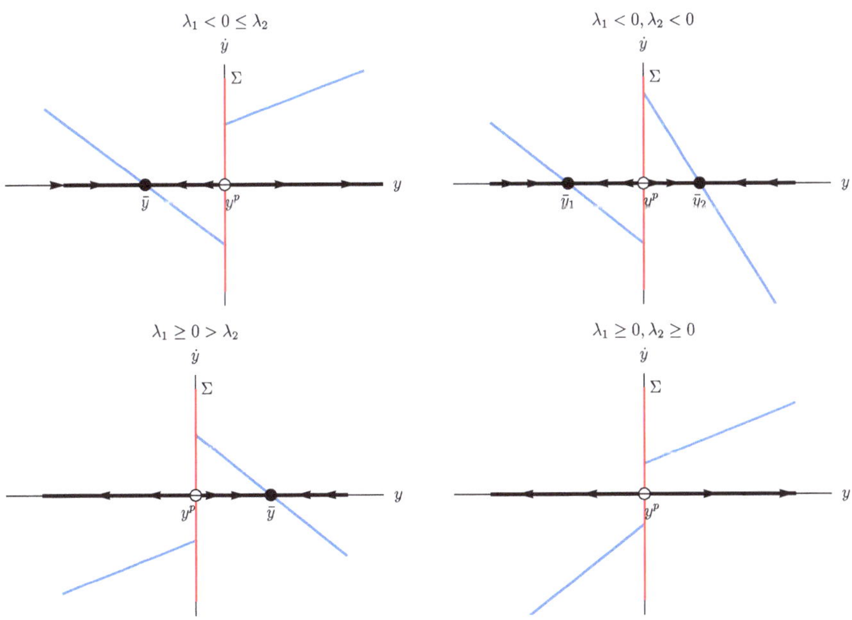

Fig. 6.6 Phase diagrams for Filippov ODE with $\beta_1 < 0$ and $\beta_2 > 0$: existence of pseudo-steady states

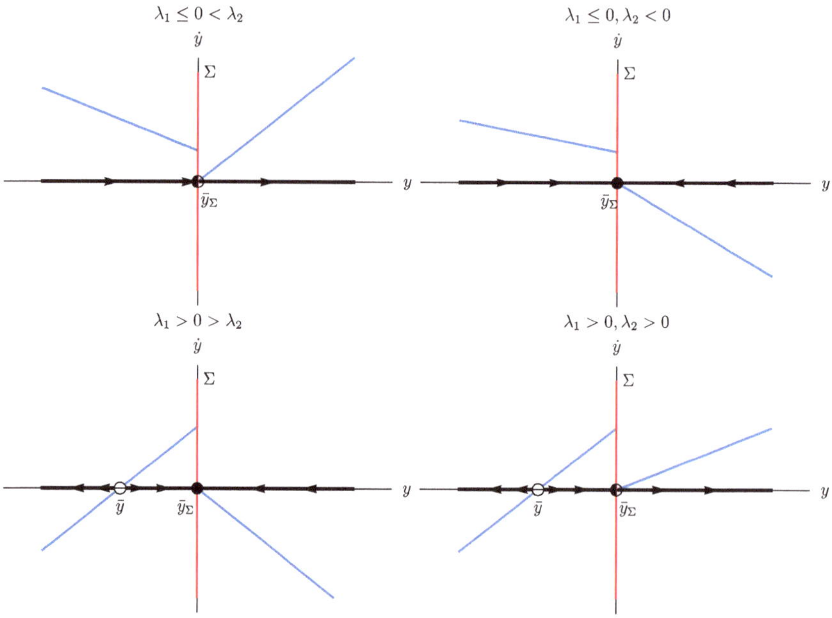

Fig. 6.7 Phase diagrams for Filippov ODE with $\beta_1 > 0$ and $\beta_2 = 0$

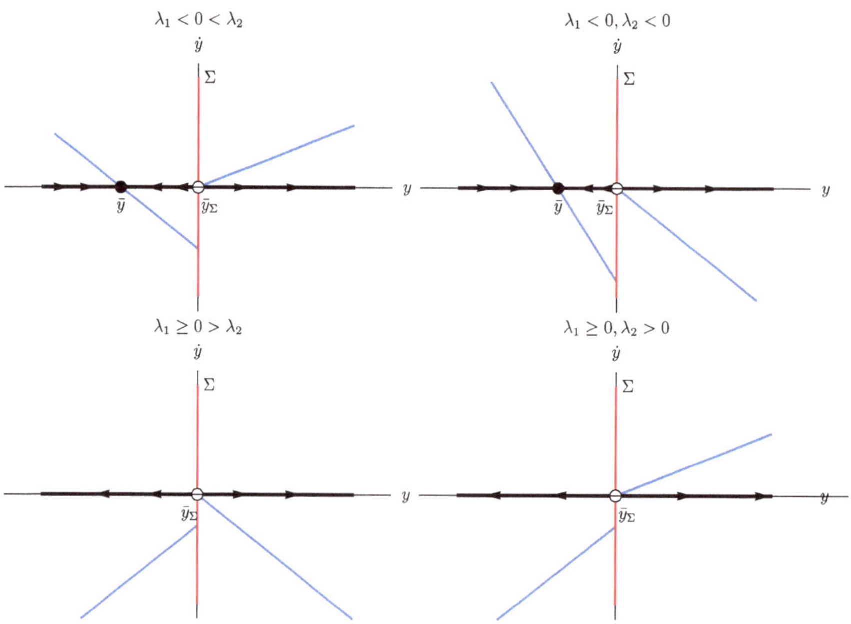

Fig. 6.8 Phase diagrams for Filippov ODE with $\beta_1 < 0$ and $\beta_2 = 0$: boundary equilibrium bifurcations

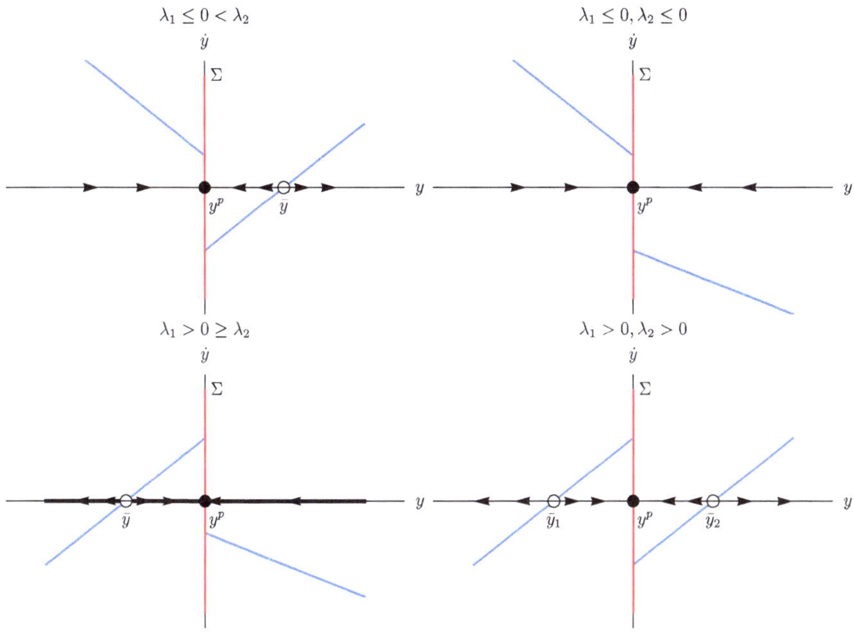

Fig. 6.9 Phase diagrams for Filippov ODE with $\beta_1 > 0$ and $\beta_2 < 0$: existence of pseudo-steady states

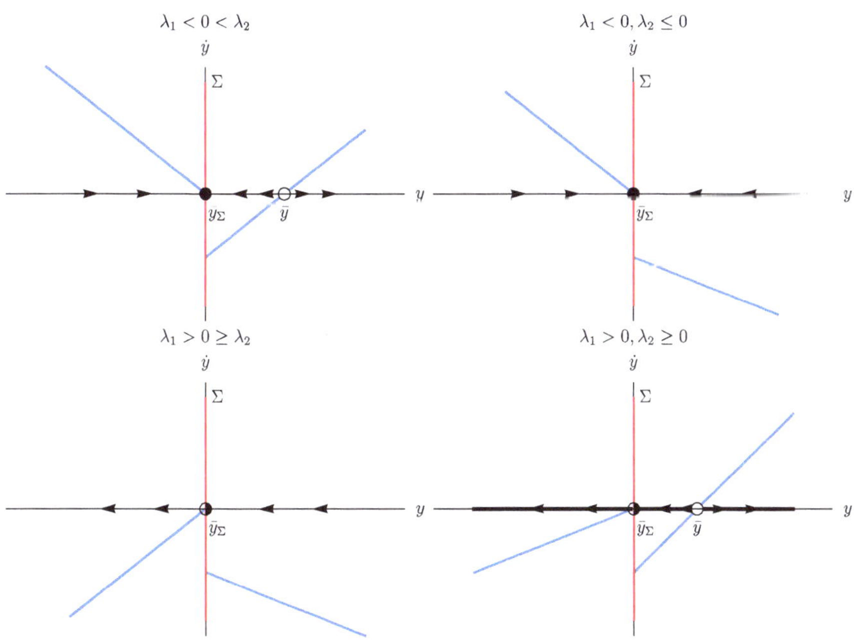

Fig. 6.10 Phase diagrams for Filippov ODE with $\beta_1 = 0$ and $\beta_2 < 0$

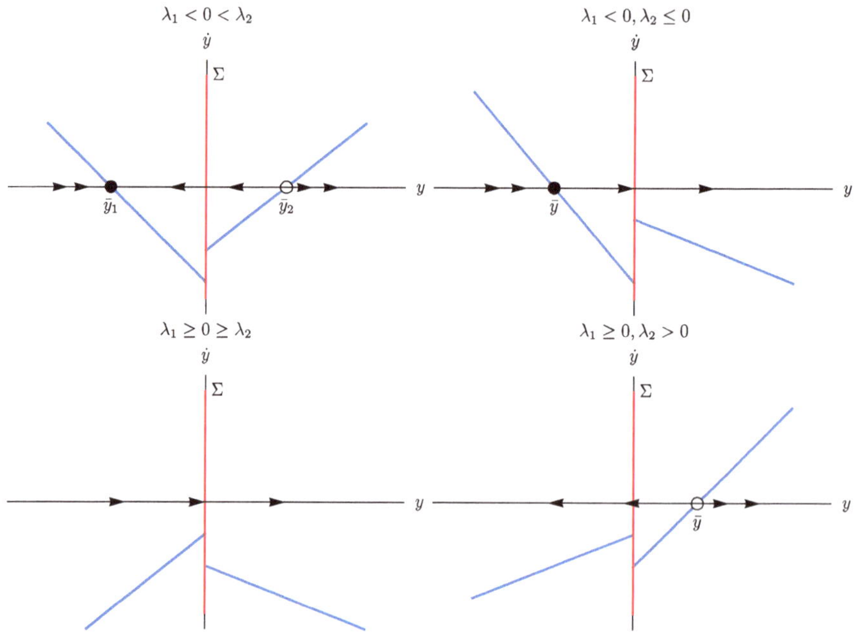

Fig. 6.11 Phase diagrams for Filippov ODE with $\beta_1 < 0$ and $\beta_2 < 0$

$$\begin{cases} \text{if } \beta_1 > 0, \beta_2 > 0, \text{ and } \lambda_1 > 0 > \lambda_2: & \text{then } \bar{y}_1 \text{ is unstable and } \bar{y}_2 \text{ is stable,} \\ \text{if } \beta_1 < 0, \beta_2 < 0, \text{ and } \lambda_1 < 0 < \lambda_2: & \text{then } \bar{y}_1 \text{ is stable and } \bar{y}_2 \text{ is unstable;} \end{cases}$$

2. or one regular steady state: $\bar{y}_1 < 0$ or $\bar{y}_2 > 0$;

$$\begin{cases} \text{if } \beta_1 > 0, \beta_2 > 0, \lambda_1 \le 0, \text{ and } \lambda_2 < 0: & \text{then } \bar{y}_2 \text{ is stable,} \\ \text{if } \beta_1 > 0, \beta_2 > 0, \lambda_1 > 0, \text{ and } \lambda_2 \ge 0: & \text{then } \bar{y}_1 \text{ is unstable,} \\ \text{if } \beta_1 < 0, \beta_2 < 0, \lambda_1 < 0, \text{ and } \lambda_2 \le 0: & \text{then } \bar{y}_1 \text{ is stable,} \\ \text{if } \beta_1 < 0, \beta_2 < 0, \lambda_1 \ge 0, \text{ and } \lambda_2 > 0: & \text{then } \bar{y}_2 \text{ is unstable;} \end{cases}$$

second, both regular and pseudo-steady states exist in the following cases:

1. two regular steady states, $\bar{y}_1 < 0$ and $\bar{y}_2 > 0$ and one pseudo-steady state $y^p = 0$:

$$\begin{cases} \text{if } \beta_1 < 0, \beta_2 > 0, \lambda_1 < 0, \text{ and } \lambda_2 < 0: & \text{then } \bar{y}_1 \text{ and } \bar{y}_2 \text{ are stable, and } y^p \text{ is unstable,} \\ \text{if } \beta_1 > 0, \beta_2 < 0, \lambda_1 > 0, \text{ and } \lambda_2 > 0: & \text{then } \bar{y}_1 \text{ and } \bar{y}_2 \text{ are unstable, and } y^p \text{ is stable;} \end{cases}$$

2. or one regular steady state, $\bar{y}_1 < 0$ or $\bar{y}_2 > 0$, and one pseudo-steady state $y^p = 0$:

$$\begin{cases} \text{if } \beta_1 < 0, \beta_2 > 0, \text{ and } \lambda_1 > 0 > \lambda_2: \text{ then } \bar{y}_2 \text{ is stable, and } y^p \text{ is unstable,} \\ \text{if } \beta_1 < 0, \beta_2 > 0, \text{ and } \lambda_1 < 0 < \lambda_2: \text{ then } \bar{y}_1 \text{ is stable, and } y^p \text{ is unstable,} \\ \text{if } \beta_1 > 0, \beta_2 < 0, \text{ and } \lambda_1 \leq 0 < \lambda_2: \text{ then } \bar{y}_2 \text{ is unstable, and } y^p \text{ is stable,} \\ \text{if } \beta_1 > 0, \beta_2 < 0, \text{ and } \lambda_1 > 0 \geq \lambda_2: \text{ then } \bar{y}_1 \text{ is unstable, and } y^p \text{ is stable.} \end{cases}$$

Assuming that $\beta_1 = 0$ or (exclusive) $\beta_2 = 0$ we have **boundary equilibrium bifurcations**.[5] There are 16 different phase diagrams (in Figs. 6.5, 6.7, 6.9 and 6.10) displaying bifurcations of several types :

1. **persistence** between regular steady states and pseudo-steady states:
   a. from one regular to one pseudo-steady state: if $\beta_1 = 0$, $\beta_2 > 0$; $\lambda_1 > 0$, $\lambda_2 \geq 0$,
   b. from two regular to one regular equilibrium and one pseudo-steady state: if $\beta_1 = 0$, $\beta_2 > 0$; $\lambda_1 > 0 > \lambda_2$
2. **non-smooth folds**:
   a. from zero regular to two steady states (one regular and one pseudo-steady state): if $\beta_1 = 0$, $\beta_2 > 0$ and $\lambda_1 < 0 < \lambda_2$
   b. from one regular to three steady states (two regular and one pseudo-steady state): if $\beta_1 = 0$, $\beta_2 > 0$; $\lambda_1 < 0$, $\lambda_2 < 0$

All the previous phase diagrams considered, we can characterize the solution path from the perspective of its contact with the switching boundary Σ. Assuming there is not a (regular or bifurcation) boundary steady state, there are three types of contacts: **no contact**, **traverse**, **collision**, and **escape**. When pseudo-steady states do not exist—this is always the case in piecewise smooth (PWS) ODE—, in the absence of boundary steady states, we observe that trajectories always **traverse** the switching boundary Σ. When pseudo-steady states exist the solution path does **not traverse** Σ, and has a behavior that looks like converging to a regular steady state. Furthermore, if a pseudo-steady state is stable (as in Fig. 6.9) the solution paths **collides** with Σ, and if the pseudo-steady state is unstable (as in Fig. 6.6) the solution paths **escape** from Σ.

6.1.2 *Scalar Non-linear PWSC ODE*

Now consider a non-linear scalar piecewise smooth or continuous ODE

$$\dot{y} = \begin{cases} f_1(y, \varphi), & y \leq 0, \\ f_2(y, \varphi), & y > 0, \end{cases} \tag{6.12}$$

[5] When $\beta_1 = \beta_2 = \beta = 0$ we have a PWS case in which $\beta = 0$.

where functions $f_j(y, \varphi)$ are linear, as in ODE (6.8) or (6.9), or are non-linear and regular as those addressed in Chap. 3. Combining the types of discontinuity with the possible non-linear forms, we conclude that the number of cases is beyond enumeration.

However, we can highlight some important definitions and facts. First, if $f_1(y, \varphi) = f_2(y, \varphi) = 0$, for $y \in \Sigma$, then ODE (6.12) is a PWS ODE, or, if $f_1(y, \varphi) \neq f_2(y, \varphi) = 0$, for $y \in \Sigma$, it is a Filippov ODE. Second, all that we have previously presented about steady states, their number and their stability properties and bifurcations, for the linear case hold. Third, but additionally, defining the set of regular steady states in sub-branch, Y_j as

$$y_j^{ss}(\varphi) = \{y \in \mathbf{y}_j : f_j(y, \varphi) = 0\},$$

we can have the following types of bifurcations: **regular (or smooth) fold bifurcation**[6] if $f_{j,y}(y) = 0$ for $y \in y_j^{ss}(\varphi)$, or a **tangent point** if $f_{j,y}(y) = 0$ for $y \in \Sigma$. Fourth, we can also have bifurcations related to pseudo-steady states if $f_{j,y}(y^p) = 0$ for $y^p \in y^{pss}$. The next example illustrates the existence of a regular fold bifurcation for a Filippov ODE

Example 6.1 Consider the ODE, having parameters $\boldsymbol{\varphi} = (\lambda_1, \varphi_1, \lambda_2, \varphi_2) \in \mathbb{R}^4$,

$$\dot{y} = \begin{cases} \lambda_1 y + \varphi_1, & \text{if } y \leq 0, \\ -y^2 + \lambda_2 y + \varphi_2, & \text{if } y > 0. \end{cases} \tag{6.13}$$

Figure 6.12 displays phase diagrams assuming positive fixed values for φ_1, λ_1 and λ_2 and for several values for φ_2. We observe that: there is a boundary equilibrium bifurcation (BEB), of the non-smooth fold type, for $\varphi_2 = 0$; there is one pseudo equilibrium for $\varphi_2 < 0$, and there is a regular fold bifurcation point for $\varphi_2 = \varphi_2^* < 0$. Observe that, at each bifurcation point the number of steady states (including the pseudo-steady state) changes. In particular, at the BEB there is a change from two regular steady states (for $\varphi_2 > 0$) to three regular steady states and one pseudo-steady state (for $0 > \varphi_2 > \varphi_2^*$). Furthermore, while the solution paths cross the switching point from branch Y_1 to branch Y_2, if a pseudo-steady state exists, and because it is stable, the solution collides to the switching boundary and does not traverse it. This would be a surprising fact if we just have determined the steady states as in a regular ODE.

Example 6.2 (*Endogenous markup with regime change*) Brito et al. (2013) postulate an economy in which the supply side is characterized by Cournotian monopolistic competition (CMC), in the production of intermediate goods, and perfect competition in the final goods' producing sector. The representative producer of the final good uses a continuous mass of input varieties, with a constant elasticity of substitution (CES) technology. The production technology in each intermediate goods' industry

---

[6] We use the notation $f_{j,y}(y) \equiv \frac{\partial f_j(y)}{\partial y}$.

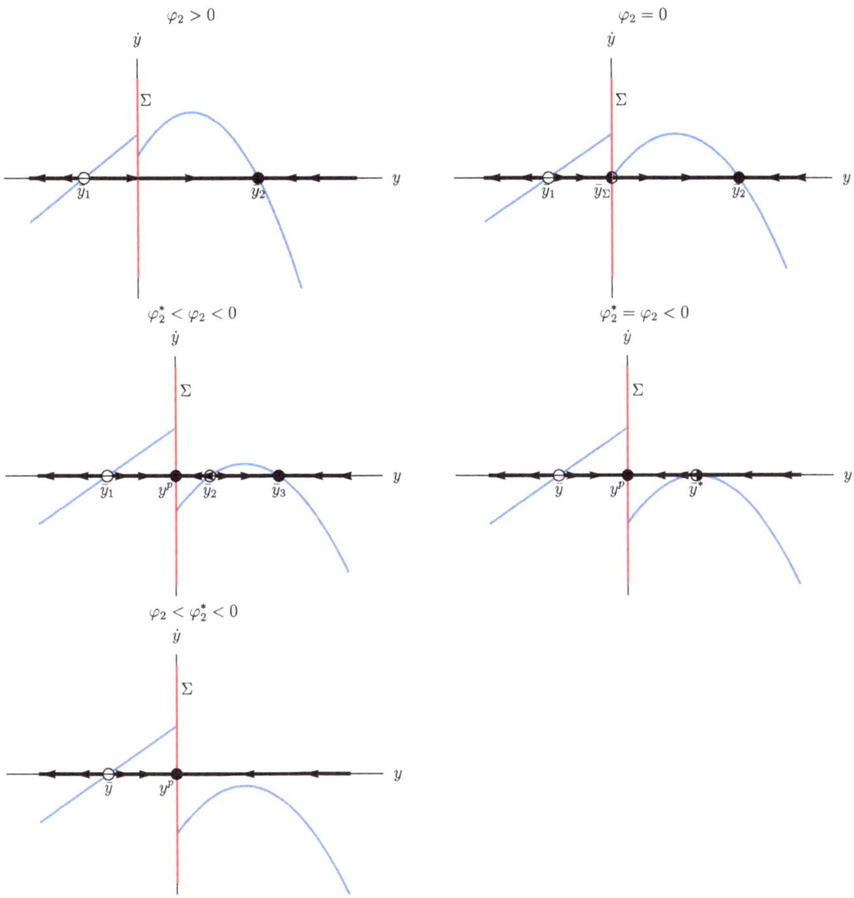

Fig. 6.12 Phase diagrams for Eq. (6.13), for $\lambda_1 > 0$, $\varphi_1 > 0$, $\lambda_2 > 0$ and several values for φ_2

is specified by a Cobb-Douglas production function, using labor and capital inputs, and have a symmetric fixed cost ϕ. In every intermediate good producing industry either there is monopoly or, when new firms enter, oligopoly. When this is the case, firms compete *à la Cournot*, but take the general price level as given. An entrant has to decide either to enter an existing industry, which makes the number of firms per industry endogenous, or has to create a new input variety in which it becomes a (temporary) monopolist.

Profit maximization, free entry, and symmetry entail the existence of a Lerner markup, $\mu = \max\{1 - \sigma^{-1}, m(k)\}$, that is endogenous and piecewise smooth. If the economy is in the (M) monopoly regime, with just one variety and one monopolist, the markup is $\mu = 1 - \sigma^{-1}$, where σ is the inverse of the elasticity of substitution between input varieties in the production of the final good (which acts as the demand elasticity for every intermediate producer). However, if it is in a CMC regime (brought about one or each of the two types of entry) the markup is $m(k) = \sqrt{\frac{\phi}{\sigma k^{\alpha}}}$, where α is the share of capital in the production of each intermediate good, and k is the per-capita capital stock. This entails a switching boundary between the two regimes that is obtained by equalizing the markups in the two regimes, $\Sigma = \{k : m(k) = \sigma^{-1}\}$.

Assuming, as in the Solow (1956) model, a "Keynesian" savings function[7] and a macroeconomic equilibrium condition between savings and (gross) investment, then the capital accumulation equation is the following PWSC ODE,[8] where we highlight the role of the fixed cost, ϕ,

$$\dot{k} = \begin{cases} F_1(k, \phi) = s\, Y_1(k) - \delta\, k, & \text{if } k \leq k_{\Sigma}(\phi), \\ F_2(k, \phi) = s\, Y_2(k) - \delta\, k, & \text{if } k \geq k_{\Sigma}(\phi), \end{cases} \tag{6.14}$$

where $k_{\Sigma} = (\phi\, \sigma)^{\frac{1}{\alpha}}$, is the switching level for capital intensity. It separates the domain $\mathrm{K} = \mathbb{R}_+$ into two subsets in which different macroeconomic regimes prevail: monopoly (M) in which $k \in \mathrm{K}_1 = [0, k_{\Sigma}]$ is relatively low, or CMC in which $k \in \mathrm{K}_2 = [k_{\Sigma}, \infty)$ is higher. The aggregate output function is a piecewise smooth function

$$Y(k) = \begin{cases} Y_1(k) = \left(1 - \dfrac{1}{\sigma}\right) k^{\alpha}, & \text{if } k \in \mathrm{K}_1, \\ Y_2(k, \phi) = \Big(1 - m(k, \phi)\Big)\, k^{\alpha}, & \text{if } k \in \mathrm{K}_2, \end{cases}$$

as is illustrated in Fig. 6.13a. We also observe that the capital accumulation equation is a piecewise smooth ODE (not Filippov) because the transition between the two states is regulated by free entry.

Comparing to Fig. 3.7a, for the Skiba model in Example 3.4, with a smooth convex-concave production function, we observe that they share a similar form. However, differently from that model we can provide an explicit microeconomic justification for that shape. In the present example, we can associate market competitiveness to the level of k: if k is low the economy tends to be in a M regime, and if k is high the economy tends to be in a CMC regime, in which both the number of firms per industry, and the number of input varieties are greater.

Comparing the bifurcation diagrams, in Figs. 3.7e and 6.13e, we observe that both display a similar type of hysteresis curve. But there are two differences: first, in this model, we can choose an economic meaningful parameter: the fixed cost in the production of an intermediate good, ϕ; and, second, while the Skiba model displays a regular hysteresis curve, in this model, we have a discontinuity induced hysteresis bifurcation curve. There are two hysteresis points, a regular hysteresis point $(k^*, \phi^*) = \{(k, \phi) : F_2(k, \phi) = F_{2,k}(k, \phi) = 0\}$, and a boundary steady state bifurcation point $(\tilde{k}, \tilde{\phi}) = \{(k, \phi) : F_1(k, \phi) = F_2(k, \phi) = 0\}$.

---

7 We are assuming also that population is constant, the labor supply is inelastic, and that there is no labor and financial market imperfections. For a model including the above-mentioned technology and product market structure within a DGE model, see Example 6.4.

8 Another way to express the ODE is $\dot{k} = s\; \max\{Y_1(k), Y_2(k)\} - \delta k$.

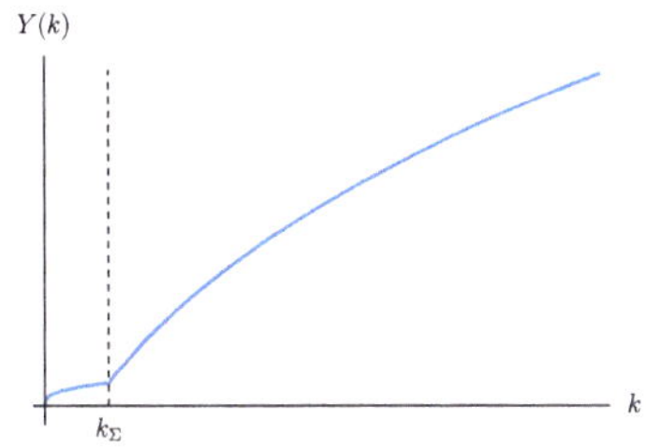

(a) Production function

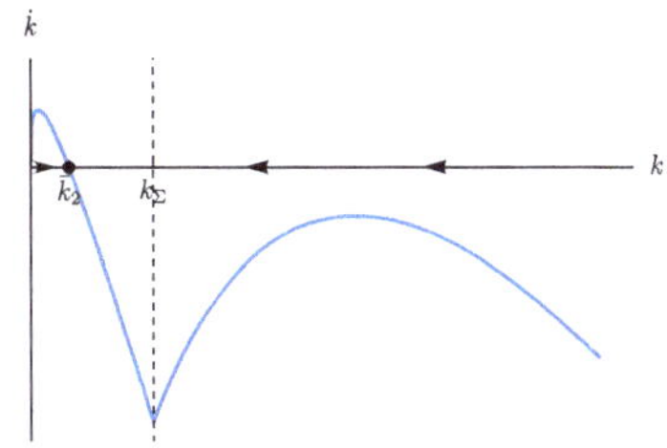

(b) Phase diagram for a high ϕ: $\phi > \phi^*$

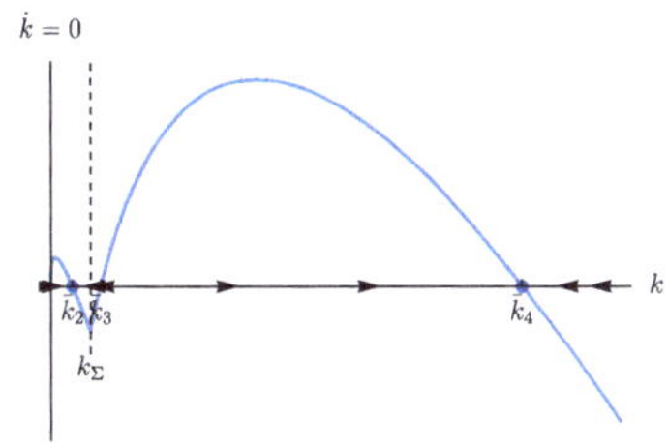

(c) Phase diagram for a intermediate ϕ: $\tilde{\phi} < \phi < \phi^*$

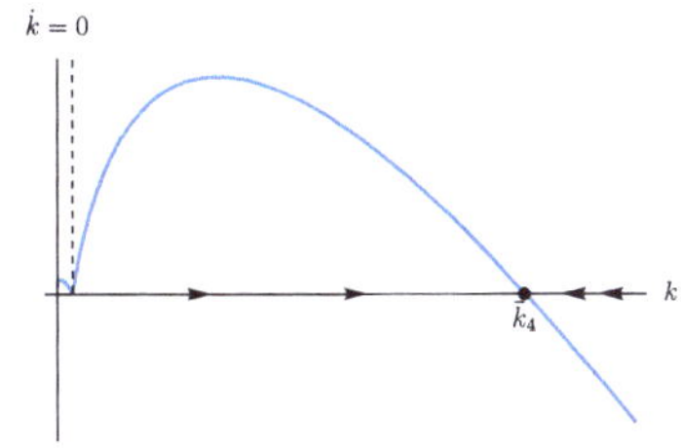

(d) Phase diagram for a low ϕ: $\phi < \tilde{\phi}$

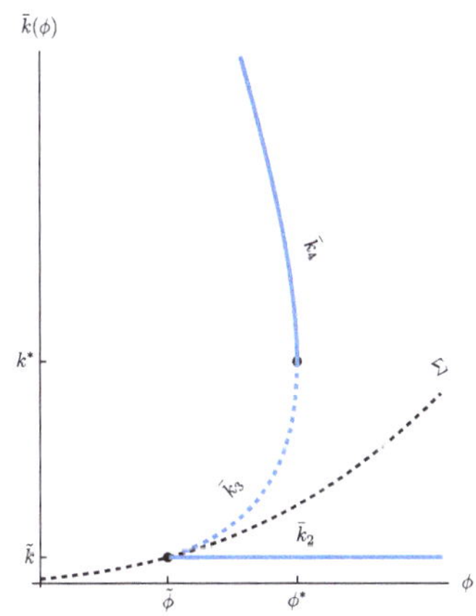

(e) Bifurcation diagram

Fig. 6.13 The "ad-hoc" version of the Brito et al. (2013) model for $0 < \alpha < 1$, $0 < s < 1$, $\delta > 0$, $\sigma > 1$, and different values for $\phi > 0$

Assuming the same values for the parameters $(\alpha, \sigma, s, \delta)$, there are three types of generic macroeconomic dynamics. First, if the fixed cost is high, $\phi > \tilde{\phi}$ there is an unique stable steady state with monopoly, $\bar{k}_2 \in K_1$ (see Fig. 6.13), and even if $k_0 \in K_2$, corresponding to a CMC regime, the stock of capital will cross, from K_2 to K_1, the switching barrier an converge to the low level $\bar{k}_2$, which is a poverty trap. Second, if the fixed cost has an intermediate value, $\tilde{\phi} < \phi < \phi^*$, then there is a stable steady state with monopoly, $\bar{k}_2 \in K_1$ (see Fig. 6.13c) and two steady states with CMC, $\bar{k}_3 < \bar{k}_4$, in which $\bar{k}_3$ is unstable and $\bar{k}_4$ is stable. Furthermore, if $k_0 < \bar{k}_3$ then the stock of capital converges to $\bar{k}_2 \in K_1$ and if $k_0 > \bar{k}_3$ it converges to $\bar{k}_4 \in K_2$. Therefore, again it is possible that starting from an initial CMC regimes, the stock of capital crosses back to a poverty trap monopoly regime, but if the economy is sufficiently far away from the switching barrier, it converges to a higher level of capital intensity, and therefore income per capita. At last, if the fixed cost is low, $\phi < \phi^*$ there is again one unique a stable CMC steady state, $\bar{k}_4 \in K_2$ (see Fig. 6.13d), and even if $k_0 \in K_1$, corresponding to a monopoly regime, the stock of capital will cross the switching barrier, from K_1 to K_2, an converge to high level $\bar{k}_4$. An interesting fact is that the width of the stable manifold of the monopoly state, and therefore the attraction of the poverty trap, increases with the fixed cost, and with a reduction of the elasticity of substitution between varieties: it can be $\mathcal{W}_1^s = K = \mathbb{R}_+$ in the first case, $\mathcal{W}_1^s = K_1 \cup [\tilde{k}, \bar{k}_2)$, where $[\tilde{k}, \bar{k}_2) \in K_2$ in the second case, and it is empty in the third case (with a low ϕ). This also means that, when there is barrier crossing, the stable and unstable submanifolds intersect over the switching barrier.

6.2 Planar PWSC ODE

In this section, we go back to the PWSC ODE (6.2), of function $\mathbf{y} : T \rightarrow \mathbf{Y} \subseteq \mathbb{R}^2$, where set $\mathbf{Y}$ is partitioned into two open subsets, as in Eq. (6.1), where the switching manifold is defined by $H(\mathbf{y}) = 0$ such that $D_{\mathbf{y}} H(\mathbf{y}) \neq 0$ for all $\mathbf{y} \in \Sigma$.

Next, we present an introduction to the qualitative approach to the dynamics of the PWSC system, called the Filippov's convex method, in Sect. 6.2.1, we characterize the types of solutions close to the switching boundary, in Sect. 6.2.2, which enables the setting up of a taxonomy of the different types of steady states in PWSC ODE, in Sect. 6.2.3. The next subsections present an introduction to bifurcation analysis of the two branch case, in Sect. 6.2.4, and of discontinuity induced bifurcations, in Sect. 6.2.5.

6.2.1 Filippov's Convex Method

As $\mathbf{f}(\mathbf{y})$ may not be defined if $\mathbf{y} \in \Sigma$, the following formalism of differential inclusions is used

$$\dot{\mathbf{y}} \in \partial \mathbf{f}(\mathbf{y}) = \begin{cases} \mathbf{f}_1(\mathbf{y}), & \text{if } \mathbf{y} \in \mathbf{Y}_1 \iff H(\mathbf{y}) \leq 0, \\ \overline{\text{co}}\{\mathbf{f}_1(\mathbf{y}), \mathbf{f}_2(\mathbf{y})\}, & \text{if } \mathbf{y} \in \Sigma \iff H(\mathbf{y}) = 0, \\ \mathbf{f}_2(\mathbf{y}), & \text{if } \mathbf{y} \in \mathbf{Y}_2 \iff H(\mathbf{y}) \geq 0, \end{cases} \tag{6.15}$$

where the convex hull containing $\mathbf{f}_1$ and $\mathbf{f}_2$ is

$$\overline{\text{co}}\{\mathbf{f}_1(\mathbf{y}), \mathbf{f}_2(\mathbf{y})\} = \{(1-\alpha)\,\mathbf{f}_1(\mathbf{y}) + \alpha\,\mathbf{f}_2(\mathbf{y}) : \; 0 \leq \alpha \leq 1\}.$$

This is called the Filippov convex specification of the PWSC ODE, and the solution to the ODE (6.15) is called the solution in the sense of Filippov.

We can apply the same method, as the one used for scalar equations, for determining a **generalized Jacobian**

$$\partial \mathbf{f}(\mathbf{y}) = \begin{cases} \mathrm{D}_{\mathbf{y}}\mathbf{f}_1(\mathbf{y}), & \text{if } \mathbf{y} \in \mathbf{Y}_1, \\ \overline{\text{co}}\{\mathrm{D}_{\mathbf{y}}\mathbf{f}_1(\mathbf{y}), \mathrm{D}_{\mathbf{y}}\mathbf{f}_2(\mathbf{y})\}, & \text{if } \mathbf{y} \in \Sigma, \\ \mathrm{D}_{\mathbf{y}}\mathbf{f}_2(\mathbf{y}), & \text{if } \mathbf{y} \in \mathbf{Y}_2, \end{cases}$$

where

$$\overline{\text{co}}\{\mathrm{D}_{\mathbf{y}}\mathbf{f}_1(\mathbf{y}), \mathrm{D}_{\mathbf{y}}\mathbf{f}_2(\mathbf{y})\} = \{(1-\alpha)\,\mathrm{D}_{\mathbf{y}}\mathbf{f}_1(\mathbf{y}) + \alpha\,\mathrm{D}_{\mathbf{y}}\mathbf{f}_2(\mathbf{y}) : \; 0 \leq \alpha \leq 1\}.$$

If we evaluate the generalized Jacobian at $\mathbf{y} \in \Sigma$ we can also determine the associated **generalized eigenvalues** $\{\lambda : \det(\partial \mathbf{f}(\mathbf{y}) - \lambda\,\mathbf{I}) = 0, \mathbf{y} \in \mathbf{Y}\}$.

This formalism is sufficiently general to encompass the two types of PWSC, both piecewise smooth and piecewise continuous, if we assume that $\mathbf{y}$ is a continuous function of time[9]: **PWS** ODE satisfy

$$\mathbf{f}_1(\mathbf{y}) = \mathbf{f}_2(\mathbf{y}), \; \mathrm{D}_{\mathbf{y}}\mathbf{f}_1(\mathbf{y}) \neq \mathrm{D}_{\mathbf{y}}\mathbf{f}_2(\mathbf{y}), \; \text{if } \mathbf{y} \in \Sigma,$$

implying that $\mathbf{f}(\mathbf{y})$ is single-valued but $\partial \mathbf{f}(\mathbf{y})$ is set-valued at Σ, and **Filippov** ODE satisfy

$$\mathbf{f}_1(\mathbf{y}) \neq \mathbf{f}_2(\mathbf{y}), \; \text{if } \mathbf{y} \in \Sigma,$$

implying $\mathcal{F}(\mathbf{y})$ is set-valued at Σ.

Solutions to PWSC. The theory of differential inclusions (see Leine and van de Wouw, 2008) guarantees existence, however, in some cases, uniqueness may not hold. As for the scalar case, there are two types of solutions: first, a solution is called **standard** if, given an initial value $\mathbf{y}_0 \in \mathbf{Y}_j$, the solution trajectory lies in the same branch $\mathbf{Y}_j$. In this case, the solution can be characterized with the methods for regular ODE presented in the previous chapters; second, a solution is called **non-standard** if it collides with Σ. In this case, several types of solution paths can take place. The

[9] That is, we are excluding hybrid ODE.

general solution to a specific PWSC ODE, if there is collision with Σ, is obtained by concatenating the two types of solution (standard and non-standard).

However, the dynamics at the switching boundary needs to be closely inspected.

6.2.2 Dynamic Behavior at the Switching Boundary

The projections of the vector fields $\mathbf{f}_1$ and $\mathbf{f}_2$ onto the normal of the surface manifold $H(\mathbf{y}) = 0$ are can be analyzed by considering the Lie derivatives

$$\mathcal{L}_{\mathbf{f}_j}(H(\mathbf{y})) = \mathrm{D}_{\mathbf{y}} H(\mathbf{y})\, \mathbf{f}_j(\mathbf{y}), \quad \text{for } \mathbf{y} \in \Sigma,$$

where $\mathrm{D}_{\mathbf{y}} H(\tilde{\mathbf{y}})$ is the vector normal to Σ.[10] Let

$$\sigma(\mathbf{y}) \equiv \mathcal{L}_{F_1}(H(\mathbf{y})).\mathcal{L}_{F_2}(H(\mathbf{y})),$$

be the product of the Lie derivatives at the two sides of the switching boundary. It can be shown that there are two main types of contacts of the generalized vector field with Σ.

First, there is **transversal crossing** at a point $\mathbf{y} \in \Sigma$ (see Fig. 6.14a) if $\mathcal{L}_{\mathbf{f}_1}(H(\mathbf{y})) \cdot \mathcal{L}_{\mathbf{f}_2}(H(\mathbf{y})) > 0$ for $\mathbf{y} \in \Sigma$. In this case, the flow exits from $\mathbf{Y}_1$ and enters into $\mathbf{Y}_2$ if $\mathcal{L}_{\mathbf{f}_1}(H(\tilde{\mathbf{y}})) > 0$ and $\mathcal{L}_{\mathbf{f}_2}(H(\tilde{\mathbf{y}})) > 0$, and exits from $\mathbf{Y}_2$ and enters into $\mathbf{Y}_1$ if $\mathcal{L}_{\mathbf{f}_1}(H(\tilde{\mathbf{y}})) < 0$ and $\mathcal{L}_{\mathbf{f}_2}(H(\tilde{\mathbf{y}})) < 0$. The **crossing set** is the, generally open, set

$$\Sigma_t = \{\mathbf{y} \in \Sigma : \sigma(\mathbf{y}) > 0\},$$

that is the set of all transversal crossing points in Σ.

Second, there is **sliding** at a point $\mathbf{y} \in \Sigma$ if $\mathcal{L}_{\mathbf{f}_1}(H(\mathbf{y})) \cdot \mathcal{L}_{\mathbf{f}_2}(H(\mathbf{y})) \leq 0$ for $\mathbf{y} \in \Sigma$, if flows starting close to Σ and coming from branch $\mathbf{Y}_1$ and/or $\mathbf{Y}_2$, enter into contact with Σ and slide along it. The **sliding set**,

$$\Sigma_s = \{\mathbf{y} \in \Sigma : \sigma(\mathbf{y}) \leq 0\},$$

is the union of closed sliding subsets and of isolated sliding points.

At last, there is a **tangent point** if there is at least one branch $\mathbf{Y}_j$, for $j \in \{1, 2\}$, such that there is a point in the sliding set satisfying $\mathcal{L}_{\mathbf{f}_j}(H(\tilde{\mathbf{y}})) = 0$, for $\tilde{\mathbf{y}} \in \Sigma_s$. At those points, the flow is tangent to Σ from side $\mathbf{Y}_j$. If $\mathcal{L}_{\mathbf{f}_j}(H(\mathbf{y})) = 0$, for $j = 1, 2$, for $\tilde{\mathbf{y}} \in \Sigma_s$ and $\mathcal{L}_{\mathbf{f}_1 - \mathbf{f}_2}(H(\tilde{\mathbf{y}})) = 0$ then this point is a **singular tangent point** where both vectors $\mathbf{f}_1$ and $\mathbf{f}_2$ are tangent to Σ, or one or both vanish.

The sliding motion, associated to each sliding point, only exists for Filippov ODE. We can further classify sliding points according to their stability properties:

1. there is an **attractive sliding mode** at a point $\tilde{\mathbf{y}} \in \Sigma_s$ (see Fig. 6.14b) if $\mathcal{L}_{\mathbf{f}_1}(H(\tilde{\mathbf{y}})) > 0$ and $\mathcal{L}_{\mathbf{f}_2}(H(\tilde{\mathbf{y}})) < 0$. It exists if a solution flow reaching Σ does

[10] We can write $\dot{h}_j = \mathcal{L}_{\mathbf{f}_j}(H(\mathbf{y}))$ as the Lie derivative represents the time variation of the vector field $\mathbf{f}_j$ in the neighborhood of the switching boundary.

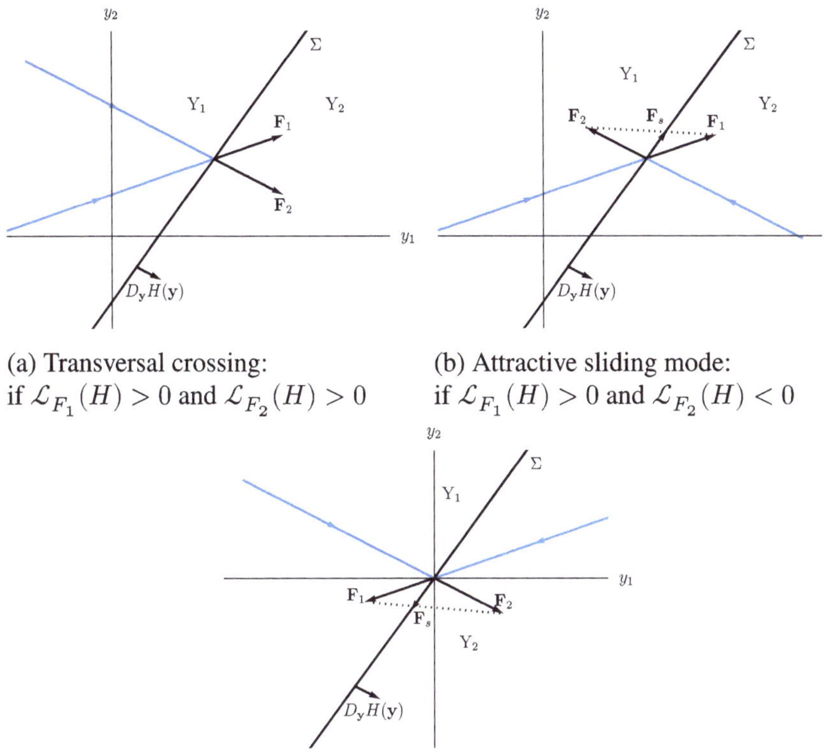

(a) Transversal crossing: if $\mathcal{L}_{F_1}(H) > 0$ and $\mathcal{L}_{F_2}(H) > 0$

(b) Attractive sliding mode: if $\mathcal{L}_{F_1}(H) > 0$ and $\mathcal{L}_{F_2}(H) < 0$

(c) Repulsive sliding mode: if $\mathcal{L}_{F_1}(H) < 0$ and $\mathcal{L}_{F_2}(H) > 0$

Fig. 6.14 Types of dynamics in the neighborhood of Σ

not leave the switching surface, and it is unique in forward time. In this case, the set of all attractive points in Σ_s is called a **stable sliding** segment

$$\Sigma_s^s = \{\mathbf{y} \in \Sigma_s : \mathcal{L}_{\mathbf{f}_1}(H(\mathbf{y})) > 0 \text{ and } \mathcal{L}_{\mathbf{f}_2}(H(\mathbf{y})) < 0\};$$

2. there is a **repulsive sliding mode** at a point $\tilde{\mathbf{y}} \in \Sigma_s$ (see Fig. 6.14c) if $\mathcal{L}_{\mathbf{f}_1}(H(\tilde{\mathbf{y}})) < 0$ and $\mathcal{L}_{\mathbf{f}_2}(H(\tilde{\mathbf{y}})) > 0$. It exists if a solution flow reaching Σ leaves the switching surface, and the solution is not unique in forward time. In this case, the set of all repulsive points in Σ_s is called a **unstable sliding** segment.

$$\Sigma_s^u = \{\mathbf{y} \in \Sigma_s : \mathcal{L}_{\mathbf{f}_1}(H(\mathbf{y})) < 0 \text{ and } \mathcal{L}_{\mathbf{f}_2}(H(\mathbf{y})) > 0\}.$$

The next proposition presents a taxonomy of the dynamics within the switching set:

Proposition 6.4 (Dynamics within set Σ) *Assume that:*

1. *there is a non-empty slidding set:* $\mathcal{L}_{\mathbf{f}_1}(H(\mathbf{y})) \cdot \mathcal{L}_{\mathbf{f}_2}(H(\mathbf{y})) < 0$*, and* $\mathcal{L}_{\mathbf{f}_1}(H(\mathbf{y})) > 0$ *and* $\mathcal{L}_{\mathbf{f}_2}(H(\mathbf{y})) < 0$ *for* $\mathbf{y} \in \Sigma$*;*
2. *and* $\mathcal{L}_{\mathbf{f}_1-\mathbf{f}_2}(H(\mathbf{y})) \neq 0$ *for* $\mathbf{y} \in \Sigma$*.*

If there is a weighting function

$$\eta(y) = \frac{\mathcal{L}_{\mathbf{f}_1}(H(\mathbf{y}))}{\mathcal{L}_{\mathbf{f}_1-\mathbf{f}_2}(H(\mathbf{y}))}, \tag{6.16}$$

such that $0 < \eta(\mathbf{y}) < 1$*, for* $\mathbf{y} \in \Sigma$*, then there is a sliding motion and the sliding vector field can be defined by*

$$\mathbf{f}_s(\mathbf{y}) = (1 - \eta(\mathbf{y}))\,\mathbf{f}_1(\mathbf{y}) + \eta(\mathbf{y})\mathbf{f}_2(\mathbf{y}),\ \ \mathbf{y} \in \Sigma,$$

and there is a ***sliding differential equation***

$$\dot{\mathbf{y}} = \mathbf{f}_s(\mathbf{y}),\ \ \mathbf{y} \in \Sigma \tag{6.17}$$

that has an unique solution in forward time, for a given initial state $\mathbf{y}_0 \in \Sigma$*.*

In fact, we have the differential-algebraic equation (DAE)

$$\begin{cases} \dot{\mathbf{y}} = \mathbf{f}_s(\mathbf{y}), \\ H(\mathbf{y}) = 0 \end{cases}$$

and along the sliding vector field (at non-isolated sliding points) we have $\mathcal{L}_{\mathbf{f}_s}(H(\mathbf{y})) = 0$. A steady state of the sliding ODE is called a **pseudo-steady state**, and belongs to the set $\mathbf{y}^{pss} = \{\mathbf{y} \in \Sigma_s : \mathbf{f}_s(\mathbf{y}) = \mathbf{0}\}$. Its stability properties can be characterized, as in smooth ODE, by the eigenvalues of the Jacobian $\mathrm{D}_{\mathbf{y}}\mathbf{f}_s(\mathbf{y})$ for $\mathbf{y} \in \mathbf{y}^{pss}$, as we will see next.

Example 6.3 Consider the following planar linear PWSC ODE for $\mathbf{y} = (y_1, y_2) \in \mathbb{R}^2$:

$$\dot{\mathbf{y}} = \begin{cases} \mathbf{f}_1(\mathbf{y}), & \text{if } y_1 \leq 1 \\ \mathbf{f}_2(\mathbf{y}), & \text{if } y_1 > 1 \end{cases} \tag{6.18}$$

where

$$\mathbf{f}_1(\mathbf{y}) = \begin{pmatrix} -y_1 - y_2 + 1 \\ y_1 - y_2 + 1 \end{pmatrix}, \text{ and } \mathbf{f}_2(\mathbf{y}) = \begin{pmatrix} y_1 - y_2 - 1 \\ y_2 - 2 \end{pmatrix}$$

Therefore $\Sigma = \{\mathbf{y} \in \mathbb{R}^2 : y_1 = 1\}$, $\mathbf{Y}_1 = \{\mathbf{y} \in \mathbb{R}^2 : y_1 < 1\}$, and $\mathbf{Y}_2 = \{\mathbf{y} \in \mathbb{R}^2 : y_1 > 1\}$. We write $\tilde{\mathbf{y}} = (1, y_2)$ as a generic element of Σ. Observe that, because the ODE is linear in the interior of both branches $\mathbf{Y}_1$ and $\mathbf{Y}_2$, the Jacobians are constant, i.e,

$$\mathrm{D_y}\mathbf{f}_1(\mathbf{y}) = \begin{pmatrix} -1 & -1 \\ 1 & -1 \end{pmatrix}, \text{ and } \mathrm{D_y}\mathbf{f}_2(\mathbf{y}) = \begin{pmatrix} 1 & -1 \\ 0 & 1 \end{pmatrix}.$$

First, we classify the ODE. As, for $\mathbf{y} = (1, y_2)$ (i.e, for values of $\mathbf{y} \in \Sigma$), we have

$$\mathbf{f}_1(1, y_2) = \begin{pmatrix} -y_2 \\ 2 - y_2 \end{pmatrix} \neq \mathbf{f}_2(1, y_2) = \begin{pmatrix} -y_2 \\ y_2 - 2 \end{pmatrix}.$$

and

$$\mathrm{D_y}\mathbf{f}_1(1, y_2) = \begin{pmatrix} -1 & -1 \\ 1 & -1 \end{pmatrix} \neq \mathrm{D_y}\mathbf{f}_2(1, y_2) = \begin{pmatrix} 1 & -1 \\ 0 & 1 \end{pmatrix}.$$

then the ODE is of the Filippov type.

Second, we address the existence and number of regular steady states. We find $\bar{\mathbf{y}}_1 = (0, 1) \in \mathbf{Y}_1$ and $\bar{\mathbf{y}}_2 = (3, 2) \in \mathbf{Y}_2$. As the eigenvalues of $\mathrm{D_y}\mathbf{f}_1(\bar{\mathbf{y}}_1)$ are $\lambda_{1,\pm} = -1 \pm i$, and $\mathrm{D_y}\mathbf{f}_2(\bar{\mathbf{y}}_2)$ are $\lambda_{2,\pm} = 1 \pm i$ then steady state $\bar{\mathbf{y}}_1$ is a stable focus and steady state $\bar{\mathbf{y}}_2$ is an unstable node with multiplicity. This indicates that there should exist transversal crossings of the switching boundary coming from $\mathbf{Y}_2$ with a continuation in set $\mathbf{Y}_1$. However, we still need to find the type of contact with Σ.

Third, to solve that problem, we need to characterize the dynamics within set Σ. To do this, we determine $\mathcal{L}_{\mathbf{f}_j}(H)(\mathbf{y}) = \mathrm{D_y}H(\mathbf{y}) \cdot \mathbf{f}_j(\mathbf{y})$ for $j = 1, 2$. As $\mathrm{D_y}H(\mathbf{y}) = (1, 0)^\top$ then $\mathcal{L}_{\mathbf{f}_1}(H)(\mathbf{y}) = -y_1 - y_2 + 1$ and $\mathcal{L}_{\mathbf{f}_2}(H)(\mathbf{y}) = y_1 - y_2 - 1$, and at $\tilde{\mathbf{y}} = (1, y_2)$, yields $\mathcal{L}_{\mathbf{f}_1}(H)(\tilde{\mathbf{y}}) = \mathcal{L}_{\mathbf{f}_2}(H)(\tilde{\mathbf{y}}) = -y_2$. Then

$$\sigma(\tilde{\mathbf{y}}) = \mathcal{L}_{F_1}(H(\tilde{\mathbf{y}}))\mathcal{L}_{F_2}(H(\tilde{\mathbf{y}})) = y_2^2 \geq 0$$

the sliding set is empty, and there is crossing almost everywhere. However, as at $y_2 = 0$ we have

$$\mathcal{L}_{\mathbf{f}_1}(H(1, 0)) = \mathcal{L}_{\mathbf{f}_2}(H(1, 0)) = 0$$

there are two trajectories tangent to Σ coming from both sides. In order to prove that point $\tilde{\mathbf{y}}_s = (1, 0)$ is a singular tangency point, we compute $\mathcal{L}_{\mathbf{f}_1 - \mathbf{f}_2}(H(\dot{\mathbf{y}}_s))$. Because

$$\mathbf{f}_1(\mathbf{y}) - \mathbf{f}_2(\mathbf{y}) = \begin{pmatrix} -2y_1 + 2 \\ y_1 - 2y_2 + 3 \end{pmatrix}$$

then $\mathcal{L}_{\mathbf{f}_1 - \mathbf{f}_2}(H(\mathbf{y})) = -2y_1 + 2$ and $\mathcal{L}_{\mathbf{f}_1 - \mathbf{f}_2}(H(\tilde{\mathbf{y}}_s)) = 0$ which proves that point $\tilde{\mathbf{y}}_s$ is indeed a singular tangency point. Summing up: there are transversal crossings almost everywhere in set Σ, except at point $\mathbf{y} = (1, 0)$ which is a singular tangency point.

Figure 6.15 illustrates the phase diagram. It displays two isoclines in subset $\mathbf{Y}_1$, $I_{\mathbf{Y}_1} = \{\mathbf{y} \in \mathbf{Y}_1 : \dot{\mathbf{y}} = \mathbf{0}\} = (y_2 = 1 - y_1, y_2 = 1 + y_1)$, and in subset $\mathbf{Y}_2$, $I_{\mathbf{Y}_2} = \{\mathbf{y} \in \mathbf{Y}_2 : \dot{\mathbf{y}} = \mathbf{0}\} = (y_2 = y_1 - 1, y_2 = 2)$, that are continuous at Σ. The two isoclines intersect at two regular steady states: a stable focus and an unstable node. There are two trajectories that are tangent to the switching surface Σ at the same point $T = (1, 0)$, which is a singular tangential point. It separates Σ into two different

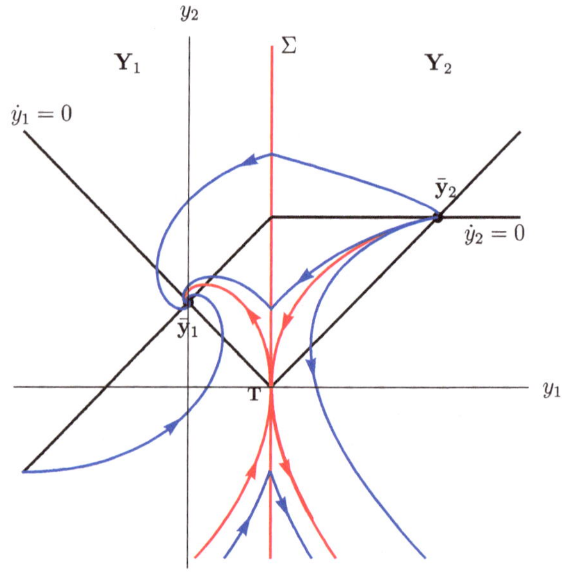

Fig. 6.15 Phase diagram for ODE (6.18)

crossing subsets. The trajectories crossing Σ at points $y_2 > 0$ converge to the regular steady state in $\bar{\mathbf{y}}^1$ and the trajectories crossing at $y_2 < 0$ become unbounded within set $\mathbf{Y}_2$.

6.2.3 *Steady States for PWSC ODE*

In this section, we present a complete taxonomy of two-branch PWSC ODE, specified in Eq. (6.2), by extending several definitions already presented, for both PWS and Filippov ODE.

Definition 6.6 (*Regular steady state*) A **regular steady state**, $\bar{\mathbf{y}}_j$, belonging to branch $j \in \{1, 2\}$, is a vector $\mathbf{y} \in \mathbf{Y}_j$ such that $\mathbf{f}_j(\mathbf{y}, \varphi) = \mathbf{0}$, for a given parameter $\varphi \in \Phi$. The **regular steady state set for branch** $j \in \{1, 2\}$ is the set of regular steady states belonging to set $\mathbf{Y}_j$,

$$\mathbf{y}_j^{ss} = \{\mathbf{y} \in \mathbf{Y}_j : \mathbf{f}_j(\mathbf{y}, \varphi) = \mathbf{0}\}, \text{ for } j = 1, 2.$$

The **regular steady state set** for ODE (6.2) is the union of the regular steady state sets for the two branches, $\mathbf{y}^{ss} = \mathbf{y}_1^{ss} \cup \mathbf{y}_2^{ss}$.

Therefore, $\bar{\mathbf{y}}_1 \in \mathbf{y}_1^{ss}$ if $\mathbf{f}_1(\bar{\mathbf{y}}_1, \varphi) = \mathbf{0}$ and $H(\bar{\mathbf{y}}_1, \varphi) < 0$, and $\bar{\mathbf{y}}_2 \in \mathbf{y}_2^{ss}$ if $\mathbf{f}_2(\bar{\mathbf{y}}_2, \varphi) = \mathbf{0}$ and $H(\bar{\mathbf{y}}_2, \varphi) > 0$. We already saw that we can have multiple regular steady in each or both branches.

Definition 6.7 (*Virtual steady state*) A **virtual steady state** belonging to branch $j \in \{1, 2\}$ is a vector $\mathbf{y} \notin \mathbf{Y}_j$ such that $\mathbf{f}_j(\mathbf{y}, \varphi) = \mathbf{0}$ for a given parameter $\varphi \in \Phi$. The **virtual steady state set for branch** $j \in \{1, 2\}$ is the set of virtual steady states belonging to set $\mathbf{Y}_{-j}$

$$\mathbf{y}_j^{vss} = \{\mathbf{y} \notin \mathbf{Y}_j : \mathbf{f}_j(\mathbf{y}, \varphi) = \mathbf{0}\}, \text{ for } j = 1, 2.$$

The **virtual steady state set** for ODE (6.2) is the union of the virtual steady state sets for the two branches $\mathbf{y}^{vss} = \mathbf{y}_1^{vss} \cup \mathbf{y}_2^{vss}$.

Therefore, $\bar{\mathbf{y}}_1^v \in \mathbf{y}_1^{vss}$ if $\mathbf{f}_1(\bar{\mathbf{y}}_1^v, \varphi) = \mathbf{0}$ and $H(\bar{\mathbf{y}}_1^v, \varphi) > 0$, and $\bar{\mathbf{y}}_2^v \in \mathbf{y}_2^{vss}$ if $\mathbf{f}_2(\bar{\mathbf{y}}_2^v, \varphi) = \mathbf{0}$ and $H(\bar{\mathbf{y}}_2^v, \varphi) < 0$. There can be multiple virtual steady states in each or both branches.

Any of the previous two types of steady states belong to the interior of the respective branch, $\mathbf{Y}_j$, which implies that they do not belong to the switching boundary set Σ.

As we already saw, there are several steady states belonging to the switching boundary: boundary steady states and pseudo steady states.

Definition 6.8 (*Boundary steady state*) A **boundary steady state**, $\bar{\mathbf{y}}_{\Sigma,j}$, is a steady state belonging to the intersection $\mathbf{Y}_j \cap \Sigma$. The set of boundary steady states in branch $\mathbf{Y}_j$ is

$$\mathbf{y}_j^{bss} = \left\{\mathbf{y} \in \Sigma : \mathbf{f}_j(\mathbf{y}, \varphi) = 0\right\}, \quad j = 1, 2.$$

The **boundary steady state set** is the union of the boundary steady states for every branch: $\mathbf{y}^{bss} = \mathbf{y}_1^{bss} \cup \mathbf{y}_2^{bss}$.

A boundary steady state is said be regular if the related branch condition is an inequality, as $H(\mathbf{y}, \varphi) \leq 0$ and/or $H(\mathbf{y}, \varphi) \geq 0$. It is said to be virtual if the related branch condition is a strict inequality, as $H(\mathbf{y}, \varphi) < 0$ and/or $H(\mathbf{y}, \varphi) > 0$. In particular:

1. for PWS ODE, as we have $\mathbf{f}_1(\mathbf{y}, \varphi) = \mathbf{f}_2(\mathbf{y}, \varphi)$ if $\mathbf{y} \in \Sigma$, if the boundary functions are not strict inequalities then $\mathbf{y}_1^{bss} = \mathbf{y}_2^{bss} = \mathbf{y}^{bss}$;
2. for Filippov ODE, as we have $\mathbf{f}_1(\mathbf{y}, \varphi) \neq \mathbf{f}_2(\mathbf{y}, \varphi)$ if $\mathbf{y} \in \Sigma$ implying $\mathbf{y}_1^{bss} \cap \mathbf{y}_2^{bss} = \emptyset$, every boundary steady state, if it exists, either belongs to $\mathbf{y}_1^{bss}$ or to $\mathbf{y}_2^{bss}$.

Definition 6.9 (*Pseudo steady state*) A **pseudo-steady state**, $\mathbf{y}^p$, is a steady state of the sliding ODE (6.17). That is, it is a vector $\mathbf{y} \in \Sigma$ such that $\mathbf{f}_s(\mathbf{y}, \varphi) = (1 - \eta)\mathbf{f}_1(\mathbf{y}, \varphi) + \eta\mathbf{f}_2(\mathbf{y}, \varphi) = \mathbf{0}$, where $\eta(\cdot)$ is defined in Eq. (6.16). The **pseudo steady state set** is the set of all pseudo steady states, for a fixed $\varphi \in \Phi$,

$$\mathbf{y}^{pss} = \left\{\mathbf{y} \in \Sigma : \mathbf{f}_s(\mathbf{y}, \varphi) = \mathbf{0}\right\}. \tag{6.19}$$

Both PWS and Filippov ODE can have interior or boundary regular and/or virtual steady states. However, pseudo steady states only exist for Filippov ODE. For Filippov ODE, a pseudo-steady state is neither a regular nor a virtual steady states for $\mathbf{f}_1$ or $\mathbf{f}_2$. As we already saw, it is a steady state for a sliding vector field, which allows us to characterize it further: (1) as a **regular pseudo-steady state** if it belongs to the set

$$\mathbf{y}^{rpss} = \Big\{\mathbf{y} \in \Sigma : \mathbf{f}_s(\mathbf{y}, \varphi) = \mathbf{0}, \text{ and } \eta(\mathbf{y}, \varphi) \in (0, 1)\Big\};$$

(2) or as a **virtual pseudo-steady state** if it belongs to the set

$$\mathbf{y}^{vpss} = \Big\{\mathbf{y} \in \Sigma : \mathbf{f}_s(\mathbf{y}, \varphi) = \mathbf{0}, \text{ and } \eta(\mathbf{y}, \varphi) \notin (0, 1)\Big\}.$$

6.2.4 *Bifurcation Analysis in the Two Branch Case*

Bifurcation theory studies the change in the dynamic properties of the system, in particular the change in the number and stability of the invariant sets, when parameters in Φ change. That it, it studies **structural stability** properties. In PWSC ODE, a general classification of possible classes of dynamics is not simple because it may involve the change in the number and other features of the switching boundary. According to the researchers in the area, the bifurcation theory for PWSC is still in its infancy. Possibly, a complete bifurcation taxonomy of all the possible types of dynamics is perhaps impossible, given the very large number of possible cases. Researchers in the field adopt a "pragmatic approach" (see di Bernardo et al. (2008a)).

In a broad view, we can classify possible bifurcations that can exist in ODE (6.2) into two groups:

1. **regular or smooth bifurcations**, if switching boundaries are not involved. That is, all the bifurcations which may exist in smooth ODE (as presented in Chap. 5 and in Kuznetsov (2023)) also hold in PWSC ODE. Those bifurcations are related to the existence of several **invariant sets** (as steady states, limit cycles, homoclinic or heteroclinic trajectories), and to local or global generic and non-generic dynamics associated to them;
2. **discontinuity induced bifurcations**, if a switching boundary is involved.

Definition 6.10 (*Discontinuity induced bifurcation*) A **discontinuity induced bifurcation** (DIB) is said to exist at a point $(\mathbf{y}, \varphi_0) \in \Sigma \times \Phi$ if, for a small change in that parameter, the ODE (6.2) ceases to be structurally stable.

The following DIBs have been reported in the literature [11]:

[11] See a textbook presentation in di Bernardo et al. (2008a) and Colombo et al. (2012). On one parameter bifurcations in planar Filippov ODE see Kuznetsov et al. (2003). On sliding bifurcations see Jeffrey and Hogan (2011). For a recent survey, see Jeffrey et al. (2022).

1. **a boundary equilibrium bifurcation** exists if a steady state is located at the switching boundary, if pseudo-steady states may appear or disappear, or if a limit cycle is generated as a consequence of the existence of switching boundaries;
2. **a grazing bifurcation of limit cycles** exists if a limit cycle or a flow becomes tangent to the switching boundary;
3. **sliding-sticking bifurcations** can exist if there are limit cycles in a Filippov ODE;
4. **bifurcations related to the existence of corners** when there are more than two switching boundaries;
5. **particular types of global bifurcations**, v.g homoclinic-like connections between pseudo-steady states.

Next we consider DIB of boundary steady states.

6.2.5 *Discontinuity Induced Bifurcations Branching from a Steady State*

This subsection is limited to presenting some useful results for boundary steady state bifurcations.

Definition 6.11 (*Boundary steady state bifurcation point*) A **boundary steady state bifurcation** (BEB) point $(\mathbf{y}^*, \varphi^*)$ is a co-dimension one bifurcation point if there is a pair $(\mathbf{y}, \varphi) \in \mathbf{Y} \times \Phi$ such that

1. $\mathbf{y}^*$ is a boundary steady state: $\mathbf{y}^* \in \mathbf{y}^{bss}$;
2. $\mathbf{y}^*$ is a non-singular boundary boundary steady state, satisfying $\det(D_{\mathbf{y}}\mathbf{f}(\mathbf{y}^*, \varphi^*)) \neq 0$;
3. $(\mathbf{y}^*, \varphi^*)$ satisfies the non-degeneracy condition:

$$\frac{\partial H(\mathbf{y}^*, \varphi^*)}{\partial \varphi} = D_{\varphi} H(\mathbf{y}^*, \varphi^*) \quad D_{\mathbf{y}} H(\mathbf{y}^*, \varphi^*)(D_{\mathbf{y}}\mathbf{f}_j(\mathbf{y}^*, \varphi^*)^{-1} D_{\varphi}\mathbf{f}_j(\mathbf{y}^*, \varphi^*)) \neq 0.$$

There are two generic types of **discontinuity induced bifurcations branching from BEB's**. First, there is **persistence (or border crossing)** if there is a regular steady state in one branch ($\mathbf{Y}_1$ or $\mathbf{Y}_2$) that, upon a small variation of the bifurcating parameter φ, becomes a boundary steady state bifurcation point, and, upon a further variation, becomes a regular steady state in the opposite branch ($\mathbf{Y}_2$ or $\mathbf{Y}_1$), or if a virtual pseudo-steady state in a Filippov system becomes regular.

Second, there is a **non-smooth fold bifurcation** or a **non-smooth Hopf bifurcation** if, at a bifurcation value of the parameter φ, two branches of steady states collide with Σ, coalesce into one boundary steady state, and split and become virtual steady states. The boundary steady state can be of a fold type or of a Hopf type.

There are two (complementary) types of approaches for determining the type of bifurcation we can have in a specific ODE.

The di Bernardo et al. (2008a) approach. The vector field on branch $\mathbf{Y}_2$ can be specified as a (vector) function of the vector field on branch $\mathbf{Y}_1$. For PWS ODE we write, for a parameter φ,

$$\mathbf{f}_1(\mathbf{y}, \varphi) = \mathbf{f}(\mathbf{y}, \varphi), \text{ and } \mathbf{f}_2(\mathbf{y}, \varphi) = \mathbf{f}(\mathbf{y}, \varphi) + G(\mathbf{y}, \varphi)\, H(\mathbf{y}, \varphi),$$

where $G(\mathbf{y}, \varphi)$ is (2×1) vector, and for Filippov ODE we have instead

$$\mathbf{f}_1(\mathbf{y}, \varphi) = \mathbf{f}(\mathbf{y}, \varphi), \text{ and } \mathbf{f}_2(\mathbf{y}, \varphi) = \mathbf{f}(\mathbf{y}, \varphi) + \mathbf{G}(\mathbf{y}, \varphi),$$

where $\mathbf{G}(\mathbf{y}, \varphi)$ is a (2×2) matrix which should be related to $H(\mathbf{y}, \varphi)$.

Theorem 6.1 *(Boundary steady state bifurcations)For a the PWSC ODE* (6.15) *let* $(\tilde{\mathbf{y}}, \varphi^*)$ *be a boundary equilibrium bifurcation point and assume that*

$$\det \mathbf{A} \neq 0,\ N - \mathbf{C}\mathbf{A}^{-1}\mathbf{M} \neq 0,\ \text{and } 1 + \mathbf{C}\mathbf{A}^{-1}\mathbf{B} \neq 0$$

where $\mathbf{A} \equiv \mathrm{D}_{\mathbf{y}}\mathbf{f}_1(\tilde{\mathbf{y}}, \varphi^*)$, $\mathbf{M} \equiv \mathrm{D}_{\varphi}\mathbf{f}_1(\tilde{\mathbf{y}}, \varphi^*)$, $\mathbf{C} \equiv \mathrm{D}_{\mathbf{y}}H(\tilde{\mathbf{y}}, \varphi^*)$, $N = \mathrm{D}_{\varphi}H(\tilde{\mathbf{y}}, \varphi^*)$, *and* $\mathbf{B} = \mathbf{G}(\tilde{\mathbf{y}}, \varphi^*)$. *Then,* $(\tilde{y}, \varphi^*)$ *is a* ***persistence or border crossing bifurcation point*** *if* $1 + \mathbf{C}\mathbf{A}^{-1}\mathbf{B} > 0$ *and it is a* ***non-smooth fold bifurcation point*** *if* $1 + \mathbf{C}\mathbf{A}^{-1}\mathbf{B} < 0$.

Proof See di Bernardo et al. (2008a, p. 223). Consider a regular steady state in $\mathbf{Y}_1$, $\bar{\mathbf{y}}_1 \in \mathbf{y}_1^{ss}$. Then $\mathbf{f}(\bar{\mathbf{y}}_1, \varphi) = 0$ and $H(\bar{\mathbf{y}}_1, \varphi) = z_1 < 0$ is a number. Linearizing about the boundary steady state bifurcation point (BEB) $(\mathbf{y}^*, \varphi^*)$ yields a linear equation system

$$\begin{cases} \mathbf{A}\bar{\mathbf{y}}_1 + \mathbf{M}\varphi = \mathbf{0}, \\ \mathbf{C}\bar{\mathbf{y}}_1 + N\varphi = z_1 < 0. \end{cases}$$

As $\det \mathbf{A} \neq 0$ (from the definition of BEB), we can solve this system at the switching boundary and obtain $\bar{\mathbf{y}}_1 = -\mathbf{A}^{-1}\,\mathbf{M}\,\varphi$ and $z_1 = (N - \mathbf{C}\mathbf{A}^{-1}\mathbf{M})\varphi$. For a regular equilibrium in $\mathbf{Y}_2$, $\bar{\mathbf{y}}_2 \in \mathbf{y}_2^{ss}$, as $\mathbf{f}(\bar{\mathbf{y}}_2, \varphi) + \mathbf{G}(\bar{\mathbf{y}}_2, \varphi)\, z_2 = 0$ and $H(\bar{\mathbf{y}}_2, \varphi) = z_2 > 0$. Linearising about the BEB point $(\mathbf{y}^*, \varphi^*)$ we get linear equation system

$$\begin{cases} \mathbf{A}\,\bar{\mathbf{y}}_2 + \mathbf{M}\varphi + \mathbf{B}\, z_2 = \mathbf{0}, \\ \mathbf{C}\,\bar{\mathbf{y}}_2 + N\varphi = z_2 > 0 \end{cases}$$

where $\mathbf{B} = \mathbf{G}(\mathbf{y}^*, \varphi^*)$. Solving this system at the switching boundary yields $z_2 = \frac{(N-\mathbf{C}\mathbf{A}^{-1}\mathbf{M})\varphi}{1+\mathbf{C}\mathbf{A}^{-1}\mathbf{B}}$. Then $z_1 = (1 + \mathbf{C}\mathbf{A}^{-1}\mathbf{B})\, z_2$, and if $\text{sign}(z_1) \neq \text{sign}(z_2)$ then the BEB displays persistence, and if $\text{sign}(z_1) = \text{sign}(z_2)$ there is a planar non-smooth fold bifurcation point.

As we already saw for the scalar case, in the phase diagram in Fig. 6.2, these are DIB that do not occur in regular ODE.

The Leine and Nijmeijer (2004) approach. Consider the generalized jacobian of $\mathbf{f}(\mathbf{y})$, in Eq. (6.10), and evaluate it at a point in $\mathbf{y}^{ss}$,

$$\begin{aligned}\partial\mathbf{f}(\mathbf{y},\varphi) &= \overline{\text{co}}\{\mathrm{D}_{\mathbf{y}}\mathbf{f}_1(\mathbf{y},\varphi), \mathbf{f}_2(\mathbf{y},\varphi)\}\\ &= \{(1-\alpha)\mathrm{D}_{\mathbf{y}}\mathbf{f}_1(\mathbf{y},\varphi)+\alpha\mathrm{D}_{\mathbf{y}}\mathbf{f}_2(\mathbf{y},\varphi),\ 0\le\alpha\le 1\}\Big|_{\mathbf{y}=\mathbf{y}^*,\varphi=\varphi^*}.\end{aligned}$$

A stronger version of their results is the following: for a regular or a pseudo-steady state

1. there is persistence if $0\notin \det(\partial\mathbf{f}(\mathbf{y}^*,\varphi^*))$;
2. there is a non-smooth fold if $0\in\det(\partial\mathbf{f}(\mathbf{y}^*,\varphi^*))$, that is if the spectrum of $\partial\mathbf{f}(\mathbf{y}^*,\varphi^*)$ crosses the imaginary axis with a zero complex part;
3. there is a non-smooth Hopf bifurcation if the spectrum of $\partial\mathbf{f}(\mathbf{y}^*,\varphi^*)$ crosses the imaginary axis elsewhere.

6.3 Applications to Economics

Next we present examples of planar PWS and Filippov ODE.

Example 6.4 (*DGE Endogenous Markups*) The Brito et al. (2013) model is a DGE model in which the technology and market environments were described in Example 6.2. If we assume a representative household with a discounted (at rate $\rho>0$) intertemporal utility functional with a logarithmic utility function, and an inelastic labor supply, the DGE is represented by the planar piecewise smooth (PWS) ODE,

$$\begin{pmatrix}\dot{k}\\ \dot{c}\end{pmatrix}=\begin{cases}\mathbf{f}_1(k,c), & \text{if } \sigma\, m(k)\le 1,\\ \mathbf{f}_2(k,c), & \text{if } \sigma\, m(k)\ge 1,\end{cases} \tag{6.20a}$$

where (see again Example 6.2 for the notation)

$$\mathbf{f}_1(k,c)=\begin{pmatrix}Y_1(k)-c-\delta\,k\\ c\,(R_1(k)-\rho-\delta)\end{pmatrix},\ \text{and}\ \mathbf{f}_2(k,c)=\begin{pmatrix}Y_2(k)-c-\delta\,k\\ c\,(R_2(k)-\rho-\delta)\end{pmatrix},$$

where k is a predetermined variable and c is non-predetermined, and there are two side conditions, the initial value for k is fixed, $k(0)=k_0$, and the transversality condition, $\lim_{t\to\infty}\frac{k(t)}{c(t)}e^{-\rho t}=0$, has to be satisfied. The output of the final good and the rate of return of capital are piecewise smooth functions

$$Y(k)=\begin{cases}Y_1(k)=(1-\sigma^{-1})\,k^{\alpha}, & \text{if } \sigma\, m(k)\le 1,\\ Y_2(k)=(1-m(k))\,k^{\alpha}, & \text{if } \sigma\, m(k)\ge 1,\end{cases}$$

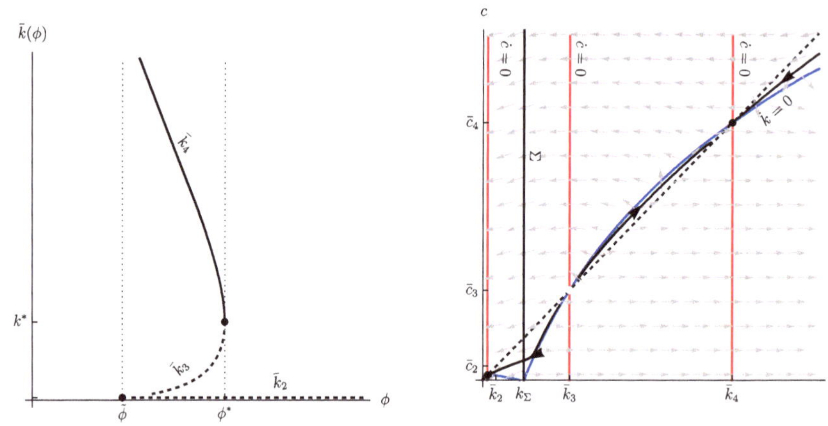

(a) Bifurcation diagram

(b) Phase diagram for an intermediate ϕ: $\tilde{\phi} < \phi < \phi^*$

Fig. 6.16 The Brito et al. (2013) model for $0 < \alpha < 1$, $\rho > 0$, $\delta > 0$, $1 < \sigma\bar{\sigma}$, and different values for $\phi > 0$

$$R(k) = \begin{cases} R_1(k) = \alpha\,(1-\sigma^{-1})\,k^{\alpha-1}, & \text{if } \sigma\,m(k) \leq 1, \\ R_2(k) = \alpha\,(1-m(k))\,k^{\alpha-1}, & \text{if } \sigma\,m(k) \geq 1. \end{cases}$$

The switching surface, $\Sigma = \{(k, c) : \sigma\,m(k) = 1\}$, splits the state space $\mathbf{Y} = \mathbb{R}^2_+$ into the monopoly set $\mathbf{Y}_1 = \{(k, c) : \sigma\,m(k) \leq 1\}$, where there is only one firm, and the CMC set $\mathbf{Y}_2 = \{(k, c) : \sigma\,m(k) \geq 1\}$ in which there are several industries and several firms per industry. Equivalently, as the stock of capital at the switching boundary is $k_\Sigma = (\phi\,\sigma)^{\frac{1}{\sigma}}$, there is monopoly if $0 < k \leq k_\Sigma$, and there is CMC if $k \geq k_\Sigma$. As $\mathbf{f}_1(k_\Sigma, c) = \mathbf{f}_2(k_\Sigma, c)$, for any $c > 0$, and $D_{(k,c)}\mathbf{f}_1(k_\Sigma, c) \neq D_{(k,c)}\mathbf{f}_2(k_\Sigma, c)$ then the ODE is a piecewise smooth not a Filippov ODE.

Furthermore, as the rate of return of capital $R(k)$ satisfies the Inada properties, then an interior steady state always exists, but may not be unique. We can have one, two, or three interior steady states.[12] While the cases in which there is one unique steady state are generic, the case in which there are two steady states are not generic and there are local bifurcation points. There is a **local regular fold bifurcation** point, branching from zero to two regular steady states, $(k^*, c^*, \phi^*) = \{(k, c, \phi) : c = Y_2(k) - \delta\,k,\ R_2(k) = \rho + \delta,\ R_{2,k}(k) = 0\}$. At this point $c^* = \beta\,k^*$, where $\beta \equiv \frac{\rho+(1-\alpha)\delta}{\alpha}$, $\phi^* = \frac{\sigma}{\bar{\sigma}^2}\,(k^*)^\alpha$, and $k^* = \left(\frac{\alpha\,(\bar{\sigma}-1)}{\bar{\sigma}\,(\rho+\delta)}\right)^{\frac{1}{1-\alpha}}$, where $\bar{\sigma} \equiv \frac{2-\alpha}{2\,(1-\alpha)}$.

There is also a **discontinuity induced bifurcation** (DIB) point $(\tilde{k}, \tilde{c}, \tilde{\phi}) = \{(k, c, \phi) : c = Y_1(k) - \delta\,k = Y_2(k) - \delta\,k,\ R_1(k) = R_2(k) = \rho + \delta\}$. At this point

[12] We have a steady state in the origin $(k, c) = (0, 0)$ and, if ϕ is sufficiently low, there is also a steady state of the ODE (6.20), for $c = 0$ and $k > \bar{k}_4$, but that does not satisfy the transversality condition, as in the RCK model.

we have $\tilde{c} = \beta\,\tilde{k}$, $\tilde{k} = \left(\frac{\alpha\,(\sigma-1)}{\sigma\,(\rho+\delta)}\right)^{\frac{1}{1-\alpha}}$, and $\tilde{\phi} = \left(\frac{1}{\sigma}\right)^{\frac{1}{1-\alpha}}\tilde{k}^{\alpha}$. We can characterize the type of DIB point that exists in this mode by applying theorem 6.1: as det $(\mathbf{A}) = \tilde{c}R_{1,k}(\tilde{k}) < 0$, $N - \mathbf{CA}^{-1}\mathbf{M} = (2\,\sigma\,\tilde{\phi})^{-1} > 0$ and $1 + \mathbf{CA}^{-1}\mathbf{B} = \frac{\sigma-\bar{\sigma}}{\sigma-1} < 0$ if $\sigma < \bar{\sigma}$ then that bifurcation point is a non-smooth fold bifurcation point.[13]

We can also prove that there are **no sliding trajectories**. Writing $H(k, c) = \sigma\,m(k)$, and as $\Sigma = (k_\sigma, c)$, for an arbitrary $c > 0$, we have $\mathcal{L}_{\mathbf{f}_j}(H(k, c)) = m'(k)(Y_j(k) - c - \delta\,k)$, for $j = 1, 2$. Applying Filippov's convex method, from Sect. 6.2.2, as, we have $\mathcal{L}_{\mathbf{f}_j}(H(k_\Sigma, c)) = m'(k_\Sigma)(Y_j(k_\Sigma) - c - \delta\,k_\Sigma)$, where $m'(k_\Sigma) = -\frac{\alpha}{2\sigma k_\Sigma} < 0$, then $\sigma(k_\Sigma, c) = (m'(k_\Sigma))^2\,(Y(k_\Sigma) - c - \delta\,k_\Sigma)^2 > 0$, because $Y_1(k_\Sigma) = Y_2(k_\Sigma)$ ruling out sliding trajectories. We conclude that, if k_Σ is not a steady state, all the trajectories cross the switching surface, and only if k_Σ is a boundary steady state, there is no crossing.

From the global dynamics perspective, Figure 6.16a displays the bifurcation diagram, taking ϕ as the bifurcating parameter, under the assumption that $1 < \sigma < \bar{\sigma}$. Again a **discontinuity induced hysteresis curve**, similar to the one appearing in Example 6.2, is obtained. It is also similar to the regular hysteresis curve that was found in the Skiba (1978) model (see Example 5.5 and Fig. 5.10c).[14] In this case, the solid segments correspond to saddle-points and the dashed segments to unstable nodes. There are two bifurcation values of the fixed cost ϕ, $\tilde{\phi} < \phi^{\star}$, such that: for $\phi > \phi^*$ there is a unique steady state with a single saddle-point monopoly firm, with low levels for the capital stock and consumption, $(\bar{k}_2, \bar{c}_2)$; for $\phi < \tilde{\phi}$ there is one unique single saddle-point steady state CMC regime, with high levels for the capital stock and consumption, $(\bar{k}_4, \bar{c}_4)$; and for intermediate values of the fixed cost, $\tilde{\phi} < \phi < \phi^*$ there are three steady states comprising a monopolist saddle-point, $(\bar{k}_2, \bar{c}_2)$, and two CMC steady states, one unstable lower level steady state $(\bar{k}_3, \bar{c}_3)$, and a saddle-point higher level steady state $(\bar{k}_4, \bar{c}_4)$.

This last case is illustrated in Fig. 6.16b, which is similar to the phase diagram 6.13c in Example 6.2, although the steady states are determined not only at a zero net investment point but also by the equality between the rate of return of capital at the "natural" level given by the sum of the rate of time preference and the depreciation rate. Comparing with the analogous phase diagram for the Skiba (1978) model, in Fig. 5.10b, we observe that there are also heteroclinic trajectories connecting the intermediate, unstable steady state, with each of two other two steady states, but differently from Skiba model, we have in this example uniqueness of transition trajectories instead of multiplicity.

Example 6.5 (*Pollution control with irreversibility*) The first contributions to optimal pollution control, in the presence of irreversibility, were Keeler et al. (1972) and

---

[13] Vector $\mathbf{G} = (-k^{\alpha}, -\alpha\,c\,k^{\alpha})$. Observe that if we assume that $\sigma > \bar{\sigma}$ the non-smooth bifurcation point is a persistence point instead because, in this case, we would have $1 + \mathbf{CA}^{-1}\mathbf{B} > 0$.

[14] Skiba (1978) is a model for a centralized economy and Brito et al. (2013) model is a representative agent equilibrium economy, in which the general equilibrium is not Pareto efficient. In the first case, the rate of return of capital is always equal to the marginal productivity of capital, i.e, $R(k) = Y_k(k)$, while in the second model, particularly in the CMC regime, we have $Y_{2,k}(k) > R_{2,k}(k)$. This can, in part, explain the difference in the bifurcation curves for the low equilibrium.

Forster (1975). A two-stage version of the model has been proposed by Tähvonen and Withagen (1996). In this example, we formalize it as an optimal control problem of a PWSC ODE. In this model, there are two variables, the stock of pollution, z, and consumption which is equal to output, y. Output generates a flow of utility $U(y)$ (an increasing and strictly concave function), and a flow of pollution $\alpha\, y$. The stock of pollution is a source of disutility, $D(z)$ (an increasing and strictly convex function). However, nature has some renewal capacity. There is irreversibility, in the sense that the rate of decay is higher (lower) if the pollution stock is below (above) a threshold $\bar{z}$. The MHDS (see Sect. 12.4) is an initial-terminal PWSC ODE

$$\begin{pmatrix} \dot{z} \\ \dot{y} \end{pmatrix} = \mathbf{f}_j(z, y) \in \mathbf{Y}_j, \text{ for } j = 1, 2, \tag{6.21}$$

where $z(0) = z_0$ is given, the transversality condition is $\lim_{t\to\infty} U'(y(t))\, z(t)\, e^{-\rho t} = 0$, and $\rho > 0$ is the rate of time preference. The switching boundary, $\Sigma = \{(z, y) \in \mathbb{R}^2_+ : H(z, y) = 0\}$ where $H(z, y) = z - \bar{z}$, partitions the state space $\mathbb{R}^2_+$ into two branches: a low pollution environment $\mathbf{Y}_1 = \{(z, y) \in \mathbb{R}^2_+ : H(z, y) < 0\}$, and a high pollution environment $\mathbf{Y}_2 = \{(z, y) \in \mathbb{R}^2_+ : H(z, y) > 0\}$. The discontinuous vector field is

$$\mathbf{f}_j(z, y) = \begin{pmatrix} \alpha\, y - \delta_j\, z \\ (\rho + \delta_j)\, \frac{U'(y)}{U''(y)} - \alpha\, \frac{D'(z)}{U''(y)} \end{pmatrix}, \text{ for each } t \in \mathbb{R}_+, \text{ for } j = 1, 2.$$

Assuming $\delta_1 > \delta_2 \geq 0$ implies $\mathbf{f}_1(\bar{z}, y) \neq \mathbf{f}_2(\bar{z}, y)$, and the ODE is of the Filippov type. As there are a large number of different dynamics, both at the interior of the two branches and at the boundary, we postulate a simple version of the model leading to a PWSC linear ODE, by assuming $U(y) = \nu\, y\, (2\, \bar{y} - y)$ and $D(z) = \frac{\gamma}{2} z^2$. Therefore: preferences display satiation at $y = \bar{y}$, and $\mu \equiv \frac{\gamma}{2\nu}$ represents the preference for goods relative to pollution.[15] Then

$$\mathbf{f}_j(z, y) = \begin{pmatrix} \alpha\, y - \delta_j\, z \\ (\rho + \delta_j)\, (y - \bar{y}) + \alpha\, \mu\, z \end{pmatrix}, \text{ for } j = 1, 2,$$

where $\mu \equiv \frac{\gamma}{2\nu}$. All parameters are strictly positive. Next, we address separately the dynamics inside the branches, from the ones at the switching boundary, with the purpose of finding the possible phase diagrams.

The potential (regular and virtual) steady states are

$$\bar{z}_j = \frac{(r + d_j)\, \bar{y}}{d_j\, (r + d_j) + \mu}, \quad \bar{y}_j = d_j\, z_j, \text{ for } j = 1, 2,$$

[15] The no satiation assumption could be incorporated by setting $\bar{y} = \infty$.

where $r \equiv \frac{\rho}{\alpha}, d_j \equiv \frac{\delta_j}{\alpha}$ are normalized time preference and nature regeneration rates. Defining $e_j \equiv (r + d_j)(\bar{x} - d_j) - \mu$, for $j = 1, 2$, where $\bar{x} \equiv \frac{\bar{y}}{\bar{z}}$ is the ratio between the satiation level of consumption and a critical level of pollution (before nature's absorption capacity starts to decline) we have the following regular steady states: (1) if $e_1 \leq 0$ and $e_2 \leq 0$ then there is one unique regular steady state $(\bar{z}_1, \bar{y}_1) \in \mathbf{Y}_1$, (2) if $e_1 > 0$ and $e_2 > 0$ then there is one unique regular steady state in $(\bar{z}_2, \bar{y}_2) \in \mathbf{Y}_2$; (3) if $e_1 \leq 0$ and $e_2 > 0$ then there are two regular steady states, $(\bar{z}_1, \bar{y}_1) \in \mathbf{Y}_1$ and $(\bar{z}_2, \bar{y}_2) \in \mathbf{Y}_2$; or if $e_1 > 0$ and $e_2 \leq 0$ then there are no steady states inside each branch.[16] As $\det(\mathrm{D}_{(z,y)}\mathbf{f}_j(z, y)) = -(d_j\,(r + d_j) + \mu) < 0$, for $j = 1, 2$, for any $(z, y) \in \mathbb{R}^2_+$, then steady states are always saddle points. The stable eigenspace passing trough steady state $(\bar{z}_j, \bar{y}_j)$, regular or virtual, is $\mathcal{E}^s_j = \{(z, y) : y = -\lambda_j\,\bar{z}_j + (d_j + \lambda_j)\,z\}$, where $\lambda_j = \frac{r}{2} - \left(\left(\frac{r}{2} + d_j\right)^2 + \mu\right)^{\frac{1}{2}} < 0$ is the negative eigenvalue of the Jacobian of $\mathbf{f}_j(z, y)$. Each eigenvector has a regular and a virtual segment, and hits the switching surface at point $(\bar{z}, -\lambda_j\,\bar{z}_j + (d_j + \lambda_j)\,\bar{z})$.

At the boundary Σ, we find $\dot{h}_j = \mathcal{L}_{\mathbf{f}_j}(H(z, y)) = \alpha\,(y - d_j\,\bar{z})$, for $j = 1, 2$ then $\dot{h}_j \gtreqless 0$ if and only if $y \gtreqless d_j\,\bar{z}$. Therefore, we can partition the switching surface Σ into a traverse subset $\Sigma_t = \{(\bar{z}, y) : y \in \left[0, d_2\bar{z}\right) \cup \left(d_1\,\bar{z}, \infty\right)\}$, a sliding subset $\Sigma_s = \{(\bar{z}, y) :\ y \in \left(d_2\,\bar{z}, d_1, \bar{z}\right)\}$, and two grazing points $\Sigma_g = \left\{\left(\bar{z}, d_1\,\bar{z}\right), \left(\bar{z}, d_2\,\bar{z}\right)\right\}$ (see Sect. 6.2.2). Furthermore, as $\theta = \left.\frac{\mathcal{L}_{\mathbf{f}_1}(H(z,y))}{\mathcal{L}_{\mathbf{f}_1-\mathbf{f}_2}(H(z,y))}\right|_{\Sigma} = \frac{y - d_1\,\bar{z}}{(d_2 - d_1)\,\bar{z}}$, then the sliding vector field is

$$\mathbf{f}_s = (1 - \theta)\mathbf{f}_1 + \theta\,\mathbf{f}_2 = \begin{pmatrix} 0 \\ 2\,\bar{z}\,\Pi(y) \end{pmatrix}, \quad \text{for } j = 1, 2,$$

where $\Pi(y) = \bar{z}\Pi(x)$, and $\Pi(x) \equiv (r + x)(x - \bar{x}) + \mu$, for $x \equiv \frac{y}{\bar{z}}$. Pseudo-steady states are the equilibrium points within the sliding subset. As $\Pi(x) = 0$ if and only if $\bar{x}^s_{1,2} = \frac{\bar{x}-r}{2} \pm \left(\left(\frac{\bar{x}+r}{2}\right)^2 - \mu\right)^{\frac{1}{2}}$, and their local stability properties can be assessed by $\Pi'(y) = \bar{z}\Pi'(x) = \bar{z}\left(2x + (r - \bar{x})\right)$, then: (1) if $\mu < \left(\frac{\bar{x}+r}{2}\right)^2$ and $e_1 < 0$ and $e_2 < 0$, we have two pseudo-steady states $\bar{x}^s_1 < \bar{x}^s_2$, where $\bar{x}^s_2$ is unstable and $\bar{x}^s_1$ is stable; (2) if $\mu > \left(\frac{\bar{x}+r}{2}\right)^2$ and $e_1 < 0$ and $e_2 > 0$ there is one unstable equilibrium point $\bar{x}^s_2$; (3) if $\mu > \left(\frac{\bar{x}+r}{2}\right)^2$ and $e_1 > 0$ and $e_2 < 0$ there is one stable pseudo-steady state $\bar{x}^s_1$; or (4) in all other cases there are no pseudo-steady state points.

Figure 6.17 illustrates four possible cases, assuming that $\mu > d_1\,d_2$,[17] unfolded by parameters $\bar{x}$ and r, illustrated by the bifurcation diagram in Fig. 6.17f, given value for parameters d_1 and d_2 such that $d_1 > d_2$. In that figure, as we have a mixed

---

[16] In the first case, there is a virtual steady state $(\bar{z}_2, \bar{y}_2) \in \mathbf{Y}_1$, in the second case, there is a virtual steady state $(\bar{z}_1, \bar{y}_1) \in \mathbf{Y}_2$, and in the last case, there are two virtual steady states $(\bar{z}_1, \bar{y}_1) \in \mathbf{Y}_2$ and $(\bar{z}_2, \bar{y}_2) \in \mathbf{Y}_1$.

[17] Under this assumption, the two lines $e_1(\bar{x}, r) = 0$ and $e_2(\bar{x}, r) = 0$ intersect at one point, and if $\mu \leq d_1\,d_2$ they do not intersect.

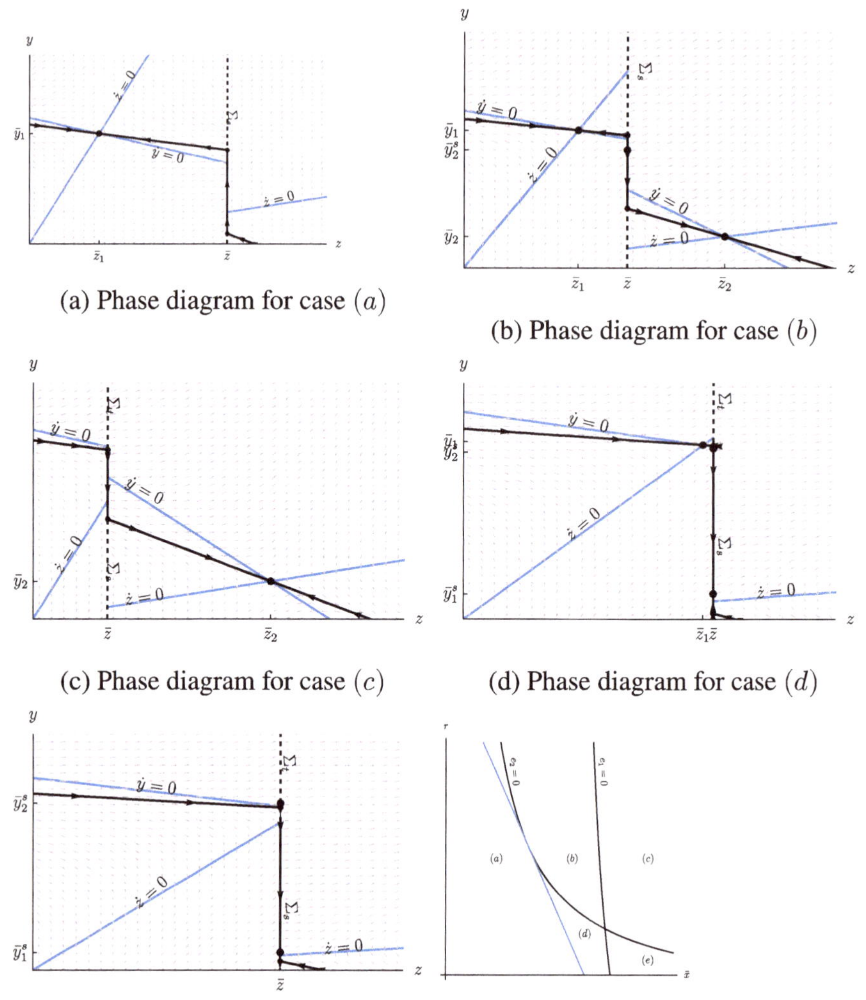

(a) Phase diagram for case (a)

(b) Phase diagram for case (b)

(c) Phase diagram for case (c)

(d) Phase diagram for case (d)

(e) Phase diagram for case (e)

(f) Bifurcation diagram

Fig. 6.17 The optimal pollution model: case (a) one steady state in $\mathbf{Y}_1$; case (b) two steady states and one unstable pseudo-steady state; case (c) one steady state in $\mathbf{Y}_2$; case (d) one steady state in $\mathbf{Y}_1$ and two pseudo-steady states (one stable and one unstable); case (e) two pseudo-steady states (one stable and one unstable) initial-terminal value problem, we only depict trajectories that satisfy the transversality constraint, that is, trajectories that converge to a steady state. We illustrate five representative phase diagrams: (a) in Fig. 6.17a, for low values of the relative satiation, there is one steady state in branch $\mathbf{Y}_1$, with high output and low pollution. Even if the economy starts in branch $\mathbf{Y}_2$ (with high pollution), the economy optimally chooses to reduce pollution. Observe that the crossing through the switching boundary involves sliding with a temporary small response of pollution but high increase in output; (b) in Fig. 6.17b, for higher levels of the relative satiation and a high relative impatience, there are steady states in both branches and one unstable pseudo-steady state, and there are multiple solutions. A pseudo-steady state divides two possible paths, one converging to a low pollution high income steady state in $\mathbf{Y}_1$ and another converging to a high pollution low income steady state in $\mathbf{Y}_2$; (c) in Fig. 6.17c, for intermediate levels of the relative satiation and a low relative impatience, there is one steady state in the high pollution branch, $\mathbf{Y}_2$. This case is a high-pollution counterpart to case (a) because the economy converges to the high pollution steady state even if it started in a low pollution branch $\mathbf{Y}_1$, and again the transition traversing the switching boundary is non-monotonic; (d) in Fig. 6.17d, for intermediate levels of relative satiation, there is one steady state in branch $\mathbf{Y}_1$, and there are two pseudo-steady states. The higher output pseudo steady state $(\bar{z}, \bar{y}_2^p)$ defines the limit to the stable manifold of the regular steady state. The other pseudo-steady state $(\bar{z}, \bar{y}_1^p)$ is stable and operates as a discontinuity induced steady state, and attracts trajectories coming form the high pollution branch $\mathbf{Y}_2$ and from the sliding subset; at last (e) in Fig. 6.17e, there are only two pseudo-steady states, and all optimal trajectories converge to the stable pseudo-steady state $(\bar{z}, \bar{y}_1^s)$.

6.4 References

Seminal mathematical contributions comprise Feigin (1970) and Filippov (1988). This presentation was made possible by the existence of recent monographs (Leine and Nijmeijer, 2004; di Bernardo et al., 2008a; Leine and van de Wouw, 2008), and survey papers (di Bernardo et al., 2008b). A recent survey for higher dimensional ODE can be found in (Jeffrey et al., 2022). There are some applications in economics: the "disequilibrium" model of Honkapohja and Ito (1983), the growth model of van Marrewijk and Verbeek (1993), a New-Keynesian DGE model in Brito et al. (2013), and an endogenous growth model with both variety expansion and quality ladders in Gil et al. (2013).

Chapter 7
Singular ODE

In the previous chapters the vector fields were assumed to be globally Lipschitz continuous function, and, in particular, they had bounded first derivatives everywhere. In this chapter, we deal with ODE in which vector fields $f(y)$, for scalar equations, and $\mathbf{f}(\mathbf{y})$, for planar equations, are not globally or locally Lipschitz continuous, that is, that may not be globally or locally bounded. We introduce time-dependent ODE, with domain $\mathrm{T} = \mathbb{R}_+$, that we call **singular ODE**.

For scalar ODE, there are two main types of differential equations: the **fast-slow ODE**

$$\epsilon\, \dot{y} = g(y), \;\; y : \mathrm{T} \rightarrow \mathrm{Y} \subseteq \mathbb{R}, \tag{7.1}$$

where $g(\cdot)$ is a Lipschitz function, and the parameter $\epsilon \in \mathrm{E} = (-\bar{\epsilon}, \bar{\epsilon})$, where $\bar{\epsilon} > 0$ is a small (close to zero) number, and the **constrained ODE** (or **impasse ODE**)

$$\delta(y)\, \dot{y} = g(y), \;\; y : \mathrm{T} \rightarrow \mathrm{Y} \subseteq \mathbb{R}, \tag{7.2}$$

if there is at least a point $y^i \in \mathrm{Y}$ such that $\delta(y^i) = 0$, and both $g(\cdot)$ and $\delta(\cdot)$ are Lipschitz functions.

In this chapter we call **scalar singular ODE in explicit form** to the functional equation over $y : \mathrm{T} \rightarrow \mathrm{Y} \subseteq \mathbb{R}$

$$\dot{y} = f(y, \epsilon) = \frac{g(y)}{\epsilon\, \delta(y)}, \;\; y : \mathrm{T} \rightarrow \mathrm{Y} \subseteq \mathbb{R}. \tag{7.3}$$

Associated to this explicit form, we specify two other forms: the **scalar singular ODE in separated form**

$$\epsilon\, \delta(y)\, \dot{y} = g(y), \;\; y : \mathrm{T} \rightarrow \mathrm{Y} \subseteq \mathbb{R}, \tag{7.4}$$

and the **scalar singular ODE in de-singularized form**

P. B. Brito, *Economic Dynamics and Distributions*, Dynamic Modeling and Econometrics in Economics and Finance 32, https://doi.org/10.1007/978-3-031-94717-9_7

$$\dot{y} = f^d(y, \epsilon)\ y : \mathrm{T} \to \mathrm{Y} \subseteq \mathbb{R}, \tag{7.5}$$

if $f^d(y, \epsilon) \neq 0$ for all $(y, \epsilon) \in \mathrm{Y} \times \mathrm{E}$.

Therefore, function $f(\cdot)$ can display two types of singularities, a **fast-slow singularity** if $\epsilon = 0$, or an **impasse singularity** if $\epsilon \neq 0$ and there is a point $y^i \in \mathrm{Y}$ such that $y = y^i$. In the first case $f(y, 0) = \infty$ for all $y \in \mathrm{Y}$, and therefore $f(\cdot)$ is not *globally* Lipschitz, and in the second case $f(y^i, \epsilon) = \infty$ and $f(y, \epsilon)$ is finite for any $y \neq y^i \in \mathrm{Y}$ and $\epsilon \neq 0$, and $f(\cdot)$ is not *locally* Lipschitz. We could consider ODE featuring both fast-slow and impasse singularities, but in this chapter, it is assumed that this is not the case. Steady states, $\bar{y}$, satisfying $f(\bar{y}, \epsilon) = 0$ can exist. A particular feature of singular ODE is related to the possible existence of singular steady states, that is to points, y^s such that $g(y^s, \epsilon) = \epsilon\, \delta(y^s) = 0$. For singularities of the impasse type, steady states can be regular or singular, in the sense that we can have $\delta(\bar{y})$ different or equal to zero.

As $f(\cdot)$ may not be, globally or locally, Lipschitz, the usual existence and uniqueness of solutions to ODE[1] may not hold for the whole domain T. In this chapter, we are particularly concerned with the properties of the solutions of ODE, and on the associated bifurcations. Therefore, we consider instead parameterized scalar singular equation ODE depending on parameters $\varphi \in \Phi \subseteq \mathbb{R}^p$, where $p \geq 1$ is the number of parameters,

$$\dot{y} = f(y, \varphi, \epsilon) = \frac{g(y, \varphi)}{\epsilon\, \delta(y, \varphi)},\quad y : \mathrm{T} \to \mathrm{Y} \subseteq \mathbb{R}. \tag{7.6}$$

To bifurcations associated to the existence of singularities, we call generically **singularity induced bifurcations**.

For planar ODE, in which $\mathbf{y} : \mathrm{T} \to \mathbf{Y} \subseteq \mathbb{R}^2$, we can represent the **explicit form** in one of the two equivalent forms

$$\dot{\mathbf{y}} = \mathbf{A}(\mathbf{y}, \varphi, \epsilon)^{-1}\, \mathbf{g}(\mathbf{y}, \varphi, \epsilon)\ \mathbf{y} : \mathrm{T} \to \mathbf{Y} \subseteq \mathbb{R}^2, \tag{7.7}$$

or

$$\mathbf{A}(\mathbf{y}, \varphi, \epsilon)\, \dot{\mathbf{y}} = \mathbf{g}(\mathbf{y}, \varphi, \epsilon),\ \ \mathbf{y} : \mathrm{T} \to \mathbf{Y} \subseteq \mathbb{R}^2, \tag{7.8}$$

where $\mathbf{y} = (y_1, y_2)^\top \in \mathbf{Y}$, $\varphi \in \Phi$, and

$$\mathbf{g}(\mathbf{y}, \varphi, \epsilon) = \begin{pmatrix} g_1(\mathbf{y}, \varphi, \epsilon) \\ g_2(\mathbf{y}, \varphi, \epsilon) \end{pmatrix}, \text{ and } \mathbf{A}(\mathbf{y}, \varphi, \epsilon) \equiv \begin{pmatrix} a_{11}(\mathbf{y}, \varphi, \epsilon)\ a_{12}(\mathbf{y}, \varphi, \epsilon) \\ a_{21}(\mathbf{y}, \varphi, \epsilon)\ a_{22}(\mathbf{y}, \varphi, \epsilon) \end{pmatrix},$$

where it is assumed that the determinant of $\mathbf{A}$ takes the form

$$\delta(\mathbf{y}, \varphi, \epsilon) = \epsilon\ \det\left(\mathbf{A}(\mathbf{y}, \varphi, \epsilon)\right).$$

[1] See e.g Hirsch and Smale (1974, Chap. 8).

It is also assumed that $g_1(.)$, $g_2(.)$, and $\delta(.)$ are continuous and continuously differentiable functions of $\mathbf{y}$. Equation (7.7) (or (7.8)) is said to be a **planar singular ODE in explicit form** if there is a one-dimensional manifold in $\mathbf{Y} \times \Phi \times \mathrm{E}$ such that $\delta(\mathbf{y}, \varphi, \epsilon) = 0$. If $\epsilon = 0$ and $\det(\mathbf{A}(\mathbf{y}, \varphi, 0)) \neq 0$, for all elements of $\mathbf{Y} \times \Phi$ then it is said to be a **fast-slow ODE in explicit form**, and if $\epsilon \neq 0$ and there are points $\mathbf{y}^i \in \mathbf{Y}$ such that $\det(\mathbf{A}(\mathbf{y}^i, \varphi, 0)) = 0$ then it is said to be a planar **constrained or impasse ODE in explicit form**.

Again there are two useful forms for a singular ODE: we call **planar singular ODE in separated form** to the functional equation

$$\delta(\mathbf{y}, \varphi, \epsilon)\, \dot{\mathbf{y}} = \mathrm{adj}(\mathbf{A}(\mathbf{y}, \varphi, \epsilon))\, \mathbf{g}(\mathbf{y}, \varphi, \epsilon) \tag{7.9}$$

where $\mathrm{adj}\mathbf{A}(.)$ is the adjoint matrix of $\mathbf{A}$, and we call **planar singular ODE in de-singularized form** to

$$\dot{\mathbf{y}} = \mathbf{f}^d(\mathbf{y}, \varphi, \epsilon) \equiv \mathrm{adj}(\mathbf{A}(\mathbf{y}, \varphi, \epsilon))\, \mathbf{g}(\mathbf{y}, \varphi, \epsilon). \tag{7.10}$$

A point $p \in \mathbf{Y} \times \Phi \times \mathrm{E}$ is said to be **regular** if $\det(\mathbf{A}(p)) \neq 0$ and we say there is a **singularity** at point p if $\det(\mathbf{A}(p)) = 0$. The set of all singularities is called the **singular set**:

$$\mathcal{S} \equiv \{(\mathbf{y}, \varphi, \epsilon) \in \mathbf{Y} \times \Phi \times \mathrm{E} : \epsilon\, \delta(\mathbf{y}, \varphi, \epsilon) = 0\}. \tag{7.11}$$

As for the scalar singular ODE, the vector field $\mathbf{f}(p)$ takes infinite values globally or locally, which can generate special properties to that ODE.

In order to understand other consequences of having singularities, consider the qualitative approaches to the dynamics of regular ODE, studied in Chaps. 4 and 5. Let $\bar{\mathbf{y}}$ be a steady state, an element of the steady state set

$$\Gamma = \{(\mathbf{y}, \varphi) \in \mathbf{Y} \times \Phi : \ \mathbf{f}(\mathbf{y}, \varphi, \epsilon) = \mathbf{0}\}$$

If set Γ is non-empty, we can compute the Jacobian of the vector field $\mathbf{f}(\mathbf{y}, \cdot)$ of ODE (7.7), at a point $\mathbf{y} \in \Gamma$,

$$\mathbf{J} \equiv \mathrm{D}_{\mathbf{y}}\mathbf{f}(\bar{\mathbf{y}}, \varphi, \epsilon) = \frac{1}{\epsilon\, \delta(\bar{\mathbf{y}}, \varphi, \epsilon)} \begin{pmatrix} j_{1,1}(\bar{\mathbf{y}}, \varphi, \epsilon) & j_{1,2}(\bar{\mathbf{y}}, \varphi, \epsilon) \\ j_{2,1}(\bar{\mathbf{y}}, \varphi, \epsilon) & j_{2,2}(\bar{\mathbf{y}}, \varphi, \epsilon) \end{pmatrix}. \tag{7.12}$$

It has trace and determinant given by

$$\begin{aligned} \mathrm{tr}(\mathbf{J}) &= \frac{j_{1,1}(\bar{\mathbf{y}}, \varphi, \epsilon) + j_{2,2}(\bar{\mathbf{y}}, \varphi, \epsilon)}{\epsilon\, \delta(\bar{\mathbf{y}}, \varphi, \epsilon)}, \\ \det(\mathbf{J}) &= \frac{j_{1,1}(\bar{\mathbf{y}}, \varphi, \epsilon)\, j_{2,2}(\bar{\mathbf{y}}, \varphi, \epsilon) - j_{1,2}(\bar{\mathbf{y}}, \varphi, \epsilon)\, j_{2,1}(\bar{\mathbf{y}}, \varphi, \epsilon)}{(\epsilon\, \delta(\bar{\mathbf{y}}, \varphi, \epsilon))^2}. \end{aligned}$$

Two consequences from the existence of singularities, at a steady state, are readily apparent. First, if functions $g_1(.)$ and $g_2(.)$ are sufficiently regular at steady state $\bar{\mathbf{y}}$, and if we evaluate them (naïvely) at a singularity then we would obtain infinitely-valued eigenvalues.[2] Regular bifurcations occur, generically, when there is a non-hyperbolic steady state, that is when the Jacobian has at least one eigenvalue with zero real part. When it is unfolded, there is at least one eigenvalue of the Jacobian whose real part changes sign. When there is a singular bifurcation, a small change in a parameter changes the sign of at least one eigenvalue not by crossing zero but crossing one asymptote (that is infinity). Second, similarly to regular bifurcation points, singularities introduce a bifurcation-like change in the local dynamics. If there is not any further degeneracy, at a singular point, the sign of the determinant $\det(\mathbf{J})$ also changes. This means that the dimensions of the stable and unstable manifolds changes when a singularity bifurcation is unfolded. Third, there are also similar global effects when the singularities do not coincide with steady states. In this case, evaluating the Jacobian $D_{\mathbf{y}}\mathbf{f}(\mathbf{y}, \varphi, \epsilon)$ at a point $(\mathbf{y}, \varphi, \epsilon) \in \mathcal{S}$, that is not a steady state, we would also obtain infinitely-value eigenvalues, and a different type of transition dynamics than those studied for regular ODE.

The facts on the qualitative dynamics for regular ODE that were studied in Chaps. 4 and 5 need to be adapted and extended in order to deal with the dynamics of the solutions to scalar or planar singular ODE, (7.6), or (7.7), that is, for dealing with local or global dynamic analysis of ODE whose vector fields are not globally (for fast-slow singularities) or locally (for impasse singularities) Lipschitz.

Given the huge dimension of the subject, we just present a brief introduction to some simple methods allowing for the characterization of the dynamics of the solution to an ODE in the presence of singularities. We introduce the following assumption:

Assumption 7.1 Singularities can only be of one of the following two types: **singular-perturbation** or **fast-slow** singularities, for which $\epsilon = 0$ and $\delta(\mathbf{y}, \varphi, 0) \neq 0$ for all $(\mathbf{y}, \varphi) \in \mathbf{Y} \times \Phi$, or **impasse** singularities, for which $\epsilon \neq 0$ and for a manifold in $\mathbf{Y}$ such that $\delta(\mathbf{y}, \varphi, \epsilon) = 0$.

The general method to deal with the two types singularities is based on the next idea: because the presence of singularities generates infinite-valued levels for the vector fields, and its eigenvalues, we transform the system in order to remove the source of the singularity, and use the same methods as for regular ODE on its de-singularized form.[3] Section 7.1 is dedicated to fast-slow (or singularly perturbed) ODE and singularities, and Sect. 7.2 to constrained ODE and impasse singularities.

[2] Even if there is a local bifurcation at a singular steady state the Jacobian evaluated in this way will hide the local dynamic behavior.

[3] A naïve intuition on this idea is the following: if we were trying to understand the trajectory of an object traveling close to the light speed, it would be better to travel close to it rather than observing its path from a fixed point.

7.1 Singular Perturbed or Fast-Slow ODE

We consider some simple cases for linear scalar and planar equations, in Sects. 7.1.1 and 7.1.2, and introduce next a general method for dealing with fast-slow singularities, in Sect. 7.1.3.

7.1.1 *Linear Scalar Equations*

The **linear scalar fast-slow ODE** in explicit form is the functional equation

$$\dot{y} = \frac{\lambda\, y + b}{\epsilon}, \text{ for } y : \mathrm{T} \to \mathrm{Y} \subseteq \mathbb{R}, \tag{7.13}$$

where $g(y) = \lambda\, y + b$, ϵ takes values in a small interval E centered at 0, and $\lambda \neq 0$ is finite.

Consider first the homogeneous ODE where $b = 0$. It is easy to see that the rate of growth of y approaches infinity, for all t, when ϵ tends to zero: $\lim_{\epsilon \to 0} \dot{y}/y = \lambda/\epsilon = \pm\infty$. This implies that, even if $\epsilon \neq 0$, the rate of growth is very high, which obfuscates the dynamics. To uncover the dynamics, we remove the source of singularity by redefining time as $\tau \equiv \frac{t}{\epsilon}$, that we call **fast time**. In contrast, t is called **slow time**. The associated de-singularized ODE is called **associated fast ODE**, or layer ODE, and it is specified by $y' = \lambda\, y$, where $y' = dy/d\tau$.

The pair of the separated and de-singularized versions of the homogeneous ODE (7.13), where $b = 0$,

$$\epsilon\, \dot{y} = \lambda\, y, \tag{7.14a}$$

$$y' = \lambda y, \tag{7.14b}$$

is called the **fast-slow system**.

If $\epsilon \neq 0$, the general solutions to the ODE in the two time scales, slow and fast, are

$$y(t, \epsilon) = y(0)\, e^{\frac{\lambda}{\epsilon} t}, \tag{7.15a}$$

$$y(\tau) = y(0)\, e^{\lambda \tau}. \tag{7.15b}$$

Both solutions are continuous and smooth at $t = 0$. If $y(0) \neq 0$, the asymptotic behavior of the trajectories is the following

$$\lim_{t \to +\infty} y(t, \epsilon) = \begin{cases} +\infty, & \text{if } \lambda/\epsilon > 0 \text{ and } y(0) > 0, \\ -\infty, & \text{if } \lambda/\epsilon > 0 \text{ and } y(0) < 0, \\ 0, & \text{if } \lambda/\epsilon < 0. \end{cases}$$

Stability properties depend on the sign of the ratio λ/ϵ. The steady state $\bar{y} = 0$ is an attractor if $\lambda/\epsilon < 0$, or it is a repeller if $\lambda/\epsilon > 0$.

If $\epsilon = 0$, the fast-slow system is instead

$$0 = \lambda\, y, \tag{7.16a}$$

$$y' = \lambda y, \tag{7.16b}$$

and the behavior of the solution for $t > 0$ is revealed by the solution to the fast (or layer) Eq. (7.15b): if $\lambda < 0$ and $\tau > 0$ ($\tau < 0$) then the solution decays monotonically to 0 (increases monotonically to $+\infty$); or if $\lambda > 0$ and $\tau > 0$ ($\tau < 0$) then the solution increases monotonically to $+\infty$ (decays monotonically to 0). Going back to the slow-time scale solution, in Eq. (7.15a), we find that if $\epsilon = 0$ the solution is $y(t) = 0$ if $y(0) = 0$, and if $y(0) \neq 0$, then the solution is **discontinuous at** $t = 0$

$$y(t, 0) = \begin{cases} y(0), & \text{for } t = 0, \\ 0, & \text{for each } t > 0, \quad \text{if } \lambda < 0, \\ \pm\infty, & \text{for each } t > 0, \quad \text{if } \lambda > 0. \end{cases}$$

Then, for a fixed λ, if we let ϵ vary within its domain E, starting from $-\bar{\epsilon}$, leads to a change in the stability properties of the solution flow. For small ϵ the solution decays monotonically (or increases monotonically), it becomes discontinuous for $\epsilon = 0$, and changes its stability properties to increase monotonically (decrease monotonically) upon crossing that value. The type of discontinuity at $\epsilon = 0$ is determined by the sign of λ, and is revealed by the dynamics of the fast or layer ODE, that can be asymptotically stable or unstable.

Consider now the non-homogeneous ODE in which $b \neq 0$. We have the **scalar linear fast-slow system**

$$\epsilon\, \dot{y} = \lambda\, y + b \tag{7.17a}$$

$$y' = \lambda\, y + b \tag{7.17b}$$

where it is assumed that $\lambda \neq 0$. Then, there is one unique steady state $\bar{y} = -b/\lambda$. The solutions for the slow time and fast time scales are

$$y(t, \epsilon) = \bar{y} + (y(0) - \bar{y})e^{\frac{\lambda}{\epsilon}t}, \text{ and } y(\tau) = \bar{y} + (y(0) - \bar{y})e^{\lambda\tau}.$$

If $\epsilon \neq 0$ the slow time solution is continuous and smooth in t, and, if $y(0) \neq \bar{y}$, it satisfies

$$\lim_{t\to+\infty} y(t, \epsilon) = \begin{cases} +\infty, & \text{if } \lambda/\epsilon > 0 \text{ and } y(0) > \bar{y}, \\ -\infty, & \text{if } \lambda/\epsilon > 0 \text{ and } y(0) < \bar{y}, \\ \bar{y}, & \text{if } \lambda/\epsilon < 0 \text{ for any } y(0). \end{cases}$$

If $\epsilon = 0$ we obtain again a discontinuous function of t, considering again the solution of the layer ODE, if $y(0) \neq \bar{y}$,

$$y(t,0) = \begin{cases} y(0), & \text{for } t = 0, \\ \bar{y}, & \text{for each } t > 0, \quad \text{if } \lambda < 0, \\ \pm\infty, & \text{for each } t > 0, \quad \text{if } \lambda > 0. \end{cases}$$

The layer equation reveals the dynamics close to $\epsilon = 0$. It has two components: a constant **outer** solution $\bar{y}$, and an initial **layer correction**, $(y(0) - \bar{y})e^{\lambda\tau}$, that decays monotonically to zero, if $\lambda < 0$, or increases monotonically if $\lambda > 0$, when $\tau \to \infty$. The monotonic dynamics of the fast equation again provides an indication on the type of the discontinuous "jump" that the slow (or original) solution has to endure.

Exercise 7.1 Solve the fast-slow ODE, in separated form, $\epsilon\,\dot{y} = b$ for $b \neq 0$

7.1.2 *Linear Planar Equations*

The **linear planar fast-slow ODE**, over mapping $\mathbf{y} : \mathrm{T} \to \mathbf{Y} \subseteq \mathbb{R}^2$ where $\mathbf{y}(t) = (y_1(t), y_2(t))$, in explicit form is

$$\dot{\mathbf{y}} = \mathbf{A}\,\mathbf{y} + \mathbf{B} = \mathbf{A}_\epsilon^{-1}\,(\mathbf{A}_s\,\mathbf{y} + \mathbf{B}_s),\ \ \mathbf{y} : \mathrm{T} \to \mathbf{Y} \subseteq \mathbb{R}^2, \tag{7.18}$$

where $\mathbf{A}_\epsilon$ and $\mathbf{A}_s$ are $\mathbb{R}^{2\times2}$ matrices, and $\mathbf{B}_s \in \mathbb{R}^2$. The de-singularized or layer ODE is the functional equation over $\mathbf{y}(\tau) = (y_1(\tau), y_2(\tau))$, where $\tau = t/\epsilon$ is the fast time scale,

$$\mathbf{y}' = \mathrm{adj}(\mathbf{A}_\epsilon)\,(\mathbf{A}_s\,\mathbf{y} + \mathbf{B}_s) = \mathbf{A}_f\,\mathbf{y} + \mathbf{B}_f. \tag{7.19}$$

We assume that matrix $\mathbf{A}_\epsilon$ is of the form $\mathbf{A}_\epsilon = \left(\begin{smallmatrix} 1 & 0 \\ 0 & \epsilon \end{smallmatrix}\right)$, and matrix $\mathbf{A}_s$ is a canonical matrix.[4] Then there is a non-singular matrix $\mathbf{P}$ such that any canonical matrix satisfies $\mathbf{A}_s = \mathbf{P}\,\boldsymbol{\Lambda}\,\mathbf{P}^{-1}$ where $\boldsymbol{\Lambda}$ is the Jordan explicit form of $\mathbf{A}_s$. Given the very large number of possible cases, we restrict to two cases in which matrix $\mathbf{A}_s$ is in one of the following two Jordan forms: the node case, in which $\mathbf{A}_s = \boldsymbol{\Lambda}_+$ has a positive discriminant, $\Delta(\mathbf{A})$, and two real distinct eigenvalues, or the focus case, in which $\mathbf{A}_s = \boldsymbol{\Lambda}_-$ has a negative discriminant and two complex conjugate eigenvalues. We also assume that the ODE is homogeneous by setting $\mathbf{B}_s = \mathbf{0}$. Those two cases are examined next.

Node case. Let the linear planar fast-slow ODE in explicit form be

$$\begin{aligned} \dot{y}_1 &= \lambda_1\,y_1, \\ \dot{y}_2 &= \frac{\lambda_2}{\epsilon}\,y_2. \end{aligned} \tag{7.20}$$

[4] See Chap. 4.

The **linear planar fast-slow system**, consisting of the ODE in both separated (slow time) and de-singularized (fast time) forms, is

$$\begin{aligned} \dot{y}_1 &= \lambda_1\, y_1, & y_1' &= \epsilon\, \lambda_1\, y_1, \\ \epsilon\, \dot{y}_2 &= \lambda_2\, y_2, & y_2' &= \lambda_2\, y_2, \end{aligned} \tag{7.21}$$

where $\epsilon \in \mathrm{E}$, and $\varphi = (\lambda_1, \lambda_2) \neq (0, 0)$ is a real-valued vector.

The ODE (7.21) has one unique steady state $\bar{\mathbf{y}} = (\bar{y}_1, \bar{y}_2) = (0, 0)$. As matrix $\mathbf{A}$, has spectrum $\sigma(\mathbf{A}) = \{\lambda_1, \frac{\lambda_2}{\epsilon}\}$, then the steady state is: (1) a stable node if $\lambda_1 < 0$ and $\lambda_2/\epsilon < 0$, (2) an unstable node if $\lambda_1 > 0$ and $\lambda_2/\epsilon > 0$; or (3) a saddle point if $\lambda_1 < 0$ and $\lambda_2/\epsilon > 0$ or $\lambda_1 > 0$ and $\lambda_2/\epsilon < 0$.

If $\epsilon \neq 0$, the solution to ODE in the slow time scale is

$$\mathbf{y}(t, \epsilon) = \begin{pmatrix} y_1(t, \epsilon) \\ y_2(t, \epsilon) \end{pmatrix} = \begin{pmatrix} y_1(0) e^{\lambda_1 t} \\ y_2(0) e^{\frac{\lambda_2}{\epsilon} t} \end{pmatrix}, \text{ for each } t \in \mathbb{R}_+,$$

and in the fast time scale is

$$\mathbf{y}(\tau, \epsilon) = \begin{pmatrix} y_1(\tau) \\ y_2(\tau) \end{pmatrix} = \begin{pmatrix} y_1(0) e^{\epsilon \lambda_1 \tau} \\ y_2(0) e^{\lambda_2 \tau} \end{pmatrix}, \text{ for each } \tau \in \mathbb{R}_+(\mathbb{R}_-), \text{ if } \epsilon > 0 (< 0).$$

If $\epsilon = 0$ the fast-slow system (7.21) becomes

$$\begin{aligned} \dot{y}_1 &= \lambda_1\, y_1, & y_1' &= 0, \\ 0 &= \lambda_2\, y_2, & y_2' &= \lambda_2\, y_2, \end{aligned} \tag{7.22}$$

If $y_2(0) \neq 0$, the solutions for the two time scales are

$$\mathbf{y}(t, 0) = \begin{cases} (y_1(0), y_2(0)), & \text{for } t = 0, \\ \left(y_1(0) e^{\lambda_1 t}, 0\right), & \text{for each } t \in (0, \infty), \quad \text{if } \lambda_2 < 0, \\ \left(y_1(0) e^{\lambda_1 t}, \infty\right), & \text{for each } t \in (0, \infty), \quad \text{if } \lambda_2 > 0, \end{cases}$$

$$\mathbf{y}(\tau, 0) = \left(y_1(0), y_2(0)\, e^{\lambda_2 \tau}\right), \text{ for each } \tau \in \mathbb{R}_+(\mathbb{R}_-), \text{ if } \epsilon > 0 (< 0).$$

While the slow time solution is discontinuous at $t = 0$, because $y_2(t)$ is discontinuous, the fast time solution is everywhere continuous in τ. If $y_2(0) \neq \bar{y}_2 = 0$, the slow-$y_2$ jumps to its asymptotic value at $t = 0^+$, the fast-y_2 has transition dynamics converging monotonically to its steady state level, $\bar{y}_2 = 0$, if $\lambda_2 < 0$, or it increases unboundedly if $\lambda_2 > 0$ for $\tau \in (0, \infty)$. If $y_2(0) = \bar{y}_2 = 0$ the fast time solution would be stationary.

We define a set of critical points in set $\Omega = \mathbf{Y} \times \Omega$,

$$\mathcal{S}_0^p = \{(\mathbf{y}, \varphi) \in \Omega : y_2 = 0\},$$

and its projection onto set $\mathbf{Y}$,

$$\Gamma_E = \{\mathbf{y} \in \mathbf{Y} : y_2 = 0\}.$$

Set Γ_E is an attractor if $\lambda_2 \epsilon < 0$, and it is a repeller if $\lambda_2 \epsilon > 0$. The dimension of the stable manifold depends also on the sign of λ_1: if $\lambda_1 < 0$ then the stable manifold has dimension two if $\epsilon\, \lambda_2 < 0$ and has dimension one if $\epsilon\, \lambda_2 > 0$.

Figure 7.1a and b illustrates the solution dynamics for $\lambda_1 < 0$ and $\lambda_2 < 0$, when ϵ tend to zero from both positive and negative values. The phase diagrams show trajectories starting from the same initial point $\mathbf{y}(0) = \mathbf{y}_0$, for different values of ϵ converging to zero from positive values in Fig. 7.1a , and from negative values in Fig. 7.1b , respectively. In the first case, if $\epsilon \downarrow 0$, the trajectory becomes discontinuous at $t = 0$, where $\mathbf{y}(0) = \mathbf{y}_0 = (1, 1)$ at $t = 0$, and $\mathbf{y}(0^+) = (1, 0)$ at $t = 0^+$, and converges over time to the steady state $\bar{\mathbf{y}} = \mathbf{0}$. If $\epsilon > 0$, the stable manifold is of dimension two and $\mathcal{W}^s = \mathbf{Y}$. In the second case, if $\epsilon \uparrow 0$, the solution also becomes discontinuous at $t = 0$, but now it jumps discontinuously from $\mathbf{y}(0) = \mathbf{y}_0 = (1, 1)$, at $t = 0$, to $\mathbf{y}(0^+) = (1, +\infty)$, not converging to the steady state. In that figure the curve $\mathcal{S}_0^p$, is the one dimensional stable manifold, that is $y_2(0) = 0$. For initial points belonging to it the solution converges, in the slow time scale, to the steady state $\bar{\mathbf{y}} = \mathbf{0}$. Therefore: first, in those graphs, the initial point $\mathbf{y}(0) = \mathbf{y}_0 = (1, 1)$ belongs to the stable subspace in the case of Fig. 7.1a and to the unstable subspace in the case of Fig. 7.1b , and therefore for a given value of λ_2 (negative in the case of the figure) the local dynamics of the solution, and the dimension of the stable and unstable subspaces is not changed when ϵ converges monotonically to zero. This illustrates the Fenichel (1979) theorem (see next subsection); second, when ϵ changes sign (for fixed values of the other parameters) it changes the dimension of the stable subspaces. Therefore, ϵ acts as a bifurcation parameter having a critical level at $\epsilon = 0$.

Focus case. The parameter ϵ operates not only as a bifurcation parameter, by determining the dimension of the local stable and unstable subspaces, but it also modifies the solution paths by generating discontinuities and changing their monotonicity properties. Consider a particular planar linear fast slow ODE having the explicit form

$$\begin{aligned} \dot{y}_1 &= \alpha\, y_1 + \beta\, y_2, \\ \dot{y}_2 &= -\frac{\beta}{\epsilon}\, y_1 + \frac{\alpha}{\epsilon}\, y_2, \end{aligned} \tag{7.23}$$

where $\mathbf{A}_s = \mathbf{\Lambda}_-$, such that $\alpha \neq 0$ and $\beta \neq 0$, and again $\mathbf{B}_s = \mathbf{0}$. The separated and de-singularized ODE constitute the following fast-slow system

$$\begin{aligned} \dot{y}_1 &= \alpha y_1 + \beta y_2 \qquad & y_1' &= \epsilon(\alpha y_1 + \beta y_2) \\ \epsilon \dot{y}_2 &= -\beta y_1 + \alpha y_2 \qquad & y_2' &= -\beta y_1 + \alpha y_2. \end{aligned} \tag{7.24}$$

ODE (7.23) has one unique steady state $\bar{\mathbf{y}} = (\bar{y}_1, \bar{y}_2) = (0, 0)$. Furthermore, as matrix $\mathbf{A}$ has $\mathrm{tr}(\mathbf{A}) = \frac{\alpha(1+\epsilon)}{\epsilon}$ and $\det(\mathbf{A}) = \frac{\alpha^2+\beta^2}{\epsilon}$, then the eigenvalues of $\mathbf{A}$ are

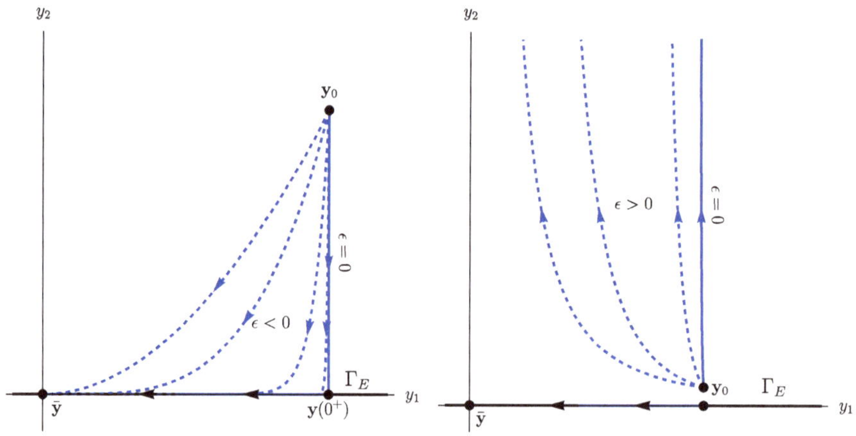

(a) Phase diagram for $\epsilon \downarrow 0$ (b) Phase diagram for $\epsilon \uparrow 0$

Fig. 7.1 Phase diagram for the ODE (7.20) where $\lambda_1 < 0$ and $\lambda_2 < 0$. We depict a sequence of paths for a sequence of ϵ converging to zero

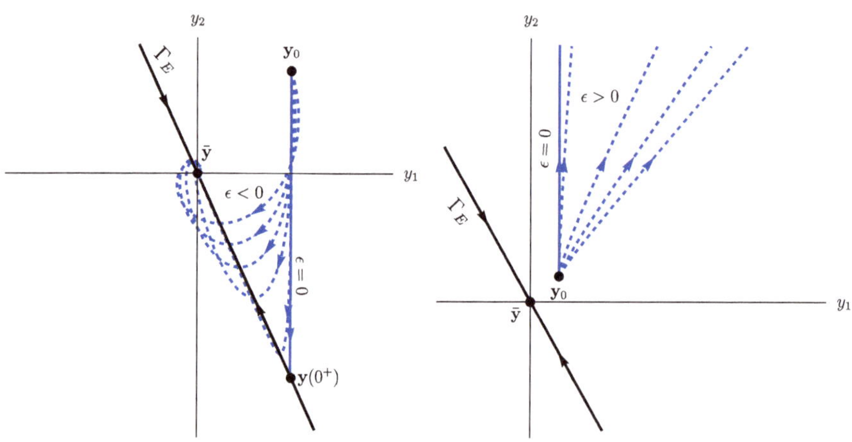

(a) Phase diagram for $\epsilon \uparrow 0$ (b) Phase diagram for $\epsilon \downarrow 0$

Fig. 7.2 Phase diagrams for Eq. (7.23) when $\alpha < 0$, $\beta > 0$ and ϵ converging to zero from positive values (**a**) and from negative values (**b**). When ϵ is closer to zero the trajectories becomes monotonic, and discontinuous in the limit

$$\lambda_{\pm} = \frac{\alpha(1+\epsilon)}{2\epsilon} \pm \left(\frac{\alpha^2(\epsilon-1)^2 - 4\epsilon\beta^2}{(2\epsilon)^2}\right)^{1/2}.$$

Then: if $\epsilon < 0$ the steady state is a saddle for any values of α, or, if $\epsilon > 0$, it is a stable unstable node or focus, if $\alpha < 0$ or $\alpha > 0$, respectively, or it is a center if $\alpha = 0$. Defining

$$\epsilon_{-}^{*} \equiv \frac{\alpha^2 + 2\beta^2 - 2\sqrt{\alpha^2\beta^2 + \beta^4}}{\alpha^2}, \text{ and } \epsilon_{+}^{*} \equiv \frac{\alpha^2 + 2\beta^2 + 2\sqrt{\alpha^2\beta^2 + \beta^4}}{\alpha^2} > 0,$$

then the equilibrium point is a focus if $\epsilon \in (\epsilon_{-}^{*}, \epsilon_{+}^{*})$ or it is a node if $\epsilon < \epsilon_{-}^{*}$ or $\epsilon > \epsilon_{+}^{*}$.

For any value of ϵ, and if $\alpha \neq 0$, we have the projected singular set

$$\Gamma_{\mathrm{E}} = \left\{\mathbf{y} \in \mathbf{Y} : y_2 = \frac{\beta}{\alpha}\, y_1\right\}.$$

This set is coincident with the stable manifold when the steady state is a saddle, and is one of the eigenspaces when the steady state is a stable node. As in the node case, for any ϵ, if an initial point is in Γ_{E} then there will be convergence to the steady state.

If $\epsilon = 0$, the general solution to the fast-time ODE, for an arbitrary initial point $\mathbf{y}(0) = (y_1(0), y_2(0))$, is

$$\mathbf{y}(\tau) = \begin{pmatrix} y_1(\tau) \\ y_2(\tau) \end{pmatrix} = \begin{pmatrix} y_1(0) \\ \frac{\beta}{\alpha}\, y_1(0) + (y_2(0) - \frac{\beta}{\alpha}\, y_1(0))e^{\alpha\tau} \end{pmatrix}, \text{ for each } \tau \in \mathbb{R}.$$

If $\tau > 0$, the solution converges to the surface $y_2 = \frac{\beta}{\alpha}\, y_1$ if $\alpha < 0$.

Assuming that $\alpha < 0$, $\beta > 0$ and ϵ is different from zero, Fig. 7.2 illustrates two families of solution paths, starting form the same initial point $\mathbf{y}(0) = \mathbf{y}_0 = (1, 1)$, for different values of ϵ converging to zero, starting from positive values in Fig. 7.2a , and from negative values in Fig. 7.2b . In Fig. 7.2a , ϵ starts from a value within the interval $(\epsilon_{-}^{*}, \epsilon_{+}^{*})$, the steady state is a stable focus, and the transition is non-monotonic. For smaller values of ϵ, the trajectory becomes monotonic, with the steady state becoming a stable node, and when $\epsilon = 0^{+}$ it jumps discontinuously to the surface Γ_{E} and converges over time to the steady state along that surface. If $\epsilon < 0$ the steady state is always a saddle point. In Fig. 7.2b, the initial point $\mathbf{y}_0$ belongs to the unstable subspace $\mathcal{W}^{u}$, for any initial value of $\epsilon < 0$, and therefore trajectories are unbounded over time. However, when $\epsilon \to 0^{+}$ the speed of departure from the surface Γ_{E} is increasing until it becomes infinite in the limit. Again, a solution path starting at the surface Γ_{E} converges over time to the steady state.

7.1.3 Non-Linear Planar Equations

This subsection contains a brief introduction to the qualitative dynamics of non-linear ODE having fast-slow singularities for $\mathbf{y} : \mathrm{T} \to \mathbf{Y} \in \mathbb{R}^2$.[5] The ODE in the slow-time explicit form is

$$\dot{\mathbf{y}} = \mathbf{f}(\mathbf{y}, \varphi, \epsilon) = \mathbf{A}_\epsilon^{-1}\, \mathbf{g}(\mathbf{y}, \varphi), \;\; \text{for } t \in \mathbb{R}_+, \tag{7.25}$$

where

$$\mathbf{f}(\mathbf{y}, \varphi, \epsilon) = \begin{pmatrix} f_1(\mathbf{y}, \varphi, \epsilon) \\ f_2(\mathbf{y}, \varphi, \epsilon) \end{pmatrix} = \begin{pmatrix} g_1(\mathbf{y}, \varphi) \\ \frac{1}{\epsilon}\, g_2(\mathbf{y}, \varphi) \end{pmatrix}.$$

The fast-slow ODE system, combining the ODE in separated (slow-time) and de-singularized (fast-time) forms is,

$$\mathbf{A}_\epsilon \dot{\mathbf{y}} = \mathbf{g}(\mathbf{y}, \varphi), \;\; \text{for } t \in \mathbb{R}_+, \tag{7.26a}$$

$$\mathbf{y}' = \mathbf{f}^f(\mathbf{y}, \varphi, \epsilon), \;\; \text{for } \tau \in \mathbb{R}, \tag{7.26b}$$

where

$$\mathbf{f}^f(\mathbf{y}, \varphi, \epsilon) = \begin{pmatrix} \epsilon\, g_1(\mathbf{y}, \varphi) \\ g_2(\mathbf{y}, \varphi) \end{pmatrix}.$$

If $\epsilon \neq 0$ the dynamics of equation (7.25) is regular and was the subject of Chap. 5. The main aspects that interest us here are the following. First, given any initial value $\mathbf{y}(0) \in \mathbf{Y}$ a solution exists, it is unique, and it is continuous for all $t \in [0, \infty)$. Second, a regular steady state $\bar{\mathbf{y}}$ is an element of the steady state set

$$\Gamma = \{(\mathbf{y}, \varphi) \in \mathbf{Y} \times \Phi : \mathbf{f}(\mathbf{y}, \varphi, \epsilon) = \mathbf{0}\}.$$

If this set is non-empty, then the local dynamics is characterized by the eigenvalues of the Jacobian $\mathbf{J}(\epsilon) \equiv \mathrm{D}_{\mathbf{y}}\mathbf{f}(\bar{\mathbf{y}}, \varphi, \epsilon)$. Clearly, regular local bifurcation points can exist in Γ, if it contains non-hyperbolic steady states, and local bifurcations, if, for instance, $\det(\mathbf{J}(\epsilon)) = 0$, for values of $\epsilon \neq 0$. Global type of trajectories, generic or not, can also exist. However, the (regular) bifurcations presented in Chap. 5 do not cover the case in which we could have $\epsilon = 0$.

If $\epsilon = 0$, we define the **fast-slow singularity set**

$$\mathcal{S}^p \equiv \{(\mathbf{y}, \varphi, \epsilon) \in \mathbf{Y} \times \Phi \times \mathrm{E} : \; \epsilon = 0\},$$

as the subset of the generic singularity set defined in (7.11). For elements to this set, the fast-slow is defined by the pair of ODE coupling the separated (slow time) form

[5] A very complete reference for these ODE is Kuehn (2015).

$$\dot{y}_1 = g_1(\mathbf{y}, \varphi), \tag{7.27a}$$
$$0 = g_2(\mathbf{y}, \varphi), \tag{7.27b}$$

and the de-singularized (fast time) form

$$y_1' = 0, \tag{7.28a}$$
$$y_2' = g_2(\mathbf{y}, \varphi). \tag{7.28b}$$

Equation (7.28b) allows us to define a the family of manifolds in set $\mathbf{Y} \times \mathbf{\Phi}$,

$$\mathcal{S}_0^p = \left\{(\mathbf{y}, \varphi) \in \mathcal{S}^p : \ g_2(\mathbf{y}, \varphi) = 0\right\},$$

called **singular-perturbed critical subset**. The projection of this subset onto the state space $\mathbf{Y}$ is

$$\Gamma_E = \{\mathbf{y} \in \mathbf{Y} : g_2(\mathbf{y}, \varphi) = 0\}.$$

The steady state set Γ has a non-empty intersection with Γ_E when ϵ converges to zero, because if $\epsilon \neq 0$ the manifold Γ_E would be very close to the isocline for y_2, and this isocline tends to Γ_E when ϵ tends to zero.

If function $g_2(\cdot)$ is a sufficiently smooth function of y_2, then the implicit function theorem allows us to write $y_2 = h(y_1)$. Upon substitution in Eq. (7.27a) we find that, within Γ_E, the dynamics is driven by the scalar differential equation for y_1,

$$\dot{y}_1 = g_1(y_1, h(y_1, \varphi), \varphi).$$

Therefore, within set Γ_E, the variable y_1 is a continuous function of the slow time scale t, while y_2 is discontinuous. In order to determine the dynamics out of the singular-perturbed critical subset, $\mathbf{Y}/\Gamma_{\mathrm{E}}$, and the type of discontinuity of y_2, we consider the ODE (7.28b) for y_2 as a function of the fast time scale,

$$y_2' = g_2(y_1, y_2, \varphi). \tag{7.29}$$

This ODE has a solution that is a continuous function of the fast time scale, τ, where y_1 is taken as a parameter.

As we already saw in Chap. 3, except for linear ODE, we usually cannot solve those two scalar ODE operating in the interior of set Γ_E, in closed form. However, a qualitative characterization of the solution to a singularly perturbed ODE can be accomplished by introducing a further stratification to sets $\mathcal{S}_0^p$ and Γ_E.

Consider a singular point $(\mathbf{y}_p, \varphi_p) \in \mathcal{S}_0^p$. That point is called a **fast-slow regular point** if $g_{2,y_2}(\mathbf{y}_p, \varphi_p) \neq 0$[6], and it is called a **fast-slow singular point** if $g_{2,y_2}(\mathbf{y}_p, \varphi) = 0$. The set of **fast-slow singular points** is defined bf

[6] We use the notation $g_{2,y_2}(\mathbf{y}, \cdot) \equiv \frac{\partial g(y_1, y_2, \cdot)}{\partial y_2}$.

$$\mathcal{S}_1^p = \left\{(\mathbf{y}, \varphi) \in \mathcal{S}_0^p : \; g_{2,y_2}(\mathbf{y}, \varphi) = 0\right\}, \tag{7.30}$$

and its projection onto set $\mathbf{Y}$ is defined by $\Gamma_{E,1} = \{\mathbf{y} \in \Gamma_E : g_{2,y_2}(\mathbf{y}, \varphi) = 0\}$.

The set of fast-slow regular points, $\mathcal{S}_0^p/\mathcal{S}_1^p$, is the set of hyperbolic points of the fast ODE (7.29), while the set of singular points, $\mathcal{S}_1^p$ is the set of non-hyperbolic points. The non-hyperbolic and hyperbolic points over set $\mathbf{Y}$ are $\Gamma_{E,1}$ and $\Gamma_E/\Gamma_{E,1}$, respectively. Those points are not necessarily steady states of the ODE.

The set $\Gamma_{E,1}$ introduces a partition of set Γ_E, that we can write $\Gamma_E = \Gamma_E^a \cup \Gamma_{E,1} \cup \Gamma_E^r$, where Γ_E^a is the set of **fast-slow regular attracting** points, satisfying $g_{2,y_2}(\mathbf{y}, \varphi) < 0$, and Γ_E^r is the set of **fast-slow regular repelling** points, satisfying $g_{2,y_2}(\mathbf{y}, \varphi) > 0$. If there are non-hyperbolic points, that is if set $\mathcal{S}_1^p$ is empty, then set $\mathcal{S}_0^p$ contains only attractor points (as in Fig. 7.1a) or only repelling points (as in Fig. 7.1b).

Furthermore, if the sets $\mathcal{S}_1^p$ and $\Gamma_{E,1}$ are empty, then one of the most important results on the fast-slow system, the Fenichel (1979) theorem, applies. It states that, in a neighborhood of Γ_E, the dimension of the stable manifold is not changed by a small variation of the parameter ϵ, as we have observed in Fig. 7.1a and b. We have also found that, when $\epsilon \to 0$ and the steady state point is stable, the trajectories will approach a monotonic behavior independently from the Jacobian $\mathbf{J}(\epsilon)$, having real or complex roots for values of ϵ sufficiently higher than ϵ. This means that there is a critical level for ϵ such that oscillatory dynamics will change to monotonic.

If the fast-slow set of singular points, $\mathcal{S}_1^p$ and $\Gamma_{E,1}$ are non-empty, then they can be partitioned further. It is important to observe that, while the set of fast-slow critical points Γ_E introduces a one-dimensional manifold (with attracting and/or repelling subsets), over the phase space $\mathbf{Y}$, the set of fast-slow singular points, $\Gamma_{E,1}$, is a zero-dimensional manifold (i.e., a point) over set $\mathbf{Y}$. This means that a further stratification on $\mathcal{S}_1^p$ involves, in general, a stratification of the parameter set Φ.

There are already some results on fast-slow singular bifurcation analysis. We can have fold-, transcritical-, and hysteresis-like fast-slow bifurcations (see Kuehn, 2015, Chaps. 3, 8). This would induce a partition over set $\mathcal{S}_1^p$, among several subsets, say $\mathcal{S}_2^p$.

For instance, a fast-slow fold bifurcation point exists if there is a point $(\mathbf{y}, \varphi) \in \mathcal{S}_0^p$ satisfying the following conditions: $\mathbf{g}(\mathbf{y}, \varphi, 0) = \mathbf{0}$, $g_{2,y_2}(\mathbf{y}, \varphi) = 0$, $g_{2,y_1}(\mathbf{y}, \varphi) \neq 0$, and $g_{2,y_2y_2}(\mathbf{y}, \varphi) \neq 0$.[7]

7.1.4 Application to Economics

Example 7.1 (*The* Benhabib and Farmer (1994) *model*) This is a New-Keynesian DGE model in which there are externalities in production leading to increasing returns at the aggregate level, and endogenous labor supply. The DGE is represented

[7] We use the notation $g_{2,y_2y_2}(\mathbf{y}, \cdot) \equiv \frac{\partial^2 g(y_1, y_2, \cdot)}{\partial y_2^2}$. See again Chap. 5 for the relevant bifurcation conditions.

by a singular perturbed ODE[8], for the stocks of capital and labor (K, L), in explicit form,

$$\dot{K} = K^{\alpha} L^{\beta} (1 - b\, L^{1+\xi}), \tag{7.31a}$$

$$\dot{L} = \frac{L}{\epsilon}\Big(K^{\alpha-1} L^{\beta} \left(a - \alpha (1 - b\, L^{1+\xi})\right) - \rho\Big), \tag{7.31b}$$

and by an initial condition for the pre-determined variable $K(0) = k_0$ and a transversality condition. The labor stock is the non-predetermined variable in our representation of the model. The parameters $\varphi = (\alpha, \beta, a, b, \xi, \rho)$ are all positive, and represent the marginal products of capital and labor in production, the income factor shares of capital and labor, the inverse of the Frisch elasticity of labor supply and the rate of time preference, respectively. Benhabib and Farmer (1994) assume that there are increasing returns to scale, $\alpha + \beta > 1$, the marginal products of both inputs differ from factor shares, $0 < \alpha < a < 1$, and $0 < \beta < b < 1$, and $a + b = 1$. Furthermore, the singularity parameter, ϵ, is defined by the difference between the productivity of capital at the private level and the Frisch elasticity, $\epsilon = \beta - (1 + \xi)$, and can have any sign: $\epsilon \gtreqless 0$ if and only if $\beta \gtreqless 1 + \xi$.

The ODE (7.31) has one unique steady state,

$$(\bar{K}, \bar{L}) = \left(\left(\frac{a}{\rho}\, b^{\frac{\beta}{1+\xi}}\right)^{\frac{1}{1-\alpha}}, b^{\frac{1}{1+\xi}}\right),$$

and its Jacobian, evaluated at the steady state, $\mathbf{J}(\epsilon)$, is a function of ϵ. The trace and the determinant of the Jacobian are

$$\operatorname{tr}(\mathbf{J}(\epsilon)) = \frac{\rho}{\epsilon}\left(\beta - \frac{\alpha}{a}(1 + \xi)\right), \text{ and } \det(\mathbf{J}(\epsilon)) = \frac{\rho^2}{\epsilon}\left(\frac{(1-\alpha)(1+\xi)}{a}\right).$$

With the previous assumptions regarding the parameters, we find that, if $\epsilon < 0$ then the steady state is a saddle point, it is a stable node if $0 < \epsilon < \tilde{\epsilon}$, or it is a stable focus if $\epsilon > \tilde{\epsilon}$, for $\tilde{\epsilon} \equiv \frac{(a\,\beta - \alpha\,(1+\xi))^2}{4a\,(1-\alpha)(1+\xi)}$.

If $\epsilon = 0$, there is a fast-slow singularity, and the de-singularized ODE, for the fast-time scale, is

$$K' = 0, \tag{7.32a}$$

$$L' = L\left(K^{\alpha-1}\left(\alpha\, b + (a - \alpha)\, L^{\beta}\right) - \rho\right). \tag{7.32b}$$

Therefore, the projected singular-perturbed critical subset,

$$\Gamma_{\mathrm{E}} = \left\{(K, L) \in \mathbb{R}^2_+ : L = h(K) \equiv \left(\frac{\rho\, K^{1-\alpha} - \alpha\, b}{a - \alpha}\right)^{\frac{1}{\beta}}\right\},$$

[8] We follow the approach in Brito et al. (2017).

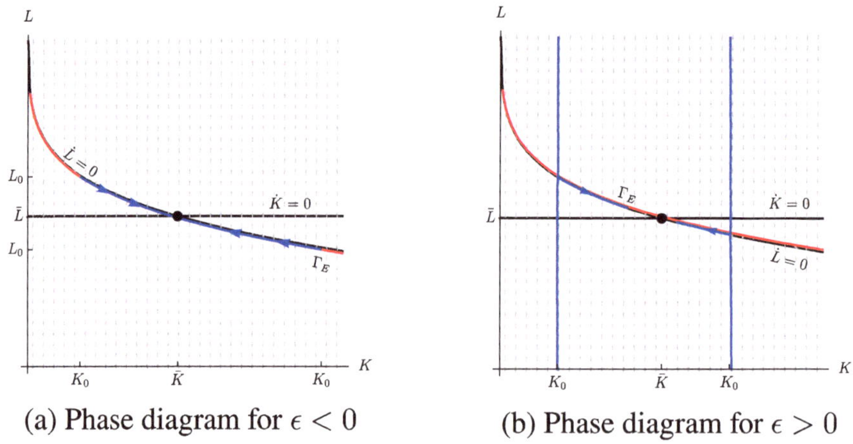

Fig. 7.3 Phase diagrams for the Benhabib and Farmer (1994) model, for $\epsilon \equiv \beta - (1+\xi)$. The DGE is determinate in **a** and indeterminate in **b**

is non-empty for $K > \left(\frac{\alpha b}{\rho}\right)^{\frac{1}{1-\alpha}}$. In order to find if there are fast-slow singular points, that is if set $\Gamma_{E,1}$, in Eq. (7.30), is non-empty, we compute $\frac{\partial L'(K,L)}{\partial L}$ for points in Γ_E. As

$$\frac{\partial L'(K,L)}{\partial L} = \frac{\rho \beta L^{\beta}(a-\alpha)}{(a-\alpha)L^{\beta} + \alpha b} > 0$$

if $a > \alpha$, then Γ_E contains only fast-slow regular attracting points. Then, the qualitative dynamics is similar to the one illustrated in Fig. 7.1 if $\epsilon < \tilde{\epsilon}$.

Thus, there are two main types of macroeconomic regimes. The first regime, in which the relative productivity of capital is low, implying $\epsilon < 0$, is characterized by having a steady state that is a saddle point, and by a determinate DGE (see Fig. 7.3a). Given any initial level for the stock of capital K_0, there is a single equilibrium level of the labor stock, $L(0) \approx H(K_0)$, that satisfies the transversality constraint. Over time the convergence to the steady state will take place along a path, staying inside set Γ_E, very close to the isocline $\dot{L} = 0$. The second regime, in which the relative productivity of capital is sufficiently high, implying $\epsilon > 0$, is characterized by having a steady state that is a asymptotically stable, and, therefore, by an indeterminate DGE (see Fig. 7.3b). Given an initial level for the stock of capital K_0 there is an infinity of equilibrium levels for the labor stock, $L(0)$. However, the existence of a singularity, for any initial beliefs regarding equilibrium labor supply, $L(0)$, will force a very fast convergence to the surface Γ_E, such that $L(t) \approx H(K_0)$, for $t \approx 0^+$, and a dynamics over time similar to the one for the first regime in which there is determinacy.

7.2 Constrained ODE and Impasse Singularities

In this subsection, we assume that $\epsilon = 1$, and study the dynamics of scalar ODE (7.6) in which $\delta(y) = 0$, and planar ODE (7.7) in which det $(\delta(\mathbf{y})) = 0$, locally.

Those ODE are called **constrained ODE**, because they can be associated to differential algebraic equations (DAE) when there is a singularity. For example, this is the case in which we have two time-dependent variables (y, z) governed by a DAE specified by the ODE $\dot{y} = f(y, z)$, where $f(\cdot)$ is a Lipschitz function, and by the algebraic equation $0 = g(y, z)$, where there is at least one point (y^i, z^i) such that $g_z(y^i, z^i) = 0$. If this is the case, we cannot use the implicit function theorem to obtain a monotonic function $z = h(y)$ such that the DAE is equivalent to the regular ODE $\dot{y} = f(y, h(y))$. Next we address the solutions of these equations by using the constrained-ODE (CODE) approach, following Zhitomirskii (1993).

The **scalar constrained ODE** (CODE) in its explicit form is the functional equation, over $y : \mathrm{T} \to \mathbf{Y} \subseteq \mathbb{R}$,

$$\dot{y} = f(y) = \frac{g(y)}{\delta(y)}, \text{ for each } t \in \mathrm{T}, \tag{7.33}$$

where functions $g(y)$ and $\delta(y)$ are continuous and continuously differentiable, and there is a non-empty set of points, $\Gamma_I = \{y \in \mathrm{Y} : \delta(y) = 0\}$. The **separated** and **de-singularized** forms of the ODE are

$$\delta(y)\,\dot{y} = g(y), \tag{7.34a}$$

$$\dot{y} = f^d(y) \equiv g(y). \tag{7.34b}$$

The **planar constrained ODE in explicit form** is the functional equation over mapping $\mathbf{y} : \mathrm{T} \to \mathbf{Y} \subseteq \mathbb{R}^2$

$$\dot{\mathbf{y}} = \mathbf{f}(\mathbf{y}, \varphi) = \mathbf{A}^{-1}(\mathbf{y}, \varphi)\,\mathbf{g}(\mathbf{y}, \varphi), \text{ for each } t \in \mathrm{T}, \tag{7.35}$$

where

$$\mathbf{g}(\mathbf{y}, \varphi) = \begin{pmatrix} g_1(\mathbf{y}, \varphi) \\ g_2(\mathbf{y}, \varphi) \end{pmatrix}, \text{ and } \mathbf{A}(\mathbf{y}, \varphi) \equiv \begin{pmatrix} a_{11}(\mathbf{y}, \varphi) & a_{12}(\mathbf{y}, \varphi) \\ a_{21}(\mathbf{y}, \varphi) & a_{22}(\mathbf{y}, \varphi) \end{pmatrix},$$

and

$$\mathbf{A}^{-1}(\mathbf{y}, \varphi) = \delta(\mathbf{A}(\mathbf{y}, \varphi))^{-1}\, \mathrm{adj}(\mathbf{A}(\mathbf{y}, \varphi)),$$

where functions $\delta(\mathbf{A}(\mathbf{y}, \cdot)) = \det(\mathbf{A}(\mathbf{y}, \cdot))$ and $\mathbf{g}(\mathbf{y}, \cdot)$ are continuous and continuously differentiable, and there is at least one point $\mathbf{y}^i \in \mathbf{Y}$, called **impasse point**, such that $\delta(\mathbf{A}(\mathbf{y}^i, \varphi)) = 0$.

The constrained ODE in its **separated form** is

$$\mathbf{A}(\mathbf{y}, \varphi)\, \dot{\mathbf{y}} = \mathbf{g}(\mathbf{y}, \varphi), \text{ for each } t \in \mathrm{T}, \tag{7.36}$$

and in its **de-singularized form** is

$$\dot{\mathbf{y}} = \mathbf{f}^d(\mathbf{y}, \varphi) = \operatorname{adj}(\mathbf{A}(\mathbf{y}, \varphi))\, \mathbf{g}(\mathbf{y}, \varphi). \tag{7.37}$$

The particular singular set for CODE, generally defined in Eq. (7.11), is called **impasse set** and is defined by

$$\mathcal{S}_0^i \equiv \{(\mathbf{y}, \varphi) \in \mathbf{Y} \times \Phi : \delta(\mathbf{y}, \varphi) = 0\}, \tag{7.38}$$

and its projection onto $\mathbf{Y}$ is **the set of impasse points**

$$\Gamma_I \equiv \{\mathbf{y} \in \mathbf{Y} : \delta(\mathbf{y}, \varphi) = 0\}. \tag{7.39}$$

Again the steady state set is $\Gamma = \{\mathbf{y} \in \mathbf{Y} : \mathbf{g}(\mathbf{y}, \varphi) = \mathbf{0}\}$.

In the rest of this section, we present a very brief introduction to constrained ODE. In Sects. 7.2.1 and 7.2.2, linear scalar and linear planar ODE, respectively, are explicitly solved. Section 7.2.3 is an introduction to the qualitative theory of the dynamics of non-linear CODE in a neighborhood of the impasse set Γ_I. One application to economics can be found in Sect. 7.2.4.

7.2.1 Linear Scalar Equations

Linear scalar constrained ODE are special cases of equation (7.33), in which functions $g(\cdot)$ and $\delta(\cdot)$ are linear functions of $y : \mathrm{T} \to \mathrm{Y} \subseteq \mathbb{R}$. In their explicit, separated and de-singularized forms, we consider

$$\dot{y} = f(y) = \frac{\lambda\, y + b}{y - y^i}, \tag{7.40a}$$

$$(y - y^i)\, \dot{y} = \lambda\, y + b, \tag{7.40b}$$

$$\dot{y} = f^d(y) = \lambda\, y + b, \tag{7.40c}$$

where λ and b are arbitrary real constants, $y^i \in \mathrm{Y}$ is fixed, $g(y) = \lambda\, y + b$ and $\delta(y) = y - y^i$. We can observe that: first, there is an impasse point for $y = y^i$; second, if $\lambda \neq 0$ there is a point $\bar{y} = -\frac{b}{\lambda}$ such that $g(\bar{y}) = 0$; third, this point is a steady state only if $\bar{y} \neq y^i$; fourth, if $\lambda \neq 0$ and $\bar{y} = y^i$ then, although $\delta(y^i) = g(\bar{y})$, a steady state does not exist because, using l'Hôpital's rule, Eq. (7.40) becomes $\dot{y} = \lambda \neq 0$; at last, if $\lambda = 0$ then $g(0) = b$ and there is not a steady state in this case.

The next Proposition presents the general solution to CODE (7.40), for the case in which there is one unique impasse point, y^i, for different values of the parameters.

Proposition 7.1 (General solutions to the linear constrained ODE (7.40)) *Assume that $y^i \in \mathbf{Y}$ is fixed. Then the general solution to Eq. (7.40) can take the following forms:*

(1) if $\lambda = 0$, then there is a constant $\tau \equiv -\frac{(y(0)-y^i)^2}{2b}$, and two functions

$$y_1(t) = y^i - \sqrt{2b\,(t-\tau)}, \tag{7.41a}$$

$$y_2(t) = y^i + \sqrt{2b\,(t-\tau)}, \tag{7.41b}$$

such that:

(a) if $b > 0$, then the general solution is

$$y(t) = \begin{cases} y_1(t) \text{ for each } t \in [0,\infty), & \text{if } y(0) < y^i, \\ y_2(t) \text{ for each } t \in [0,\infty), & \text{if } y(0) > y^i, \end{cases}$$

(b) if $b < 0$ then $\tau > 0$, and the general solution is

$$y(t) = \begin{cases} y_1(t) \text{ for each } t \in [0,\tau], & \text{if } y(0) \le y^i, \\ y_2(t) \text{ for each } t \in [0,\tau], & \text{if } y(0) \ge y^i, \end{cases}$$

(2) if $\lambda \neq 0$, and $b + \lambda\, y^i \neq 0$, or equivalently, if $\bar{y} \neq y^i$, then there is a constant $\tau \equiv \frac{1}{\lambda}\left(y^i - y(0) + (\bar{y} - y^i)\log\left(\frac{\bar{y}-y(0)}{\bar{y}-y^i}\right)\right)$, and two functions

$$y_1(t) = \bar{y} + \left(\bar{y} - y^i\right) W_0(\eta(t)), \text{ for } \eta(t) \in [-e^{-1}, \infty), \tag{7.42a}$$

$$y_2(t) = \bar{y} + \left(\bar{y} - y^i\right) W_{-1}(\eta(t)), \text{ for } \eta(t) \in [-e^{-1}, 0), \tag{7.42b}$$

where $\eta(t) = -\exp\left(\frac{\lambda(t-\tau)}{\bar{y}-y^i} - 1\right)$, and $W_0(\cdot)$ and $W_{-1}(\cdot)$ are the principal and secondary branches of the Lambert-W function, such that:

(a) if $\lambda < 0$, $b > 0$ and $\bar{y} > y^i$, then the general solution is

$$y(t) = \begin{cases} y_2(t) \text{ for each } t \in [0,\infty), & \text{if } y(0) < y^i, \\ y_1(t) \text{ for each } t \in [0,\infty), & \text{if } y(0) > y^i, \end{cases}$$

(b) if $\lambda < 0$, $b \le 0$, and $\bar{y} < y^i$, then the general solution is

$$y(t) = \begin{cases} y_1(t) \text{ for each } t \in [0,\infty), & \text{if } y(0) < \bar{y}, \\ y_1(t) \text{ for each } t \in [0,\tau], & \text{if } \bar{y} < y(0) < y^i, \\ y_2(t) \text{ for each } t \in [0,\tau], & \text{if } y(0) > y^i, \end{cases}$$

(c) if $\lambda > 0$, $b \ge 0$, and $\bar{y} < y^i$, then the general solution is

$$y(t) = \begin{cases} y_1(t) \text{ for each } t \in [0, \infty), & \text{if } y(0) < y^i, \\ y_2(t) \text{ for each } t \in [0, \infty), & \text{if } y(0) > y^i, \end{cases}$$

(d) if $\lambda > 0$, $b < 0$, and $\bar{y} > y^i$, then the general solution is

$$y(t) = \begin{cases} y_2(t) \text{ for each } t \in [0, \tau], & \text{if } y(0) < y^i, \\ y_1(t) \text{ for each } t \in [0, \tau], & \text{if } y^i < y(0) < \bar{y}, \\ y_1(t) \text{ for each } t \in [0, \infty), & \text{if } y(0) > \bar{y}; \end{cases}$$

(3) if $\lambda \neq 0$, and $b + \lambda\, y^i = 0$, or equivalently, $\bar{y} = y^i$, then the general solution is

$$y(t) = y(0) + \lambda\, t, \text{ for each } t \in [0, \infty). \tag{7.43}$$

Proof Assume there is one impasse point $y^i \in \mathbf{Y}$. (1) If $\lambda = 0$ then the CODE in the separated form is $(y - y^i)\dot{y} = b$. Using the method of separation of variables and integrating yields $\frac{y(t)^2}{2} - y^i\, y(t) = \frac{y(0)^2}{2} - y^i\, y(0) + b\, t$ for $t \geq 0$. This is equivalent to $(y(t) - y^i)^2 = (y(0) - y^i)^2 + 2\, b\, t$. Solving for $y(t)$ yields, formally $y(t) = y^i \pm \sqrt{(y(0) - y^i)^2 + 2\, b\, t}$. Defining $\tau = -\frac{(y(0)-y^i)^2}{2\, b}$, we have the two branches (7.41a) and (7.41a). If $b > 0$, then $\tau < 0$, and the solution exists for all $t \in [0, \infty)$. If $b > 0$, then $\tau > 0$ and the solution only exists for $t \in [0, \tau]$. In both cases, the solution satisfies $y(t = 0) = y(0)$, that requires

$$y(0) = \begin{cases} y^i - |y(0) - 1| = y(0) & \text{only if } y(0) < y^i \\ y^i + |y(0) - 1| = y(0) & \text{only if } y(0) > y^i. \end{cases}$$

(2) Let $\lambda \neq 0$ and $\bar{y} \neq y^i$, which is equivalent to assuming that $b + \lambda\, y^i \neq 0$. Then the CODE in the form (7.40b) is $(y - y^i)\dot{y} = \lambda\,(y^i - \bar{y})$. Using the method of separation of variables and integrating yields $y(t) + (\bar{y} - y^i)\, \log\big(y(t) - \bar{y}\big) = y(0) + (\bar{y} - y^i)\, \log\big(y(0) - \bar{y}\big) + \lambda\, t$ for $t \geq 0$. Dividing by $\bar{y} - y^i$, taking exponentials on both sides, and dividing the outcome again by $\bar{y} - y^i$, yields $\frac{y(t)-y^i}{\bar{y}-y^i}\, \exp\left(\dfrac{y(t) - y^i}{\bar{y} - y^i}\right)$ where

$$\eta(t) \equiv \left(\frac{y(0) - \bar{y}}{\bar{y} - y^i}\right) \exp\left[\frac{y(0) - \bar{y} + \lambda\, t}{\bar{y} - y^i}\right]. \tag{7.44}$$

The solution to the CODE is, formally, $y(t) = \bar{y} + (\bar{y} - y^i)\, W(\eta(t))$, where $W(\cdot)$ is the Lambert-W function.[9] Given the properties of the $W(\cdot)$ function, then two

[9] The Lambert-W function is the solution to equation $xe^x = a, x = W(a)$. The Lambert function has the following useful properties: (a) $W(x)$ has domain $x \in [-e^{-1}, +\infty)$; (b) it is single-valued and increasing function for $x > 0$, and it has double-valued for $x \in [-e^{-1}, 0)$; (c) it has two branches: the

branches for the solution to the CODE, depending on the sign of $\eta(t)$, as in Eqs. (7.42a) and (7.42b). In order to specify under which conditions the solution follows a particular branch, we apply the properties of the W function to our specific equation: (a) defining $\tau \equiv \frac{1}{\lambda}\left[y^i - y(0) + (\bar{y} - y^i)\log\left(\frac{\bar{y}-y(0)}{\bar{y}-y^i}\right)\right]$, we have $\eta(t) = -e^{-1}$ if and only if $t = \tau$, which implies that $W_0(\eta(\tau)) = W_{-1}(\eta(\tau)) = -1$ and $y_1(\tau) = y_2(\tau) = y^i$; (b) if $\text{sign}(\lambda) = \text{sign}(\bar{y} - y^i)$ then $\eta(t) \gtreqless 0$ if and only if $\left(\frac{y(0)-\bar{y}}{\bar{y}-y^i}\right) \gtreqless 0$; (c) if $\text{sign}(\lambda) \neq \text{sign}(\bar{y} - y^i)$ then $\eta(t) \gtreqless 0$ if and only if $\left(\frac{y(0)-\bar{y}}{\bar{y}-y^i}\right) \lesseqgtr 0$. Therefore, we have the following cases: (i) if $\text{sign}(\lambda) = \text{sign}(\bar{y} - y^i) = \text{sign}(y(0) - \bar{y})$ or if $\text{sign}(\lambda) \neq \text{sign}(\bar{y} - y^i) = \text{sign}(y(0) - \bar{y})$ then $\eta(t) > 0$ for $t \in [0, \infty)$ and $y(t) = y_1(t)$ for $t \in [0, \infty)$; (ii) if $\text{sign}(\lambda) = \text{sign}(\bar{y} - y^i) \neq \text{sign}(y(0) - \bar{y})$ then $\eta(t) < 0$ and as $\eta'(t) < 0$, then $\eta(t) \in (0, -1)$ and $W_0(t) \in [-e^{-1}, 0]$ and $y(t) = y_1(t)$ for $t \in [0, \tau]$; or (iii) if $\text{sign}(\lambda) \neq \text{sign}(\bar{y} - y^i) \neq \text{sign}(y(0) - \bar{y})$ then $\eta(t) < 0$ and, as $\eta'(t) > 0$, then $\eta(t) \in (0, -1)$ and $W_{-1}(t) \in [-e^{-1}, 0]$ and $y(t) = y_2(t)$ for $t \in [0, \infty)$.
(3) Let $\lambda \neq 0$ and $\bar{y} = y^i$. Then ODE (7.40) becomes $\dot{y} = \lambda$, a regular linear ODE with solution $y(t) = y(0) + \lambda t$.

Figure 7.4 illustrates the possible paths and the phase diagrams if $\lambda = 0$, that is for the two cases (1) in Proposition 7.1. If $b > 0$ Fig. 7.6a shows that the solution increases over time for any time $t \in [0, \infty)$ for any initial $y(0) \neq y^i$, while if $b < 0$ Fig. 7.6b shows that the solution decreases over time until it collides, at time $t = \tau$, with the impasse point, $y(\tau) = y^i$, and there is no continuation for $t > \tau$. Therefore, we call point $y = y^i$ an **impasse repeller** point when solutions diverge from y^i over time, as in the first case, and we call $y = y^i$ an **impasse attractor** point when solutions collide with y^i in finite time and have no continuation afterward. Figure 7.5 illustrates the possible paths and the phase diagrams for the four cases (2) in Proposition 7.1, in which $\lambda \neq 0$. In every case, in addition to an impasse point (that can be an attractor or a repeller) we have a steady state, that can be stable or unstable. Observe the type of "stability" of the impasse and the steady state points are the opposite when the impasse point is attractor and the steady state is unstable, and when the impasse point is repeller and the steady state is asymptotically stable. Furthermore, when the impasse point is attractor there is no continuation of the solution after a finite critical time τ. In Fig. 7.5a , corresponding to solution (2.a), we have $y^i < \bar{y}$, the impasse point is repeller and the steady state is stable, and therefore for any initial point the solution, although having two branches, exists for $t \in [0, \infty)$. In Fig. 7.5b , corresponding to solution (2.b), we have $y^i > \bar{y}$, and the impasse point is an attractor and the steady state is unstable, and for initial points in the interval $(\bar{y}, \infty)$ the solution only exists for $t \in [0, \tau)$, as trajectories collide with the impasse point in finite time, while for initial points $y(0) < \bar{y}$, solutions exist for $t \in [0, \infty)$. Fig. 7.5c and d , corresponding to cases (2.c), and (2.d), respectively, have a similar interpretation.

---

principal branch $W_0(x) \in [-1, \infty)$ for $x \in [-e^{-1}0)$, and the secondary branch $W_{-1}(x) \in [-1, -\infty)$ for $x \in [-e^{-1}, 0)$; and (d) at point $x = -e^{-1}$ we have $W_0(-e^{-1}) = W_{-1}(-e^{-1}) = -1$. See Corless et al. (1996) or Brito et al. (2008).

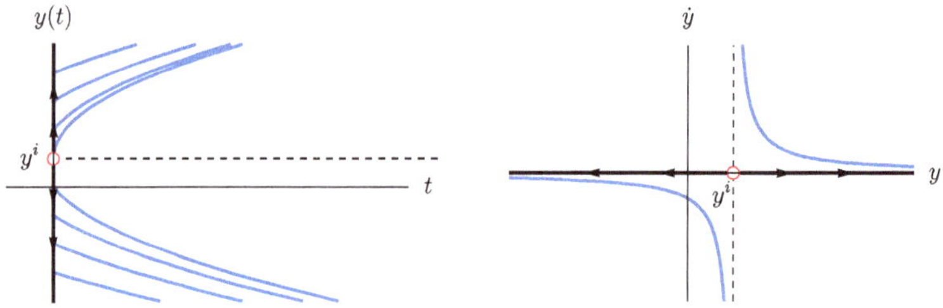

(a) Case (1,a): $\lambda = 0$, $b > 0$

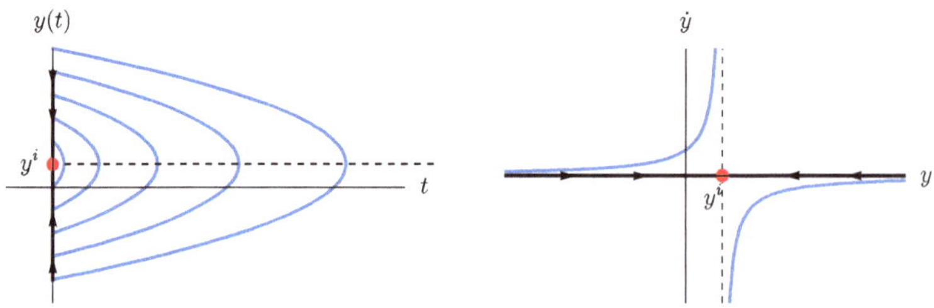

(b) Case (1,b): $\lambda = 0$, $b < 0$

Fig. 7.4 Phase diagram and trajectories ODE (7.40) for $y^i > 0$, $\lambda = 0$, and $b \neq 0$, from Proposition 7.1

Figure 7.6 illustrates two cases (3) in Proposition 7.1, for $\lambda \neq 0$ and when the steady state is an impasse point. This is a non-generic case in which trajectories traverse Γ_I through the impasse point, where $y(t)$ increases (decreases) over time after crossing, if $\lambda > 0$ ($\lambda < 0$).

Example 7.2 (*Interest rate rule policy, and singularity of government debt*) Let a central bank use an interest rate rule depending on the government budget deficit $r = \bar{r} + \rho\,\dot{b}$ where $\bar{r} > 0$ is the exogenous component, and $\rho > 0$ is the proportional response to the deficit, $\dot{b}$, where b is the net public debt. The government budget constraint is $\dot{b} = g + r\,b$, where g is the primary deficit, assumed to be constant. Substituting the interest rate in the government budget constraint, yields the constrained ODE, in separated form,

$$\left(b - \frac{1}{\rho}\right)\dot{b} = -\frac{g}{\rho} - \frac{\bar{r}}{\rho}\,b.$$

As $\bar{r} > 0$, if there is a primary deficit, $g > 0$, then $\bar{r} + \rho\,g > 0$ and we have case (2.*b*) of Proposition 7.1. If there is a primary surplus, $g < 0$ such that $\bar{r} + \rho\,g < 0$ we have case (2.*a*). From our previous results, we know that the dynamics of debt depends on the initial level of the government net position, $b(0)$. We can conclude that this type of rule can lead to the demise of the interest rate rule, in the sense that

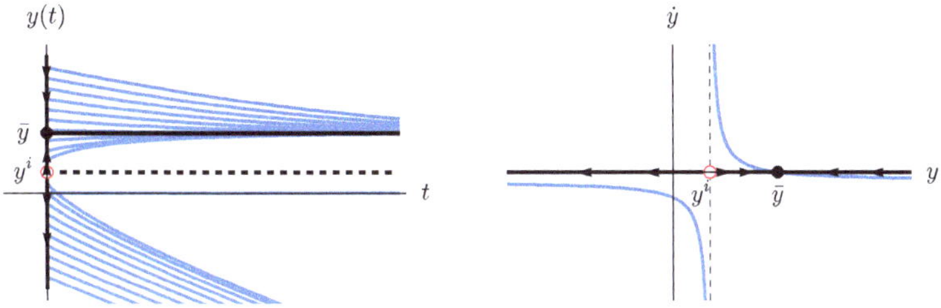

(a) Case (2,a): $\lambda < 0, b > 0$

(b) Case (2,b): $\lambda < 0, b \leq 0$

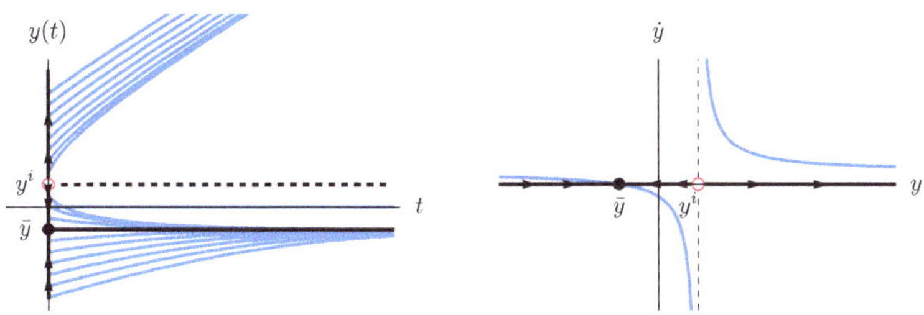

(c) Case (2,c): $\lambda > 0, b \geq 0$

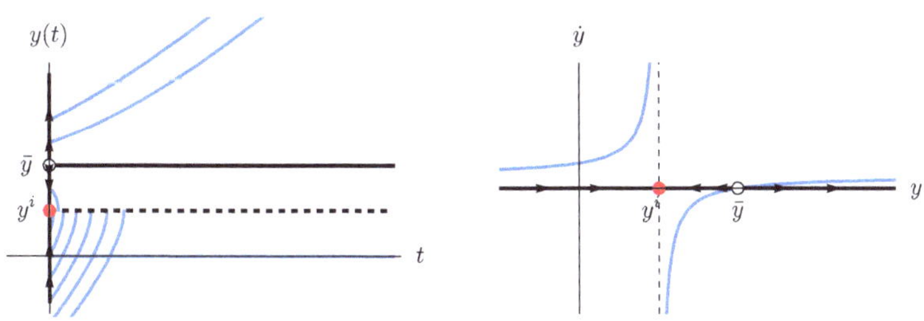

(d) Case (2,d): $\lambda > 0, b < 0$

Fig. 7.5 Phase diagram and trajectories for ODE (7.40), for $y^i \neq \bar{y}$, $y^i > 0$, and $\lambda \neq 0$, from Proposition 7.1

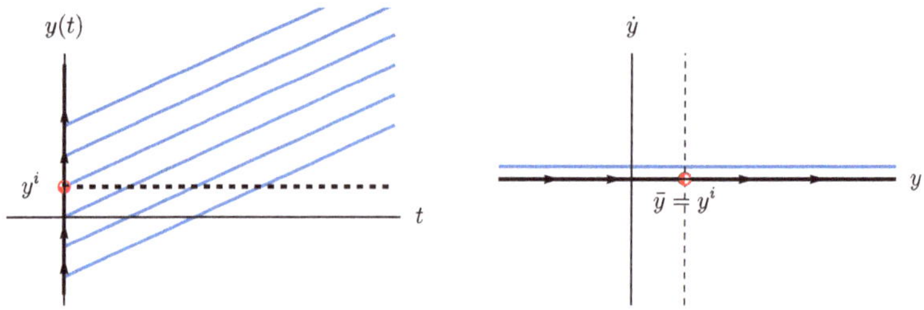

(a) Case (3) for $\lambda > 0$, $b < 0$

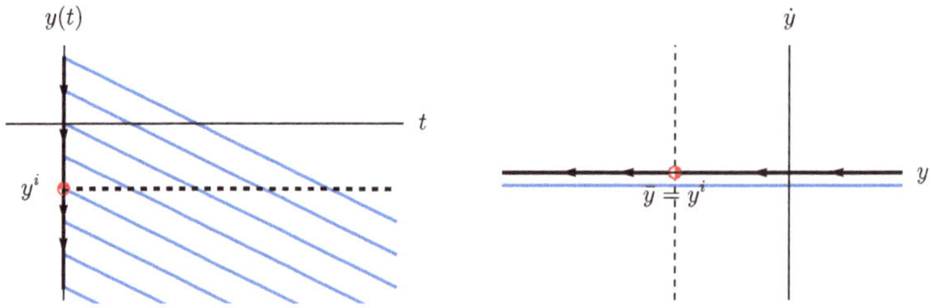

(b) Case (3) for $\lambda > 0$, $b > 0$

Fig. 7.6 Phase diagram and trajectories for ODE (7.40) if $y^i = \bar{y}$

there is no continuation for this policy if $b(0) > 0$, and there is a permanent primary deficit.

We considered the linear scalar CODE for which the singular set has always only one element, $\Gamma_I = \{y^i\}$, for all $\varphi = (b, \lambda) \in \mathbb{R}^2$, with the projection of **impasse set** $\mathcal{S}_0^i = \{(y, \varphi) \in \mathbf{Y} \times \Phi : y = y^i\}$. The impasse point introduces a partition into set $\mathbf{Y} = \mathbf{Y}_- \cup \mathcal{S}_0^i \cup \mathbf{Y}_+$, where set $\mathbf{Y}_- = \{y \in \mathbf{Y} : y < y^i\}$ and $\mathbf{Y}_+ = \{y \in \mathbf{Y} : y > y^i\}$ have dimension one, and the set Γ_I has dimension zero. The steady state set $\Gamma = \{y \in \mathbf{Y} : \lambda y + b = 0\}$ is not empty if $\lambda \neq 0$, and has at most one element. Sets Γ_I and Γ have a intersection if and only if $b + \lambda y^i = 0$. This defines a **singular impasse point**, and a singular impasse set $\Gamma_{I,1} = \{\mathbf{y} \in \mathbf{Y} : b + \lambda y^i = 0\}$, that is a projection of the **singular impasse set** $\mathcal{S}_1^i = \{(y, \varphi) \in \mathbf{Y} \times \Phi : \bar{y} = y^i\}$.

7.2.2 *Planar Equations*

Zhitomirskii (1993) presents the following normal forms for planar constrained ODE, defined on functions mapping $\mathbf{y} : \mathrm{T} \to \mathbf{Y} \subseteq \mathbb{R}^2$, where $\mathbf{y} = (y_1, y_2)$, displaying impasse singularities.

Example 7.3 (*Impasse-repeller points*) Consider the ODE in separated form

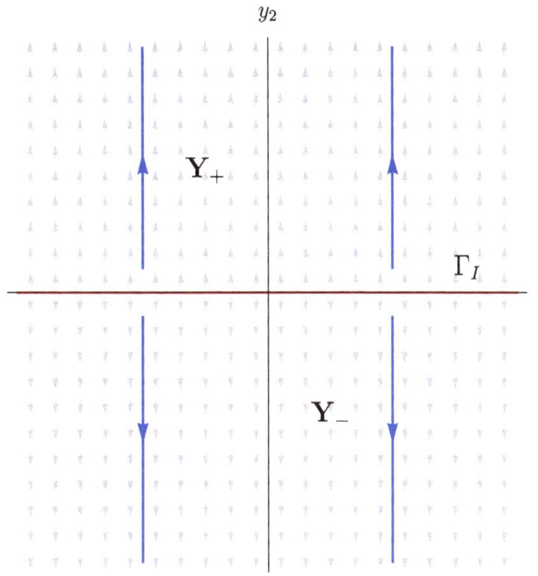

Fig. 7.7 Phase diagram and trajectories for impasse ODE (7.45)

$$\begin{aligned} \dot{y}_1 &= 0, \\ y_2\,\dot{y}_2 &= 1, \end{aligned} \tag{7.45}$$

where $\delta(\mathbf{y}) = y_2$, and $\mathbf{g}(\mathbf{y}) = (0, 1)^\top$. This CODE has no steady state and the impasse set is $\Gamma_I = \{\mathbf{y} \in \mathbf{Y} : y_2 = 0\}$, a horizontal curve in a (y_1, y_2)-diagram, see Fig. 7.7. Using the previous methods it is easy to see that its has a closed form solution

$$\begin{pmatrix} y_1(t) \\ y_2(t) \end{pmatrix} = \begin{pmatrix} y_1(0) \\ \pm\left(\frac{t}{2} + y_2(0)^2\right)^{1/2} \end{pmatrix}, \text{ for each } t \in [0, \infty).$$

Therefore, the solution exists for $t \in [0, \infty)$ and if $y_2(0) > 0$ then $\lim_{t\to\infty} y_2(t) = +\infty$ and if $y_2(0) < 0$ then $\lim_{t\to\infty} y_2(t) = -\infty$. Therefore, the impasse surface Γ_I partitions $\mathbf{Y}$ into $\mathbf{Y}_- = \{\mathbf{y} \in \mathbf{Y} : y_2 < 0\}$ and $\mathbf{Y}_+ = \{\mathbf{y} \in \mathbf{Y} : y_2 > 0\}$, in which all trajectories become asymptotically unbounded. To assess analytically the dynamics in the neighborhood of Γ_I, we use the Lie derivative of the de-singularized vector field for a change in $\delta(\cdot)$

$$\mathcal{L}_{\mathbf{g}}(\delta(\mathbf{y})) = \mathrm{D}_{\mathbf{y}}\delta(\mathbf{y}) \cdot \mathbf{g}(\mathbf{y}) = \frac{\partial \delta(\mathbf{y})}{\partial y_1}\, g_1(\mathbf{y}) + \frac{\partial \delta(\mathbf{y})}{\partial y_2}\, g_2(\mathbf{y}),$$

that in this case yields $\mathcal{L}_{\mathbf{g}}(\delta(\mathbf{y})) = 1 > 0$. Then impasse set is only composed of impasse-repeller regular points, and there are no impasse singular points.

Example 7.4 (*Impasse-attractor points*) Consider the CODE in separated form

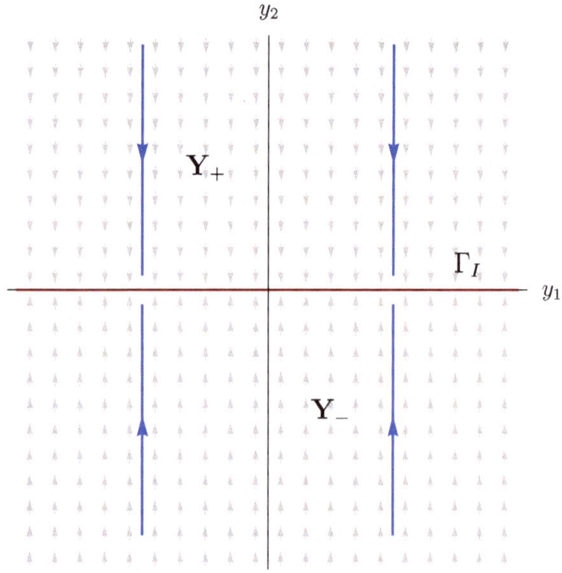

Fig. 7.8 Phase diagram and trajectories for Eq. (7.46)

$$\begin{aligned} \dot{y}_1 &= 0, \\ y_2\,\dot{y}_2 &= -1, \end{aligned} \tag{7.46}$$

where $\delta(\mathbf{y}) = y_2$, and $\mathbf{g}(\mathbf{y}) = (0, -1)^\top$. The solution is

$$\begin{pmatrix} y_1(t) \\ y_2(t) \end{pmatrix} = \begin{pmatrix} y_1(0) \\ \pm\left(y_2(0)^2 - \frac{t}{2}\right)^{1/2} \end{pmatrix}, \quad \text{for each } t \in [0, \tau],$$

where $\tau \equiv 2\,y_2(0)^2$. It has no continuation for $t > \tau$. The singular set is again $\Gamma_I = \{\mathbf{y} \in \mathbf{Y} : y_2 = 0\}$, and, as $\mathcal{L}_{\mathbf{g}}(\delta(\mathbf{y})) = -1 < 0$, there are only impasse-attractor points, as can be observed in Fig. (7.8).

Example 7.5 (*Impasse-tangent point*) Consider the CODE in separated form

$$\begin{aligned} \dot{y}_1 &= 0, \\ (y_1 + y_2^2)\, \dot{y}_2 &= 1, \end{aligned} \tag{7.47}$$

where $\delta(\mathbf{y}) = y_1 + y_2^2$ is non-linear, and $\mathbf{g}(\mathbf{y}) = (0, 1)^\top$. We find again that $y_1(t) = y_1(0)$, is an arbitrary initial point for any t, but if we use the method of separation of variables for solving the second equation, we find that $y_2(t)$ is the solution to the cubic equation

$$y_2(t)^3 + 3y_1(0)y_2(t) + 3(t - \tau) = 0$$

where $\tau \equiv -y_2(0)(y_1(0) + (1/3)y_2(0)^2)$. This polynomial equation can have one or three real roots in the generic case, or two real roots in the non-generic case. In the first case the solution has three branches and is only defined for finite time intervals $t \in [0, \tau_i]$, where τ_i is different for each different branch. In the second case, there is only one branch for the solution that exists for $t \in [0, \infty)$.

However, we can apply a qualitative analysis to characterize the solution in the neighborhood of the impasse set $\Gamma_I = \{\mathbf{y} \in \mathbf{Y} : y_1 + y_2^2 = 0\}$, a parabola in the orthant $y_1 \leq 0$ of space $\mathbf{Y}$ (see Fig. 7.9). As $\mathcal{L}_\mathbf{g}(\delta(\mathbf{y})) = 2\, y_2$, then there is a non-empty impasse attractor set $\Gamma_I^a = \{\mathbf{y} \in \mathbf{Y} : y_1 + y_2^2 = 0,\ \text{and}\ y_2 < 0\}$, a non-empty impasse repeller set $\Gamma_I^r = \{\mathbf{y} \in \mathbf{Y} : y_1 + y_2^2 = 0,\ \text{and}\ y_2 > 0\}$, and an impasse singular set $\Gamma_{I,1} = \{\mathbf{y} \in \mathbf{Y} : y_1 + y_2^2 = 0,\ \text{and}\ y_2 = 0\} = \{(0, 0)\}$. As $\mathrm{D}_\mathbf{y}\delta(\mathbf{y}) = (1, 2y_2)$, and $\mathrm{D}_\mathbf{y}\delta(\bar{\mathbf{y}}) = (1, 0)$, we have $\mathcal{L}_\mathbf{g}(\delta(\bar{\mathbf{y}})) = 0$ because $\delta(\mathbf{y})$ and $\mathbf{g}(\mathbf{y})$ are orthogonal at the impasse singular point, which is called **impasse-tangent point**. Observe that trajectories starting from $y_1(0) = 0$ passing through this point, coming from set $\mathbf{Y}_+$ (outside the parabola curve), are continued over time. This is not the case for trajectories starting from $y_1(0) < 0$, and coming from set $\mathbf{Y}_+$.

In the previous examples, the planar CODE did not depend on parameters, and it was assumed that $\mathbf{A}(\mathbf{y}, \varphi) = \left(\begin{smallmatrix} 1 & 0 \\ 0 & \delta(\mathbf{y}) \end{smallmatrix}\right)$, and therefore that $\det(\mathbf{A}(\mathbf{y}, \varphi)) = \delta(\mathbf{y})$. In the next examples, matrices $\mathbf{A}(\cdot)$ have different forms.

Example 7.6 (*Impasse transversal points*) Consider the planar ODE in separated form that is dependent on a parameter $\lambda \in \mathbb{R}$,

$$\begin{aligned} \frac{1}{\lambda}\dot{y}_2 - \dot{y}_1 &= 1, \\ (y_2 - y_1)\, \dot{y}_2 &= \lambda\, y_2, \end{aligned} \tag{7.48}$$

where $\mathbf{A}(\mathbf{y}, \lambda) = \left(\begin{smallmatrix} -1 & \lambda^{-1} \\ 0 & y_2 - y_1 \end{smallmatrix}\right)$, and $\mathbf{g}(\mathbf{y}, \lambda) = (1, \lambda\, y_2)^\top$. Then $\delta(\mathbf{y}, \lambda) = y_1 - y_2$. Again there are no steady states, $\Gamma = \emptyset$, and the impasse set, $\Gamma_I = \{\mathbf{y} \in \mathbf{Y} : y_1 - y_2 = 0\}$, generates subsets of set $\mathbf{Y}$ into $\mathbf{Y}_- = \{\mathbf{y} \in \mathbf{Y} : y_1 < y_2\}$ and $\mathbf{Y}_+ = \{\mathbf{y} \in \mathbf{Y} : y_1 > y_2\}$.

The explicit and de-singularized forms of this CODE are

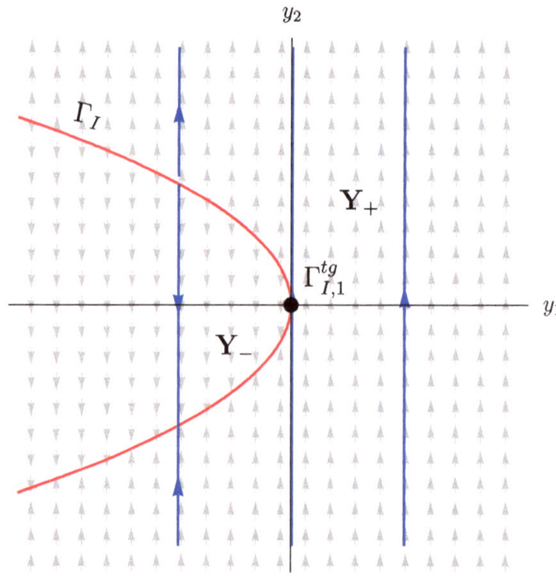

Fig. 7.9 Phase diagram and trajectories for Eq. (7.47)

$$\dot{\mathbf{y}} = \mathbf{f}(\mathbf{y}) = \Big(\frac{y_1}{y_2 - y_1}, \frac{\lambda\, y_2}{y_2 - y_1}\Big)^\top,$$
$$\dot{\mathbf{y}} = \mathbf{f}^d(\mathbf{y}) = \Big(- y_1, -\lambda\, y_2\Big)^\top.$$

There is an impasse singular point,$\mathbf{y}^{is} = (0, 0)$. The local dynamics at the impasse singular point can be characterized from the eigenvalues of the Jacobian $\mathbf{J}^f(\mathbf{y}) = \mathrm{D}_\mathbf{y}\mathbf{f}^d(\mathbf{y}) = \left(\begin{smallmatrix} -1 & 0 \\ 0 & -\lambda \end{smallmatrix}\right)$. Therefore, as $\sigma(\mathbf{J}^f(\mathbf{y})) = \{-1, -\lambda\}$, then, if $\lambda < 0$ the impasse singular point is a **transversal saddle point**, or if $\lambda > 0$ it is a **transversal node point**. Furthermore, the solution to DE $\dot{\mathbf{y}} = \mathbf{f}^d(\mathbf{y})$, $y_1(t) = y_1(0)\, e^{-t}$, and $y_2(t) = y_2(0)\, e^{-\lambda t}$, allows for characterizing the dynamics for both types of ODE.

Figure 7.10a illustrates the phase diagram when the impasse singular point is an **impasse-transversal saddle**, which is the case if $\lambda < 0$. We observe that, as regards the impasse singular point, $\mathbf{y}^{is} = (0, 0)$, the manifold $y_2 = 0$ is a stable eigenvector within $\mathbf{Y}_+$ (where $y_1 > y_2$) and it is an unstable eigenvector within set $\mathbf{Y}_-$ (where $y_1 < y_2$), and the manifold $y_1 = 0$ is an unstable eigenvector within $\mathbf{Y}_+$ and it is a stable eigenvector within set $\mathbf{Y}_-$. We also observe that the impasse set Γ_I is separated into an attractor subset $\Gamma_I^a = \{\mathbf{y} \in \Gamma_I : y_1 > 0\}$, and a repeller subset $\Gamma_I^r = \{\mathbf{y} \in \Gamma_I : y_1 < 0\}$. We can prove this by computing (see Eq. (7.53)

$$\mathcal{L}_{\mathbf{f}^d}(\delta(\mathbf{y})) = -y_1 + \lambda\, y_2. \tag{7.49}$$

Evaluated at any point in Γ_I, where $y_1 = y_2$, if we assume that $\lambda < 0$ then $\mathcal{L}_{\mathbf{f}^d}(\delta(\mathbf{y}) < 0$ if $y_1 = y_2 > 0$ and $\mathcal{L}_{\mathbf{f}^d}(\delta(\mathbf{y})) > 0$ if $y_1 = y_2 < 0$.

Therefore, there are three types of trajectories starting from subset $\mathbf{Y}_+$ ($\mathbf{Y}_-$): if $y_2(0) > 0$ they collide in finite time with Γ_I^a, or, if $y_2(0) = 0$, they cross the impasse set at point $\mathbf{y}^{is} = (0, 0)$ and continue into subspace $\mathbf{Y}_-$ ($\mathbf{Y}_+$) and have $\lim_{t\to\infty} \mathbf{y}(t) = (-\infty, 0)$ ($\lim_{t\to\infty} \mathbf{y}(t) = (0, -\infty)$); or, if $y_2(0) < 0$, they evolve within subset $\mathbf{Y}_+$ ($\mathbf{Y}_-$) and are repelled away from the impasse repeller set Γ_I^r and have $\lim_{t\to\infty} \mathbf{y}(t) = (0, -\infty)$ ($\lim_{t\to\infty} \mathbf{y}(t) = (-\infty, 0)$). Summing up, there are two paths that cross this singular saddle point, but differently from regular saddle points, it is reached done in finite time, with a continuation in the other subset, and the stable eigenvectors change direction.

Figure 7.10b illustrates the phase diagram when the impasse singular point is an **impasse-transversal node**, which is the case if $\lambda > 0$. We observe that, as regards the impasse singular point $\mathbf{y}^{is} = (0, 0)$, while the stable manifold within subset $\mathbf{Y}_+$ (where $y_1 > y_2$) is of dimension two and it is empty within subset $\mathbf{Y}_-$, the unstable manifold within set $\mathbf{Y}_+$ is empty and has dimension two within subset $\mathbf{Y}_-$. We also observe that the non-singular subsets of the impasse set, i.e., $\Gamma_I/\mathbf{y}^{is}$, is an attractor as a boundary to subset $\mathbf{Y}_+$ and is a repeller as a boundary of subset $\mathbf{Y}_-$. To prove this, we consider again Eq. (7.49) and find that, if $\lambda > 0$, $\mathcal{L}_{\mathbf{f}^d}(\delta(\mathbf{y}), < 0$ if $y_1 > \lambda y_2 > 0$ and $\mathcal{L}_{\mathbf{f}^d}(\delta(\mathbf{y}), < 0$ if $y_1 < \lambda y_2 < 0$.

Therefore, there are three types of trajectories: if $\mathbf{y}(0) \in \mathbf{Y}_+$ either they collide with Γ_I or they cross it and become asymptotically unbounded, or, if $\mathbf{y}(0) \in \mathbf{Y}_-$ they become asymptotically unbounded. Thus, we have an infinite number of flows crossing from $\mathbf{Y}_+$ to set $\mathbf{Y}_-$, as in a regular stable node, but differently from these, the attraction and crossing of the impasse singular point is done in finite time, with a continuation in the subset $\mathbf{Y}_-$, where the flow becomes asymptotically unbounded. In this sense, stability property is only temporary and is not permanent as in regular ODE.

Example 7.7 (*Impasse focus points*) Consider the planar ODE in separated form depending on a parameter $\lambda \neq 0$,

$$\begin{aligned} \lambda \dot{y}_1 - \dot{y}_2 &= 1 + \lambda^2, \\ y_2\, \dot{y}_2 &= -\lambda y_1 - y_2, \end{aligned} \tag{7.50}$$

where $\mathbf{A}(\mathbf{y}, \lambda) = \left(\begin{smallmatrix} \lambda & -1 \\ 0 & y_2 \end{smallmatrix}\right)$, and $\mathbf{g}(\mathbf{y}, \lambda) = (1 + \lambda^2, \text{-}\lambda\, y_1 - y_2)^\top$. Then $\delta(\mathbf{y}, \lambda) = \lambda\, y_2$. Again there are no steady states, $\Gamma = \emptyset$. The impasse set $\Gamma_I = \{\mathbf{y} \in \mathbf{Y} : y_2 = 0\}$, introduces a partition of set $\mathbf{Y}$ into $\mathbf{Y}_- = \{\mathbf{y} \in \mathbf{Y} : \lambda\, y_2 < 0\}$ and $\mathbf{Y}_+ = \{\mathbf{y} \in \mathbf{Y} : \lambda\, y_2 > 0\}$.

The explicit and de-singularized forms of this CODE are

$$\begin{aligned} \dot{\mathbf{y}} &= \mathbf{f}(\mathbf{y}) = \left(\frac{\lambda\, y_2 - y_1}{y_2}, \text{-}\frac{\lambda\, y_1 + y_2}{y_2}\right)^\top, \\ \dot{\mathbf{y}} &= \mathbf{f}^d(\mathbf{y}) = \Big(\lambda\, (\lambda\, y_2 - y_1), \text{-}\lambda\, (\lambda\, y_1 + y_2)\Big)^\top. \end{aligned}$$

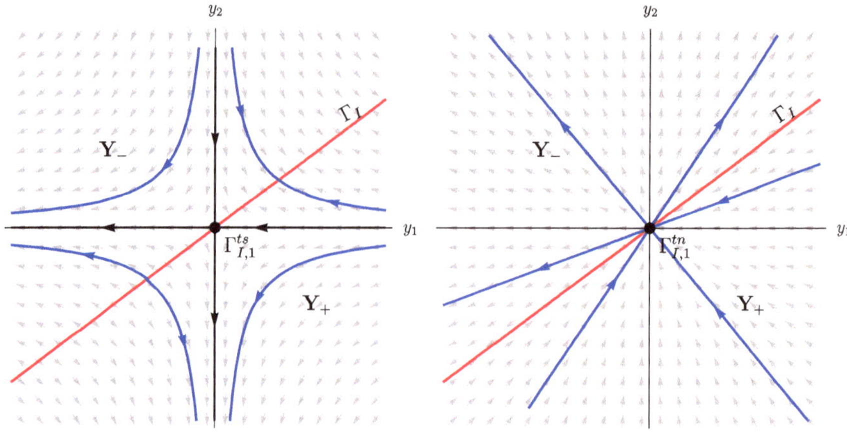

(a) Phase diagram for $\lambda < 0$ (b) Phase diagram for $\lambda > 0$

Fig. 7.10 Phase diagrams of CODE (7.48) for different λ

There is an impasse singular point $\mathbf{y}^{is} = (0, 0)$. The local dynamics at the impasse singular point can be characterized from the eigenvalues of the Jacobian $\mathbf{J}^f(\mathbf{y}) = \mathrm{D}_{\mathbf{y}}\mathbf{f}^d(\mathbf{y}) = \begin{pmatrix} -\lambda & \lambda^2 \\ -\lambda^2 & -\lambda \end{pmatrix}$. Therefore, as $\sigma(\mathbf{J}^f(\mathbf{y})) = \{-\lambda \pm \lambda^2 i\}$, the eigenvalues are complex conjugate and the impasse singular point is a **transversal focus point**, for any $\lambda \neq 0$.

As $\mathcal{L}_{\mathbf{f}^d}(\delta(\mathbf{y})) = -\lambda^2(\lambda\, y_1 + y_2)$ then, for points $\mathbf{y}^i \in \Gamma_I$, we have $\mathcal{L}_{\mathbf{f}^d}(\delta(\mathbf{y}^i)) = -\lambda^3\, y_1$, if $\lambda^3\, y_1 > 0$ ($\lambda^3\, y_1 < 0$), and therefore, point $\mathbf{y}^i$ is an attractor (a repeller), and $\mathbf{y}^{is} = (0, 0)$ is an **impasse focus point**.

The phase diagram in Fig. 7.11 displays the phase diagrams for $\lambda < 0$ and $\lambda > 0$. In both phase diagrams, the impasse focus point is at the origin. Although there are trajectories that reach this point in finite time from both subsets ($\mathbf{Y}_+$ or $\mathbf{Y}_-$) and trajectories that leave this point to the other subset ($\mathbf{Y}_-$ or $\mathbf{Y}_+$), it can be proved that there are no crossing trajectories. That is the solution has no continuation after a finite time, for any initial $\mathbf{y}(0) \in \mathbf{Y}$. This is justified by the fact that the de-singularized ODE in polar coordinates is the same as for the regular stable focus case ($\dot{r} = -\lambda\, r$ and $\dot{\theta} = \lambda^2$).

7.2.3 CODE Dynamics: An Introduction

This subsection is a brief introduction to the qualitative dynamics of non-linear constrained ODE taking stock from what we have already learned from some normal forms.

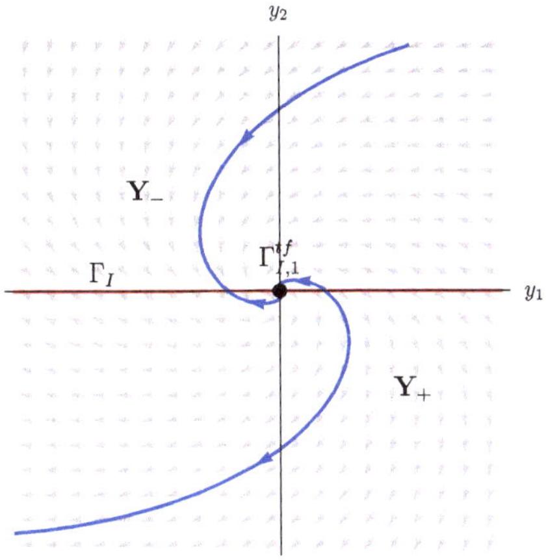

(a) Phase diagram for $\lambda < 0$

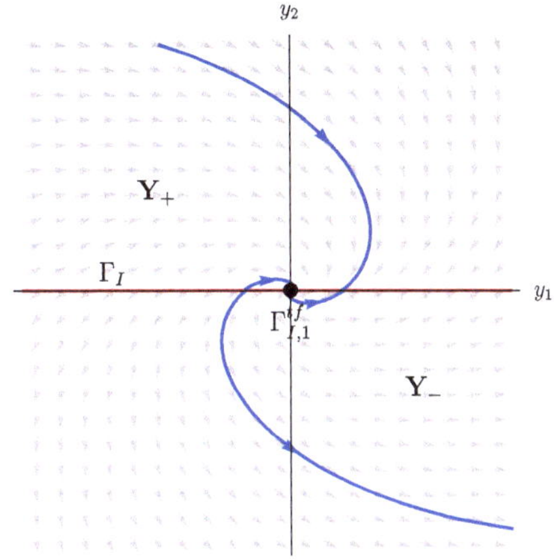

(b) Phase diagram for $\lambda > 0$

Fig. 7.11 Phase diagram for CODE (7.50) for different λ

Scalar CODE. We start with a parameterized scalar equation over $y : \mathrm{T} \rightarrow \mathrm{Y} \subseteq \mathbb{R}$, in its explicit form

$$\dot{y} = \frac{g(y, \varphi)}{\delta(y, \varphi)}, \tag{7.51}$$

extending Eq. (7.33), and the associated separated and de-singularized forms

$$\delta(y, \varphi)\, \dot{y} = g(y, \varphi), \tag{7.52a}$$

$$\dot{y} = f^d(y, \varphi) = g(y, \varphi), \tag{7.52b}$$

extending Eqs. (7.34a) and (7.34b). In this subsection we assume that $g(y, \cdot)$ and $\delta(y, \cdot)$ can be non-linear, but are always continuous and continuously differentiable functions. We define the set of impasse points, and its projection onto set Y, by

$$\mathcal{S}_0^i = \{(y, \varphi) \in \mathrm{Y} \times \Phi : \delta(y, \varphi) = 0\}, \text{ and } \Gamma_I = \{y \in \mathrm{Y} : \delta(y, \varphi) = 0\}$$

respectively, and the set of steady states by $\Gamma = \{y \in \mathrm{Y} : g(y, \varphi) = 0\}$. In Sect. 7.2.1, we dealt with the case in which the constitutive functions, $\delta(y)$ and $g(y)$, were linear, and showed that generic equations had at most one impasse point, and at most one steady state. For the non-linear case, the number of impasse points and/or the number of steady states can be greater than one but finite, that is Γ_I has at least one element, and Γ can be empty or have a finite, but greater than one, number of elements. For regular points, that is for $y \in \mathbf{Y}/\Gamma_I$, the qualitative analysis of the dynamics can be done by applying the theory that was exposed in Chap. 3.

To simplify, assume that the impasse set has only one element $\Gamma_I = \{y^i\}$, such that $\delta(y^i, \varphi) = 0$. As we already saw, there are several consequences from the existence of impasse points. First, each impasse point introduces a partition over the image set Y between subsets $Y_- = \{y \in Y : \delta(y, \varphi) < 0\}$, and $Y_+ = \{y \in Y : \delta(y, \varphi) > 0\}$. Second, it introduces a distinction between regular steady states, if $\Gamma_I \cap \Gamma$ is empty, and impasse steady states, if $P_I = \Gamma_I \cap \Gamma = \{y \in \Gamma_I : g(y, \varphi) = 0\}$ is not empty. Third, it allows us to locate steady states as regards each impasse point, by splitting the steady state set into $\Gamma_- = \{y \in Y_- : g(y, \varphi) = 0\}$ and $\Gamma_+ = \{y \in Y_+ : g(y, \varphi) = 0\}$. This location has an effect on the local dynamics at a regular steady state. Differentiating the vector field in CODE (7.33) yields

$$f'(y) = \frac{\partial f(y)}{\partial y} = \frac{g'(y)\,\delta(y) - g(y)\delta'(y)}{(\delta(y))^2}$$

At a regular steady state $\bar{y} \in \Gamma$, we have $f'(\bar{y}) = \frac{g'(\bar{y})\,\delta(\bar{y})}{\delta(\bar{y})^2}$, for $y^i \neq \bar{y}$. Then the steady state is asymptotically stable if $g'(\bar{y})\,\delta(\bar{y}) < 0$, and it is unstable if $g'(\bar{y})\,\delta(\bar{y}) > 0$. If there are two steady states located in two different branches of set Γ, e.g, one in branch Γ_- and another in branch Γ_+, it is possible that they have both the same stability properties, e.g., they can be both asymptotically stable or unstable. Those cases cannot exist in a non-linear regular ODE. Fourth, this fact implies that the existence of impasse singularities introduces different types of global dynamics, when comparing with regular ODE, because they can change the nature of the invariant (stable or unstable) manifolds, and they introduce the problem of existence of continuation of the solutions.

At last, the local dynamics in the neighborhood of set Γ_I needs to be addressed. In order to understand the dynamics at an impasse point, $y^i \in \Gamma_I$, we differentiate function $\delta(y, \cdot)$ along the direction defined by the de-singularized vector field (7.34b)

$$\mathcal{L}_{f^d}(\delta(y, \varphi)) = \delta'(y)\,g(y), \text{ for } y \in \Gamma_I.$$

This allows us to establish a taxonomy of impasse points. We say an impasse point $y^i \in \Gamma_I$ is a **regular impasse point** if $\delta'(y^i)\,g(y^i) \neq 0$, and it is a **singular impasse point** if $\delta'(y^i)\,g(y^i) = 0$; a regular impasse point is an **attractor impasse point** if $\delta'(y^i)\,g(y^i) < 0$ and it is a **repeller impasse point** if $\delta'(y^i)\,g(y^i) > 0$.

Therefore, we can define the set of impasse attractor points $\Gamma_I^a = \{y \in \Gamma_I : \delta'(y)\,g(y) < 0\}$, the set of impasse-repeller points $\Gamma_I^r = \{y \in \Gamma_I : \delta'(y)\,g(y) > 0\}$, and the set of impasse singular points $\Gamma_{I,1} = \{y \in \Gamma_I : \delta'(y)\,g(y) = 0\}$, and the associated set $\mathcal{S}_1^i = \{(y, \varphi) \in \mathbf{Y} \times \Phi : \delta'(y)\,g(y) = 0\}$.

The set of singular impasse points, $\mathcal{S}_1^i$, can be further stratified among the set of **impasse-tangent points**, satisfying $\delta'(y^i) = 0$ and $g(y^i) \neq 0$, and the set of **impasse-traverse points**, satisfying $\delta'(y^i) \neq 0$ and $g(y^i) = 0$. Observe that this is a set containing singular steady states. While tangent points cannot exist if $\delta(y)$ is linear, traverse points exist if $\delta(y)$ is linear, as in case (3) of Proposition 7.1. Indeed, solution trajectories of a constrained ODE can only cross the impasse surface if there is a traverse impasse point.

Planar CODE. Next, consider planar non-linear CODE. They are represented in the explicit form, in Eq. (7.35), in the separated form, in Eq. (7.36), and, for purposes of qualitative dynamics analysis, in the de-singularized form in Eq. (7.37), in which vector function $\mathbf{g}(\mathbf{y}, \cdot)$ and function $\delta(\mathbf{y}, \cdot)$ can be non-linear, and, if this is the case, they are continuous and continuously differentiable.

The **set of impasse points**, and its projection onto set $\mathbf{Y}$ are

$$\mathcal{S}_0^i = \{(\mathbf{y}, \varphi) \in \mathbf{Y} \times \Phi : \delta(\mathbf{y}, \varphi) = 0\}, \text{ and } \Gamma_I = \{\mathbf{y} \in \mathbf{Y} : \delta(\mathbf{y}, \varphi) = 0\},$$

respectively, where $\delta(\mathbf{y}, \varphi) \equiv \det(\mathbf{A}(\mathbf{y}, \varphi))$ are assumed to be non-empty. If $\delta(\mathbf{y}, \varphi)$ is linear, as a function of $\mathbf{y}$, then Γ_I has only one element. However, if it is non-linear then more than one impasse surface can exist.

The set of steady states is denoted by

$$\Gamma = \{\mathbf{y} \in \mathbf{Y} : \mathbf{g}(\mathbf{y}, \varphi) = \mathbf{0}\}.$$

The impasse set Γ_I introduces a partition into the image set, $\mathbf{Y}$, between subsets $\mathbf{Y}_- = \{\mathbf{y} \in \mathbf{Y} : \delta(\mathbf{y}, \varphi) < 0\}$ and $\mathbf{Y}_+ = \{\mathbf{y} \in \mathbf{Y} : \delta(\mathbf{y}, \varphi) > 0\}$. If there are steady states, and set Γ is non-empty, they will be located in set $\mathbf{Y}_-$, or in set $\mathbf{Y}_+$ if the functions in the CODE are linear, or in one or both sets, $\mathbf{Y}_-$ and $\mathbf{Y}_+$, if the vector function $\mathbf{g}(\mathbf{y}, \cdot)$ and/or matrix function $\mathbf{A}(\mathbf{y}, \cdot)$ are non-linear. We define a **regular steady state** as a steady state that does not belong to Γ_I, and an **impasse-singular steady state** as a steady state in Γ_I. We can define two sets of regular steady states according to their location relative to the partition introduced by Γ_I,

$$\Gamma_- = \{\mathbf{y} \in \mathbf{Y}_- : \mathbf{g}(\mathbf{y}, \varphi) = \mathbf{0}\}, \text{ and } \Gamma_+ = \{\mathbf{y} \in \mathbf{Y}_+ : , \mathbf{g}(\mathbf{y}, \varphi) = \mathbf{0}\}.$$

If there is a non-empty intersection between the steady state and the impasse sets, there is a non-empty set of impasse-steady states

$$P_I = \{\mathbf{y} \in \Gamma_I : \mathbf{g}(\mathbf{y}, \varphi) = \mathbf{0}\}.$$

Analogously to scalar ODE, we have to separate the local analysis in the neighborhood of a steady state $\bar{y} \in \Gamma$, as in regular ODE, from the local analysis in the neighborhood of an impasse point $\mathbf{y}^i \in \Gamma_I$. The dynamics in the neighborhood of a regular point, belonging to each of the two subsets $\mathbf{Y}_-$ and $\mathbf{Y}_+$ is well known, but the dynamics in the neighborhood of a singular point is not. A non-empty impasse set introduces some possible local dynamics, which do not occur in equations in which it is empty. For instance, if there are two equilibrium points in a regular ODE, the dimension of the stable manifolds is different at every equilibrium point. If the impasse surface is non-empty and there is one steady state in every subset, Γ_- and Γ_+, then it is possible that the dimension of the local stable manifolds can be the same (for instance two saddle points). This cannot occur in regular ODE. The dynamics in the neighborhood of an impasse point also introduces new types of stable and unstable manifolds, as regards regular ODE. They can partition stable manifolds associated to a regular steady state into subsets with different dimensions. The dynamics for P_I points is associated to local or global bifurcation points.

A complete taxonomy, even if it exists, is beyond our intention for this book.[10] We will only focus next on two problems related to: (1) the type of dynamics we should expect in the neighborhood of Γ_I; (2) finding the conditions under which there are trajectories that traverse set Γ_I.

Local analysis close to a regular steady state. Assume that set Γ is non-empty, and consider a regular steady state $\bar{\mathbf{y}}$ such that $\bar{\mathbf{y}} \notin \Gamma_I$, or equivalently such that $\delta(\bar{\mathbf{y}}, \varphi) \neq 0$. As the impasse surface changes the local dynamics, we have to study local dynamics by using the CODE in explicit form. Therefore, the eigenvalues of the Jacobian

$$\mathbf{J}(\bar{\mathbf{y}}) = \mathrm{D}_{\mathbf{y}}\mathbf{f}(\bar{\mathbf{y}}, \varphi) = \mathrm{D}_{\mathbf{y}}\Big(\mathbf{A}^{-1}(\bar{\mathbf{y}}, \varphi)\, \mathbf{g}(\bar{\mathbf{y}}, \varphi)\Big).$$

will allow us to find the type of steady state, the dimension of the local stable and unstable manifolds, and the possible existence of local bifurcation, by using the results from Chaps. 4 and 5.

Local analysis close to an impasse point. We extend and adapt the approach by Zhitomirskii (1993) and Llibre et al. (2002),[11] is based on the determining the existence of the kernel and image line fields generated by matrix $\mathbf{A}$ and their relationship with $\mathcal{S}_0^i$. They define two functions $K(\mathbf{y}, \varphi) = \mathrm{D}_{\mathbf{y}} \det(\mathbf{A}(\mathbf{y}, \varphi)) \cdot v$ where $v \in \mathrm{kern}(\mathbf{A}(\mathbf{y}, \varphi)) - \{\mathbf{0}\}$ and $I(\mathbf{y}, \varphi) = \mathbf{f}^d(\mathbf{y}, \varphi) = \mathrm{adj}(\mathbf{A}(\mathbf{y}, \varphi))\, \mathbf{g}(\mathbf{y}, \varphi)$.

There are two types of impasse-singular points: a point $\mathbf{y}_k^i$ such that $K(\mathbf{y}_k^i, \varphi) = 0$ is called **kernel-singular point**, and a point $\mathbf{y}_i^i$ such that $I(\mathbf{y}_i^i, \varphi) = 0$ is called **image-singular point**. A point which is not kernel singular is called **kernel-regular** and a point which is not image singular is called **image-regular**.

Using the Lie derivative of $\mathbf{f}^d$ in the direction of the manifold defining the impasse surface $\delta(\mathbf{y}, \varphi) = 0$, we have

$$\mathcal{L}_{\mathbf{f}^d}(\delta(\mathbf{y}, \varphi)) = \mathrm{D}_{\mathbf{y}}\delta(\mathbf{y}, \varphi) \cdot \mathbf{f}^d(\mathbf{y}, \varphi) = \mathrm{D}_{\mathbf{y}}\delta(\mathbf{y}, \varphi) \cdot (\mathrm{adj}(\mathbf{A}(\mathbf{y}, \varphi))\mathbf{g}(\mathbf{y}, \varphi)). \quad (7.53)$$

We find $K(\mathbf{y}, \varphi) = \mathrm{D}_{\mathbf{y}}\delta(\mathbf{y}, \varphi) \cdot \mathbf{f}^d(\mathbf{y}, \varphi)$ if both $\mathrm{D}_{\mathbf{y}}\delta(\mathbf{y}, \varphi) \neq \mathbf{0}$ and $\mathbf{f}^d(\mathbf{y}, \varphi) \neq \mathbf{0}$, and again $I(\mathbf{y}, \varphi) = \mathbf{f}^d(\mathbf{y}, \varphi)$.[12] Then

$$\Gamma_I^k = \{(\mathbf{y}, \varphi) \in \Gamma_I : \ \mathcal{L}_{\mathbf{f}^d}(\delta(\mathbf{y}, \varphi)) = 0, \ \text{and } \mathrm{D}_{\mathbf{y}}\delta(\mathbf{y}, \varphi) \neq \mathbf{0} \text{ and } \mathbf{f}^d(\mathbf{y}, \varphi) \neq \mathbf{0}\}$$

is the set of kernel-singular impasse points, and

$$\Gamma_I^i = \{(\mathbf{y}, \varphi) \in \Gamma_I : \ \mathbf{f}^d(\mathbf{y}, \varphi) = \mathbf{0}\}$$

---

[10] The dimension of the task could be even greatly increased if one would assume that the constitutive functions could be piecewise smooth, which would require considering the material studied in Chap. 6.

[11] They assume that matrix $\mathbf{A} = \begin{pmatrix} 1 & 0 \\ 0 & \delta(\mathbf{y}, \varphi) \end{pmatrix}$, and we assume a general square matrix function.

[12] The cited authors defined $K(\mathbf{y}, \varphi) = \delta_{y_2}(\mathbf{y}, \varphi)$ and $I(\mathbf{y}, \varphi) = (0, f_2(\mathbf{y}, \varphi))^\top$ that cannot be applied to our CODE (7.47), for example. See Cardin et al. (2012), for our more general case.

is the set of image-singular impasse points. An **impasse-regular** point is an impasse point that is neither kernel-singular nor image-singular; that is , it is a point satisfying $\mathcal{L}_{\mathbf{f}^d}(\delta(\mathbf{y},\varphi)) \neq 0$.[13] We can define the set of **impasse-singular** points by

$$\mathcal{S}_1^i = \left\{(\mathbf{y},\varphi) \in \mathcal{S}_0^i \,:\, \mathcal{L}_{\mathbf{f}^d}(\delta(\mathbf{y},\varphi)) = 0 \right\},$$

and its projection onto set $\mathbf{Y}$ by

$$\Gamma_{I,1} = \{\mathbf{y} \in \Gamma_I \,:\, \mathcal{L}_{\mathbf{f}^d}(\delta(\mathbf{y},\varphi)) = 0 \}.$$

If $\Gamma_{I,1}$ is non-empty, it introduces a partition into the set of **impasse-regular** points, $\Gamma_I / \Gamma_{I,1}$ between two subsets: the sets of **attractor impasse points** $\Gamma_{I,1}^a$, and of **repeller impasse points** $\Gamma_{I,1}^r$, where

$$\Gamma_{I,1}^a = \{\mathbf{y} \in \Gamma_I \,:\, \mathcal{L}_{\mathbf{f}^d}(\delta(\mathbf{y},\varphi)) < 0 \}, \text{ and } \Gamma_{I,1}^r = \{\mathbf{y} \in \Gamma_I \,:\, \mathcal{L}_{\mathbf{f}^d}(\delta(\mathbf{y},\varphi)) > 0 \}.$$

The geometrical intuition of the several impasse points that were defined is the following: first, flows starting close to impasse repeller point in Γ_I continue by moving away from the starting point inside the same branch ($\mathbf{Y}_+$ or $\mathbf{Y}_-$), while a flow starting close to an attractor point in Γ_I terminate in **finite time** in Γ_I and does not continue; second, if the set $\Gamma_{I,1}$ is empty there are only impasse-repeller points, as in Fig. 7.7, or impasse-attractor points, as in Fig. 7.8; third, if a kernel-singular point exists then there is one flow that is tangent to Γ_I at that point (see Fig. 7.9); fourth, if an image-singular point exists then there are trajectories that collide, in finite time, with Γ_I, and they can traverse it, or not, in one or two directions depending on the type of the image-impasse point being a transversal node, a saddle, or a focus (as in Figs. 7.10 or 7.11).

In order to characterize image-singular and kernel-regular impasse points, $\mathbf{y}^i \in \Gamma_I^i$, we can apply a similar approach as for regular steady states of the de-singularized CODE, in Eq. (7.37), by finding the variational ODE $\dot{y} = \mathbf{J}^d(\mathbf{y}^i)\,(\mathbf{y} - \mathbf{y}^i)$, where

$$\mathbf{J}^d(\mathbf{y}^i) = \mathrm{D}_{\mathbf{y}}\mathbf{f}^d(\mathbf{y}^i,\varphi) = \mathrm{D}_{\mathbf{y}}\Big(\mathrm{adj}(\mathbf{A}(\mathbf{y}^i,\varphi))\,\mathbf{g}(\mathbf{y}^i,\varphi)\Big),$$

at point $\mathbf{f}^d(\mathbf{y}^i,\varphi) = \mathbf{0}$. Then $\mathbf{y}^i$ is a **transversal saddle** point if $\det(\mathbf{J}^d(\mathbf{y}^i)) < 0$, it is a **transversal node** point if $\det(\mathbf{J}^d(\mathbf{y}^i)) > 0$ and $\mathrm{tr}(\mathbf{J}^d(\mathbf{y}^i)) > 0$, or it is a **transversal focus** if $\Delta(\mathbf{J}^d(\mathbf{y}^i)) < 0$. There are degenerate impasse transversal points if the Jacobian $\mathbf{J}^d(\mathbf{y}^i)$ has eigenvalues with zero real parts. This opens the way for introducing further partitions into set $\mathcal{S}_0^i$ and for investigating the existence of singularity induced bifurcations.

If the image-singular point is a transversal saddle or node, we can study the crossing paths, or the tangent to the crossing paths, by determining the eigenspaces associated to the approximate (or exact) solutions of the de-singularized variational

[13] (Riaza, 2008, p. 164) shows that the condition $\mathrm{D}_{\mathbf{y}}\det(\mathbf{A}(\mathbf{y},\varphi)\cdot I(\mathbf{y},\varphi,) \neq 0$ is equivalently satisfied.

ODE, which has a Jacobian matrix with a diagonal Jordan canonical form,

$$\mathbf{y}(t) = \mathbf{y}^i + h_- \mathbf{P}^- e^{\lambda^i_- t} + h_+ \mathbf{P}^+ e^{\lambda^i_+ t}$$

where $\lambda^i_- < \lambda^i_+$ are the eigenvalues of $\mathbf{J}^d(\mathbf{y}^i)$ and $\mathbf{P}^-$ and $\mathbf{P}^+$ are the associated eigenvectors, and $\mathbf{h} = (h_-, h_+) = \mathbf{P}^{-1}(\mathbf{y}(0) - \mathbf{y}^i)$. For instance, the stable and unstable eigenspaces at a transversl saddle $\mathbf{y}^i$, for paths coming from $\mathbf{Y}_+$, such that $\mathbf{y}(0) \in \mathbf{Y}_+$, can have slopes given by $\mathbf{P}^-$ and $\mathbf{P}^+$, respectively, while the stable and unstable eigenspaces at $\mathbf{y}^i$, for paths coming from $\mathbf{Y}_-$, such that $\mathbf{y}(0) \in \mathbf{Y}_-$, can have slopes given by $\mathbf{P}^+$ and $\mathbf{P}^-$, respectively. This explains the shape of phase diagram Fig. 7.10a. For a transversal node, the eigenspaces associated to $\mathbf{P}^-$ and $\mathbf{P}^+$, are stable eigenspaces for only one subset, $\mathbf{Y}_-$ or $\mathbf{Y}_+$, and are unstable eigenspaces for the other subset, $\mathbf{Y}_+$ or $\mathbf{Y}_-$. This explains the shape of phase diagram Fig. 7.10b, where the transversal node is analogous to a stable node from side $\mathbf{Y}_+$ of Γ_I and to an unstable node from side $\mathbf{Y}_-$. Furthermore, as the impasse set is attractive from side $\mathbf{Y}_+$, this means that the segment of the local stable manifold associated to $\mathbf{y}^i$, say $\mathcal{W}^{s,loc}_{\mathbf{y}^i} \subset \mathbf{Y}_+$, is a strict subset. This means that there is a subset of points in $\mathbf{Y}_+ / \mathcal{W}^{sloc}_{\mathbf{y}^i}$ whose continuation paths collide with Γ_I and do not continue by crossing at the transversal-singular point.

Therefore, the set $\mathcal{S}^i_1$, and its projection Γ_I, can be stratified further into three types of sets: (1) several types of non-degenerate image-singular and kernel regular points, (2) several types of non-degenerate image-regular and kernel-singular points, and (3) several degenerate image-singular or kernel regular points.

Impasse steady states and other non-generic points. As for the scalar case (see Proposition 7.1), an impasse or singular steady state is a non-generic point, requiring a particular value for the parameters. We saw that for the scalar case, it reduced the CODE to a regular ODE in which there is no steady state, and therefore, the impasse set is "transparent" to the solution trajectories. This is not necessarily the case for planar CODE

Together with singular steady states, degenerate image-singular or kernel-singular points can be assembled into set $\mathcal{S}^i_2$, associated to **non-generic impasse-singular points**. This set gathers four types of elements: (1) degenerate image-singular points, (2) degenerate kernel-singular points, (3) kernel-singular points and (4) impasse steady states. Observe that set $\mathcal{S}^i_2$ consists of points in the impasse-singular surface (i.e., manifolds of dimension zero in $\mathbf{Y}$) and manifolds of co-dimension one or two in the parameter space Φ. We call **impasse bifurcation point** to an element of set $\mathcal{S}^i_2$ that is not a degenerate steady state. A degenerate singular steady state point is called a **singularity induced bifurcation point**.[14]

[14] See Llibre et al. (2002) for a complete presentation of local bifurcations for the planar constrained ODE case.

7.2.4 Application to Economics

The consequences of the existence of impasse points on the existence of macroeconomic equilibrium in DGE models, is discussed in Brito et al. (2017). Assuming that there is a non-empty impasse surface Γ_I, there are several implications for the type of impasse points that can be consistent with DGE paths, because a necessary condition for a trajectory to be an equilibrium path, when representative agents have infinity horizons, is that trajectories should be continued for $t \to \infty$. First, a **necessary** condition for a candidate DGE path $\mathbf{y}(t)$ to traverse Γ_I, at a finite time t_A, is that there is at least one image-singular point, $\mathbf{y}_i^i$, and $\mathbf{y}(t_A) = \mathbf{y}_i^i$. Second, another necessary condition is that the image singular point is a transversal saddle or a transversal node. There is no crossing at transversal foci points. Third, there are other conditions as well, basically associated to the need for the crossing to be in the "right" direction, meaning that crossing trajectories should enter subspaces of $\mathbf{Y}$ having a steady state with an appropriate dimension of the local stable manifold. Fourth, if there are singular steady states, i.e, steady states located in Γ_I, satisfying $\bar{y} = y^i$, that are degenerate image-singular points satisfying $\mathbf{g}(\bar{y}^i, \varphi) = 0$ and det $(\mathbf{J}(\bar{y}^i) = 0$, they can be associated to a steady state equilibrium path.

As the next example shows, if there is an image-singular transversal node point, **temporary indeterminacy** can exist, in the sense that an equilibrium path crossing the impasse point and converging to a steady state can display temporary indeterminacy before (after) crossing the impasse point and can display determinacy after (before) crossing it. This type of dynamics can never occur in DGE represented by non-linear regular ODE, which can only display either **permanent determinacy** or **permanent determinacy**.

Example 7.8 (*Brit et al. (2017) model*) This is a new-Keynesian model in which there is one competitive market for a final good using a continuum of intermediate inputs, and every producer of an intermediate input uses capital and labor with a Cobb-Douglas technology. Intermediate good firms compete in a Cournotian monopolistic competition market structure. This implies there is an endogenous markup driving a wedge between factor returns and their marginal productivities. The representative household is characterized by a benchmark utility function with endogenous labor supply. The government follows a balanced budget policy, and adopts a distortionary countercyclical tax rule. The DGE depends on the paths of the stocks of capital and labor, (k, ℓ), that are governed by the CODE in separated form, representing the macroeconomic equilibrium, and the intertemporal arbitrage condition,

$$\dot{k} = (1 - T(k, \ell))\, Y(k, \ell)(1 - L(\ell)), \tag{7.54a}$$

$$\delta(k, \ell)\, \dot{\ell} = \ell\,(1 - T(k, \ell))\,(R(k, \ell) - \rho), \tag{7.54b}$$

where $Y(k, \ell) = k^{\alpha}\, \ell^{1-\alpha}$, is the production function, $T(k, \ell) = \phi\, Y(k, \ell)^{\mu}$ is the distortionary tax rule, $L(\ell) = (1 - \alpha)\, \ell^{-\chi-1}$ is the household's net labor income ratio, $R(k, \ell) = r(k, \ell)\left[\mu\, T(k, \ell) + \left(1 - (1 + \mu)\, T(k, \ell)\right) L(\ell)\right]$, is the rate of return for

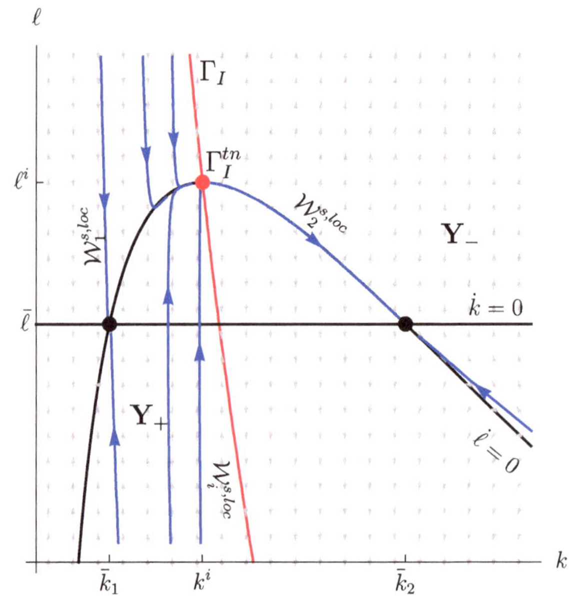

Fig. 7.12 Temporary indeterminacy in the CMC model

capital, $r(k, \ell) = \alpha Y(k, \ell)/k$ is the marginal productivity of capital, and $\delta(k, \ell) = \epsilon (T(k, \ell) - T_s)$ where $\epsilon \equiv \alpha + \chi - (1 - \alpha)\mu$, captures the difference between the observed tax rate and the critical level $T_s \equiv \frac{\alpha+\chi}{\epsilon}$. The following parameters $\boldsymbol{\varphi} = (\alpha, \chi, \rho, \phi, -\mu) \in (0, 1) \times \mathbb{R}^4_{++}$ entered the previous functions: the share of capital, α, the inverse of the Frisch elasticity, χ, the rate of time preference, ρ, and the parameters of the tax rule, ϕ and μ. We also assume that $\epsilon > 0$ to rule out a fast-slow singularity. Furthermore, it is assumed that $\mathbf{Y} = \{(k, \ell) \in \mathbb{R}^2_{++} : 0 < T(k, \ell) < 1\}$.

The steady state set can be simplified to $\Gamma = \{(k, \ell) : L(\ell) = 1, \ (1 - T(k, \ell))r(k, \ell) = \rho\}$, and the impasse set is $\Gamma_I = \{(k, \ell) \in \mathbf{Y} : T(k, \ell) = \tau_s\}$. The reason for the existence of an impasse singularity is associated to the concavity of the tax rule. We can thus partition the image set between the subset of low tax rates $\mathbf{Y}_- = \{(k, \ell) \in \mathbf{Y} : T(k, \ell) < T_s\}$, and high tax rates $\mathbf{Y}_+ = \{(k, \ell) \in \mathbf{Y} : T(k, \ell) > T_s\}$. For a given ℓ, the low (high) tax rate corresponds to a high (low) level for k. The same qualitative relationship holds for the endogenous equilibrium markups: high (low) markups are associated to low (high) levels of capital, for a given ℓ.

Several cases are possible, by taking ϕ and χ as bifurcating parameters. There is one or there are two steady states, that can be located in one or both subsets $\mathbf{Y}_-$ and $\mathbf{Y}_+$ The steady states also have different types of local stability (determinacy) properties. However, when two steady states are located in the same branch the dimension of their local stable manifolds is different, while when they are located in different branches, we can have a case in which they are both saddle points (i.e, determinate).

It can also be shown that there is one or there are two impasse points, and they are of the impasse image-singular type, in particular they can be impasse-transversal saddles or nodes.

Figure 7.12 illustrates the phase diagram for a particular case in which there are two steady states, $(\bar{k}_1, \bar{\ell})$, and $(\bar{k}_2, \bar{\ell})$, the first in the high markup branch and the second in the low markup branch, and they are both saddle points, and thus locally determinate. There are also two image-singular points. The one that is illustrated in that phase diagram is a transversal node, that is attracting from the high markup side and is repelling from the low markup side. We also observe some global dynamic features of the model: the stable manifold associated to the high markup steady state, $\mathcal{W}_1^{s,loc}$, constitutes the boundary to the basin of attraction of the impasse point (k^i, ℓ^i); the other boundary is set by two trajectories, $\mathcal{W}_i^{s,loc}$, that separates the points in the low markup subset that are attracted to the impasse point from those that collide with (for low ℓ) or are repelled (for high ℓ) from the impasse surface, Γ_I; and from the infinite number of trajectories that cross that surface, there is only one that converges to the high markup steady state $(\bar{k}_2, \bar{\ell})$. A sufficient condition for a solution to CODE (7.54a)–(7.54b) to be DGE (in which the transversality condition is satisfied) is that it converges to a steady state, given any initial level for the capital stock, which places restrictions on the initial level for the labor stock. Therefore, there are only three types of solution paths: first, if $k(0)$ is in the high markup branch, $\mathbf{Y}_+$, the DGE displays convergence over time to the high markup steady state $(\bar{k}_1, \bar{\ell})$ along $\mathcal{W}_1^{s,loc}$; second, if the initial capital is $k(0) > k^i$ then the DGE converges to low markup steady state $(\bar{k}_2, \bar{\ell})$, along $\mathcal{W}_2^{s,loc}$; third, if $k(0)$ is low and the initial point is inside the basin of attraction of Γ_I^{tn}, then there is infinite number of DGE trajectories that eventually cross Γ_I and converge to the low markup steady state $(\bar{k}_2, \bar{\ell})$. In the first two cases, there is **permanent determinacy** and in the third there is **temporary indeterminacy, and asymptotic determinacy**. The last type of DGE paths can only exist in the presence of impasse singularities, and suggests the intuition that short-run macroeconomic volatility, due to the presence of self-fulfilling prophecies, does not entail necessarily long-run volatility.

7.3 References

The main reference on fast-slow ODE is Kuehn (2015). For constrained ODE (or the associate DAE), we follow the approach by Zhitomirskii (1993). See also, Venkatasubramanian et al. (1995a), Llibre et al. (2002) and Riaza (2008). For a similar approach for DAE see, e.g., Venkatasubramanian et al. (1995b), Beardmore et al. (2004), Riaza (2008), and Marszalek (2009).

Applications to economics, explicitly using the general theory, are very rare. For an application to DGE theory, see Brito et al. (2017). For an endogenous growth model with government debt dynamics and fiscal rules leading to impasse points see Brito (2017). For other economic models displaying impasse singularities, see Barnett and He (2002) and He and Barnett (2006).

Part II
Functional Calculus and Calculus of Variations

Chapter 8
Introduction to Functional Calculus

In this chapter, we provide a heuristic and pragmatic approach to functional calculus in economics. As we will see in this and the next chapters, functionals are pervasive in both theoretical and applied economic concepts and problems.

We introduce spaces of functions mapping one continuum in another continuum, $\mathcal{Y}$, in which several additional properties (concerning their level, continuity, differentiability, and boundedness properties) are imposed. A **functional**, denoted by $\mathsf{F}[y]$, is a mapping of a function space into a real number $\mathsf{F} : \mathcal{Y} \rightarrow \mathbb{R}$.

There are three main components in the definition of a functional: the space of functions describing the object under study $\mathcal{Y}$, the domain of those functions, denoted by X, or **X** if it is multi-dimensional, and a functional introducing a valuation over those functions, F. In economics, support spaces can be classified according to their internal hierarchy, or according to the continuity of their elements. First, according to their internal hierarchy, we can classify them as **spaces of varieties**, or **spaces of variations**. Spaces of varieties do not have an order structure, and can be considered as just a tagging device, like the listing of economic sectors into primary, secondary or tertiary, or listings of different consumer goods belonging to a consumption basket, or different inputs in a production function. In this sense, spaces of functions over spaces of varieties are not normed vector spaces. We call spaces of variations to domains of variables having an order structure, as is the case, for instance, when they refer to size, distance, skill level, income, or time. Spaces of functions over spaces of variations are normed vector spaces, and the functional is an operator.[1] Second, according to their continuity properties, we can distinguish between **discrete support sets**, if $\mathrm{X} \subseteq \mathbb{N}$, and **continuous support sets**, if $\mathrm{X} \subseteq \mathbb{R}$. In the first case, the domain of functions is finite-dimensional or countable, and in the second it is infinite-dimensional. Next

[1] A normed vector space is a vector space (v.g, a set of numbers, continuous functions, continuous differentiable functions, bounded functions, or distributions) over which a norm (that is a measure) is defined. An operator is a mapping between two normed vector spaces. In this chapter, we are interested in operators between spaces of continuous functions and the space of real numbers.

P. B. Brito, *Economic Dynamics and Distributions*, Dynamic Modeling and Econometrics in Economics and Finance 32, https://doi.org/10.1007/978-3-031-94717-9_8

we will mostly deal with the continuous case, and with sets that can be ordered or not. When they are ordered, it is assumed that $\mathsf{X} \subseteq \mathbb{R}$.

The following two functionals over functions $y : \mathsf{X} \to \mathsf{Y} \subseteq \mathbb{R}$, where X can have or not an order structure, cover most cases we are interested in: the **linear functional**

$$\mathsf{F}[y] = \int_{\mathsf{X}} f(y(x), x)\, dx, \tag{8.1}$$

and the **transformed linear functional**

$$\mathsf{G}[y] = g\Big(\int_{\mathsf{X}} f(y(x), x)\, dx\Big), \tag{8.2}$$

where f and g are both functions with appropriate properties (usually regarding continuity, differentiability, boundedness, and measurability).[2]

The behavioral or performance function $f(\cdot)$ provides a measure for point-wise levels, $y = y(x)$, for each $x \in \mathsf{X}$, while the functional functional $\mathsf{F}[y]$ provides a measure for the whole distribution $\mathbf{y}_{\mathsf{X}} = \big(y(x)\big)_{x \in \mathsf{X}}$, by a weighted sum of the point-wise valuations in functional (8.1). In functional (8.2), $\mathsf{G}[y]$ assesses the distribution $\mathbf{y}_{\mathsf{X}}$ not only by its point-wise weighted sum but also by a direct valuation of their weighted sum. In functional (8.1), the direct dependence on the independent variable x, in $f(\cdot, x)$, is introduced to capture support specific valuation effects not directly associated to the size of $y(\cdot)$, as discounting or distributional effects, for instance.

There are several extensions to functionals (8.1), and analog extensions to functional (8.2), that have applications in economics. First, we can have a n-dimensional space of functions depending on a single independent variable, $\mathbf{y} \in \mathcal{Y}$, where $\mathbf{y} : \mathsf{X} \to \mathbf{Y} \subseteq \mathbb{R}^n$, and

$$\mathsf{F}[\mathbf{y}] = \int_{\mathsf{X}} f(\mathbf{y}(x), x)\, dx. \tag{8.3}$$

For instance, for $n = 2$, we have $\mathsf{F}[y_1, y_2] = \int_{\mathsf{X}} f(y_1(x), y_2(x), x)\, dx$.

Second, we can have a one-dimensional function of a vector of dependent variables, $y \in \mathcal{Y}$, where $y : \mathbf{X} \to \mathsf{Y} \subseteq \mathbb{R}$, for $\mathbf{X} = \otimes_{i=1}^{m} \mathsf{X}_i \subseteq \mathbb{R}^m$, and

$$\mathsf{F}[y] = \int_{\mathbf{X}} f(y(\mathbf{x}), \mathbf{x})\, d\mathbf{x} \tag{8.4}$$

where $d\mathbf{x} = \prod_{i=1}^{m} dx_i$. For instance, if $m = 2$, we have

$$\mathsf{F}[y] = \int_{\mathsf{X}_1} \int_{\mathsf{X}_2} f(y(x_1, x_2), x_1, x_2)\, dx_2\, dx_1.$$

[2] The finite-dimensional discrete analog functionals over $y : \mathsf{X} \to \mathbb{R}$ are $\mathsf{F}[y] = \sum_{\mathsf{X}} f(y_x, x)$ or $\mathsf{G}[y] = g\Big(\sum_{\mathsf{X}} f(y_x, x)\Big)$.

Third, we can combine the previous two cases, considering vector functions of a vector of independent variables, $\mathbf{y} \in \mathcal{Y}$, where $\mathbf{y} : \mathbf{X} \subseteq \mathbb{R}^m \to \mathbf{Y} \subseteq \mathbb{R}^n$, and

$$\mathsf{F}[\mathbf{y}] = \int_{\mathbf{X}} f(\mathbf{y}(\mathbf{x}), \mathbf{x})\, d\mathbf{x}. \tag{8.5}$$

For instance, for $m = n = 2$, we have

$$\mathsf{F}[y_1, y_2] = \int_{X_1} \int_{X_2} f(y_1(x_1, x_2), y_2(x_1, x_2), x_1, x_2)\, dx_2\, dx_1.$$

In calculus of variations problems, there is another type of extension to functional (8.1), involving first derivatives as arguments in the performance function,

$$\mathsf{F}[y] = \int_{X} f\big(y'(x), y(x), x\big)\, dx, \tag{8.6}$$

or first and second derivatives, as

$$\mathsf{F}[y] = \int_{X} f\big(y''(x), y'(x), y(x), x\big)\, dx. \tag{8.7}$$

If both the support and function $\mathbf{y}$ are higher dimensional, and the set function $f(\cdot)$ involves partial derivatives in the performance function, we have

$$\mathsf{F}[\mathbf{y}] = \int_{\mathbf{X}} f\big(\mathrm{D}_{\mathbf{xx}}\mathbf{y}(\mathbf{x}), \mathrm{D}_{\mathbf{x}}\mathbf{y}(\mathbf{x}), \mathbf{y}(\mathbf{x}), \mathbf{x}\big)\, d\mathbf{x}. \tag{8.8}$$

Functional $\mathsf{G}[\cdot]$ can also be analogously extended.

Several economic models involve optimizing particular (qualitative or explicit) specifications of one of the above functionals, that can be constrained or not by algebraic or functional side conditions. In order to address them, and to characterize the properties of functionals, we need some results from generalized calculus or, when the space of functions can be a normed vector space, from calculus and optimization over normed vector spaces. The purpose of this chapter is to present an introduction to those topics, with a view to economic applications. In particular, we provide the generalized calculus facts allowing to set up and solve some infinite-dimensional optimization problems. Section 8.1 presents some results on calculus of functionals (or generalized calculus). Section 8.2 presents problems involving maximization of functionals. In this chapter, we consider problems in which the variation of the state function $y(\cdot)$ is not an argument of the performance criterium directly, as in calculus of variations problems, or indirectly, as in optimal control problems. We leave those problems to the following chapters of this Part and to Parts III and IV.

8.1 Functional or Variational Calculus

In this section, we first present the concept of functional, or Gâteaux, derivative, and apply it to linear functionals in Sect. 8.1.1. Section 8.1.2 presents linear functionals. Linear functionals and generalized differentials for spike perturbations are the object of Sect. 8.1.3. Extensions to functionals involving derivatives can be found in Sect. 8.1.4. Section 8.1.5 contains several applications of functional calculus in economics.

8.1.1 Functionals and Functional Derivatives

Let $\mathcal{Y}$ be the set of functions y mapping a continuous index set X to a subset of the set of real numbers Y, that is $y \in \mathcal{Y}$ for $y : \mathrm{X} \to \mathrm{Y} \subseteq \mathbb{R}$. In particular, we consider $\mathcal{Y} \in C^1(\mathrm{X}, \mathrm{Y})$, the set of continuous functions with domain X and range in a subset of $\mathbb{R}$.[3] Recall our definition of a **functional** F as the mapping, $\mathsf{F} : \mathcal{Y} \to \mathbb{R}$.

A **variation of functional** F is defined as the change in the functional if function $y \in \mathcal{Y}$ is perturbed to $y + dy \in \mathcal{Y}$, $\Delta\mathsf{F}[y] = \mathsf{F}[y + dy] - \mathsf{F}[y]$. As the functional range is a real number, this means that $\Delta\mathsf{F}[y]$ is a real number as well.

The **variation of the functional along direction** $\eta \in \mathcal{Y}$ is the variation of the functional defined as

$$\mathrm{D}\,\mathsf{F}[y] = \mathsf{F}[y + \epsilon\,\eta] - \mathsf{F}[y],$$

where ϵ is a constant.

Definition 8.1 The **Gâteaux differential** is defined as the variation of the functional in the direction $\eta \in \mathcal{Y}$ when the constant ε is infinitesimal

$$\mathrm{D}_{\eta(\cdot)}\mathsf{F}[y] = \lim_{\varepsilon \to 0} \frac{\mathsf{F}[y + \varepsilon\,\eta] - \mathsf{F}[y]}{\varepsilon}. \tag{8.9}$$

The Gâteaux differential is a weak definition of a functional differential that can be defined even for spaces without a norm. In practice, it is denoted as

$$\mathrm{D}_{\eta(\cdot)}\mathsf{F}[y] = \frac{d}{d\varepsilon}\mathsf{F}[y + \varepsilon\,\eta]\Big|_{\varepsilon=0}.$$

---

[3] Recall that a function $y(\cdot)$ is continuous at point $x_0 \in \mathrm{X}$ if $\lim_{x \to x_0} y(x) = y(x_0)$ for $x \in \mathrm{X}$ in a neighborhood of x_0. In particular, this means that the right and left limits are equal: that is $y(x_0^+) = y(x_0^-) = y(x_0)$ where $y(x_0^-) = \lim_{x \uparrow x_0}$ and $y(x_0^+) = \lim_{x \downarrow x_0}$. Function $y(\cdot)$ is continuous if it is continuous for all $x \in \mathrm{X}$.

Equivalently, the perturbed functional is obtained from the Taylor expansion,[4]

$$\mathsf{F}[y+\varepsilon\,\eta] = \mathsf{F}[y] + \mathrm{D}_{\eta(\cdot)}\mathsf{F}[y]\,\varepsilon + O(\varepsilon^2).$$

In analogy to the case of finite-dimensional analysis,[5] it can be shown that the Gâteaux differential is a linear functional of η. If y is a linear operator on a normed vector space, and it is bounded, we can invoke the Riesz-Frechet theorem Riesz and Sz.-Nagy (1955, p. 61) which states that if $\mathsf{G}[\eta]$ is a bounded operator in a inner-product space, then there is a function ζ such that $\mathsf{G}[\eta]$ can be represented by the inner product $\mathsf{G}[\eta] = \langle \zeta, \eta \rangle$ and by the measure $\|\mathsf{G}\| = \|\zeta\|$. This means that the Gâteaux differential of a linear functional admits the representation

$$\mathrm{D}_{\eta(\cdot)}\mathsf{F}[y] = \int_{\mathrm{X}} dy(x)\,\eta(x)\,dx.$$

This representation can be extended even to spaces of functions without a norm.

Definition 8.2 Let $y \in \mathcal{Y}$ be a space of functions. We call **Gâteaux derivative** of the functional $\mathsf{F}[y]$ at a point $x \in \mathrm{X}$, to the Gâteaux differential in which the perturbation is a Dirac-δ generalized function at an isolated point x,

$$\frac{\delta \mathsf{F}[y]}{\delta y(x)} \equiv dy(x) = \int_{\mathrm{X}} dy(s)\,\delta(s-x)\,ds.$$

Therefore, we can **represent the Gâteaux differential** for the perturbation by function $\eta \in \mathcal{Y}$ by the linear functional[6]

$$\mathrm{D}_{\eta(\cdot)}\mathsf{F}[y] = \int_{\mathrm{X}} \frac{\delta \mathsf{F}[y]}{\delta y(x)}\,\eta(x)\,dx. \tag{8.10}$$

We can extend the previous definitions to second-order variations. The **second-order variation** of the functional $\mathsf{F}[y]$,

---

[4] To see this observe that Eq. (8.9) is equivalent to $\lim_{\varepsilon\to 0} \frac{\mathsf{F}[y+\varepsilon\,\eta]-\mathsf{F}[y]}{\varepsilon} - \mathrm{D}_{\eta(\cdot)}\mathsf{F}[y] = 0$, and therefore, to $\lim_{\varepsilon\to 0} \frac{\mathsf{F}[y+\epsilon\,\eta]-\mathsf{F}[y]-\mathrm{D}_{\eta(\cdot)}\mathsf{F}[y]\,\varepsilon}{\varepsilon} = 0$. Writing the numerator as $g(\varepsilon)$, which is possible because a functional has its range in set $\mathbb{R}$, we can use the "little-o" notation where $o(\varepsilon) = \lim_{\varepsilon\to 0} \frac{g(\varepsilon)}{\varepsilon} = 0$.

[5] The Gâteaux differential generalizes the concept of directional derivative in elementary calculus. Let $f(\mathbf{y}) = f(y_1, \ldots, y_n)$. The directional derivative in the direction given by the vector $\mathbf{h} = (h_1, \ldots, h_n)^\top$ is

$$\mathrm{D}_{\mathbf{h}} f(\mathbf{y}) \equiv \lim_{\varepsilon\to 0} \frac{f(\mathbf{y}+\varepsilon\mathbf{h}) - f(\mathbf{y})}{\varepsilon} = \sum_i \frac{\partial f(\mathbf{y})}{\partial y_i}\,h_i$$

is a linear functional of $\mathbf{h} = (h_1, \ldots, h_n)^\top$.

[6] This is the analog to a partial derivative in classical calculus. In applications, it allows for the determination of marginal utilities and elasticities of substitution in a continuum setting. See Sect. 8.1.5.

$$\Delta^2 \mathsf{F}[y] = \Delta\big(\Delta\mathsf{F}[y + d_1 y] - \Delta\mathsf{F}[y]\big) = \mathsf{F}[y + d_1 y + d_2 y] - \mathsf{F}[y + d_2 y] - \big(\mathsf{F}[y + d_1 y] - \mathsf{F}[y]\big),$$

is the perturbation by $d_2\, y \in \mathcal{Y}$, of the perturbation of the functional by $d_1\, y \in \mathcal{Y}$.

The second-order Gâteaux differential of functional $\mathsf{F}[y]$ is defined as the Gâteaux differential along direction $\eta_2 \in \mathcal{Y}$, of the Gâteaux differential along direction $\eta_1 \in \mathcal{Y}$,

$$\mathrm{D}_{\eta_1(\cdot),\eta_2(\cdot)}\mathsf{F}[y] = \lim_{\varepsilon_2\to 0}\lim_{\varepsilon_1\to 0} \frac{\mathsf{F}[y + \varepsilon_1\,\eta_1 + \varepsilon_2\,\eta_2] - \mathsf{F}[y + \varepsilon_2\,\eta_2] - \big(\mathsf{F}[y + \varepsilon_1\,\eta_1] - \mathsf{F}[y]\big)}{\varepsilon_1\,\varepsilon_2}.$$

We can extend the representation of (8.10) to the second-order "crossed" Gâteaux differential

$$\mathrm{D}_{\eta_1(\cdot),\eta_2(\cdot)}\mathsf{F}[y] = \int_{\mathrm{X}}\int_{\mathrm{X}} \frac{\delta^2\mathsf{F}[y]}{\delta y(x)\,\delta y(x')}\,\eta_1(x)\,\eta_2(x')\,dx\,dx', \tag{8.11}$$

where

$$\frac{\delta^2\mathsf{F}[y]}{\delta y(x)\,\delta y(x')}, \quad \text{for each } x, x' \in \mathrm{X}$$

is the second-order "crossed" Gâteaux derivative,associated to perturbations $\eta_1(x)$ and $\eta_2(x)$.

If $\frac{\delta^2\mathsf{F}[y]}{\delta y(x)\,\delta y(x')}$ is bounded, then by Fubini's theorem, the crossed derivatives satisfy

$$\mathrm{D}_{\eta_1(\cdot),\eta_2(\cdot)}\mathsf{F}[y] = \mathrm{D}_{\eta_2(\cdot),\eta_1(\cdot)}\mathsf{F}[y].$$

The second-order "own" Gâteaux differential, associated to perturbation $\eta(\cdot)$, is

$$\mathrm{D}_{\eta(\cdot)^2}\mathsf{F}[y] = \int_{\mathrm{X}\times\mathrm{X}} \frac{\delta^2\mathsf{F}[y]}{\delta y(x)^2}\,\eta(x)^2\,dx^2, \tag{8.12}$$

and the second-order "own" Gâteaux derivative is

$$\frac{\delta^2\mathsf{F}[y]}{\delta y(x)^2}, \quad \text{for } x \in \mathrm{X}.$$

Therefore, a generalization of the second-order Taylor expansion for a "own" perturbation $y(x) + \varepsilon\,\eta(x) \to y(x) + \varepsilon^2\,\eta(x)^2$ is

$$\mathsf{F}[y + \varepsilon\,\eta] = \mathsf{F}[y] + \mathrm{D}_{\eta(\cdot)}\mathsf{F}[y]\,\varepsilon + \frac{1}{2}\,\mathrm{D}_{\eta(\cdot)^2}\mathsf{F}[y]\,\varepsilon^2 + O(\varepsilon^3),$$

where $\mathrm{D}_{\eta(\cdot)^2}\mathsf{F}[y] \equiv \mathrm{D}_{\eta(\cdot),\eta(\cdot)}\mathsf{F}[y]$.

Some **general properties of the Gâteaux differential** of functionals $\mathsf{F} : \mathcal{Y} \to \mathbb{R}$, of one-dimensional functions of one variable, $y \in \mathcal{Y}$, where $y : \mathrm{X} \to \mathrm{Y} \subseteq \mathbb{R}$, are:

1. multiplication by a constant: let $a \in \mathbb{R}$, then $\mathrm{D}_{\eta(\cdot)}\{a\,\mathsf{F}[y]\} = a\mathrm{D}_{\eta(\cdot)}\mathsf{F}[y]$
2. sum of functionals: let $\mathsf{F}_1[y]$ and $\mathsf{F}_2[y]$ be two functionals, then $\mathrm{D}_{\eta(\cdot)}\{\mathsf{F}_1[y] + \mathsf{F}_2[y]\} = \mathrm{D}_{\eta(\cdot)}\mathsf{F}_1[y] + \mathrm{D}_{\eta(\cdot)}\mathsf{F}_2[y]$
3. product rule: let $\mathsf{F}_1[y]$ and $\mathsf{F}_2[y]$ be two functionals, then $\mathrm{D}_{\eta(\cdot)}\{\mathsf{F}_1[y]\cdot\mathsf{F}_2[y]\} = \{\mathrm{D}_{\eta(\cdot)}\mathsf{F}_1[y]\}\cdot\mathsf{F}_2[y] + \{\mathrm{D}_{\eta(\cdot)}\mathsf{F}_2[y]\}\cdot\mathsf{F}_1[y]$
4. chain rule: let $f : \mathbb{R} \to \mathbb{R}$ be a monotonic function and $\mathsf{F}[y]$ be a functional, then $\mathrm{D}_{\eta(\cdot)}\{f(\mathsf{F}[y])\} = f'(\mathsf{F}[y])\,\{\mathrm{D}_{\eta(\cdot)}\mathsf{F}[y]\}$

The last property can be applied to find generalized differentials and derivatives of the transformed functional (8.2), that can be written as

$$\mathsf{G}[y] = g(\mathsf{F}[y]), \tag{8.13}$$

where $\mathsf{F} : \mathcal{Y} \to \mathbb{R}$ is a functional over function $y \in \mathcal{Y}$, and $g(\cdot)$ is a $C^2(\mathbb{R})$ function. The variation of the functional $\mathsf{G}[y + \varepsilon\, y] - \mathsf{G}[y]$, associated to a perturbation $y(x) \to y(x) + \varepsilon\,\eta(x)$, for $\eta \in \mathcal{Y}$, has the Gâteaux representation of the generalized differential

$$\mathrm{D}_{\eta(\cdot)}\mathsf{G}[y] = \lim_{\varepsilon\to 0}\frac{\mathsf{G}[y + \varepsilon\, y] - \mathsf{G}[y]}{\varepsilon} = \int_{\mathrm{X}} \frac{\delta\mathsf{G}[y]}{\delta y(x)}\,\eta(x)\,dx,$$

where the Gâteaux derivative, for a general perturbation $\eta \in \mathcal{Y}$, is

$$\frac{\delta\mathsf{G}[y]}{\delta y(x)} = g'(\mathsf{F}[y])\frac{\delta\mathsf{F}[y]}{\delta y(x)}, \text{ for each } x \in \mathrm{X}. \tag{8.14}$$

The second-order generalized own differential is

$$\mathrm{D}_{\eta(\cdot)^2}\mathsf{G}[y] = \int_{\mathrm{X}\times\mathrm{X}} \frac{\delta^2\mathsf{G}[y]}{\delta y(x)^2}\,\eta(x)^2\,dx^2,$$

where the generalized own derivative is

$$\frac{\delta^2\mathsf{G}[y]}{\delta y(x)^2} = g''(\mathsf{F}[y])\left(\frac{\delta\mathsf{F}[y]}{\delta y(x)}\right)^2 + g'(\mathsf{F}[y])\frac{\delta^2\mathsf{F}[y]}{\delta y(x)^2}, \text{ for each } x \in \mathrm{X}. \tag{8.15}$$

The second-order generalized differential for a "crossed" perturbation is

$$\mathrm{D}_{\eta(\cdot)\eta'(\cdot)}\mathsf{G}[y] = \int_{\mathrm{X}\times\mathrm{X}} \frac{\delta^2\mathsf{G}[y]}{\delta y(x)\delta y(x')}\,\eta(x)\,\eta'(x')\,dx\,dx',$$

where the generalized crossed derivative is

$$\frac{\delta^2 \mathsf{G}[y]}{\delta y(x)\delta y(')} = g''(\mathsf{F}[y])\, \frac{\delta \mathsf{F}[y]}{\delta y(x)}\, \frac{\delta \mathsf{F}[y]}{\delta y(x')} + g'(\mathsf{F}[y]) \frac{\delta^2 \mathsf{F}[y]}{\delta y(x)\, \delta y(x')}, \text{ for each } x, x' \in \mathrm{X}. \tag{8.16}$$

8.1.2 Linear Functionals and Their Generalized Differentials for General Perturbations

In this subsection, we apply the previous concepts of functional differentiation to particular versions of linear functionals (8.1)–(8.5), where function $f : \mathbf{Y} \times \mathbf{X} \rightarrow \mathbb{R}$ is a smooth function, i.e., $f \in C^2(\mathbf{X} \times \mathbf{Y})$, for $\mathbf{X} \subseteq \mathbb{R}^m$ and $\mathbf{Y} \subseteq \mathbb{R}^n$, where $m \in \{1, 2\}$ and $n \in \{1, 2\}$, and the functionals are assumed to be well defined.

Consider, first, the linear functional (8.1). This functional has the generalized differential (in the Gâteaux sense), for a general perturbation $y(x) \rightarrow y(x) + \varepsilon\, \eta(x)$ where $\eta \in \mathcal{Y}$, given by Eq. (8.10), where the Gâteaux derivative is

$$\frac{\delta \mathsf{F}[y]}{\delta y(x)} = \frac{\partial f(y(x), x)}{\partial y(x)}, \text{ for each } x \in \mathrm{X}. \tag{8.17}$$

It has the second-order crossed generalized differential given by Eq. (8.11), where the second-order crossed derivative is

$$\frac{\delta^2 \mathsf{F}[y]}{\delta y(x)\delta y(x')} = \frac{\partial}{\partial y(x')}\left(\frac{\partial f(y(x), x)}{\partial y(x)}\right) = 0, \text{ for each } x, x' \in \mathrm{X}. \tag{8.18}$$

and the second-order own generalized differential given by Eq. (8.12), where the second-order "own" derivative is

$$\frac{\delta^2 \mathsf{F}[y]}{\delta y(x)^2} = \frac{\partial^2 f(y(x), x)}{\partial y(x)^2}, \text{ for each } x \in \mathrm{X}. \tag{8.19}$$

Example 8.1 For a linear functionals of the form (8.1), that are common in economics, the generalized derivatives are:

1. if $\mathsf{F}[y] = \int_{\mathrm{X}} a\, dx$, where $a \in \mathbb{R}$, then $\dfrac{\delta \mathsf{F}[y]}{\delta y(x)} = \dfrac{\delta^2 \mathsf{F}[y]}{\delta y(x)^2} = 0$, for each $x \in \mathrm{X}$;
2. if $\mathsf{F}[y] = \int_{\mathrm{X}} a\, y(x)\, dx$, then $\dfrac{\delta \mathsf{F}[y]}{\delta y(x)} = a$, and $\dfrac{\delta^2 \mathsf{F}[y]}{\delta y(x)^2} = 0$, for each $x \in \mathrm{X}$;
3. if $\mathsf{F}[y] = \int_{\mathrm{X}} y(x)^2\, dx$, then $\dfrac{\delta \mathsf{F}[y]}{\delta y(x)} = 2\, y(x)$, and $\dfrac{\delta^2 \mathsf{F}[y]}{\delta y(x)^2} = 2$, for each $x \in \mathrm{X}$;
4. if $\mathsf{F}[y] = \int_{\mathrm{X}} e^{y(x)}\, dx$, then $\dfrac{\delta \mathsf{F}[y]}{\delta y(x)} = \dfrac{\delta^2 \mathsf{F}[y]}{\delta y(x)^2} = e^{y(x)}$, for each $x \in \mathrm{X}$;

5. if $\mathsf{F}[y] = \int_{\mathrm{X}} y(x)^{\theta}\, dx$, then $\dfrac{\delta \mathsf{F}[y]}{\delta y(x)} = \theta\, y(x)^{\theta-1}$ and $\dfrac{\delta^2 \mathsf{F}[y]}{\delta y(x)^2} = \theta\,(\theta-1)\, y(x)^{\theta-2}$, for each $x \in \mathrm{X}$;
6. and if $\mathsf{F}[y] = \int_{\mathrm{X}} e^{g(y(x))}\, dx$, then $\dfrac{\delta \mathsf{F}[y]}{\delta y(x)} = g'(y(x))\, e^{g(y(x))}$ and $\dfrac{\delta^2 \mathsf{F}[y]}{\delta y(x)^2} = g''(y(x)) e^{g(y(x))} + (g'(y(x)))^2\, e^{g(y(x))}$, for each $x \in \mathrm{X}$.

In all the above cases $\dfrac{\delta^2 \mathsf{F}[y]}{\delta y(x) y(x')} = 0$, for $x, x' \in \mathrm{X}$.

Consider now the linear functional (8.2) and introduce a general perturbation $y(x) \to y(x) + \varepsilon\, \eta(x)$. This functional has the first-order generalized derivative (in the Gâteaux sense) given by Eq. (8.14). As functional F is the linear functional (8.1), then the generalized derivative takes the form

$$\frac{\delta \mathsf{G}[y]}{\delta y(x)} = g'\big(\mathsf{F}[y]\big)\, \frac{\partial f(y(x), x)}{\partial y(x)}, \quad \text{for each } x \in \mathrm{X}. \tag{8.20}$$

For the same reason, the second-order crossed derivative (8.16) is

$$\begin{aligned} \frac{\delta^2 \mathsf{G}[y]}{\delta y(x) \delta y(x')} &= g''\big(\mathsf{F}[y]\big) \left(\frac{\partial f(y(x'), x')}{\partial y(x')}\right) \left(\frac{\partial f(y(x), x)}{\partial y(x)}\right) + \\ &\quad + g'\big(\mathsf{F}[y]\big)\, \frac{\partial}{\partial y(x')} \left(\frac{\partial f(y(x), x)}{\partial y(x)}\right) \text{ for each } x, x' \in \mathrm{X}, \end{aligned} \tag{8.21}$$

and the second-order own generalized derivative (8.15) becomes

$$\frac{\delta^2 \mathsf{G}[y]}{\delta y(x)^2} = g''\big(\mathsf{F}[y]\big) \left(\frac{\partial f(y(x), x)}{\partial y(x)}\right)^2 + g'\big(\mathsf{F}[y]\big) \frac{\partial^2 f(y(x), x)}{\partial y(x)^2}, \quad \text{for each } x \in \mathrm{X}. \tag{8.22}$$

Comparing the second-order crossed derivatives of the linear and transformed functionals, in Eqs. (8.18) and (8.21), we observe that the transformed functional incorporates an aggregate effect, if $g(\cdot)$ is non-linear.

Example 8.2 Consider the transformed functional $\mathsf{G}[y] = \left(\int_{\mathrm{X}} \phi(y(x))^{\theta}\, dx\right)^{\frac{1}{\theta}}$, where $\phi(\cdot)$ is a smooth function, and θ is a real number. If we denote $\mathsf{F}[y] = \int_{\mathrm{X}} \phi(y(x))^{\theta}\, dx$ and $g(\mathsf{F}[y]) = \mathsf{F}[y]^{\frac{1}{\theta}}$, we obtain the first-order generalized derivative $\dfrac{\delta \mathsf{G}[y]}{\delta y(x)} = \phi'(y(x)) \left(\dfrac{\mathsf{G}[y]}{y(x)}\right)^{1-\theta}$, for each $x \in \mathrm{X}$, the second-order crossed generalized derivative

$$\frac{\delta^2 \mathsf{G}[y]}{\delta y(x) \delta y(x')} = (1-\theta) \left(\frac{\mathsf{G}[y]^2}{\phi(y(x))\, \phi(y(x'))}\right)^{1-\theta} \frac{\phi'(y(x))\, \phi'(y(x'))}{\mathsf{G}[y]}$$

and the second-order own generalized derivative

$$\frac{\delta^2 \mathsf{G}[y]}{\delta y(x)^2} = (1-\theta)\left(\frac{\mathsf{G}[y]}{\phi(y(x))}\right)^{2(1-\theta)} \frac{(\phi'(y(x)))^2}{\mathsf{G}[y]} + \\ + \left(\frac{\mathsf{G}[y]}{\phi(y(x))}\right)^{1-\theta}\left(\phi''(y(x)) - (1-\theta)\frac{\phi'(y(x))}{\phi(y(x))}\right)$$

Prove this.

Functionals of vector functions. For linear functionals of vector functions, (8.3) $\mathbf{y}(x) = (y_1(x), \ldots, y_n(x)) \in \mathcal{Y}$, where $\mathbf{y} : \mathrm{X} \subseteq \mathbb{R} \to \mathbf{Y} \subset \mathbb{R}^n$, and for a perturbation $\boldsymbol{\eta}(x) \in \mathcal{Y}$, the Gâteaux differential is

$$\mathrm{D}_{\boldsymbol{\eta}(\cdot)}\mathsf{F}[\mathbf{y}] = \int_{\mathrm{X}} \left\langle \frac{\delta \mathsf{F}[\mathbf{y}]}{\delta \mathbf{y}(x)}, \boldsymbol{\eta}(x) \right\rangle dx, \tag{8.23}$$

where the Gâteaux gradient is $\frac{\delta \mathsf{F}[\mathbf{y}]}{\delta \mathbf{y}(x)} = \left(\frac{\delta \mathsf{F}[\mathbf{y}]}{\delta y_1(x)}, \ldots, \frac{\delta \mathsf{F}[\mathbf{y}]}{\delta y_n(x)}\right)$.

For $n = 2$ we have the generalized differential

$$\mathrm{D}_{\boldsymbol{\eta}(\cdot)}\mathsf{F}[\mathbf{y}] = \int_{\mathrm{X}} \left(\frac{\delta \mathsf{F}[\mathbf{y}]}{\delta y_1(x)} \eta_1(x) + \frac{\delta \mathsf{F}[\mathbf{y}]}{\delta y_2(x)} \eta_2(x) \right) dx, \tag{8.24}$$

and the generalized gradient

$$\frac{\delta \mathsf{F}[\mathbf{y}]}{\delta \mathbf{y}(x)} = \begin{pmatrix} \frac{\delta \mathsf{F}[\mathbf{y}]}{\delta y_1(x)} \\ \frac{\delta \mathsf{F}[\mathbf{y}]}{\delta y_2(x)} \end{pmatrix} = \begin{pmatrix} \frac{\partial f(y_1(x), y_2(x), x)}{\partial y_1(x)} \\ \frac{\partial f(y_1(x), y_2(x), x)}{\partial y_2(x)} \end{pmatrix}, \text{ for each } x \in \mathrm{X}.$$

Example 8.3 Let $\mathsf{F}[\mathbf{y}] = \int_{\mathrm{X}} y_1(x)^{\alpha}\, y_2(x)^{1-\alpha} dx$, for a number α. Then the generalized gradient is

$$\frac{\delta \mathsf{F}[\mathbf{y}]}{\delta \mathbf{y}(x)} = \begin{pmatrix} \alpha y_1(x)^{\alpha-1}\, y_2(x)^{1-\alpha} \\ (1-\alpha) y_1(x)^{\alpha}\, y_2(x)^{-\alpha} \end{pmatrix}.$$

Exercise 8.1 Find the generalized own and crossed Hesseans $\frac{\delta^2 \mathsf{F}[\mathbf{y}]}{\delta \mathbf{y}(x)^2}$, and $\frac{\delta^2 \mathsf{F}[\mathbf{y}]}{\delta \mathbf{y}(x) \delta \mathbf{y}(x')}$ for the $n = 2$ case.

Functionals with independent vector variables. For the linear functional, in Eq. (8.4), of functions of vector variables $y(\mathbf{x}) \in \mathcal{Y}$, where $y : \mathbf{X} \subseteq \mathbb{R}^m \to \mathrm{Y} \subset \mathbb{R}$ and $\mathbf{X} = \otimes_{i=1}^m \mathrm{X}_i$, and for a perturbation $y(\mathbf{x}) \to y(\mathbf{x}) + \varepsilon\, \eta(\mathbf{x})$ for $\eta(\mathbf{x}) \in \mathcal{Y}$, the Gâteaux differential is

$$\mathrm{D}_{\eta(\cdot)}\mathsf{F}[y] = \int_{\mathbf{X}} \frac{\delta \mathsf{F}[y]}{\delta y(\mathbf{x})} \eta(\mathbf{x})\, \mathbf{dx} \tag{8.25}$$

where $\mathbf{dx} = \prod_{i=m}^{1} dx_i$, and the Gâteaux derivative is

$$\frac{\delta \mathsf{F}[y]}{\delta y(\mathbf{x})} = \frac{\partial f(y(\mathbf{x}), \mathbf{x})}{\partial y(\mathbf{x})}, \text{ for each } \mathbf{x} \in \mathbf{X}. \tag{8.26}$$

For the two-dimensional $m = 2$ case, where $y(\mathbf{x}) = y(x_1, x_2)$, the linear functional is

$$\mathsf{F}[y] = \int_{X_1} \int_{X_2} f\big(y(x_1, x_2), x_1, x_2\big)\, dx_2\, dx_1 \tag{8.27}$$

where $f(y, \mathbf{x})$ is continuously differentiable function of the mapping $y : \mathrm{Y} \to \mathbf{X} = \mathrm{X}_1 \times \mathrm{X}_2 \subseteq \mathbb{R}^2$, and its Gâteaux differential is

$$\mathrm{D}_{\eta(\cdot)}\mathsf{F}[y] = \int_{X_1} \int_{X_2} \frac{\delta \mathsf{F}[y]}{\delta y(\mathbf{x})} \eta(x_1, x_2)\, dx_2\, dx_1, \tag{8.28}$$

where the Gâteaux derivative is

$$\frac{\delta \mathsf{F}[y]}{\delta y(\mathbf{x})} = \frac{\partial f(y(x_1, x_2), x_1, x_2)}{\partial y(x_1, x_2)}, \text{ for each } (x_1, x_2) \in \mathrm{X}_1 \times \mathrm{X}_2. \tag{8.29}$$

Exercise 8.2 Find the generalized own and crossed Hesseans $\frac{\delta^2 \mathsf{F}[y]}{\delta y(\mathbf{x})^2}$, and $\frac{\delta^2 \mathsf{F}[y]}{\delta y(\mathbf{x}) \delta y(\mathbf{x}')}$ for the $m = 2$ case.

Functionals with both dependent and independent vector variables. For the linear functional, in Eq. (8.5), of vector functions of vector variables $\mathbf{y}(\mathbf{x}) \in \mathcal{Y}$, where $\mathbf{y} : \mathbf{X} \subseteq \mathbb{R}^m \to \mathbf{Y} \subset \mathbb{R}^n$, for $\mathbf{X} = \otimes_{i=1}^m \mathrm{X}$ and $\mathbf{Y} = \otimes_{i=1}^n \mathrm{Y}_i$, introducing a perturbation $\mathbf{y}(\mathbf{x}) \to \mathbf{y}(\mathbf{x}) + \varepsilon \boldsymbol{\eta}(\mathbf{x})$ for $\boldsymbol{\eta}(\mathbf{x}) \in \boldsymbol{\mathcal{Y}}$, yields the Gâteaux differential

$$\mathrm{D}_{\boldsymbol{\eta}(\cdot)}\mathsf{F}[\mathbf{y}] = \int_{\mathbf{X}} \Big\langle \frac{\delta \mathsf{F}[\mathbf{y}]}{\delta \mathbf{y}(\mathbf{x})}, \boldsymbol{\eta}(\mathbf{x}) \Big\rangle\, \mathbf{dx}. \tag{8.30}$$

For the double two-dimensional case in which $m = n = 2$, in which $\mathbf{y}(\mathbf{x}) = (y_1(x_1, x_2), y_2(x_1, x_2))$ is a mapping $\mathbf{y} : \mathbf{X} \subseteq \mathbb{R}^2 \to \mathbf{Y} \subseteq \mathbb{R}^2$, where $\mathbf{Y} = \mathrm{Y}_1 \times \mathrm{Y}_2 \subseteq \mathbb{R}^2$, and $\mathbf{X} = \mathrm{X}_1 \times \mathrm{X}_2 \subseteq \mathbb{R}^2$, the linear functional

$$\mathsf{F}[\mathbf{y}] = \int_{X_1} \int_{X_2} f\big(y_1(x_1, x_2), y_2(x_1, x_2), x_1, x_2\big)\, dx_2\, dx_1, \tag{8.31}$$

has the Gâteaux differential, for a general perturbation $\mathbf{y}(\mathbf{x}) \to \mathbf{y}(\mathbf{x}) + \varepsilon\, \boldsymbol{\eta}(\mathbf{x}) \in \boldsymbol{\mathcal{Y}}$,

$$\mathrm{D}_{\boldsymbol{\eta}(\cdot)}\mathsf{F}[\mathbf{y}] = \int_{X_1} \int_{X_2} \Big\langle \frac{\delta \mathsf{F}[\mathbf{y}]}{\delta \mathbf{y}(x_1, x_2)}, \boldsymbol{\eta}(x_1, x_2) \Big\rangle\, dx_2\, dx_1, \tag{8.32}$$

where the Gâteaux derivative is

$$\frac{\delta \mathsf{F}[\mathbf{y}]}{\delta \mathbf{y}(\mathbf{x})} = \begin{pmatrix} \frac{\partial f(y_1(\mathbf{x}), y_2(\mathbf{x})\mathbf{x}}{\partial y_1(\mathbf{x})} \\ \frac{\partial f(y_1(\mathbf{x}), y_2(\mathbf{x})\mathbf{x})}{\partial y_2(\mathbf{x})} \end{pmatrix}, \text{ for each } \mathbf{x} = (x_1, x_2) \in \mathrm{X}_1 \times \mathrm{X}_2. \tag{8.33}$$

8.1.3 Generalized Differentials for Spike Perturbations

We have considered general perturbations $\eta : \mathrm{X} \to \mathbb{R}$. A degenerate perturbation concentrated at a specific point $x = x_0$, also called "spike" perturbation, is an important particular case. It can be represented by a Dirac-delta generalized function $\delta(x - x_0)$ (or degenerate distribution).[7] For the linear functional (8.1), we have the Gâteaux derivative[8]

$$\mathrm{D}_{\delta(x_0)}\mathsf{F}[y] = \int_{\mathrm{X}} \frac{\delta \mathsf{F}[y]}{\delta y(x)} \delta(x - x_0)\, dx = \frac{\delta \mathsf{F}[y]}{\delta y(x_0)} = \frac{\partial f(y(x_0), x_0)}{\partial y(x_0)}. \tag{8.34}$$

and for functional (8.2) we have the Gâteaux derivative

$$\mathrm{D}_{\delta(x_0)}\mathsf{G}[y] = g'(\mathsf{F}[y]) \int_{\mathrm{X}} \frac{\delta \mathsf{F}[y]}{\delta y(x)} \delta(x - x_0)\, dx = \frac{\delta \mathsf{G}[y]}{\delta y(x_0)} = g'(\mathsf{F}[y]) \frac{\partial f(y(x_0), x_0)}{\partial y(x_0)}. \tag{8.35}$$

This perturbation provides an infinite-dimensional analog to the finite-dimensional partial derivative of a function. Therefore, it also provides a generalization of the concepts of marginal utility and marginal productivity, if functionals $\mathsf{F}[\cdot]$ or $\mathsf{G}[\cdot]$ represent an utility functional over a continuum of goods, or a production technology with a continuum of inputs, as we will see next.

8.1.4 Functionals for Functions Depending on Derivatives

In this subsection, we extend the linear functional (8.1) by assuming that the integrand function $f(.)$ is a continuous and continuously differentiable function not only of y but also of its first and second derivatives, $y'(\cdot)$ and $y''(\cdot)$. These functionals are explicitly or implicitly used in calculus of variations and optimal control problems, and their functional derivatives appear in the optimality conditions of those problems.

---

[7] Dirac-δ is a distribution, not a function, and has the following properties:

$$\delta(x - x_0) = \begin{cases} 0 & \text{if } x \neq x_0 \\ \infty & \text{if } x = x_0 \end{cases},$$

$\int_{-\infty}^{\infty} \delta(x) dx = 1$ and $\int_{-\infty}^{\infty} \delta(x - y_0)\, y(x)\, dx = y(x_0)$.

[8] In mathematical physics, this is sometimes called as a Volterra derivative.

8.1.4.1 First-Order Derivatives

The following linear functional is frequently found in economics,

$$\mathsf{F}[y] = \int_{\mathrm{X}} f\big(y'(x), y(x), x\big)\, dx, \tag{8.36}$$

where function $f(\cdot) : \mathrm{Y} \times \mathrm{Y} \times \mathrm{X} \to \mathbb{R}$ is assumed to be continuous and continuous differentiable in (y', y). We denote $F(x) \equiv f(y'(x), y(x), x)$.

Lemma 8.1 *The (first-order) Gâteaux differential of functional* (8.36)*, for an arbitrary perturbation* $\eta \in \mathcal{Y}$*, is*

$$D_{\eta(\cdot)}\mathsf{F}[y] = \int_{\mathrm{X}} \frac{\delta \mathsf{F}[y]}{\delta y(x)}\eta(x)\, dx + \int_{\mathrm{X}} \frac{\partial F(x)}{\partial y'(x)}\eta(x) \tag{8.37}$$

where the generalized derivative is

$$\frac{\delta \mathsf{F}[y]}{\delta y(x)} = \frac{\partial F(x)}{\partial y(x)} - \frac{d}{dx}\Big(\frac{\partial F(x)}{\partial y'(x)}\Big), \text{ for each } x \in \mathrm{X}, \tag{8.38}$$

and, if $\mathrm{X} = [x_0, x_1]$ *is a closed set,*

$$\int_{\mathrm{X}} \frac{\partial F(x)}{\partial y'(x)}\eta(x) = \frac{\partial F(x_1)}{\partial y'(x_1)}\eta(x_1) - \frac{\partial F(x_0)}{\partial y'(x_0)}\eta(x_0),$$

or, if $\mathrm{X} = (x_0, x_1)$ *is an open set,*

$$\int_{\mathrm{X}} \frac{\partial F(x)}{\partial y'}\eta(x) = \lim_{x \uparrow x_1} \frac{\partial F(x)}{\partial y'(x)}\eta(x) - \lim_{x \downarrow x_0} \frac{\partial F(x)}{\partial y'(x)}\eta(x).$$

Proof We have to introduce not only the perturbation $y(x) \to y(x) + \varepsilon\eta(x)$ but also the perturbation of the first derivative $y'(x) \to y'(x) + \varepsilon\eta'(x)$. From Eq. (8.24) we have

$$\begin{aligned} \mathrm{D}_{\eta(\cdot)}\mathsf{F}[y] &= \int_{\mathrm{X}} \Big(\frac{\delta \mathsf{F}[y]}{\delta y(x)}\eta(x) + \frac{\delta \mathsf{F}[y]}{\delta y'(x)}\eta'(x)\Big)\, dx \\ &= \int_{\mathrm{X}} \frac{\partial f(y'(x), y(x), x)}{\partial y(x)}\eta(x)\, dx + \int_{\mathrm{X}} \frac{\partial f(y'(x), y(x), x)}{\partial y'(x)}\eta'(x)\, dx, \end{aligned}$$

where, integrating the second integral by parts

$$\begin{aligned} \int_{\mathrm{X}} \frac{\partial f(y'(x), y(x), x)}{\partial y'(x)}\eta'(x)\, dx &= \int_{\mathrm{X}} \frac{\partial f(y'(x), y(x), x)}{\partial y'(x)}\eta(x) - \\ &\quad - \int_{\mathrm{X}} \frac{d}{dx}\Big(\frac{\partial f(y'(x), y(x), x)}{\partial y'(x)}\Big)\eta(x)\, dx. \end{aligned}$$

Upon substitution, yields the differential (8.37). □

Example 8.4 Let $\mathsf{F}[y] = \int_{x_0}^{x_1} F(x)\,dx$ where $F(x) = \ln\big(y(x)^{\alpha}\, y'(x)^{\beta}\big)$. As $F_y(x) = \frac{\alpha}{y(x)}$, $F_{y'}(x) = \frac{\beta}{y'(x)}$ and $\frac{dF_{y'}(x)}{dx} = -\frac{\beta\, y''(x)}{(y'(x))^2}$ then $\frac{\delta \mathsf{F}[y]}{\delta y(x)} = \frac{\alpha}{y(x)} + \frac{\beta\, y''(x)}{(y'(x))^2}$, and the Gâteaux differential for a perturbation $\big(\eta(x)\big)_{x\in[x_0,x_1)}$ is

$$\mathrm{D}_{\eta(\cdot)}\mathsf{F}[y] = \int_{x_0}^{x_1} \Big(\frac{\alpha}{y(x)} + \frac{\beta\, y''(x)}{(y'(x))^2}\Big)\,\eta(x)\,dx + \beta\Big(\frac{\eta(x_1)}{y'(x_1)} - \frac{\eta(x_0)}{y'(x_0)}\Big).$$

8.1.4.2 Second-Order Derivatives

Consider now the linear functional

$$\mathsf{F}[y] = \int_{\mathrm{X}} f(y''(x), y'(x), y(x), x)\,dx, \tag{8.39}$$

where function $f(\cdot)$ is assumed to be continuous and continuous differentiable, at least up to the second order, in (y'', y', y). Next, we denote $F(x) = f(y''(x), y'(x), y(x), x)$.

Lemma 8.2 *The (first-order) Gâteaux differential of functional* (8.39), *for an arbitrary perturbation* $\eta \in \mathcal{Y}$, *is*

$$\begin{aligned}\mathrm{D}_{\eta(\cdot)}\mathsf{F}[y] = &\int_{\mathrm{X}} \frac{\delta \mathsf{F}[y]}{\delta y(x)}\,\eta(x)\,dx + \\ &+ \int_{\mathrm{X}} \Big(\frac{\partial F(x)}{\partial y'(x)} - \frac{d}{dx}\Big(\frac{\partial F(x)}{\partial y'(x)}\Big)\Big)\,\eta(x) + \int_{\mathrm{X}} \frac{\partial F(x)}{\partial y''(x)}\,\eta'(x),\end{aligned} \tag{8.40}$$

where the Gâteaux derivative is

$$\frac{\delta \mathsf{F}[y]}{\delta y(x)} = \frac{\partial F(x)}{\partial y(x)} - \frac{d}{dx}\Big(\frac{\partial F(x)}{\partial y'(x)}\Big) + \frac{d^2}{dx^2}\Big(\frac{\partial F(x)}{\partial y''(x)}\Big), \text{ for each } x \in \mathrm{X} \tag{8.41}$$

and, in the case in which set $\mathrm{X} = [x_0, x_1]$ *is closed*

$$\begin{aligned}\int_{\mathrm{X}} \frac{\partial F(x)}{\partial y'(x)}\,\eta(x) &= \frac{\partial F(x_1)}{\partial y'(x_1)}\,\eta(x_1) - \frac{\partial F(x_0)}{\partial y'(x_0)}\,\eta(x_0)\\ \int_{\mathrm{X}} \frac{\partial F(x)}{\partial y''(x)}\,\eta'(x) &= \frac{\partial F(x_1)}{\partial y''(x_1)}\,\eta'(x_1) - \frac{\partial F(x_0)}{\partial y''(x_0)}\,\eta'(x_0)\\ \int_{\mathrm{X}} \frac{d}{dx}\Big(\frac{\partial F(x)}{\partial y''(x)}\Big)\,\eta(x) &= \frac{d}{dx}\Big(\frac{\partial F(x_1)}{\partial y''(x_1)}\Big)\,\eta(x_1) - \frac{d}{dx}\Big(\frac{\partial F(x_0)}{\partial y''(x_0)}\Big)\,\eta(x_0).\end{aligned}$$

Proof In this case, we have

$$\begin{aligned} \mathsf{D}_{\eta(\cdot)}\mathsf{F}[y] &= \int_{\mathrm{X}} \left(\frac{\delta \mathsf{F}[y]}{\delta y(x)} \eta(x) + \frac{\delta \mathsf{F}[y]}{\delta y'(x)} \eta'(x) + \frac{\delta \mathsf{F}[y]}{\delta y''(x)} \eta''(x) \right) dx \\ &= \int_{\mathrm{X}} \frac{\delta \mathsf{F}[y]}{\delta y(x)} \eta(x)\, dx + \int_{\mathrm{X}} \frac{\delta \mathsf{F}[y]}{\delta y'(x)} \eta'(x)\, dx + \int_{\mathrm{X}} \frac{\delta \mathsf{F}[y]}{\delta y''(x)} \eta''(x)\, dx \\ &= \int_{\mathrm{X}} \frac{\partial F(x)}{\partial y(x)} \eta(x)\, dx + \int_{\mathrm{X}} \frac{\partial F(x)}{\partial y'(x)} \eta'(x)\, dx + \int_{\mathrm{X}} \frac{\partial F(x)}{\partial y''(x)} \eta''(x)\, dx. \end{aligned}$$

Integrating by parts the second integral, we have

$$\int_{\mathrm{X}} \frac{\delta F(x)}{\delta y'(x)} \eta'(x)\, dx = \int_{\mathrm{X}} \frac{\partial F(x)}{\partial y'(x)} \eta(x) - \int_{\mathrm{X}} \frac{d}{dx} \left(\frac{\partial F(x)}{\partial y'(x)} \right) \eta(x)\, dx.$$

Integrating by parts the third integral,[9] we have

$$\begin{aligned} \int_{\mathrm{X}} \frac{\delta F(x)}{\delta y''(x)} \eta''(x)\, dx = & \int_{\mathrm{X}} \frac{\partial F(x)}{\partial y''(x)} \eta'(x) - \int_{\mathrm{X}} \frac{d}{dx} \frac{\partial F(x)}{\partial y''(x)} \eta'(x)\, dx \\ & + \int_{\mathrm{X}} \frac{d^2}{dx^2} \frac{\partial F(x)}{\partial y''(x)} \eta(x)\, dx. \end{aligned}$$

□

Extensions to Gâteaux functionals as

$$\mathsf{F}[y] = \int_{\mathrm{X}} f(\mathrm{D}_{\mathbf{x}} y(\mathbf{x}), y(\mathbf{x}), \mathbf{x})\, d\mathbf{x}$$

or

$$\mathsf{F}[y] = \int_{\mathrm{X}} f(\mathrm{D}_{\mathbf{xx}} y(\mathbf{x}), \mathrm{D}_{\mathbf{x}} y(\mathbf{x}), y(\mathbf{x}), \mathbf{x})\, d\mathbf{x}$$

can be found in Chaps. 14 and 16.

8.1.5 Functionals in Economics

In this section, we present several examples of functionals that are relatively common in economics, and characterize their properties relative to analog finite-dimensional cases.

[9] This is because $\int uv''\,dx = uv' - \int u'v'dx = uv' - \left(u'v - \int u''v dx\right)$.

8.1.5.1 Generalized Means

The generalized mean is frequently used in economics. It can be used for independent variables belonging to a variety or a variation space (i.e., without or with an order structure).[10] Let there be different varieties or levels belonging to a continuum $\mathrm{I} \subseteq \mathbb{R}$, and let $x(i)$ be a quantity for the variety or the level $i \in \mathrm{I}$, where $x : \mathrm{I} \to \mathrm{X} \subseteq \mathbb{R}$. The distribution of the quantities for all varieties or levels is denoted by $\mathbf{x}_\mathrm{I} = \big(x(i)\big)_{i\in\mathrm{I}}$.

Definition 8.3 (*Generalized mean*) The **generalized mean** of the distribution $\mathbf{x}_\mathrm{I}$ is the functional[11]

$$\mathsf{M}_\rho[x] = \left(\int_\mathrm{I} w(i)\, x(i)^\rho\, di\right)^{\frac{1}{\rho}}, \text{ for } \rho \in [-\infty, \infty] \tag{8.42}$$

where $w : \mathrm{I} \to (0, 1)$ is a weighting function such that $\int_\mathrm{I} w(i)\, di = 1$.

Proposition 1 (Generalized means) *The generalized mean, defined in Eq.* (8.42), *takes several special forms for different values of* ρ:

1. *if* $\rho = 0$ *then* $\mathsf{M}_0[x] = \exp\left(\int_\mathrm{I} \ln\left(x(i)^{w(i)}\right) di\right) = \int_\mathrm{I} x(i)^{w(i)}\, di$ *is a geometric mean of* $\mathbf{x}_\mathrm{I}$;
2. *if* $\rho = 1$ *then* $\mathsf{M}_1[x] = \int_\mathrm{I} w(i)\, x(i)\, di$ *is a generalized arithmetic mean of* $\mathbf{x}_\mathrm{I}$;
3. *if* $\rho = -1$ *then* $\mathsf{M}_{-1}[x] = \dfrac{1}{\int_\mathrm{I} \frac{w(i)}{x(i)}\, di}$ *is a harmonic mean of* $\mathbf{x}_\mathrm{I}$;
4. *if* $\rho = \infty$ *then* $\mathsf{M}_\infty[x] = \max\left\{\mathbf{x}_\mathrm{I}\right\}$ *is the mode of* $\mathbf{x}_\mathrm{I}$;
5. *if* $\rho = -\infty$ *then* $\mathsf{M}_{-\infty}[x] = \min\left\{\mathbf{x}_\mathrm{I}\right\}$ *is the minimum of* $\mathbf{x}_\mathrm{I}$.

In general $\mathsf{M}_{-\infty}[x] \leq \mathsf{M}_\rho[x] \leq \mathsf{M}_\infty[x]$.

Proof We can write Eq. (8.42) equivalently as

$$\mathsf{M}_\rho[x] = \exp\left\{\ln\left[\left(\int_\mathrm{I} w(i)\, x(i)^\rho\, di\right)^{\frac{1}{\rho}}\right]\right\} = \exp\left\{\frac{1}{\rho}\ln\left(\int_\mathrm{I} w(i)\, x(i)^\rho\, di\right)\right\}.$$

For the case $\rho = 0$ we use l'Hôpital's rule as regards the exponent

[10] In the last case, if a generalized mean of the absolute value of a function is bounded, it defines a norm in a L^ρ Lebesgue function space.

[11] See Bullen (2003).

$$\lim_{\rho\to 0} \frac{\ln\left(\int_{\mathrm{I}} w(i)\,x(i)^{\rho}\,di\right)}{\rho} = \lim_{\rho\to 0} \frac{\int_{\mathrm{I}} w(i)\,\ln(x(i))\,x(i)^{\rho}\,di}{1}$$
$$= \int_{\mathrm{I}} w(i)\,\ln(x(i))\,di$$
$$= \int_{\mathrm{I}} \ln\left(x(i)^{w(i)}\right) di.$$

Then $\mathsf{M}_0[x] = \exp\left(\int_{\mathrm{I}} \ln\left(x(i)^{w(i)}\right)\right) di$. The generalized means for $\rho = 1$ and $\rho = -1$ are obtained by direct substitution.

If function $x \in \mathcal{X} \in L^{\infty}(\mathbb{R})$ then there is a maximum $x^* = ||x||_{\infty}$, and $\mathsf{M}_{\infty}[x] = \lim_{\rho\to\infty} x^* \left(\int_{\mathrm{I}} w(i) \left(\frac{x(i)}{x^*}\right)^{\rho} di\right)^{\frac{1}{\rho}} = x^* = \max\{\mathbf{x}_{\mathrm{I}}\}$.

As we have $\mathsf{M}_{\rho}[x] = \mathsf{M}_{-\rho}\left[\frac{1}{x}\right]$, for any ρ, and as $\min\{\mathbf{x}_{\mathrm{I}}\} = \max\left\{\mathbf{x}_{\mathrm{I}}^{-1}\right\}$, if $x \in L^{\infty}(\mathbb{R})$ then $\mathsf{M}_{-\infty}[x] = \lim_{\rho\to\infty} \mathsf{M}_{\rho}\left[x^{-1}\right] = \min\{\mathbf{x}_{\mathrm{I}}\}$. □

The generalized (Gâteaux) first-order derivative of the generalized mean (8.42), for a finite ρ, is

$$\frac{\delta \mathsf{M}_{\rho}[x]}{\delta x(i)} = w(i)\left(\frac{\mathsf{M}_{\rho}[x]}{x(i)}\right)^{1-\rho}, \text{ for each } i \in \mathrm{I}. \tag{8.43}$$

The generalized second-order "own" derivative is

$$\frac{\delta^2 \mathsf{M}_{\rho}[x]}{\delta x(i)^2} = -(1-\rho)\,\frac{w(i)}{x(i)}\left(\frac{\mathsf{M}_{\rho}[x]}{x(i)}\right)^{1-\rho}\left(1 - w(i)\left(\frac{\mathsf{M}_{\rho}[x]}{x(i)}\right)^{-\rho}\right), \text{ for each } i \in \mathrm{I}, \tag{8.44}$$

and the generalized (Gâteaux) second-order "crossed" derivative is

$$\frac{\delta^2 \mathsf{M}_{\rho}[x]}{\delta x(i)\delta x(j)} = (1-\rho)\,\frac{w(i)\,w(j)}{x(i)}\left(\frac{\mathsf{M}_{\rho}[x]}{x(j)}\right)^{1-\rho}\left(\frac{\mathsf{M}_{\rho}[x]}{x(i)}\right)^{-\rho}, \text{ for each } i, j \in \mathrm{I}. \tag{8.45}$$

Exercise 8.3 Prove the previous results.

Next we show that there are some well-known special cases corresponding to Cobb-Douglas (if $\rho = 0$), generalized linear (if $\rho = 1$) or to the Leontieff (if $\rho = -\infty$) production functions when there are continua of inputs.

8.1.5.2 Varieties in Consumption Theory

Let $c(i) \geq 0$ be the quantity of good of variety $i \in \mathrm{I}$. We define a consumption basket as the ensemble of quantities of all varieties, $\mathbf{c}_{\mathrm{I}} = \big(c(i)\big)_{i\in \mathrm{I}}$. The support set I is a non-ordered continuum of varieties. Sometimes the index set is specified as $\mathrm{I} = [0, N]$, where N measures the span of varieties. The consumption of a specific variety $\iota \in \mathrm{I}$, $c(\iota)$, is equivalent to the product of the distribution $\mathbf{c}_{\mathrm{I}}$ by a ι-centered Dirac-delta generalized function, $c(\iota) = \int_{\mathrm{I}} \delta(i - \iota)\, c(i)\, di$. The set of variety distributions is the function space with non-negative components $\mathcal{C} = \{c(i) : \mathrm{I} \to \mathbb{R} : c(i) \geq 0 \text{ for all } i \in \mathrm{I}\}$.

A utility functional $\mathsf{U}[c]$ is a measure of the value of a consumption basket $\mathbf{c}_{\mathrm{I}}$. It maps the distribution of varieties into a real number, i.e., $\mathsf{U} : \mathcal{C} \to \mathbb{R}$, which allows for ordering different consumption baskets.

There are two main types of utility functionals in the literature, the additive utility functional $\mathsf{U}[c] = \int_{\mathrm{I}} u(c(i))\, di$, where $u(c)$ is increasing and strictly concave (assuming no satiation), $u''(c) < 0 < u'(c)$ for $c \in (0, \infty)$,[12] and the Dixit and Stiglitz (1977) (DS)[13] utility functional, or aggregator,

$$\mathsf{U}[c] = \left(\int_{\mathrm{I}} c(i)^{1-\gamma} di\right)^{\frac{1}{1-\gamma}}, \tag{8.46}$$

where $\gamma \geq 0$. Observe that if $\gamma = 0$, the DS utility functional reduces to the additive functional. We assume that $\gamma > 0$ from now on. As this functional can be taken as an infinite-dimensional generalization of the CES aggregator it is called a CES (constant elasticity of substitution) utility functional, for a reason that will be clear bellow. Observe that although it is similar to the generalized mean (8.42), it is not a generalized mean because the utility of every variety is not weighted.

As the DS utility of a consumption basket is a transformed functional, of type (8.13), of a functional included in Example 8.2, then a change in the consumption basket, such as $c \to c + \epsilon\, dc \in \mathcal{C}$, generates a variation in the utility of the basket (8.46) given by the Gâteaux differential

$$\mathrm{D}_{dc(\cdot)} \mathsf{U}[c] = \int_{\mathrm{I}} \frac{\delta \mathsf{U}[c]}{\delta c(i)}\, dc(i)\, di,$$

where the Gâteaux derivative

$$\mathsf{U}_i \equiv \frac{\delta \mathsf{U}[c]}{\delta c(i)} = \left(\frac{\mathsf{U}[c]}{c(i)}\right)^{\gamma}, \text{ for each } i \in \mathrm{I},$$

corresponds to the **marginal utility** of variety $i \in \mathrm{I}$. To see this, observe that a variation in the quantity of variety $\iota \in \mathrm{I}$, represented by a perturbation $dc(\iota) =$

[12] See the recent contribution by Zhelobodko et al. (2012) or Dhingra and Morrow (2019).

[13] See also Krugman (1980) or Melitz (2003).

$\delta(i-\iota)\,dc(i)$, increases the utility of the consumption basket $\mathbf{c}_{\mathrm{I}}$ by $\mathrm{D}_{\delta(\iota)}\mathsf{U}[c] = \int_{\mathrm{I}} \frac{\delta \mathsf{U}[c]}{\delta c(i)} \delta(i-\iota)\,dc(i)\,di = \mathsf{U}_{\iota}\,dc(\iota)$.

The marginal rate of substitution between varieties ι and ι', denoted by $MRS_{\iota,\iota'}$ is defined by the change in the quantities of two components of a basket, say ι and $\iota' \neq \iota$, such that the level of utility is unchanged. Formally $MRS_{\iota,\iota'} = -\frac{dc(\iota')}{dc(\iota)}\Big|_{\mathrm{D}_{dc}\mathsf{U}[c]=0}$ where $dc(i) = 0$ for $i \notin \{\iota, \iota'\}$. As $\mathrm{D}_{dc}\mathsf{U}[c] = \frac{\delta \mathsf{U}[c]}{\delta c(\iota)}\,dc(\iota) + \frac{\delta \mathsf{U}[c]}{\delta c(\iota')}\,dc(\iota') = 0$ then the marginal rate of substitution between varieties ι and ι' for the DS aggregator is

$$MRS_{\iota,\iota'} = -\frac{dc(\iota')}{dc(\iota)}\Big|_{\mathsf{U}=\text{const}} = \frac{\dfrac{\delta \mathsf{U}[c]}{\delta c(\iota)}}{\dfrac{\delta \mathsf{U}[c]}{\delta c(\iota')}} = \left(\frac{c(\iota)}{c(\iota')}\right)^{-\gamma}, \text{ for any } \iota, \iota' \in \mathrm{I}.$$

We can also use the second-order Gâteaux derivatives to derive the implicit properties of the utility functional (8.46) as regards to the change in marginal utility and the substitution between varieties. The own and crossed Allen-Uzawa elasticities are

$$\varepsilon_{\iota,\iota} = -\frac{\dfrac{\delta^2 \mathsf{U}[c]}{\delta c(\iota)^2}\,c(\iota)}{\dfrac{\delta \mathsf{U}[c]}{\delta c(\iota)}} = \gamma\left(1 - \left(\frac{\mathsf{U}[c]}{c(\iota)}\right)^{\gamma-1}\right), \text{ for each } \iota \in \mathrm{I},$$

and

$$\varepsilon_{\iota,\iota'} = -\frac{\dfrac{\delta^2 \mathsf{U}[c]}{\delta c(\iota)\,\delta c(\iota')}\,c(\iota')}{\dfrac{\delta \mathsf{U}[c]}{\delta c(\iota)}} = -\gamma\left(\frac{\mathsf{U}[c]}{c(\iota')}\right)^{\gamma-1}, \text{ for each } \iota, \iota' \in \mathrm{I}.$$

Therefore, there is decreasing marginal utility for any variety ι if $\gamma \geq 1$, that is the case if the utility function $u(\cdot)$ is concave. In addition, as $\varepsilon_{\iota,\iota'} < 0$ then any two varieties ι and ι' are Edgeworth-Pareto (EP) complements. The fact that there is no EP independence between every pair of varieties, is related to the fact that the utility functional is a transformation of a linear functional, as we already saw in Eq. (8.21).

We can also determine a generalized elasticity of variety substitution between any two varieties

$$\begin{aligned} EVS_{\iota,\iota'} &= \frac{d\ln\left(c(\iota)/c(\iota')\right)}{d\ln\left(MRS_{\iota,\iota'}\right)} = \\ &= \frac{c(\iota)\,\dfrac{\delta \mathsf{U}[c]}{\delta c(\iota)} + c(\iota')\,\dfrac{\delta \mathsf{U}[c]}{\delta c(\iota')}}{c(\iota)\,\dfrac{\delta \mathsf{U}[c]}{\delta c(\iota)}\,\varepsilon_{\iota',\iota'} - 2\,c(\iota)\,\dfrac{\delta \mathsf{U}[c]}{\delta c(\iota)}\,\varepsilon_{\iota,\iota'} + c(\iota')\,\dfrac{\delta \mathsf{U}[c]}{\delta c(\iota')}\,\varepsilon_{\iota,\iota}}, \text{ for each } \iota \in \mathrm{I} \end{aligned}$$

Substituting the previous formulas and simplifying we find $EVS_{\iota,\iota'} = \gamma^{-1}$, for any two varieties. Therefore, γ is routinely called the inverse of the elasticity of variety substitution, and this justifies calling the utility functional (8.46) a CES (constant elasticity of substitution) aggregator.

Let there be a continuum but finite span of varieties $\mathrm{I} = [0, N]$, where N is finite. The relative love for variety is usually defined by the elasticity of $\mathsf{U}[c]$ as regards N.[14] Then we find

$$LV = \frac{\frac{\partial \mathsf{U}[c]}{\partial N} N}{\mathsf{U}[c]} = \frac{N}{1-\gamma}\left(\frac{c(N)}{\mathsf{U}[c]}\right)^{1-\gamma}.$$

For a homogeneous distribution $c(i) = \bar{c}$ for all $i \in [0, N]$, then $\mathsf{U}[\bar{c}] = \bar{c}\,N$ and $LV = \frac{1}{1-\gamma} = \frac{EVS}{ESV-1}$. As $\frac{\partial LV}{\partial ESV} = -\frac{1}{(ESV-1)^2}$ then, for a CES aggregator, love-for-variety decreases with the elasticity of substitution between varieties, and it is independent from the number of varieties.

Exercise 8.4 Prove all the previous results.

8.1.5.3 Intertemporal Consumption Theory

Consider instead consumption over time as a mapping $c : \mathrm{T} \to \mathrm{C} \subseteq \mathbb{R}_+$, where $c(t)$ is the consumed quantity of one good at time t. In this case, the support set is the ordered set $\mathrm{T} = [0, T) \subseteq \mathbb{R}_+$, where T can be finite or infinite. A consumption path over time is $\mathbf{c}_{\mathrm{T}} = (c(t))_{t\in\mathrm{T}}$, and an intertemporal utility functional provides a measure of the value of consumption paths. A benchmark functional is the time additive discounted intertemporal utility functional

$$\mathsf{U}[c] = \int_0^T u(c(t))\, e^{-\rho t} dt \tag{8.47}$$

where $\rho > 0$ is the rate of time preference, and the benchmark utility function is strictly increasing and concave, $u''(c) < 0 < u'(c)$. Therefore, the utility functional is a mapping $\mathsf{U} : \mathcal{C} \to \mathbb{R}$, where $\mathcal{C} = \{c : \mathrm{T} \to \mathrm{C} \subseteq \mathbb{R}_+ : c(t) > 0\}$. The variation of the utility for a general perturbation in the consumption path $\mathbf{c}_{\mathrm{T}}$, where $c(t) \to c(t) + \epsilon\, dc(t)$, for each $t \in \mathrm{T}$, is

$$\mathrm{D}_{dc(\cdot)}\mathsf{U}[c] = \int_0^T \frac{\delta \mathsf{U}[c]}{\delta c(t)}\, dc(t)\, dt$$

where the Gâteaux derivative is

$$\frac{\delta \mathsf{U}[c]}{\delta c(t)} = u'(c(t))\, e^{-\rho t}, \text{ for each } t \in \mathrm{T}.$$

[14] See Matsuyama and Ushchev (2023), for a recent reassessment.

The marginal utility for consumption at a specific time $t \in \mathrm{T}$, can, again, be modeled by a spike change in consumption at time τ, $dc(\tau) = \delta(t - \tau)dc(t)$,

$$\mathsf{U}_\tau \equiv \frac{\mathrm{D}_{dc(\tau)}\mathsf{U}[c]}{dc(\tau)} = \frac{\delta \mathsf{U}[c]}{\delta c(\tau)} = u'(c(\tau))\, e^{-\rho\tau}.$$

The **intertemporal marginal rate of substitution** between consumption at time t_0 and $t_1 > t_0$ is

$$IMRS_{t_0,t_1} = \frac{\mathsf{U}_{t_0}}{\mathsf{U}_{t_1}} = \frac{u'(c(t_0))}{u'(c(t_1))}\, e^{\rho\,(t_1 - t_0)}.$$

And the own and the crossed Allen-Uzawa elasticities are

$$\epsilon_{t_i,t_i} = -\frac{\mathsf{U}_{t_i,t_i} c(t_i)}{\mathsf{U}_{t_i}} = -\frac{u''(c(t_i))\, c(t_i)}{u'(c(t_i))},\quad \epsilon_{t_i,t_j} = -\frac{\mathsf{U}_{t_i,t_j} c(t_j)}{\mathsf{U}_{t_i}} = 0, \text{ for any, } t_i \neq t_j,$$

there is decreasing marginal utility, in the sense that an increase in consumption at time t_i decreases marginal intertemporal utility $\mathsf{U}[c]$, and that consumption in two different times are EP independent. This last property if a consequence of the linearity of the functional (8.47), as was showed for Eq. (8.18).

Assuming a stationary consumption path, such that $c(t) = \bar{c} > 0$, for all $t \in \mathrm{T}$, then we can show that the utility functional (8.47) displays: impatience, because $IMRS_{t_0,t_1} = e^{\beta(t_1 - t_0)} > 1$ if $t_1 > t_0$, and a constant elasticity of intertemporal substitution

$$EIS_{t_0,t_1} = \frac{d \ln\left(c(t_0)/c(t_1)\right)}{d \ln\left(MRS_{t_0,t_1}\right)} = -\frac{u''(\bar{c})\, \bar{c}}{u'(\bar{c})},\quad \text{for any } t_0 \neq t_1 \in \mathrm{T}.$$

For an isoelastic utility function (see (2.17)) $u(c) = \frac{c^{1-\theta}-1}{1-\theta}$ we have $EIS_{t_0,t_1} = \frac{1}{\theta}$.

Exercise 8.5 Prove all the previous results.

8.1.5.4 Intertemporal Choice of Consumption Baskets

We can combine the two previous cases, by considering the path of consumption baskets over time, $\mathbf{c}_{\mathrm{I,T}} = \left(c(i,t)\right)_{(i,t)\in \mathrm{I}\times\mathrm{T}}$, and the utility functional over paths of consumption baskets

$$\mathsf{U}[c] = \int_0^T u(\mathsf{V}[c(t)])\, e^{-\rho t}\, dt, \text{ where } \mathsf{V}[c(t)] = v^{-1}\Big(\int_0^N v(c(i,t))\, di\Big)$$

where $\mathsf{U}[c]$ is an intertemporally additive discounted utility functional, and $\mathsf{V}[c(t)]$ is a temporary variety aggregator. In this case, the Gâteaux differential for a consumption variation $c(i,t) \to c(i,t) + \epsilon\, dc(i,t)$ is

$$\mathrm{D}_{dc(\cdot)}\mathsf{U}[c] = \int_{\mathrm{T}}\int_{\mathrm{I}} \frac{\delta \mathsf{U}[c]}{\delta c(i,t)}\, dc(i,t)\, di\, dt.$$

where the Gâteaux derivative is $\frac{\delta \mathsf{U}[c]}{\delta c(i,t)}$. Again, introducing a Dirac-delta perturbation centered in a particular pair variety-time (i_0, t_0), $dc(i_0, t_0) = \delta(i - i_0, t - t_0)\, d(c(i,t))$, the Gâteaux differential is $\mathrm{D}_{\delta(i_0,t_0)}\mathsf{U}[c] = \int_{\mathrm{T}}\int_{\mathrm{I}} \frac{\delta \mathsf{U}[c]}{\delta c(i,t)} \delta(i - i_0, t - t_0)\, dc(i,t) di\, dt = \frac{\delta \mathsf{U}[c]}{\delta c(i_0,t_0)} dc(i_0, t_0)$, and, again, the marginal utility is represented by the Gâteaux derivative

$$\mathsf{U}_{(i_0,t_0)} = \frac{\mathrm{D}_{\delta(i_0,t_0)}\mathsf{U}[c]}{dc(i_0,t_0)} = \frac{\delta \mathsf{U}[c]}{\delta c(i_0,t_0)}, \quad \text{for } (i_0,t_0) \in \mathrm{I} \times \mathrm{T}.$$

We can obtain several types of marginal rates of substitution: inter-varieties, from variety i_0 to variety i_1 at time t_0; intertemporal, from time t_0 to time t_1 for variety i_0; or from both variety i_0 at time t_0 to variety t_1 at time t_1. For the most general case, we have

$$MRS_{(i_0,t_0),(i_1,t_1)} = \frac{\mathsf{U}_{(i_0,t_0)}}{\mathsf{U}_{(i_1,t_1)}} = \frac{\dfrac{\delta \mathsf{U}[c]}{\delta c(i_0,t_0)}}{\dfrac{\delta \mathsf{U}[c]}{\delta c(i_1,t_1)}}.$$

We can also find that the EP complementary/substitutability properties are different from the previous inter-variety- and intertemporal- only cases by using the second-order crossed generalized derivative

$$\mathsf{U}_{(i_0,t_0),(i_1,t_1)} = \frac{\delta^2 \mathsf{U}[c]}{\delta c(i_1,t_1)\delta c(i_0,t_0)}.$$

Example 8.5 Consider the benchmark case in which the utility function is a generalized logarithm (CRRA) $u(\cdot) = \ln_\theta(c)$ and the variety aggregator is DS, as in Eq. (8.46), with N varieties. Then, we have the utility functional

$$\mathsf{U}[c] = \int_0^T \frac{1}{1-\theta}\Big[\Big(\int_0^N c(i,t)^{1-\gamma} di\Big)^{\frac{1-\theta}{1-\gamma}} - 1\Big] e^{-\rho t}\, dt$$

that has the generalized first-order derivative

$$\mathsf{U}_{(i,t)} = e^{-\rho t}\mathsf{V}[c(t)]^{\gamma-\theta} c(i,t)^{-\gamma}, \quad \text{for each } (i,t) \in \mathrm{I} \times \mathrm{T}$$

and the generalized second-order derivatives

$$\mathsf{U}_{(i,t),(i,t')} = 0, \quad \text{for any } t \neq t' \in \mathrm{T}, \quad \text{for each } i \in \mathrm{I},$$

$$\mathsf{U}_{(i,t),(i',t')} = 0, \text{ for any } (i,t) \neq (i',t') \in \mathrm{I} \times \mathrm{T},$$

and

$$\mathsf{U}_{(i,t),(i',t)} = (\gamma - \theta)\, e^{-\rho t} \left(\frac{\mathsf{V}[c(t)]^{-\theta-1}}{c(i,t)} \frac{\mathsf{V}[c(t)]}{c(i',t)} \right)^{\gamma}, \ (i,t) \neq (i',t) \in \mathrm{I} \times \mathrm{T}.$$

We saw, in the static case, that varieties i and i' were EP complements, and, in the single variety intertemporal case, that consumption at time t and t' was EP independent. In this more general model, there is still EP independence when intertemporal choices between the same or different varieties are at stake. However, two varieties can only be temporary EP complements if $\gamma > \theta$. It is possible that they might be temporary EP substitutes if $\gamma < \theta$.

8.1.5.5 Consumption Choice Under Uncertainty

Consider a stochastic environment represented by a probability space $(\Omega, \mathbb{P})$, in which Ω is a continuum of events or states of nature, ω, and $\mathbb{P} = (p(\omega))_{\omega \in \Omega}$ is a probability measure, such that $0 < p(\omega) < 1$ and $\int_{\Omega} p(\omega) d\omega = 1$. The consumption of the household is, ex-ante, a random variable $C = \big(c(\omega)\big)_{\omega \in \Omega}$, in which $c(\omega) > 0$, is the consumption realization in state of nature ω. Let the value of every consumption realization be assessed by the Bernoulli utility function, $u(c)$ that is assumed to be strictly increasing and concave $u''(c) \leq 0 < u'(c)$.

It is assumed that the household has imperfect information: it knows *ex-ante* the consumption realizations and the probability for every state of nature, but does not know which state will materialize, and therefore, the *ex-post* realization of consumption. Therefore, as the only relevant decision variable for households, in an uncertainty environment, are contingent consumption profiles, C, the theory of choice under uncertainty addresses the choice among consumption profiles, and not isolated consumption realizations.

The benchmark theory of choice under utility[15] states that contingent consumption profiles, C, can be ranked by means of an expected utility functional, or **von-Neumann-Morgenstern utility functional**,

$$\mathsf{U}[c] = \mathbb{E}[u(C)] = \int_{\Omega} u(c(\omega))\, p(\omega)\, d\omega, \tag{8.48}$$

that is a linear functional over states of nature.

We call **certainty equivalent** to the certain consumption realization that has the same utility as the contingent consumption profile C,

$$\bar{c} = u^{-1}\Big(\mathbb{E}[u(C)]\Big). \tag{8.49}$$

[15] For the history of the theory of choice under uncertainty, see Szpiro (2020) or Moscati (2023).

This is a transformed functional, and can be conceived as providing an alternative valuation for contingent consumption profiles.

Observe that the average consumption is also a linear functional,

$$\mathbb{E}[C] = \mathsf{M}[c] = \int_{\Omega} c(\omega)\, p(\omega)\, d\omega.$$

We say there is **risk aversion** (neutrality) if the consumer is willing to pay a positive (zero) premium $\mathbb{E}[C] - \bar{c} > 0$ ($\mathbb{E}[C] - \bar{c} = 0$), in terms of diminished consumption, in order to insure against uncertainty. If $u(\cdot)$ is strictly concave then the Jensen's inequality implies $\mathbb{E}[u(C)] < u(\mathbb{E}[C])$, meaning that the utility functional that is specified with a concave utility function displays risk aversion. There is risk neutrality if the utility function is linear, implying $\mathbb{E}[u(C)] = u(\mathbb{E}[C])$

Assuming the expected utility functional in Eq. (8.48), the marginal utility for consumption in state ω is positive

$$\frac{\delta \mathbb{E}[u(C)]}{\delta c(\omega)} = u'(c(\omega))\, p(\omega) > 0.$$

The marginal rate of substitution for consumption between two states of nature ω and ω' is

$$MRS_{\omega,\omega'} = \frac{u'(c(\omega))\, p(\omega)}{u'(c(\omega'))\, p(\omega')}, \text{ for any } \omega \neq \omega' \in \Omega$$

weights both the consumption levels and the likelihoods of different states of nature. We find that if $c(\omega) = c(\omega')$ and $p(\omega) > p(\omega')$ then consumption in state ω is more valuable that at state ω', because likelihood of its realization is higher. The change in the marginal utility of consumption in the state of nature ω, $c(\omega)$, for variations in consumption in the same state

$$\frac{\delta^2 \mathbb{E}[u(C)]}{\delta c(\omega)^2} = u''(c(\omega))\, p(\omega) < 0, \text{ for each } \omega \in \Omega$$

displays decreasing marginal utility, as another implication of the presence of risk aversion. Furthermore, as a consequence of the linearity of the von-Neumann-Morgenstern utility functional, the change in the marginal utility of consumption in state ω for a change in consumption in any other state of nature ω', we have

$$\frac{\delta^2 \mathbb{E}[u(C)]}{\delta c(\omega)\, \delta c(\omega')} = 0, \text{ for each } \omega' \neq \omega \in \Omega,$$

means that consumption in any two different states of nature is EP independent.[16]

However, if the household assesses the value of contingent consumption profiles by their certainty equivalent, we can find some differences in its implicit behavior.

[16] This is a cardinal consequence of he independence axiom.

The marginal utility for consumption in state ω is proportional to the marginal utility for the vNM utility functional,

$$\frac{\delta \bar{c}}{\delta c(\omega)} = \frac{\partial \bar{c}}{\partial \mathbb{E}[u(C)]} \frac{\delta \mathbb{E}[u(c)]}{\delta c(\omega)} = \frac{\partial \bar{c}}{\partial \mathbb{E}[u(C)]} u'(c(\omega))\, p(\omega) > 0, \text{ for each } \omega \in \Omega, \tag{8.50}$$

because $\frac{\partial \bar{c}}{\partial \mathbb{E}[u(C)]} > 0$. The marginal rate of substitution between consumption in any two different states of nature is equal to the one for the vNM utility functional. The second-order own Gâteaux derivative for consumption in state ω $c(\omega)$ is, from Eq. (8.22),

$$\frac{\delta^2 \bar{c}}{\delta c(\omega)^2} = \frac{\partial^2 \bar{c}}{\partial \mathbb{E}[u(C)]^2} (u'(c(\omega))\, p(\omega))^2 + \frac{\partial \bar{c}}{\partial \mathbb{E}[u(C)]} u''(c(\omega))\, p(\omega) \gtreqless 0, \tag{8.51}$$

is ambiguous and has a sign that is possibly state-dependent, because the first term is positive, given that $\frac{\partial^2 \bar{c}}{\partial \mathbb{E}[u(C)]^2} > 0$ as $u''(c) < 0$, and the second term is negative. From Eq. (8.21), the second-order crossed Gâteaux derivative for consumption in state ω, for a change in consumption in state ω', is

$$\frac{\delta^2 \bar{c}}{\delta c(\omega)\, \delta c(\omega')} = \frac{\partial^2 \bar{c}}{\partial \mathbb{E}[u(C)]^2} u'(c(\omega))\, p(\omega)\, u'(c(\omega'))\, p(\omega') > 0, \text{ for } \omega \neq \omega' \in \Omega, \tag{8.52}$$

if there is risk aversion. Therefore, if there is risk aversion, the certainty equivalent criterium introduces two main changes as regards the vNM benchmark: this utility can be decreasing or increasing depending on the state of nature, it is decreasing independently of the state of nature, and realizations in two different states of nature are EP complementary, and not EP independent independent as in the vNM case.

Example 8.6 Assuming a CRRA Bernoulli utility function, $u(c) = \frac{c^{1-\varrho}-1}{1-\varrho}$, with a coefficient of relative risk aversion $\varrho > 0$ and a vNM utility functional

$$\mathbb{E}[u(C)] = \int_\Omega \frac{c(\omega)^{1-\varrho} - 1}{1-\varrho} p(\omega)\, d\omega,$$

then the marginal utility for consumption in state ω is $\frac{\delta \mathbb{E}[u(C)]}{\delta c(\omega)} = c(\omega)^{-\varrho}\, p(\omega) > 0$, and the second-order derivatives are $\frac{\delta^2 \mathbb{E}[u(c)]}{\delta c(\omega)^2} = -\theta\, c(\omega)^{-\varrho-1}\, p(\omega) < 0$, for each $\omega \in \Omega$, and $\frac{\delta^2 \mathbb{E}[u(c)]}{\delta c(\omega)\, \delta c(\omega')} = 0$. The certainty equivalent is

$$\bar{c} = \left(1 + (1-\varrho)\, \mathbb{E}[u(C)]\right)^{\frac{1}{1-\varrho}}.$$

For this utility functional, we have $\frac{\partial \bar{c}}{\partial \mathbb{E}[u(C)]} = \bar{c}^{\varrho}$ and $\frac{\partial^2 \bar{c}}{\partial \mathbb{E}[u(C)]^2} = \varrho\, \bar{c}^{2\,\varrho-1}$ that are both positive if $\bar{c} > 0$. Therefore, the marginal utility is positive for every state of nature

$$\frac{\delta \bar{c}}{\delta c(\omega)} = p(\omega)\left(\frac{\bar{c}}{c(\omega)}\right)^{\varrho} > 0, \text{ for each } \omega \in \Omega.$$

The second-order own Gâteaux derivative,

$$\frac{\delta^2 \bar{c}}{\delta c(\omega)^2} = \varrho \frac{p(\omega)}{c(\omega)}\left(\frac{\bar{c}}{c(\omega)}\right)^{\varrho}\left[p(\omega)\left(\frac{\bar{c}}{c(\omega)}\right)^{\varrho-1} - 1\right],$$

is negative (positive) if $p(\omega) < \pi(\omega)$ ($p(\omega) > \pi(\omega)$) where $\pi(\omega) \equiv \left(\frac{c(\omega)}{\bar{c}}\right)^{\varrho-1}$. The crossed derivative is always positive as proved before,

$$\frac{\delta^2 \bar{c}}{\delta c(\omega)\,\delta c(\omega')} = \varrho \frac{p(\omega)\,p(\omega')}{c(\omega)}\left(\frac{\bar{c}}{c(\omega)}\right)^{\varrho}\left(\frac{\bar{c}}{c(\omega')}\right)^{\varrho} > 0, \text{ for each } \omega \in \Omega.$$

We can show that, if $\varrho = 1$ (the utility function is logarithmic) that then the marginal utility is decreasing for consumption realizations in all states of nature. However, if $\varrho > 1$ they can be decreasing or increasing depending on the probability and the ratio $c(\omega)/\bar{c}$: for a state of nature with a low (high) probability or a high (low) relative realization as regards the certainty equivalent then the marginal utility is decreasing (increasing). Therefore, the marginal utility for an increase in the realization of consumption at a particular state of nature is decreasing if the own decreasing marginal utility effect dominates the distribution-wide effect, via the certainty equivalent.

8.1.5.6 Intertemporal Consumption Decision Under Uncertainty

We can combine time and state of nature heterogeneities. Consider a stochastic environment represented a continuum of states Ω and by a flow of probabilities $\left(\mathbb{P}(t)\right)_{t\in \mathrm{T}}$ where the probability measure at any time $\mathbb{P}(t) = (p(t,\omega))_{\omega\in\Omega}$, satisfies $0 < p(t,\omega) < 1$ and $\int_{\Omega} p(t,\omega)d\omega = 1$, for every $t \in \mathrm{T} = \mathbb{R}_{+}$. The consumption of the household is a stochastic process $\mathbf{C}_{\mathrm{T}} = \left(C(t)\right)_{t\in\mathrm{T}} = \left(c(t,\omega)\right)_{(t,\omega)\in\Omega\times\mathrm{T}}$, in which the consumption realization in state of nature ω at time t is positive $c(t,\omega) > 0$.

There are two main discounted intertemporal additive utility functionals. They can be distinguished by the uncertainty aggregation assumption: the benchmark functional aggregates temporary expected utilities of consumption, and the second aggregates certainty equivalents.

The first is the **intertemporal von-Neumann-Morgenstern** utility functional[17]

$$\mathsf{U}[c] = \int_{\mathrm{T}} \mathbb{E}[u(C(t))]\, e^{-\rho} dt = \int_{\mathrm{T}\times\Omega} u(c(t,\omega))\, p(t,\omega)\, e^{-\rho}\, d\omega\, dt, \tag{8.53}$$

[17] This utility functional is the benchmark in the finance literature starting with Merton (1971, 1973).

where $u''(\cdot) < 0 < u'(\cdot)$ and $\rho > 0$. This utility functional is jointly linear in both time and state dimensions. The Gâteaux marginal utility is positive for every state of nature and time

$$\frac{\delta \mathsf{U}[c]}{\delta c(t,\omega)} = u'(c(t,\omega))\, p(t,\omega)\, e^{-\rho t} > 0, \text{ for each } (t,\omega) \in \mathrm{T} \times \Omega.$$

The second-order generalized derivatives are

$$\frac{\delta^2 \mathsf{U}[c]}{\delta c(t,\omega)^2} = u''(c(t,\omega))\, p(t,\omega)\, e^{-\rho t} < 0, \quad \text{for any } (t,\omega) \in \mathrm{T} \times \Omega$$

and

$$\frac{\delta^2 \mathsf{U}[c]}{\delta c(t,\omega)\delta c(t',\omega)} = \frac{\delta^2 \mathsf{U}[c]}{\delta c(t,\omega)\delta c(t,\omega')} = 0, \text{ for any } (t,\omega) \neq (t',\omega') \in \mathrm{T} \times \Omega,$$

display the same properties, for each time, as the static counterpart in Sect. 8.1.5.5: functional (8.53) displays decreasing own marginal utility, EP independence across states of nature, and over time. This is a consequence of the joint linearity.

The second intertemporal utility functional we consider is a version of the Epstein and Zin (1989) (EZ) utility functional, in which utility functional is the present value of the path of utilities of certainty equivalents over time $\bar{c}_{\mathrm{T}} = \big(\bar{c}(t)\big)_{t\in\mathrm{T}}$,

$$\mathsf{U}[c] = \int_{\mathrm{T}} u(\bar{c}(t))\, e^{-\rho}\, dt, \tag{8.54}$$

where the utility function $u(\cdot)$ is increasing and concave, $u''(c) < 0 < u'(c)$, as a function of the temporary certainty equivalent.

If we assume the temporary utility aggregator, at each time $t \in \mathrm{T}$, is von-Neumann-Morgenstern functional

$$\mathbb{E}[v(C(t))] = \int_{\Omega} v(c(t,\omega))\, p(t,\omega)\, d\omega$$

where the Bernoulli utility function is again strictly increasing and concave $v''(c) < 0 < v'(c)$. The temporary certainty equivalent is

$$\bar{c}(t) = v^{-1}\Big(\mathbb{E}[v(C(t))]\Big) = v^{-1}\Big(\int_{\Omega} v(c(t,\omega)) p(t,\omega)\, d\omega\Big),$$

it satisfies, by Jensen's inequality, $\bar{c}(t) < \mathbb{E}[C(t)] = \int_{\Omega} c(t,\omega) p(t,\omega)\, d\omega$, where the concavity of $v(\cdot)$ incorporates an implicit measure of risk aversion. Therefore, the certainty equivalent is a non-linear functional. Therefore, the functional (8.54) can be expanded to

$$\begin{aligned} \mathsf{U}[c] &= \int_{\mathrm{T}} u\Big(v^{-1}\Big(\mathbb{E}[v(C(t))]\Big)\Big)\, e^{-\rho t} dt = \\ &= \int_{\mathrm{T}} u\Big(v^{-1}\Big(\int_{\Omega} v(c(t,\omega))p(t,\omega)\, d\omega\Big)\Big)\, e^{-\rho}\, dt. \end{aligned} \tag{8.55}$$

The Gâteaux derivative for the temporary certainty equivalent of a temporary realization of consumption in state of nature ω, at time t, is also positive,

$$\frac{\delta \bar{c}(t)}{\delta c(t,\omega)} = \frac{\partial \bar{c}(t)}{\partial \mathbb{E}[v(C(t))]} v'(c(t,\omega))\, p(t,\omega) > 0, \text{ for each } \omega \in \Omega.$$

which implies that the intertemporal marginal utility for a spike change in consumption at time t and for state of nature ω associated to the whole process $\mathbf{C}_{\mathrm{T}}$, is also positive

$$\begin{aligned} \frac{\delta \mathsf{U}[c]}{\delta c(t,\omega)} &= u'(\bar{c}(t))\, \frac{\delta \bar{c}(t)}{\delta c(t,\omega)} e^{-\rho t} = \\ &= u'(\bar{c}(t))\, \frac{\partial \bar{c}(t)}{\partial \mathbb{E}[v(C(t))]} v'(c(t,\omega))\, p(t,\omega)\, e^{-\rho t}, \text{ for each } (t,\omega) \in \Omega \times \mathrm{T}. \end{aligned}$$

The second own generalized derivative,

$$\begin{aligned} \frac{\delta^2 \mathsf{U}[c]}{\delta c(t,\omega)^2} &= e^{-\rho t}\, p(t,\omega) \left(u''(\bar{c}(t))\, v'(c(t,\omega)) \left(\frac{\partial \bar{c}(t)}{\partial \mathbb{E}[v(C(t))]} \right)^2 + \right. \\ &\left. + u'(\bar{c}(t)) \left(v'(c(t,\omega)) \frac{\partial^2 \bar{c}(t)}{\partial \mathbb{E}[v(C(t))]^2} + v''(c(t,\omega)) \frac{\partial \bar{c}(t)}{\partial \mathbb{E}[v(C(t))]} \right) \right) \end{aligned} \tag{8.56}$$

can have, again, any sign. Although the first component, related to an intertemporal effect, is negative, there is a second component, related to a "static" effect already found in Eq. (8.51), that is ambiguous.

There are two crossed second-order effects, one related with time and another related to the states of nature. The second crossed generalized derivatives regarding time displays EP independence,

$$\frac{\delta^2 \mathsf{U}[c]}{\delta c(t,\omega)\delta c(t',\omega)} = 0, \text{ for each } t \neq t' \in \mathrm{T}, \text{ for each} \omega \in \Omega \tag{8.57}$$

because the utility functional is time-additive. The second crossed generalized derivative regarding states of nature

$$\frac{\delta^2 \mathsf{U}[c]}{\delta c(t,\omega)\delta c(t,\omega')} = e^{-\rho t}\, p(t,\omega)\left(u''(\bar{c}(t))\Big(\frac{\partial \bar{c}(t)}{\partial \mathbb{E}[v(C(t))]}\Big)^2 + \right.$$
$$\left. + u'(\bar{c}(t))\frac{\partial^2 \bar{c}(t)}{\partial \mathbb{E}[v(C(t))]^2}\right) v'(c(t,\omega))\, v'(c(t,\omega')), \quad (8.58)$$

does not display again EP independence in general, although the static utility functional for choices over states of nature, (8.48), is EP independent. If both utility functions $u(\cdot)$ and $v(\cdot)$ are concave, then we will have EP complementarity, meaning that an increase in the consumption in any state of nature will increase the certainty equivalent level of consumption. This is a consequence of the fact that taking the certainty equivalent as a measure of the temporary consumption entails a form of non-linearity. Therefore, if functions $v(\cdot)$ and $u(\cdot)$ are monotonic, and $v(\cdot) = u(\cdot)$ then utility functional (8.53) is a special case of utility functional (8.55).

Example 8.7 Assume constant EIS and CRRA utility functions $u(c) = \frac{c^{1-\theta}-1}{1-\theta}$ and $v(c) = \frac{c^{1-\varrho}-1}{1-\varrho}$, where $\theta > 0$ and $\varrho > 0$. The temporary generalized derivative for the certainty equivalent at time t as regards realization $c(t,\omega)$ is positive

$$\frac{\delta \bar{c}(t)}{\delta c(t,\omega)} = p(t,\omega)\Big(\frac{\bar{c}(t)}{c(t,\omega)}\Big)^{\varrho} > 0, \ \text{for each } (t,\omega) \in \mathrm{T}\times\Omega.$$

The temporary second-order own generalized derivative,

$$\frac{\delta^2 \bar{c}(t)}{\delta c(t,\omega)^2} = \varrho\,\frac{p(t,\omega)}{c(t,\omega)}\Big(\frac{\bar{c}(t)}{c(t,\omega)}\Big)^{\varrho}\left(p(t,\omega)\Big(\frac{\bar{c}(t)}{c(t,\omega)}\Big)^{\varrho-1} - 1\right),$$

is negative (positive) $p(t,\omega) < \pi(t,\omega)$ $(p(t,\omega) > \pi(t,\omega))$, where $\pi(t,\omega) \equiv \left(\frac{\bar{c}(t)}{c(t,\omega)}\right)^{1-\varrho}$, and the crossed derivative

$$\frac{\delta^2 \bar{c}(t)}{\delta c(t,\omega)\,\delta c(t,\omega')} = \varrho\, p(t,\omega)p(t,\omega')\Big(\frac{\bar{c}(t)}{c(t,\omega)}\,\frac{\bar{c}(t)}{c(t,\omega')}\Big)^{\varrho}\frac{1}{\bar{c}(t)} > 0,$$

for $\omega \neq \omega' \in \Omega$ and for each $t \in \mathrm{T}$, displays EP complementarity. Therefore, an increase in the consumption realization in state of nature ω at time t increases intertemporal welfare

$$\frac{\delta \mathsf{U}[c]}{\delta c(t,\omega)} = e^{-\rho t}\,\bar{c}^{-\theta}\,\frac{\delta \bar{c}(t)}{\delta c(t,\omega)} > 0.$$

But this intertemporal marginal utility is decreasing in $c(t,\omega)$, because,

$$\frac{\delta^2 \mathsf{U}[c]}{\delta c(t,\omega)^2} = -e^{-\rho t}\,\bar{c}^{-\theta}\left(\frac{1}{\bar{c}(t)}\Big(\frac{\delta \bar{c}(t)}{\delta c(t,\omega)}\Big)^2 - \frac{\delta^2 \bar{c}(t)}{\delta c(t,\omega)^2}\right) < 0,$$

there is intertemporally EP complementary in the consumption realization at any other state of nature, $\omega' \neq \omega$, at the same time because,

$$\frac{\delta^2 \mathsf{U}[c]}{\delta c(t,\omega)\,\delta c(t,\omega')} = e^{-\rho t}\,\bar{c}^{-\theta}\,\frac{\delta^2 \bar{c}(t)}{\delta c(t,\omega)\,\delta c(t,\omega')} > 0,$$

and, as we already saw, there is EP intertemporal independence between $c(t,\omega)$ and $c(t',\omega)$ or $c(t',\omega')$.

8.1.5.7 Infinite-Dimensional Production Functions

Functionals also allow for dealing with technologies of production using a continuum of inputs, that could be intermediate goods or heterogeneous labor or capital inputs. In the first case, the support set is usually non-ordered, but in the two last cases, it can be ordered by skill or productivity levels.

Let $x(i)$, the quantity of input of variety $i \in \mathrm{I}$, be a mapping $x : \mathrm{I} \to \mathrm{X} \subseteq \mathbb{R}_+$. Inputs belong to the function space $\mathcal{X} = \{x : \mathrm{I} \to \mathbb{R}_+\}$. A distribution of the quantities of all varieties is denoted by $\mathbf{x}_\mathrm{I} = \big(x(i)\big)_{i\in\mathrm{I}}$. Again, the quantity of input of a particular variety $\iota \in \mathrm{I}$, $x(\iota)$, can be obtained as $x(\iota) = \int_\mathrm{I} \delta(i-\iota)\,x(i)\,di$.

A production functional $\mathsf{F}[x]$ maps the distribution of inputs from all varieties into real production, a non-negative number, $\mathsf{F} : \mathcal{X} \to \mathbb{R}_+$. The level of production using a continuum of varieties is set to be $y = \mathsf{F}[x]$. There are several production functionals in the literature[18]: (1) the generalized AK production function $\mathsf{F}[x] = \int_\mathrm{I} A(i)\,x(i)\,di$, in which $A(i)$ is the specific marginal productivity of input $i \in \mathrm{I}$; (2) the generalized CES production function

$$\mathsf{F}[x] = \Big(\int_\mathrm{I} A(i)\,x(i)^{\frac{\epsilon-1}{\epsilon}}\,di\Big)^{\frac{\epsilon}{\epsilon-1}} \tag{8.59}$$

which is the infinite-dimensional analog of $F(\{x\}) = \Big(\sum_{i\in\mathrm{I}} A_i\,x_i^{\frac{\epsilon-1}{\epsilon}}\Big)^{\frac{\epsilon}{\epsilon-1}}$, for $\mathrm{I} = \{i_1,\ldots,i_n\}$; (3) the O-ring production functional (see Kremer, 1993) $\mathsf{F}[x] = \exp\Big(\int_\mathrm{I} \gamma(i)\log\,(x(i))\,di\Big)$ where $\int_\mathrm{I} \gamma(i) = 1$. This generalized function is the infinite-dimensional analog to the finite-dimensional geometric average $F(\{x\}) = \prod_{i\in\mathrm{I}} x_i^{\gamma_i}$ for $\sum_\mathrm{I} \gamma_i = 1$, which makes the O-ring production functional an infinite-dimensional analog to the Cobb-Douglas production function; or (4) the translog production functional $\mathrm{F}[x] = \exp\big(\int_\mathrm{I} \gamma(i)\log\,(x(i)) + \gamma(i)\log\,(x(i)^2)\,di\big)$.

For the CES production functional (8.59), we have the marginal productivity of a particular input $i \in \mathrm{I}$,

$$\frac{\delta \mathrm{F}[x]}{\partial x(i)} = A(i)\left(\frac{\mathrm{F}[x]}{x(i)}\right)^{\frac{1}{\varepsilon}} = A(i)\left(\frac{y}{x(i)}\right)^{\frac{1}{\varepsilon}},\ \text{for each } i \in \mathrm{I},$$

[18] See, for instance, Parenti et al. (2017) and Bucci and Ushchev (2016).

which implies that the marginal rate of transformation between inputs ι and ι' is

$$MRT_{\iota,\iota'} = \frac{\dfrac{\delta \mathsf{F}[x]}{\partial x(\iota)}}{\dfrac{\delta \mathsf{F}[x]}{\partial x(\iota')}} = \frac{A(\iota)}{A(\iota')}\left(\frac{x(\iota')}{x(\iota)}\right)^{\frac{1}{\varepsilon}}, \text{ for any } \iota, \iota' \in \mathrm{I}.$$

The (generalized) Allen-Uzawa elasticities are, for variety ι is

$$\varepsilon_{\iota,\iota} = \frac{1}{\varepsilon}\left(1 - A(i)\left(\frac{y}{x(\iota)}\right)^{\frac{1-\varepsilon}{\varepsilon}}\right), \text{ for any } \iota \in \mathrm{I}$$

and for varieties ι and ι' is

$$\varepsilon_{\iota,\iota'} = -\frac{A(\iota')}{\varepsilon}\left(\frac{y}{x(\iota')}\right)^{\frac{1-\varepsilon}{\varepsilon}} \text{ for any } \iota \neq \iota' \in \mathrm{I},$$

meaning that varieties ι and ι' are (generalized) EP complements. The elasticity of variety substitution is constant $EVS_{\iota,\iota'} = \varepsilon$.index[subj]Production functionals!properties!elasticity of variety substitution Input variety expansion can have an effect on production. If we assume $\mathrm{I} = [0, N]$, and we measure the effect of variety expansion by the elasticity of output as regards the number of varieties, calling it elasticity of variety expansion,

$$EVE = \frac{\dfrac{\partial \mathsf{F}[x]}{\partial N} N}{\mathsf{F}[x]} = \frac{\epsilon\, A(N)\, N}{\epsilon - 1}\left(\frac{y}{x(N)}\right)^{\frac{1-\epsilon}{\epsilon}},$$

we find that, if an increase in the elasticity of variety substitution satisfies $\epsilon > 1 (\epsilon < 1)$, output increases as $EVE > 0$ (reduces as $EVE < 0$),

Exercise 8.6 Prove the previous results.

Exercise 8.7 Find the MRT, the Allen-Uzawa elasticities, and the elasticity of variety substitution for the other production functionals.

8.1.5.8 Inequality Measures

Let $y \in \mathrm{Y} \subseteq \mathbb{R}_+$ be the income of a household, and let households be ordered by income. Let $F(y)$ be the proportion of the households having an income smaller than y. The distribution of income is a mapping $F : \mathrm{Y} \to [0, 1]$, and, if it is differentiable, $f(y) = F'(y)$ is the density function. The distribution of households across all levels of income is thus $\mathbf{F}_{\mathrm{Y}} = \big(F(y)\big)_{y \in \mathrm{Y}}$. Social welfare literature addresses the issue of measuring inequality, for comparing different income distributions, by a **social welfare functional** $\mathsf{W}[F]$.

There are two main approaches to measuring economic inequality: the ethical and the axiomatic approaches. The ethical approach starts from the principle of transfer (attributed to Dalton, 1920) and measures inequality by means of a social welfare function. The most well-known is the KAS social welfare function (from Kolm, 1969, Atkinson, 1970 and Sen, 1973) which ranks income distributions F by a linear social welfare functional

$$\mathsf{W}[F] = \int_{\mathrm{Y}} W(y)\, dF(y) \tag{8.60}$$

where the welfare function is increasing and concave $W''(\cdot) < 0 < W'(\cdot)$. The axiomatic approach is based upon setting some general properties that an inequality measurement should satisfy. Assuming anonimity, continuity, differentiability, aggregative decomposition, and scale indifference, several authors[19] proved that the unique inequality measure satisfying those properties is of the generalized entropy functional[20]

$$GE[f] = \frac{1}{c\,(1-c)} \int_{\mathrm{Y}} \left(\left(\frac{y}{\mathbb{E}[y]} \right)^{c} - 1 \right) f(y)\, dy.$$

If $c = 1$ or $c = 0$ we have the Theil T or the Theil L indices, and $\mathbb{E}[y] = \int_{\mathrm{Y}} y\, f(y)\, dy$.

Let us consider the functional (8.60). It has been shown (by Kolm, 1969, Atkinson, 1970, Sen, 1973, and also by Dasgupta et al., 1973) that $\mathsf{W}[F]$ displays inequality aversion if and only the social welfare function $W(y)$ is concave. If the social welfare function is concave then $\mathbf{F}^1$ is better than $\mathbf{F}^2$ if and only if $F^1(y) \le F^2(y)$ for every $y \in \mathbb{R}_+$. This is similar to assuming first-order stochastic dominance.[21] However, if we cannot establish first-order stochastic dominance, which occurs when Lorenz curves intersect, we need to introduce a stronger criterium beyond concavity. Defining, the equality distributed income equivalent by $y^{ede} = W^{-1}\Big(\int_{\mathrm{Y}} W(y)\, dF(y) \Big)$, and the mean income by $m_1 = \int_{\mathrm{Y}} dF(y)$, the Kolm-Atkinson-Sen (KAS) **relative inequality** index $I(F) = 1 - y^{ede}/m_1$ is a benchmark measure of income inequality, where $I \in (0, 1)$. A higher I is associated to a higher measured level of income inequality. The KAS social welfare function is

$$W(y) = \frac{y^{1-\xi} - 1}{1 - \xi}, \tag{8.61}$$

where $\xi > 0$ is the coefficient **relative social income inequality aversion.**[22] The measure of social welfare associated to income distribution $\mathbf{F} = (F(y))_{y \in \mathrm{Y}}$, in the presence of inequality aversion and the KAS social welfare functional is,

---

[19] For the discrete case see Bourguignon 1979, Cowell, 1980, Shorrocks, 1980; 1984.

[20] Do not confuse with the Shannon (1948) entropy.

[21] See Müller and Stoyan (2002, p.16).

[22] This welfare function has been introduced in Atkinson (1970). Absolute inequality aversion can be formalized by $W(y) = -\frac{e^{-\lambda y}}{\lambda}$.

$$\mathsf{W}[F;\xi] = \int_{y\in \mathrm{Y}} W(y)\,dF(y) = \frac{m_{1-\xi}-1}{1-\xi}, \tag{8.62}$$

where $m_{1-\xi} = \int_{\mathrm{Y}} y^{1-\xi}\,dF(y)$ is the $1-\xi$ moment associated to distribution $\mathbf{F}$. In this case, $y^{ede} = (m_{1-\xi})^{\frac{1}{1-\xi}}$ and $I(F) = 1-(m_{1-\xi})^{\frac{1}{1-\xi}}/m_1$. As $W(y)$ is strictly concave then the functional $\mathsf{W}[F;\xi]$ introduces a complete ordering on the space of income distributions. We can also say that distribution $\mathbf{F}^1$ is better than distribution $\mathbf{F}^2$ if and only if $\mathsf{W}[F^1;\xi] > \mathsf{W}[F^2;\xi]$, for a given inequality aversion parameter ξ.

Exercise 8.8 Find the Gâteaux derivatives for the KAS social welfare functional.

8.1.5.9 In Information Theory and Econometrics

Let $X \in \mathrm{X} \subseteq \mathbb{R}$ be a random variable, and let the probability of realization of x to be $f(x) = \mathbb{P}[X = x]$. The probability density function is a mapping $f : \mathrm{X} \to (0,1)$ such that $\int_{\mathrm{X}} f(x)\,dx = 1$. **Entropy** is a measure of the degree of information conveyed by variable X. The higher the entropy is the "more random" variable X is. Shannon (1948) introduces a measure of information content in the density f by the entropy functional

$$\mathrm{H}[f] = -\int_{\mathrm{X}} f(x)\ln(f(x))dx. \tag{8.63}$$

The higher $\mathrm{H}[f]$ is the less informative density f is.[23] To see this, we can compute the first-order functional derivative,

$$\frac{\delta\mathsf{H}[f]}{\delta f(x)} = -(1+\ln(f(x)), \text{ for each } x \in \mathrm{X},$$

and the second-order own functional derivative

$$\frac{\delta^2\mathsf{H}[f]}{\delta f(x)^2} = -\frac{f'(x)}{f(x))} \text{ for each } x \in \mathrm{X}.$$

Remark: if a given realization x has density $f(x) \to 0^+$ then $\dfrac{\delta\mathsf{H}[f]}{\delta f(x)} \to \infty$. This means that a rare event is highly informative.[24]

[23] Observe that $\mathrm{H}[f]$ is a positive number because $\ln(f(x)) < 0$ as $f(x) \in (0,1)$ for every $x \in \mathrm{X}$.

[24] For an entropy-based econometric approach, see Judge and Mittelhammer (2012),

8.2 Problems Involving Functionals

Several (static) problems in economics, and in other sciences, are formalized as maximization of functionals with or without constraints. The first-order conditions of those problems involve solving functional equations, as will be seen in Sect. 8.2.1. In Sect. 8.2.2, several applications are presented.

8.2.1 Optimization of Functionals

The function $y^* \in \mathcal{Y}$ is an **extremum** for functional $\mathsf{F}[y]$ if the associated value of functional $\mathsf{F}[y^*]$ does not change for any small arbitrary perturbation $\eta \in \mathcal{Y}$, such that $y = y^* + \varepsilon\,\eta \in \mathcal{Y}$. Therefore, y^* is an extremum only if

$$D_{\eta(\cdot)}\mathsf{F}[y^*] = 0, \text{ for } y^* \in \mathcal{Y}.$$

An extremum $y^* \in \mathcal{Y}$ is a **maximum**, only if, for an arbitrary perturbation η, such that $y = y^* + \varepsilon\,\eta$, we have $\mathsf{F}[y^*] \geq \mathsf{F}[y]$, and it is a minimum only if $\mathsf{F}[y^*] \leq \mathsf{F}[y]$. From the generalized Taylor expansion in Eq. (8.1.1) a necessary condition of second order for a maximum is $D^2_{\eta(\cdot)}\mathsf{F}[y^*] \leq 0$, for $y^* \in \mathcal{Y}$.

Next, we consider three different problems involving maximization of a functional without constraints, with a functional constraint, and with a point-wise constraint. We assume throughout that the functional exists.

The **unconstrained maximization problem** is: find function $y \in \mathcal{Y}$, mapping $y : \mathrm{X} \to \mathrm{Y} \subseteq \mathbb{R}$, that solves the problem

$$\max_{y(\cdot)} \mathsf{F}[y]. \tag{8.64}$$

Proposition 8.1 (Necessary optimality condition for problem (8.64)) *Let function* $y^* : \mathrm{X} \to \mathbb{R}$ *solve problem* (8.64). *Then it satisfies the condition*

$$\frac{\delta \mathsf{F}[y^*]}{\delta y(x)} = \left.\frac{\delta \mathsf{F}[y]}{\delta y(x)}\right|_{y(x)=y^*(x)} = 0, \text{ for each } x \in \mathrm{X}. \tag{8.65}$$

Proof Assume we know the maximum y^*. The objective functional evaluated at the optimum is $\mathsf{F}[y^*]$, which is a number. Introducing an arbitrary admissible perturbation $\eta \in \mathcal{Y}$ at the optimum, such that $y(x) = y^*(x) + \varepsilon\,\eta(x)$ for every $x \in \mathrm{X}$, yields the functional $\mathsf{F}[y] = \mathsf{F}[y^* + \varepsilon\,\eta]$. If y^* is an optimum then we should have $\mathsf{F}[y^*] \geq \mathsf{F}[y]$ for any $y \in \mathcal{Y}$. Expanding $\mathsf{F}[y]$ in a neighborhood of the optimum we have

$$\mathsf{F}[y] = \mathsf{F}[y^*] + \mathrm{D}_{\eta(\cdot)}\mathsf{F}[y^*]\,\varepsilon + O(\varepsilon^2)$$

Therefore, we can have $\mathsf{F}[y^*] \geq \mathsf{F}[y]$ only if $\mathrm{D}_{\eta(\cdot)}\mathsf{F}[y^*] = 0$ for an arbitrary perturbation. As $\mathrm{D}_{\eta(\cdot)}\mathsf{F}[y^*] = \int_{\mathrm{X}} \frac{\delta \mathbf{F}[y^*}{\delta y(x)} \eta(x)\, dx = 0$ then y^* is a maximum if and only if condition (8.65) is satisfied. □

Proposition 8.2 (Sufficient optimality condition for problem (8.64)) *Assume that: first, the functional is concave in the sense that it satisfies the condition*

$$\frac{\delta^2 \mathsf{F}[y]}{\delta y(x)^2} \leq 0, \textit{ for any } y \in \mathcal{Y},$$

and if there a function $y^* \in \mathcal{Y}$ *such that* $\frac{\delta \mathsf{F}[y^*]}{\delta y(x)} = 0$ *then* $y^* \in \mathcal{Y}$ *is a solution to Problem* (8.64).

Proof A Taylor expansion of the functional $\mathsf{F}[y]$ at a function $y^* \in \mathcal{Y}$, for an arbitrary admissible perturbation $\eta(x) \in \mathcal{Y}$ up to the second order yields

$$\mathsf{F}[y] = \mathsf{F}[y^*] + \mathrm{D}_{\eta(\cdot)}\mathsf{F}[y^*]\,\varepsilon + \mathrm{D}_{\eta(\cdot)\eta(\cdot)}\mathsf{F}[y^*]\,\varepsilon^2 + O(\varepsilon^3).$$

If $\mathrm{D}_{\eta(\cdot)\eta(\cdot)}\mathsf{F}[y] \leq 0$ for any $y \in \mathcal{Y}$ and $\mathrm{D}_{\eta(\cdot)}\mathsf{F}[y] = 0$ if $y = y^*$ then $\mathsf{F}[y^*] \geq \mathsf{F}[y]$ and y^* is an optimum. But $\mathrm{D}_{\eta(\cdot)^2}\mathsf{F}[y] \leq 0$ if and only if $\frac{\delta^2 \mathsf{F}[y]}{\delta y(x)^2} \leq 0$. □

Example 8.8 Consider the linear functional (8.1) and assume $y^*(x)$ is known. For any admissible perturbation $y(x) = y^*(x) + \varepsilon\eta(x)$ we have the variation $\mathsf{F}[y] - \mathsf{F}[y^*] = \int_{\mathrm{X}} f\big(y^*(x) + \varepsilon\eta(x)\big) - f\big(y^*(x)\big)\, dx$. Therefore, the Gâteaux differential is $\mathrm{D}_{\eta(\cdot)}\mathsf{F}[y^*] = \int_{\mathrm{X}} f'(y^*(x))\, \eta(x)\, dx$ which implies that, for any admissible perturbation, $\eta(\cdot)$ the first-order conditions for a maximum is

$$f'(y^*(x), x) = 0 \text{ for each } x \in \mathrm{X}.$$

Exercise 8.9 Prove that the maximum for the transformed functional (8.2) is

$$g'(\mathsf{F}[y^*])\, f'(y^*(x), x) = 0 \text{ for each } x \in \mathrm{X}.$$

The **maximization problem with a functional constraint** is: find function $y \in \mathcal{Y}$, where $y : \mathrm{X} \to \mathbb{R}$, that maximizes a functional subject to a functional constraint, involving the whole distribution of y:

$$\max_{y(\cdot)} \{\mathsf{F}[y] : \mathsf{G}[y] = 0\}. \tag{8.66}$$

Proposition 8.3 (Necessary optimality conditions for problem (8.66)) *Let function* $y^* : \mathrm{X} \to \mathbb{R}$ *solve problem* (8.66). *Then there is a number* $\lambda \in \mathbb{R}$ *such that the following conditions are satisfied*

$$\frac{\delta \mathsf{F}[y^*]}{\delta y(x)} = \lambda \frac{\delta \mathsf{G}[y^*]}{\delta y(x)}, \textit{ for each } x \in \mathrm{X}. \tag{8.67a}$$

$$\mathsf{G}[y^*] = 0. \tag{8.67b}$$

where the first condition involves Gâteaux derivatives, and the second is a functional equation.

Proof We define a generalized Lagrange functional, parameterized by the Lagrange multiplier $\lambda \in \mathbb{R}$, $\mathsf{L}[y] = \mathsf{F}[y] - \lambda\,\mathsf{G}[y]$. At the optimum, (y^*, λ^*), we have $\mathsf{L}[y^*] = \mathsf{F}[y^*]$, because $\mathsf{G}[y^*] = 0$. Introducing admissible perturbations on $y^*(x)$ and λ^*, $y(x) = y^*(x) + \varepsilon\,\eta(x) \in \mathcal{Y}$ and $\lambda = \lambda^* + \varepsilon\,\psi$, and making a Taylor expansion in the neighborhood of the optimum, we have

$$\mathsf{L}[y] = \mathsf{L}[y^*] + \mathrm{D}_{\eta(\cdot)}\mathsf{L}[y^*]\,\varepsilon + \frac{\partial \mathsf{L}[y^*]}{\partial \lambda}\,\psi\,\varepsilon + O(\varepsilon^2),$$

where $\mathrm{D}_{\eta(\cdot)}\mathsf{L}[y^*] = \mathrm{D}_{\eta(\cdot)}\mathsf{F}[y^*] - \lambda^*\,\mathrm{D}_{\eta(\cdot)}\mathsf{G}[y^*]$, and $\dfrac{\partial \mathsf{L}[y^*]}{\partial \lambda} = \mathsf{G}[y^*]$. Function y^* is an optimum only if $\mathsf{L}[y^*](\lambda^*) \geq \mathsf{L}[y](\lambda)$, that only holds if $\mathrm{D}_{\eta(\cdot)}\mathsf{L}[y^*] = 0$ and $\dfrac{\partial \mathsf{L}[y^*]}{\partial \lambda} = 0$. As those conditions should hold in the optimum for any arbitrary admissible perturbations, we obtain conditions (8.67b) and (8.69a). □

Exercise 8.10 Assume that both functionals in Problem (8.66) are linear functionals, as in

$$\max_{y(\cdot)} \int_X f(y(x), x)\,dx, \tag{8.68a}$$

$$\text{subject to} \int_X g(y(x), x)dx = 0.$$

From Proposition 8.3, we obtain the necessary conditions for an optimum

$$\frac{\partial f(y^*(x), x)}{\partial y(x)} = \lambda\,\frac{\partial g(y^*(x), x)}{\partial y(x)}, \text{ for each } x \in X, \tag{8.69a}$$

$$\int_X g(y^*(x), x)dx = 0, \tag{8.69b}$$

Exercise 8.11 Assume that both functionals in Problem (8.66) are transformed functionals of type 1 (8.2)

$$\max_{y(\cdot)} h\Big(\int_X f(y(x), x)\,dx\Big), \tag{8.70a}$$

$$\text{subject to } l\Big(\int_X g(y(x), x)dx\Big) = 0.$$

Show that the necessary conditions for an optimum are

$$h'(\mathsf{F}[y^*])\frac{\partial f(y^*(x), x)}{\partial y(x)} = \lambda\,l'(\mathsf{G}[y^*])\,\frac{\partial g(y^*(x), x)}{\partial y}, \text{ for each } x \in X, \tag{8.71a}$$

$$l\Big(\int_X g(y(x), x)dx\Big) = 0. \tag{8.71b}$$

where $\mathsf{F}[y^*] = \int_{\mathsf{X}} f(y^*(x), x)\, dx$ and $\mathsf{G}[y^*] = \int_{\mathsf{X}} g(y^*(x), x)\, dx$.

Exercise 8.12 Modify problem (8.66) by assuming constraint $\mathsf{G}[y] \le 0$. Find the Karush-Kuhn-Tucker conditions for optimality.

Example 8.9 (Static cake eating problem) Consider the consumption of a cake that can be eaten by bites of size $c(x)$, where the number of bites belongs to the continuum $x \in \mathsf{X} = [0, \bar{x}]$, where $\bar{x}$ is a fixed number of bites. As the size of the cake is $\bar{c} > 0$, then $\int_0^{\bar{x}} c(x)\, dx \le \bar{c}$. The size of the flow of bites is determined by maximizing the utility functional $\mathsf{U}[c] = \int_0^{\bar{x}} u(c(x))\, dx$, where the utility function is increasing and concave $u''(c) < 0 < u'(c)$. The problem is to find a function $c^* \in \{c : [0, \bar{x}] \to \mathbb{R}_+\}$ that maximizes the utility functional subject to the cake size constraint. The first-order conditions, using the solution to exercise 8.12, are

$$u'(c(x))\left(\bar{c} - \int_0^{\bar{x}} c(x)\, dx\right) = 0,\ \ u'(c(x)) \ge 0,\ \text{ and } \bar{c} \ge \int_0^{\bar{x}} c(x)\, dx.$$

The solution depends on the utility function. First, consider a utility function $u(c) = \frac{c^{1-\theta}-1}{1-\theta}$ with constant elasticity of substitution (CES). As $u'(c) = c^{-\theta} > 0$, for every $c > 0$, then the solution is $c^*(x) = \frac{\bar{c}}{\bar{x}}$, for each $x \in [0, \bar{x}]$. Second, consider the quadratic utility function $u(c) = c - \frac{c^2}{2\tilde{c}}$, where $\tilde{c} > 0$. As $u'(c) = 1 - \frac{c}{\tilde{c}}$ then $u'(c) \ge 0$ if and only if $c \le \tilde{c}$, which allows us to interpret $\tilde{c}$ as a satiation level. The solution of the problem depends on the relationship between $\bar{c}$ and $\bar{x}\,\tilde{c}$: (i) if $\bar{c} \ge \bar{x}\,\tilde{c}$ then the solution is the same as for the CES case; but (ii) if $\bar{c} < \bar{x}\,\tilde{c}$ then the solution is $c^*(x) = \tilde{c}$, for each, $x \in [0, \bar{x}]$, meaning that the proportion $\frac{\bar{c}-\tilde{c}\bar{x}}{\bar{c}}$ of the cake is left uneaten. Prove those results.

The **maximization problem with an algebraic constraint** is: find function $y \in \mathcal{Y}$, where $y : \mathsf{X} \to \mathbb{R}$, that maximizes a functional subject to a continuum of constraints for each value of $x \in \mathsf{X}$:

$$\max_{y(\cdot)}\{\mathsf{F}[y] : g(y(x), x) = 0,\ \text{ for each } x \in \mathsf{X}\}. \tag{8.72}$$

Proposition 8.4 (Necessary optimality conditions for problem (8.72)) *The pair of functions y^* and λ, where $\lambda : \mathsf{X} \to \mathbb{R}$, are an optimum to problem* (8.72) *if they satisfy the two continuum of constraints*

$$\frac{\delta \mathsf{F}[y^*]}{\delta y(x)} = \lambda(x)\,\frac{\partial g(y^*(x), x)}{\partial y(x)},\ \textit{for each } x \in \mathsf{X}, \tag{8.73a}$$

$$g(y^*(x), x) = 0,\ \textit{for each } x \in \mathsf{X}. \tag{8.73b}$$

Proof While in problem (8.66) we had one constraint, in this case, we have an infinity of constraints. Therefore, we have to introduce a Lagrange function $\lambda : \mathsf{X} \to \mathbb{R}$ (instead of a Lagrange multiplier $\lambda \in \mathbb{R}$ as in problem (8.66)). The Lagrange functional is now

$$\mathsf{L}[y,\lambda] = \mathsf{F}[y] - \mathsf{G}[y,\lambda], \text{ where } \mathsf{G}[y,\lambda] = \int_{\mathrm{X}} \lambda(x)\, g(y(x),x)dx,$$

and, at the optimum we have $\mathsf{L}[y^*,\lambda^*]$. Introducing perturbations in y^* and λ^*, such that $y(x) = y^*(x) + \varepsilon\,\eta(x)$, and $\lambda(x) = \lambda^*(x) + \varepsilon\,\psi(x)$, for each $x \in \mathrm{X}$, and expanding $\mathsf{L}[y,\lambda]$ in a neighborhood of $\mathsf{L}[y^*,\lambda^*]$ yields,

$$\mathsf{L}[y,\lambda] = \mathsf{L}[y^*,\lambda^*] + \mathrm{D}_{\eta(\cdot)}\mathsf{L}[y^*,\lambda^*]\,\varepsilon + \mathrm{D}_{\psi(\cdot)}\mathsf{L}[y^*,\lambda^*]\,\varepsilon + O(\varepsilon^2),$$

where the Gâteaux differentials are

$$\mathrm{D}_{\eta(\cdot)}\mathsf{L}[y^*,\lambda^*] = \int_{\mathrm{X}} \left(\frac{\delta \mathsf{F}[y^*]}{\partial y(x)} - \frac{\delta \mathsf{G}[y^*,\lambda^*]}{\partial y(x)}\right) \eta(x)\,dx,$$

and

$$\mathrm{D}_{\psi(\cdot)}\mathsf{L}[y^*,\lambda^*] = -\int_{\mathrm{X}} \frac{\delta \mathsf{G}[y^*,\lambda^*]}{\partial \lambda(x)}\,\psi(x)\,dx.$$

Necessary conditions for the optimum are $\mathrm{D}_{\eta(\cdot)}\mathsf{L}[y^*,\lambda^*] = 0$ and $\mathrm{D}_{\psi(\cdot)}\mathsf{L}[y^*,\lambda^*] = 0$ for arbitrary perturbations, $\eta(\cdot)$ and $\psi(\cdot)$. As $\frac{\delta \mathsf{G}[y^*,\lambda^*]}{\partial y(x)} = \lambda^*(x)\,\frac{\partial g(y^*(x),x)}{\partial y(x)}$, and $\frac{\delta \mathsf{G}[y^*,\lambda^*]}{\partial \lambda(x)} = g(y^*(x),x)$, conditions (8.73a) and (8.73b) are obtained. □

Exercise 8.13 Assume that the objective functional in Problem (8.72) is a linear functional. Show that the optimality conditions in Proposition 8.4 are

$$\frac{\partial f(y^*(x),x)}{\partial y(x)} = \lambda(x)\,\frac{\partial g(y^*(x),x)}{\partial y(x)}, \quad \text{for each } x \in \mathrm{X}, \tag{8.74a}$$

$$g(y^*(x),x) = 0, \quad \text{for each } x \in \mathrm{X}. \tag{8.74b}$$

Exercise 8.14 Assume that the objective functional in Problem (8.72) is a transformed functional $\mathsf{F}[y] = h\Big(\int_{\mathrm{X}} f(y(x),x)\,dx\Big)$. Show that the optimality conditions from Proposition 8.4 are

$$h'(\mathsf{F}[y^*])\frac{\partial f(y^*(x),x)}{\partial y(x)} = \lambda(x)\,\frac{\partial g(y^*(x),x)}{\partial y}, \quad \text{for each } x \in \mathrm{X}, \tag{8.75a}$$

$$g(y^*(x),x) = 0, \quad \text{for each } x \in \mathrm{X}. \tag{8.75b}$$

Exercise 8.15 Modify problem (8.72) by assuming constraints $g(y(x),x) \leq 0$. Find the Karush-Kuhn-Tucker conditions for optimality.

8.2.2 Applications

8.2.2.1 Household Optimal Choice of Varieties

Consider a household with an utility functional as in Sect. 8.1.5.2, valuing the utility of a consuming basket $\mathbf{c}_\mathrm{I} = \big(c(i)\big)_{i\in \mathrm{I}}$, where $c(i)$ is the consumption of variety $i \in \mathrm{I}$, of a continuum of varieties I. The household nominal income is $Y > 0$. The basket nominal cost is the functional $\mathsf{E}[c] = \int_\mathrm{I} p(i)\, c(i)\, di$, where $p(i)$ is the relative price of variety i relative to the income deflator. The set of prices for all varieties, $(p(i))_{i\in\mathrm{I}}$, where $p(i) > 0$, are given to the household. Assuming that there are no savings, the budget constraint is $\mathsf{E}[c] = Y$.

The household problem is to find the optimal basket $\mathbf{c}_\mathrm{I}^*$ that solves the problem

$$\begin{aligned} &\max_{c(\cdot)} \mathsf{U}[c] = \left(\int_\mathrm{I} c(i)^{1-\gamma} di\right)^{\frac{1}{1-\gamma}}, \\ &\text{subject to } \mathsf{E}[c] = \int_\mathrm{I} p(i)\, c(i)\, di = Y, \end{aligned} \tag{8.76}$$

where $\gamma > 0$ is the inverse of the elasticity of variety substitution (see Sect. 8.1.5.2).

Proposition 8.5 *The solution to the household problem* (8.76) *is the basket* $\mathbf{c}^* = \big(c^*(i)\big)_{i\in\mathrm{I}}$, *where the optimal quantity of variety* $i \in \mathrm{I}$, *is*

$$c^*(i) = \frac{Y}{\mathsf{P}}\left(\frac{\mathsf{P}}{p(i)}\right)^{\frac{1}{\gamma}}, \text{ for each } i \in \mathrm{I}. \tag{8.77}$$

where $\mathsf{P} = \left(\int_\mathrm{I} p(i)^{\frac{\gamma-1}{\gamma}}\right)^{\frac{\gamma}{\gamma-1}}$ *is the household's true cost of living index.*

Proof This model is analogous to Problem (8.66), displaying a transformed objective functional and a linear constraint functional (see Exercises 8.10 and 8.11). The Lagrange functional can be interpreted as the indirect utility functional

$$\mathsf{L}[c;\lambda] = \mathsf{U}[c] + \lambda\left(Y - \mathsf{E}[c]\right) = \left(\int_\mathrm{I} c(i)^{1-\gamma} di\right)^{\frac{1}{1-\gamma}} - \lambda \int_\mathrm{I} p(i)\, c(i)\, di + \lambda\, Y.$$

The first-order conditions for problem (8.76) are

$$\frac{\delta \mathsf{L}[c^*]}{\delta c(i)} = \frac{\delta \mathsf{U}[c^*]}{\delta c(i)} - \lambda \frac{\delta \mathsf{E}[c^*]}{\delta c(i)} = 0, \text{ for each } i \in \mathrm{I}, \tag{8.78a}$$

$$\mathsf{E}[c^*] = Y. \tag{8.78b}$$

The optimality condition (8.78a) allows us to find $c^*(i) = \mathsf{U}[c^*]\big(\lambda\, p(i)\big)^{-\frac{1}{\gamma}}$. Substituting in the utility functional, and canceling $\mathsf{U}[c^*]$ yields $1 =$

$\left(\int_{\mathrm{I}} \left(\lambda\, p(i)\right)^{\frac{\gamma-1}{\gamma}} di\right)^{\frac{1}{1-\gamma}}$. Then

$$\frac{1}{\lambda} = \mathsf{P} \equiv \left(\int_{\mathrm{I}} p(i)^{\frac{\gamma-1}{\gamma}}\, di\right)^{\frac{\gamma}{\gamma-1}},$$

where P is interpreted as the consumer price deflator (or the true cost of living index) using as numéraire the income deflator. Substuting in the constraint (8.78b), and simplifying, yields $\mathsf{P}\,\mathsf{U}[c^*] = Y$. Substituting again in the expression for $c(i)$ yields the solution to the consumer problem in Eq. (8.77). □

We can write the demand for variety as

$$c^*(i) = Y\, C(p(i), \mathsf{P}), \quad \text{for each } i \in \mathrm{I},$$

where Y is real income, deflated by the GDP deflator. As P is functional of the distribution of prices across all varieties, $\mathsf{P} = \mathsf{P}[p] = \left(\int_{\mathrm{I}} p(i)^{\frac{\gamma-1}{\gamma}}\right)^{\frac{\gamma}{\gamma-1}}$, then the marginal change of P for a change in any price $p(i)$ is the Gâteaux derivative,

$$\frac{\delta \mathsf{P}[p]}{\delta p(i)} = \left(\frac{\mathsf{P}}{p(i)}\right)^{\frac{1}{\gamma}}, \quad \text{for any } i \in \mathrm{I}.$$

Comparative statics results can be obtained from Eq. (8.77).

First, consumption for any variety i is linearly increasing with real income,

$$\frac{\partial c(i)}{\partial Y} = C(p(i), \mathsf{P}) > 0, \quad \text{for each } i \in \mathrm{I}.$$

This is an income effect. Second, the change in consumption of variety i for changes in its "own" price is

$$\begin{aligned}\frac{\partial c(i)}{\partial p(i)} &= Y\left(\frac{\partial C(p(i), \mathsf{P})}{\partial p(i)} + \frac{\partial C(p(i), \mathsf{P})}{\partial \mathsf{P}}\,\frac{\delta \mathsf{P}[p]}{\delta p(i)}\right) \\ &= -\frac{c^*(i)}{\gamma\, p(i)}\left(1 + (\gamma - 1)\left(\frac{\mathsf{P}}{p(i)}\right)^{\frac{1-\gamma}{\gamma}}\right), \quad \text{for each } i \in \mathrm{I},\end{aligned}$$

and in the price of any other variety, $j \neq i$,

$$\frac{\partial c^*(i)}{\partial p(j)} = \frac{\delta \mathsf{C}^*[p]}{\delta p(j)} = \frac{(1-\gamma)}{\gamma}\,\frac{c^*(i}{\mathsf{P}}\left(\frac{\mathsf{P}}{p(j)}\right)^{\frac{1}{\gamma}}, \quad \text{for any } j \neq i \in \mathrm{I}$$

is ambiguous in both cases. However, the responses to both own and other price changes depend on the elasticity of variety substitution $EVS_{i,j} = 1/\gamma$ (see Sect. 8.1.5.2). If $0 < \gamma < 1$ then $EVS_{i,j} > 1$ and an increase in the own price, $p(i)$, has an ambiguous effect on the consumption of variety i, and an increase in the price of any other variety, $p(j)$ for $j \neq i$ increases its consumption. In this case, this means that although different varieties are EP complements, they are Hicks substitutes. The ambiguity of the response to the "own" price changes depends on the level of $p(i)$ relative to P. If $p(i) > (1-\gamma)^{\frac{\gamma}{1-\gamma}}\,\mathsf{P}$ then $\dfrac{\partial c(i)}{\partial p(i)} < 0$. Observe that both multipliers display both an income effect and a substitution effect, and if $0 < \gamma < 1$ the substitution effect dominates.

8.2.2.2 Household Decision in a Static Stochastic Arrow-Debreu Economy

Consider the household problem in a stochastic environment as in Sect. 8.1.5.5. The household is entitled to stochastic endowment that is represented by a random variable $Y = \big(y(\omega)\big)_{\omega\in\Omega}$, where $y(\omega)$ is the endowment that will be received if state of nature ω materializes. The consumption of the household is also a random variable $C = \big(c(\omega)\big)_{\omega\in\Omega}$, in which $c(\omega) > 0$, is the consumption realization in state of nature ω. In an autarkic economy, $c(\omega) = y(\omega)$, that is consumption would be, *ex-post*, equal to the endowment drawn by nature. Instead, let us assume that there are **Arrow-Debreu (AD) contracts** that allow for delivery of one unit of the good, for a price $q(\omega)$, if state of nature $\omega \in \Omega$ is drawn. Assume that there is no extra source of income, in addition to selling the endowment in the AD markets associated to every specific state of nature $\omega \in \Omega$ is drawn. The existence of these contracts allows for having $c(\omega) \neq y(\omega)$: households can insure consumption is bad states of nature and sell the excess endowment in good states of nature, that is, it allows for partial or complete insurance. We say there is partial insurance if the variance of C is lower than the variance of Y, and we say there is complete insurance if the variance of C is equal to zero. It is assumed that the household has imperfect information-it knows, for every state, their probability and the associated endowment realization, but does not know, *ex-ante* which state will materialize-and that it values consumption by a von-Neumann-Morgenstern utility functional as in Sect. 8.1.5.5, and a Bernoulli utility function satisfying Pratt (1964)'s constant relative risk aversion, with a coefficient $\varrho > 0$, The household (insurance) problem is

$$\begin{aligned} &\max_{c(\cdot)} \mathrm{U}[c] = \int_0^\infty \frac{c(\omega)^{1-\varrho}-1}{1-\varrho}\, p(\omega)\, d\omega, \\ &\text{subject to } \int_\Omega q(\omega)\, c(\omega)\, d\omega = y = \int_\Omega q(\omega)\, y(\omega)\, d\omega, \end{aligned} \tag{8.79}$$

where y is the market value of the *ex-ante* endowments to be received by the household.[25] We call the ratio between the market state price and its probability, $\mu(\omega) = \frac{q(\omega)}{p(\omega)}$, the **stochastic kernel** for state $\omega \in \Omega$. Therefore, $\mu = \big(\mu(\omega)\big)_{\omega\in\Omega}$ is the distribution of the stochastic kernel.

[25] A discrete version of this model is in Gollier (2001, Chap. 13).

Proposition 8.6 *The solution to the household problem* (8.79) *is a distribution of optimal AD contracts allowing the ex-ante distribution of consumption across states of nature be* $C^* = \left(c^*(\omega)\right)_{\omega\in\Omega}$, *where the optimal consumption in state* ω *is*

$$c^*(\omega) = \mu(t,\omega)^{-\frac{1}{\varrho}} \frac{y}{\mathbb{E}\left[\mu^{\frac{\varrho-1}{\varrho}}\right]} \text{ for each } t \in [0,\infty). \tag{8.80}$$

where

$$\mathbb{E}\left[\mu^{\frac{\varrho-1}{\varrho}}\right] = \int_{\Omega} \mu(\omega)^{\frac{\varrho-1}{\varrho}} p(\omega)\, d\omega.$$

The solution for consumption is state-dependent, except if there are complete markets in which $\mu(\omega) = \mu$ for all states of nature. In this case, the household can perform AD contracts such that it totally insures consumption at level $c^*(\omega) = \frac{y}{\mu}$ for every state of nature, although the endowment is stochastic. If markets are incomplete, then the stochastic kernel is state dependent and complete insurance is not possible.

We can have (generalized) comparative statics results for the change in $c^*(\omega)$ resulting from changes in the AD price for the contract insuring the same state or any other state of nature. The change in consumption $c^*(\omega)$ is in general ambiguous for changes in the own price,

$$\frac{\delta c^*(\omega)}{\delta q(\omega)} = \frac{c^*(\omega)}{q(\omega)\, y}\left(q(\omega)\left(y(\omega) - c^*(\omega)\right) - \frac{y - q(\omega)\, c^*(\omega)}{\varrho}\right), \text{ for each } \omega \in \Omega,$$

but, because $y > q(\omega)\, c^*(\omega)$, for every ω then, if the household does not trade in the AD market for that state, or is a net buyer, then $\frac{\delta c^*(\omega)}{\delta q(\omega)} < 0$. We can only have a reversed sign if ϱ is very large and the household is a seller in the AD market for state ω. The change in consumption $c^*(\omega)$ is in general ambiguous for changes in the price of state $\omega' \neq \omega$,

$$\frac{\delta c^*(\omega)}{\delta q(\omega')} = \frac{c^*(\omega)}{y}\left(y(\omega') - c^*(\omega') + \frac{c^*(\omega')}{\varrho}\right), \text{ for each } \omega \neq \omega' \in \Omega.$$

Although there is EP independence, there can be Hicksian complementarity if $\varrho \leq 1$, or Hicksian substitutability if $\varrho > 1$, and the household should be a buyer in the ω'-market if $y(\omega') < \frac{\varrho-1}{\varrho} c(\omega')$.

Exercise 8.16 Prove all the previous results.

Exercise 8.17 Solve a modified version of the Problem (8.79) in which the household maximizes the certainty equivalent of the contingent consumption distribution C.

8.2.2.3 Intertemporal Consumption Choice in a Deterministic Arrow-Debreu Economy

Let a household have a flow of endowments of a good over time $\left(y(t)\right)_{t\in \mathrm{T}}$, where $y(t) \geq 0$ is the quantity of the good to be delivered at time $t \geq 0$. In a Robinson Crusoe's economy, the household has the autarkic consumption constraint $c(t) = y(t)$, at every time t. Alternatively, assume that, at time $t = 0$, there are markets for Arrow-Debreu contracts in which the household can sell or buy the quantity $|y(t) - c(t)|$ at time $t = 0$, for delivery at every time $t > 0$ in the future, at price $q(t)$. In this case, the household can be free from the autarkic constraint, but it still has a budget constraint.

Assume that there is perfect foresight and the household has the utility functional that was specified in Sect. 8.1.5.3. The household's problem is to find the optimal path of $\mathbf{c}_{\mathrm{T}} = \left(c(t)\right)_{t\in \mathrm{T}}$, given its beliefs at time $t = 0$ over the paths $\left(y(t), q(t)\right)_{t\in \mathrm{T}}$, that solves the problem (assuming an infinite horizon)

$$\begin{aligned} &\max_{c(\cdot)} \mathrm{U}[c] = \int_0^{\infty} \frac{c(t)^{1-\theta} - 1}{1-\theta} e^{-\rho t}\, dt, \\ &\text{subject to } \int_0^{\infty} q(t)\, c(t)\, dt = W, \end{aligned} \tag{8.81}$$

where we assume that the household income for selling its endowment in the AD markets, $W = \int_0^{\infty} q(t)\, y(t)\, dt$, is positive. A sufficient condition for this is that there is at least one time $t = \tau$ such that $y(\tau) > 0$ and $y(t) = 0$ for all $t \neq \tau \in \mathrm{T}$.

Proposition 8.7 (Consumption in an intertemporal AD economy) *The solution to the household problem* (8.81) *is a distribution of optimal AD contracts at time* $t = 0$ *allowing for the path of consumption* $\mathbf{c}^* = \left(c^*(t)\right)_{t\in \mathrm{T}}$, *where optimal consumption at time t is*

$$c^*(t) = \left(e^{-\rho t} q(t)\right)^{\frac{1}{\theta}} \mathsf{h}[q], \text{ for each } t \in [0, \infty), \tag{8.82}$$

where

$$\mathsf{h}[q] = \frac{W}{\mathsf{Q}[q]} = \frac{\int_0^{\infty} q(t) y(t) dt}{\int_0^{\infty} \left(e^{-\rho t} q(t)^{\theta-1}\right)^{\frac{1}{\theta}} dt}$$

is the household's net wealth, deflated by a discounted price deflator, $\mathsf{Q}[q]$.

Proof The Lagrange functional is $\mathsf{L}[c] = \int_0^{\infty} \frac{c(t)^{1-\theta}-1}{1-\theta} e^{-\rho t} dt + \lambda \left(W - \int_0^{\infty} q(t)\, c(t)\, dt\right)$. The first-order conditions for an optimum are

$$\begin{aligned} \frac{\delta \mathsf{L}[c]}{\delta c(t)} &= e^{-\rho t} c(t)^{-\theta} - \lambda q(t) = 0, \text{ for each } t \in [0, \infty) \\ \frac{\partial \mathsf{L}[c]}{\partial \lambda} &= W - \int_0^{\infty} q(t)\, c(t) = 0 \end{aligned}$$

Solving the first condition for $c(t)$ and substituting in the constraint yields $\lambda^{-\frac{1}{\theta}} = \frac{W}{\mathsf{Q}[q]}$, which allows us to obtain the optimal consumption at time t in Eq. (8.82). □

If $q(t) > 0$, for any t, and as $W > 0$, then consumption is strictly positive for all $t \in \mathrm{T}$. The existence of AD contracts allows for positive consumption even at times in which the household is not getting an endowment, that is $y(t) = 0$.

In Eq. (8.82), we observe that there is a positive wealth effect, as consumption increases with W. In order to study the effect on consumption at time t from an increase in the own price $q(t)$, or in the price for delivery at a different date $q(t')$, we need to find the effect of an increase of any price in the real net wealth of the household, $\mathsf{h}[q]$, that is a functional in q,

$$\frac{\delta \mathsf{h}[q]}{\delta q(t)} = \frac{1}{\mathsf{Q}[q]}\Big(y(t) - \frac{\theta-1}{\theta}c^*(t)\Big) = \frac{1}{\mathsf{Q}[q]}\Big(y(t) - c^*(t) + \frac{c^*(t)}{\theta}\Big).$$

If the utility function is logarithmic (that is the case if $\theta = 1$) or if the household is a net seller for AD contracts to time t, such that $y(t) \geq c^*(t)$, and increase in the price will increase net wealth. Then, the multipliers are

$$\frac{\partial c^*(t)}{\partial q(t)} = \frac{c^*(t)}{q(t)\,W}\Big(q(t)\,(y(t) - c^*(t)) - \frac{W - q(t)c^*(t)}{\theta}\Big), \tag{8.83a}$$

$$\frac{\partial c^*(t)}{\partial q(t')} = \frac{c^*(t)}{W}\Big(y(t') - c^*(t') + \frac{c^*(t')}{\theta}\Big). \tag{8.83b}$$

The signs of the multipliers for spike shocks in AD prices depend on the size of endowments in any particular time, and on the $EIS = 1/\theta$. If $c^*(t) \approx y(t)$ then $\frac{\partial c^*}{\partial q(t)} < 0$ and $\frac{\partial c^*}{\partial q(t')} > 0$. In this case, although there is EP independence, there is Hiscksian substitutability. If $y(t)$ is very large, then the sign of $\frac{\partial c^*}{\partial q(t)}$ can change, and if $y(t')$ is close to zero the sign of $\frac{\partial c^*}{\partial q(t')}$ can change as well. In the logarithmic case, we have

$$\frac{\partial c^*(t)}{\partial q(t)} = \frac{c^*(t)}{q(t)}\Big(\frac{q(t)\,y(t)}{W} - 1\Big) \leq 0, \text{ and} \frac{\partial c^*(t)}{\partial q(t')} = c^*(t)\frac{y(t')}{W} \geq 0,$$

if $y(t) \geq 0$ for each $t \in [0, \infty)$, and because $q(t)y(t) \leq W$.

8.2.2.4 Intertemporal Consumption Choice in a Stochastic Arrow-Debreu Economy

The information set is the same as in Sect. 8.1.5.6. The household is entitled to a flow of (real) endowments given by the stochastic process $\mathbf{Y}_{\mathrm{T}} = \big(Y(t)\big)_{t\in \mathrm{T}}$ where $Y(t) = \Big(y(t,\omega)\Big)_{\omega\in\Omega}$, and $y(t,\omega) \geq 0$ is the endowment at time t conditional on the realization of state of nature ω. At time $t = 0$ there is a continuum of AD markets for delivery of one unit of the good at every time t and for each state of nature ω, at price $q(t,\omega)$ (relative to the price of the spot market, for immediate delivery at time $t=0$). The **stochastic discount factor** is $\mu(t,\omega)=\frac{q(t,\omega)}{p(t,\omega)}$, and the stochastic process for the stochastic discount factor (SDF) is denoted by $\mathbf{M}_{\mathrm{T}}=\big(M(t)\big)_{t\in\mathrm{T}}$, where $M(t)=\big(\mu(t,\omega)\big)_{\omega\in\Omega}$.

Assuming again imperfect information and that the household preferences are represented by an intertemporal EZ utility functional (see Sect. 8.1.5.6), the household problem is now

$$\begin{aligned}&\max_{c(\cdot)}\mathsf{U}[c]=\int_{\mathrm{T}}u\Big(v^{-1}\Big(\int_{\Omega}v(c(t,\omega))p(t,\omega)\,d\omega\Big)\Big)\,e^{-\rho}\,dt,\\ &\text{subject to }\int_{\mathrm{T}\times\Omega}q(t,\omega)\,c(t,\omega)d\omega\,dt=W,\end{aligned}\tag{8.84}$$

where household wealth, deflated by the price of the spot AD contracts is $W=\int_{\mathrm{T}\times\Omega}q(t,\omega)\,y(t,\omega)d\omega\,dt$. We assume further that $u(\cdot)$ and $v(\cdot)$ display constant EIS $\theta^{-1>0}$, and constant CRRA $\varrho>0$, respectively (see Sect. 8.1.5.6).

Proposition 8.8 *The solution to the household problem* (8.84) *is a distribution of optimal AD contracts allowing the path of consumption* $\mathbf{C}^*_{\mathrm{T}}=\big(C^*(t)\big)_{t\in\mathrm{T}}$, *where optimal consumption at time t, contingent on the state of nature* ω, *is*

$$c^*(t,\omega)=\mu(t,\omega)^{-\frac{1}{\varrho}}\,\frac{W}{\mathsf{M}(t)}\left(\frac{e^{-\frac{\rho}{\theta}t}\mathsf{M}(t)^{\frac{\varrho(1-\theta)}{\theta(1-\varrho)}}}{\int_{\mathrm{T}}e^{-\frac{\rho}{\theta}t}\mathsf{M}(t)^{\frac{\varrho(1-\theta)}{\theta(1-\varrho)}}\,dt}\right),\ \textit{for each } t\in[0,\infty).\tag{8.85}$$

where there is a utility-based wealth deflator,

$$\mathsf{M}(t)=\mathbb{E}\Big[M(t)^{\frac{\varrho-1}{\varrho}}\Big]=\int_{\Omega}\mu(t,\omega)^{\frac{\varrho-1}{\varrho}}\;p(t,\omega)\,d\omega,\textit{for each } t\in\mathrm{T}.\tag{8.86}$$

Proof Solving the optimality condition $\frac{\delta\mathsf{U}[c]}{\delta c(t,\omega)}=\lambda\,q(t,\omega)$ for $c(t,\omega)$, and solving together with the certainty equivalent $\bar{c}(t)$ yields

$$c(t,\omega)=\mu(t,\omega)^{-\frac{1}{\varrho}}\left(\lambda\,e^{\rho t}\right)^{-\frac{1}{\theta}}\mathsf{M}(t)^{\frac{\theta-\varrho}{(\varrho-1)\theta}},$$

for $\mathsf{M}(t)$ in Eq. (8.86). The Lagrange multiplier, using the budget constraint, is $\lambda^{-\frac{1}{\theta}}=\frac{W}{R}$ where $R=\int_{\mathrm{T}}e^{-\frac{\rho}{\theta}t}\mathsf{M}(t)^{\frac{\varrho(1-\theta)}{\theta(1-\varrho)}}\,dt$, and substituting back in the optimal consumption we obtain (8.85). □

Again, if $c^*(t,\omega)$ is state independent and the variance of $C^*(t)=\big(c^*(t,\omega)\big)_{\omega\in\Omega}$ is smaller than the variance of $Y(t)=\big(y(t,\omega)\big)_{\omega\in\Omega}$ we say there is incomplete insurance, and if $c^*(t,\omega)=\bar{c}^*(t)$ is state independent we say there is complete insurance.

In this general case, we expect that optimal consumption at time t, contingent on the state of nature ω, $c^*(t,\omega)$, depend on the SDF for the same time and state of nature, $\mu(t,\omega)$, but also on the SDF on both different states, $\mu(t,\omega')$, and different times, $\mu(t',\omega)$. The channel through which intertemporal and distributional effects operate involves the deflator $\mathsf{M}(t)$, for the state shocks, and involves the integral over time of $\mathsf{M}(t)$, for the intertemporal effect. Therefore, we can have both Hicksian substitutability or complementarity over time and/or over states of nature.

Several simple cases are possible. In the additive case, in Eq. (8.53), that is when $\varrho=\theta$, we find that

$$c^*(t,\omega)=\mu(t,\omega)^{-\frac{1}{\theta}}\frac{W}{\mathsf{M}(t)}\left(\frac{e^{-\frac{\varrho}{\theta}t}\mathsf{M}(t)}{\int_{\mathrm{T}}e^{-\frac{\varrho}{\theta}t}\mathsf{M}(t)\,dt}\right),$$

and we can still have Hicskian state and time substitutability or complementarity. If the utility function $u(\cdot)$ is logarithmic and $T=\infty$, yields $c^*(t,\omega)=\rho\,\mu(t,\omega)^{-\frac{1}{\varrho}}\frac{W}{\mathsf{M}(t)}$ and there is only state dependent Hicksian effects. If the function $v(\cdot)$ is logarithmic we obtain

$$c^*(t,\omega)=\frac{W}{\mu(t,\omega)}\frac{e^{-\frac{\varrho}{\theta}t}\mathsf{M}(t)^{\frac{\theta-1}{\theta}}}{\int_{\mathrm{T}}e^{-\frac{\varrho}{\theta}t}\mathsf{M}(t)^{\frac{\theta-1}{\theta}}\,dt},$$

where $\mathsf{M}(t)=\exp\left(\int_{\Omega}\ln\left(\mu(t,\omega)\right)p(t,\omega)\,d\omega\right)$, and there are again state- and time-dependent effects. In all the previous cases, consumption is state dependent. If markets are complete at every time, such that $\mu(t,\omega)=\mu(t)=e^{-rt}$ where $r>0$ is a risk-free interest rate, then consumption becomes state-independent

$$c^*(t,\omega)=\bar{c}^*(t)=\frac{\rho+r\,(\theta-1)}{\theta}\,\mathbb{E}[y(t)]\,e^{-\frac{\rho+r\,(\theta-1)}{\theta}t},\ \text{for each } (t,\omega)\in\mathrm{T}\times\Omega,$$

and there is complete insurance.

8.2.2.5 Single-Product Firm with a Continuum of Intermediate Inputs

A component of new-Keynesian macro models is the problem of a producer of a final good using a continuum of inputs $\mathbf{x}_{\mathrm{I}}=(x(i))_{i\in\mathrm{I}}$, where $x(i)$ is the quantity of the input of variety i. When bundles of inputs are infinite-dimensional, that is when I is a continuum, those inputs are assumed to be intermediary goods not factors of production. Assume that the firm has a generalized constant elasticity of substitution (CES) production function

$$y=\mathsf{F}[x]\equiv\left(\int_{\mathrm{I}}A(i)\,x(i)^{\frac{\epsilon-1}{\epsilon}}\,di\right)^{\frac{\epsilon}{\epsilon-1}}$$

with specific input productivity $A(i)$, where ϵ is the elasticity of substitution between inputs.

Total sales are $S = p\,y$, where p is the final good price and is given to the firm, and y is the output. Therefore, nominal sales are a functional $\mathsf{S}[x] = p\,\mathsf{F}[x]$ where p is the price of the output. The total cost is also a functional $\mathsf{C}[x] = \int_{\mathrm{I}} w(i)x(i)di$, where $w(i)$ denotes the price of input $i \in \mathrm{I}$. We can denote by $\mathbf{A}_{\mathrm{I}} = \big(A(i)\big)_{i\in\mathrm{I}}$ and $\mathbf{w}_{\mathrm{I}} = \big(w(i)\big)_{i\in\mathrm{I}}$ the profiles of productivity and costs associated to the input profile $\mathbf{x}_{\mathrm{I}} = \big(x(i)\big)_{i\in\mathrm{I}}$. Therefore, firm profits are also a functional over the x, $\Pi[x] = p\,\mathsf{F}[x] - \mathsf{C}[x]$. Assuming that the firm is a price-taker both in the product market and in every intermediate good market, $i \in \mathrm{I}$, the firm's problem is to find the optimal input bundle $\mathbf{x}^*_{\mathrm{I}} = \big(x^*\big)_{i\in\mathrm{I}}$ that solves the problem

$$\begin{aligned}
&\max_{x(\cdot)} \Pi[x] = p\left(\int_{\mathrm{I}} A(i)\,x(i)^{\frac{\epsilon-1}{\epsilon}}\,di\right)^{\frac{\epsilon}{\epsilon-1}} - \int_{\mathrm{I}} w(i)x(i)di,\\
&\text{subject to} \left(\int_{\mathrm{I}} A(i)\,x(i)^{\frac{\epsilon-1}{\epsilon}}\,di\right)^{\frac{\epsilon}{\epsilon-1}} = d,
\end{aligned} \tag{8.87}$$

where d is the market demand. In equilibrium $y = \mathsf{F}[x] = d$.

Exercise 8.18 Prove that the solution to problem (8.87) is

$$x^*(i) = d\left(\frac{W\,A(i)}{w(i)}\right)^{\epsilon}, \text{ for each } i \in \mathrm{I}$$

where $W = \mathsf{W}[A, w] \equiv \left(\int_{\mathrm{I}} A(i)^{\epsilon}\,w(i)^{1-\epsilon}\,di\right)^{\frac{1}{1-\epsilon}}$ is a measure of the aggregate marginal costs. Perform the comparative statics exercise for changes in demand and in the prices of inputs. Study the comparative statics for a change in the price of own input i and j another arbitrary input $\neq i$ of the demand of input i by the firm.

Exercise 8.19 Extend the previous example to a multi-product firm using only one factor of production.

Exercise 8.20 Extend the previous example to a monopolist firm.

Exercise 8.21 Extend the previous example to a monopsonist firm.

8.2.2.6 Optimal Taxation

We present next a model for the optimal taxation of heterogeneous households by a social planner that wants to maximize social welfare, while keeping the government budget balanced. We assume that households have heterogeneous skill levels and have homogeneous preferences. The government has perfect information on both the distribution of income and the skill level, for every household, and has a fixed level of aggregate expenditures.

Order a population by the skill level, θ, where $\theta \in \Theta \subseteq \mathbb{R}_+$, and let $f(\theta) \in (0, 1)$ be the population density. Therefore $\int_{\Theta} f(\theta)\,d\theta = 1$. Let $\ell(\theta)$, $y(\theta)$, and $c(\theta)$ denote the work effort, income and consumption, respectively, by people with skill level θ. Let income be dependent on the level of skill and on the work effort: $y(\theta) = \theta\, \ell(\theta)$. At last, let the utility for a household with skill level θ be a function of consumption and work effort, $u(\theta) = U(c(\theta), \ell(\theta))$.

The government has an exogenous expenditure level G and wants to find an optimal tax schedule $T(\theta) = \tau(\theta)\, y(\theta)$ which implements a social optimum. Assume that the social optimal criterium is the average social welfare

$$\mathsf{W}[u] = \int_{\Theta} W\big(u(\theta)\big)\, f(\theta)\, d\theta.$$

Assume that the social planner not only observes the distribution of income of household for any skill level θ, $y(\theta)$ but also has complete information on its components, the work effort $\ell(\theta)$, and its productivity θ.[26]

The problem is simplified if we use the implicit function theorem to solve $u(\theta) = U(c(\theta), \ell(\theta))$ for $c(\theta) = C\big(u(\theta), \ell(\theta)\big)$. We can find the optimal allocation of utility and work effort, u^* and ℓ^*, by solving the functional problem

$$\max_{u(\cdot),\ell(\cdot)} \int_{\Theta} W\big(u(\theta)\big)\, f(\theta)\, d\theta,$$
$$\text{subject to } \int_{\Theta} \Big(\theta\, \ell(\theta) - C\big(u(\theta), \ell(\theta)\big)\Big)\, f(\theta)\, d\theta = G.$$

The Lagrange functional is

$$\mathsf{L}[u, \ell] = \int_{\Theta} \Big(W(u(\theta)) - \lambda\, \big(\theta\, \ell(\theta) - C\big(u(\theta), \ell(\theta)\big)\Big)\, f(\theta)\, d\theta + G.$$

The first-order conditions are

$$\frac{\delta \mathsf{L}}{\delta u(\theta)} = \Big(W'(u(\theta)) + \lambda\, C_u\big(u(\theta), \ell(\theta)\big)\Big)\, f(\theta) = 0, \text{ for each } \theta \in \Theta \tag{8.88a}$$
$$\frac{\delta \mathsf{L}}{\delta \ell(\theta)} = -\lambda\, \Big(\theta - C_\ell\big(u(\theta), \ell(\theta)\big)\Big)\, f(\theta) = 0, \text{ for each } \theta \in \Theta \tag{8.88b}$$
$$\int_{\Theta} \Big(\theta\, \ell(\theta) - C\big(u(\theta), \ell(\theta)\big)\Big)\, f(\theta)\, d\theta = G. \tag{8.88c}$$

Equations (8.88a) and (8.88b) hold if and only if

$$W'(u(\theta)) + \lambda\, C_u\big(u(\theta), \ell(\theta)\big) = 0$$
$$C_\ell\big(u(\theta), \ell(\theta)\big) = \theta.$$

[26] This makes this model a simpler version of the Mirrlees (1971) model in which the government only observes the distribution of income but cannot separate the skill level from effort.

If this system has a unique solution, we can write it as $u(\theta) = U^*(\theta, \lambda)$, and $\ell(\theta) = L^*(\theta, \lambda)$ then, substituting in the budget constraint (8.88c) yields, a hopefully unique, value for λ

$$\lambda^* = \left\{\lambda : \int_\Theta \left(\theta\, L(\theta, \lambda) - C\big(U^*(\theta, \lambda), L^*(\theta, \lambda)\big)\right) f(\theta)\, d\theta = G\right\}.$$

If this is the case, the solution to the optimal tax problem is the optimal distribution of utility and labor effort $u^*(\theta) = U^*(\theta, \lambda^*)$ and $\ell^*(\theta) = L^*(\theta, \lambda^*)$. If we can find the consumption and income distributions, $c^*(\theta) = C\big(u^*(\theta), \ell^*(\theta)\big)$ and $y^*(\theta) = \theta\, \ell^*(\theta)$, then the optimal optimal tax-transfer policy that implements the social optimum, and finances the government expenditure G, can be found by finding

$$T^*(\theta) = y^*(\theta) - c^*(\theta),$$

that represents net tax for a household of skill level θ. Therefore, we can separate households that should receive transfers, having $T^*(\theta) < 0$, from those that should pay taxes, having $T^*(\theta) > 0$. The net tax distribution depends on the level of government expenditure, and on the properties of the social welfare, household utility, and population density functions.

Exercise 8.22 Assume there is no inequality aversion (see Sect. 8.1.5.8), that is $W(u) = u$ and that the household utility function is $U(c, \ell) = \ln(c) + \alpha \ln(1 - \ell)$, for $\alpha > 0$. Prove that, if $\frac{1+\alpha}{\alpha}\,\theta > \mathbb{E}[\theta] - G > 0$, the tax policy that implements the optimal distribution is $T^*(\theta) = \theta - \theta_c$ where $\theta_c = \mathbb{E}[\theta] - G$. The tax schedule is linear in the skill level, and skill levels below (above) θ_c receive a transfer (pay taxes).

Exercise 8.23 Assume that the social welfare function is $W(u) = -\frac{e^{-\beta u}}{\beta}$, for $\beta > 0$, and that the household utility function is $U(c, \ell) = c + (1 - \ell)^\xi$, for $\xi > 0$.[27] Is there inequality aversion in this case? Defining $v^*(\theta)\left(\frac{\theta}{\xi}\right)^{\frac{\xi}{\xi-1}}$ and $V^* = \mathbb{E}[v^*]$, prove that the tax that implements the optimal distribution is $T^*(\theta) = \theta + \mathbb{E}[\theta] + (1 - \xi)\,(V^* + v^*(\theta) - G$. Although the tax schedule is increasing in the level of skill $(T^*)'(\theta) = 1 - v'(\theta) > 0$, show that it may not be progressive (hint, show that $(T^*)''(\theta) < 0$ for $\xi > 1$).

Exercise 8.24 Find the optimal tax policy when the private utility function is $U(c, \ell) = \log(c) - \frac{\ell^{1+\xi}}{1+\xi}$, for $\xi > 0$, and the social welfare function displays no inequality aversion, $W(u) = u$.

[27] This example is in Diamond (1998).

8.2.2.7 Optimal Derivation of Statistical Distributions

The **principle of maximum entropy**, where entropy is defined in Sect. 8.1.5.9, allows for finding the statistic distributions that satisfy some constraints.[28] A one-dimensional continuous probability density function is a mapping $f : \mathsf{X} \subseteq \mathbb{R} \to (0, 1)$, such that it satisfies, at least, a normalization constraint $\int_{\mathsf{X}} f(x)\,dx = 1$. Additional restrictions involving moments or other type of functionals (ex: the Gini index) can also be required. And, as we will see next, we obtain different optimal statistic distributions if the support set X, and the additional constraint are different.

Maximum entropy with the simplest constraint. The problem is: using the Shannon specification of entropy, in Eq. (8.63), find the continuous density function $f : \mathsf{X} \to (0, 1)$, with support $\mathsf{X} = [a, b]$ where $b > a$, such that it maximizes entropy and satisfies the normalization constraint $\int_a^b f(x)dx = 1$. Formally,

$$\max_{f(\cdot)} \mathsf{H}[f] = -\int_a^b f(x)\ln(f(x))\,dx,$$
$$\text{subject to} \int_a^b f(x)dx = 1.$$

This is a problem of maximizing a linear functional subject to a linear functional over function $f(\cdot)$. The solution is the **uniform probability density function** $f^*(x) = \dfrac{1}{b-a}$, for each $x \in \mathsf{X} = [a, b]$. To prove this result, we write the Lagrange functional

$$\mathsf{L}[f] = \int_a^b L(f(x), \lambda)dx = -\int_a^b \Big(\ln(f(x)) + \lambda\Big) f(x)\,dx + \lambda,$$

where λ is a Lagrange multiplier. The first-order conditions are, using the functional derivative from Sect. 8.1.5.9,

$$\frac{\delta\mathsf{L}[f^*]}{\delta f(x)} = -\big(\ln(f(x)) + 1\big) - \lambda = 0, \text{ for each } x \in \mathsf{X}$$
$$\frac{\mathsf{L}[f^*]}{\partial\lambda} = \int_a^b f(x)\,dx = 1.$$

As $f(x) = e^{-(1+\lambda)}$ from the optimum condition, the constraint $\int_a^b e^{-(1+\lambda)}\,dx = 1$ allows us to find $1 + \lambda = \ln(b - a)$. Substituting back, yields $f^*(x) = 1/(b - a)$.[29]

[28] First formulated by Jaynes (1957a, 1957b). See Jaynes (2003). A concise presentation is in Cover and Thomas (2006, Chap. 12). From a statistical mechanics perspective, with historical remarks on the Boltzmann-Gibbs entropy see Tsallis (2009). For a maximum entropy characterization of income distributions, see Park and Bera (2018).

[29] This is indeed a maximum because, as we saw in Sect. 8.1.5.9 the entropy functional is concave in a generalized sense, $\frac{\delta^2\mathsf{H}[f]}{\delta f(x)^2} < 0$.

Maximum entropy with constraints for the first and second moments. The problem is now: find the continuous maximum entropy density distribution $\mathbf{f} = (f(x))_{x \in \mathrm{X}}$ with support in $\mathrm{X} = (-\infty, \infty)$, such that, in addition to the normalization constraint, it has mean μ and variance σ^2, $\mathbb{E}[x] = \mu$, and $\mathbb{E}[(x-\mu)^2] = \sigma^2$, where μ and σ are real numbers.[30] Recall that $\mathbb{E}[x] = \int_{-\infty}^{\infty} x f(x) dx$, and $\mathbb{V}[x] = \int_{-\infty}^{\infty} (x-\mu)^2 f(x) dx$. Formally, the problem is: find $f : \mathrm{X} \to (0, 1)$, where $\mathrm{X} = (-\infty, \infty)$ that solves the problem[31] of maximizing a linear functional subject to three linear functionals,

$$\max_{f(\cdot)} \mathsf{H}[f] = -\int_{-\infty}^{\infty} f(x) \ln(f(x))\, dx,$$

subject to

$$\int_{-\infty}^{\infty} f(x)dx = 1,$$

$$\int_{-\infty}^{\infty} x f(x)dx = \mu,$$

$$\int_{-\infty}^{\infty} x^2 f(x)dx = \sigma^2 + \mu^2.$$

The solution is: the optimal PDF is the Gaussian or normal distribution, $f^* \sim N(\mu, \sigma^2)$, where

$$f^*(x) = \frac{1}{\sqrt{2\pi\sigma^2}} e^{-\frac{(x-\mu)^2}{2\sigma^2}}, \text{ for each } x \in (-\infty, \infty).$$

To prove this, write the Lagrange functional

$$\begin{aligned} \mathsf{L}[f] = &-\int_{-\infty}^{\infty} \left(\ln(f(x)) + \lambda_0 + \lambda_1 x + \lambda_2 x^2 \right) f(x)\, dx \\ &+ \lambda_0 + \lambda_1 \mu + \lambda_2 (\sigma^2 + \mu^2), \end{aligned}$$

where λ_0, λ_1 and λ_2 are Lagrange multipliers (all real numbers). The first-order condition for $f(x)$ is,

$$\frac{\delta \mathsf{L}[f^*]}{\delta f(x)} = -\ln(f(x)) - 1 - \lambda_0 - \lambda_1 x - \lambda_2 x^2 = 0$$

yields

$$f(x) = e^{-(1+\lambda_0+\lambda_1 x+\lambda_2 x^2)}.$$

---

[30] See Shannon (1948, p. 36), or Cover and Thomas (2006, Chap. 12).

[31] Prove that $\mathbb{E}[x^2] = \sigma^2 + \mu^2$ if and only if $\mathbb{V}[x] = \sigma^2$.

There are two alternative methods for finding the solution. The first method is: conjecture that the solution is a Gaussian function $g(x) = a\,e^{-\frac{(x-b)^2}{2c^2}}$ where a, b, and $c > 0$ are undetermined coefficients. If the conjecture is correct, they can be obtained by substituting this function in the three constraints. Using the properties of the Gaussian integral (see Sect. 15.7.2), we have a algebraic system of three equations and three unknowns

$$\int_{-\infty}^{\infty} g(x)dx = a\sqrt{2\pi c} = 1,$$
$$\int_{-\infty}^{\infty} xg(x)dx = ba\sqrt{2\pi c} = \mu,$$
$$\int_{-\infty}^{\infty} x^2 g(x)dx = a\sqrt{2\pi c}\left(b^2 + c\right) = \sigma^2.$$

Solving this system yields $a = \frac{1}{2\pi\sigma^2}$, $b = \mu$ and $c = \sigma^2$, implying function $g(x)$ becomes $g(x) = e^{\log(a) - \frac{1}{2c}(x^2 - 2bx + b^2)}$. Matching the exponent with $f(x)$, we find the Lagrange multipliers

$$\lambda_0 = -1 - \ln(2\pi\sigma^2)^{-\frac{1}{2}} + \frac{\mu^2}{2\sigma^2},\ \lambda_1 = -\frac{\mu}{\sigma^2},\ \lambda_2 = \frac{1}{2\sigma^2}.$$

The second method is: substitute function $f(x)$ in the constraints and determine the Lagrange multipliers. Assuming that $\mathrm{Re}(\lambda_2) > 0$, we find

$$\int_{-\infty}^{\infty} f(x)dx = \sqrt{\frac{\pi}{\lambda_2}} e^{-1-\lambda_0+\frac{\lambda_1^2}{4\lambda_2}} = 1,$$
$$\int_{-\infty}^{\infty} xf(x)dx = -\frac{\lambda_1}{2\lambda_2} = \mu,$$
$$\int_{-\infty}^{\infty} x^2 f(x)dx = \left(\frac{\lambda_1^2 + 2\lambda_2}{(2\lambda_2)^2}\right) = \sigma^2 + \mu^2.$$

Solving for λ_0, λ_1 and λ_2, we obtain the same result.

8.3 References

For generalized calculus and its application in economics, see Ok (2007). For generalized calculus in applied mathematics, see Smith (1974), Drabek and Milota (2013, Chap. 3), and Siddiqi (2018, Chap. 5). For functional analysis applied to optimization, but with an emphasis in non-differentiable functions, see Clarke (2013).

Chapter 9
Introduction to Calculus of Variations

A **calculus of variations problem** (CV) has the purpose of finding an extremum of a functional of the form

$$\mathsf{F}[y] \equiv \int_{\mathrm{X}} F\big(y'(x), y(x), x\big)\, dx, \tag{9.1}$$

with or without side constraints, where $F(.)$ is a known objective function, and the mapping $y : \mathrm{X} \to \mathrm{Y} \subseteq \mathbb{R}$ is unknown. A calculus of variations problem is distinguished from non-variational optimization problems, already presented in Sect. 8.2, because the objective functional contains derivatives of y. That is, in Eq. (9.1), it is assumed that $F_{y'}(y'(x), y(x), x) \neq 0$,[1] except in a subset of measure zero.

To understand the effect of the derivative on the optimum, consider instead the objective functional $\mathsf{F}[y] \equiv \int_{\mathrm{X}} F\big(y(x), x\big)\, dx$. If there are no other constraints, and $y^*(x)$ is the optimum, a necessary condition is $\frac{\delta \mathsf{F}[y^*]}{\delta y(x)} = F_y\big(y^*(x), x\big) = 0$. This condition is a point-wise optimality criterium, tracing out the solution function $y^*(x)$ for every point $x \in \mathrm{X}$. The dependency of the optimum on any other point, say $x' \in \mathrm{X}$, only exists via the Lagrange multiplier when there are side conditions involving a functional, as we saw in Sect. 8.2. However, if the objective function depends on the derivative $y'(\cdot)$, as in functional (9.1) then the optimum condition involves derivatives of $y^*(x)$, thus depending on the local interaction of of the solution function between different x values. This has two consequences: first, the optimum cannot be just determined by a point wise extremum, even in the absence of functional side conditions, and, second, any side condition imply a dependence of the solution of the CV problem on its distribution over the entire support X.

Consequently, a CV problem can only be consistently defined for sets of independent variables with an order structure, that is $\mathrm{X} \subseteq \mathbb{R}$. It cannot be formulated for optimization upon varieties, as was the case for non-variational problems formulated

[1] The notation $F_{y'}(y'(x), y(x), x) \equiv \frac{\partial F(y'(x), y(x), x)}{\partial y'}$ is used.

P. B. Brito, *Economic Dynamics and Distributions*, Dynamic Modeling and Econometrics in Economics and Finance 32, https://doi.org/10.1007/978-3-031-94717-9_9

in Sect. 8.2. This also implies that the solution belongs to the space of differentiable functions, i.e., $y \in \mathcal{Y} \in C^1(\mathrm{X})$, or $C^2(\mathrm{X})$.

In this chapter, we are also concerned with the differences between problems in which the independent variable is time or is not time. While CV problems in which the independent variable is time seek to find optimal trajectories, CV problems in which the independent variable is not time seek to find optimal distributions.[2]

When the independent variable is not time, denoted generically by $x \in \mathrm{X} \subseteq \mathbb{R}$, CV problems formalize microeconomic decisions over a continuum, or static macroeconomic problems with heterogeneity among agents, in which x is related, v.g., to information or searching frictions, skills, space locations. For example, in models in which x refers to the skill levels, and the skill domain is bounded as $\mathrm{X} = [x_0, x_1]$, x_0 refers to the lowest skill in the distribution and x_1 to the highest skill. The boundary values for the independent variable, x_0 and/or x_1, can be fixed or can be determined optimally. Furthermore, the boundary values for the dependent variable, $y(x_0)$ and/or $y(x_1)$, can be known or not, meaning that they can also be fixed or determined optimally. This type of problems, in which the independent variable has a finite domain, is the object of Sect. 9.1.

However, a very large (possibly the biggest) part of the calculus of variations problems in economics has time as the independent, denoted by $t \in \mathrm{T} \subseteq \mathbb{R}_+$. This is the case in macroeconomic, investment and finance, and growth models. In general the initial time is assumed to be known, and is set to $t_0 = 0$, and the terminal time can be finite or infinite, as can be seen next in Sect. 9.2.

9.1 Calculus Variations for a Generic Independent Variable

In this section, we specify and present the optimum conditions for several types of CV problems in which the independent variable, x, belongs to an ordered set $\mathrm{X} \subseteq \mathbb{R}$. Section 9.1.1 is dedicated to the simplest problems in which the boundaries of set X, and the associated values of $y(x)$ are fixed or free, and Sect. 9.1.2 presents several extensions to those problems.

9.1.1 Problems with Bounded Domains

Throughout this section, we assume that there is only one independent variable with a closed or open domain $\mathrm{X} = [x_0, x_1] \subset \mathbb{R}$, or $\mathrm{X} = (x_0, x_1) \subseteq \mathbb{R}$, where $x_0 < x_1$, and there is a single independent variable $y : \mathrm{X} \to \mathrm{Y} \subseteq \mathbb{R}$.

[2] In Chaps. 14 and 16, we briefly address problems in which the goal is to find optimal distribution trajectories.

The simplest CV problem. Let the two bounds of set X, x_0 and x_1, and two elements of set Y, $y_0 \neq y_1$, be fixed. The simplest CV problem objective is to find function $y : \mathrm{X} \to \mathrm{Y} \subseteq \mathbb{R}$, that solves the problem:

$$\max_{y(\cdot)} \int_{x_0}^{x_1} F(y'(x), y(x), x)\, dx \tag{9.2a}$$

subject to

$$y(x_0) = y_0, \text{ and } y(x_1) = y_1 \text{ fixed,} \tag{9.2b}$$

$$x_0, \text{ and } x_1 \text{fixed.} \tag{9.2c}$$

We call **admissibility set**, $\mathcal{Y}$, to the set

$$\mathcal{Y} \equiv \left\{ y \in C^2(\mathrm{X}; \mathrm{Y}) : y(x_0) = y_0 \text{ and } y(x_1) = y_1, \text{ and } x_0, x_1, y_0, y_1 \text{fixed} \right\},$$

which is the set of continuous and continuously differentiable functions satisfying the side conditions of the CV problem. The solution to the problem, y^*, if it exists, is an element of set $\mathcal{Y}$ that maximizes the objective functional (9.2a).

Let $\zeta = (x_0, x_1, y_0, y_1)$ denote the data of the problem. The optimum value of the objective functional is a function depending on the data of the problem, and is called **value function**,[3]

$$V(\zeta) = \mathsf{F}[y^*] \equiv \max_{y \in \mathcal{Y}} \int_{x_0}^{x_1} F\big(y^{*\prime}(x), y^*(x), x\big)\, dx.$$

Proposition 9.1 (First-order necessary conditions for the simplest CV problem) *Let $y^* \in \mathcal{Y}$ be the solution to the simplest CV problem* (9.2). *Then it satisfies the* ***Euler-Lagrange equation***[4]

$$F_y\big(y^{*\prime}(x), y^*(x), x\big) = \frac{d}{dx}\Big(F_{y'}\big(y^{*\prime}(x), y^*(x), x\big)\Big), \text{ for each } x \in \mathrm{X}, \tag{9.3}$$

together with the boundary conditions

$$y^*(x_0) = y_0, \text{ and } y^*(x_1) = y_1. \tag{9.4}$$

Proof Assume that $y^* \in \mathcal{Y}$ is known. Then $\mathsf{F}[y^*]$ is also known. Function y^* is an optimum only if $\mathsf{F}[y^*] \geq \mathsf{F}[y]$ for any other admissible function $y \in \mathcal{Y}$. Take an admissible variation over y^*, $y = y^* + \delta y$ such that the variation is a *parameterized perturbation* of y^*, satisfying $\delta y = \varepsilon\, \eta$ where $\eta : \mathrm{X} \to \mathrm{Y}$ and ε is a number. A variation to be admissible has to satisfy $y(x_1) = y^*(x_1) = y_1$ and $y(x_0) = y^*(x_0) =$

---

[3] The functional $\mathsf{F}[y^*]$ is a function depending on a particular value of the parameter set ζ. Therefore, we can take it as a mapping $V : \mathrm{X} \times \mathrm{Y} \to \mathbb{R}$.

[4] We use the notation $F_y(y', y, x) = \frac{\partial F(y', y, x)}{\partial y}$ and $F_{y'}(y', y, x) = \frac{\partial F(y', y, x)}{\partial y'}$.

y_0. Therefore, an admissible perturbation has to satisfy $\eta(x_0) = \eta(x_1) = 0$, and it can take arbitrary values $\eta(x) \in \mathrm{Y}$ for each x in the interior of the domain X. The value functional for the perturbed function y is

$$\mathsf{F}[y] = \mathsf{F}[y^* + \varepsilon\eta] = \int_{x_0}^{x_1} F\big(y^{*\prime}(x) + \varepsilon\,\eta'(x), y^*(x) + \varepsilon\,\eta(x), x\big)\,dx.$$

A first-order expansion of the functional $\mathsf{F}[y]$ in a neighborhood of y^*, is $\mathsf{F}[y] = \mathsf{F}[y^*] + \mathrm{D}_{\eta(\cdot)}\mathsf{F}[y^*]\,\varepsilon + O(\varepsilon^2)$. Then $\mathsf{F}[y^*] \geq \mathsf{F}[y]$ only if the first Gâteaux differential of $\mathsf{F}[y]$ evaluated at the optimum is zero: $\mathrm{D}_{\eta(\cdot)}\mathsf{F}[y^*] = 0$. From Lemma 8.1 the Gâteaux differential, evaluated at y^*, for an arbitrary perturbation η, is

$$\mathrm{D}_{\eta(\cdot)}\mathsf{F}[y^*] = \int_{x_0}^{x_1} \Big(F_y\big(y^{*\prime}(x), y^*(x), x\big)\,\eta(x) + F_{y'}\big(x, y^*(x), y^{*\prime}(x)\big)\,\eta'(x)\Big)\,dx.$$

Integrating by parts the second integral yields

$$\begin{aligned}\mathrm{D}_{\eta(\cdot)}\mathsf{F}[y^*] = &\int_{x_0}^{x_1} \Big(F_y\big(y^{*\prime}(x), y^*(x), x\big) - \frac{d}{dx}\Big(F_{y'}\big(y^{*\prime}(x), y^*(x), x\big)\Big)\Big)\,\eta(x)\,dx \\ &+ F_{y'}(y^{*\prime}(x_1), y^*(x_1), x_1)\,\eta(x_1) - F_{y'}(y^{*\prime}(x_0), y^*(x_0), x_0)\,\eta(x_0).\end{aligned} \tag{9.5}$$

As $\eta(x_1) = \eta(x_0) = 0$ for the perturbation to be admissible, then Eq. (9.5) reduces to

$$\mathrm{D}_{\eta(\cdot)}\mathsf{F}[y^*] = \int_{x_0}^{x_1} \Big(F_y\big(y^{*\prime}(x), y^*(x), x\big) - \frac{d}{dx}\Big(F_{y'}\big(y^{*\prime}(x), y^*(x), x\big)\Big)\Big)\,\eta(x)\,dx.$$

If $F(.)$ is a continuous function, we can use the following result: if $h := [x_0, x_1] \to \mathbb{R}$ is a continuous function and we specify $\int_{x_0}^{x_1} h(x)\eta(x)dx$ for all C^1 functions η such that $\eta(x_0) = \eta(x_1) = 0$ then $\int_{x_0}^{x_1} h(x)\eta(x)dx = 0$ if and only if $h(x) = 0$ for all $x \in (x_0, x_1)$. Therefore $\mathrm{D}_{\eta(\cdot)}\mathsf{F}[y^*] = 0$ if an only if condition (9.3) holds for every $x \in \mathrm{X} = [x_0, x_1]$.

Expanding the derivatives, we find that the Euler-Lagrange (EL) equation is the second-order ODE, if $F_{y'y'} \neq 0$,

$$F^*_{y'y'}(x)\,y''(x) + F^*_{y'y}(x)\,y'(x) + F^*_{y'x}(x) - F^*_y(x) = 0,\ \text{ for each } x_0 < x < x_1,$$

where $F^*(x) = F(y^{*\prime}(x), y^*(x), x)$, and using a similar notation for the other derivatives.[5]

Alternatively, we can transform the EL into a system of first-order ODE. If we define $y_1 = y$ and $y_2 = y'$ then

---

[5] Specifically, $F^*_{y'}(x) = \frac{\partial F(y^{*\prime}(x), y^*(x), x)}{\partial y'(x)}$, $F^*_y(x) = \frac{\partial F(y^{*\prime}(x), y^*(x), x)}{\partial y(x)}$, $F^*_x(x) = \frac{\partial F(y^{*\prime}(x), y^*(x), x)}{\partial x}$, $F^*_{y'y'}(x) = \frac{\partial^2 F(y^{*\prime}(x), y^*(x), x)}{\partial y'(x)^2}$, et caetera.

$$y_1' = y_2$$
$$F^*_{y_2y_2}(x, y_1, y_2)\, y_2' = F^*_{y_1}(x, y_1, y_2) - F^*_x(x, y_1, y_2) - F^* F_{y_2y_1}(x, y_1, y_2) y_2.$$

To find the particular solution to the EL equation, we need to have side conditions allowing for the explicit determination of the value of y at the two boundaries for $x = x_0$ and $x = x_1$. In the simplest problem's case, both the boundary values of the independent variable and of the dependent variable are given. The **existence of solutions to the CV** problem is thus reduced to the existence of solutions to a mixed initial-terminal ODE problem.

The previous first-order necessary condition only refers to an extremum. To be a maximand, it has to satisfy the following second-order necessary condition:

Proposition 9.2 (Second-order necessary conditions for problem (9.2)) *Let $y^* \in \mathcal{Y}$ be the solution to the CV problem (9.2). The solution is a maximand only if it satisfies the Legendre-Clebsch condition*

$$F_{y'y'}(y^{*\prime}(x), y^*(x), x) \leq 0, \; \textit{for every } x \in \mathrm{X} \tag{9.6}$$

Proof A second-order expansion of the functional $\mathsf{F}[x]$ in a neighborhood of y^*, yields $\mathsf{F}[y] = \mathsf{F}[y^*] + \mathrm{D}_{\eta(\cdot)}\mathsf{F}[y^*]\,\varepsilon + \frac{1}{2}\mathrm{D}_{\eta(\cdot)\eta(\cdot)}\mathsf{F}[y^*]\,\varepsilon^2 + O(\varepsilon^3)$. Because at the optimum for any admissible perturbation η we have $\mathrm{D}_{\eta(\cdot)}\mathsf{F}[y^*] = 0$, and at a maximum $\mathsf{F}[y] \leq \mathsf{F}[y^*]$, a necessary condition is $\mathrm{D}_{\eta(\cdot)\eta(\cdot)}\mathsf{F}[y^*] \leq 0$, that is

$$\begin{aligned}
\mathrm{D}_{\eta(\cdot)\eta(\cdot)}\mathsf{F}[y^*] &= \int_{x_0}^{x_1} \left(F^*_{yy}(x)\, \eta(x)^2 + 2\, F^*_{yy'}(x)\, \eta(x)\, \eta'(x) + F^*_{y'y'}(x)\, (\eta'(x))^2 \right) dx \\
&= \int_{x_0}^{x_1} \left(F^*_{yy}(x)\, \eta(x)^2 + F^*_{yy'}(x) \frac{d}{dx}\left(\eta(x)^2\right) + F^*_{y'y'}(x)\, (\eta'(x))^2 \right) dx \\
&= \int_{x_0}^{x_1} \left(\left(F^*_{yy}(x) - \frac{d}{dx} F^*_{yy'}(x) \right) \eta(x)^2 + F^*_{y'y'}(x) (\eta'(x))^2 \right) dx + \\
&\quad + F^*_{yy'}(x) \eta(x)^2 \Big|_{x_0}^{x_1}.
\end{aligned}$$

From the admissibility conditions $\eta(x_0) = \eta(x_1) = 0$, and following Liberzon (2012, pp. 59–60), it can be shown that $\mathrm{D}_{\eta(\cdot)\eta(\cdot)}\mathsf{F}[y^*] \leq 0$ only if condition (9.6) holds.

Proposition 9.3 (Sufficient conditions for problem (9.2)) *Assume that $F_{y'y'}(y', y, x) \leq 0$, for each $x \in \mathrm{X}$, and let $y^* \in \mathcal{Y}$ satisfy the EL equation (9.3). Then y^* is the solution to problem (9.2).*

Proof See Liberzon (2012, pp. 62–68)

Proposition 9.4 ((Necessary and sufficient conditions of problem (9.2)) *Consider the simplest calculus of variations problem (9.2) and assume that $F_{y'y'}(y', y, x) \leq 0$ for every $x \in \mathrm{X}$. Then the EL equation (9.3), and the boundary conditions (9.4) are necessary and sufficient conditions for the solution to problem* (9.2).

Free boundary values for the dependent variable. Let the boundaries of the set of independent variables be fixed, and assume that at least one of the associated values for the dependent variable, $y(x_0)$ or $y(x_1)$, is free. The CV problems with fixed boundary values for the independent variable, and several alternative free boundary values for the dependent variable can be formally stated as follows

$$\begin{aligned}
\max_{y(\cdot)} \int_{x_0}^{x_1} & F(y'(x), y(x), x)\, dx \\
& \text{subject to } x_0 \text{ and } x_1 \text{ fixed,} \\
& \text{and } y(x_0) \text{ free, and } y(x_1) \text{ free,} && (9.7a) \\
& \text{or } y(x_0) = y_0 \text{ fixed, and } y(x_1) \text{ free,} && (9.7b) \\
& \text{or } y(x_0) \text{ free, and } y(x_1) = y_1 \text{ fixed.} && (9.7c)
\end{aligned}$$

The data and the value function for problem with constraint (9.7a) is $\zeta = (x_0, x_1)$, and the value function is

$$V(\zeta) = \max_{y(\cdot)} \left\{ \int_{x_0}^{x_1} F(y'(x), y(x), x)\, dx : x_0, \text{ and } x_1 \text{ fixed} \right\}.$$

The value functions for problems with constraint (9.7b), or constraint (9.7c), are $V(x_0, x_1, y_0)$, or $V(x_0, x_1, y_1)$, respectively, and the function domain $\mathcal{Y}$ is defined accordingly.

Proposition 9.5 (First-order necessary conditions for CV problems with fixed boundary independent variables) *Let y^* be the solution to a CV problem with fixed boundary values for the independent variable, x_0 and x_1, and with at least one free boundary value for the dependent variable* (9.7). *Then it satisfies the EL equation* (9.3), *and one of the following conditions regarding the values of the dependent variable for boundary values of the independent variable:*

1. *if both boundary values are free (as in constraint* (9.7a)*)*

$$F_{y'}(y^{*\prime}(x_0), y^*(x_0), x_0) = 0, \text{ and } F_{y'}(y^{*\prime}(x_1), y^*(x_1), x_1) = 0 \tag{9.8}$$

2. *if the lower boundary value is given by $y(x_0) = y_0$, and the upper boundary value is free (as in constraint* (9.7b)*)*

$$y^*(x_0) = y_0, \text{ and } F_{y'}(y^{*\prime}(x_1), y^*(x_1), x_1) = 0 \tag{9.9}$$

3. *if the upper boundary value is given by $y(x_1) = y_1$, and the lower boundary value is free (as in constraint* (9.7c)*)*

$$F_{y'}(y^{*\prime}(x_0), y^*(x_0), x_0) = 0, \text{ and } y^*(x_1) = y_1. \tag{9.10}$$

Proof The proof follows the same steps as the proof of Proposition (9.1) until Eq. (9.5). However, in the present case, the boundary values for the perturbation function, $\eta(x_0)$ and/or $\eta(x_1)$, are different from zero when the associated value for the dependent variable, $y(x_0)$ and/or $y(x_1)$, is free. In particular, we have: for problem the problem with constraint (9.7a) $\eta(x_0)$ and $\eta(x_1)$ can take any value, for the problem with constraint(9.7b) $\eta(x_0) = 0$ and $\eta(x_1)$ can take any value, and for problem with constraint (9.7c) $\eta(x_0)$ can take any value and $\eta(x_1) = 0$. Then to have $\mathsf{D}_{\eta(\cdot)}\mathsf{F}[y^*] = 0$ in Eq. (9.5), in addition to the Euler-Lagrange condition, we should also have

$$F_{y'}\left(y^{*\prime}(x_j), y^*(x_j), x_j\right)\eta(x_j) = 0, \text{ for } j = 0, 1. \tag{9.11}$$

Thus, the two adjoint conditions for an optimum at the boundary values x_j, for $j = 0, 1$, are: First, if $y(x_j)$ is fixed, i.e., $y(x_j) = y_j$, an admissible perturbation has to satisfy $\eta(x_j) = 0$, implying that condition (9.11) is satisfied. This is the case in Proposition 9.1. Second, if $y(x_j)$, for any $j \in \{0, 1\}$, is free, then the related perturbation value is arbitrary and $\eta(x_j) \neq 0$ in general. The optimally condition (9.11) holds if and only if $F_{y'}\left(y^{*\prime}(x_j), y^*(x_j), x_j\right) = 0$, for the associated j. Therefore, the adjoint conditions (9.8), (9.9), or (9.10) are obtained.

In time-varying models in which the value of the state variable is known at time $t = 0$ and the terminal value of the state variable is endogenous, we supplement the Euler-Lagrange with condition (9.9).

Free boundary values for the independent variable. There are models in which the initial value of the independent variable is unknown. This is the case, for instance, in optimal taxation models of the Mirrlees (1971) type in which the independent variable is the skill level and the initial condition refers to a cutoff level of skill bellow which taxes should be zero.

We consider the following problems: find function $y \in \mathcal{Y} \in C^2(\mathrm{X}^*)$, that is the continuously differentiable functions $y : \mathrm{X}'' \to \mathrm{Y} \subseteq \mathbb{R}$, where $y(x_j^*) = y_j$, for $j \in \{0, 1\}$, are known, and where X^* has at least one boundary value, x_0^* and/or x_1^*, that is free. We can have the following cases: $\mathrm{X}^* = [x_0^*, x_1]$, or $\mathrm{X}^* = [x_0, x_1^*]$ or $\mathrm{X}^* = [x_0^*, x_1^*]$, where x_j is fixed and x_j^* is free. The objective functional is again (9.1).

Formally, the three possible CV problems, with alternative constraints, are :

$$\max_{y(\cdot), x_0, x_1} \int_{x_0}^{x_1} F(y'(x), y(x), x)\, dx$$

$$\begin{aligned} &\text{subject to } y(x_0) = y_0 \text{ and } y(x_1) = y_1 \text{ fixed,} \\ &\text{and } x_0 \text{ and } x_1 \text{ free,} && (9.12a) \\ &\text{or } x_0 \text{ fixed, and } x_1 \text{ free,} && (9.12b) \\ &\text{or } x_0 \text{ free, and } x_1 \text{ fixed.} && (9.12c) \end{aligned}$$

In the case of the CV problem with constraint (9.12a), the data of the problem are $\zeta = (y_0, y_1)$, and the value functional is

$$V(y_0, y_1) = \max_{y(\cdot), x_0, x_1} \int_{x_0}^{x_1} F(y'(x), y(x), x)\, dx = \int_{x_0^*}^{x_1^*} F(y^{*\prime}(x), y^*(x), x)\, dx,$$

where x_0^* and/or x_1^* are determined at the optimum. The value functional for CV problems with constraints (9.12b) or (9.12c) are defined analogously.

Proposition 9.6 (First-order necessary conditions for free boundary exogenous variable CV problems) *Let y^* be the solution to a CV problem with known boundary values for the state variable, y_0 and y_1, and free terminal values for the independent variable, as in constraints* (9.12a), (9.12b), *or* (9.12c). *Then it satisfies the EL equation* (9.3), *and one of the following boundary conditions:*

1. *if both boundary values for the independent variable are free (in problem with constraint* (9.12a)*)*

$$\begin{aligned} F(y^{*\prime}(x_0^*), y_0, x_0^*) - F_{y'}(y^{*\prime}(x_0^*), y_0, x_0^*)\, y^{*\prime}(x_0^*) &= 0, \\ F(y^{*\prime}(x_1^*), y_1, x_1^*) - F_{y'}(y^{*\prime}(x_1^*), y_1, x_1^*)\, y^{*\prime}(x_1^*) &= 0; \end{aligned} \tag{9.13}$$

2. *if the lower boundary value for the independent variable is fixed, $x_0^* = x_0$, and the upper boundary for the independent variable is free (problem with constraint* (9.12b)*)*

$$x_0^* = x_0, \text{ and } F(y^{*\prime}(x_1^*), y_1, x_1^*) - F_{y'}(y^{*\prime}(x_1^*), y_1, x_1^*)\, y^{*\prime}(x_1^*) = 0; \tag{9.14}$$

3. *if the upper boundary value for the independent variable is fixed, $x_1^* = x_1$, and the lower boundary for the independent variable is free (problem with constraint* (9.12c)*)*

$$F(y^{*\prime}(x_0^*), y_0, x_0^*) - F_{y'}(y^{*\prime}(x_0^*), y_0, x_0^*)\, y^{*\prime}(x_0^*) = 0, \text{ and } x_1^* = x_1. \tag{9.15}$$

Proof Let $x_0 = x_0^*$ and/or $x_1 = x_1^*$ be fixed. The solution to the dependent function $y^*(x)$ for $x \in [x_0^*, x_1^*]$, can be free or can be fixed. In the last case, we write $y^*(x_j^*) = y_j$, where y_j is fixed, for $j \in \{0, 1\}$.Two types of perturbations are introduced: a perturbation to the dependent variable $y(x) = y^*(x) + \varepsilon\eta(x)$, and another perturbation to the independent variable $x = x^* + \varepsilon\chi(x)$. If we denote $y_j^* = y^*(x_j^*)$, for $j = 0, 1$, the two boundary values for the independent and dependent variables are $P_j^* \equiv (x_j^*, y_j^*)$, for $j = 0, 1$, at the optimum. The related terminal points for the perturbed solution are $P_j = (x_j^* + \varepsilon\chi_j, y_j^* + \varepsilon\eta_j)$, for $j = 0, 1$. At the optimum, the objective functional $\mathsf{F}[y^*; x^*] = \int_{x_0^*}^{x_1^*} F(y^{*\prime}(x), y^*(x), x)\, dx$, is integrated between two optimum boundary values. The perturbed functional is

$$\mathsf{F}[y^* + \varepsilon\eta; x^* + \varepsilon\chi] = \int_{x_0^* + \varepsilon\chi_0}^{x_1^* + \varepsilon\chi_1} F(y^{*\prime}(x) + \varepsilon\eta'(x), y^*(x) + \varepsilon\,\eta(x), x)\, dx.$$

The variation of the functional $\Delta \mathsf{F} = \mathsf{F}[y^* + \varepsilon\eta; x^* + \varepsilon\chi] - \mathsf{F}[y^*; x^*]$ is

$$\begin{aligned}
\Delta \mathsf{F} = & \int_{x_0^*+\varepsilon\,\chi_0}^{x_1^*+\varepsilon\,\chi_1} F\big(y^{*\prime}(x) + \varepsilon\,\eta'(x), y^*(x) + \varepsilon\,\eta(x), x\big)\,dx \\
& - \int_{x_0^*}^{x_1^*} F(y^{*\prime}(x), y^*(x), x)\,dx \\
= & \int_{x_0^*}^{x_1^*} \Big(F(y^{*\prime}(x) + \varepsilon\,\eta'(x), y^*(x) + \varepsilon\,\eta(x), x) - F(y^{*\prime}(x), y^*(x), x)\Big)dx \\
& + \int_{x_1^*}^{x_1^*+\varepsilon\chi_1} F(y^{*\prime}(x) + \varepsilon\,\eta'(x), y^*(x) + \varepsilon\,\eta(x), x)\,dx \\
& - \int_{x_0^*+\varepsilon\chi_0}^{x_0^*} F(y^{*\prime}(x) + \varepsilon\,\eta'(x), y^*(x) + \varepsilon\,\eta(x), x)\,dx
\end{aligned}$$

Denoting $F^*\big(x\big) = F\big(y^{*\prime}(x), y^*(x), x\big)$ and using the mean-value theorem, yields

$$\Delta \mathsf{F} = \varepsilon\Bigg(\int_{x_0^*}^{x_1^*} \big(F_y^*(x)\,\eta(x) + F_{y'}^*(x)\,\eta'(x)\big)\,dx + F(\tilde{x}_1)\,\chi_1 - F(\tilde{x}_0)\,\chi_0\Bigg),$$

where $\tilde{x}_1 \in (x_1^*, x_1^* + \varepsilon\,\chi_1)$ and $\tilde{x}_0 \in (x_0^*, x_0^* + \varepsilon\,\chi_0)$. The Gâteaux differential $\mathsf{D}_{(\eta,\chi)}\mathsf{F}[y^*; x^*] = \lim_{\varepsilon\to 0} \frac{\Delta\mathsf{F}(\varepsilon)}{\varepsilon}$, is

$$\mathsf{D}_{(\eta,\chi)}\mathsf{F}[y^*; x^*] = \int_{x_0^*}^{x_1^*} \Big(F_y^*(x)\eta(x) + F_{y'}^*(x)\,\eta'(x)\Big)\,dx + F^*(x_1^*)\,\chi_1 - F^*(x_0^*)\,\chi_0.$$

Integrating by parts yields

$$\begin{aligned}
\mathsf{D}_{(\eta,\chi)}\mathsf{F}[y^*; x^*] = & \int_{x_0^*}^{x_1^*} \Big(F_y^*(x) - \frac{d}{dx}F_{y'}^*(x)\Big)\,\eta(x)dx + \\
& + F_{y'}^*(x_1^*)\,\eta(x_1^*) - F_{y'}^*(x_0^*)\,\eta(x_0^*) + F^*(x_1^*)\,\chi_1 - F^*(x_0^*)\,\chi_0.
\end{aligned}$$

We only know the perturbations for the state variables at the perturbed boundaries x_0 and x_1, and not at the optimum, x_0^* and x_1^*, which inhibits the computation of the integral in the last equation. In order to find $\eta(x_j^*)$, we approximate $\eta(x_j^*) \approx \eta_j - y'(x_j^*)\chi_j$ for $j \in \{0, 1\}$. Therefore,

$$\begin{aligned}
D_{(\eta,\chi)}\mathsf{F}[y^*; x^*] = & \int_{x_0^*}^{x_1^*} \Big(F_y^*(x) - \frac{d}{dx}F_{y'}^*(x)\Big)\,\eta(x)dx + F_{y'}^*(x_1^*)\eta_1 - F_{y'}^*(x_0^*)\eta_0 + \\
& + \big(F^*(x) - F_{y'}^*(x)\,y'(x)\big)\big|_{x=x_1^*}\,\chi_1 - \big(F^*(x) - F_{y'}^*(x)\,y'(x)\big)\big|_{x=x_0^*}\,\chi_0.
\end{aligned} \tag{9.16}$$

As the terminal values of the state variables, $y(x_0^*) = y_0$ and $y(x_1^*) = y_1$, are fixed, then the terminal perturbation for the independent variable should satisfy $\eta_0 = \eta_1 = 0$. Therefore, $\mathsf{D}_{(\eta,\chi)}\mathsf{F}[y^*; x^*] = 0$ if and only if the EL equation (9.3) is satisfied, and $\left(F^*(x) - F^*_{y'}(x)y'(x)\right)\big|_{x=x_1^*}\chi_1 = 0$ if x_1 is free, and/or $\left(F^*(x) - F^*_{y'}(x)y'(x)\right)\big|_{x=_0^*}\chi_0 = 0$ if x_0 is free. From those conditions we can find the three constraints (9.13), (9.14) and (9.15).

Free boundaries for both independent and dependent variables. The most general problem is: find function $y^* \in \mathcal{Y}$ among continuously differentiable functions $y : \mathrm{X}^* \to \mathrm{Y} \subseteq \mathbb{R}$, where X^* has at least one unknown boundary, x_0^* and/or x_1^*, as in the previous subsection, and the terminal values for the state variables, $y(x_0^*)$ and/or $y(x_1^*)$, can also be free. The objective functional is again (9.1).

Formally, the following problem contais all the alternative cases :

$$
\begin{aligned}
&\max_{y(\cdot)} \int_{x_0}^{x_1} F(y'(x), y(x), x)\,dx \\
&\quad \text{subject to} \\
&\quad y(x_0) \text{ and } x_0 \text{ free, and } y(x_1) = y_1 \text{ and } x_1 \text{ fixed,} && \text{(9.17a)} \\
&\quad \text{or } y(x_0) = y_0 \text{ and } x_0 \text{ fixed, and } y(x_1) \text{ and } x_1 \text{ free,} && \text{(9.17b)} \\
&\quad \text{or } y(x_0),\ x_0,\ y(x_1), \text{ and } x_1 \text{ free.} && \text{(9.17c)}
\end{aligned}
$$

In case of problem (9.17c), the data of the problem, ζ, is empty, which means that it only involves parameters that may be present in function $F(.)$. The value functional is

$$
V = \max_{y(\cdot), x_0, x_1} \int_{x_0}^{x_1} F(y'(x), y(x), x)dx = \int_{x_0^*}^{x_1^*} F(x, y^*(x), y^{*\prime}(x))dx
$$

where x_0^* and/or x_1^* and $y^*(x_0^*)$ and/or $y^*(x_1^*)$ are determined endogenously.

Proposition 9.7 (First-order necessary conditions for the free boundary independent and state variable problems) *Let y^* be the solution of the CV problem with free boundary values for the state variable and for the independent variables, problems* (9.17a), (9.17b), *or* (9.17c).. *Then it satisfies the Euler-Lagrange equation* (9.3), *and one of the boundary conditions:*

1. *if both values for lower boundaries are free (for a problem with constraint* (9.17a)),

$$
\begin{aligned}
&F_y(y^{*\prime}(x_0^*), y^*(x_0^*), x_0^*) = F_{y'}(y^{*\prime}(x_0^*), y^*(x_0^*), x_0^*) = 0, \\
&y^*(x_1) = y_1;
\end{aligned}
\tag{9.18}
$$

2. *if both values for upper boundaries are free (for a problem with constraint* (9.17b))

$$
\begin{aligned}
&y^*(x_0) = y_0, \\
&F_y(y^{*\prime}(x_1^*), y^*(x_1^*), x_1^*) = F_{y'}(y^{*\prime}(x_1^*), y^*(x_1^*), x_1^*) = 0;
\end{aligned}
\tag{9.19}
$$

Table 9.1 Adjoint conditions for bounded domain CV problems

Data		Side conditions at the optimum	
x_j	$y(x_j)$	x_j^*	$y^*(x_j^*)$
Fixed	Fixed	x_j	y_j
Fixed	Free	x_j	$F_{y'}(y^{*\prime}(x_j), y^*(x_j), x_j) = 0$
Free	Fixed	$F(y^{*\prime}(x_j^*), y_j, x_j^*) -$ $y^{*\prime}(x_j^*)\, F_{y'}(y^{*\prime}(x_j^*), y_j, x_j^*) = 0$	y_j
Free	Free	$F(y^{*\prime}(x_j^*), y^*(x_j^*), x_j^*) = 0$	$F_{y'}(y^{*\prime}(x_j^*), y^*(x_j^*), x_j^*) = 0$

3. *if all terminal values for x and $y(x)$ are free (for a problem with constraint (9.17c))*

$$F_y(y^{*\prime}(x_0^*), y^*(x_0^*), x_0^*) = F_{y'}(y^{*\prime}(x_0^*), y^*(x_0^*), x_0^*) = 0, \tag{9.20a}$$

$$F_y(y^{*\prime}(x_1^*), y^*(x_1^*), x_1^*) = F_{y'}(y^{*\prime}(x_1^*), y^*(x_1^*), x_1^*) = 0. \tag{9.20b}$$

Proof We can take condition Eq. (9.16), and consider the perturbations $\eta_0 \neq 0$, $\eta_1 \neq 0$, $\chi_0 \neq 0$ and/or $\chi_1 \neq 0$, when $y(x_0)$, $y(x_1)$, x_0, and/or x_0, respectively, are free.

Table 9.1 assembles the side conditions for all the previous CV problems. Observe that if we consider all the possible combinations of the information on both boundaries, we have **16 possible cases**.

Example 9.1 (*Spatial sale distribution with adjustment costs*) Assume a Salop (1979) circle with a fixed size $L > 0$, and consider the problem of a firm selling a product across that region, and facing adjustment costs for relocating its product. It is assumed that the firm is a price taker, and the price, $p > 0$, is the same across all locations. Denoting the firm's sales at location x by $y(x)$, the firm's problem is to maximize the average profit

$$\Pi[y] = \max_{y(\cdot)} \frac{1}{L} \int_{-L/2}^{L/2} p\, y(x) - \frac{1}{2} \left(y'(x)\right)^2 dx \tag{9.21a}$$

$$\text{subject to } y(-L/2) = y(L/2) = \bar{y}, \tag{9.21b}$$

where $\bar{y}$ can be thought as the sales on the most distant location. This is a simplest CV problem in which both the two boundary values for the independent and the dependent variable are known. The EL condition is $y''(x) + p = 0$, for each $-\frac{L}{2} \leq p \leq \frac{L}{2}$, and has solution $y(x) = c_1 + c_2\, x - \frac{p}{2} x^2$. Evaluating it at the two boundary conditions (9.21b) yields $c_1 = \bar{y} + \frac{p}{2} \left(\frac{L}{2}\right)^2$ and $c_2 = 0$. Therefore, the solution is

$$y^*(x) = \bar{y} + \frac{p}{2}\left(\left(\frac{L}{2}\right)^2 - x^2\right), \text{ for each } -\frac{L}{2} \leq p \leq \frac{L}{2}.$$

The profit is $\Pi[y^*] = p\left(\bar{y} + \frac{p}{3}\left(1 - \frac{p}{2}\right)\left(\frac{L}{2}\right)^2\right)$.

Observe that, if the firm would like to find $\bar{y}$ optimally, the problem would not have a solution in the reals for $L > 0$, because the boundary conditions would be incompatible as $c_2 = -\frac{p}{2}L$ and $c_2 = \frac{p}{2}L$.

9.1.2 Extensions

Several extensions to the previous CV problems are possible, including the introduction of second-order derivatives in the objective function $F(\cdot)$, and the addition of of functional constraints, as in Chap. 8, or more than one independent variable, as in Chaps. 14 and 16. In this section, we consider a problem, that is frequent in economics, in which there are inequality boundary constraints.

Let the two limits for the independent variable, x_0 and x_1, be fixed, but $y(x_0)$ and $y(x_1)$ be constrained by $R_0(x_0, y(x_0)) \geq 0$ and $R_1(x_1, y(x_1)) \geq 0$. Formally, the problem is:

$$\max_{y(\cdot)} \int_{x_0}^{x_1} F(y'(x), y(x), x)\, dx$$
$$\text{subject to } x_0, \text{ and } x_1 \text{ fixed,}$$
$$R_0(y(x_0), x_0) \geq 0, \tag{9.22a}$$
$$R_1(y(x_1), x_1) \geq 0. \tag{9.22b}$$

Proposition 9.8 (First-order necessary conditions for the inequality constrained problem) *Let y^* be the solution to the CV problem (9.13). Then it satisfies the Euler-Lagrange equation* (9.3)*, the boundary conditions*

$$F_{y'}(y^{*\prime}(x_0), y^*(x_0), x_0)\, R_0(y^*(x_0), x_0) = 0, \tag{9.23a}$$
$$F_{y'}(y^{*\prime}(x_1), y^*(x_1), x_1)\, R_1(y^*(x_1), x_1) = 0, \tag{9.23b}$$

and the constraints (9.22a) *and* (9.22b) *evaluated at the optimum, $R_0(y^*(x_0), x_0) \geq 0$, and $R_0(y^*(x_1), x_1) \geq 0$.*

Proof The Lagrange functional, with Lagrange multipliers associated to the two boundary inequality constraints (9.22a) and (9.22b), is

$$\mathsf{L}[y] = \int_{x_0}^{x_1} F(y'(x), y(x), x)\, dx + \mu_0\, R_0(y(x_0), x_0) + \mu_1\, R_1(y(x_1), x_1).$$

Perturbing the functional at the optimum from $\mathsf{L}[y^*]$ to $\mathsf{L}[y] = \mathsf{L}[y^* + \varepsilon\,\eta]$, yields the Gâteaux differential

$$\begin{aligned} \mathrm{D}_\eta \mathsf{L}[y^*] &= \int_{x_0}^{x_1} \left(F_y^*(x) - \frac{d}{dx} F_{y'}^*(x)\right) \eta(x)\, dx + \\ &+ \left(F_{y'}^*(x_0) + \mu_0\, R_{0,y}^*(x_0)\right) \eta(x_0) + \left(F_{y'}^*(x_1) + \mu_1\, R_{1,y}^*(x_1)\right) \eta(x_1) \end{aligned}$$

where $F^*(x) = F(y^{*\prime}(x), y^*(x), x)$ and $R_{j,y}^*(x_j) = \frac{\partial R_j(y^*(x_j), x_j)}{\partial y(x)}$, for $j = 0, 1$. As the values of the dependent variable for the boundary values of the independent variable are implicitly free, the associated admissible perturbation levels are $\eta(x_j) \neq 0$, for $j = 0, 1$. Therefore, the necessary condition $\mathrm{D}_\eta L[y^*] = 0$ is satisfied if and only if the adjoint conditions $F_{y'}^*(x_j) + \mu_j\, R_{j,y}^*(x_j) = 0$ hold for each $j = 0, 1$. Furthermore, due to the existence of two static inequality constraints at the boundaries x_0 and x_1, the Karush-Kuhn-Tucker (KKT) complementarity slackness conditions,

$$\mu_j\, R_j^*(x_j) = 0,\;\; \mu \geq 0 \text{ and } R_j^*(x_j) \geq 0,\;\; \text{for } j = 0, 1,$$

where $R_j^*(x_j) \equiv R_j(y^*(x_j), x_j)$, are also necessary. Multiplying the adjoint conditions by $R_j^*(x_j)$ we obtain equivalent conditions

$$R_j^*(x_j)\, F_{y'}^*(x_j) + \mu_j\, R_j^*(x_j)\, R_{j,y}^*(x_j) = 0,\;\; \text{for } j = 0, 1,$$

that, from the KKT complementary slackness conditions, is equivalent to $F_{y'}^*(x_j)$ $R_j^*(x_j) = 0$. Therefore $\mathrm{D}_\eta \mathsf{L}[y^*] = 0$ if the EL equation (9.3), and the adjoint boundary conditions (9.23a) and (9.23b), are satisfied.

Exercise 9.1 Find the first-order necessary conditions for CV problems with the same boundary constraints, (9.22a) and (9.22b), but in which $y(x_j)$ are fixed but x_j are free, for $j = 0, 1$.

Exercise 9.2 Find the first-order necessary conditions for CV problems with the same boundary constraints, (9.22a) and (9.22b), but in which both $y(x_j)$ and x_j are free, for $j = 0, 1$.

Exercise 9.3 Find the first-order necessary conditions for the CV problems in Table 9.1 when there is an additional functional side condition $\int_{x_0}^{x_1} G(y'(x), y(x), x)$ $dx \leq g_0$, where g_0 is a known constant and function $G(\cdot)$ is continuously differentiable in (y, y').

9.2 Calculus of Variations for Intertemporal Problems

For time-dependent, or intertemporal, CV problems we can apply the previous results, for formally similar problems, taking into consideration that the independent variable usually belongs to the unbounded domain $\mathbb{R}_+$. This fact introduces some differences that we detail next.

9.2.1 Discounted Infinite Horizon CV Problems

The benchmark problem in macroeconomics and growth theory has three main features. First, time is the independent variable, the initial time and the level of the state variable are both known, usually denoted by $t_0 = 0$ and $y(0) = y_0$, and the terminal time, or horizon, is infinite. Therefore, the domain of the independent variable is $\mathrm{T} = \mathbb{R}_+ = [0, \infty)$. Second, the objective function is a discounted function of type $F(t, y, \dot{y}) = f(y, \dot{y})\, e^{-\rho t}$, where $e^{-\rho t}$ is a discount factor with a time-independent rate of discount $\rho \geq 0$, and the **current value objective function** $f(y, \dot{y})$ is time-independent. Third, there are two main versions of the problem depending on the terminal constraint for the state variable: it can be free or it can be constrained.

Free asymptotic state problem. The discounted infinite horizon problem is to find a function $y \in \mathcal{Y} = \{y \in C^2(\mathrm{T}) : y(0) = y_0,\ \lim_{t\to\infty} y(t) = \text{free}\}$ that solves the problem[6]:

$$\begin{aligned} \max_{y(\cdot)} \mathsf{F}[y] \equiv \int_0^\infty & f(\dot{y}(t), y(t))\, e^{-\rho t} dt,\ \ \rho \geq 0 \\ & \text{subject to } y(0) = y_0,\ \text{fixed} \\ & \lim_{t\to\infty} y(t) \text{ free} \end{aligned} \tag{9.24}$$

This can be considered a problem with a fixed initial time and value for the state variable, a fixed terminal time but a free terminal value for the state variable (see Table 9.1):

Proposition 9.9 (First-order necessary conditions for the discounted infinite horizon problem with free terminal state) *Let y^* be the solution to problem* (9.24). *Then it satisfies the Euler-Lagrange equation*

$$\frac{d}{dt}\left(f_{\dot{y}}(\dot{y}^*(t), y^*(t))\right) = f_y(\dot{y}^*(t), y^*(t)) + \rho f_{\dot{y}}(\dot{y}^*(t), y^*(t)),\ \textit{for each}\ t \in [0, \infty), \tag{9.25}$$

the (so-called) transversality condition

$$\lim_{t\to\infty} f_{\dot{y}}(\dot{y}^*(t), y^*(t))\, e^{-\rho t} = 0 \tag{9.26}$$

and the initial condition $y^*(0) = y_0$.

Proof In the proof for the free boundaries value problem with constraint (9.7b), we set $t_1 \to \infty$ and take it as fixed but let $\lim_{t\to\infty} y^*(t)$ be free. In this discounted problem, the Euler-Lagrange equation (9.3) starting $F_y^*(t) = \frac{d}{dt} F_{\dot{y}}^*(t)$, is equivalent to $e^{-\rho t} f_y(\dot{y}^*, y^*) = \frac{d}{dt}\left(e^{-\rho t} f_{\dot{y}}(\dot{y}^*, y^*)\right)$, and the terminal condition $\lim_{t\to\infty} F_{\dot{y}}^*(t) = 0$, is satisfied if (9.26) holds.

[6] We use again the Newtonian notation for the time-dependent derivative $\dot{y} = \frac{dy(t)}{dt}$.

Observe that the Euler-Lagrange equation (9.25) is again a second-order non-linear autonomous ODE

$$f_y(\dot{y}^*, y^*) + \rho f_{\dot{y}}(\dot{y}^*, y^*) - f_{\dot{y}y}(\dot{y}^*, y^*)\,\dot{y} - f_{\dot{y}\dot{y}}(\dot{y}^*, y^*)\,\ddot{y} = 0.$$

The inequality constrained terminal state problem. In several problems, in economics, the former condition can lead to an asymptotic state that does not make economic sense (v.g, a negative level for a capital stock). The discounted infinite horizon problem is to find $y \in \mathcal{Y} = \{y \in C^2(\mathrm{T}) : y(0) = y_0,\ \lim_{t\to\infty} R(y(t), t) \geq 0\}$ that solves the benchmark discounted infinite horizon model

$$\begin{aligned} \max_{y(\cdot)} \mathsf{F}[y] \equiv \int_0^\infty & f(\dot{y}(t), y(t))e^{-\rho t}dt,\ \rho \geq 0 \\ & \text{subject to } y(0) = y_0,\ \text{fixed} \\ & \lim_{t\to\infty} R(y(t), t) \geq 0. \end{aligned} \tag{9.27}$$

Proposition 9.10 (First-order necessary conditions for the discounted infinite horizon problem with constrained terminal state) *Let y^* be the solution to problem* (9.27). *Then it satisfies the Euler-Lagrange equation* (9.25), *the initial condition $y^*(0) = y_0$, and the (so-called) transversality condition*

$$\lim_{t\to\infty} f_{\dot{y}}(\dot{y}^*(t), y^*(t))\, R(t, y^*(t))e^{-\rho t} = 0. \tag{9.28}$$

Exercise 9.4 Prove this. Observe that, as the terminal constraint is analogous to $\lim_{t\to\infty} y(t) \geq 0$, we have to introduce a Lagrange multiplier associated to the terminal time, as in problem (9.22).

9.2.2 Applications to Economics

Example 9.2 (*The resource depletion, or cake eating, problem*) Instead of the static cake-eating problem in Exercise 8.9, this example considers that activity takes place over time. Let the size of the cake (or of the resource) at time $t \in \mathrm{T} = [0, T]$ be denoted by the non-negative variable $w(t)$. The initial size of the cake is fixed by $w_0 > 0$. The cake is instantaneously consumed by the amount $c(t) = -\dot{w}(t) \geq 0$. The value of the cake is represented by the utility functional,

$$\mathsf{U}[W] = \int_0^T u(c(t))\, e^{-\rho t}dt,$$

where the utility function is increasing and concave $u''(c) < 0 < u'(c)$. Several properties are implicit in this functional: (1) the consumer is impatient, because time is discounted at a positive rate $\rho > 0$; (2) the felicity at every point in time is only a function of the instantaneous consumption (preferences are inter temporally additive); (3) more consumption means more felicity but at a decreasing rate (the increase in utility for big bites is smaller than for small bites); and (4) there is no satiation (there is not a bite with a zero or negative marginal utility): consumption is always good. If we assume that the cake has to be completely consumed by date $T > 0$, the necessary conditions for an optimum are

$$\frac{d}{dt}u'(-\dot{w}^*) = 0, \text{ for each } t \in [0, T] \tag{9.29a}$$

$$w^*(0) = w_0 > 0 \text{ fixed} \tag{9.29b}$$

Next, we present several versions of this problem for different utility functions and terminal conditions.

Example 9.3 (*Cake eating problem with the terminal state given*) Let the utility functional be a particular case of the utility functional already presented in Sect. 8.1.5.3, $\mathrm{U}[c] = \int_0^T \ln(c(t))e^{-\rho t}dt$, and let the cake be completely consumed until time $T > 0$, where T is fixed and finite. This is a CV problem analogous to problem (9.2), with value functional

$$V(w_0, T, \rho) = \max_w \mathsf{U}[w] = \max_w \left\{ \int_0^T \ln(-\dot{w}(t))e^{-\rho t}dt : w \in \mathcal{W} \right\}.$$

where $\mathcal{W} = \{w \in C^2(\mathrm{T}; \mathbb{R}_+) : w(0) = w_0 > 0, \text{ and } w(T) = 0\}$. The necessary (and in this example also sufficient) conditions for an optimum are

$$\begin{cases} \ddot{w}^* + \rho\dot{w}^* = 0, & \text{for each } 0 < t < T, \\ w^*(0) = w_0, & t = 0, \\ w^*(t) = 0, & t = T. \end{cases}$$

The solution to the EL equation is $w(t) = w(0) - \frac{\dot{w}(0)}{\rho}\left(1 - e^{-\rho t}\right)$. Substituting the two constraints, yields the solution to the problem

$$w^*(t) = \frac{e^{-\rho t} - e^{-\rho T}}{1 - e^{-\rho T}} w_0, \text{ for } t \in [0, T].$$

The value of the cake depends on the initial size, the horizon, and the rate of time preference

$$\begin{aligned} V(w_0, T, \rho) &= \int_0^T \ln(-\dot{w}^*(t))e^{-\rho t}dt = \\ &= \frac{1}{\rho}\left[\left(1 + \ln\left(\frac{1 - e^{-\rho T}}{\rho w_0}\left(e^{-\rho T} - 1\right)\right)\right)\right] + T\,e^{-\rho T}, \end{aligned}$$

that, if the consumer is rational, this should be its reservation price for the cake. If $\rho = 0.01$ and the cake lasts for 1 week and the calorie content is $w_0 = 1000$ then the reservation price should be $V(10, 0.01, 1/52) \approx 0.12$ per 100 calories.

Example 9.4 (*Cake eating problem: infinite horizon*) If we assume an infinite horizon and introduce the terminal condition $\lim_{t\to\infty} w(t) \geq 0$, meaning that we cannot have a negative size or the cake (or level of resource) asymptotically, the problem is an example of the ((9.27)) problem. The first-order conditions are now:

$$\begin{cases} \ddot{w}^*(t) + \rho\dot{w}^*(t) = 0, & \text{for each } t \in (0,\infty), \\ w^*(0) = w_0, \\ \lim_{t\to\infty} -e^{-\rho t}\frac{w^*(t)}{\dot{w}^*(t)} = 0. \end{cases}$$

We already found the solution to the EL equation, $w(t) = w_0 - \frac{\dot{w}(0)}{\rho}\left(1 - e^{-\rho t}\right)$ for $t \in [0,\infty)$. Then $\dot{w}(t) = -\dot{w}(0)\,e^{-\rho t}$ and $\lim_{t\to\infty} -e^{-\rho t}\frac{w^*(t)}{\dot{w}^*(t)} = \frac{w_0}{\dot{w}(0)} - \frac{1}{\rho}$. Imposing the transversality condition, we find $\dot{w}(0) = \rho w_0$, yielding the solution

$$w^*(t) = w_0 e^{-\rho t},\ \ t \in [0,\infty).$$

Again $\lim_{t\to\infty} w^*(t) = 0$.

Example 9.5 (*The benchmark representative household problem*) The benchmark representative household model in a finance economy is analog to problem presented in Sect. 8.2.2.3, but in which the household can decide over consumption over time, by participating in spot financial markets that trade a risk-free financial asset. In this simple example, it is assumed that there are no financial frictions, and there is perfect information. Again, at every point in time, the household receives a non-financial endowment, y, that we assume to be constant. Therefore, the household budget constraint is $\dot{a} = y + ra(t) - c(t)$ for each $t \in [0,\infty)$, where $a(t)$ is the net wealth at time t. The financial asset has a rate of return r, assumed to be constant over time, and the household can be in a short or a long position. The household is a net debtor if $a(t) < 0$ and is a net creditor if $a(t) > 0$. There are two more constraints to the household problem: the initial asset position, $a(0) = a_0$, where a_0 is given, and a non-Ponzi-game condition $\lim_{t\to\infty} a(t)e^{-rt} = 0$. All the variables are measured in real terms. We define the non-financial wealth as $h(t) = \int_t^\infty ye^{r(t-s)}ds$.

Assuming the utility functional in Sect. 8.1.5.3, satisfying the benchmark conditions $u''(c) < 0 < u'(c)$, and taking the consumption at time t than can be financed by the endowment and the participation in the financial market, the household problem is to find the optimal asset position $a \in \mathcal{A} = \{a \in C^2(\mathrm{T};\mathbb{R}) : a(0) = a_0,$ and $\lim_{t\to\infty} a(t)e^{-rt} = 0\}$, such that it solves the problem:

$$\begin{aligned} &\max_{a(\cdot)} \int_0^\infty u\Big(y + r\,a(t) - \dot{a}(t)\Big)\,e^{-\rho t}\,dt, \\ &\quad \text{subject to } a(0) = 0, \\ &\quad \lim_{t\to\infty} a(t)\,e^{-rt} = 0. \end{aligned} \tag{9.30}$$

Proposition 9.11 (Solution to the household consumption-finance benchmark problem) *Consider the household consumption problem in a frictionless finance economy (9.30), in which the utility function has $EIS = \theta^{-1}$ and assume that $\rho > 0$, $\theta > 0$ and $\rho + r(\theta - 1) > 0$. Then the optimal consumption and net asset position are*

$$c^*(t) = (r - \gamma)\left(a_0 + h_0\right) e^{\gamma t}, \text{ for each } t \in [0, \infty), \tag{9.31a}$$

$$a^*(t) = -h_0 + (a_0 + h_0)\, e^{\gamma t}, \text{ for each } t \in [0, \infty). \tag{9.31b}$$

where $h_0 = \frac{y}{r}$ is the initial non-financial wealth.

Proof The first-order conditions - necessary and sufficient in this case - are

$$(r - \rho)\, u'\big(y + r\,a - \dot{a}\big) + (r\,\dot{a} - \ddot{a})\, u''\big(y + r\,a - \dot{a}\big) = 0, \text{ for } t \in \mathrm{T}, \tag{9.32a}$$

$$a(0) = a_0, \tag{9.32b}$$

$$-\lim_{t\to\infty} u'\big(y + r\,a - \dot{a}\big)\, a(t)\, e^{-\rho t} = 0. \tag{9.32c}$$

As $\dot{c} = r\,\dot{a} - \ddot{a}$, and assuming an utility function with $EIS = \theta^{-1}$, the EL equation is equivalent to

$$\dot{c} = \gamma\, c, \text{ where } \gamma \equiv \frac{r - \rho}{\theta}. \tag{9.33}$$

Solving this ODE, to get $c(t) = c(0)\, e^{\gamma t}$, substituting it in the budget constraint, $\dot{a} = y + r\,a - c(t)$, and solving it for a, yields $a(t) = -h_0 + (a_0 + h_0)\, e^{r t} + \frac{c(0)}{r-\gamma}\left(e^{r t} - e^{\gamma t}\right)$, for $t \in [0, \infty)$, where $h_0 = \frac{y}{r}$ and $a(0) = a_0$. Substituting this solution, depending on the unknown $c(0)$, in the transversality condition (9.32c), we have

$$\begin{aligned}
&\lim_{t\to\infty} c(0)^{-\theta}\, e^{-rt}\Big[- h_0 + (a_0 + h_0)\, e^{r t} + \frac{c(0)}{r - \gamma}\left(e^{r t} - e^{\gamma t}\right)\Big] = \\
&= \lim_{t\to\infty} c(0)^{-\theta}\left[a_0 + h_0 + \frac{c(0)}{r - \gamma} - \frac{c(0)}{r - \gamma}\, e^{(\gamma - r)t}\right] \\
&= c(0)^{-\theta}\left(a_0 + h_0 + \frac{c(0)}{r - \gamma}\right)
\end{aligned}$$

because $\gamma - r < 0$ if and only if $\rho + r(\theta - 1) > 0$. The transversality condition (9.32c) is satisfied if and only if $c(0) = (r - \gamma)(a_0 + h_0)$. Substituting back in the solutions to $c(t)$ and $a(t)$ yields the solutions to the CV problem in Eqs. (9.31a) and (9.31b). The transversality condition, and therefore the solution to Problem (9.30), exists only if $r > 0$ and $r > \gamma$. Therefore, observing the solutions for consumption (9.31a), and for the net asset position (9.31b), the time behavior of the solution depends on the sign of γ, and therefore on the difference between the market rate of return, r and the rate of time preference ρ (a measure of household's patience relative to the market): (i) if $r > \rho$ then both consumption and the net asset position converge to infinity in infinite time; (ii) if $r = \rho$ consumption and net wealth are constant over time $c^*(t) = r\,a_0 + y$, and $a^*(t) = a_0$ for each $t \in [0, \infty)$, or (iii) if $r < \rho$ consumption converges over time to zero and the net asset position converges to the "natural debt limit" , $\lim_{t\to\infty} c^*(t) = 0$, and $\lim_{t\to\infty} a^*(t) = -h_0 < 0$.

Example 9.6 (*The* Rotemberg (1982)*'s firm problem*) This example presents the firm's problem in Rotemberg (1982) model of sticky prices. The firm produces one good and is a monopolist in its market, although it operates in a monopolistic competition environment. This means that it faces a downward sloping demand curve for its output $y(p) = p^{-\sigma}$, where $\sigma > 1$. It faces two costs, a production cost $C(p) = w\,y$, where $w > 0$, and an adjustment menu cost for changing its price $\frac{\zeta}{2}\left(\frac{\dot{p}}{p}\right)^2$. The firm's problem is to find the optimal price policy that maximizes the present value of profits, using the market interest rate, r, assumed to be constant and positive, as the discount factor. The firms optimal price policy, $\mathbf{p}_{\mathrm{T}} = \left(p^*(t)\right)_{t\in\mathrm{T}}$, solves the following CV problem:

$$\max_{p(\cdot)} \int_0^{\infty} \left(\left(p(t) - w\right) p(t)^{-\sigma} - \frac{\zeta}{2}\left(\frac{\dot{p}(t)}{p(t)}\right)^2\right) e^{-r\,t}\,dt, \tag{9.34a}$$

$$\text{subject to } p(0) = p_0,$$

$$\lim_{t\to\infty} e^{-r\,t}\,p(t) = 0, \tag{9.34b}$$

where the constraint (9.34b) forces prices to be bounded in present value terms. The first-order conditions for an optimum (that are both necessary and sufficient) are

$$\begin{cases} \dfrac{\ddot{p}}{p} - \left(\dfrac{\dot{p}}{p}\right)^2 - r\,\dfrac{\dot{p}}{p} + \left((1-\sigma)\,p + \sigma\,w\right)\dfrac{p^{-\sigma}}{\zeta} = 0, \text{ for each } t \in \mathbb{R}_+, \\ p(0) = p_0, \\ \lim_{t\to\infty} e^{-r\,t}\,p(t) = 0, \end{cases} \tag{9.35a}$$

where (9.35a) is the Euler-Lagrance equation. There is an unique steady state, $\bar{p}^* = m\,w$, where $m \equiv \frac{\sigma}{\sigma-1}$ is the markup. Although there is no closed form solution to the problem, we can show that the optimal price level converges to the linear approximation $p^*(t) = \bar{p}^* + \frac{p_0-\bar{p}^*}{\bar{p}^*}\,e^{\lambda\,t}$, for $t \in [0, \infty)$, where $\lambda = \frac{r}{2} - \left(\left(\frac{r}{2}\right)^2 + m\,\frac{\sigma}{\zeta}\,(\bar{p}^*)^{1-\sigma}\right)^{\frac{1}{2}} < 0$. If the initial price is not at the steady state level, i.e., $p_0 \neq \bar{p}^*$, and because the relative half life of the adjustment is $\tau = \frac{\ln(0.5)}{\lambda}$, the price adjustment is faster if the markup m is high, and ζ is low.

9.3 References

For the history of the calculus of variations, see Goldstine (1980). A general and important reference is Gel'fand and Fomin (1963). See also van Brunt (2010). For applications to economics, see Liberzon (2012). On the cake-eating problem, see Romer (1986a).

Chapter 10
Introduction to Optimal Control: The Maximum Principle Approach

Pontryagin et al. (1962) Maximum Principle (PMP) provides first-order necessary conditions for the optimal control problem (OCP), from a slightly different approach to functional optimization problems than the Calculus of Variations approach. Although the two approaches lead to the same solution, when applied to the same simple problems, the PMP can be more convenient when the constraints take the form of non-elementary ordinary differential equations. As can be seen next in this chapter, the first-order conditions are a system of ordinary differential equations (and not an implicit second-order ODE as in the calculus of variations approach). This allows for a direct application of the qualitative theory of ODE and the use geometrical methods, to obtain a characterization of the solution, to an ODE problem even when the problem is not explicitly specified and/or when explicit solutions cannot be obtained.

An OCP consists in finding, or characterizing the behavior, of a state variable whose **variation** depends on a control variable, in order to maximize (or minimize) a functional. This separation of the control and state variables is the main reason why the PMP can be more flexible as regards the CV approach.

In this chapter, we provide an introduction to simple OCP, in which there is only one state variable, starting with the case of an arbitrary independent variable, in Sect. 10.1, and proceeding with OCP over time, and the dynamics of their solution paths, in Sect. 10.2.

10.1 The Maximum Principle in an Arbitrary Domain

In this section, we consider OCP of a function of an arbitrary independent variable, x, with domain $\mathrm{X} \subseteq \mathbb{R}$. We write $\mathrm{X} = [x_0, x_1]$ if it is a closed set, or $\mathrm{X} = (x_0, x_1)$ if it is open, where $x_0 < x_1$ are fixed or can be free and determined optimally, and can be finite or infinite.

P. B. Brito, *Economic Dynamics and Distributions*, Dynamic Modeling and Econometrics in Economics and Finance 32, https://doi.org/10.1007/978-3-031-94717-9_10

An OCP involves two variables, the **state variable**, $y(\cdot)$ and the **control variable**, $u(\cdot)$. As, in this chapter, we consider only problems in which the state variable is of dimension one, the state variable is a mapping $y : \mathrm{X} \subseteq \mathbb{R} \to \mathrm{Y} \subseteq \mathbb{R}$, and the control variable is a mapping $u : \mathrm{X} \subseteq \mathbb{R} \to \mathrm{U} \subseteq \mathbb{R}^m$, where $m \geq 1$. That is, there can be more than one control variable. Again it is important to distinguish between the **point-wise level** of variables, i.e., $y(x)$ and $u(x)$, for a specific x, from **curves or profiles traced out in the image sets**, Y and U, of variables , i.e., $\mathbf{y}_{\mathrm{X}} = \big(y(x)\big)_{x \in \mathrm{X}}$, and $\mathbf{u} = \big(u(x)\big)_{x \in \mathrm{X}}$.

Definition 10.1 (*Optimal control problem*) An OCP consists in finding the continuously differential function $y \in \mathcal{Y} \in C^1(\mathrm{X}; \mathrm{Y})$, mapping $y : \mathrm{X} \subseteq \mathbb{R} \to \mathrm{Y} \subseteq \mathbb{R}$, and the piecewise continuously differentiable (vector) function $u \in \mathcal{U} \in PC^1(\mathrm{X}; \mathrm{U})$, mapping $u : \mathrm{X} \subseteq \mathbb{R} \to \mathrm{U} \subseteq \mathbb{R}^m$, where $m \geq 1$, that maximize the **objective functional**

$$\mathsf{J}[y, u] \equiv \int_{x_0}^{x_1} F(y(x), u(x), x)\, dx \tag{10.1}$$

subject to an ODE constraint,

$$y'(x) = G(y(x), u(x), x), \ \text{ for } x \in \mathrm{X}, \tag{10.2}$$

and to other side conditions, that can be functionals or point-wise conditions for the boundaries of set X, and/or their associated levels in Y.

A **solution** to an OCP is represented by two optimal curves $\mathbf{y}^* = \big(y^*(x)\big)_{x \in \mathrm{X}}$ and $\mathbf{u}^* = \big(u^*(x)\big)_{x \in \mathrm{X}}$, traced out in the specified or optimized domains, Y and U, by state and control variables, $y^*(x)$ and $u^*(x)$, for every $x \in \mathrm{X}$.

The **Pontriyagin's maximum principle** (PMP) provides necessary (and sometimes also sufficient) conditions for an optimum, involving the **Hamiltonian** function,

$$H(y, u, \lambda, x) = F(y, u, x) + \lambda(x)\, G(y, u, x).$$

where λ, called the **co-state variable**, is a piecewise continuous mapping $\lambda : \mathrm{X} \to \mathbb{R}$. At the optimum, if it exists, the maximized functional is a function, sometimes called **value function**

$$V(\boldsymbol{\varphi}) = \mathsf{J}[y^*, u^*]$$

depending on the data of the problem, denoted by $\boldsymbol{\varphi}$, that includes information on the boundary values for set X and the associated elements in Y,

Next, we present the optimality conditions, according to the PMP, when there are point-wise side constraints, in Sect. 10.1.1, and when there are functional side constraints, in Sect. 10.1.2.

10.1.1 Boundary Constraints

Fixed boundary values for the independent variable We assume that the boundary values for set $\mathrm{X} = [x_0, x_1]$, for $x_0 < x_1$, x_0 and x_1 are fixed. The OCP is to find an optimal control curve $\mathbf{u}^* = (u^*(x))_{x \in [x_0, x_1]}$ that maximizes the functional (10.1) subject to ODE constraint (10.2), and additional side conditions for $y(x_0)$ and/or $y(x_1)$, can be fixed or be chosen optimally.[1]

Four variants of the simplest problem are considered:

$$\max_{u(\cdot)} \int_{x_0}^{x_1} F(y(x), u(x), x)\, dx, \tag{10.3a}$$

$$\text{subject to } x_0 \text{ and } x_1 \text{ fixed}, \tag{10.3b}$$

$$y' = G(y(x), u(x), x), \quad \text{for } x \in [x_0, x_1], \tag{10.3c}$$

$$\text{one of the next boundary conditions on } y(x_0) \text{ and/or } y(x_1),$$

$$y(x_0) = y_0, \quad \text{and } y(x_1) = y_1 \text{ fixed}; \tag{10.3d}$$

$$y(x_0) = y_0 \text{ fixed}, \quad \text{and } y(x_1) \text{ free}; \tag{10.3e}$$

$$y(x_0) \text{ free}, \quad \text{and } y(x_1) = y_1 \text{ fixed}; \tag{10.3f}$$

$$y(x_0), \quad \text{and } y(x_1) \text{ free}. \tag{10.3g}$$

The data referring to the four variants of problem are: $\boldsymbol{\varphi} = (x_0, x_1, y_0, y_1)$ for problem with constraint (10.3d), $\boldsymbol{\varphi} = (x_0, x_1, y_0)$ for problem with constraint (10.3e), $\boldsymbol{\varphi} = (x_0, x_1, y_1)$ for problem with constraint (10.3f), and $\boldsymbol{\varphi} = (x_0, x_1)$ for problem with constraint (10.3g).

Proposition 10.1 (PMP for fixed boundary values of the independent variable) *Let $(\mathbf{y}^*, \mathbf{u}^*)$ be a solution to the OC problem (10.3) in which one of the constraints (10.3d), or (10.3e), or (10.3f) or (10.3g) is be satisfied. Then there is a piecewise continuous function $\lambda : [x_0, x_1] \to \mathbb{R}$, called co-state variable, such that the curves $(\mathbf{y}^*, \mathbf{u}^*, \boldsymbol{\lambda})$ satisfy the following conditions:*

- *the optimality condition*[2]*:*

$$H_u(y^*(x), u^*(x), \lambda(x), x) = 0, \; \textit{for each } x \in [x_0, x_1], \tag{10.4}$$

- *the adjoint equation*

$$\lambda'(x) = -H_y(y^*(x), u^*(x), \lambda(x), x), \; \textit{for each } x \in (x_0, x_1), \tag{10.5}$$

---

[1] This is similar to the case dealt in the CV problems (9.2) and (9.7).

[2] The following notation is used: $H_u(y(x), u(x), \lambda(x), x) \equiv \frac{\partial}{\partial u(x)} H(y(x), u(x), \lambda(x), x)$ is the derivative evaluated at $x \in \mathrm{X}$ for any functions (y, u, λ) and $H_u(y^*(x), u^*(x), \lambda(x), x)$ is the derivative evaluated at the optimal levels (y^*, u^*). The other derivatives, v.g. for y, use the same notation.

- *the constraint of the problem:*

$$y^{*\prime}(x) = G(y^*(x), u^*(x), x),\ \textit{for each } x \in (x_0, x_1), \tag{10.6}$$

- *and the adjoint conditions associated to the boundary conditions (10.3d) to (10.3g)*
  - *for a problem with constraint (10.3d)*

$$y^*(x_0) = y_0 \textit{ for } x = x_0,\ \textit{and } y^*(x_1) = y_1 \textit{ for } x = x_1, \tag{10.7}$$

  - *for a problem with constraint (10.3e)*

$$y^*(x_0) = y_0 \textit{ for } x = x_0,\ \textit{and } \lambda(x_1) = 0 \textit{ for } x = x_1, \tag{10.8}$$

  - *for a problem with constraint (10.3f)*

$$\lambda(x_0) = 0 \textit{ for } x = x_0,\ \textit{and } y^*(x_1) = y_1 \textit{ for } x = x_1, \tag{10.9}$$

  - *for a problem with constraint (10.3g)*

$$\lambda(x_0) = 0 \textit{ for } x = x_0,\ \textit{and } \lambda(x_1) = 0 \textit{ for } x = x_1. \tag{10.10}$$

Proof Let $\mathbf{u}^* = \big(u^*(x)\big)_{x\in X}$ be an optimal control curve and let $\mathbf{y}^* = \big(y^*(x)\big)_{x\in X}$ be the associated optimal state curve. The value of the problem is $\mathsf{J}[y^*, u^*] = \int_{x_0}^{x_1} F^*(x)\,dx$, where $F^*(x) \equiv F(y^*(x), u^*(x), x)$. Then

$$\begin{aligned}\mathsf{J}[y^*, u^*] &= \int_{x_0}^{x_1} F^*(x)dx = \int_{x_0}^{x_1} F^*(x) + \lambda(x)\left(G^*(x) - y^{*\prime}(x)\right)\Big]dx = \\ &= \int_{x_0}^{x_1} \Big[H^*(x) - y^{*\prime}(x)\,\lambda(x)\Big]\,dx,\end{aligned}$$

where $G^*(x) \equiv G(y^*(x), u^*(x), x)$ and $H^*(x) \equiv H\Big(y^*(x), u^*(x), \lambda(x), x\Big)$. A perturbation is introduced in the optimal state-control pair $(y, u) = (y^*, u^*) + \varepsilon\,\boldsymbol{\eta}$, where ε is a constant and $\boldsymbol{\eta} = (\eta_y, \eta_u)$. A perturbation is admissible if, for constraint (10.3d) it satisfies $\eta_y(x_0) = \eta_y(x_1) = 0$, for constraint (10.3e) it satisfies $\eta_y(x_0) = 0$ and $\eta_y(x_1) \neq 0$, for constraint (10.3f) it satisfies $\eta_y(x_0) \neq 0$ and $\eta_y(x_1) = 0$, or for constraint (10.3g) it satisfies $\eta_y(x_0) \neq 0$ and $\eta_y(x_1) \neq 0$. The pair (y^*, u^*) is an extremum if $\mathsf{J}[y^*, u^*] \geq \mathsf{J}[y, u]$ for any other admissible pair (u, y) of perturbed functions. The first-order Taylor approximation of the value functional at (y^*, u^*) is

$$\mathsf{J}[y, u] = \mathsf{J}[y^*, u^*] + \mathrm{D}_{\boldsymbol{\eta}(\cdot)}\mathsf{J}[y^*, u^*]\,\epsilon + O(\epsilon^2)$$

where the Gâteaux differential is[3]

$$\mathsf{D}_{\eta(\cdot)}\mathsf{J}[y^*, u^*] = \int_{x_0}^{x_1} \Big[H_u^*(x)\,\eta_u(x) + H_y^*(x)\,\eta_y(x) - \lambda(x)\,\eta_y'(x) \Big] dx =$$

$$= \int_{x_0}^{x_1} \Big[H_u^*(x)\,\eta_u(x) + \big(H_y^*(x) + \lambda'(x)\big)\,\eta_y(x) \Big] dx + \lambda(x_0)\,\eta_y(x_0) - \lambda(x_1)\,\eta_y(x_1),$$

using integration by parts.

Then $\mathsf{J}[y, u] \leq \mathsf{J}[y^*, u^*]$ only if $\mathsf{D}_{\eta(\cdot)}\mathsf{J}[y^*, u^*] = 0$, which, using similar arguments as in the case of the calculus of variations problem (see Chap. 9), is equivalent to the Pontriyagin's conditions: $H_u^*(x) = \lambda'(x) - H_y^*(x) = 0$, for each $x \in \mathrm{X}$. The adjoint constraints have to satisfy the following conditions: $\lambda(x_0)\,\eta_y(x_0) = \lambda(x_1)\,\eta_y(x_1) = 0$. From this and the admissible value for $\eta_y(x_0)$ and/or $\eta_y(x_1)$ then the associated adjoint constraint (10.7), (10.8), (10.9) or (10.10) are obtained.

Constraints on the boundary values of set. X Let the boundary values for the state variables, y, be fixed, and the boundary values of the independent variable be free, and therefore be chosen optimally, i.e., x_0^* and/or x_1^* are free, and $y(x_0^*) = y_0$ and $y(x_1^*) = y_1$ are fixed. This implies that the boundaries of the domain of the independent variable become endogenous, leading us to denote it by X^*. The OCP is to find the optimal cut-off values for the independent variable, x_0^* and/or x_1^*, and optimal curves $\big(u^*, y^*\big)_{x\in \mathrm{X}^*}$ that maximize the functional (10.1) subject to ODE constraint (10.2).

Four variants of problems with potential decisions on the cut-off values for X are considered:

$$\max_{u(\cdot)} \int_{x_0}^{x_1} F(y(x), u(x), x)\,dx, \tag{10.11a}$$

$$\text{subject to } y(x_0) = y_0 \text{ and } y(x_1) = y_1 \text{ fixed}, \tag{10.11b}$$

$$y' = G(y(x), u(x), x), \ \text{ for } x \in [x_0, x_1], \tag{10.11c}$$

one of the next boundary conditions on x_0 and/or x_1,

$$x_0 \text{ and } x_1 \text{ fixed;} \tag{10.11d}$$

$$x_0 \text{ fixed and } x_1 \text{ free;} \tag{10.11e}$$

$$x_0 \text{ free and } x_1 \text{ fixed;} \tag{10.11f}$$

$$x_0 \text{ and } x_1 \text{ free..} \tag{10.11g}$$

Proposition 10.2 (PMP for free domain and fixed boundary state variable optimal control problems) *Let* $(\mathbf{y}^*, \mathbf{u}^*)$ *be a solution curve to the OC problem (10.11) where* $y(x_0) = y_0$ *and* $y(x_1) = y_1$ *are fixed. Then there is an optimal domain for the independent variable* $\mathrm{X}^* = [x_0^*, x_1^*] \subset \mathbb{R}$*, a piecewise continuous function* $\lambda :, x^* \to \mathbb{R}$,

[3] We use the notation $H_x^*(x) \equiv \frac{\partial H\left(y^*(x), u^*(x), \lambda(x), x\right)}{\partial u(x)}$, and $H_y^*(x) \equiv \frac{\partial H\left(y^*(x), u^*(x), \lambda(x), x\right)}{\partial y(x)}$.

called co-state variable, such that the curves $(\mathbf{y}^*, \mathbf{u}^*, \boldsymbol{\lambda})$ *satisfy the optimality condition (10.4), the adjoint equation (10.5) and the ODE constraint of the problem (10.6), all for* $x \in Int(X^*)$*, and the adjoint condition associated to one of the boundary conditions (10.11d) to (10.11g):*

- *for a problem with constraint (10.11d)* $y^*(x_0) = y_0$ *and* $y^*(x_1) = y_1$ *and* x_0 *and* x_1 *are fixed;*
- *for a problem with constraint (10.11e)* $y^*(x_0) = y_0$ *and* $y^*(x_1^*) = y_1$ *and*

$$x_0^* = x_0 \text{ and } H(y_1, u^*(x_1^*), x_1^*) - y^{*\prime}(x_1^*)\,\lambda(x_1^*) = 0; \tag{10.12}$$

- *for a problem with constraint (10.11f)* $y^*(x_0^*) = y_0$ *and* $y^*(x_1) = y_1$ *and*

$$H(y_0, u^*(x_0^*), x_0^*) - y^{*\prime}(x_0^*)\lambda(x_0^*) = 0 \text{ and } x_1^* = x_1; \tag{10.13}$$

- *for a problem with constraint (10.11g)* $y^*(x_0^*) = y_0$ *and* $y^*(x_1^*) = y_1$ *and*

$$\begin{aligned} &H(y_0, u^*(x_0^*), x_0^*) - y^{*\prime}(x_0^*)\,\lambda(x_0^*) = 0 \\ &\qquad\qquad \text{and } H(y_1, u^*(x_1^*), x_1^*) - y^{*\prime}(x_1^*)\,\lambda(x_1^*) = 0. \end{aligned} \tag{10.14}$$

Proof In this case, in addition to the optimal pair (y^*, u^*), we have to find optimal boundary values for the independent variables, x_0^* and x_1^*. Using the same method as in the previous proof, but introducing additional perturbations for the terminal values of the independent variables, $x_j = x_j^* + \varepsilon\,\chi_j$ for $j \in \{0, 1\}$,[4] the Gâteaux differential of the value functional becomes

$$\begin{aligned} \mathrm{D}_{(\eta(\cdot),\chi)}\mathrm{J}[y^*, u^*; x^*] = &\int_{x_0^*}^{x_1^*} \Big[H_u^*(x)\,\eta_u(x) + H_y^*(x)\,\eta_y(x) - \lambda(x)\,\eta_y'(x)\Big]\,dx + \\ &+ H^*(x_1^*)\,\chi_1 - H^*(x_0^*)\,\chi_0. \end{aligned}$$

where $H^*(x_j^*) = H(y^*(x_j^*), u^*(x_j^*, x_j^*), \lambda(x_j^*))$ for $j = 0, 1$. Integrating by parts, yields

$$\begin{aligned} \mathrm{D}_{(\eta(\cdot),\chi)}\mathrm{J}[y^*, u^*; x^*] = &\int_{x_0^*}^{x_1^*} \Big[H_u^*(x)\,\eta_u(x) + \big(H_y^*(x) + \lambda'(x)\big)\,\eta_y(x)\Big]\,dx + \\ &+ \lambda(x_0^*)\,\eta_y(x_0) - \lambda(x_1^*)\,\eta_y(x_1) + H^*(x_1^*)\,\chi_1 - H^*(x_0^*)\,\chi_0. \end{aligned}$$

Using the approximation $\eta(x_j) \approx \eta(x_j^*) + y'(x_j^*)\,\chi_j$, for $j = 0, 1$,[5] yields

---

[4] For more details, see the proofs of propositions 9.6, and 10.1.

[5] As in the proof of Proposition 9.6 in Chap. 9, and analogously to Eq. (9.16).

$$\mathrm{D}_{(\eta(\cdot),\chi)}\mathrm{J}[y^*, u^*; x^*] = \int_{x_0^*}^{x_1^*} \left[H_u^*(x)\eta_u(x) + \left(H_y^*(x) + \lambda'(x)\right)\eta_y(x)\right] dx$$
$$+ \left(H^*(x_1^*) - y^{*\prime}(x_1^*)\lambda(x_1^*)\right)\chi_1 - \left(H^*(x_0^*) - y^{*\prime}(x_0^*)\lambda(x_0^*)\right)\chi_0 +$$
$$+ \lambda(x_0^*)\,\eta_y(x_0^*) - \lambda(x_1^*)\,\eta_y(x_1^*) \quad (10.15)$$

As the terminal values for the state variable are fixed, then $\eta_y(x_0^*) = \eta_y(x_1^*) = 0$, and we obtain the adjoint necessary conditions for the optimum, for the different versions of the problem, in Eqs. (10.12)–(10.14).

General boundary constraints. The most general problem with bounded sets X and Y can be specified if we allow for x_0 and/or x_1 to be fixed or free, and for $y(x_0)$ and/or $y(x_1)$ to be fixed or free:

$$\max_{u(\cdot),x_0,x_1,y(x_0),y(x_1)} \int_{x_0}^{x_1} F(y(x), u(x), x)\, dx \quad (10.16a)$$

subject to

$$y' = G(y(x), u(x), x), \text{ for } x \in [x_0, x_1] \quad (10.16b)$$

$$x_0 \text{ and/or } x_1, \text{ and/or } y(x_0) \text{ and/or } y(x_1) \text{ are free.} \quad (10.16c)$$

When there are free boundary conditions for the independent variable x, or for the associated state variable $y(x)$, those boundary values become decision variables and should be optimized. In the first case, the OCP is to find the optimal cut-off values for the independent variable, x_0^* and/or x_1^*, and an optimal control $(u^*(x))_{t\in[x_0^*,x_1^*]}$ that maximizes the functional (10.16a) subject to ODE constraint (10.16b) and having fixed or free boundary values for the state variable. In the second case, the OCP is to find the optimal boundary values for the state variable, $y^*(x_0)$ and/or $y^*(x_1)$ and an optimal control $(u^*(x))_{t\in[x_0,x_1]}$ that maximizes the functional (10.16a) subject to ODE constraint (10.16b) and having fixed or free cut-off values for the independent variable.

The necessary conditions include the optimality condition (10.4), the adjoint equation (10.5) and the ODE constraint of the problem (10.6), all for $x \in \text{int}(\mathrm{X}^*)$. To obtain the adjoint conditions associated to the terminal values of the state variable, when they need to be optimized, we set, in Eq. (10.15), $\eta_y(x_0^*) \neq 0$ and/or $\eta_y(x_1^*) \neq 0$. Therefore, the adjoint condition associated to $y^*(x_j^*)$ and $\lambda(x_j^*) = 0$, imply that the adjoint condition associated to the optimal boundary value of the independent variable x_j^* is $H^*(x_j^*) = 0$, for $j = 0, 1$.

The adjoint conditions presented in Table 10.1 cover all the possible sixteen cases, and can be compared to Table 9.1:

The index j refers to the lower boundary when $j = 0$ and to the upper boundary when $j = 1$, y_j and x_j refer to fixed values, and y_j^* and x_j^* refer to free values at the optimum.

Example 10.1 (*Optimal spatial distribution*) Let $\mathrm{X} = [\underline{x}, \overline{x}]$ represent a region and consider the problem of a firm that seeks to sell a product across that region but faces

Table 10.1 Adjoint conditions for bounded domain OC problems

Data		Boundary conditions for an optimum	
x_j	$y(x_j)$	x_j^*	$y^*(x_j^*)$
Fixed	Fixed	x_j	y_j
Fixed	Free	x_j	$\lambda(x_j) = 0$
Free	Fixed	$H(y_j, u^*(x_j^*), x_j^*) - y^{*\prime}(x_j^*)\lambda(x_j^*) = 0$	y_j
Free	Free	$H(y^*(x_j^*), u^*(x_j^*), x_j^*) = 0$	$\lambda(x_j^*) = 0$

adjustment costs when it tries to relocate sales, and a transportation cost. It is assumed that the firm is price taker, and cannot discriminate prices across locations. The problem of the firm is to maximize profits subject to a spatial relocation constraint[6]:

$$\max_{u(\cdot)} \int_{\underline{x}}^{\overline{x}} p\, y(x) - \frac{1}{2}\left(u(x)\right)^2 dx,$$
$$\text{subject to} \tag{10.17}$$
$$y'(x) = u(x) - \delta\, y(x),$$
$$\text{other side conditions,}$$

where $y(x)$ represents quantities of a product sold in location x, $u(x)$ is the net increase in sales in location x, $p > 0$ is the price, and $\delta > 0$ is a cost of transportation. The Hamiltonian for this control problem is

$$H(y, u, \lambda) = p\, y - \frac{u^2}{2} + \lambda\, (u - \delta\, y).$$

The static optimality and the canonical conditions are $u(x) = \lambda(x)$ and $\lambda'(x) = \delta\, \lambda(x) - p$, respectively, for each $x \in \mathrm{X}$. Therefore, an optimal distribution $(\mathbf{y}^*, \mathbf{u}^*)$, satisfies the linear planar ODE

$$y'(x) = u(x) - \delta\, y(x), \ \text{ for each } x \in \mathrm{X}, \tag{10.18a}$$
$$u'(x) = \delta\, u(x) - p, \ \text{ for each } x \in \mathrm{X}, \tag{10.18b}$$

plus the side conditions and their duals. Solving the linear ODE system (10.18a)–(10.18b) yields a general solution for $x \in \mathrm{X}$

$$y(x) = \frac{p+\delta}{\delta^2} + w_-\, e^{-\delta\, x} + w_+\, e^{\delta\, x}, \ \text{ for each } x \in \mathrm{X}, \tag{10.19a}$$
$$u(x) = \frac{p}{\delta} + 2\,\delta\, w_+\, e^{\delta\, x}, \ \text{ for each } x \in \mathrm{X}, \tag{10.19b}$$

[6] Compare to the CV Example 9.1.

where w_- and w_+ are two arbitrary constants. Next, we consider some cases differing by the boundary conditions.

Fixed boundary values for the independent and state variable. Let $\underline{x}$ and $\overline{x}$ be fixed, and $y(\underline{x}) = \underline{y}$ and $y(\overline{x}) = \overline{y}$ be also fixed. This is the case in the first row of Table 10.1 for both $j = 0$ and $j = 1$. Defining $z(x) = y(x) - \frac{p+\delta}{\delta^2}$, and applying the constraints to the general solution for the state variable y, in Eq. (10.19a), yields a linear system on the two arbitrary constants w_- and w_+,

$$\underline{z} = w_- \, e^{-\delta \, \underline{x}} + w_+ \, e^{\delta \, \underline{x}}, \; \overline{z} = w_- \, e^{-\delta \, \overline{x}} + w_+ \, e^{\delta \, \overline{x}}.$$

Solving this system and substituting back in Eq. (10.19a) and (10.19b), yields the solution to the problem

$$y^*(x) = \frac{p+\delta}{\delta^2} + \underline{z} \left(\frac{e^{\delta\,(\overline{x}-x)} - e^{\delta\,(x-\overline{x})}}{e^{\delta\,(\overline{x}-\underline{x})} - e^{\delta\,(\underline{x}-\overline{x})}} \right) + \overline{z} \left(\frac{e^{\delta\,(x-\underline{x})} - e^{\delta\,(\underline{x}-x)}}{e^{\delta\,(\overline{x}-\underline{x})} - e^{\delta\,(\underline{x}-\overline{x})}} \right), \text{ for } x \in \mathrm{X},$$

and

$$u^*(x) = \frac{p}{\delta} + 2\delta \left(\frac{\overline{z}\, e^{\delta\,(x-\underline{x})} - \underline{z}\, e^{\delta\,(x-\overline{x})}}{e^{\delta\,(\overline{x}-\underline{x})} - e^{\delta\,(\underline{x}-\overline{x})}} \right), \text{ for } x \in \mathrm{X}.$$

Free boundary values for the state variable. If $y(\overline{x})$ and $y(\underline{x})$are both free, independently from the fact that $\overline{x}$ and $\underline{x}$ are fixed, a solution exists only if $\overline{x} = \underline{x}$. This is because we should have (see the second and fourth rows of Table 10.1) $u(\overline{x}) = u(\underline{x}) = 0$, that is $e^{\delta\overline{x}} = e^{\delta\underline{x}}$. If $\overline{x} \neq \underline{x}$, there is no solution to the problem.

Fixed boundary values for the independent variable and free boundary state variable. Let $\underline{x}$ and $\overline{x}$ be fixed, and $y(\underline{x}) = \underline{y}^*$ and $y(\overline{x}) = \overline{y}^*$ be free, that corresponds to the third row of Table 10.1. As $H^*(\underline{x}) = y^{*\prime}(\underline{x})\, u(\underline{x})$ and $H^*(\overline{x}) = y^{*\prime}(\overline{x})\, u(\overline{x})$ if and only if $u(\underline{x})^2 = 2\, p\, \underline{y}^*$ and $u(\overline{x})^2 = 2\, p\, \overline{y}^*$. There is a solution to this version of the problem if solving those two last equations together with equations $y(\underline{x}) - \underline{y}^*$ and $y(\overline{x}) = \overline{y}^*$ (using general solutions (10.19a) and (10.19b)), allow us to find w_-, w_+, $\overline{y}^*$, and $\underline{y}^*$.

10.1.2 Extension: Integral Constraints

In this subsection, OCP subject to integral constraints of the iso-perimetric type are studied. These are functional constraints specified by the integration of a known function of the state, and/or of controls, over the whole domain of the independent variable. There are other types of integral constraints, of non-iso perimetric type, in which the integration is done over a sub-domain of the independent variable. While an iso-perimetric constraint has dimension zero (it is a scalar), and thus requires the introduction of one adjoint variable, the other types of integral constraints require the introduction of adjoint functions. Next we introduce an iso-perimetric constraint in the specification of OCP.

Let the independent variable be $x \in \mathrm{X} \subseteq \mathbb{R}$, where $\mathrm{X} = [x_0, x_1]$, the state variable be $y : \mathrm{X} \to \mathrm{Y} \subseteq \mathbb{R}$, and the control variable be $u : \mathrm{X} \to \mathrm{U} \subseteq \mathbb{R}$. The OCP is

$$\max_{x_0, x_1, y(x_0), y(x_1), u(\cdot)} \int_{x_0}^{x_1} F(y(x), u(x), x)\, dx,$$

subject to

$$\int_{x_0}^{x_1} G_0(y(x), u(x), x)\, dx \leq \bar{G}, \tag{10.20a}$$

$$\frac{dy(x)}{dx} = G_1(y(x), u(x), x) \text{ for each } x \in \mathrm{X}, \tag{10.20b}$$

$$x_0, x_1, y(x_0), y(x_1) \text{ free}, \tag{10.20c}$$

where $\bar{G}$ is a constant. This OCP has one functional constraint of the iso-perimetric type, (10.20a), one ordinary differential equation constraint, (10.20b), and has free boundary values for the independent variable, and for the associated values for the state variable,.

There are several versions for this problem. For instance: (1) the simplest problem when x_0, x_1, $y(x_0)$ and $y(x_1)$ are fixed; (2) the free terminal problem when x_0 and $y(x_0)$ are fixed and x_1 and $y(x_1)$ are free; (3) a problem in which the limit values of the indices, x_0 and x_1, are fixed and the state values, and $y(x_0)$ and $y(x_1)$, are free; or (4) a problem in which the limit values of the indices, x_0 and x_1, are free and the state values, and $y(x_0)$ and $y(x_1)$, are fixed.

For this problem, we need to respecify the Hamiltonian function as

$$H(x) = H(y, u, \lambda_0, \lambda_1, x) = F(y, u, x) - \lambda_0\, G_0(y, u, x) + \lambda_1(x)\, G_1(y, u, x)$$

where λ_0 is a constant, and λ_1 is a mapping $\lambda_1 : \mathrm{X} \to \mathbb{R}$. The maximized Hamiltonian, is

$$H^*(x) = \max_{u(\cdot)} H(y(x), u(x), \lambda_0, \lambda_1(x), x), \text{ for each } x \in \mathrm{X}.$$

where $H^*(x) = H(y^*(x), u^*(x), \lambda_0, \lambda_1(x), x)$.

Proposition 10.3 (PMP for the integral-constrained optimal control problem) *Let (y^*, u^*) be a solution to the OC problem (10.20). Then there is a variable λ_0 and a piecewise continuous function $\lambda : \mathrm{X} \to \mathbb{R}$, such that $(y^*, u^*, \lambda_0, \lambda_1)$ satisfy the following conditions:*

- *the optimality condition:*

$$H_u(y^*(x), u^*(x), \lambda_0, \lambda_1(x), x) = 0, \text{ for } x \in \left[x_0^*, x_1^*\right]; \tag{10.21}$$

- *the adjoint equation*

$$\lambda_1'(x) + H_y(y^*(x), u^*(x), \lambda_0, \lambda_1(x), x) = 0, \text{ for } x \in \left[x_0^*, x_1^*\right]; \tag{10.22}$$

- *the initial and terminal conditions associated to the independent and state variables*

$$\lambda_1\left(x_j^*\right) = 0, \text{ for } j = 0, 1 \tag{10.23a}$$

$$H^*\left(x_j^*\right) - \lambda_j\left(x_j^*\right) y^{*\prime}\left(x_j^*\right) = 0, \text{ for } j = 0, 1 \tag{10.23b}$$

- *for admissible solutions, i.e., satisfying*

$$\int_{x_0^*}^{x_1^*} G_0(y^*(x), u^*(x), x)dx = \bar{G} \tag{10.24a}$$

$$y^{*\prime}(x) = G_1(y^*(x), u^*(x), x) \; x \in \left(x_0^*, x_1^*\right) \tag{10.24b}$$

Proof The value functional at the optimum is

$$\mathsf{J}[y^*, u^*] = \int_{x_0^*}^{x_1^*} F(y^*(x), u^*(x), x)\, dx. \tag{10.25}$$

Substituting the definition of the Hamiltonian and integrating by parts yields

$$\begin{aligned}\mathsf{J}[y^*, u^*] &= \int_{x_0^*}^{x_1^*} \left(H\left(y^*(x), u^*(x), \lambda_0, \lambda_1(x), x\right) - y^{*\prime}(x)\right) dx + \lambda_0 \bar{G} \\ &= \int_{x_0^*}^{x_1^*} \left(H(y^*(x), u^*(x), \lambda_0, \lambda_1(x), x) + \lambda_1'(x)\, y^*(x)\right) dx + \\ &\quad + \lambda_1(x_1^*)\, y^*\left(x_1^*\right) - \lambda_1\left(x_0^*\right) y^*\left(x_0^*\right) + \lambda_0\, \bar{G}\end{aligned}$$

Introduce the arbitrary (functional) perturbations $y^*(x) \to y(x) = y^*(x) + \varepsilon\, \eta_y(x)$, $u^*(x) \to u(x) = u^*(x) + \varepsilon\, \eta_u(x)$, and the (point) perturbations $x_j^* \to x_j = x_t^* + \varepsilon\, \chi_t$, for $j = 0, 1$ and $y_j^* \to y_j = y_j^* + \varepsilon\, \iota_j$, for $j = 0, 1$, such that

$$\iota(x_j^*) = \iota_j - y'(x_j^*)\, \chi_j, \;\; j = 0, 1 \tag{10.26}$$

The Gâteaux functional derivative for the perturbed functional, in an admissible neighborhood of the optimum, is

$$
\begin{aligned}
\mathrm{D}_{\boldsymbol{\eta}(\cdot),\iota_0,\iota_1}\mathsf{J}[y^*,u^*] = &\int_{x_0^*}^{x_1^*}\Big[H\big(y^*(x)+\varepsilon\eta_y(x),u^*(x)+\varepsilon\eta_u(x),\lambda_0,\lambda_1(x),x\big)-\\
&-H\big(y^*(x),u^*(x),\lambda_0,\lambda_1(x),x\big)+\lambda_1'(x)\big(y^*(x)+\varepsilon\eta_y(x)-y^*(x)\big)\Big]dx+\\
&+\lambda_1(x_1^*)\left(y^*(x_1^*)+\varepsilon\eta_y(x_1^*)\right)-\lambda_1(x_0^*)\left(y^*(x_0^*)+\varepsilon\eta_y(x_0^*)\right)-\\
&-\lambda_1(x_1^*)\,y^*(x_1^*)+\lambda_1(x_0^*)\,y^*(x_0^*)+\\
&+\left(H\left(y^*(x),u^*(x),\lambda_0,\lambda_1(x),x\right)\Big|_{x=x_1^*}\right)\chi_1-\\
&-\left(H\left(y^*(x),u^*(x),\lambda_0,\lambda_1(x),x\right)\Big|_{x=x_0^*}\right)\chi_0
\end{aligned}
$$

Using a first-order Taylor approximation, considering Eq. (10.26), collecting terms, factoring out, and simplifying the notation we have,

$$
\begin{aligned}
\mathrm{D}_{\boldsymbol{\eta}(\cdot),\iota_0,\iota_1}\mathsf{J}[y^*,u^*] = &\int_{x_0^*}^{x_1^*}\left[H_u^*(x)\,\eta_u(x)+\left(H_y^*(x)+\lambda_1'(x)\right)\,\eta_y(x)\right]dx+\\
&+\lambda_1(x_1^*)\,\eta_y(x_1^*)-\lambda_1(x_0^*)\,\eta_y(x_0^*)+H^*(x_1^*)\,\chi_1-H^*(x_0^*)\,\chi_0\\
=&\int_{x_0^*}^{x_1^*}\left[H_u^*(x)\,\eta_u(x)+\left(H_y^*(x)+\lambda_1'(x)\right)\,\eta_y(x)\right]dx+\\
&+\lambda_1(x_1^*)\,\iota_1-\lambda_1(x_0^*)\,\iota_0+\left(H^*(x_1^*)-\lambda_1(x_1^*)\,(y^*)'(x_1^*)\right)\,\chi_1-\\
&-\left(H^*(x_0^*)-\lambda_1(x_0^*)\,(y^*)'(x_0^*)\right)\,\chi_0.
\end{aligned}
$$

As at the optimum $\mathrm{D}_{\boldsymbol{\eta}(\cdot),\iota_0,\iota_1}\mathsf{J}[y^*,u^*]=0$, we derive Eqs. (10.21)–(10.23b).

Example 10.2 (*The* Mirrlees (1971) *model*) This is an optimal tax policy model, in which a social planner seeks to maximize the social welfare of a population of households with heterogeneous skills, $\theta\in\Theta=[\underline{\theta},\bar{\theta}]$. However, differently from the case in Sect. 8.2.2.6, the social planner knows the distribution of income, $y(\theta)$, but cannot separate the contribution of effort, $\ell(\theta)$, and the skill level, θ. Recall that $y(\theta)=\theta\,\ell(\theta)$, and that $u(\theta)=U(c(\theta),\ell(\theta))$ is the utility of an household with skill level θ, in which the contribution of consumption is positive and the contribution of labor effort is negative, $U_\ell(c,\ell)<0<U_c(c,\ell)$. The skill distribution of households is exogenous and is denoted by $f(\theta)$, satisfying $0<f(\theta)<1$ and $\int_\Theta f(\theta)\,d\theta=1$. The government has two constraints: a budget constraint, and an incentive compatibility constraint. The tax paid by a household having skill level θ is $\tau(\theta)=T(u(\theta),\ell(\theta))=y(\theta)-C\big(u(\theta),\ell(\theta)\big)$, where $C(\cdot)$ is consumption. As there is no savings then $c(\theta)=y(\theta)$, then $u(\theta)=u(\ell(\theta))$. The incentive compatibility constraint is the seminal contribution in Mirrlees (1971), to address the existence of the imperfect information by the social planner: it incorporates into the design of the tax schedule a mechanism seeking to incentivize higher skilled households to exert higher effort instead of acting as a low skilled workers in order to reduce their tax payments.

When there is inequality aversion (see Sect. 8.1.5.8), the social welfare problem of finding an optimal tax policy can be equivalently specified as a problem of finding the optimal distribution of effort among heterogeneously skilled households that solves the problem

$$\max_{l(\theta)} \int_{\underline{\theta}}^{\bar{\theta}} W[u(\theta)]\, f(\theta)\, d\theta, \tag{10.27a}$$

subject to

$$\int_{\underline{\theta}}^{\overline{\theta}} T\big(u(\theta), \ell(\theta)\big)\, f(\theta)\, d\theta = G, \tag{10.27b}$$

$$u'(\theta) = -\frac{\ell(\theta) u_\ell(\ell(\theta))}{\theta}, \text{ for each } \theta \in \Theta, \tag{10.27c}$$

$$\underline{\theta}, \bar{\theta} \text{ free}, \tag{10.27d}$$

$$u(\underline{\theta}), u(\bar{\theta}) \text{ free}. \tag{10.27e}$$

Equation (10.27b) is the **resource constraint**, Eq. (10.27c) is the **incentive compatibility constraint** (ICP). The constraints (10.27d) and (10.27e) are introduced to account for the fact that the tax authority limits and levels of taxes at both ends of the skill distribution should be optimally derived. This means that there can be upper or lower extremes of the skill distribution that are not taxed. The Hamiltonian function is

$$H(\theta) = W(u(\theta))\, f(\theta) + \lambda_0\, T\big(u(\theta), \ell(\theta)\big)\, f(\theta) - \\ - \lambda_1(\theta)\, \frac{\ell(\theta)\, u_\ell(\ell(\theta))}{\theta}, \quad \text{for each } \theta \in \Theta,$$

where λ_0 is the adjoint constant associated to the budget constraint (10.27b), and $\lambda_1(\theta)$ is the co-state variable associated to the ICP (10.27c). The first-order conditions for an optimum, applying Proposition 10.3), comprise: first, a static optimality condition equating the private costs and social benefits of a change in the labor effort

$$\lambda_1(\theta)\, u_\ell(\ell(\theta))\, (1 - \sigma_\ell(\theta)) = \lambda_0\, \theta\, f(\theta)\, T_\ell\big(u(\theta), \ell(\theta)\big), \quad \text{for each } \theta \in \Theta, \tag{10.28}$$

where the elasticity labor supply is $\sigma_\ell(\theta) \equiv -\frac{u_{\ell\ell}(\ell(\theta))\,\theta}{u_\ell(\ell(\theta)}$, which implies $\lambda_0 < 0$; second, the adjoint ODE

$$\lambda_1'(\theta) = -f(\theta)\left(W'(u(\theta)) + \lambda_0\, T_u\big(u(\theta), \ell(\theta)\big)\right), \quad \text{for each } \theta \in \Theta, \tag{10.29}$$

is also an arbitrage condition, in utility terms, between the social benefit of an increase in the progressiveness of the tax schedule and the private cost of higher taxation, for each household of skill level θ; third, the ICP constraint (10.27c) and the budget constraint (10.27b), that jointly constitute the admissibility conditions; fourth, the boundary constraints that allows us to determine the cut-off households that pay taxes, and their optimal tax levels (see Table 10.1)

$$\lambda_1(\underline{\theta}^*) = \lambda_1(\overline{\theta}^*) = 0, \tag{10.30}$$

and

$$\begin{aligned} f(\underline{\theta}^*) \left(W(u(\underline{\theta}^*)) + \lambda_0 \, T\big(u(\underline{\theta}^*), \ell(\underline{\theta}^*)\big) \right) &= \\ = f(\overline{\theta}^*) \left(W(u(\overline{\theta}^*)) + \lambda_0 \, T\big(u(\underline{\theta}^*), \ell(\underline{\theta}^*)\big) \right) &= 0. \end{aligned} \tag{10.31}$$

In principle, we can find general solutions to the optimal distribution across skill levels of utility $u^*(\theta)$ and the co-state variable $\lambda_1(\theta)$, by solving ODE's (10.29) and (10.27c). Then we can obtain the optimal labor effort distribution across skill levels $\ell^*(\theta)$, obtained from Eq. (10.28), and the Lagrange adjoint λ_0 from Eq. (10.27b). Those distributions depend on arbitrary values for the boundary values of θ, λ_1 and u, at the optimum. Those optimal boundary values $\underline{\theta}^*$, $\overline{\theta}^*$, $\lambda_1(\underline{\theta}^*)$, $\lambda_1(\overline{\theta}^*)$, $u(\underline{\theta}^*)$, and $u(\overline{\theta}^*)$ can be obtained from the general solution for $\lambda_1(\theta)$ and from the optimum conditions (10.30) and (10.31). Therefore, in principle, we can obtain a complete distribution for the optimum utility and labor effort $\mathbf{u}^* = (u^*(\theta))_{\theta\in\Theta^*}$ and $\boldsymbol{\ell}^* = (\ell^*(\theta))_{\theta\in\Theta^*}$, where $\Theta^* = [\underline{\theta}^*, \overline{\theta}^*]$, and the optimal tax structure $\boldsymbol{\tau}^* = \big(T(u^*(\theta), \ell^*(\theta)\big)_{\theta\in\Theta^*}$.

Exercise 10.1 Find the optimal effort and tax policy that solves problem (10.27) assuming the same behavioral functions as in Exercise 8.22. Try to provide an intuition on the optimal progressiveness of taxation.

Exercise 10.2 Find the optimal effort and tax policy that solves problem (10.27) assuming the same behavioral functions as in Exercise 8.23. Try to provide an intuition on the optimal progressiveness of taxation.

Exercise 10.3 Find the optimal effort and tax policy that solves problem (10.27) assuming the same behavioral functions as in Exercise 8.24. Try to provide an intuition on the optimal progressiveness of taxation.

10.2 The Maximum Principle Approach in the Time Domain

In most applications in macroeconomics and growth theory, the independent variable is time, which allows for the application of the well-developed dynamic systems theory for setting up and solving dynamic optimization problems. Regarding the time domain, T, there are two main types of problems, finite horizon problems, in which $\mathrm{T} = [t_0, t_1]$, and infinite horizon problems, in which $\mathrm{T} = \mathbb{R}_+$. The optimal control problem (OCP) involves two variables, the **state variable**, y, and the **control variable**, u. In this chapter, we consider again only problems in which the state variable is a dimension one mapping $y : \mathrm{T} \to \mathrm{Y} \subseteq \mathbb{R}$, and the control variable is a $m-$dimensional mapping $u : \mathrm{T} \to \mathrm{U} \subseteq \mathbb{R}^m$, where $m \geq 1$. It is important to distinguish between the **temporary level** of variables, i.e., $y(t')$ and $u(t')$, for a specific $t = t'$, from the **paths or trajectories** over the some time-domain $\mathbf{y} = \big(y(t)\big)_{t\in\mathrm{T}}$ and $\mathbf{u} = \big(u(t)\big)_{t\in\mathrm{T}}$. A solution to an OCP allows for finding optimal trajectories, $\mathbf{y}^* = \big(y^*(t)\big)_{t\in\mathrm{T}}$ and $\mathbf{u}^*_{\mathrm{T}} = \big(u^*(t)\big)_{t\in\mathrm{T}}$ that maximize (or minimize) an objective functional, subject to an ODE, and some other constraints. The constraints can be functionals specified for the whole time domain, point-wise temporal constraints (usually for the initial and the terminal times), or constraints limiting the trajectory paths of the state and/or control variables (the next chapter addresses this case).

In this section, finite and infinite horizon OCPs are presented in Sects. 10.2.1, and 10.2.2, respectively. The dynamics of the solution to infinite horizon problems and their relationship with the qualitative theory of ODE is the subject of Sect. 10.2.3. Section 10.2.4 contains some benchmark applications to economics.

10.2.1 Finite Horizon Problems

Formally, a time-dependent OCP with a bounded time domain $\mathrm{T} = [t_0, t_1] \subset \mathbb{R}_+$, consists in finding a continuous and continuously differentiable state variable $y \in \mathcal{Y} \subset C^1(\mathrm{T}; \mathrm{Y})$, mapping $y : \mathrm{T} \to \mathrm{Y} \subseteq \mathbb{R}$, and a piecewise continuous control variable $u \in \mathcal{U} \subset PC^1(\mathrm{T}; \mathrm{U})$, mapping $u : \mathrm{T} \to U \subseteq \mathbb{R}^m$, where $m \geq 1$, that maximizes the functional

$$\mathsf{J}[y, u] \equiv \int_{t_0}^{t_1} F(y(t), u(t), t)\, dt \tag{10.32}$$

subject to the ODE

$$\dot{y} = G(y(t), u(t), t), \ \text{ for } t \in [t_0, t_1], \tag{10.33}$$

and to some additional constraints referring to the boundaries of set T, and the associated values for the state variable $y \in \mathrm{Y}$, that we specify next. The **value function** is the optimal level of the objective functional

$$V(\boldsymbol{\varphi}) = \mathsf{J}[y^*, u^*] = \max_{u(\cdot)} \mathsf{J}[y, u],$$

and depends on a set of parameters $\boldsymbol{\varphi}$, containing the data of the problem, v.g., fixed boundary values for the time domain, and/or the associated values of the state variable. The **Pontriyagin's maximum principle** (PMP) involves the **Hamiltonian** function,

$$H(y(t), u(t), \lambda(t), t) = F(y(t), u(t), t) + \lambda(t)\, G(y(t), u(t), t), \ \text{ for each } t \in \mathrm{T},$$

where the piecewise continuous function, mapping $\lambda : \mathrm{T} \to \mathbb{R}$, is called **co-state**variable. The path of the co-state variable is denoted by $\boldsymbol{\lambda} = (\lambda(t))_{t \in \mathrm{T}}$.

We start with the cases in which the two boundary times, t_0 and t_1, are fixed, and the values of the state variables at those times, $y(t_0)$ and/or $y(t_1)$, can be fixed or can be chosen optimally, drawing on the formally identical problems dealt with in Sect. 10.1.1, with an appropriate change in the notation.

Proposition 10.4 (PMP for fixed boundary values of the independent variable) *Let $(y^*(t), u^*(t))$ be functions with domain $\mathrm{T} = [t_0, t_1]$, where t_0 and t_1 are fixed, that maximize the objective functional (10.32) subject to the constraint ODE (10.33), and some other terminal conditions on the state variable at the boundary times, $y(t_0)$ and/or $y(t_1)$. Then there is a piecewise continuous function $\lambda : \mathrm{T} \to \mathbb{R}$, called co-state variable, such that the paths $(\mathbf{y}^*, \mathbf{u}^*, \boldsymbol{\lambda})$ satisfy the following conditions:*

- *the optimality condition*

$$H_u(y^*(t), u^*(t), \lambda(t), t) = 0, \text{ for each } t \in [t_0, t_1], \tag{10.34}$$

- *the adjoint equation*

$$\dot{\lambda} = -H_y(y^*(t), u^*(t), \lambda(t), t), \text{ for each } t \in (t_0, t_1), \tag{10.35}$$

- *the constraint of the problem*

$$\dot{y}^* = G(y^*(t), u^*(t), t), \text{ for each } t \in (t_0, t_1), \tag{10.36}$$

- *and one of the following terminal conditions:*

$$y^*(t_0) = y_0 \text{ and } y^*(t_1) = y_1, \text{ if } y_0 \text{ and } y_1 \text{ are fixed}; \tag{10.37a}$$
$$y^*(t_0) = y_0 \text{ and } \lambda(t_1) = 0, \text{ if } y_0 \text{ is fixed and } y(t_1) \text{ is free}; \tag{10.37b}$$
$$\lambda(t_0) = 0, \text{ and } y^*(t_1) = y_1, \text{ if } y(t_0) \text{ is free and } y_1 \text{ is fixed}; \tag{10.37c}$$
$$\lambda(t_0) = 0, \text{ and } \lambda(t_1) = 0, \text{ if } y(t_0) \text{ and} y(t_1) \text{ are free}. \tag{10.37d}$$

Next we have the cases in which the two boundary values for the state variable, $y(t_0)$ and $y(t_1)$, are fixed, and the terminal times, t_0 and t_1, are fixed or are free and can be chosen optimally.

Proposition 10.5 (PMP for free domain and fixed boundary state variable optimal control problems) *Let $(y^*(t), u^*(t))$ be functions with domain $\mathrm{T} = [t_0, t_1]$, where $y^*(t_0) = y_0$ and/or $y^*(t_1) = y_1$ are fixed, that maximize the objective functional (10.32) subject to the constraint ODE (10.33), with or without fixed values for the boundary values of the state variable, $y(t_0)$ and/or $y(t_1)$. Then there is an optimal domain for the independent variable $\mathrm{T}^* = [t_0^*, t_1^*] \subset \mathbb{R}$, a piecewise continuous function $\lambda : \mathrm{T}^* \to \mathbb{R}$, called co-state variable, such that the paths $(\mathbf{y}^*, \mathbf{u}^*, \boldsymbol{\lambda})$ satisfy the*

Table 10.2 Adjoint conditions for bounded domain OC problems

data		Boundary conditions for an optimum	
t_j	$y(t_j)$	t_j^*	$y^*(t_j^*)$
Fixed	Fixed	t_j	y_j
Fixed	Free	t_j	$\lambda(t_j) = 0$
Free	Fixed	$H(y_j, u^*(t_j^*), \lambda(t_j^*), t_j^*,) - \dot{y}^*(t_j^*)\lambda(t_j^*) = 0$	y_j
Free	Free	$H(y^*(t_j^*), u^*(t_j^*), \lambda(t_j^*), t_j^*,) = 0$	$\lambda(t_j^*) = 0$

optimality condition (10.34), the adjoint equation (10.35), and the ODE constraint of the problem (10.36), for each $t \in Int(\mathrm{T}^*)$, and one of the following terminal conditions:

$$y^*(t_0) = y_0 \text{ and } y^*(t_1) = y_1, \text{ if } t_0 \text{ and } t_1 \text{ fixed;} \tag{10.38a}$$
$$y^*(t_0) = y_0 \text{ and } H^*(t_1^*) = \dot{y}^*(t_1^*)\lambda(t_1^*) \text{ if } t_0 \text{ fixed and } t_1^* \text{ free;} \tag{10.38b}$$
$$H^*(t_0^*) = \dot{y}^*(t_0^*)\lambda(t_0^*) \text{ and } y^*(t_1) = y_1, \text{ if } t_0^* \text{ free and } t_1 \text{ fixed;} \tag{10.38c}$$
$$H^*(t_0^*) = \dot{y}^*(t_0^*)\lambda(t_0^*) \text{ and } H^*(t_1^*) = \dot{y}^*(t_1^*)\lambda(t_1^*), \text{ if } t_0^* \text{ and } t_1^* \text{ free.} \tag{10.38d}$$

where $H^*(t_j^*) = H(y_j, u^*(t_j^*), \lambda(t_j^*), t_j^*)$, for $j = 0, 1$.

In Proposition 10.4, we assumed that t_0 and t_1 where fixed, $y(t_0)$ and $y(t_1)$ could be fixed or free, and in Proposition 10.5, we assumed that y_0 and y_1 are fixed, and t_0 and t_1 could be fixed or free. In the more general case several other possibilities exist: for instance $y(t_0)$ and t_0 are fixed and $y(t_1)$ and t_1 are free, or $y(t_0)$ and t_0 are free and $y(t_1)$ and t_1 are fixed, or $y(t_0)$ and t_0 are free and $y(t_1)$ and t_1 are free. In fact, there are 16 possible combinations of boundary conditions.

The necessary conditions for the more general case include the optimality condition (10.34), the adjoint equation (10.35), and the ODE constraint of the problem (10.36), for each $t \in \mathrm{int}(\mathrm{T}^*)$, and one of the adjoint conditions presented in Table 10.2 for each $j = 0, 1$:

The index refers to the lower boundary when $j = 0$ and to the upper boundary when $j = 1$. The necessary conditions for a particular OCP include only one entry for each $j = 0$ and $j = 1$.

Inequality terminal constraints. Consider an OCP in which set T has fixed limits, t_0 and t_1, the initial value of the state variable is known, $y(t_0) = y_0$, and the terminal value of the state variable satisfies the inequality constraint

$$R(y(t_1), t_1) \geq 0. \tag{10.39}$$

We call this problem a **constrained terminal value problem**.

Proposition 10.6 (PMP for the constrained terminal value problem) *Let $(y^*(t), u^*(t))$ be functions with domain $\mathrm{T} = [t_0, t_1]$, where $y^*(t_0) = y_0$ and/or $y^*(t_1) = y_1$ are fixed,* *that maximize the objective functional (10.32) subject to the constraint ODE (10.33), to a fixed initial value $y(t_0) = y_0$, and to the terminal constraint (10.39). Then there is a piecewise continuous function $\lambda : \mathrm{T} \to \mathbb{R}$, called co-state variable, such that the paths $(\mathbf{y}^*, \mathbf{u}^*, \boldsymbol{\lambda})$ satisfy the optimality condition (10.34), the adjoint equation (10.35) for each $t \in Int(\mathrm{T})$, and the transversality condition*

$$\lambda(t_1)\, R(y^*(t_1), t_1) = 0. \tag{10.40}$$

for admissible paths, satisfying the admissibility constraints, the ODE constraint (10.36), and the initial condition $y^*(t_0) = y_0$.

Proof We follow similar steps as in the proof of Proposition 10.1. At the optimum the value functional is

$$\begin{aligned} \mathsf{J}[y^*, u^*] &= \int_{t_0}^{t_1} F(y^*(t), u^*(t), t)\, dt \\ &= \int_{t_0}^{t_1} \big(H(y^*(t), u^*(t), \lambda(t), t) - \dot{y}^*(t)\lambda(t)\big)\, dt + \psi\, R(y(t_1), t_1), \end{aligned}$$

where ψ is a Lagrange adjoint multiplier associated to constraint (10.39), and the KKT conditions

$$R(y^*(t_1), t_1) \geq 0,\ \ \psi \geq 0,\ \ \psi\, R(y^*(t_1), t_1) = 0,$$

are satisfied. The Gâteaux differential for an arbitrary admissible perturbation $(\delta y, \delta u) = \varepsilon(\eta_y, \eta_u)$ around (y^*, u^*), satisfying $\eta_y(t_0) = 0$ and $\eta_y(t_1) \neq 0$, is now

$$\begin{aligned} \mathrm{D}_{(\eta_y(\cdot), \eta_u(\cdot))}\mathsf{J}[y^*, u^*] = \int_{t_0}^{t_1} & \big[H_u^*(t)\, \eta_u(t) + (H_y^*(t) + \dot{\lambda}(t))\, \eta_y(t)\big]\, dt + \\ & + \big(\psi\, R_y(t_1, y^*(t_1)) - \lambda(t_1)\big)\, \eta_y(t_1). \end{aligned}$$

Then $\mathrm{D}_{(\eta_y(\cdot), \eta_u(\cdot))}\mathsf{J}[y^*, u^*] = 0$ only if $H_u^*(t) = \dot{\lambda}(t) - H_y^*(t) = 0$, that is if the optimality condition (10.34), and the adjoint equation (10.35) are satisfied for each $t_0 < t < t_1$. At last, because $\eta_y(t_1) \neq 0$, the remaining necessary condition for an optimum is $\psi\, R_y(y^*(t_1), t_1) = \lambda(t_1)$. Multiplying both terms by $R(y^*(t_1), t_1)$, and using the KKT condition, yields condition (10.40).

10.2.2 Infinite Horizon Problems

The benchmark problem in macroeconomics and growth theory is the **discounted infinite horizon problem** in which the domain of the independent variable is $\mathrm{T} = R^+$,

$$\max_{u(\cdot)} \int_0^\infty e^{-\rho t} f(y(t), u(t))dt, \tag{10.41a}$$

$$\text{subject to } y(0) = y_0 \text{ fixed,}$$

$$\dot{y} = g(y, u), \quad \text{for } t \in \mathrm{T}, \tag{10.41b}$$

$$\lim_{t\to\infty} y(t) \text{ is free} \tag{10.41c}$$

$$\text{or } \lim_{t\to\infty} R(y(t), t) \geq 0, \tag{10.41d}$$

where the discount rate is $\rho > 0$, and function $R(t, y)$ is known and takes the form of a solvability or sustainability condition. The objective function $F(y(t), u(t), t) \equiv e^{-\rho t} f(y(t), u(t))$, has a time-dependent discount factor, that is a bounded function of time, and the constraint (10.41b) is an autonomous ODE.

It is convenient to define the **current-value Hamiltonian** function

$$\begin{aligned} h(y(t), u(t), q(t)) &= f(y(t), u(t)) + q(t)\, g(y(t), u(t)) = \\ &= e^{\rho t}\, H(y(t), u(t), \lambda(t), t), \text{ for each } t \in \mathrm{T} \end{aligned} \tag{10.42}$$

where $q(t) = e^{\rho t}\lambda(t)$ is the **current-value co-state variable**. Then, $H(t, y, u, \lambda)$ is called **discounted Hamiltonian** and and λ is called the **discounted co-state variable**. The co-state variable q is a piecewise continuous function of time $q : \mathrm{T} \to \mathrm{Q} \subseteq \mathbb{R}$.

The current-value Hamiltonian (10.42) is time-independent, in the sense that it does not depend directly on time as the discounted Hamiltonian. If, alternatively, the constraint is a non-autonomous ODE, i.e., if $\dot{y} = g(t, y, u)$, then Hamiltonian is also explicitly time dependent, $h(y(t), u(t), q(t)t) = f(y(t), u(t)) + q(t)\, g(t, y(t), u(t))$.

Proposition 10.7 (PMP for infinite horizon problems) *Let (y^*, u^*) be the optimal state and control trajectory pair. Then there is a PC^1 continuous co-state variable q such that the following conditions hold:*

- *the optimality condition*

$$h_u(y^*(t), u^*(t), q(t)) = 0, \textit{ for each } t \in [0, \infty), \tag{10.43}$$

- *the adjoint equation for the current value co-state variable*

$$\dot{q} = \rho q - h_y(y^*(t), u^*(t), q(t)), \text{ for each } t \in [0, \infty), \tag{10.44}$$

- *one of the two following transversality conditions:*

$$\lim_{t\to\infty} e^{-\rho t} q(t) = 0, \text{ if the problem has constraint (10.40c)}, \tag{10.45a}$$

$$\lim_{t\to\infty} e^{-\rho t} q(t)\, R(y^*(t), t) = 0, \text{ if the problem has constraint (10.41d)}, \tag{10.45b}$$

- *and the admissibility conditions*

$$\dot{y}^* = g(y^*(t), u^*(t)), \text{ for each } t \in [0, \infty), \tag{10.46a}$$

$$y^*(0) = y_0, \text{ for } t = 0. \tag{10.46b}$$

Proof See the proof of Proposition 10.4, for the case in which t_0, t_1, $y(t_0)$ are fixed and $y(t_1)$ is free, and the proof of Proposition 10.6, taking $t_1 \to \infty$.

If the constraint ODE is non-autonomous, $\dot{y} = g(y(t), u(t), t)$, first-order conditions (10.44) and (10.46a) are also non-autonomous ODE, because the current value Hamiltonian becomes (directly) time-dependent, i.e., $h(t) = h(y(t), u(t), q(t), t) = f(y(t), u(t)) + q(t)\, g(y(t), u(t), t)$.

Sufficient conditions for an optimum. There is a large literature on the nature of the transversality condition for infinite horizon problems. In particular, it has been discussed if it is a necessary or sufficient condition for an optimum, and this issue is related to the boundedness of the objective functional.[7]

In the rest of this chapter, we assume that the **Arrow sufficiency condition**, $h^*_{uu} = h(y^*, u^*, q) \leq 0$, holds. This condition guarantees that the first-order necessary conditions for an extremum, presented in Proposition 10.7, are also sufficient.

10.2.3 The Dynamics of Optimal Control Problems

The infinite-horizon discounted OCP is at the core of growth theory since the 1960s, and of macroeconomics since the early 1980s. In several applications, the constitutive functions, $f(y, u)$ and $g(y, u)$, are non-linear (sometimes not even specified explicitly), implying that explicit solutions cannot be obtained. However, a general qualitative characterization of the dynamics of the solution to an OCP is possible, at least in regular cases.

Let functions $f(y, u)$ and $g(y, u)$ be regular, that is, continuous and smooth $C^2(\mathrm{Y} \times \mathrm{U})$ functions.[8] The necessary (and sometimes also sufficient) conditions for

---

[7] Halkin (1974)showed that it may not be satisfied. For infinite horizon problems, this issue has been discussed and settled in Michel (1982) and Kamihigashi (2001).

[8] In Chap. 12, we consider non-regular cases.

the infinite-horizon discounted OCP, presented in Proposition 10.7, can be compactly presented as a differential-algebraic system, that we call **modified Hamiltonian dynamic system** (MHDS) in extended form,

$$\dot{y} = h_q(y, u, q) = g(y, u), \tag{10.47a}$$

$$\dot{q} = \rho\, q - h_y(y, u, q), \tag{10.47b}$$

$$0 = h_u(y, u, q). \tag{10.47c}$$

We write $\mathbf{F} = \Big(h_q(y, u, q), \rho\, q - h_y(y, u, q), h_u(y, u, q)\Big)^\top$. The Jacobian of MHDS (10.47) is

$$\mathbf{J} \equiv \mathbf{DF}(y, u, q) = \begin{pmatrix} g_y(y, u) & g_u(y, u) & 0 \\ -h_{yy}(y, u, q) & -h_{uy}(y, u, q) & \rho - g_y(y, u) \\ h_{yu}(y, u, q) & h_{uu}(y, u, q) & g_u(y, u) \end{pmatrix}, \tag{10.48}$$

where all the partial derivatives are evaluated at an arbitrary point (y, u, q). The second-order partial derivatives are $h_{yy}(y, u, q) = f_{yy}(y, u) + q\, g_{yy}(y, u)$, and if the functions are continuous $h_{yu}(y, u, q) = f_{yu}(y, u) + q\, g_{yu}(y, u) = h_{uy}(y, u, q) = f_{uy}(y, u) + q\, g_{uy}(y, u)$ and $h_{uu}(y, u, q) = f_{uu}(y, u) + q\, g_{uu}(y, u)$, $h_{qy}(y, u, q) = h_{yq}(y, u, q) = g_y(y, u)$, and $h_{qu}(y, u, q) = h_{uq}(y, u, q) = g_u(y, u)$. The determinant of $\mathbf{J}(y, u, q)$ is

$$\det(\mathbf{J}) = (g_y - \rho)\left(g_y\, h_{uu} - g_u\, h_{uy}\right) + g_u\left(g_u\, h_{yy} - g_y\, h_{yu}\right) \tag{10.49}$$

The (local) existence, uniqueness and stability properties of a steady state, $(\bar{y}, \bar{u}, \bar{q}) \in \mathrm{Y} \times \mathrm{U} \times \mathrm{Q}$ such that $\dot{y} = \dot{q} = 0$, can be assessed by the value of $\det(\bar{\mathbf{J}}) = \det(\mathbf{DF}(\bar{y}, \bar{u}, \bar{q}))$. The following cases are possible. First, if $f(y, u)$ and $g(y, u)$ are linear functions of both the state and the control variables, all the second derivatives of $h(\cdot, q)$ are equal to zero, then $\det \mathbf{J}(y, u, q) = 0$, which implies that either a steady state does not exist, or there is an infinite number of steady states, or a solution to the OCP may not exist. Second, if $h(\cdot)$ is linear in the control variable, then $h_{uu} = h_{yu} = h_{uy} = 0$ and $\det \mathbf{J}(y, u, q) = (g_u)^2\, h_{yy}$, where g_u is a constant. A steady state exists if $h_{yy} \neq 0$, and if $h_{yy}(\bar{y}, \bar{u}, \bar{q}) < 0$ then the steady state is a saddle point. Third, if the objective function is independent of the state variable, as in several macroeconomic models, that is $f = f(u)$ and the constraint is linear in y then $h_{uy} = q\, g_{uy}$ and $h_{yy} = 0$. In this case, we have

$$\det(\mathbf{J}) = (g_y - \rho)\left(g_y\, h_{uu} - g_u\, q\, g_{uy}\right) - g_u\, q\, g_y\, g_{yu},$$

which does not rule out the existence of a steady state. However, if the constraint function is also linear in u we cannot rule out that $\det(\mathbf{F}) = 0$, and that either a steady state does not exist or there is an infinite number of steady states. Indeed, this is the case in simple endogenous growth models and simple models for the representative household, as can be seen next. At last, if functions $f(y, u)$ and $g(y, u)$ are non-linear, such that we have locally det $(\mathbf{J}) = 0$, a necessary condition for the existence of a local bifurcation can be satisfied. This can generate complex dynamics solutions to OCP. Although the number of possible cases is very large, we can say, in abstract, that any bounded trajectory converging to a steady state or to a limit cycle is a candidate solution to the problem, but we cannot rule out the possibility of existence of multiple solutions.

If functions $f(.)$ and $g(.)$ are sufficiently smooth, we can characterize qualitatively the optimal dynamics of (y, q) (or (y, u)), by using the algebraic equation (10.47c). There are two general cases.

First, if there is at least a point $(y, u, q) \in \mathrm{Y} \times \mathrm{U} \times \mathrm{Q}$ such that $h_{uu}(y, u, q) = 0$, then solving algebraic equation (10.47c) for u and substituting in the ODEs (10.47a) and (10.47b) yields a **singular** ODE. Using our approach from Chap. 7,[9] we can generally say that candidate solutions to the OCP should have continuation towards infinite time and converge to a steady state or to a trajectory that satisfies the transversality condition. Trajectories that are attracted to the singular surface, at the point in which $h_{uu}(y, u, q) = 0$, cannot be candidate trajectories.

Second, if instead $h_{uu}(y, u, q) \neq 0$ for all $(y, u, q) \in \mathrm{Y} \times \mathrm{U} \times \mathrm{Q}$, then the DAE system (10.47) is **regular**, and, invoking the implicit function theorem, solving Eq. (10.47c) for u allows us to find this control variable as a monotonic function of the state and the co-state variables,

$$u = U(y, q), \text{ where } U_y = -\frac{h_{uy}}{h_{uu}}, \text{ and } U_q = -\frac{g_u}{h_{uu}}.$$

If we substitute this control representation in the ODE (10.47a) and (10.47b) we obtain the **modified Hamiltonian dynamic system** (MHDS), in reduced form, as a non-linear planar ODE,

$$\begin{pmatrix} \dot{y} \\ \dot{q} \end{pmatrix} = \mathbf{M}(y, q) \equiv \begin{pmatrix} g\big(y, U(y, q)\big) \\ \rho q - h_y\big(y, U(y, q), q\big) \end{pmatrix}. \tag{10.50}$$

The Jacobian of the MHDS is

$$\mathrm{D}\mathbf{M}(y, q) = \begin{pmatrix} g_y(y, q) + g_u(y, q)\, U_y(y, q) & g_u(y, q)\, U_q(y, q) \\ -h_{yy}(y, q) - h_{yu}(y, q)\, U_y(y, q) & \rho - h_{yu}(y, q)\, U_q(y, q) - h_{yq}(y, q) \end{pmatrix}.$$

Lemma 10.1 *Let $\rho > 0$ and assume that $h_{uu}(y, u, q) \neq 0$ for every $(y, u, q) \in \mathrm{Y} \times \mathrm{U} \times \mathrm{Q}$. Then the Jacobian* $\mathrm{D}\mathbf{M}(y, q)$ *has* $\mathrm{tr}(\mathrm{D}\mathbf{M}(y, q)) = \rho > 0$ *for any point $(y, q) \in \mathrm{Y} \times \mathrm{Q}$.*

---

[9] In Brito et al. (2017), we deal with the case in which we can have locally $h_{uu} = 0$, which implies that the solution can be non-unique locally for a subset of the space $\mathrm{Y} \times \mathrm{Q}$.

Proof As $g_u\, U_y = -g_u\, \frac{h_{uy}}{h_{uu}} = h_{uy}\, U_q = -h_{uy}\, \frac{g_u}{h_{uu}}$, we readily observe that the Jacobian matrix has the form $\mathrm{D}\mathbf{M}(y,q) = \begin{pmatrix} M_{11}(y,q) & M_{12}(y,q) \\ M_{21}(y,q) & \rho - M_{11}(y,q) \end{pmatrix}$. Therefore $\mathrm{tr}(\mathrm{D}\mathbf{M}(y,q)) = M_{11} + \rho - M_{11} = \rho > 0$.

This is a global property, depending on the condition $h_{uu} \neq 0$ being globally satisfied. This implies that there is a steady state and that there are no singularities. Assume the MHDS has, at least, one steady state, $(\bar{y}, \bar{q}) = \{(y,q) : \dot{y} = \dot{q} = 0\}$. In the neighborhood of $(\bar{y}, \bar{q})$, we can approximate the non-linear MHDS (10.50) by the linear ODE

$$\begin{pmatrix} \dot{y}(t) \\ \dot{q}(t) \end{pmatrix} = \mathrm{D}\mathbf{M}(\bar{y}, \bar{q}) \begin{pmatrix} y(t) - \bar{y} \\ q(t) - \bar{q} \end{pmatrix},$$

where the Jacobian $\mathrm{D}_{(y,q)}\mathbf{M}(y,q)$ is evaluated at the steady state $(\bar{y}, \bar{q})$. If functions $f(.)$ and $g(.)$ have no singularities, then we can obtain a generic characterization of the dynamics of the MHDS, and, therefore, of the solution to the OCP.

Proposition 10.8 *Let there be a steady state $(\bar{y}, \bar{q})$ for the MHDS system (10.50). The steady state can never be a stable node or a stable focus. The solution to the MHDS system is time dependent and displays transitional dynamics converging to the steady state only if it is a saddle point, that is if* $\det(\mathrm{D}_{(q,y)}\mathbf{M}(\bar{y}, \bar{q})) < 0$.[10]

Proof Evaluating the Jacobian at a steady state $(\bar{y}, \bar{q})$, with $\bar{u} = U(\bar{y}, \bar{q})$, we find that the Jacobian matrix is

$$\mathrm{D}\mathbf{M}(\bar{y}, \bar{q}) = \begin{pmatrix} \bar{m}_{11} & \bar{m}_{12} \\ \bar{m}_{21} & \rho - \bar{m}_{11} \end{pmatrix} = \begin{pmatrix} \bar{g}_y - \dfrac{\bar{g}_u \bar{h}_{uy}}{\bar{h}_{uu}} & -\dfrac{(\bar{g}_u)^2}{\bar{h}_{uu}} \\ -\bar{h}_{yy} + \dfrac{(\bar{h}_{uy})^2}{\bar{h}_{uu}} & \rho - \bar{g}_y + \dfrac{\bar{g}_u \bar{h}_{uy}}{\bar{h}_{uu}}, \end{pmatrix} \tag{10.51}$$

where $\bar{g}_y = g_y\big(U(\bar{y}, \bar{q}), \bar{y}\big)$, etc. Then

$$\mathrm{tr}(\mathrm{D}_{(q,y)}\mathbf{M}(\bar{y}, \bar{q})) = \rho > 0, \tag{10.52}$$

implies that if a steady state exists it cannot be a stable node or focus. Furthermore, as

$$\det(\mathrm{D}\mathbf{M}(\bar{y}, \bar{q})) = \bar{m}_{11}\left(\rho - \bar{m}_{11}\right) - \bar{m}_{12}\,\bar{m}_{21}, \tag{10.53}$$

the sign is ambiguous. However, a steady state can be an unstable node or an unstable focus if $\det(\mathrm{D}_{(q,y)}\mathbf{M}(\bar{y}, \bar{q})) > 0$, a saddle-point if $\det(\mathrm{D}\mathbf{M}(\bar{y}, \bar{q})) < 0$ or a degenerate saddle node if $\det(\mathrm{D}\mathbf{M}(\bar{y}, \bar{q})) = 0$. Transitional dynamics converging to a steady state only exists if the steady state is a saddle-point.

Then we have the following possible cases for a regular OCP. First, if there is a steady state $(\bar{y}, \bar{q})$ that is a saddle point, satisfying $\det(\mathrm{D}\mathbf{M}(\bar{y}, \bar{q})) < 0$, with an

[10] This is a stronger version of the Brock and Scheinkman (1976) condition. See Carlson et al. (1996).

associated stable manifold $\mathcal{W}^{s,loc} = \{(y, q) \in \mathrm{Y} \times \mathrm{Q} \subseteq \mathbb{R}^2 : \lim_{t\to}(y(t), q(t)) = (\bar{y}, \bar{q})\}$, then a solution exists only if $y(0) = y_0 \in \mathcal{W}^{s,loc}$. In this case, the solution to the OC problem is (at least locally) unique, and the optimal trajectory is asymptotically tangent to the stable eigenspace $\mathcal{E}^s$ associated to Jacobian $\mathbf{DM}(\bar{y}, \bar{q})$. Second, if the steady state is unstable, satisfying det $(\mathbf{DM}(\bar{y}, \bar{q})) > 0$, and the stable manifold is empty, then candidate solutions diverge away from the steady state. The existence of a solution to the OCP depends on the transversality condition (Eq. (10.46b)) being satisfied. If this is the case, the solution to the OCP is **non-stationary**, and asymptotically unbounded. Third, if det $(\mathbf{DM}(y, q)) = 0$ globally there is an infinite number of candidate steady states.[11] However, the MHDS can be "anchored" by the initial condition $y(0) = y_0$, which means that the solution to the OCP exists, it is unique, and it is **stationary** (time independent).

Conclusion. Therefore, if there is a unique solution to the infinite-horizon discounted OCP, with a single state variable y, and if the constitutive functions are monotonic, three types of solutions can occur: (1) a unique time varying solution converging to a steady state, belonging to the stable manifold associated to that steady state, if the determinant of the MHDS evaluated at the steady state is negative; (2) a time-varying unbounded solution which grows over time if the determinant of the MHDS evaluated at the steady state is positive and the solution satisfies the transversality condition; or (3) a stationary solution if the determinant of the MHDS evaluated at the steady state is equal to zero.[12]

10.2.4 Applications to Economics

10.2.4.1 Simplest Problems

Some of the following problems have been solved by using the calculus of variations approach in Chap. 9.

Example 10.3 (*The resource depletion or cake-eating problem*) The (non-renewable) resource depletion problem can be considered an infinite horizon version of the cake-eating problem, already solved in Example 9.4 using a CV approach. Next, we solve it by using the PMP approach. Recall that the problem is

$$\begin{aligned} \max_{c(\cdot)} & \int_0^\infty e^{-\rho t} \ln (c(t))dt, \ \rho > 0, \\ & \text{subject to } w(0) = w_0 > 0, \text{ fixed}, \\ & \dot{w}(t) = -c(t), \text{ for } t \in [0, \infty), \\ & \lim_{t\to\infty} w(t) \geq 0. \end{aligned} \tag{10.54}$$

[11] If the determinant is equal to zero locally and not globally, we need to study the global solutions to the OCP to be able to characterize it.

[12] However, the existence of linear functions $f(\cdot)$ and $g(\cdot)$ calls for a slightly different version of the PMP (see singular problems in Chap. 12).

In this problem, the control variable is consumption, c, and the state variable is the level of the resource, w. The current-value Hamiltonian is $h = \ln(c) - q\,c$. Applying the PMP, in Proposition 10.7, and solving for c, we have $c(t) = q^{-1}(t)$. Then the MHDS, jointly with the initial and the transversality conditions, is

$$\dot{w} = -q(t)^{-1}, \tag{10.55a}$$

$$\dot{q} = \rho\, q(t), \tag{10.55b}$$

$$\lim_{t\to\infty} e^{-\rho t}\, q(t)\, w(t) = 0, \tag{10.55c}$$

$$w(0) = w_0. \tag{10.55d}$$

The Jacobian of the MHDS has tr $= \rho > 0$ and det $= 0$. In this case, there is no steady state.[13] The MHDS has a closed form solution: ODE (10.55b) has the solution $q(t) = q(0)\, e^{-\rho\, t}$, where $q(0)$ is arbitrary; substituting in ODE (10.55a), using the method of separation of variables, and the initial condition, yields $w(t) = w_0 + \frac{1}{\rho\, q(0)}\left(e^{\rho\, t} - 1\right)$; and, at last, substituting in the transversality condition, Eq. (10.55c), and solving for $q(0)$ yields $1 = \rho\, w_0\, q(0)$. Then the solution to the problem is (compare with the solution in Example 9.4)

$$w^*(t) = w_0\, e^{-\rho t},\;\; c^*(t) = \frac{1}{q(t)} = \rho\, w_0\, e^{-\rho t},\;\; \text{for each } t \in [0, \infty).$$

It is optimal to consume the resource over time (eat the cake) until it is completely depleted asymptotically. The half like of this process is close to 34.7 years if $\rho = 2\%$.

Example 10.4 (*The consumption-saving problem*) The benchmark household problem in macroeconomics was also specified and solved as a CV problem in Example 9.5. Now we specify and solve it as a OCP using the PMP. The problem is

$$\begin{aligned} \max_{c(\cdot)} \int_0^{\infty} & e^{-\rho t}\frac{c(t)^{1-\theta} - 1}{1-\theta} dt,\;\; \rho > 0, \\ & \text{subject to } a(0) = a_0,\;\; \text{fixed}, \\ & \dot{a} = y - c + r\,a,\;\; \text{for } t \in [0, \infty), \\ & \lim_{t\to\infty} a(t) \geq 0, \end{aligned} \tag{10.56}$$

where y and r denote the (constant) non-financial income, and the rate of return of the risk-free financial asset. The control variable is consumption, c, and the state variable, a is the net asset position of the household. It is assumed that $y > 0$, $\theta > 0, r > 0$, and $\rho + r\,(\theta - 1) > 0$. Defining $\gamma \equiv \frac{r-\rho}{\theta}$, the last assumption implies the relationship $r > \gamma$ is satisfied. The current value Hamiltonian is $h(a, c, q) =$

---

[13] This is the case for the MHDS in variables (w, q). If we specify the MHDS in variables (w, c), the steady states define a manifold $c = \rho\, w$, and the related MHDS has trace equal to zero and determinant equal to $-\rho < 0$. Thus, the analysis in Sect. 10.2.3 can be slightly modified when the determination of the solution to an OCP is conducted in the space (y, u).

$\frac{c^{1-\theta}-1}{1-\theta} + q\,(y - c + r\,a)$, applying the PMP in Proposition 10.7 yields the MHDS, in variables (a, c), together with the initial and the transversality condition:

$$\dot{a} = y - c + r\,a, \tag{10.57a}$$

$$\dot{c} = \gamma\,c, \tag{10.57b}$$

$$a(0) = a_0, \tag{10.57c}$$

$$0 = \lim_{t\to\infty} c(t)^{-\theta}\,a(t)\,e^{-\rho t}. \tag{10.57d}$$

The MHDS has the solution

$$a^*(t) = \bar{a} + (a_0 - \bar{a})\,e^{\gamma\,t}, \text{ for } t \in [0, \infty), \tag{10.58a}$$

$$c^*(t) = (r - \gamma)\,(a_0 - \bar{a})\,e^{\gamma\,t}, \text{ for } t \in [0, \infty), \tag{10.58b}$$

where $\bar{a} = -\frac{y}{r}$ is the steady state net asset position. As the model is linear then the MHDS is also a linear planar ODE, with eigenvalues $\lambda_- = \gamma$ and $\lambda_+ = r$. The dynamic properties of the solution provide an illustration of the several types of the dynamics of the solution for regular OCP that were referred to in Sect. 10.2.3. There are three possible cases depending on the sign of γ: first, if $\gamma < 0$ then there is one steady state $(\bar{a}, 0)$ that is a saddle point, and the solution satisfying the transversality condition belongs to the stable manifold passing through that point; second, if $\gamma = 0$ the Jacobian of the MHDS has one zero and one positive eigenvalue, there is an infinity of steady states, and the solution to the problem is stationary over time, with constant a and c, that is "anchored" by the initial condition at a_0; at last, if $\gamma > 0$ then the steady state $(\bar{a}, 0)$ is an unstable node, and the solution to the problem, if $a_0 > \bar{a}$, is unbounded. See the discussion of the possible solutions in Example 9.5.

10.2.4.2 Qualitatively Specified Models: The RCK Model

For most OCP explicit solutions cannot be obtained, when the objective function and the constraint vector field are not explicitly specified, or if they are specified and are non-linear. In those cases, we can still characterize the qualitative properties of the solution to OCP by using the method exposed in Sect. 10.2.3, together with the qualitative theory of ODE in Chap. 5.

A well-known case is the Ramsey (1928) model, called, in the Cass (1965), Koopmans (1965) version, the RCK model. This is a model for a (benevolent) social planner that seeks to maximize a social welfare functional subject to a macroeconomic constraint, and initial and boundary conditions. For a utilitarian social planner, the problem is

$$V(k_0) = \max_{c()} \int_0^{\infty} u(c(t))\, e^{-\rho t}\, dt, \tag{10.59a}$$

$$\text{subject to } k(0) = k_0, \text{ fixed,}$$

$$\dot{k} = f(k) - c - \delta\, k, \text{ for each } t \in \mathbb{R}_+, \tag{10.59b}$$

$$\lim_{t\to\infty} k(t) \geq 0, \tag{10.59c}$$

where $\rho > 0$ is the social rate of time preference. The solution to this model was studied in Example 5.4, where it was shown that the MHDS has three steady states but only two steady states satisfy the transversality condition. Furthermore, that example shows that if the initial capital stock, k_0, is positive there is convergence to a steady state that is a saddle point, in which $f'(\bar{k}) = \rho + \delta$ and $c = f(\bar{k})$. The stable manifold passing through this steady state is a heteroclinic curve connecting it to the unstable steady state $(k, c) = (0, 0)$. As this steady state is hyperbolic, the Grobman-Hartmann theorem justifies the common approach in economics to characterize the dynamics, and to do comparative dynamics exercises, by using a linear variational ODE.

10.2.4.3 Unbounded Solutions and Endogenous Growth Models

The qualitative characterization of the solution, appealing to the Grobman-Hartman theorem, and the quantitative determination of its approximation to a steady state requires the existence of an hyperbolic steady state, usually a saddle point. However, in some cases, in particular in endogenous growth models, either the steady state is a unstable node of focus, and solutions do not converge to it, or the solution which interests us can be unbounded over time. In this case, the existence of solutions depends on the transversality condition being satisfied. We need to find a global solution not a local approximation. Next we show the approach to this problems that has been adopted in the endogenous growth literature.[14]

Example 10.5 (*Benchmark endogenous growth model*) Consider a simplified version of the Rebelo (1991)AK-model, where $\mathrm{T} = [0, \infty)$:

$$\begin{aligned} &\max_{c(\cdot)} \int_0^{\infty} e^{-\rho t} \frac{C(t)^{1-\theta} - 1}{1-\theta}\, dt, \quad \rho > 0, \ \theta > 0, \\ &\text{subject to } K(0) = K_0 > 0 \text{ fixed,} \\ &\dot{K} = A\, K(t) - C, \text{ for each } t \in \mathrm{T}, \\ &\lim_{t\to\infty} K(t) \geq 0, \end{aligned} \tag{10.60}$$

where $A > 0$ is the total factor productivity, and the stock of capital includes all types of capital, physical and human capital, under the assumption that they are perfect EP

[14] See Barro and Sala-i-Martin (2004) and Acemoglu (2009).

substitutes.It is assumed that $A(\theta - 1) + \rho > 0$. The MHDS in the (K, C)-space, where the rate of growth is $\gamma \equiv \frac{A-\rho}{\theta} > 0$, together with the initial and the transversality conditions, is

$$\dot{K} = A\,K - C, \text{ for each } t \in \mathrm{T}, \tag{10.61a}$$
$$\dot{C} = \gamma C, \text{ for each } t \in \mathrm{T}, \tag{10.61b}$$
$$K(0) = k_0, \tag{10.61c}$$
$$\lim_{t\to\infty} C(t)^{-\theta}\,K(t)\,e^{-\rho t} = 0. \tag{10.61d}$$

The MHDS has a steady state $(\bar{K}, \bar{C}) = (0, 0)$ that is an unstable node. The local stable manifold is therefore empty, and all solutions diverge over time from the steady state. On the other hand, as the MHDS is linear, we can find a general explicit solution to the MHDS, and find the particular solution that satisfies the transversality condition. We already proceed in this way to find the solution to the household problem, in Example 10.4, when $r > \rho$.

The economic growth literature has developed an alternative method to find the solution to those problems, that is particularly useful in higher dimensional models. It consists in solving a variational system in the neighborhood of a **balanced growth path**, and not in the neighborhood of a steady state as in the RCK model, such that the initial and the transversality conditions hold.

The method proceeds along five steps. First, define the variables using multiplicative deviations along exponential trends with proportional growth rates. In our case, we try the case in which the rates of growth are equal $K(t) = k(t)\,e^{gt}$ and $C(t) = c(t)\,e^{gt}$. Second, we specify the MHDS in the detrended variables (k, c), $\dot{k} = (A - g)\,k - c$ and $\dot{c} = (\gamma - g)c$. Third, obtain g from a stationary solution to the system in detrended variables. In our case, setting $g = \gamma$ transforms the MHDS (10.61) to

$$\dot{k} = (A - \gamma)\,k - c, \;\; \text{for each } t \in \mathrm{T}, \tag{10.62a}$$
$$\dot{c} = 0, \;\; \text{for each } t \in \mathrm{T}, \tag{10.62b}$$
$$k(0) = k_0, \tag{10.62c}$$
$$\lim_{t\to\infty} c(t)^{-\theta}\,k(t)\,e^{-(\gamma\,(\theta-1)+\rho)\,t} = 0. \tag{10.62d}$$

Solving (10.62b) yields $c(t) = c(0)$, for every $t \in \mathrm{T}$, where $c(0)$ is an arbitrary constant. Substituting c, and as $k(0) = k_0$, the solution to ODE (10.62a) is $k(t) = \frac{c(0)}{A-\gamma} + \left(k_0 - \frac{c(0)}{A-\gamma}\right) e^{(A-\gamma)\,t}$. Then the transversality condition (10.62d) is satisfied if and only if $A > \gamma$ and $c(0) = (A - \gamma)\,k_0$. As $A > \gamma$, if $A\,(\theta - 1) + \rho > 0$, then the solution to the problem (10.60) is

$$K^*(t) = k_0\,e^{\gamma\,t} \text{ and } C^*(t) = (A - \gamma)\,k_0\,e^{\gamma\,t} \text{for each } t \in \mathrm{T}. \tag{10.63}$$

This solution represents a **balanced growth path** in which capital, output, and consumption evolve exponentially, at the same rate of growth, to infinity in infinite time. The difference to the RCK model is related to technology: in the Ramsey model, there are decreasing marginal return to capital, and in this case there are constant marginal returns. Substituting the optimal consumption function in the utility functional yields the value function

$$V(k_0) = \int_0^\infty e^{-\rho t} \frac{C^*(t)^{1-\theta} - 1}{1-\theta} dt = \frac{1}{1-\theta}\left[(A-\gamma)^{-\theta} k_0^{1-\theta} - \frac{1}{\rho}\right], \tag{10.64}$$

as a measure of the utility of the initial capital, given the behavioral (ρ and θ) and technological (A) parameters, and assuming an utilitarian social planner.

10.3 References

The seminal contribution to optimal control theory is Pontryagin et al. (1962). Brief histories of optimal control theory can be found in Sargeant (2000) or Peschl and Plail (2009). Introductory textbooks include Kamien and Schwartz (1991), and Chiang (1992). More advanced references are Fleming and Rishel (1975), Seierstad and Sydsæter (1987), and Weber (2011). Important early applications to economics are Arrow and Kurz (1970) and Intriligator (2002). Complete references with application to economics and management science are Tu (1984), Grass et al. (2008), and Sethi (2019). For infinite-horizon optimal control problems, see Carlson et al. (1996). For economic growth textbooks, intensively using the PMP, see Barro and Sala-i-Martin (2004) and Acemoglu (2009). On the relationship between optimal control and nonlinear dynamics, see Brock and Malliaris (1989).

Chapter 11
Introduction to the Dynamic Programming Principle

The **dynamic programming principle** (DPP) states informally that, in order to find a solution to an optimal control problem (OCP), we have to find an optimal increment at the next point in time, assuming that we proceed in the same way at all the other times. In other words, finding a solution to the OCP is achieved by finding the **policy function**, that is the best feedback rule for the control variable. This rule is provided by the **Hamilton-Jacobi-Bellman** (HJB) equation. This equation encodes the optimal variational mechanism of the DPP approach, and it is either a partial differential equation for the finite horizon problem, or an implicit ODE for the infinite horizon problem. This is a counterpart to the PMP in which the optimal variational mechanism is provided by the adjoint ODE equation together with a terminal condition, for the case in which the initial data is fixed.

The DPP approach has some pros and cons when compared to the PMP approach. On the plus side, it provides a clearer intuition to dynamic optimization, it allows for an easier extension to stochastic OCP, and, it is preferred by researchers that use numerical methods. On the minus side, the terminal condition, particularly for infinite horizon problems, that should complement the HJB differential equation is not always clear, studying the differentiability properties of the HJB equation requires advanced functional analysis, and it does not lead immediately to the application of the qualitative theory of ODE for finding and characterizing the solution to optimal control problems. However, we will see that this shortcoming can be eliminated by the invoking the **envelope theorem**.

In this chapter, we consider only time-dependent OCP. The HJB equation, and its role in solving OCP, is the object of Sect. 11.1. Section 11.2 contains several applications to economics. In Sect. 11.3, we relate the DPP with the PMP approaches via a simple version of the envelope theorem.

P. B. Brito, Economic Dynamics and Distributions, Dynamic Modeling and Econometrics in Economics and Finance 32, https://doi.org/10.1007/978-3-031-94717-9_11

11.1 HJB Equations and the Solution to OCP

Sections 11.1.1 and 11.1.2 are dedicated to present HJB equations for finite and infinite horizon problems, respectively, and Sect. 11.1.3 shows how to use them to obtain the solution to OCP.

11.1.1 The Finite Horizon Case

Consider again the free terminal state optimal control problem: among functions $y \in \mathcal{Y} \subset C^1(\mathrm{T}; \mathrm{Y})$ and $u \in \mathcal{U} \subset PC^1(\mathrm{T}; \mathrm{U})$, where $\mathrm{T} = [0, \mathrm{T}]$ and $\mathrm{Y} \subseteq \mathbb{R}$ and $\mathrm{U} \subseteq \mathbb{R}$, satisfying

$$\dot{y} = G(y(t), u(t), t), \text{ for each } t \in [0, T], \tag{11.1}$$

and $y(0) = y_0$ is fixed, T is fixed, and $y(T)$ is free. The problem is to find the pair of functions $(y^*(t), u^*(t))$ that maximize the functional

$$\mathsf{J}[y, u] \equiv \int_0^T F(y(t), u(t), t)\, dt, \tag{11.2}$$

subject to the previous constraints.

Proposition 11.1 (DPP: first-order necessary conditions for the finite horizon problem) Let $(y^*(t), u^*(t))_{t\in\mathrm{T}}$ be the solution to the optimal control problem. Then it satisfies the **Hamilton-Jacobi-Bellman equation**, if $V(t, y)$ is a $C^{1,2}(\mathrm{T}, \mathrm{Y})$ function, such that

$$-V_t(t, y) = \max_{u\in\mathcal{U}} \left\{F(t, y, u) + V_y(t, y)G(t, y, u)\right\}, \text{ for each } (t, y) \in \mathrm{T} \times \mathrm{Y}. \tag{11.3}$$

Proof Let the functional over the state and control functions starting from an arbitrary time $t \geq 0$ be $(y, u) : [t, T] \to \mathrm{Y} \times \mathrm{U} \subseteq \mathbb{R}^2$, and let the objective functional starting at an arbitrary time $t \in \mathrm{T}$ be denoted by $\mathsf{J}[y, u; t] = \int_t^T F(y(s), u(s), s)ds$. The **value function** is

$$V(t, y(t)) \equiv \max_{(u(s)|s\in[t,T])} \mathsf{J}[y, u; t]$$

for $y(t) \in Y$. The **principle of dynamic programming** states the following: for every $(t, y) \in \mathrm{T} \times \mathrm{Y}$ and every $\Delta t \in (0, T - t]$, the value function satisfies

$$V(t, y(t)) = \max_{(u(s)|s\in[t,t+\Delta t])} \left\{\int_t^{t+\Delta t} F(y(s), u(s), s)\, ds + V(t + \Delta t, y(t + \Delta t))\right\}$$

where $y(t+\Delta t) = y(t) + G(t, y(t), u(t))\Delta t + O(\Delta t^2)$. A first-order Taylor expansion of the value function, if V is $C^{1,2}$ function, yields[1]

$$V(t+\Delta t, y(t+\Delta t)) = V(t, y(t)) + V_t(t, y(t))\,\Delta t + \\ + V_y(t, y(t))\,G(t, y(t), u(t))\,\Delta t + O(\Delta t^2).$$

If the interval Δt is sufficiently small, the mean-value theorem allows us to write $\int_t^{t+\Delta t} F(y(s), u(s), s)ds = F(y(t), u(t), t)\,\Delta t$. Then

$$V(t, y(t)) = \max_{(u(s)|s\in[t,t+\Delta t])} \Big\{ F(y(t), u(t), t)\Delta t + V(t, y(t)) + V_t(t, y(t))\Delta t + \\ + V_y(t, y)g(t, y(t), u(t))\Delta t \Big\}.$$

Cancelling out $V(t, y(t))$, dividing by Δt, taking $\Delta t \to 0$ and observing that the pair $(t, y(t))$ is an arbitrary element of $T \times Y$, the solution to the finite horizon OCP satisfies the HJB equation (11.3). □

The HJB equation (11.3) for the finite horizon optimal control problem is a PDE (partial differential equation).

11.1.2 Infinite Horizon Discounted Optimal Control Problem

Now let T $= [0, \infty)$. The infinite horizon discounted optimal control problem is to find functions $y^* \in \mathcal{Y} \subset C^1(\mathrm{T}; \mathrm{Y})$ and $u^* \in \mathcal{U} \subset C^1(\mathrm{T}; \mathrm{U})$, for $\mathrm{Y} \subseteq \mathbb{R}$ and $\mathrm{U} \subseteq \mathbb{R}$, that solve the problem

$$\begin{aligned} &\max_{u(\cdot)} \int_0^\infty e^{-\rho t} f(y(t), u(t))dt, \\ &\text{subject to } y(0) = y_0 \text{ fixed}, \\ &\dot{y} = g(y(t), u(t)), \text{ for each } t \in [0, \infty). \end{aligned} \tag{11.4}$$

Proposition 11.2 (DPP: first-order necessary condition for the infinite horizon problem) Let $(y^*(t), u^*(t))_{t\in\mathrm{T}}$ be the solution to the discounted infinite horizon problem (11.4). Then it satisfies the **Hamilton-Jacobi-Bellman equation**, if $v(y)$ is a $C^2(\mathrm{Y})$ function, such that

$$\rho\, v(y) = \max_u \left\{ f(y, u) + v'(y)\, g(y, u) \right\}. \tag{11.5}$$

for each $t \in \mathrm{T}$, where $y(t) = y$ and $u(t) = u$.

[1] We use the notation $V_t(t, y) = \frac{\partial V(t,y)}{\partial t}$, and $V_y(t, y) = \frac{\partial V(t,y)}{\partial y}$.

Proof Let $y(t) = y$. The discounted value function, for the continuation of the program after $t \geq 0$, is $V(t, y) \equiv \int_t^\infty e^{-\rho s} f(y^*(s), u^*(s))ds$. The current value value function, $v(y) = e^{\rho t} V(t, y) = \int_t^\infty e^{-\rho(s-t)} f(y^*(s), u^*(s))ds$, is independent of time. Therefore, $V(t, y) = e^{-\rho t} v(y)$, and $V_t(t, y) = -\rho\, e^{-\rho t} v(y)$ and $V_y(t, y) = e^{-\rho t} v'(y)$. Substituting those derivatives in HJB equation (11.3) yields HJB equation (11.5). □

In the case of the discounted infinite horizon, the HJB equation (11.5) is not a PDE but an ODE in implicit form. Again the optimum is obtained not by finding the whole optimal path to the state and control variables, but by obtaining the functions that maximize the value of the problem.

11.1.3 The HJB and the Solution to OCP

In applications, the DPP approach to solving the optimal control problem, that is to find the paths $\left(u^*(t), u^*(t)\right)_{t \in \mathrm{T}}$, proceeds in the following steps, for the infinite horizon OCP (11.4).

First, we need to find the **policy function**. Let $h(u, y) \equiv f(y, u) + v'(y)g(y, u)$. Then HJB equation (11.5) can be written as $\rho v(y) = \max_u h(u, v)$. The first-order condition for an optimum is

$$\frac{\partial h(u, y)}{\partial u} = \frac{\partial f(y, u)}{\partial u} + v'(y)\, \frac{\partial g(y, u)}{\partial u} = 0.$$

If $h(\cdot)$ is continuously differentiable as a function of u, the implicit function theorem can be invoked to obtain $u^* = U(y)$. Function $U(.)$ in the DPP literature is called **policy function**. It gives the optimal control as a function of the state variable. This is why it is called a **feedback control** problem. If we substitute the policy function in the HJB equation (11.5), we obtain the **maximized HJB** equation as ODE in implicit form

$$\rho v(y) = f(y, U(y)) + v'(y)\, g(y, U(y)), \tag{11.6}$$

where the state variable y is the independent variable and the value function, $v(y)$, is the unknown function. This ODE is usually highly non-linear. Assuming we can solve it and find $v(y)$, then we can obtain the optimal policy function, $U(y)$, because it is a function of $v'(y)$.

Second, substituting the policy function in the ODE constraint to the OCP problem (11.4), yields an initial-value ODE problem over time,

$$\dot{y} = g(y, U(y)),\ \ t \in [0, \infty)$$

which, together with the initial condition $y(0) = y_0$, would, hopefully, allow for the determination of the solution to the state variable, $y^*(t)$, and $u^*(t) = U(y^*(t))$.

If we can find the policy function, then obtaining the optimal dynamics for y reduces to solving an initial-value ODE problem instead of a mixed initial-terminal value problem (or two-point boundary value problem) as is the case for the CV and for the PMP approaches.

However, only in a very small number of models, we can obtain closed form solutions to the HJB equation. Next we show some simple economic models in which this is possible, to illustrate the reasoning behind the DPP approach.

11.2 Applications

Example 11.1 (*The resource depletion problem*) The resource depletion problem (or the infinite horizon cake eating problem), that was as a CV problem in Example 9.4, or PMP problem in Example 10.3, can also be solved by using the DPP. Assume that the utility function is logarithmic. The HJB equation is the implicit ODE for function $v(w)$,

$$\rho v(w) = \max_{c}\{\ln(c) + v'(w)(-c)\}.$$

Let $h(c, w) \equiv \ln(c) - v'(w)\,c$. Then the optimality condition $\frac{\partial h(c,w)}{\partial c} = 0$ allows us to find the policy function $c^* = C(w) = \left(v'(w)\right)^{-1}$. Substituting in the HJB equation, we obtain the maximized HJB equation as the mplicit ODE

$$\ln(v'(w)) = -(1 + \rho v(w)). \tag{11.7}$$

A **frequently used method** for solving this ODE is based upon guessing a solution and using the **method of undetermined coefficients**. This method consists in introducing an affine function (by guessing or by using a symmetry approach see Chang, 2004) depending on arbitrary constants, that can be determined after substituting the guessed function in Eq. (11.7). As the constraint of the problem is a linear, satisfying $dw(t) = -c(t)\,dt$, this suggests a trial function that is an affine function of $\ln(w)$, $v(w) = a + b\ln(w)$, for $w > 0$, where parameters a and b are arbitrary. If the conjecture is right we would be able to determine the undetermined coefficients from Eq. (11.7).

As $v'(w) = \frac{b}{w}$, and substituting back in Eq. (11.7), yields $\rho a + 1 + \ln(b) = \ln(w)(1 - \rho b)$. Setting both sides of the equation equal to zero allows us to find $b = 1/\rho$ and $a = (\ln\rho - 1)/\rho$. Therefore,

$$v(w) = \frac{\ln\rho\, w - 1}{\rho}$$

and the optimal policy function is , $c^* = (v'(w))^{-1} = \rho\, w$.

A less used, but, in our opinion better, alternative method is to **directly solve the ODE**(11.7), noting that it is equivalent to the non-linear ODE in explicit form $v'(w) = e^{-(1+\rho\, v(w))}$. This ODE has the explicit general solution,[2] provided that $\rho\,(w+k) > 0$,

$$v(w) = -\frac{1}{\rho}\,(\ln\,(\rho(w+k) - 1)\,,$$

where k is an arbitrary constant. Therefore, $c^* = \left(v'(w)\right)^{-1} = \rho\,(w+k)$, and substituting in the constraint $\dot{w} = -c = -\rho\,(w+k)$, we have the general time-dependent solution, depending on an arbitrary constraint, k, $w(t) = -k + (w(0)+k)e^{-\rho t}$. The fact that this solution is incompletely specified, reveals a potential problem when using the first trial method: we are implicitly introducing an integrability assumption without being aware of it.

In this example, it is natural to assume that $\lim_{t\to\infty} w(t) = 0$, which allows us to find the arbitrary constant from $w(\infty) = -k = 0$. Then we obtain the same solution as in Examples 9.4, and 10.3:

$$c^*(t) = \rho\, w_0 e^{-\rho t},\;\; w^*(t) = w_0 e^{-\rho t},\;\; \text{for each } t \in [0,\infty).$$

Exercise 11.1 (*Cake eating with a constant EIS utility function*) In the Example 11.1 assume instead the utility function is $u(c) = \frac{c^{1-\theta}-1}{1-\theta}$ for $\theta > 0$. Show that in this case the solution is $c^*(t) = \frac{\rho}{\theta}\, w_0 e^{-\frac{\rho}{\theta}t}$ and $w^*(t) = w_0 e^{-\frac{\rho}{\theta}t}$ for $t \in [0,\infty)$.

Example 11.2 (*The consumption-saving problem*) Consider again the problem specified in Example 10.4 and solved using the PMP. Applying the HJB equation (11.5) to our problem we have

$$\rho\, v(a) = \max_{c}\left\{\frac{c^{1-\theta}-1}{1-\theta} + v'(a)\,(y - c + r\, a)\right\}. \tag{11.8}$$

The policy function is $c^* = C(a) \equiv \left(v'(a)\right)^{-\frac{1}{\theta}}$. Then the maximized HJB equation is

$$\rho\, v(a) = \frac{\theta}{1-\theta}\left(v'(a)\right)^{\frac{\theta-1}{\theta}} - \frac{1}{1-\theta} + r\, v'(a)\, w(a) \tag{11.9}$$

where $w(a) = \frac{y}{r} + a$, is the sum of non-financial and financial net wealth, or total net wealth. To solve this (implicit) ODE, we use again the trial method, conjecturing that $v(a) = a + b\, w(a)^{1-\theta}$, with arbitrary parameters a and b to be determined. After substitution in Eq. (11.9), we get

$$a\,\rho + \frac{1}{1-\theta} = w(a)^{1-\theta}\, b\,\theta\left[\left(b\,(1-\theta)\right)^{-1/\theta} - (r-\gamma)\right]$$

[2] The ODE $y'(x) = e^{-(a+b\,y(x))}$ has the closed form solution $y(x) = \frac{\ln\,(b(k+x))-a}{b}$, if $b\,(x+k) - a > 0$, for an arbitrary k.

where $\gamma \equiv (r-\rho)/\theta$. Setting both sides to zero, yields $a = \frac{1}{\rho(\theta-1)}$ and $b = \frac{(r-\gamma)^{-\theta}}{1-\theta}$, the value function is

$$v(a) = \frac{1}{1-\theta}\Big[(r-\gamma)^{-\theta}\left(\frac{y}{r}+a\right)^{1-\theta} - \frac{1}{\rho}\Big].$$

Then the policy function is $C^*(a) = (r-\gamma)\left(\frac{y}{r}+a\right) > 0$ only if $r > \gamma$. The net asset dynamics, at the optimum, is the solution to the initial value ODE problem

$$\dot{a}^* = y + r\,a - C^*(a) = \gamma\left(\frac{y}{r}+a\right), \text{ for each } t \in \mathrm{T},$$
$$a(0) = a_0.$$

Then the optimal net asset position is

$$a^*(t) = -\frac{y}{r} + \left(\frac{y}{r}+a_0\right)e^{\gamma t}, \text{ for } t \in [0,\infty),$$

and optimal consumption is

$$c^*(t) = (r-\gamma)\left(\frac{y}{r}+a_0\right)e^{\gamma t}, \text{ for } t \in [0,\infty).$$

Exercise 11.2 Prove that if the utility function is $u(c) = \ln(c)$, then that the value function is

$$v(a) = \frac{1}{\rho}\Big[\frac{r-\rho}{\rho} + \ln\big(\rho\, w(a)\big)\Big]. \tag{11.10}$$

Hint: this corresponds to the case $\theta = 1$. Use the property $f(x) = \exp(\ln f(x))$ and the l'Hôpital theorem to obtain the value function from the one derived in Example 11.2.

Observation. The utility function is a generalized logarithm $u(C) = \ln_\theta(C) = \frac{C^{1-\theta}-1}{1-\theta}$, which is surprisingly little acknowledge in the literature. The constant EIS utility function is frequently specified in the literature as

$$u(c) = \begin{cases} \dfrac{c^{1-\theta}}{1-\theta}, & \text{if } \theta \neq 1, \\ \ln(c), & \textit{if } \theta = 1. \end{cases}$$

In this case, the value function is instead $v(a) = \dfrac{(r-\gamma)^{-\theta}}{1-\theta} w(a)^{1-\theta}$. However, we cannot obtain value function (11.10) when we set $\theta = 1$. If we use this formulation of the constant EIS case we should set $u(c) = \ln(c) - 1$ instead of $u(c) = \ln(c)$, when $\theta = 1$.

Example 11.3 (*The* Ramsey (1928) *model*) Consider again the model specified in Eq. (10.59). The HJB for this model is

$$\rho v(k) = \max_{c} \left\{ u(c) + v'(k)\,(F(k) - c) \right\}$$

The optimality condition is $u'(c) = v'(k)$. If $u(\cdot)$ is sufficiently smooth then we obtain the policy function $c = C(k) = (u')^{-1}\big(v'(k)\big)$. Substituting back in the HJB equation yields the implicit ODE in $v(k)$

$$\rho v(k) = u\big(C(k)\big) + v'(k)\big(F(k) - C(k)\big)$$

which does not have a closed form solution in general.

Exercise 11.3 For the case in which $u(c) = \frac{c^{1-\theta}-1}{1-\theta}$ and $F(k) = k^{\alpha}$, such that if $\theta = \alpha$ the policy function is $c^* = C(k) = \frac{\rho}{\alpha}k$, and, therefore we can find a closed form solution to the RCK model Prove this.

Example 11.4 (*Benchmark endogenous growth model*) Recall Example 10.5, and the AK model specified in problem (10.60). The associated HJB equation is

$$\rho v(K) = \max_{c} \left\{ \frac{C^{1-\theta} - 1}{1-\theta} + v'(K)\,(A\,K - C) \right\},$$

yielding the policy function $C = (v'(K))^{-\frac{1}{\theta}}$. Therefore, the HJB equation at the optimum is the implicit ODE

$$\rho\, v(K) = v'(K)\left(\frac{\theta}{1-\theta}\, v'(K)^{-\frac{1}{\theta}} + A\,K\right) - \frac{1}{1-\theta}$$

Assuming a trial function $v(K) = \alpha + \beta\, K^{1-\theta}$, and substituting, yields $\alpha\,\rho\,(1-\theta) = 1$ and $(1-\theta)\,\beta = (A-\gamma)^{-\theta}$, for $\gamma \equiv \frac{A-\rho}{\theta}$, yields the value function

$$v(K) = \frac{1}{1-\theta}\left((A-\gamma)^{-\theta}\, K^{1-\theta} - \frac{1}{\rho}\right),$$

that allows us to obtain the optimum consumption function $C^*(K) = (A-\gamma)K$, and the budget constraint

$$\dot{K}^* = A\,K^* - C^*(K) = \gamma\, K^* \text{ for each } t \in \mathrm{T}.$$

From the fixed initial level of the capital stock, $K(0) = k_0$, we finally get the optimal paths for capital, output, and consumption

$$K^*(t) = k_0\, e^{\gamma t},\;\; Y^*(t) = A\, k_0\, e^{\gamma t},\;\; C^*(t) = (A-\gamma)\, k_0\, e^{\gamma t} \text{ for } t \in [0, \infty).$$

11.3 The Envelope Theorem

The following proposition establishes a relationship between the optimality conditions of the PMP and of the DPP, for regular discounted infinite horizon problems. It has two useful consequences: it provides one interpretation of the co-state variable of the PMP, and allows for using the qualitative theory of ODE to study the dynamics of the solution satisfying the HJB equation.[3]

Proposition 11.3 (Envelope theorem) Consider the optimal control problem (11.4), and its HJB equation (11.5). If function $v(\cdot)$ is $C^2(\mathbb{R})$, and we define $q = v'(y)$, then the optimum and adjoint equations of the Pontriyagin's maximum principle, Eqs. (10.43) and (10.44), are obtained.

Proof Consider the HJB equation (11.5), and write $h(y, u) = f(y, u) + v'(y)\, g(y, u)$. If functions $f(\cdot)$ and $g(\cdot)$ are differentiable, the optimality condition, $h_u(u, y) = f_u(y, u) + v'(y)\, g_u(y, u) = 0$ allows us to obtain the policy function $u^* = U(y)$.

Write $q = v'(y)$, and therefore, we can write $h(y, u) = f(y, u) + q\, g(y, u)$.[4] With this change of variables we obtain the optimality conditions according to the PMP.[5] First, the static optimality condition (10.43), according to the PMP, is formally equivalent to $h_u(y, u^*) = f_u(y, u^*) + q\, g_u(y, u^*) = 0$. Second, taking the derivative to the HJB equation (11.5), as regards the state variable, evaluated at the optimum yields

$$\begin{aligned}\rho\, v'(y) =& f_y(y, u^*) + f_u(y, u^*)\, U'(y) + v''(y)\, g(y, u^*) + \\ &+ v'(y)\left(g_y(y, u^*) + g_u(y, u^*)\, U'(y)\right) \\ =& f_y(y, u^*) + v''(y)\, g(y, u^*) + v'(y)\, g_y(y, u^*) + \\ &+ \Big(f_u(y, u^*) + v'(y)\, g_u(y, u^*)\Big)\, U'(y) \\ =& f_y(y, u^*) + v''(y)\, g(y, u^*) + v'(y)\, g_y(y, u^*),\end{aligned}$$

if we apply the envelope theorem. Therefore, if function $v(\cdot)$ is smooth

$$g(y, u^*) = \frac{\rho\, v'(y) - f_y(y, u^*) - v'(y)\, g_y(y, u^*)}{v''(y)} = \frac{\rho\, q - h_y(y, u^*)}{v''(y)} \tag{11.11}$$

using the previous definition. The constraint to the problem evaluated at the optimum is $\frac{dy}{dt} = g(y, u^*)$. Taking the time derivative of $q = v'(y)$ implies

$$\frac{dq}{dt} = v''(y)\, \frac{dy}{dt} = v''(y) g(y, u^*) = \rho\, q - h_y(y, u^*)$$

[3] The relationship with the CV approach can be seen in Beckmann (1968, p. 29).

[4] See Beckmann (1968, p. 33).

[5] Although the DPP approach is silent to what to do with the transversality condition. However, Ekeland (2010) shows that the solution to the problem exists if there is a steady state value for the state variable and at the steady state $\bar{y}$ we should have $\rho\, v(\bar{y}) = v'(\bar{y})$.

which is the adjoint ODE (10.44) of the PMP.[6] □

Example 11.5 (*The* Ramsey (1928) *model*)Consider the Ramsey problem with a constant EIS utility function and a Cobb-Douglas production function. The HJB equation is

$$\rho\, v(k) = \max_{c} \left\{ \frac{c^{1-\theta} - 1}{1-\theta} + v'(k) \left(k^{\alpha} - c\right) \right\}$$

where $\theta > 0$ and $0 < \alpha < 1$. The policy function is $c^* = C(k) = \left(v'(k)\right)^{-\frac{1}{\theta}}$. Then

$$\frac{dc}{dt} = -\frac{c^*\, v''(k)}{\theta\, v'(k)} \frac{dk}{dt}.$$

From the envelope theorem, we obtain

$$\rho\, v'(k) = v''(k) \left(k^{\alpha} - c^*\right) + \alpha\, k^{\alpha-1}\, v'(k)$$

as

$$\dot{k} \equiv \frac{dk}{dt} = k^{\alpha} - c^* = \frac{v'(k) \left(\rho - \alpha\, k^{\alpha-1}\right)}{v''(k)}$$

then, we obtain the Ramsey-Keynes rule associated to the PMP,

$$\dot{c} = \frac{c}{\theta} (\alpha\, k^{\alpha-1} - \rho).$$

Although this model has no closed form solution, we can characterize it by invoking the qualitative theory of the ODE, as was done in Example 5.4.

11.4 References

The seminal contribution is Bellman (1957). Beckmann (1968) is a classical reference, that influenced the approach in this chapter. A recent textbook from an important contributor to this area is Sethi (2019).

[6] This requires that the value function is a $C^2(\mathrm{Y})$ function with $v''(y) \neq 0$, at least locally.

Chapter 12
Optimal Control of ODE: Extensions

This chapter contains several extensions to simple optimal control problems dealt in Chap. 10. Section 12.1 introduces singular control problems. Optimal control problems with several types of discontinuities are the subject of the next three sections: discontinuities introduced by instantaneous constraints to state and control variables in Sect. 12.2, discontinuities in two-stage optimal control problems in Sect. 12.3, and piecewise continuous or smooth ODE constraints in Sect. 12.4. A brief presentation of two-state optimal control problems is done in Sect. 12.5.

12.1 Singular Optimal Control

In Chap. 10, it was assumed that the Hamiltonian function was smooth (continuously differentiable up at least to the second order) and globally concave as regards the control variable, implying that the Arrow sufficiency condition holds. We saw that this property guarantees, in most cases, existence and uniqueness of solutions to an OCP. We say an optimal control problem (OCP) is **singular** if that property does not hold. In the literature,[1] this denomination is reserved to OCP in which the Hamiltonian function is linear in the control variable. As we will see next, this property implies that a solution to a singular OCP may not exist, or it can exist and be non-unique. However, we can conceive a much less studied case in which the maximized Hamiltonian dynamic system (MHDS), for time-dependent problems, is singular in the sense studied in Chap. 7. This can be the case for OCP in which the Hamiltonian is non-monotonic or is convex-concave, in the control variable. In this case the singularity of the MHDS is local, and not global as in the linear case. The examples in Sect. 7.2.4 suggest that, again, solutions to an OCP may not exist, or they can exist but not be unique.

[1] See Grass et al. (2008, Chap. 3.5) for a complete reference.

P. B. Brito, *Economic Dynamics and Distributions*, Dynamic Modeling and Econometrics in Economics and Finance 32, https://doi.org/10.1007/978-3-031-94717-9_12

In this section, we consider an optimal control problem in which both the objective function and the ODE constraint are linear in the control variable. This is equivalent to stating that the Hamiltonian has the following properties $H_u(y, u, t)$ does not depend on u, and $H_{uu}(y, u, t) = 0$, for all $u \in \mathrm{U}$. This poses two problems: first, it does not allow for obtaining the optimal control from the static first-order condition, and sufficiency—from the perspective of the Arrow sufficiency condition—may not be satisfied. We use the same notation for teh domain sets and the variables as in the previous two chapters.

We consider the intertemporal problem with $\mathrm{T} = [0, T) \subseteq \mathbb{R}_+$:

$$\begin{aligned}\max_{u(\cdot)} \mathrm{J}[y, u] = \int_{\mathrm{T}} & F(y(t), u(t), t)\, dt,\\ &\text{subject to } y(0) = y_0 \text{ fixed},\\ &\dot{y} = G(y(t), u(t), t),\\ &R(y(T), T) \geq 0.\end{aligned} \tag{12.1}$$

In this case, the Hamiltonian function is

$$\begin{aligned}H(t) &\equiv H(y(t), u(t), \lambda_0, \lambda(t), t)\\ &= \lambda_0\, F(y(t), u(t), t) + \lambda(t)\, G(y(t), u(t), t), \quad \text{for each; } t \in \mathrm{T},\end{aligned}$$

where a new scalar Lagrange multiplier λ_0, multiplying the objective function, is introduced.

An informal version of the Pontriyagin's principle states the following: If $(\mathbf{y}^*, \mathbf{u}^*)$ is an optimum path, then there is a scalar $\lambda_0 \in \{0, 1\}$ and a piecewise continuous function $\lambda : \mathrm{T} \to \mathbb{R}$ such that the optimum satisfies, for each $t \in \mathrm{T}$,

$$\begin{aligned}&H_u^*(t) = H_u(y^*, u^*, t) = 0, \quad \text{for each } t \in \mathrm{T},\\ &\dot{\lambda} = -H_y^*(t) = -H_y(y^*, u^*, t), \quad \text{for each } t \in \mathrm{T},\\ &\dot{y} = H_\lambda^*(t) = G(t, y^*(t), u^*(t)), \quad \text{for each } t \in \mathrm{T},\\ &\lambda(T)\, R_y^*(y(T), T) = 0, \quad \text{for } t = T,\\ &y^*(0) = y_0, \quad \text{for } t = 0.\end{aligned}$$

To guarantee existence of solutions, we may have to enlarge the domain of the domain of the state variable y to the set of piecewise-continuous functions $y : \mathrm{T} \to \mathrm{Y} \subseteq \mathbb{R}$.

Example 12.1 (*Household problem with a linear utility function*) Here, we are concerned with the household problem with a linear utility function, an isoelastic utility function with an infinitely-valued elasticity of substitution. The household problem is

$$
\begin{aligned}
\max_{c(\cdot)} \int_0^\infty & c(t)\, e^{-\rho t} dt, \\
& \text{subject to } a(0) = a_0, \ \text{fixed}, \\
& \dot{a} = r\, a + y - c, \\
& \lim_{t\to\infty} a(t) \geq 0.
\end{aligned}
$$

The current-value Hamiltonian function is

$$
h(a, c, \lambda_0, \lambda) = \lambda_0\, c + \lambda\, (r\, a + y - c)
$$

where λ_0 is a number equal to 0 or 1, and λ is a function of time. The Hamiltonian function is linear in (c, a). The first-order conditions are

$$
\begin{aligned}
& \lambda_0 = \lambda(t), \ \text{ for each } t \in \mathrm{T}, \\
& \dot{\lambda} = \lambda\, (\rho - r), \ \text{ for each } t \in \mathrm{T}, \\
& \dot{a} = r\, a + y - c, \ \text{ for each } t \in \mathrm{T}, \\
& \lim_{t\to\infty} \lambda(t)\, a(t) e^{-\rho t} = 0, \\
& a(0) = a_0.
\end{aligned}
$$

Setting $\lambda_0 = 1$ then $\lambda(t) = 1$ for every $t \in [0, \infty)$, which implies $\dot{\lambda} = 0$. A solution only exists if $r = \rho$. If we assume this to be the case, solve the budget constraint and substitute the transversality condition, then we have $\int_0^\infty e^{-\rho s}\, c(s)\, ds = a_0 + \frac{y}{r}$. As there are no more constraints on the functional form for $c(t)$, as in the constant EIS case, the solution for c is indeterminate. That is, there is an infinite number of consumption trajectories that satisfy that constraint. In particular, if we let $c(t) = c_0\, e^{\eta t}$, where $\eta < \rho = r$ yields $c^*(t) = (\rho - \eta)\left(a_0 + \frac{y}{r}\right) e^{\eta t}$, meaning that the solution can be increasing, constant, or decreasing over time if $\eta \gtreqless 0$, respectively.

Setting $\lambda_0 = 0$, instead, implies $\lambda(t) = 0$ and $\dot{\lambda} = 0$, for every $t \in [0, \infty)$. In this case, we do not need the existence condition $r = \rho$. However, for any r, and $\rho > 0$, the transversality condition is satisfied for any trajectories of a and c. In this case, the solution always exists but is indeterminate, in the sense that it holds for any path of c.

From the economic point of view, the problem is misspecified. In order to overcome this misspecification, either we introduce more curvature in the utility function, or a satiation point, or some adjustment costs in consumption, or bounds in the net asset position of the household.

12.2 Static Constraints on the State and Control Variables

Several dynamic optimization problems involve instantaneous constraints that introduce time-wise limits on the levels of the control variable, or on the state variable, or on both variables.

For example, these constraints can take the form of a lower bound on variables, formalized by inequality $Q(t, y(t), u(t)) \geq 0$. This type of constraint, if it is binding, introduces discontinuities in the co-state and the control variables, and non-differentiability in the state variable, at the points in time at which it becomes binding. A general OCP with that constraint is, for $\mathrm{T} = [0, T) \subseteq \mathbb{R}_+$:

$$\max_{u(\cdot)} \mathrm{J}[y, u] = \int_{\mathrm{T}} F(y(t), u(t), t)\, dt, \tag{12.2a}$$

$$\text{subject to } y(0) = y_0 \text{ fixed}, \tag{12.2b}$$

$$\dot{y} = G(y(t), u(t), t), \;\; \text{for each } t \in \mathrm{T}, \tag{12.2c}$$

$$Q(y(t), u(t), t) \geq 0, \;\; \text{for each } t \in \mathrm{T}, \tag{12.2d}$$

$$R(y(T), T) \geq 0, \tag{12.2e}$$

where $Q(y, u, t) \geq 0$ is a joint constraint on the state and control variable. It encompasses the case of pure control constraint, in which $Q_y(t) = 0$, and of pure state constraint, in which $Q_u(t) = 0$. The Hamiltonian function is[2]

$$H(y(t), u(t), \lambda_0, \lambda(t), t) = \lambda_0\, F(y(t), u(t), t) + \lambda(t)\, G(y(t), u(t), t), \;\; \text{for each } t \in \mathrm{T}$$

and the Lagrange is

$$L(y(t), u(t), \lambda_0, \lambda(t), t) = H(y(t), u(t), \lambda_0, \lambda(t), t) + \\ + \nu(t)\, Q(y(t), u(t), t) \text{ for each } t \in \mathrm{T},$$

where $\nu : \mathrm{T} \to \mathbb{R}$ is the Lagrange multiplier associated to the instantaneous constraint. An informal version[3] of the PMP states the following: if (y^*, u^*) is an optimum, then there is a scalar $\lambda_0 \in \{0, 1\}$, a piecewise continuous function co-state variable, $\lambda : \mathrm{T} \to \mathbb{R}$, a Lagrange multiplier $\nu : \mathrm{T} \to \mathbb{R}$, a variable $\zeta(t_i)$, at the times of discontinuity t_i for $\lambda(\cdot)$, and constant KKT multipliers α, and β that satisfy the following conditions: for each $t \in [0, T)$

$$L_u^*(t) = H_u^*(t) + \nu(t)\, Q_u^*(t) = 0,$$
$$\dot{\lambda} = -L_y^*(t) = -H_y^*(t) - \nu(t)\, Q_y^*(t),$$
$$\dot{y} = L_\lambda^*(t) = G(t, y^*(t), u^*(t)),$$
$$\nu(t) \geq 0, \;\; Q^*(t) \geq 0, \;\; \text{and } \nu(t)\, Q^*(t) = 0,$$

---

[2] This includes the case in which $F(\cdot)$ is linear in u.

[3] See Hartl et al. (1995).

at the terminal time $t = T$,

$$\lambda(T^-) = \alpha\, Q_y^*(T) + \beta\, R_y^*(T),$$
$$\alpha \geq 0,\ \ Q_y^*(T) \geq 0,\ \text{ and } \alpha\, Q_y^*(T) = 0,$$
$$\beta \geq 0,\ \text{ and } \beta\, R^*(T) = 0,$$

and, for any binding time t_i, in which the static constraint (12.2d) becomes binding, the co-state variable $\lambda(\cdot)$ can have a discontinuity given by

$$\lambda(t_i^-) = \lambda(t_i^+) + \zeta(t_i)\, Q_y^*(t_i),$$
$$H^*(t_i^-) = H^*(t_i^+) + \zeta(t_i)\, Q^*(t_i),$$
$$\zeta(t_i) \geq 0,\ \ Q^*(t_i),\ \text{ and } \zeta(t_i)\, Q^*(t_i) = 0,$$

where $\lambda(t_i^-) = \lim_{t \uparrow t_i} \lambda(t)$ and $\lambda(t_i^+) = \lim_{t \downarrow t_i} \lambda(t)$, and an analogous notation for $H^*(t)$.

Exercise 12.1 (*Lower and upper static constraints*) Find the optimality conditions for an OCP with the same specification as problem (12.2), but in which the constraint (12.2d) is substituted by constraint $\underline{Q} \leq Q(t, y(t), u(t)) \leq \overline{Q}$, for each $t \in \mathrm{T}$, in which $\underline{Q} < \overline{Q}$ are two real numbers.

Exercise 12.2 (*Household problem with no debt constraint*) Consider the household consumption-saving problem in Example 10.4 but in which the no-debt constraint $a(t) \geq 0$, for each $t \in \mathrm{T}$, is introduced. Find the solution to the problem. In particular, study the case in which $\rho > r$.

12.3 Two-Stage OCP

Two-stage OCP are deterministic dynamic optimization problems in which there is a **switching time,** $t_1 \in \mathrm{T}$, such that there are two different sub-problems, before and after it. We call initial problem to the first stage sub-problem and continuation problem to the second stage sub-problem. This problem can be extended to multi-stage problems, in which there are several switching times. There are several versions to the two-stage problem. In this section, we consider the case in which both the objective function and the constraint vector field switch at the same time, and in which there is a cost to switch. However, we introduce two versions to that problem depending on the assumption regarding the switching time: in Sect. 12.3.1 the switching time is fixed (exogenous), and in Sect. 12.3.2, it is free (endogenous).

12.3.1 Fixed Switching Time

In this subsection, we consider the two-stage problems in which the switching time, t_1, is fixed. We present two problems: a first problem with a fixed and finite terminal time, and a second problem with a fixed terminal time but with a terminal constraint on the state variable, which allows us to deal with the infinite horizon problems, which are the benchmark in economics.

Finite terminal time. Let $\mathrm{T} = [t_0, t_2] \subset \mathbb{R}_+$, where $t_0 < t_2$ are fixed and assume there is a fixed switching time, t_1, such that $t_0 < t_1 < t_2$. We assume that $y(t_0) = y_0$ is fixed, and the terminal value of the state variable is constrained by the inequality $R(y(t_2), t_2) \geq 0$. The value of the state variable at the switching time, $y(t_1)$, is determined endogenously by a particular match between the first stage and the second stage sub-problems. We consider the two-stage problem:

$$
\begin{aligned}
&\max_{u(\cdot)} \left\{ \int_{t_0}^{t_1} F_1(y(t), u(t), t)\, dt + \int_{t_1}^{t_2} F_2(y(t), u(t), t)\, dt \right\} - S(y(t_1), t_1),\\
&\qquad \text{subject to } t_0, t_1, t_2 \text{ fixed},\\
&\qquad y(t_0) = y_0 \text{ fixed},\\
&\qquad y(t_1) = y_1 \text{ free},\\
&\qquad \dot{y} = \begin{cases} G_1(y(t), u(t), t) & \text{for each } t_0 \leq t < t_1,\\ G_2(y(t), u(t), t) & \text{for each } t_1 < t < t_2, \end{cases}\\
&\qquad R(t_2, y(t_2)) \geq 0,
\end{aligned}
\tag{12.3}
$$

where $S(y(t_1), t_1)$ is a switching cost function. The usual requirement that functions $F_1(\cdot)$, $F_2(\cdot)$, $G_1(\cdot)$, $G_2(\cdot)$ and $S(\cdot)$ are continuous and continuously differentiable are assumed to hold. The Hamiltonian function is now a piecewise function of time

$$
H(t) = \begin{cases} H_1(t) & \text{for } t_0 \leq t < t_1,\\ H_2(t) & \text{for } t_1 < t < t_2, \end{cases}
$$

where

$$
H_j(t) = H_j(y(t), u(t), \lambda(t), t) = F_j(y(t), u(t), t) + \lambda(t)\, G_j(y(t), u(t), t), \;\; j = 1, 2,
$$

where the co-state variable is a piecewise-continuous function of t. We assume that function $H(\cdot)$ is a continuous and continuously differentiable (indeed regular) function of u.

Proposition 12.1 (Two-stage optimal control for a fixed switching time) *Let* $(\mathbf{y}^*, \mathbf{u}^*) = \left(y^*(t), u^*(t)\right)_{t\in\mathrm{T}}$ *be a solution to the optimal control problem* (12.3). *Then there is a piecewise continuous function* $\lambda : \mathrm{T} \to \mathbb{R}$ *such that the following conditions are satisfied:*

- *the optimality, multiplier, and terminal condition*

$$H^*_{1,u}(t) = 0, \textit{ for each } t_0 \le t < t_1, \tag{12.4a}$$
$$H^*_{2,u}(t) = 0, \textit{ for each } t_1 < t \le t_2, \tag{12.4b}$$
$$\dot{\lambda} + H^*_{1,y}(t) = 0, \textit{ for each } t_0 \le t < t_1, \tag{12.4c}$$
$$\dot{\lambda} + H^*_{2,y}(t) = 0, \textit{ for each } t_1 < t \le t_2, \tag{12.4d}$$
$$\lambda(t_2)\, R^*(t_2) = 0; \tag{12.4e}$$

- *the admissibility conditions*

$$\dot{y} = G^*_1(t), \textit{ for each } t_0 \le t < t_1, \tag{12.5a}$$
$$\dot{y} = G^*_2(t), \textit{ for each } t_1 < t \le t_2, \tag{12.5b}$$
$$y^*(t_0) = y_0; \tag{12.5c}$$

- *and the matching condition*

$$\lambda(t_1^+) = \lambda(t_1^-) + S^*_y(t_1). \tag{12.6}$$

Proof Let $(\mathbf{y}^*, \mathbf{u}^*)$ be the solution to the OCP, and write $F^*_j(t) = F_j\big(y^*(t), u^*(t), t\big)$ and $G^*_j(t) = G_j\big(y^*(t), u^*(t), t\big)$ for $j = 1, 2$, and the associated maximized Hamitonian functions $H^*_j(t) = H_j\big(y^*(t), u^*(t), t\big) = F^*_j(t) + \lambda(t)\, G^*_j(t)$, for $j = 1, 2$. The objective functional at the optimum is

$$\mathsf{J}^* = \mathsf{J}[y^*, u^*] = \int_{t_0}^{t_1} F^*_1(t)\, dt + \int_{t_1}^{t_2} F^*_2(t)\, dt - S^*(t_1),$$

where $R^*(t_2) = R(y^*(t_2), t_2)$ and $S^*(t_1) = S(y^*(t_1), t_1)$. Defining the Lagrange functional,

$$\mathsf{L}[y, u; \psi] = \int_{t_0}^{t_1} F_1(t)\, dt + \int_{t_1}^{t_2} F_2(t)\, dt - S(t_1) + \psi\, R(t_2)$$

where $\psi \ge 0$ is the Lagrange multiplier associated to the terminal condition, we have, at the optimum $\mathsf{L}^* = \mathsf{J}^*$ if the Karush-Kuhn-Tucker conditions, $\psi \ge 0$, $R^*(t_2) \ge 0$ and $\psi\, R^*(t_2) = 0$, hold. As $F^*_j(t) = H^*_j(t) - \lambda(t)\, \dot{y}^*(t)$, we have

$$\mathsf{L}^* = \int_{t_0}^{t_1} \Big(H^*_1(t) - \dot{y}^*(t)\,\lambda(t)\Big)\, dt + \int_{t_1}^{t_2} \Big(H^*_2(t) - \dot{y}^*(t)\,\lambda(t)\Big)\, dt - S^*(t_1) + \psi\, R^*(t_2),$$

integrating by parts yields

$$\mathsf{L}^* = \int_{t_0}^{t_1} \left(H_1^*(t) + \dot{\lambda}(t)\, y^*(t) \right) dt + \int_{t_1}^{t_2} \left(H_2^*(t) + \dot{\lambda}(t)\, y^*(t) \right) dt - S^*(t_1) +$$
$$+ \psi\, R^*(t_2) + \lambda(t_0)\, y^*(t_0) - \lambda(t_1^-)\, y^*(t_1) + \lambda(t_1^+)\, y^*(t_1) - \lambda(t_2)\, y^*(t_2),$$

where the left and right limits of λ at point $t = t_1$ are $\lim_{t\uparrow t_1} \lambda(t) = \lambda(t_1^-)$ and $\lim_{t\downarrow t_1} \lambda(t) = \lambda(t_1^+)$. Using a variational approach, we introduce an arbitrary admissible perturbed trajectory $(\mathbf{y}, \mathbf{u})$ in a neighborhood of the optimum $(\mathbf{y}^*, \mathbf{u}^*)$ such that $y(t) = y^*(t) + \delta y(t)$ and $u(t) = u^*(t) + \delta u(t)$, for $t \in [t_0, t_2]$, where $(\delta y, \delta u) = \varepsilon\, (\eta_y, \eta_u)$, and $\eta_y(t_0) = 0$.

The functional derivative, of the Lagrange functional, evaluated in a neighborhood of the optimum is

$$\begin{aligned} \mathrm{D}_{\eta_y,\eta_u} \mathsf{L}[y^*, u^*] = & \int_{t_0}^{t_1} \left(H_{1,u}^*(t)\, \eta_u(t) + \left(H_{1,y}^*(t) + \dot{\lambda}(t) \right) \eta_y(t) \right) dt + \\ & + \int_{t_1}^{t_2} \left(H_{2,u}^*(t)\, \eta_u(t) + \left(H_{2,y}^*(t) + \dot{\lambda}(t) \right) \eta_y(t) \right) dt + \\ & + \left(\lambda(t_1^+) - \lambda(t_1^-) - S_y^*(t_1) \right) \eta_y(t_1) + \left(\psi\, R_y^*(t_2) - \lambda(t_2) \right) \end{aligned}$$

A necessary condition for an optimum is that the Gâteaux differential is equal to zero, $\mathrm{D}_{\eta_y,\eta_u} \mathsf{L}[y^*, u^*] = 0$, for any admissible arbitrary perturbation η_y and η_u. Together with KKT conditions, $\psi \geq 0$, $R^*(t_2) \geq 0$ and $\psi\, R^*(t_2) = 0$, yields the necessary conditions in Eqs. (12.4a)–(12.6). □

In applications, the two-stage problem can be solved in three steps. In the first step, we solve the second-stage problem by assuming that the initial time, t_1, is fixed but the initial value of the state variable, $y(t_1)$, is free. That is, we solve the sub-problem.

$$\begin{aligned} & \max_{u(\cdot)} \int_{t_1}^{t_2} F_2(y(t), u(t), t)\, dt, \\ & \quad \text{subject to } y(t_1) \text{ free}, \\ & \quad \dot{y} = G_2(y(t), u(t), t), \\ & \quad R(t_2, y(t_2)) \geq 0. \end{aligned}$$

Solving backwards, we can find, in principle, the initial value for the co-state variable, $\lambda(t_1^+)$. In a second step, we solve the first-stage sub-problem,

$$\begin{aligned} & \max_{u(\cdot)} \int_{t_0}^{t_1} F_1(y(t), u(t), t)\, dt, \\ & \quad \text{subject to} \\ & \quad \dot{y} = G_1(y(t), u(t), t), \\ & \quad y(t_0) = y_0 \text{ given}, \end{aligned}$$

to obtain a solution for $u(t)$ and $y(t)$ depending on the terminal value for the co-state variable $\lambda(t_1^-)$. In the final step, knowing $\lambda(t_1^-)$ and $\lambda(t_1^+)$, and using the matching condition (12.6) we can find the optimum value of the state variable at the switching point t_1, that is $y^*(t_1) = y_1$, after substituting in the solution for the state variable evaluated at $t = t_1$. This allows us to trace out the solution to the control and state variables for the whole domain T. Observe that the solution to the initial value of the control variable at the optimum, $u^*(0)$, depends on the complete information introduced by the switch. As it is non-predetermined, and without introducing further constraints, the solution to the control variable, $\mathbf{u}^*$, can be continuous and continuously differentiable for all $t \in \mathrm{T}$, without discontinuities at the switching point. However, because the state variable is pre-determined, specially if it is present in the switching constraint, the state variable is continuous but non-differentiable at $t = t_1$.

Discounted infinite horizon. To solve the problem

$$\begin{aligned} \max_{u(\cdot)} \int_{t_0}^{t_1} & F_1(y(t), u(t), t)\, dt + \int_{t_1}^{\infty} F_2(y(t), u(t), t)\, dt, \\ & \text{subject to } t_0, t_1 \text{ fixed}, \\ & \dot{y} = \begin{cases} G_1(y(t), u(t), t) & \text{for each } t_0 \le t < t_1, \\ G_2(y(t), u(t), t) & \text{for each } t_1 < t < \infty, \end{cases} \\ & y(t_0) = y_0 \text{ fixed}, \\ & y(t_1) = y_1 \text{ free}, \\ & \lim_{t\to\infty} R(y(t), t) \ge 0, \end{aligned} \tag{12.7}$$

we can adapt Proposition 12.1to this case by setting $t_2 = \infty$.

Example 12.2 Consider the dynastic household problem in Example 10.4 in which the non-financial income has a switch at a known time $\tau \in (0, \infty)$ such that

$$y(t) = \begin{cases} y_1, & \text{for } 0 \le t < \tau, \\ y_2, & \text{for } \tau \le t < \infty, \end{cases}$$

where $y_1 \neq y_2$. This example refers to life-cycle changes in income related to employed/unemployed stages or work/retirement stages. The household problem is to maximize the utility functional $\int_{\tau}^{\infty} \frac{c(t)^{1-\theta}-1}{1-\theta} e^{-\rho t}\, dt$ subject to the budget constraint $\dot{a} = r\,a + y(t) - c$ for each $t \in [0, \infty)$, where $y(t)$ is time-dependent, and to the non-Ponzi game constraint $\lim_{t\to\infty} a(t) \ge 0$, given the initial net wealth $a(0) = a_0$. where $r > 0$ is the interest rate, $\rho > 0$ is the rate of time preference, and the parameters satisfy the conditions $\theta > 0$, and $(\theta - 1)\,r + \rho > 0$. As referred, we solve the problem backwards in three steps.

First step: terminal stage. We solve the problem for the interval $t \in [\tau, \infty)$ assuming we know the initial value of the state variable $a(\tau) = a_\tau$, and that $y(t) = y_2$ for $t \in [\tau, \infty)$. Denoting the co-state variable for the second phase by $\lambda_2 : [\tau, \infty) \to$

$\mathbb{R}$, the necessary (and in this case sufficient) conditions for an optimum are

$$\begin{aligned} &c(t)^{-\theta}\, e^{-\rho t} = \lambda_2(t), \text{ for each } t \in [\tau, \infty),\\ &\dot{\lambda}_2 = -r\,\lambda_2, \text{ for each } t \in [\tau, \infty),\\ &\dot{a} = r\,a + y_2 - c, \text{ for each } t \in [\tau, \infty),\\ &a(\tau) = a_1, \text{ for } t = \tau,\\ &\lim_{t\to\infty} \lambda_2(t)\, a(t) = 0, \end{aligned}$$

Solving the Euler equation yields $\lambda_2(t) = c(\tau)^{-\theta}\, e^{-\rho(t-\tau)}$. Substituting in the optimality condition we find $c(t) = c(\tau)\, e^{\gamma(t-\tau)}$, and, solving the budget constraint, we obtain $e^{-r(t-\tau)}\, a(t) = a_1 + e^{r\,\tau} \left(\int_\tau^t \left(y_2 - c(\tau)\, e^{\gamma(s-\tau)}\right) ds \right)$, where $c(\tau)$ is unknown in both candidate solutions. Multiplying by the solution by $\lambda_2(t)$, using the transversality condition, and assuming that $r - \gamma > 0$, as before, we find the optimal consumption at the time of the switch to be $c^*(\tau) = (r-\gamma)\left(a_1 + \frac{y_2}{r}\right)$. Therefore, the solution to the second-stage sub-problem is

$$\begin{aligned} c^*(t) &= (r-\gamma)\left(a_1 + \frac{y_2}{r}\right) e^{\gamma(t-\tau)}, \quad \tau \le t < \infty,\\ a^*(t) &= -\frac{y_2}{r} + \left(a_1 + \frac{y_2}{r}\right) e^{\gamma(t-\tau)}, \quad \tau \le t < \infty, \end{aligned}$$

and the value of the co-state variable at the time of the switch is

$$\lambda_2(\tau) = \left((r-\gamma)\left(a_1 + \frac{y_2}{r}\right)\right)^{-\theta} e^{-\rho\tau},$$

where $a^*(\tau) = a_1$ is unknown at this stage.

Second step: initial stage. Now we solve the first-stage problem, for the period $t \in [0, \tau]$, taking $a(0) = a_0$ and $a(\tau^-) = a_1$ as given. The necessary (and in this case sufficient) conditions for an optimum are

$$\begin{aligned} &c(t)^{-\theta}\, e^{-\rho t} = \lambda_1(t), \text{ for each } t \in [0, \tau),\\ &\dot{\lambda}_1 = -r\,\lambda_1, \text{ for each } t \in (0, \tau),\\ &\dot{a} = r\,a + y_1 - c, \text{ for each } t \in (0, \tau). \end{aligned}$$

where $\lambda_1 : [0, \tau) \to \mathbb{R}$ is the co-state variable for the first period. Solving the Euler equation, substituting for consumption, and solving the budget constraint, subject to the initial net asset position a_0, we find $c(t) = c(0)\, e^{\gamma t}$ and $a(t) = e^{r\,t}\Big(a_0 + \int_0^t e^{-rs}\left(y_1 - c(0)e^{\gamma\,s}\right) ds\Big)$, for each $t \in [0, \tau)$, where $c(0)$ is unknown, and $\lambda_1(\tau) = c(0)^{-\sigma}\, r^{-r\,\tau}$ at the terminal time.

Final step. We still have two unknowns: $c(0)$ and a_1. We can determine them by using the matching condition for the co-state variable and by using the solution for the net asset position obtained in the last step. That is, we can find them by solving jointly

$$\begin{cases} \lambda_1(\tau^-) = \lambda_2(\tau), \\ a(\tau^-) = a_1. \end{cases}$$

Solving the first equation yields $c^*(0) = (r - \gamma)\left(a_1 + \frac{y_2}{r}\right) e^{-\gamma\tau}$, that, upon substitution in the second equation yields

$$a_1^* = e^{\gamma\tau}\left(a_0 + \frac{y_1}{r}\left(1 - e^{-r\tau}\right) + \frac{y_2}{r}\left(e^{-r\tau} - e^{-\gamma\tau}\right)\right).$$

Solution. The optimal consumption path is $\mathbf{c}^* = \left(c^*(t)\right)_{t\in[0,\infty)}$, where

$$c^*(t) = (r - \gamma)\left(a_1^* + \frac{y_2}{r}\right) e^{\gamma(t-\tau)}, \quad \text{for } t \in [0, \infty).$$

Substituting a_1^* in the solution for the first stage, we obtain the optimal path for the net asset position $\mathbf{a}^* = \left(a^*(t)\right)_{t\in[0,\infty)}$, where

$$a^*(t) = \begin{cases} -\frac{y_1}{r} + \left(a_1^* + \frac{y_2}{r}\right) e^{\gamma(t-\tau)} + \left(\frac{y_1}{r} - \frac{y_2}{r}\right) e^{r(t-\tau)}, & \text{for } t \in [0, \tau) \\ -\frac{y_2}{r} + \left(a_1^* + \frac{y_2}{r}\right) e^{\gamma(t-\tau)}, & \text{for } t \in [\tau, \infty) \end{cases}$$

We can draw several conclusions: first, although there is a jump in y, both paths for consumption and net asset position are continuous at τ, i.e., $c^*(\tau^-) = c^*(\tau)$ and $a^*(\tau^-) = a^*(\tau)$; second, while consumption is continuously differentiable at $t = \tau$, the net asset position is not, because $\frac{d}{dt}a^*(\tau^-) \neq \frac{d}{dt}a^*(\tau)$; third, the initial consumption level should be somewhere between the initial consumption level for cases in which the non-financial income is permanently equal to y_1 or to y_2, say $c_1(0)$ and $c_2(0)$. If we assume that $y_1 > y_2$ then

$$c_2(0) < c^*(0) < c_1(0) \iff (r - \gamma)\left(a_0 + \frac{y_2}{r}\right) < c^*(0) < (r - \gamma)\left(a_0 + \frac{y_1}{r}\right);$$

at last, the solution will converge asymptotically to the solution of a problem in which the non-financial income would be permanently equal to y_2, that is $\lim_{t\to\infty} a^*(t) = -\frac{y_2}{r}$.

12.3.2 Free Switching Time

In this subsection, we let the optimizer have the choice to switch optimally from the first to the second stage by choosing t_1 within set $\mathrm{T} = [t_0, t_2]$. The **optimal switching time** is denoted by t_1^*. There are three possibilities: the switching time is interior $t_0 < t_1^* < t_2$; there is an immediate switch in the initial time of the planning time $t_1^* = t_0$; or the switch is only optimal at the terminal time $t_1^* = t_2$. The last case is only possible if the time period $\mathrm{T} = [t_0, t_2]$ is closed. It is open at the terminal time, as in infinite-horizon problems $\mathrm{T} = [t_0, t_2)$, the last option is not available.

We consider the following problem:

$$\max_{u(\cdot),t_1} \int_{t_0}^{t_1} F_1(y(t), u(t), t)\, dt + \int_{t_1}^{t_2} F_2(y(t), u(t), t)\, dt - S(t_1, y(t_1)),$$

$$\begin{aligned}
&\text{subject to } t_0, t_2 \text{ fixed and } t_1 \text{ free,}\\
&\dot{y} = \begin{cases} G_1(y(t), u(t), t) & \text{for } t_0 \le t < t_1, \\ G_2(y(t), u(t), t) & \text{for } t_1 < t < t_2, \end{cases}\\
&y(t_0) = y_0 \text{ fixed,}\\
&y(t_1) = y_1 \text{ free,}\\
&R(t_2, y(t_2)) \ge 0.
\end{aligned} \tag{12.8}$$

The optimality conditions are formally similar to the fixed-switching time case but a new timing condition, associated to the switching time, t_1 is added:

Proposition 12.2 (Two-stage optimal control for a fixed switching time) *Let* $(\mathbf{y}^*, \mathbf{u}^*, t_1^*)$ *be a solution to the optimal control problem* (12.3). *Then there is a piecewise continuous function* $\lambda : \mathrm{T} \to \mathbb{R}$ *such that the following conditions are satisfied:*

- *the optimality, multiplier, and terminal condition*

$$H_{1,u}^*(t) = 0, \text{ for each } t_0 \le t < t_1^*, \tag{12.9a}$$

$$H_{2,u}^*(t) = 0, \text{ for each } t_1^* < t \le t_2, \tag{12.9b}$$

$$\dot{\lambda} + H_{1,y}^*(t) = 0, \text{ for each } t_0 \le t < t_1^*, \tag{12.9c}$$

$$\dot{\lambda} + H_{2,y}^*(t) = 0, \text{ for each } t_1^* < t \le t_2, \tag{12.9d}$$

$$\lambda(t_2)\, R^*(t_2) = 0; \tag{12.9e}$$

- *the admissibility conditions*

$$\dot{y} = G_1^*(t), \text{ for each } t_0 \le t < t_1^*, \tag{12.10a}$$

$$\dot{y} = G_2^*(t), \text{ for each } t_1^* < t \le t_2, \tag{12.10b}$$

$$y^*(t_0) = y_0; \tag{12.10c}$$

- *the matching condition*

$$\lambda(t_1^{*+}) = \lambda(t_1^{*-}) + S_y^*(t_1^*); \tag{12.11}$$

- *and one of the following three timing conditions*[4]

$$H_1^*(t_0) - S_t^*(t_0) < H_2^*(t_0),\ \text{if}\ t_0 = t_1^* < t_2, \tag{12.12a}$$
$$H_1^*(t_1^*) - S_t^*(t_1^*) = H_2^*(t_1^*),\ \text{if}\ t_0 < t_1^* < t_2, \tag{12.12b}$$
$$H_1^*(t_2) - S_t^*(t_2) > H_2^*(t_2),\ \text{if}\ t_0 < t_1^* = t_2. \tag{12.12c}$$

Proof We extend the proof of Proposition 12.1 to the case in which the switching time is optimally chosen. This entails changes in the optimal value

$$\mathsf{J}^* = \mathsf{J}[y^*, u^*, t_1^*] = \int_{t_0}^{t_1^*} F_1^*(t)\,dt + \int_{t_1^*}^{t_2} F_2^*(t)\,dt - S^*(t_1^*),$$

where the switching time, t_1^*, is optimal. Because $t_1^* \in [t_0, t_2]$ or $t_1^* \in [t_0, t_2)$, the generalized Lagrange functional is now

$$\mathsf{L}[y, u] = \int_{t_0}^{t_1} F_1(t)\,dt + \int_{t_1}^{t_2} F_2(t)\,dt - S(t_1) + \psi\, R(t_2) + \mu_0\,(t_1 - t_0) + \mu_2\,(t_2 - t_1)$$

where μ_0 and μ_2 are auxiliary variables such that the KKT conditions

$$\mu_0 \geq 0,\ t_1 \geq t_0,\ \text{and}\ \mu_0\,(t_1 - t_0) = 0,$$
$$\mu_2 \geq 0,\ t_1 \leq t_2,\ \text{and}\ \mu_2\,(t_2 - t_1) = 0,$$

are satisfied. To apply the variational approach, in addition to the perturbations in y and u we have to introduce a perturbation in the terminal time (a constant) $t_1 = t_1^* + \epsilon\,\tau_1$ where τ_1 is an arbitrary number. To find the necessary conditions, we introduce a variation not only in the state and control variables but also in the switching time. □

12.4 Optimal Control of PWSC ODE

The two-stage problem introduces discontinuities in the specification of the constraint regarding the independent variable. A different type of discontinuity exists if the vector field of the ODE constraint is piecewise continuous or piecewise smooth, as a function of the state and/or the control variable, of the type studied in Chap. 6. We consider next the following OCP in which the constraint is a PWSC ODE with two branches:

[4] We denote $S_y(t) = \frac{\partial S(y(t),t)}{\partial y(t)}$, and $S_t(t) = \frac{\partial S(y(t),t)}{\partial t}$.

$$
\begin{aligned}
&\max_{u(\cdot)} \int_0^{\infty} f(y(t), u(t))\, e^{-\rho t} dt, \\
&\quad \text{subject to } y(0) = y_0 \text{ fixed}, \\
&\quad \dot{y} = \begin{cases} g_1(y(t), u(t)) & \text{if } s(y(t), u(t)) \leq 0, \\ g_2(y(t), u(t)) & \text{if } s(y(t), u(t)) > 0, \end{cases} \\
&\quad \lim_{t\to\infty} R(y(t), t) \geq 0.
\end{aligned}
\tag{12.13}
$$

As seen in Chap. 6 there is a switching surface within space $\Omega = \mathrm{Y} \times \mathrm{U} \subseteq \mathbb{R}^2$, $\Sigma = \{(y, u) \in \Omega : s(y, u) = 0\}$, that partitions the space Ω into two branches $\Omega_1 = \{(y, u) \in \Omega : s(y, u) < 0\}$ and $\Omega_2 = \{(y, u) \in \Omega : s(y, u) > 0\}$. Therefore, the initial point belongs to one of the following subsets Ω_1, Ω_2, or Σ.

In order to apply the PMP we have to define the (piecewise) current-value Hamiltonian

$$
h(t) = h(y(t), u(t), q(t)) = \begin{cases} h_1(y(t), u(t), q(t)), & \text{if } (y(t), u(t)) \in \Omega_1, \\ h_2(y(t), u(t), q(t)), & \text{if } (y(t), u(t)) \in \Omega_2, \end{cases}
$$

where $h_i(y(t), u(t), q(t)) = f(y(t), u(t)) + q(t)\; g_i(y(t), u(t))$ for $i = 1, 2$.

Informally, the PMP states that if $(\mathbf{y}^*, \mathbf{u}^*)$ is a solution to the OCP (12.13) then there is current time co state variable $q : \mathrm{T} \to \mathbb{R}$, such that the following conditions are satisfied: the static optimality condition

$$
h_{i,u}(y^*(t), u^*(t), q(t)) = 0, \;\text{ if } (y^*(t), u^*(t)) \in \Omega_i, \;\text{ for } i = 1, 2, \tag{12.14}
$$

the adjoint equation, which is the PWSC ODE

$$
\dot{q} = \begin{cases} \rho q(t) - h_{1,y}(y^*(t), u^*(t), q(t)), & \text{if } (y^*(t), u^*(t)) \in \Omega_1, \\ \rho q(t) - h_{2,y}(y^*(t), u^*(t), q(t)), & \text{if } (y^*(t), u^*(t)) \in \Omega_2, \end{cases} \tag{12.15}
$$

and the transversality condition,

$$
\lim_{t\to\infty} q(t)\, e^{-\rho t}\, R(y^*(t), t) = 0,
$$

for admissible paths $\left(y^*(t), u^*(t)\right)_{t\in\mathrm{T}}$ that satisfy the constraints of problem (12.13). If the two branches of the current value Hamiltonian function are sufficiently smooth as regards the control variable, then we can define a PWSC control function $U_i(y(t), q(t))$, by invoking the implicit function theorem, from Eq. (12.14), and the MHDS becomes a PWSC system, at the optimum,

$$
\begin{aligned}
\dot{y} &= \begin{cases} g_1(y(t), U_1(y(t), q(t)) & \text{if } (y(t), u(t)) \in \Omega_1, \\ g_2(y(t), U_2(y(t), q(t)) & \text{if } (y(t), u(t)) \in \Omega_2 \end{cases} \\
\dot{q} &= \begin{cases} \rho q(t) - h_{1,y}(y(t), U_1(y(t), q(t)), q(t)) & \text{if } (y(t), u(t)) \in \Omega_1, \\ \rho q(t) - h_{2,y}(y(t), U_2(y(t), q(t)), q(t)) & \text{if } (y(t), u(t)) \in \Omega_2, \end{cases}
\end{aligned}
\tag{12.16}
$$

that can be of the piecewise-smooth or Filippov type, depending on the type of discontinuity. The number of possible cases is enormous. However, from our analysis in Chap. 6, there are some issues that should be taken into account. Let us assume that we start from an initial state $y(0) = y_0$ belonging to one of the two branches (or to surface Σ). First, the solution to the MHDS stays inside one branch, an in that case it behaves as in the simple infinite horizon optimal control problem studied in Sect. 10.2.3. Second, if the solution to the MHDS collides with the surface Σ and traverses the surface, it does it at a specific time, and tends to converge to a steady state in the other branch. In this case, it looks like the solution to a two-stage optimal control problem with free switching time. Third, the solution to the MHDS can collide with the surface Σ and can either converge to a pseudo-equilibrium belonging to it, or can exit the surface to the same branch or to the other branch and converge to a steady state. In higher dimensional problems, other types of dynamics are possible. Next, we illustrate with one of the very rare examples in the literature.

Example 12.3 (Tähvonen and Withagen (1996) *model*) The Tähvonen and Withagen (1996) is a model for the optimal control of pollution in which production, y, generates an increase in the stock of pollution, z. The natural resource is renewable, absorbing part of the pollution, but its capacity of renewal depends on the size of the pollution stock. If the pollution level is higher than a critical level $\bar{z} > 0$, it is assumed that the renewal capacity is lower than if the pollution level is lower than that critical level. It is supposed that all the output is consumed by households, and households have preferences not only for consumption but also dislike pollution. The optimal control problem is

$$
\begin{aligned}
&\max_{y(\cdot)} \int_0^{\infty} U(y(t)) - D(z(t))\, e^{-\rho t} dt, \\
&\quad \text{subject to } z(0) = z_0 \text{ fixed}, \\
&\quad \dot{z} = \begin{cases} \alpha\, y(t) - \delta_1 z(t), & \text{if } z(t) \le \bar{z}, \\ \alpha\, y(t) - \delta_2 z(t), & \text{if } z(t) > \bar{z}, \end{cases} \\
&\quad \lim_{t \to \infty} z(t) \ge 0,
\end{aligned}
\tag{12.17}
$$

where $\alpha > 0$ is the increase of pollution for every unit level of output, and $\delta_1 > \delta_2 > 0$ are the natural regeneration capacities if the pollution is below or above the critical level $\bar{z}$. The utility function is separable in consumption and in the amenity associated to pollution, $U(c) - D(z)$, where $U(y)$, for $c = y$, is increasing and concave, and $D(z)$ is increasing and convex. The solution to this problem is studied in Example 6.5.

12.5 Optimal Control with Two State Variables

In this section, we briefly address the dynamics of discounted infinite-horizon OCP with two state variables, in which we have a vector state function $\mathbf{y} \in \mathcal{Y} \subset C^2(\mathrm{T}; \mathbb{R}^2)$, mapping $\mathbf{y} : \mathrm{T} \to \mathbf{Y} \subseteq \mathbb{R}^2$, where $\mathrm{T} = \mathbb{R}_+$, and a control vector function $\mathbf{u} \in \mathcal{U} \subset PC^1(\mathrm{T}; \mathrm{U})$, mapping $\mathbf{u} : \mathrm{T} \to \mathbf{U} \subseteq \mathbb{R}^m$, for $m \geq 1$. The problem is

$$\begin{aligned} &\max_{\mathbf{u}(\cdot)} \int_0^\infty f(\mathbf{y}(t), \mathbf{u}(t))\, e^{-\rho t}\, dt, \\ &\qquad \text{subject to } \mathbf{y}(0) = \mathbf{y}_0 \text{ fixed}, \\ &\qquad \dot{\mathbf{y}} = \mathbf{g}(\mathbf{y}(t), \mathbf{u}(t)), \;\; \text{for each } t \in \mathrm{T}, \\ &\qquad \lim_{t \to \infty} \mathbf{R}(\mathbf{y}(t), t) = \mathbf{0}, \end{aligned} \tag{12.18}$$

where $\mathbf{g} : \mathbf{Y} \times \mathbf{U} \to \mathbb{R}^2$, is a vector function.[5] The current-value Hamiltonian function is now

$$h(\mathbf{y}, \mathbf{u}, \mathbf{q}) = f(\mathbf{y}, \mathbf{u}) + \mathbf{q} \cdot \mathbf{g}(\mathbf{y}, \mathbf{u}).$$

If functions $f(\cdot)$ and $\mathbf{g}(\cdot)$ are sufficiently smooth, then we obtain the optimal control as a vector function $\mathbf{u}^* = \mathbf{U}(\mathbf{y}, \mathbf{q})$ and write the maximized Hamiltonian as a function of the state and co-state variables $h^*(\mathbf{y}, \mathbf{q})$. Then the MHDS is a four-dimensional ODE system

$$\begin{aligned} \dot{\mathbf{y}} &= \mathbf{g}(\mathbf{y}, \mathbf{U}(\mathbf{y}, \mathbf{q})) = \mathrm{D}_{\mathbf{q}} h^*(\mathbf{y}, \mathbf{q}), \\ \dot{\mathbf{q}} &= \rho\, \mathbf{q} - \mathrm{D}_{\mathbf{y}} h^*(\mathbf{y}, \mathbf{q}). \end{aligned}$$

Assuming a steady state $(\bar{\mathbf{y}}, \bar{\mathbf{q}})$ exists, then the Jacobian of the MHDS evaluated at the steady state has the following form

$$\mathbf{J} = \begin{pmatrix} \mathrm{D}_{\mathbf{qy}} h^*(\mathbf{y}, \mathbf{q}) & \mathrm{D}_{\mathbf{qq}} h^*(\mathbf{y}, \mathbf{q}) \\ -\mathrm{D}_{\mathbf{yy}} H^*(\mathbf{y}, \mathbf{q}) & \rho\, \mathbf{I}_2 - \mathrm{D}_{\mathbf{yq}} h^*(\mathbf{y}, \mathbf{q}) \end{pmatrix},$$

where, if all the primitive functions are continuous, $\mathrm{D}_{\mathbf{qy}} h^*(\mathbf{y}, \mathbf{q}) = \mathrm{D}_{\mathbf{yq}} h^*(\mathbf{y}, \mathbf{q})^\top$ and $\mathrm{D}_{\mathbf{qq}} h^*(\mathbf{y}, \mathbf{q})$ and $\mathrm{D}_{\mathbf{yy}} h^*(\mathbf{y}, \mathbf{q})$ are symmetric. Then the characteristic equation of the (4×4) Jacobian matrix $\mathbf{J}$ is the fourth-order polynomial

$$c(\lambda) = \lambda^2 - 2\rho\, \lambda^3 + M_2\, \lambda^2 - \rho\, (M_2 - \rho^2)\, \lambda + M_0$$

where M_2 is the sum of the principal minors of order 2, and M_0 is the determinant, of $\mathbf{J}$. The four eigenvalues can be separated into two groups

[5] For the necessary conditions for discounted infinite horizon OCP with three and four state variables see Brito (1999) and Brito (2000), respectively. As the Galois theory asserts that polynomials of degree five or higher cannot be solved by elementary functions, then the highest dimension of the state variable vector allowing for a general characterization of the qualitative dynamics of the MHDS through linearization is four.

$$\lambda_{s,i} = \frac{\rho}{2}\left(1 - \sqrt{w_i}\right), \text{ and } \lambda_{u,i} = \frac{\rho}{2}\left(1 + \sqrt{w_i}\right), \text{ for } i = 1, 2$$

where w_1 and w_2 are the roots of the polynomial equation $c(w) = w^2 - \varphi_1\, w + \varphi_0$, where $\varphi_1 \left(\frac{\rho}{2}\right)^2 = 6\left(\frac{\rho}{2}\right)^2 - M_2$ and $w_0 \left(\frac{\rho}{2}\right)^4 = 5\left(\frac{\rho}{2}\right)^4 - \left(\frac{\rho}{2}\right)^2 M_2 + M_0$. As $\lambda_{u,i} > 0$, for all $i = 1, 2$, then the dimension of the stable manifold of the MHDS is less than or equal to two, and an appropriate two-dimensional projection of the four-dimensional system $(\mathbf{y}, \mathbf{q})$ can locally display any phase diagram that is qualitatively similar to the generic phase portraits for planar linear ODE in Fig. 4.3. Although the solution is four dimensional vector function $(\mathbf{y}, \mathbf{q}) : \mathrm{T} \to \mathbb{R}^4$, we can find two-dimensional projections that are stable nodes, stable-foci.[6] It is also possible to have local projections that are saddles, which in endogenous growth models relates to the existence of transition dynamics towards a balanced growth path.

Exercise 12.3 (*Endogenous growth with transitional dynamics*) A simple version of the Uzawa-Lucas model without externalities[7] is

$$\max_{C, K_1, H_1, H_2} \int_0^\infty \frac{C(t)^{1-\theta}}{1-\theta} e^{-\rho t} dt,$$

subject to the accumulation equations for stocks of physical and human capital

$$\dot{K} = Y_1(t) - C(t), \text{ and } \dot{H} = Y_2(t),$$

and to the allocation constraints of the stocks between the two sectors

$$K(t) = K_1(t), \text{ and } H(t) = H_1(t) + H_2(t),$$

where the production functions for manufacturing and education are

$$Y_1(t) = A_1\, K_1(t)^\alpha\, H_1(t)^{1-\alpha}, \text{ and } Y_2(t) = A_2\, H_2(t),$$

where Y_j is the output of manufacturing ($j = 1$) and education ($j = 2$), K_j and H_j are the physical and human capital allocated to sector $j = 1, 2$ H_j human capital allocated to sector $j = 1, 2$, A_j is the TFP in sector $j = 1, 2$, K and H are the aggregate stocks of physical and human capital, $\alpha \in (0, 1)$ share of capital in manufacturing, $\theta > 0$ is the inverse of the households' EIS in consumption, $\rho > 0$ is the rate of time preference. We assume that $\rho + A_2(\theta - 1) > 0$.

Introducing an admissible detrending, by a common growth rate γ, to all variables,

$$K_j(t) = k_j(t)e^{\gamma t},\ \ H_j(t) = h_j(t)e^{\gamma t},\ \ j = 1, 2$$

[6] An illustration of this projection can be seen in Brito and Dixon (2013).

[7] The original papers are Uzawa (1965) and Lucas (1988). There are several textbook presentation of this model, for instance in Barro and Sala-i-Martin (2004, Chap. 5) and Acemoglu (2009, Chap. 10). For a general version of this model with externalities see Brito and Venditti (2010).

allows us to transform the UL model in detrended variables

$$\max_{c,k_1,h_1,h_2} \int_0^\infty \frac{c(t)^{1-\theta}}{1-\theta} e^{-(\rho-\gamma(1-\theta))t} dt,$$

$$\begin{aligned}
&\text{subject to}\\
&\dot{k}_1 = A_1 k_1(t)^\alpha h_1(t)^{1-\alpha} - c(t) - \gamma k(t),\\
&\dot{h}_2 = A_2 h_2(t) - \gamma h(t),\\
&k(t) = k_1(t),\\
&h(t) = h_1(t) + h_2(t),
\end{aligned}$$

and also subject to the initial $k(0) = k_0$, and $h(0) = h_0$, and to the terminal $\lim_{t\to\infty} k(t) \geq 0$ and $\lim_{t\to\infty} h(t) \geq 0$ conditions. At the steady state, we can obtain the long run growth rate $\bar{\gamma} = \frac{A_2-\rho}{\theta}$, which allows for obtaining the MHDS

$$\begin{aligned}
\dot{k} &= (a_1(\pi) - \bar{\gamma})\, k - c(p_k),\\
\dot{h} &= a_2(\pi)k + (A_2 - \bar{\gamma})\, h,\\
\dot{q}_k &= q_k(A_2 - r(q_k/q_h)),\\
\dot{q}_h &= 0,
\end{aligned}$$

in which q_k and q_h are the co-state variables, and the optimal consumption is $c(q_k) = q_k^{-\frac{1}{\theta}}$, $\pi = \frac{q_k}{q_h}$, $a_1 = \frac{r(\pi)}{\alpha} > 0$, and $a_2 = -A_2 \left(\frac{r(\pi)}{\alpha A_1}\right)^{\frac{1}{1-\alpha}} < 0$, where the rate of return of physical capital is $r(\pi) = \left(\alpha^\alpha (1-\alpha)^{(1-\alpha)} A_1 \left(\frac{\pi}{A_2}\right)^{1-\alpha}\right)^{\frac{1}{\alpha}}$.

If the initial conditions satisfies $k_0 = \eta\, h_0$, where $\eta = \bar{\pi}\left(\frac{A_2-\bar{\gamma}}{A_2}\right)\left(\frac{\alpha}{1-\alpha}\right) > 0$, and $\bar{\pi} = \left(\frac{\alpha_0 A_1}{A_2}\right)^{\frac{1}{1-\alpha}}$, then the economy will evolve along the BGP

$$\bar{K}(t) = \eta\, h_0\, e^{\bar{\gamma}t},\ \bar{H}(t) = h_0\, e^{\bar{\gamma}t},\ \text{ for each } t \in \mathbb{R}_+.$$

Alternatively, if $k_0 \neq \eta\, h_0$ then there is transitional dynamics. The variational MHDS is

$$\begin{pmatrix} \dot{k} \\ \dot{h} \\ \dot{q}_k \\ \dot{q}_h \end{pmatrix} = \bar{\mathbf{J}} \begin{pmatrix} k - \bar{k} \\ h - \bar{h} \\ q_k - \bar{q}_k \\ q_h - \bar{q}_h \end{pmatrix}, \text{ where } \bar{\mathbf{J}} = \begin{pmatrix} \beta & 0 & \frac{\beta(\theta-\alpha)-\alpha\mu)\bar{k}}{\alpha\theta\,\bar{p}_k} & -\frac{(\beta-\mu)\bar{k}}{\alpha\bar{p}_h} \\ -\frac{\mu}{\eta} & \mu & -\frac{\mu\bar{h}}{\alpha\bar{p}_k} & \frac{\mu\bar{h}}{\alpha\bar{p}_h} \\ 0 & 0 & \mu-\beta & \bar{\pi}(\beta-\mu) \\ 0 & 0 & 0 & 0 \end{pmatrix}$$

where $\beta = \frac{\bar{c}}{\bar{k}} = \frac{A_2}{\alpha} - \bar{\gamma} > 0$, and $\mu = A_2 - \bar{\gamma} < \beta$. The eigenvalues of the Jacobian $\bar{\mathbf{J}}$ are $\lambda_1 = \mu - \beta < 0$, $\lambda_2 = 0$, $\lambda_3 = \beta > 0$ and $\lambda_4 = \mu > 0$, and the therefore stable manifold in the neiborhood of the BGP has dimension one. Differently from the

AK one-sector endogenous growth model, in Example 10.5, there is **transitional dynamics**. The linearized solution to the detrended variables can be found to be

$$
\begin{aligned}
h(t) &= \bar{h} + (h_0 - \bar{h})\, e^{(\mu-\beta)t},\\
k(t) &= \eta\, \bar{h} + (k_0 - \eta\, \bar{h})\, e^{(\mu-\beta)t},\\
p_k(t) &= \bar{p}_k \left(1 + \frac{\alpha\, \theta\, (2\beta - \mu)}{\mu\, (\theta - \alpha)} \left(\frac{h_0 - \bar{h}}{\bar{h}}\right) e^{(\mu-\beta)t}\right),\\
p_h(t) &= \bar{p}_h,
\end{aligned}
$$

where $\bar{h} = \frac{k_0\, \mu\, (\theta-\alpha)+\eta\, h_0\, (\beta(\alpha+\theta)-\mu\theta)}{\eta(\beta(\alpha+\theta)-\alpha\, \mu)}$.

12.6 References

Comprehensive textbooks on optimal control applied to economics and management science are Grass et al. (2008), and also Sethi (2019). On optimal control problems with instantaneous constraints, see Köhler (1980), and Hartl et al. (1995). On two-stage optimal control, see Tomiyama (1985), or Amit (1986). For applications on the timing for liberalizing international capital markets or on the adoption of new technologies, see Makris (2001) and Boucekkine et al. (2004), respectively.

Part III
Partial Differential Equations

Chapter 13
First-Order PDE

This chapter presents a short introduction to first-order partial differential equations (PDE), sometimes called hyperbolic PDE.[1] A first-order (or hyperbolic) PDE is a functional equation containing one or more functions of **more than one independent variable** together with their first-order derivatives. A first-order PDE is called scalar (planar) if there is only one unknown function (two unknown functions).

In economic applications, usually one of the independent variables is time and the other independent variable is the support of some distribution. First-order PDE have been applied in continuous time overlapping generations models, vintage capital models, interest rate term-structure models, for instance. They also provide an elegant and effective way of modeling the dynamics of distribution in heterogeneous agent economies, in the absence of pure social mobility. In physics, they model advection, traveling, or transportation phenomena. Mathematical demography is also an important area of application, in particular for modeling age-dependent dynamics of population. We already saw in Chap. 11 that the Hamilton-Jacobi equation, for deterministic finite horizon optimal control problems, is a non-linear first-order PDE.

There are two benefits from studying PDE: first, they provide a powerful modeling framework for setting up, solving, and characterizing dynamic models with heterogeneity, and second, they provide a better understanding of the implicit assumptions that are introduced when using ODE (or difference equations) models for studying the dynamics of heterogeneity. The field is very large in terms of equations studied and methods involved, and is not generally in the toolbox of economists. We will only present a very brief introduction allowing to study simple linear scalar models.

Definition 13.1 (*Scalar first-order PDE*) A **scalar first-order PDE** in two independent variables $\mathbf{x} = (x_1, x_2) \in \mathbf{X} \subseteq \mathbb{R}^2$ is a functional equation defined by a known function $F : D \to \mathbb{R}$, where $D \subset \mathbb{R}^5$,

$$F(\mathrm{D}_{\mathbf{x}} y(\mathbf{x}), y(\mathbf{x}), \mathbf{x}) = 0 \tag{13.1}$$

[1] Using a classification for second-order PDE.

P. B. Brito, *Economic Dynamics and Distributions*, Dynamic Modeling and Econometrics in Economics and Finance 32, https://doi.org/10.1007/978-3-031-94717-9_13

depending on the function $y : \mathbf{X} \to \mathrm{Y} \subseteq \mathbb{R}$, and its gradient

$$\mathrm{D}_{\mathbf{x}} y(\mathbf{x}) = \left(\partial_{x_1} y(\mathbf{x}), \partial_{x_2} y(\mathbf{x})\right)^\top,$$

where we use the notation $\partial_{x_i} y(\mathbf{x}) = \frac{\partial y(\mathbf{x})}{\partial x_i}$, for $i = 1, 2$.

We say Eq. (13.1) is non-linear if $F(\cdot)$ features a non-linear function of $\mathrm{D}_{\mathbf{x}} y(\cdot)$. If $F(\cdot)$ is a linear function of the gradient, there are four types of scalar first-order PDE:

Definition 13.2 (*Scalar quasi-linear first-order PDE*) A scalar first-order PDE is quasi-linear if $F(\cdot)$ is linear in the derivatives in $\mathrm{D}_{\mathbf{x}} y(\mathbf{x})$, with the form

$$a_1(y, \mathbf{x})\, \partial_{x_1} y + a_2(y, \mathbf{x})\, \partial_{x_2} y = b(y, \mathbf{x}), \tag{13.2}$$

where the coefficients $a_1(\cdot)$, $a_2(\cdot)$, and $b(\cdot)$ are possibly non-linear function of y, and are constant or differentiable functions of $\mathbf{x}$.

Definition 13.3 (*Scalar semi-linear first-order PDE*) A scalar first-order PDE is semi-linear if $F(\cdot)$ is linear in the derivatives in $\mathrm{D}_{\mathbf{x}} y(\mathbf{x})$, with the form

$$a_1(\mathbf{x})\, \partial_{x_1} y + a_2(\mathbf{x})\, \partial_{x_2} y = b(y, \mathbf{x}), \tag{13.3}$$

where the coefficients $a_1(\cdot)$ and $a_2(\cdot)$, are constant or differentiable functions of $\mathbf{x}$, and $b(\cdot)$ is a non-linear function of y, and a constant or differentiable function of $\mathbf{x}$.

Definition 13.4 (*Scalar linear first-order PDE*) A scalar first-order PDE is linear if $F(\cdot)$ is linear in the derivatives in $\mathrm{D}_{\mathbf{x}} y(\mathbf{x})$ and y, and takes the form

$$a_1(\mathbf{x})\, \partial_{x_1} y + a_2(\mathbf{x})\, \partial_{x_2} y = b(\mathbf{x})\, y + b_0(\mathbf{x}), \tag{13.4}$$

where the coefficients $a_1(\cdot)$, $a_2(\cdot)$, $b(\cdot)$ and $b_0(\cdot)$ are constant or differentiable functions of $\mathbf{x}$.

Linear equations can be classified further as **autonomous** or **non-autonomous**, if coefficients a_1, a_2, b and b_0 are constants, or if they are functions of one or both independent variables, $\mathbf{x}$, respectively. Semi-linear PDE are called **homogeneous** or **non-homogeneous**, if function $b(y, \mathbf{x})$ is a homogeneous function of y, or if it is independent from y, respectively.

Higher-dimensional first-order PDE, also called systems hyperbolic equations, are equations of form $\mathbf{F}(D_{\mathbf{x}}\mathbf{y}(\mathbf{x}), \mathbf{y}(\mathbf{x}), \mathbf{x}) = \mathbf{0}$ where $\mathbf{x} \in \mathbb{R}^q$, for $q > 1$, and $\mathbf{y} : \mathbb{R}^q \to \mathbb{R}^n$, and $\mathrm{D}_{\mathbf{x}}\mathbf{y}$ is the Jacobian. For instance, a planar linear first-order PDE in two independent variables can be written as

$$\mathbf{A}\, \mathrm{D}_{\mathbf{x}}\mathbf{y}(\mathbf{x}) = \mathbf{B}\, \mathbf{y}(\mathbf{x})$$

where

$$\mathbf{y}(\mathbf{x}) = \begin{pmatrix} y_1(\mathbf{x}) \\ y_2(\mathbf{x}) \end{pmatrix} \text{ and } \mathrm{D}_{\mathbf{x}}\mathbf{y}(\mathbf{x}) = \begin{pmatrix} \partial_{x_1} y_1(\mathbf{x}) & \partial_{x_2} y_1(\mathbf{x}) \\ \partial_{x_1} y_2(\mathbf{x}) & \partial_{x_2} y_2(\mathbf{x}) \end{pmatrix}$$

and $\mathbf{A} \in \mathbb{R}^{2\times 2}$ and $\mathbf{B} \in \mathbb{R}^2$. Next we consider scalar equations.

A **solution to a first-order PDE** is a differentiable function $y(\mathbf{x})$ that satisfies the associated PDE. Existence and uniqueness of solutions for first-order PDE, and for problems involving them, are not guaranteed. If a solution is continuous and continuously differentiable, it is called a **classic solution**. Otherwise, if a solution is non-differentiable or it is discontinuous it is called a **generalized or weak solution**.

There are several **methods for obtaining solutions**, depending on the type of PDE. The classical methods for solving simpler equations, and that we apply next, are the **method of characteristics**, and **transformation methods** (in particular application of Laplace transforms). As we will see, the usefulness of a particular method depends on the domain of the independent variables, $\mathbf{X}$.[2] These methods simplify the first-order PDE, by making it equivalent to ODE, redefining the independent variables or transforming one of the independent variables into a parameter. In this chapter, we solve PDE by finding explicit or analytical solutions, using those two previous methods.

PDE problems. There are two main types of **problems involving first-order PDE**. We have a Cauchy problem whenever there is a single constraint on $\mathbf{x}$ along a surface $\Gamma \in \mathbf{X}$

$$\begin{cases} F(\mathrm{D}_{\mathbf{x}} y(\mathbf{x}), y(\mathbf{x}), \mathbf{x}) = 0, & \mathbf{x} \in \mathbf{X}, \\ y \mid_{\Gamma} = \phi, & \mathbf{x} \in \Gamma \subset \mathbf{X}, \end{cases} \tag{13.5}$$

where ϕ is a constant. We have a Dirichlet or **boundary-value problem** it there are two side conditions, one for each independent variable,

$$\begin{cases} F(\mathrm{D}_{\mathbf{x}} y(\mathbf{x}), y(\mathbf{x}), \mathbf{x}) = 0, & \mathbf{x} \in \mathbf{X}, \\ y(\mathbf{x}) \mid_{x_2=0} = \phi_1(x_1), & (x_1, 0) \in \mathbf{X}, \\ y(\mathbf{x}) \mid_{x_1=0} = \phi_2(x_2), & (0, x_2) \in \mathbf{X}. \end{cases} \tag{13.6}$$

In general, we call Cauchy-Dirichlet to a PDE problem in which there are side conditions over the boundary, or a segment of the boundary, of set $\mathbf{X}$, denoted by $\partial\mathbf{X}$.

Well- and ill-posed problems. Existence, uniqueness, and the properties of solutions vary widely. Again, we have to distinguish existence properties of the PDE and of an associated PDE problem. A problem is **ill-posed** if, for instance, although the PDE has a solution the problem involving the PDE may not have a solution. A problem is **well-posed** if the general solution to the PDE has a particular solution satisfying the constraints of the problem. In most cases, linear PDE and well-posed problems involving liner PDE have explicit solutions.

---

[2] This is one of the features of PDE that increases the difficulty of finding their solutions relative to ODE.

In the rest of the chapter, in Sect. 13.1, we solve linear scalar equations with an infinite domain by the method of characteristics. In Sect. 13.2, we solve some linear equations in the semi-infinite domain by Laplace transform methods. In Sect. 13.3, we briefly present time-dependent first-order PDE and the qualitative properties of the associated distributional dynamics. Section 13.4 includes several applications to economics and demography.

13.1 Equations in the Infinite Domain and the Method of Characteristics

In this section, we address solutions to several types of first-order PDE where $y(\cdot)$ has a **double-infinite domain** $\mathbf{X} = \mathbb{R}^2$, with an emphasis on the method of characteristics that is the simplest and more intuitive method for solving these first-order PDE.[3] In Sects. 13.1.1 and 13.1.2 we solve free boundary and Cauchy-Dirichlet problems for linear PDE, in Sects. 13.1.3 and 13.1.4 we address the solution to semi-linear and quasi-linear Cauchy-Dirichlet problems.

General solutions to first-order PDE generate a surface $S = \{(x_1, x_2, y(x_1, x_2)) \in \mathbf{X} \times \mathrm{Y} \subseteq \mathbb{R}^3\}$. The **method of charactetistics** is based on finding curves over surface S that correspond to general solutions of an ODE, called **characteristic curves**, that are consistent with the PDE that generates the surface S.

To apply that method, we need to recall the definition of directional derivative. Consider a function $f(\mathbf{x})$, where $\mathbf{x} \in \mathbb{R}^2$, the **directional derivative** of function $f(\cdot)$ in the direction specified by vector $\mathbf{v} = (v_1, v_2)^\top$ is defined by the limit of the variation $f(\mathbf{x} + \epsilon\, \mathbf{v}) - f(\mathbf{x})$ when $\epsilon \to 0$,

$$\mathrm{D}_{\mathbf{v}} f(\mathbf{x}) = \lim_{\epsilon \to 0} \frac{f(x_1 + \epsilon\, v_1, x_2 + \epsilon\, v_2) - f(\mathbf{x})}{\epsilon},$$

if the limit exists. If function $f(\mathbf{x})$ is differentiable, in the direction $\mathbf{v} = (v_1, v_2)^\top$, the directional derivative is equal to the dot product[4]

$$\begin{aligned} \mathrm{D}_{\mathbf{v}} f(\mathbf{x}) &= \mathrm{D}_{\mathbf{x}} f(\mathbf{x}) \cdot \mathbf{v} = \left(\partial_{x_1} f(\mathbf{x}), \partial_{x_2} f(\mathbf{x})\right) \cdot (v_1, v_2) = \\ &= \partial_{x_1} f(\mathbf{x}) v_1 + \partial_{x_2} f(\mathbf{x}) v_2. \end{aligned}$$

[3] There is not an agreement on the paternity of the method, with attributions to Paul Charpit de Villecourt (?-1794), Euler (1707–1783), Lagrange (1736–1813), Gaspard Monge (1746-1818), or others.

[4] Observe there is a relationship with the total differential. Let $z = f(x, y)$, where $f(.)$ is differentiable. The total differential is $dz = f_x(x, y)dx + f_y(x, y)dy$. If we write $dx = v_x h$ and $dy = v_y h$ then $\mathrm{D}_{\mathbf{x}} f(x, y) \cdot \mathbf{v} = \lim_{h \to 0} \frac{dz}{h}$.

13.1.1 General Solutions to Free Boundary Problems

Assume there are no side constraints, and let a direction in the domain $\mathbf{X}$ be parameterized by a variable $s \in \mathbb{R}_+$ such that $\mathbf{x} = \mathbf{X}(s)$. Then the dependent variable y becomes also function of s, $y = y(\mathbf{X}(s)) = Y(s)$. Then the derivative of $Y(s)$ as regards s,

$$Y'(s) = \frac{d}{ds} y(\mathbf{X}(s)) = \mathrm{D}_{\mathbf{x}} y(\mathbf{x}) \cdot \mathbf{X}'(s), \tag{13.7}$$

is equal to the inner product of the directional derivative of $y(\mathbf{x})$, $\mathrm{D}_{\mathbf{x}} y(\mathbf{x})$, with the direction of change of $\mathbf{x}$ relative to s, $\mathbf{X}'(s) = (X_1'(s), X_2'(s))^\top$. Therefore, a characteristic is a curve on surface S,

$$C = \{(X_1(s), X_2(s), Y(s)) \in S : Y'(s) = \mathrm{D}_{\mathbf{x}} y(\mathbf{x}) \cdot \mathbf{X}'(s), \text{ for } s \in \mathbb{R}_+\},$$

that is associated to the directional derivative implicit in a specific PDE.

First, using the definition of directional derivative, the linear PDE (13.4) can be writen as $\mathrm{D}_{\mathbf{x}} y(\mathbf{x}) \cdot (a_1(\mathbf{X}(s), a_2(\mathbf{X}(s)) = b(\mathbf{X}(s))$. It can match equation (13.7) if the following ODE are satisfied,

$$X_1'(s) = a_1(X_1(s), X_2(s)), \ s \in \mathbb{R}_+, \tag{13.8a}$$
$$X_2'(s) = a_2(X_1(s), X_2(s)), \ s \in \mathbb{R}_+, \tag{13.8b}$$

and

$$Y'(s) = b(X_1(s), X_2(s)), \ s \in \mathbb{R}_+. \tag{13.9}$$

ODE system (13.8) is called **characteristic system** and ODE (13.9) is called **compatibility condition**. Observe that in the case of a linear PDE, the characteristic system is independent of $Y(s)$, which means that the projection of curve C onto set $\mathbf{X}$ is independent of Y. Solving the characteristic ODE system to $\mathbf{X}(s)$, finding the characteristic projections in the set $\mathbf{X}$, and substituting in the compatibility condition yields a non-autonomous linear ODE $Y'(s) = B(s)$, with solution $Y(s) = Y(0) + \int_0^s B(z)dz$, for $Y(0) = f(X_1(0), X_2(0))$, where $f(\cdot)$ is an arbitrary function. From the solution to the characteristic system (13.8), we may obtain $s = S(\mathbf{x})$, implying that $X_1(0)$ and $X_2(0)$ can also be obtained as functions of $\mathbf{x}$. Transforming back to the original coordinates, yields a general solution to the PDE, $y(\mathbf{x})$, as a function of an arbitrary function $f(\cdot)$.[5]

Second, for a semi-linear equation, (13.3) the characteristic system is also (13.8). However, we have now the **compatibility condition**

$$Y'(s) = b(X_1(s), X_2(s), Y(s)), \ s \in \mathbb{R}_+, \tag{13.10}$$

[5] Analogously to ODE whose solution depends on an arbitrary constant.

that depends on $Y(s)$. This implies that the solution to the semi-linear PDE, although having similar characteristics as the linear PDE, has a different distribution. That is, although the projections onto space $\mathbf{X}$ are similar to the linear case, the surface S is different.

At last, a quasi-linear equation, (13.2), has characteristic and compatibility conditions

$$
\begin{aligned}
X_1'(s) &= a_1(X_1(s), X_2(s), Y(s)), \; s \in \mathbb{R}_+, \\
X_2'(s) &= a_2(X_1(s), X_2(s), Y(s)), s \in \mathbb{R}_+, \\
Y'(s) &= b(X_1(s), X_2(s), Y(s)), \; s \in \mathbb{R}_+.
\end{aligned}
\tag{13.11}
$$

Differently from the previous linear and semi-linear cases, there is a joint determination of the characteristics and of the dependent variable. We will see next that this fact fundamentally changes the properties of the solution to quasi-linear PDE. Next, we apply this approach to some types of PDE, starting with the linear PDE (13.4).

Proposition 13.1 (Linear autonomous PDE) *Consider the autonomous linear first-order PDE, with a zero right-hand side,*

$$\partial_{x_1} y(\mathbf{x}) + a\, \partial_{x_2} y(\mathbf{x}) = 0, \; \textit{for each } \mathbf{x} \in \mathbf{X} = \mathbb{R}^2, \tag{13.12}$$

where $a \neq 0$ is an arbitrary constant. The general solution to this equation is

$$y(\mathbf{x}) = f(x_2 - a\, x_1), \; \textit{for each } \mathbf{x} \in \mathbf{X} = \mathbb{R}^2, \tag{13.13}$$

where $f \in C^1(\mathbb{R})$ is a $\mathbb{R}^2 \to \mathbb{R}$ arbitrary differentiable function.

Proof Using the method of characteristics, the characteristic system and the consistency conditions are the three ODE

$$X_1'(s) = 1, \; X_2'(s) = a, \; \text{and } Y'(s) = 0, \; \text{for } s \in \mathbb{R}_+.$$

Observe that everywhere along each characteristic curve $C = \{(X_1(s), X_2(s), Y(s)) : Y'(s) = 0\}$, the variable Y is constant. Solving the previous ODE system yields the general solutions

$$X_1(s) = s + X_1(0), \; X_2(s) = a\, s + X_2(0), \; \text{and } Y(s) = Y(0), \; \text{for } s \in \mathbb{R}_+.$$

Along a characteristic we have $x_1 = X_1(s)$, $x_2 = X_2(s)$, and $y = y(\mathbf{x}) = Y(X_1(s), X_2(s))$. Eliminating s in the solutions to the characteristic system, yields $a\, x_1 - x_2 = a\, X_1(0) - X_2(0)$. Setting, arbitrarily $X_1(0) = 0$ yields $X_2(0) = x_2 - a\, x_1$, along a characteristic. From the solution to the compatibility condition $Y(s) = Y(0) = Y(X_1(0), X_2(0)) = f(x_2 - a\, x_1)$, we obtain function (13.13), where f is an arbitrary $C^1(\mathbb{R})$ function. To verify that (13.13) is indeed a general solution, we substitute it into Eq. (13.12).

We call **projected characteristic** to the line $x_2 = X_2(0) + a\, x_1$, where $X_2(0) \in \mathbb{R}$ is arbitrary, and we call $f(x_2 - a\, x_1)$ the **first integral** of the PDE. Figure 13.1 depicts the projected characteristic lines for $a > 0$ and $a < 0$. These curves correspond to the projection onto space (x_1, x_2) of the solution curves of the PDE (13.12) that are, in this case, $y(\mathbf{x}) = \text{constant}$.

Exercise 13.1 Prove that the solution to the PDE $a_1\, \partial_{x_1} y + a_2\, \partial_{x_2} y = 0$, for $\mathbf{x} \in \mathbb{R}^2$, is

$$y(x_1, x_2) = f(a_1\, x_2) \text{ if } a_1 \neq 0 \text{ and } a_2 = 0,$$
$$y(x_1, x_2) = f(a_2\, x_1) \text{ if } a_1 = 0 \text{ and } a_2 \neq 0.$$

Draw the projected characteristics.

Proposition 13.2 (Linear autonomous PDE with linear right-hand side) *Consider the autonomous linear first-order PDE, with a linear right hand side,*

$$\partial_{x_1} y(\mathbf{x}) + a\, \partial_{x_2} y(\mathbf{x}) = b_0 + b\, y(\mathbf{x}), \text{ for each } \mathbf{x} \in \mathbb{R}^2, \tag{13.14}$$

where $a \neq 0$, and $b \neq 0$ are constants. The general solution to this PDE is the function

$$y(\mathbf{x}) = -\frac{b_0}{b}\left(1 - e^{b\, x_1}\right) + f(x_2 - a\, x_1)\, e^{b\, x_1}, \text{ for each } \mathbf{x} \in \mathbf{X} = \mathbb{R}^2, \tag{13.15}$$

where $f(.)$ is an arbitrary C^1 function.

Proof Using the method of characteristics, we have the same characteristic system as PDE (13.12), but the compatibility condition is now $Y'(s) = b_0 + b\, Y(s)$, and has solution $Y(s) = -\frac{b_0}{b} + (Y(0) + \frac{b_0}{b})\, e^{b\, s}$. Identically to PDE (13.12), we set $X_1(0) = 0$, $s = x_1$ and $X_2(0) = x_2 - a\, x_1$, and substitute $Y(0) = y(X_1(0), X_2(0)) = f(x_2 - a\, x_1)$ and $s = x_1$ in the solution to the compatibility condition, to obtain the general solution (13.15).

Comparing PDE (13.14) with PDE (13.12), we conclude that both PDE have the same projected characteristics, depicted in Fig. 13.1, but the value of $y(.)$ will not remain constant along each characteristic (it grows if $b > 0$, or decays if $b < 0$) while in the second PDE it remains constant.

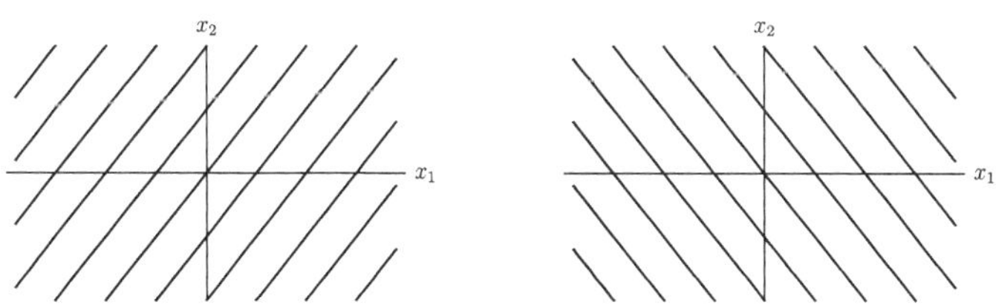

(a) $a > 0$ (b) $a < 0$

Fig. 13.1 Projected characteristics for PDE (13.12) and PDE (13.14)

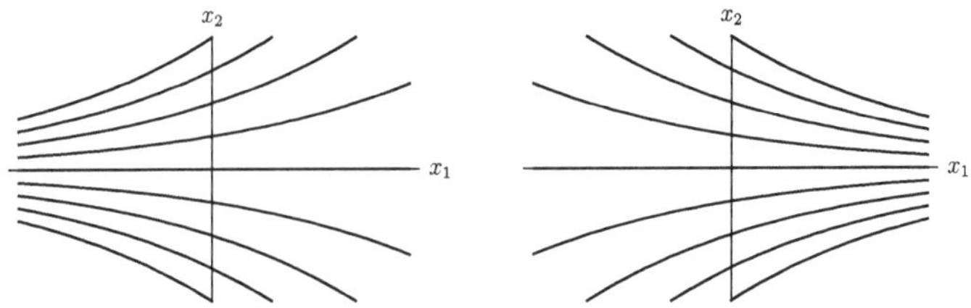

Fig. 13.2 Projected characteristic lines for PDE (13.16) with $a_0 = 0$

Proposition 13.3 (Non-autonomous linear PDE) *Consider the non-autonomous linear first-order PDE, with a zero right-hand side,*

$$\partial_{x_1} y(\mathbf{x}) + (a_0 + a\, x_2)\, \partial_{x_2} y(\mathbf{x}) = 0, \textit{ for each } \mathbf{x} \in \mathbf{X} = \mathbb{R}^2, \tag{13.16}$$

where $a \neq 0$ is an arbitrary constant. The general solution to this equation is

$$y(\mathbf{x}) = f\Big(-\frac{a_0}{a} + \big(x_2 + \frac{a_0}{a}\big)\, e^{-a\, x_1} \Big), \textit{ for each } \mathbf{x} \in \mathbf{X} = \mathbb{R}^2, \tag{13.17}$$

where $f \in C^1(\mathbb{R}^2; \mathbb{R})$ is an arbitrary differentiable function.

Proof In this case, the characteristic ODE for X_1, and the consistency condition for Y, are the same as for PDE (13.12), and the characteristic ODE is $X_2'(s) = a_0 + a X_2(s)$ for X_2. Solving this ODE yields $X_2(s) = -\frac{a_0}{a} + \big(X_2(0) + \frac{a_0}{a}\big)\, e^{a\, s}$. Setting again $s = x_1$, obtaining $x_2 = -\frac{a_0}{a} + \big(X_2(0) + \frac{a_0}{a}\big)\, e^{a\, x_1}$, and substituting in the solution to the consistency ODE $y = Y(s) = Y(0) = f(X_1(0), X_2(0))$, where $X_1(0) = 0$, and as $\big(X_2(0) + \frac{a_0}{a}\big)\, e^{a\, x_1} = x_2 + \frac{a_0}{a}$, yields solution (13.17).

Figure 13.2 depicts the projected characteristic lines for $a_0 = 0$, and for the cases in which $a > 0$ or $a < 0$. We observe that projected characteristics converge to $x_2 = 0$ when $x_1 \downarrow -\infty$, in the first case, and when $x_1 \uparrow \infty$, in the second case. When $a_0 \neq 0$ then the projected characteristics converge to $x_2 = -\frac{a_0}{a}$ instead. These curves correspond to the projection onto space (x_1, x_2) of the solution curves of the PDE (13.16) corresponding again to $y(\mathbf{x}) =$ constant.

Exercise 13.2 Show that the solution to the PDE $\partial_{x_1} y(\mathbf{x}) + a\, \partial_{x_2} y(\mathbf{x}) = b_0 + b\, y(\mathbf{x})$, for $\mathbf{x} \in \mathbb{R}^2$ is

$$y(x_1, x_2) = -\frac{b_0}{b} \left(1 - e^{b\, x_1}\right) + f(x_2 - a\, x_1)\, e^{b\, x_1}, \text{ for each } \mathbf{x} \in \mathbb{R}^2.$$

13.1.2 Cauchy-Dirichlet Problems

We say we have a Cauchy problem if we add a side condition to a PDE, not necessarily on the boundary of set $\mathbf{X}$, and Cauchy-Dirichlet problem if the side condition is on the boundary of set $\mathbf{X}$.

Autonomous PDE A widespread particular case consists in fixing a distribution for one independent variable for a given particular value of the other variable.

Proposition 13.4 *Consider the following Cauchy-Dirichlet problem for PDE* (13.12)

$$\begin{cases} \partial_{x_1} y(\mathbf{x}) + a\,\partial_{x_2} y(\mathbf{x}) = 0, & \mathbf{x} \in \mathbf{X} = \mathbb{R}^2, \\ y(0, x_2) = \phi(x_2), \ \textit{fixed} & \mathbf{x} \in \{x_1 = 0\} \times \mathbb{R}, \end{cases} \tag{13.18}$$

*where $\phi(\cdot)$ is a **known** C^1 function. The solution to this Cauchy problem is the function*

$$y(x_1, x_2) = \phi(x_2 - a\,x_1), \ \textit{for each } \mathbf{x} \in \mathbf{X} = \mathbb{R}^2. \tag{13.19}$$

Proof In the three-dimensional surface S, the constraint defines a curve $(0, x_2, \phi(x_2))$ that has a projection onto the (x_1, x_2)-space characterized by a curve passing through point $\{(0, x_2)\}$. Using the method of characteristics, that was used to determine the characteristic curve C, we parameterize the constraint by a new variable r, such that it defines a direction $\Gamma = \{(0, r) \in \mathbb{R}^2\}$. Introducing now the two parameterizations (associated to the characteristic curve and the initial condition), we have $x_1 = X_1(s, r)$, $x_2 = X_2(s, r)$ and $y = y(X_1(s, r), X_2(s, r)) = Y(s, r)$. The characteristic system and the compatibility condition become the (parameterized by r) three-dimensional initial-value ODE problem

$$\frac{\partial X_1(s, r)}{\partial s} = 1, \ \text{for } s \neq 0, \text{ and } X_1(0, r) = 0,$$
$$\frac{\partial X_1(s, r)}{\partial s} = a, \ \text{for } s \neq 0, \text{ and } X_2(0, r) = r,$$
$$\frac{\partial Y(s, r)}{\partial s} = 0, \ \text{for } s \neq 0, \text{ and } Y(0, r) = \phi(r).$$

The solution to the three independent initial value ODE problems allows us to obtain a relationship between the initial independent variables and the pair (s, r)

$$x_1 = X_1(s, r) = s, \tag{13.20}$$
$$x_2 = X_2(s, r) = a\,s + r, \tag{13.21}$$

and

$$y = Y(s, r) = Y(0, r) = \phi(r), \ \text{for any } s \in \mathbb{R}.$$

To obtain the solution in the original independent variables, $y(\mathbf{x})$, we have reverse the relationships, $s = S(x_1, x_2)$ and $r = R(x_1, x_2)$. Write the solution for the character-

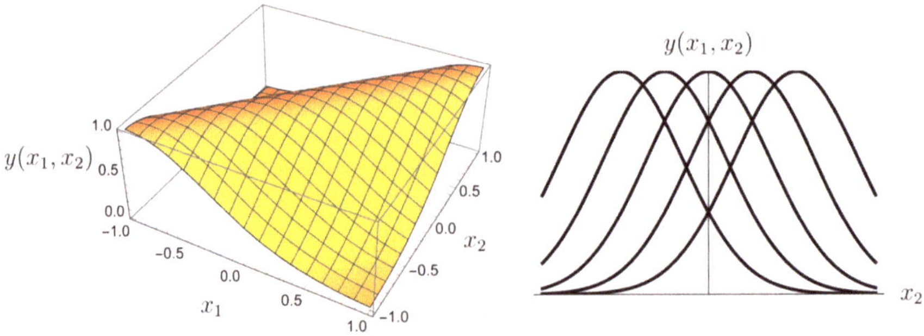

Fig. 13.3 Solution to Example 13.1: 3d plot, and 2d plot in (x_2, y) for $x_1 \in \{-1, -0.5, 0, 0.5, 1\}$

istic system, (13.20)–(13.21), as $(x_1, x_2) = \mathbf{G}(s, r) = (s, a\,s + r)$. Then inverting it yields $(s, r) = \mathbf{G}^{-1}(x_1, x_2) = (x_1, x_2 - a\,x_1)$. Then $y(x_1, x_2) = Y(s, r) = \phi(r) = \phi(x_2 - a\,x_1)$

Example 13.1 If the initial distribution is $y(0, x_2) = \phi(x_2) = e^{-x_2^2}$, then the solution to the Cauchy-Dirichlet problem (13.18) is

$$y(x_1, x_2) = e^{-(x_1 - x_2)^2}, \text{ for } \mathbf{x} \in \mathbf{X} \subseteq \mathbb{R}^2,$$

as illustrated in Fig. 13.3. The projected characteristics are the same as in Fig. 13.1.

Next, we add a growth/decay term to the linear PDE.

Proposition 13.5 *Consider the Cauchy-Dirichlet problem*

$$\begin{cases} \partial_{x_1} y(\mathbf{x}) + a\,\partial_{x_2} y(\mathbf{x}) = b\,y(\mathbf{x}), & \mathbf{x} \in \mathbb{R}^2, \\ y(0, x_2) = \phi(x_2), & \mathbf{x} \in \{x_1 = 0\} \times \mathbb{R}, \end{cases} \tag{13.22}$$

where a and b are constants, and $\phi(\cdot)$ is a ***known*** *C^1 function. Then the solution to this problem is the function*

$$y(\mathbf{x}) = \phi(x_2 - a\,x_1)\,e^{b\,x_1}, \textit{ for each } \mathbf{x} \in \mathbf{X} = \mathbb{R}^2. \tag{13.23}$$

Exercise 13.3 Prove this.

Figure 13.4 illustrates the solution to the problem in Proposition 13.5 when $b > 0$, in Fig. 13.4a, and when $b < 0$, in Fig. 13.4b. In both cases, the solution displays both advection (i.e., transport), and growing when $b > 0$, and transport and decay when $b < 0$..

Non-autonomous PDE and general boundaries Now, as an introduction to the to the method of characteristics to more general PDE, we consider non-autonomous linear PDE and Cauchy-Dirichlet problems with more general side conditions. Let us consider the non-autonomous PDE Cauchy-Dirichlet problem:

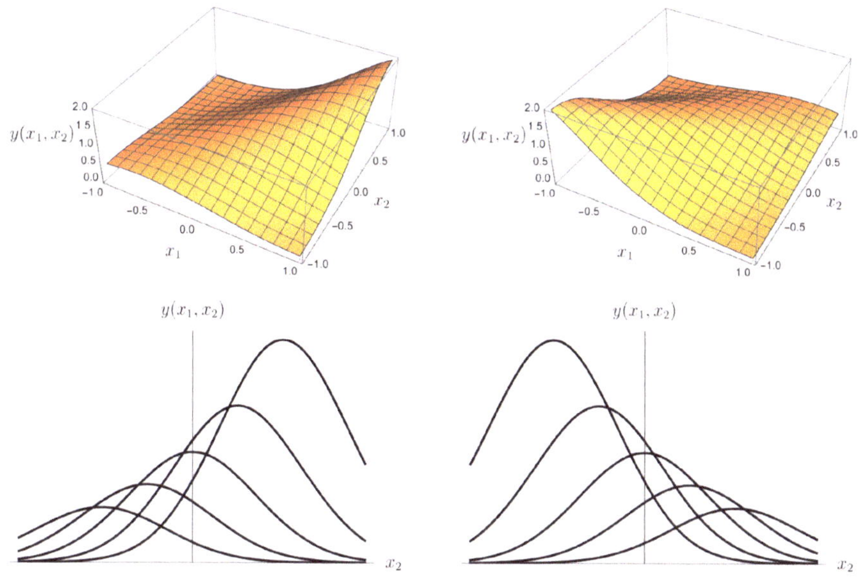

(a) $b > 0$ (b) $b < 0$

Fig. 13.4 Solution for the problem $\partial_{x_1} y + \partial_{x_2} y = b\, y$ and $y(0, x_2) = e^{-x_2^2}$, 3d plot, and 2d plot in (x_2, y) for $x_1 \in \{-1, -0.5, 0, 0.5, 1\}$, for different signs of b

$$\begin{cases} a_1(\mathbf{x})\partial_{x_1} y + a_2(\mathbf{x})\partial_{x_2} y = 0, & \mathbf{x} \in \mathbb{R}^2, \\ y \mid_\Gamma = \phi, & \mathbf{x} \in \Gamma \subset \mathbb{R}^2, \end{cases} \tag{13.24}$$

where $a_1(\mathbf{x})$ and $a_2(\mathbf{x})$ are $C^1(\mathbb{R}^2; \mathbb{R})$ functions, and there is a constraint specified by curve Γ. The solution to this problem depends on the form of the constraint surface Γ.

Let us consider the points in the constrained set parameterised by r, $\Gamma = \{(x_1, x_2) = (\gamma_1(r), \gamma_2(r))\}$ and define

$$A_1(r) \equiv a_1(\gamma_1(r), \gamma_2(r)),$$
$$A_2(r) \equiv a_2(\gamma_1(r), \gamma_2(r)).$$

Definition 13.5 We say that the constraint Γ **is characteristic** if it is tangent to the projected characteristic, and Γ is **non-characteristic** if it is not tangent to the projected characteristic.

For PDE in problem (13.24), Γ is characteristic if

$$\frac{A_1(r)}{A_2(r)} = \frac{\gamma_1'(r),}{\gamma_2'(r)},$$

and Γ is non-characteristic if

$$\frac{A_1(r)}{A_2(r)} \neq \frac{\gamma_1'(r)}{\gamma_2'(r)}. \tag{13.25}$$

Proposition 13.6 (Existence and uniqueness of solutions to problem (13.24)) *Consider the Cauchy-Dirichlet problem* (13.24). *A unique solution exists if* Γ *is non-characteristic in all its domain. If* $\det(\mathbf{G}|_\Gamma) \neq 0$ *then a local solution to the problem exist and is unique, and can be written as*

$$y(x_1, x_2) = \phi(\mathbf{G}|_\Gamma^{-1}(x_1, x_2)). \tag{13.26}$$

Proof We proceed in two step. First step: Apply the same method as in the proof of Proposition 13.4, by introducing the change in coordinates $x_1 = X_1(s, r)$, $x_2 = X_2(s, r)$, that imply $y = y(x_1, x_2) = y(X_1(s, r), X_2(s, r)) = Y(s, r)$, and specifying the characteristic initial value ODE problem

$$\frac{\partial X_1(s, r)}{\partial s} = a_1(X_1(s, r), X_2(s, r)), \text{ for } s \neq 0, \text{ and } X_1(0, r) = \gamma_1(r), \tag{13.27a}$$

$$\frac{\partial X_1(s, r)}{\partial s} = a_2(X_1(s, r), X_2(s, r)), \text{ for } s \neq 0, \text{ and } X_2(0, r) = \gamma_2(r), \tag{13.27b}$$

and the compatibility condition

$$\frac{\partial Y(s, r)}{\partial s} = 0, \text{ for } s \neq 0, \text{ and } Y(0, r) = \phi(r). \tag{13.28}$$

If we solve the characteristic ODE problem, we obtain $(x_1, x_2) = (X_1(s, r), X_2(s, r)) = \mathbf{G}(s, r)$. The solution to the compatibility condition is $Y(s, r) = Y(0, r) = \phi(r)$. It can only be transformed back to the original coordinates, $\mathbf{x}$, if we can obtain r as a function of $\mathbf{x}$.

Second step: the solution to the characteristic ODE problem is locally invertible to $s = S(\mathbf{x})$ and $r = R(\mathbf{x})$ if, invoking the inverse function theorem, we can find $(s, r) = \mathbf{G}^{-1}(\mathbf{x})$ uniquely. This is the case if the Jacobian of $\mathbf{G}$ has a non-zero determinant evaluated at point $(0, r)$. As the Jacobian is

$$D(\mathbf{G})|_\Gamma = \begin{pmatrix} \frac{\partial X_1}{\partial s}(0, r) & \frac{\partial X_1}{\partial r}(0, r) \\ \frac{\partial X_2}{\partial s}(0, r) & \frac{\partial X_2}{\partial r}0, r) \end{pmatrix} = \begin{pmatrix} A_1(r) & \gamma_1'(r) \\ A_2(r) & \gamma_2'(r), \end{pmatrix}$$

then $\det(\mathrm{D}(\mathbf{G}|_\Gamma)) \neq 0$ if condition (13.25) holds. Then we can (at least locally) determine $(s, r)|_{s=0} = \mathbf{G}^{-1}(x_1, x_2)$, and the solution to the PDE problem has the generic form $y(x_1, x_2) = \phi(\mathbf{G}^{-1}\Big|_\Gamma (x_1, x_2))$

Example 13.2 Consider the special case of problem (13.14)

$$\begin{cases} \partial_{x_1} y(\mathbf{x}) + a\, x_2\, \partial_{x_2} y(\mathbf{x}) = 0, & \text{for each } \mathbf{x} \in \mathbb{R}^2, \\ y(0, x_2) = \phi(x_2), & \text{for } \mathbf{x} \in \{x_1 = 0\} \times \mathbb{R}, \end{cases} \tag{13.29}$$

where a is a real number, and $\phi(\cdot)$ is an arbitrary $C^1(\mathbb{R})$ function. Show that the solution exists, is unique, and is

$$y(\mathbf{x}) = \phi\big(x_2\, e^{-a\, x_1}\big), \quad \text{for each } \mathbf{x} \in \mathbb{R}^2. \tag{13.30}$$

The projected characteristics of the problem in example 13.2 are analogous to the ones showed in Fig. 13.2. They satisfy $x_2 \propto e^{a\, x_1}$, and their slope depends on the sign of a: if $a > 0$ then x_2 diverges exponentially for increasing values of x_1; if $a < 0$ then x_2 converges exponentially to zero for increasing values of x_1, and if $a = 0$ then they are parallel. In particular, if the constraint is specified by $y(0, x_2) = \phi(x_2) = e^{-x_2^2}$ then the solution to the Cauchy-Dirichlet problem is

$$y(x_1, x_2) = e^{-(x_2\, e^{-a\, x_1})^2}, \quad \mathbf{x} \in \mathbb{R}^2.$$

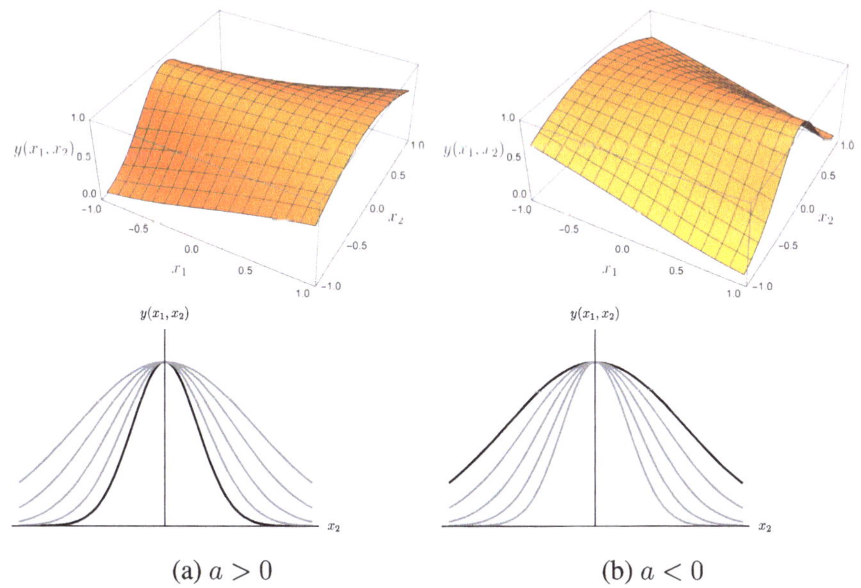

Fig. 13.5 Solution to the problem $\partial_{x_1} y + a\, x_2\, \partial_{x_2} y = 0$, and $y(0, x_2) = e^{-x_2^2}$, 3d plots and 2d plots for $x_1 \in \{-1, -0.5, 0, 0.5, 1\}$. The darker curves illustrate the case $x_1 = -1$.

Again the behavior of the solution depends on the sign of a: if $a > 0$, as illustrated in Fig. 13.5a, the solution displays rarefaction, and if $a < 0$, as illustrated in Fig. 13.5b for it displays displaying concentration, along the x_2 axis.

Example 13.3 Consider the special case of problem (13.14)

$$\begin{cases} \partial_{x_1} y(\mathbf{x}) + a\, x_2\, \partial_{x_2} y(\mathbf{x}) = b\, y(\mathbf{x}), & \text{for each } \mathbf{x} \in \mathbb{R}^2, \\ y(0, x_2) = \phi(x_2), & \text{for } \mathbf{x} \in \{x_1 = 0\} \times \mathbb{R}, \end{cases} \tag{13.31}$$

where a and b are real numbers, and $\phi(\cdot)$ is an arbitrary $C^1(\mathbb{R})$ function. Show that the solution exists, is unique, and is

$$y(\mathbf{x}) = \phi\big(x_2\, e^{-a\, x_1}\big)\, e^{b\, x_1}, \quad \text{for each } \mathbf{x} \in \mathbb{R}^2. \tag{13.32}$$

13.1.3 Cauchy-Dirichlet Problems for Semi-linear PDE

The Cauchy-Dirichlet problem for a semi-linear equation (13.3), and an associated side-condition in a surface $\Gamma \in \mathbb{R}^2$ is

$$\begin{cases} a_1(\mathbf{x})\, \partial_{x_1} y(\mathbf{x}) + a_2(\mathbf{x})\, \partial_{x_2} y(\mathbf{x}) = b\big(\mathbf{x}, y(\mathbf{x})\big), & \mathbf{x} \in \mathbb{R}^2, \\ y\,|_\Gamma = \phi, & \mathbf{x} \in \Gamma \subset \mathbb{R}^2, \end{cases} \tag{13.33}$$

where $a_1(\mathbf{x})$, and $a_2(\mathbf{x})$ are known $C^1(\mathbb{R}^2; \mathbb{R})$ functions, and $b(y, \mathbf{x})$ is a known $C^1(\mathbb{R}^3; \mathbb{R})$ function that is non-linear in y. To apply the method of characteristics, we introduce again a parameterisation associated with the characteristic surface and the constraint surface by a pair (s, r) and set $x_1 = X_1(s, r)$, $x_2 = X_2(s, r)$, and $y = Y(s, r) = y(X_1(s, r), X_2(s, r))$. The characteristic system is again (13.27a)–(13.27b). However, the consistency condition is now

$$\frac{\partial Y(s, r)}{\partial s} = b\,(s, r, Y(s, r))\,, \quad \text{for each } s \neq 0, \text{ and } Y(0, r) = \phi(r), \quad \text{for } s = 0,$$

that, differently from the linear case, in Eq. (13.28), features a non-linear ODE.

Assuming that curve Γ is non-characteristic, from the solution to the characteristic ODE, we can obtain a relationship $(x_1, x_2) = \mathbf{G}(s, r)$. If det $(\mathbf{G}\big|_\Gamma) \neq 0$, then we can obtain uniquely $(s, r) = \mathbf{G}^{-1}\Big|_\Gamma(\mathbf{x})$. In this case, the solution to the PDE problem (13.33) exists and is unique.

The main difference of semi-linear as regards linear first-order PDE, consists in the fact that the right-hand side of the compatibility condition is a non-linear ODE over

Y and not a linear ODE. This can induce different types of phenomena, in particular **blow-up** behavior, the possibility of $y(\cdot)$ be infinite for specific finite values of $\mathbf{x}$.[6]

Example 13.4 Consider the Cauchy-Dirichlet problem

$$\begin{cases} \partial_{x_1} y(\mathbf{x}) + a\,\partial_{x_2} y(\mathbf{x}) = y(\mathbf{x})^2, & \text{for each } \mathbf{x} \in \mathbb{R}^2 \\ y(0, x_2) = \phi(x_2), & \text{for } \mathbf{x} \in \{x_1 = 0\} \times \mathbb{R}. \end{cases} \tag{13.34}$$

Solving the characteristic ODE yields $s = x_1$, and $r = x_2 - a\,x_1$. The solution to the consistency condition $\frac{\partial Y(s,r)}{\partial s} = Y(s,r)^2$, for $Y(0,r) = \phi(r)$, is $Y(s,r) = \frac{\phi(r)}{1 - s\,\phi(r)}$. Making the transformation from (s, r) to $\mathbf{x}$ yields the solution to the Cauchy-Dirichlet problem

$$y(\mathbf{x}) = \frac{\phi(x_2 - a\,x_1)}{1 - x_1\,\phi(x_2 - a\,x_1)}, \text{ for each } \mathbf{x} \in \mathbb{R}^2.$$

In this case, we have blow-up if there are $\mathbf{x}$ such that $x_1\,\phi(x_2 - a\,x_1) = 1$. Assuming that $\phi(x_2) = e^{-x_2^2}$ the solution is unbounded whenever $x_1 = e^{(x_2 - a\,x_1)^2}$. This is illustrated in projected characteristics in Fig. 13.7b, and in the solution drawn in Fig. 13.7a.

13.1.4 Cauchy-Dirichlet Problems for Quasi-linear PDE

Quasi-linear PDE present some new types of features of solutions to PDE that are very different from non-linear ODE. To address this issue, consider the Cauchy-Dirichlet problem for a quasi-linear PDE (13.2),

$$\begin{cases} a_1(\mathbf{x}, y)\partial_{x_1} y + a_2(\mathbf{x}, y)\partial_{x_2} y = b(y, \mathbf{x}), & \mathbf{x} \in \mathbb{R}^2, / \Gamma \\ y \mid_\Gamma = \phi, & \mathbf{x} \in \Gamma \subset \mathbb{R}^2, \end{cases} \tag{13.35}$$

where $a_1(.)$, and $a_2(.)$ and $b(.)$ are $C^1(\mathbb{R}^3; \mathbb{R})$ functions. Introducing again a transformation of variables from $\mathbf{x}$ to (s, r), yields the characteristic and the compatibility ODE problems

[6] Comparing the types of dynamics for scalar linear when compared non-linear ODE, dealt with in Chaps. 2, 3, 6, and 7, can convince us that there is a much larger number of different types of solutions.

$$\frac{\partial X_1(s,r)}{\partial s} = a_1\left(X_1(s,r), X_2(s,r), Y(s,r)\right), \text{ if } s > 0 \text{ and } X_1(0,r) = \gamma_1(r), \tag{13.36a}$$

$$\frac{\partial X_2(s,r)}{\partial s} = a_2\left(X_1(s,r), X_2(s,r), Y(s,r)\right), \text{ if } s > 0 \text{ and } X_2(0,r) = \gamma_2(r), \tag{13.36b}$$

$$\frac{\partial Y(s,r)}{\partial s} = b\left(X_1(s,r), X_2(s,r), Y(s,r)\right), \text{ if } s > 0 \text{ and } Y(0,r) = \phi(r). \tag{13.36c}$$

This system, differently from the previous cases, lost their recursive structure, in the sense that we cannot separate the determination of the solutions for $X_1(.)$ and $X_2(.)$ from $Y(.)$: the two independent variables, X_1 and X_2, and the dependent variable, Y, are coupled.

Now the non-characteristic conditions for Γ are more involved because all three differential equations depend on (X_1, X_2, Y) and the conditions for the application of the non-characteristic condition may not hold.

The geometric meaning is the following: while for linear and semi-linear PDE, the characteristic lines are parallel and do not cross, for the quasi-linear PDE the characteristic lines can cross. This is the existence of singular points, and at singularity points, the uniqueness and even the existence of solutions are problematic.

Example 13.5 A well-known quasi-linear first-order PDE is the inviscid Burgers (1948) equation. The following Cauchy-Dirichlet problem is a particular version with our previously used side condition

$$\begin{cases} \partial_{x_1} y(\mathbf{x}) + y(\mathbf{x})\, \partial_{x_2} y(\mathbf{x}) = 0, & \mathbf{x} \in \mathbb{R}^2/\Gamma, \\ y(0, x_2) = \phi(x_2), & \mathbf{x} \in \Gamma = \{x_1 = 0\} \times \mathbb{R}. \end{cases} \tag{13.37}$$

We can prove that Γ is non-characteristic: as $A_1(r) = 0$, $A_2(r) = \phi(r)$, $\gamma_1'(r) = 0$, and $\gamma_2'(r) = 1$, then $A_1(r)\,\gamma_2'(r) - A_2(r)\,\gamma_1'(r) = 1 - 0 \times \phi(r) = 1 \neq 0$. The solution to the characteristic ODE problem, (13.36a)–(13.36b), allows for finding s explicitly. However, r can only be implicitly obtained from equation $h(r, x_1, x_2) \equiv r + x_1\,\phi(r) - x_2 = 0$. As the solution to the consistency condition is $Y(s,r) = Y(0,r) = \phi(r)$, then, the solution to the Burgers problem (13.37) is implicitly given by

$$y(\mathbf{x}) = \phi(r), \; r = R(\mathbf{x})$$

where $r = R(\mathbf{x}) \iff h(r, \mathbf{x}) = 0$. There are three types of solutions:[7] first, if the projected characteristics are parallel, then the solution will be of type $y(\mathbf{x}) = \text{const}$ along every characteristic; second, if $\phi(r)$ is locally increasing, then for two different values of r, $r_1 > r_2$, then $\phi(r_1) > \phi(r_2)$ and the solution $y(\mathbf{x})$ will display rarefaction waves along direction x_2 for increasing (or decreasing) levels of x_1, similarly to the

[7] For a thorough discussion, see Olver (2014, Sect. 2.3).

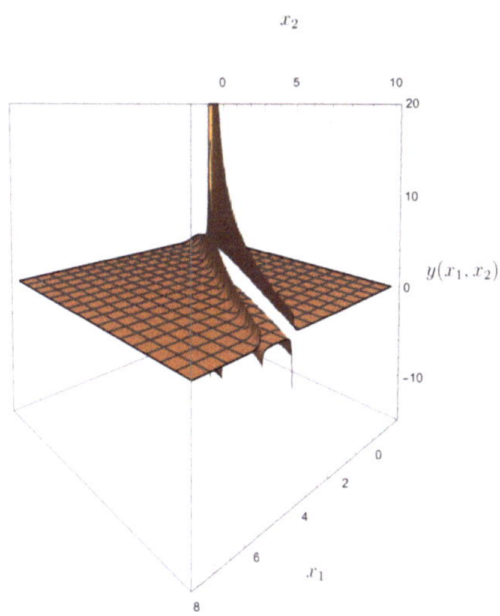

(a) Solution

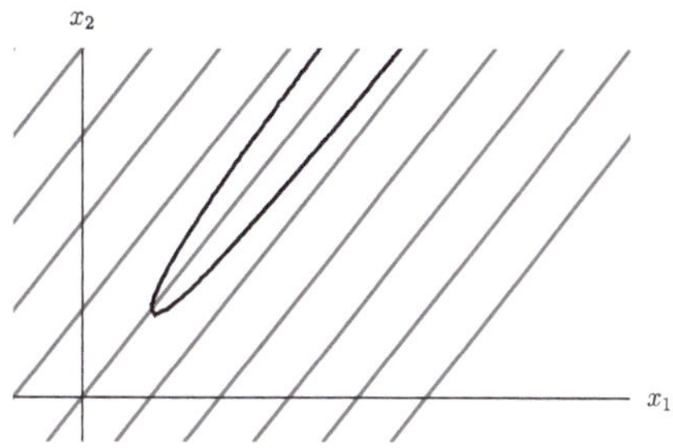

(b) Projected characteristics and blow-up curve

Fig. 13.6 Solution for the problem $\partial_{x_1} y + a\, \partial_{x_2} y = y^2$, and $y(0, x_2) = e^{-x_2^2}$, 3d plots and projected characteristics

projected characteristics in Fig. 13.2; third, if $\phi(r)$ is locally decreasing, then $r_1 > r_2$ imply $\phi(r_1) < \phi(r_2)$ and the solution $y(\mathbf{x})$ will display compression waves along direction x_2 for increasing (or decreasing) levels of x_1, and the solution to $y(\mathbf{x})$ will not be unique. In the last case, introducing some solvability conditions, gives birth to **shock waves**, which is a type of behavior not presented in linear hyperbolic PDE.

We can find singular points, separating the previous types of solutions by solving system $h(r, \mathbf{x}) = \partial_r h(r, \mathbf{x}) = 0$, that is

$$\begin{cases} x_2 = r + x_1\, \phi(r), \\ 1 + x_1\, \phi'(r) = 0. \end{cases}$$

If the side condition is $\phi(r) = e^{-r^2}$, then $\phi'(r) = -2\, r\, \phi(r)$ and, as $r^2 - x_2\, r + \frac{1}{2} = 0$. We find two critical values $r^-(x_2) = \frac{x_2}{2} - \sqrt{\left(\frac{x_2}{2}\right)^2 - \frac{1}{2}}$ and $r^+(x_2) = \frac{x_2}{2} + \sqrt{\left(\frac{x_2}{2}\right)^2 - \frac{1}{2}}$. Therefore, there are two loci $X_1^+(x_2) = r^+(x_2) e^{r^-(x_2)^2}$ and $X_1^-(x_2) = r^-(x_2) e^{r^+(x_2)^2}$, such that: if $X_1^-(x_2) < x_1 < X_1^+(x_2)$ then the solution to the Cauchy-Dirichlet problem is unique; if $x_1 < X_1^-(x_2)$ or $x_1 > X_1^+(x_2)$ then there are multiple solutions to the Cauchy-Dirichlet problem, and $(x_1, x_2) \in \left\{(x_1^-, x_2^-), (x_1^+, x_2^+)\right\}$, where $x_1^\pm = \pm\frac{\sqrt{2}\, e^{\frac{1}{2}}}{2}$ and $x_2^\pm = \pm\sqrt{2}$, are critical points separating the two previous cases.

(a) Projected characteristic, and $\mathbf{x}^*$

(b) Solution curves

Fig. 13.7 Solution to Burgers PDE with $y(0, x_2) = e^{-x_2^2}$, projected characteristics, and solution curves in a $x_2, y)$-diagram for $x_1 \in \{-2.5, x_1^-, -0.5, 0.5, x_1^+, 2.5\}$

Figure 13.7a shows the projected characteristics, illustrating the two previous types of solutions: in the first case, the solutions are similar to the previous linear and non-linear cases, in which the projected characteristics do not cross, but they are compressing for higher values of x_1; in the second case, given a value of x_1 there are values of x_2 for which there are multiple solutions for y. The curve in red displays the set of critical points in set $\mathbf{X}$ separating unique from multiple solutions to y. The two dots correspond to the onset of shock waves, as illustrated by the thick curves in Fig. 13.7b.

13.2 The Linear Equation in the Semi-Infinite Domain and Laplace Transform Methods

In the previous cases, we assumed that the independent variables were defined in the space $\mathbf{X} = \mathbb{R}^2$. Next, we show that if $\mathbf{X} \subset \mathbb{R}^2$ the solution to the first-order PDE, and its associated problems, varies both in terms of the existence of solutions and of the methods of solution, if they exist. In this case, we can have solutions that are not elementary functions (single-valued continuous mappings) but generalized functions (also called weak solutions).

For applications in economics, the semi-infinite domain is of interest. A convenient way to solve first-order PDE over the semi-infinite domain, where $\mathbf{X} = \mathbb{R}_+^2$, is to use the **Laplace transform method** instead of the method of characteristics. In order to apply it, we take one independent variable as a parameter (for instance x_1) and keep the other as an independent variable (for instance x_2).[8] Laplace transforms are a convenient tool in this case because the domain of the transformation is the semi-infinite space $\mathbb{R}_+ = [0, \infty)$. The steps for obtaining a solution are the following: first, we apply Laplace transforms to the original dependent variable, $y(\mathbf{x})$, to transform both the PDE into a parameterized ODE over $Y(\xi)$, and the side-conditions; second,

[8] The choice can be done as a mean to simplify the solution to the problem, given the constraints.

we solve the ODE problem for $Y(\xi)$; and, at last, we apply the inverse Laplace transforms to obtain the solution for $y(\mathbf{x})$.

To illustrate this, assume that $\mathbf{X} = \mathbb{R}^2_{++}$, that is $x_1 > 0$ and $x_2 > 0$, and consider the boundary-value problem (see Eq. (13.6) for the general definition)

$$\begin{cases} \partial_{x_1} y(\mathbf{x}) + a\, \partial_{x_2} y(\mathbf{x}) = 0, & \mathbf{x} \in \mathbb{R}^2_{++}, \\ y(x_1, 0) = \phi_1(x_1), & \mathbf{x} \in \mathbb{R}_{++} \times \{x_2 = 0\}, \\ y(0, x_2) = \phi_2(x_2), & \mathbf{x} \in \{x_1 = 0\} \times \mathbb{R}_{++}, \end{cases} \tag{13.38}$$

where functions $\phi_1(\cdot)$ and $\phi_2(\cdot)$ are known.

Proposition 13.7 *The solution to boundary-value problem* (13.38) *is*

$$y(\mathbf{x}) = \phi_2(x_2 - a\, x_1)\, H(x_2 - a\, x_1) + a\, \mathcal{L}^{-1} \left[\int_0^{x_1} \phi_1(s) e^{-a\xi(x_1 - s)} ds \right] (x_2) \tag{13.39}$$

where[9]

$$\phi_2(x_2 - a\, x_1)\, H(x_2 - a\, x_1) = \begin{cases} 0, & \text{if } x_2 - a\, x_1 \leq 0, \\ \phi_2(x_2 - a\, x_1), & \text{if } x_2 - a\, x_1 > 0, \end{cases}$$

and $\mathcal{L}^{-1}[f(x)](y)$ *is the inverse Laplace transform.*

Proof Let $Y(x_1, \xi)$ be the Laplace transform (see Appendix 13.6) of $y(\mathbf{x})$ taking variable x_1 as a parameter, that is $Y(x_1, \xi) \equiv \mathcal{L}[y(\mathbf{x})](\xi) = \int_0^\infty e^{-\xi\, x_2} y(x_1, x_2)\, dx_2$, where $\xi > 0$. The Laplace transform of the left-hand side of the PDE is

$$\begin{aligned} \mathcal{L}[\partial_{x_1} y(\mathbf{x}) + a\, \partial_{x_2} y(\mathbf{x})](\xi) &= \int_0^\infty e^{-\xi\, x_2} \left(\partial_{x_1} y(\mathbf{x}) + a\, \partial_{x_2} y(\mathbf{x}) \right) dx_2 \\ &= Y'(x_1, \xi) + a \left(\xi\, Y(x_1, \xi) - y(x_1, 0) \right), \\ &- Y'(x_1, \xi) + a \left(\xi\, Y(x_1, \xi) - \phi_1(x_1) \right), \end{aligned}$$

where $Y'(x_1, \xi) \equiv \frac{\partial Y(x_1, \xi)}{\partial \xi}$, and the Laplace transform of the side-condition dependent on x_2 is[10]

$$\mathcal{L}[y(0, x_2)](\xi) = \int_0^\infty e^{-\xi\, x_2} \phi_2(x_2)\, dx_2 = \Phi_2(\xi).$$

[9] The Heaviside generalized function is

$$H(z) = \begin{cases} 0, & \text{if } z \leq 0, \\ 1, & \text{if } z > 0. \end{cases}$$

[10] See Appendix 13.6 for the Laplace transform of a derivative.

Then we obtain an equivalent initial-value ODE for the transformed variable

$$\begin{cases} Y'(x_1, \xi) = -a\,\xi\, Y(x_1, \xi) + a\,\phi_1(x_1), & \text{for } x_1 > 0 \\ Y(0, \xi) = \Phi_2(\xi), & \text{for } x_1 = 0. \end{cases}$$

The solution is

$$Y(x_1, \xi) = \Phi_2(\xi) e^{-a\,\xi\, x_1} + a \int_0^{x_1} \phi_1(s) e^{-a\,\xi\,(x_1 - s)} ds.$$

The solution, as a function of the primitive variables, is obtained by transforming back taking the inverse Laplace transform (see again Appendix 13.6), $y(\mathbf{x}) = \mathcal{L}^{-1}\big[Y(x_1, \xi)\big](x_2)$. Therefore,

$$\begin{aligned} y(x_1, x_2) &= \mathcal{L}^{-1}\big[e^{-a\,\xi\, x_1}\, \Phi_2(\xi)\big](x_2) + a\,\mathcal{L}^{-1}\Big[\int_0^{x_1} \phi_1(s)\, e^{-a\,\xi\,(x_1 - s)} ds\Big] \\ &= \phi_2(x_2 - a\,x_1)\, H(x_2 - a\,x_1) + a\,\mathcal{L}^{-1}\Big[\int_0^{x_1} \phi_1(s)\, e^{-a\,\xi\,(x_1 - s)} ds\Big], \end{aligned}$$

where, the following fact was used

$$\begin{aligned} e^{-a\,x_1\,\xi}\, \Phi_2(\xi) = e^{-a\,x_1\,\xi} \int_0^{\infty} e^{-\xi s}\, \phi_2(s)\, ds &= \int_0^{\infty} e^{-\xi\,(s + a\,x_1)}\, \phi_2(s)\, ds \\ &= \int_{a\,x_1}^{\infty} e^{-\xi\, x_2} \phi_2(x_2 - a\,x_1) dx_2, \ \text{setting}(s + a\,x_1 = x_2) \\ &= \int_0^{\infty} e^{-\xi\, x_2} \phi_2(x_2 - a\,x_1)\, H(x_2 - a\,x_1)\, dx_2 = \\ &= \mathcal{L}\big[\phi_2(x_2 - a\,x_1)\, H(x_2 - a\,x_1)\big](\xi). \end{aligned}$$

Example 13.6 To obtain an illustration of the solution consider the case: $a > 0$, $\phi_2(x_2) = 0$, and $\phi_1(x_1) = \phi_{1,0}$ is a constant. In this case

$$Y(x_1, \xi) = a\,\phi_{1,0} \int_0^{x_1} e^{-a\,\xi\,(x_1 - s)}\, ds = \phi_{1,0} \left(\frac{1 - e^{-a\,\xi\, x_1}}{\xi}\right),$$

and

$$y(x_1, x_2) = \phi_{1,0}\, \mathcal{L}^{-1}\left[\frac{1 - e^{-a\,\xi\, x_1}}{\xi}\right](x_2) = \phi_{1,0}\, H(a\,x_1 - x_2).$$

Then the solution (13.39) is the function

$$y(x_1, x_2) = \begin{cases} \phi_{1,0}, & \text{for } 0 < x_2 < a\,x_1, \\ 0, & \text{for } x_2 \geq a\,x_1, \end{cases}$$

that is constant for $\mathbf{x} \in \{\mathbf{x} \in \mathbb{R}^2_{++} : 0 < x_2 < a\,x_1\}$, and it is equal to zero elsewhere.

Example 13.7 If, instead, we had the side condition $\phi_2(x_2) = e^{b\,x_2}$ and $\phi_1(x_1) = 0$, the solution (13.39) would satisfy $y(x_1, x_2) = \phi(x_2 - a\,x_1)\,H(x_2 - a\,x_1) = e^{b\,(x_2 - a\,x_1)}$ $H(x_2 - a\,x_1)$, and therefore,

$$y(x_1, x_2) = \begin{cases} 0 & \text{for } 0 < x_2 \le a\,x_1 \\ e^{b\,(x_2 - a\,x_1)} & \text{for } x_2 > a\,x_1. \end{cases}$$

In both cases, the projected characteristics are illustrated in Fig. 13.3a.

Next, we consider a Dirichlet PDE problem with a homogeneous right-hand side

$$\begin{cases} \partial_{x_1} y(\mathbf{x}) + a\,\partial_{x_2} y(\mathbf{x}) = b\,y(\mathbf{x}), & \text{for each } \mathbf{x} \in \mathbb{R}^2_{++}, \\ y(x_1, 0) = e^{\beta\,x_1}, & \text{for } x_1 > 0, \\ y(0, x_2) = 0, & \text{for } x_2 > 0. \end{cases} \tag{13.40}$$

Using the Laplace transform method, the transformed ODE problem

$$\begin{cases} Y'(x_1, \xi) = (b - a\,\xi)\,Y(x_1, \xi) + a\,e^{\beta\,x_1}, & \text{for } x_1 > 0, \\ Y(0, \xi) = 0, & \text{for } x_1 = 0, \end{cases}$$

has solution

$$Y(x_1, \xi) = a\,e^{(b - a\,\xi)\,x_1} \int_0^{x_1} e^{(\beta - b + a\,\xi)\,s}\,ds = a\,e^{b\,x_1}\,\frac{e^{(\beta - b)\,x_1} - e^{-a\,x_1\,\xi\,\xi}}{\beta - b + a\,\xi}.$$

Applying the inverse Laplace transform, we find

$$y(x_1, x_2) = \mathcal{L}^{-1}\big[Y(x_1, \xi)\big](x_2) = e^{\frac{a\,\beta\,x_1 + (b - \beta)\,x_2}{a}}\,(1 - H(x_2 - a\,x_1)).$$

The solution to problem (13.40) is

$$y(\mathbf{x}) = e^{\frac{b}{a}\,(a\,x_1 - x_2)}(1 - H(x_2 - a\,x_1)) = \begin{cases} e^{\frac{a\,\beta\,x_1 + (b - \beta)\,x_2}{a}}, & \text{if } 0 < x_2 < a\,x_1, \\ 0, & \text{if } x_2 > a\,x_1. \end{cases} \tag{13.41}$$

A graphical representation of the solution for $a = 1$, $b = 1$ and $\beta = 2$ is presented in Fig. 13.8b. The projected characteristics are shown in Fig. 13.6a but, differently from that case where y is constant, now the solution displays growth at the rate a along the characteristic lines. We can also observe in Fig. 13.6a the characteristic line that represents the boundary between the two branches of the solution.

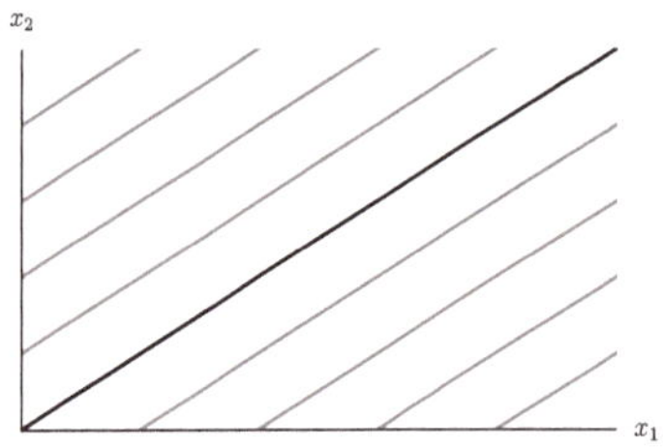

(a) Projected characteristics

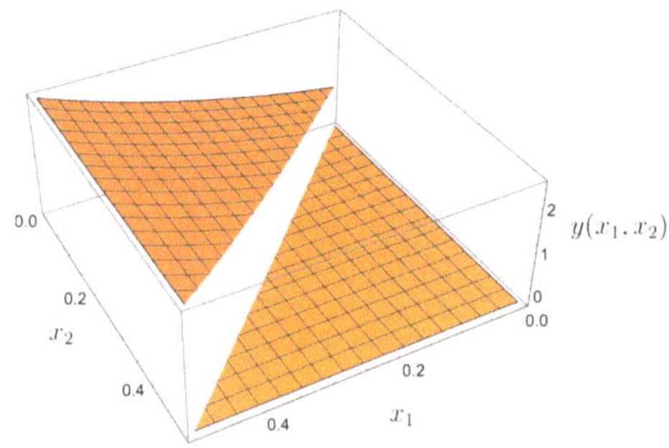

(b) Solution

Fig. 13.8 Boundary-value problem $\partial_{x_1} y + \partial_{x_2} y = 2\, y$, $y(0, x_2) = 0$, $y(x_1, 0) = e^{2\,x_1}$, for $\mathbf{x} > \mathbf{0}$

13.3 Evolution Equations

Let the vector of independent variables be $(t, x) \in \mathrm{T} \times \mathrm{X} \subseteq \mathbb{R}_+ \times \mathbb{R}$, where t denotes time, and x denotes another independent variable belonging to an ordered set X. A time-evolving distribution, $y(t, x)$, is generated by the **evolution equation**

$$\partial_t y(t, x) + \partial_x \big(g(y(t, x))\big) = 0, \text{ for each } (t, x) \in \mathrm{T} \times \mathrm{X}. \tag{13.42}$$

The general solution to the evolution equation describes the evolution of a distribution, or wave, with support in set X, traveling over time. A very common side-condition is

$$\int_{\mathrm{X}} y(t, x)\, dx = \bar{y}, \text{ for each } t \in \mathrm{T} \subseteq \mathbb{R}_+,$$

where $\bar{y}$ is a constant. In this case, we say a **conservation law** is satisfied. Given an initial function $y(0, x) = \phi(x)$, for $t = 0$, such that $\int_{\mathrm{X}} \phi(x)\, dx = \bar{y}$, the solution $y(t, x)$ displays the same mass over time. In particular, this is a model for a distribution if $\bar{y} = 1$.[11]

Assuming that a solution exists, we may be interested in characterizing the asymptotic time behavior of the distribution $\lim_{t\to\infty} y(t, x)$. This is the analog to studying the long-run behavior of an ODE. However, while the solution of an ODE can converge asymptotically to a steady state (a point) the solution of a PDE can converge asymptotically to a function (or a steady state distribution), can converge to a degenerate distribution (a Dirac delta generalized function), or can be asymptotically unbounded.

Qualitative theory of the dynamics for first-order PDE. We are unaware of the existence of a Grobmann-Hartmann theorem for quasi-linear and non-linear PDE, that would allow for a general qualitative theory for PDE as the one that exists

---

[11] We could extend the model to a case in which an aggregate is constrained over time a moving mass, $\int_{\mathrm{X}} y(t, x)\, dx = \bar{y}(t)$, for $t \in \mathrm{T}$.

for ODE.[12] However, according to Dafermos (2000, p. 13) we can can expect the following types of dynamics for generic **non-linear evolution** first-order scalar PDE

(1) blow-up if the solution becomes infinite when a certain surface of X is reached in finite time;
(2) globally bounded solutions even in the case in which x goes to infinite in infinite time;
(3) progressive concentration over time, tending asymptotically to a degenerate distribution concentrated at a finite value for x, x^*, in infinite time;
(4) shock-waves such that he solution becomes non-smooth and multivalued, if there is a collision to a particular surface;
(5) or rarefaction waves such that the distribution becomes increasingly dispersed.

For linear, semi-linear, and well-posed PDE, the distributional dynamics that characterize their solution can be explicitly obtained, and therefore, the cross-section dynamics can be explicitly described. Therefore, for the case of the **linear evolution** first-order scalar ODE

$$\partial_t y(t,x) + b(x)\,\partial_x y(t,x) = c\,y(t,x), \text{ for each } (t,x) \in \mathrm{T} \times \mathrm{X}, \tag{13.43}$$

we can be more specific. According to Olver (2014, p. 29), the characteristic curves can have one of the following generic properties:

(1) for each point $(t, x) \in \mathrm{T} \times \mathrm{X}$, there is one unique characteristic curve passing through that point;
(2) characteristic curves cannot cross;
(3) if $c = 0$ and $x = h(t)$ is a characteristic curve, then $x = h(t) + k$ is also a characteristic curve, or, if $c \neq 0$, the characteristic curves satisfy $x = k\,e^{-ct}$, for $k \in \mathbb{R}$;
(4) the path traced out by a characteristic curve $x = h(t)$, for increasing values of t, always moves in the same direction, and cannot change the direction of propagation;
(5) as $t \to \infty$ the characteristic curve either converges to a fixed point, $x(t) \to x^*$ where $c(x^*) = 0$, or go to $\pm\infty$ either in finite or infinite time. The first is a case of progressive concentration, and the second is a case of rarefaction.

To illustrate the last point, consider the problem

$$\begin{cases} \partial_t y + \beta\left(x - x^*\right)\partial_x y = 0, \text{ for each } (t,x) \in \mathrm{T} \times \mathbb{R}, \\ y(0,x) = \phi(x), \text{ for } (t,x) \in \{t = 0\} \times \mathbb{R}. \end{cases} \tag{13.44}$$

The solution to this Cauchy-Dirichlet problem is

$$y(t,x) = \phi\Big(x^* + (x - x^*)\,e^{-\beta t}\Big), \text{ for each } (t,x) \in \mathrm{T} \times \mathbb{R}. \tag{13.45}$$

[12] See Sect. 5.2.

The asymptotic dynamics of function (13.44) depends on the sign of β: if $\beta > 0$ there is a rarefaction wave, i.e., $\lim_{t\to\infty} y(t,x) = \phi(x^*)$ which is a constant for all $x \in X$, or, if $\beta < 0$ there is a progressive concentration, i, e. $\lim_{t\to\infty} y(t,x) = \delta_y(x^*)$ is a degenerate distribution whose mass is concentrated into point $x = x^*$.

13.4 Applications to Economics

Example 13.8 (*The transport equation*) An important first-order PDE is the **transport or continuity equation**, of a function $y : \mathrm{T} \times \mathrm{X} \to \mathrm{Y} \subseteq \mathbb{R}$,

$$\partial_t y(t,x) + v(t,x)\,\partial_x y(t,x) = 0, \text{ for each } (t,x) \in \mathrm{T} \times \mathbb{R}, \tag{13.46}$$

where X can be the domain of geographical locations or any other continuum of ordered elements. The **transport equation in conservative form** is

$$\partial_t y(t,x) + \partial_x \left(u(t,x)\,y(t,x)\right) = 0, \text{ for each } (t,x) \in \mathrm{T} \times \mathrm{X}, \tag{13.47}$$

satisfies the property $\int_{-\infty}^{\infty} y(t,x)dx = \bar{y}$ for each $t \in \mathrm{T}$. If $\bar{y} = 1$ and $0 < y(t,x) < 1$, then $y(\cdot)$ is analogous to a probability distribution that evolves over time.

If function $u(\cdot)$ is time independent and linear, $u = \mu\,x$, then the Cauchy-Dirichlet problem for a conservative PDE is

$$\begin{cases} \partial_t y(t,x) + \partial_x \left(\mu\,x\,y(t,x)\right) = 0, & \text{for each } (t,x) \in \mathrm{T} \times \mathrm{X}, \\ y(0,x) = \phi(x), & \text{for each } (t,x) \in \{t=0\} \times \mathrm{X}, \end{cases} \tag{13.48}$$

where $\mu \neq 0$, and $\int_{\mathrm{X}} \phi(x)dx = 1$, has the solution

$$y(t,x) = e^{-\mu\,t}\,\phi\Big(x\,e^{-\mu\,t}\Big). \text{ for each } (t,x) \in \mathrm{T} \times \mathbb{R}. \tag{13.49}$$

We can obtain this solution by several different methods. First, we can directly apply Eq. (13.32), obtained by the method of characteristics. Second, we can also find it by applying a Laplace transformation method. Third, we can reduce it to a simpler linear PDE by transforming both $x = X(z) = e^{\mu\,z}$ and $y(t, X(z)) = e^{-\mu\,t}\,v(t,z)$, if $\mu \neq 0$, yielding an equivalent Cauchy problem

$$\begin{cases} \partial_t v(t,z) + \partial_y v(t,z) = 0, & (t,z) \in \mathbb{R}_+ \times \mathbb{R}, \\ v(0,z) = \phi\big(X(z)\big), & (t,z) \in \{t=0\} \times \mathbb{R}. \end{cases}$$

The PDE has the same form of (13.3), with $a = 1$, and, from (13.18), the solution to the transformed Cauchy-Dirichlet problem is $v(t,z) = \phi(X(z-t)) = \phi\Big(e^{\mu(z-t)}\Big)$. To obtain the solution in the original variables, in Eq. (13.49), we need to transform back, using the inverse transformation $z = Z(x) = \frac{\ln x}{\mu}$. Then $y(t, x) = e^{-\mu t}\,\phi\Big(e^{\mu\,(Z(x)-t)}\Big) = e^{-\mu t}\,\phi\left(x\,e^{-\mu t}\right)$.

Example 13.9 (*Growth and inequality dynamics*) Consider an economy in which there are two types of capital, financial capital and human capital and to simplify assume they are perfect EP substitutes. It also assumed that households are heterogeneous as regards their net wealth. Let the net wealth belong to a continuum $\mathrm{K} \in \mathbb{R}$ and let $H(t, k)$ be the number of households with net wealth $k \in \mathrm{K}$, at time $t \in \mathrm{T} = [0, \infty)$. If we order households by their net wealth, the number of households at time t is $H(t) = \int_{\mathrm{K}} H(t, k)dk$. The proportion of households with wealth smaller that k at time t is $N(t, k) = \frac{H(t,k)}{H(t)}$, is an integrand for the cumulative distribution function satisfying $\int_{\mathrm{K}} dN(t, k) = 1$, for each time $t \in \mathrm{T}$. Then, setting $dN(t, k) = n(t, k)\,dk$, the household density function at time t, $n(t, k)$, satisfies $0 < n(t, k) < 1$, for each $(t, k) \in \mathrm{T} \times \mathrm{K}$, and

$$\int_{\mathrm{K}} n(t, k)dk = 1, \;\; \text{for each}\; t \in \mathrm{T}.$$

The dynamics of the density $n(\cdot)$ is driven by **social mobility**. Let a household with net wealth $k(t)$ at time t, have net wealth $k(t + \Delta t)$ at time $t + \Delta t$. If there is aggregate social mobility (but not pure social mobility)[13], we have $k(t + \Delta) = k(t) + \frac{dk(t)}{dt}\Delta t$, the rate of change of the household net wealth is

$$\frac{dk(t)}{dt} = s(t, k(t)), \;\; \text{for each}\; t \in \mathrm{T} \tag{13.50}$$

where $s(t, k(t))$ is the household saving's rate. Let the mass of the population in the net wealth segment $[\underline{k}(t), \overline{k}(t)]$ at time t be the same as the mass of population in the net wealth segment $[\underline{k}(t + \Delta t), \overline{k}(t + \Delta t)]$, at time $t + \Delta t$. Formally, this is equivalent to setting

$$\int_{\underline{k}(t+\Delta t)}^{\overline{k}(t+\Delta t)} n(t + \Delta t, k)\,dk - \int_{\underline{k}(t)}^{\overline{k}(t)} n(t, k)\,dk = 0.$$

Therefore,

$$\lim_{\Delta t \to 0}\left(\int_{\underline{k}(t+\Delta t)}^{\overline{k}(t+\Delta t)} n(t + \Delta t, k)\,dk - \int_{\underline{k}(t)}^{\overline{k}(t)} n(t, k)\,dk\right) = \frac{d}{dt}\int_{\underline{k}(t)}^{\overline{k}(t)} n(t, k)\,dk = 0.$$

invoking the Leibniz rule,

$$\frac{d}{dt}\int_{\underline{k}(t)}^{\overline{k}(t)} n(t, k)dk = \int_{\underline{k}(t)}^{\overline{k}(t)} \partial_t n(t, k)\,dk + n(t, \overline{k}(t))\,\frac{d\overline{k}(t)}{dt} - n(t, \underline{k}(t))\,\frac{d\underline{k}(t)}{dt},$$

[13] See Fields (2021) for those concepts.

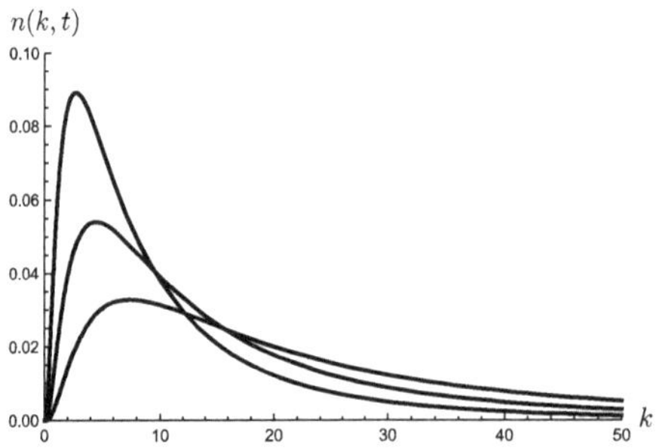

(a) Density dynamics

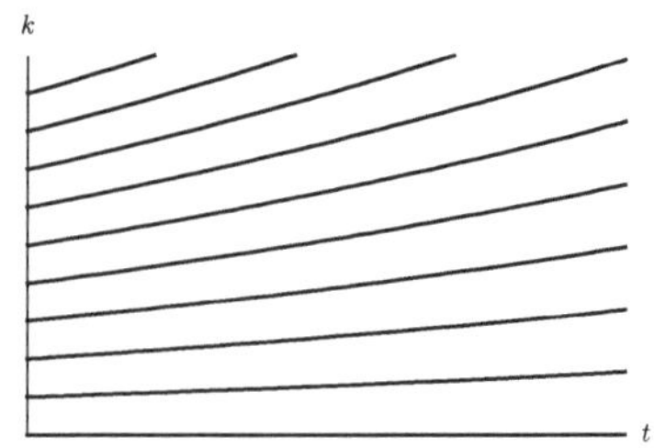

(b) Characteristic lines

Fig. 13.9 Distribution dynamics from the linear model (13.52) for $\gamma > 0$

considering Eq. (13.50), then

$$\frac{d}{dt}\int_{\underline{k}(t)}^{\overline{k}(t)} n(t,k)dk = \int_{\underline{k}(t)}^{\overline{k}(t)} \partial_t n(t,k)\,dk + n(t,\overline{k}(t))\,s(t,\overline{k}(t)) - n(t,\underline{k}(t))\,s(t,\underline{k}(t)),$$

and, from the mean-value theorem, we obtain

$$\frac{d}{dt}\int_{\underline{k}(t)}^{\overline{k}(t)} n(t,k)dk = \int_{\underline{k}(t)}^{\overline{k}(t)} \left(\partial_t n(t,k) + \frac{\partial\,(s(t,k)\,n(t,k))}{\partial k}\right) dk = 0,$$

that is satisfied and only if the following transport equation

$$\partial_t n(t,k) + \partial_k\big(s(t,k)\,n(t,k)\big) = 0 \tag{13.51}$$

holds for each $(t,k) \in \mathrm{T} \times \mathrm{K}$.

Example 13.10 (*Linear distributional model of aggregate mobility*) Let us assume that the savings function is linear and constant over time $s(t,k) = \gamma\, k$. Then we have a model for the inequality dynamics, when there is aggregate social mobility,

$$\begin{cases} \partial_t n(t,k) + \gamma k \partial_k n(t,k) + \gamma\, n(t,k) = 0, & \text{for each } (t,k) \in \mathbb{R}^2_+, \\ n(0,k) = n_0(k), & \text{for } k \in \mathbb{R}_+, \end{cases} \tag{13.52}$$

for a given initial density $n_0(k)$ such that $\int_{\mathrm{K}} n_0(k)\,dk = 1$.

As this is a version of the transport equation (13.46), the solution is

$$n(t,k) = e^{-\gamma t}\, n_0\Big(k\, e^{-\gamma t}\Big), \text{ for each } (t,k) \in \mathbb{R}^2_+. \tag{13.53}$$

Assuming an initial log-normal distribution, $n_0(k) \sim LN(\mu, \sigma^2)$, then the solution (13.54) is also a long-normal distribution with a time-dependent drift coefficient

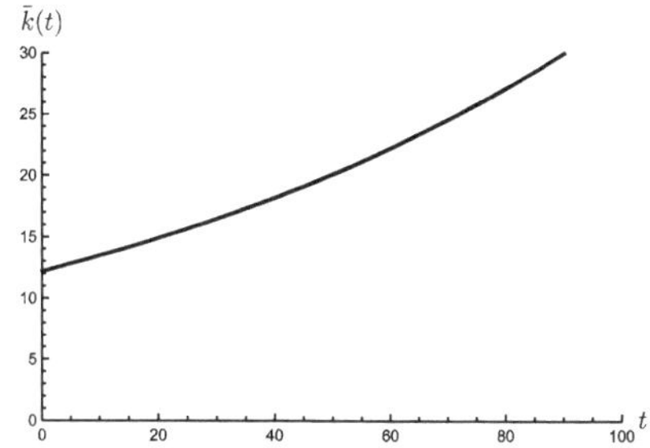

(a) Increasing average

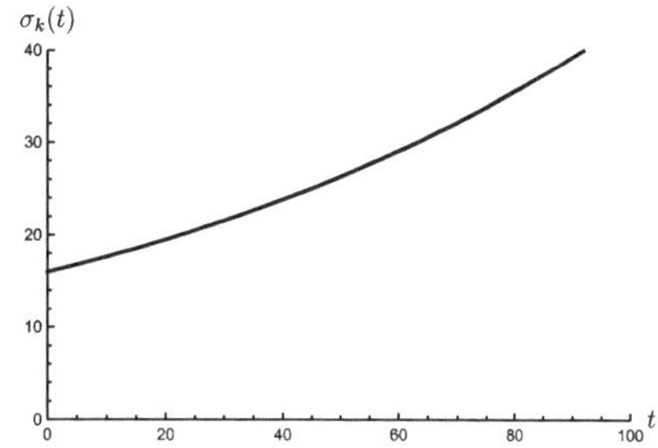

(b) Increasing st. dev.

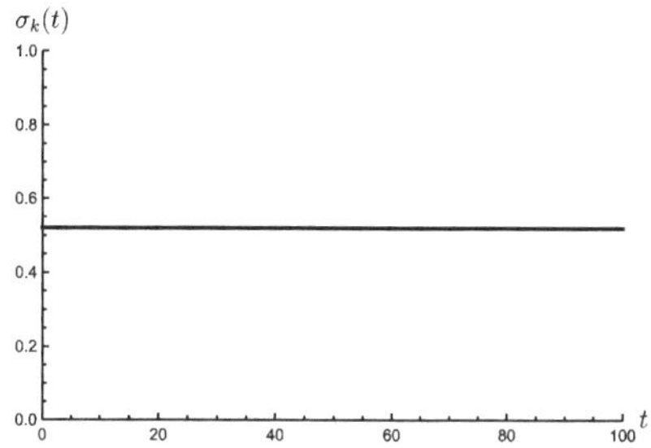

(c) Constant Gini coefficient

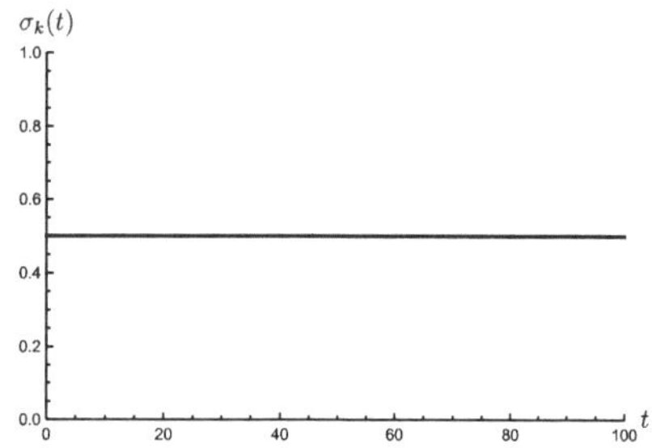

(d) Constant Theil index

Fig. 13.10 Linear accumulation function for $\gamma > 0$ and an initial log-normal distribution

$$n(t,k) \sim LN(\mu + \gamma\, t, \sigma^2),$$

$$n(t,k) = \left(2\,\pi\, k^2\, \sigma^2\right)^{-\frac{1}{2}} \exp\left(-\frac{(\log(k) - \gamma\, t - \mu)^2}{2\,\sigma^2}\right), \quad \text{for each } (t,k) \in \mathbb{R}^2_+. \tag{13.54}$$

Figure 13.9a displays the net wealth distribution evolution for three different moments in time, and Fig. 13.9b shows the projected characteristics, which can be interpreted as the dynamics of net wealth for representative households having different levels of net wealth at time $t = 0$.

This model introduces a theory of economic growth in the presence of inequality heterogeneity: economic growth can be considered a process of transportation of the mass of the households along a wealth (or income) support in the direction of higher levels of net wealth (or income). To have an aggregate characterization of the distributional dynamics, we can calculate several statistics. Figure 13.10 presents several statistics, showing that: first, there is unbounded per capita growth, with a constant growth rate $\gamma(t) = \gamma$,

$$\mu_k(t) \equiv \mathsf{M}[k(t)] = \int_0^\infty k\, n(t,k)\, dk = \mathsf{M}[k(0)]\, e^{\gamma t},$$

second, the distribution dynamics is non-ergodic, as both the average and the variance

$$\mathsf{V}[k(t)] = \int_0^\infty (k - \mu_k(t))^2\, n(t,k)\, dk = (\mu_k(t))^2 \left(e^{\sigma^2} - 1\right)$$

become unbounded over time, third, and surprisingly, although the difference between rich and poor people increases over time, the benchmark inequality measures are constant, as both the Gini index

$$\mathsf{G}[k(t)] = 1 - \frac{1}{\mathsf{M}[k(t)]} \int_0^\infty \left(1 - \int_0^k n(t,s)\, ds\right)^2 dk = \mathrm{erf}\left(\frac{\sigma}{2}\right),$$

and the Theil index

$$\mathsf{T}[k(t)](t) = \int_0^\infty \frac{k}{\mu_k(t)} \ln\left(\frac{k}{\mu_k(t)}\right) n(t,k)\, dk = \frac{\sigma^2}{2}$$

are constant over time. Therefore, there is **pure social immobility**: one household starting in a given quantile will remain (statistically) in the same quantile. This is corroborated by the quantile ratio between the wealth of households in the 90% quantile relative to households in the 10% quantile,

$$\frac{k(t)_{90}}{k(t)_{10}} = -\sigma\sqrt{2}\left[\mathrm{erf}^{-1}\left(1 - 2\frac{9}{10}\right) - \mathrm{erf}^{-1}\left(1 - 2\frac{1}{10}\right)\right],$$

that is constant over time. Seen as a model for growth among heterogeneous economies, endowed with different levels of capital, we can say that there is neither $\sigma-$ convergence nor divergence.

Example 13.11 (*Solow distributional model of aggregate mobility*) Let us assume now that instead of a linear AK model for the dynamics of capital at the household level we have a Solow (1956) capital accumulation equation

$$dk(t) = F(k(t))dt = (s\, A\, k(t)^\alpha - \delta\, k(t))\, dt$$

where $0 < s < 1$, $A > 0$, and $\delta > 0$ denote the savings rate, the total factor productivity, and the rate of depreciation of capital respectively, and $0 < \alpha < 1$ is the share of capital. There is a steady state level of capital $\bar{k} = \left(\frac{sA}{\delta}\right)^{\frac{1}{1-\alpha}}$. This is a distributional extension of the Solow (1956) model in Example 3.2. The dynamics of the household wealth density dynamics is the Cauchy problem

$$\begin{cases} \partial_t n(t,k) + \partial_k\big(F(k)\, n(t,k)\big) = 0, & \text{for each } (t,k) \in \mathrm{T} \times \mathrm{K} = R^2_{++}, \\ n(0,k) = n_0(k), & \text{for each } (t,k) \in \{t=0\} \times \mathrm{K}. \end{cases}$$

The solution is

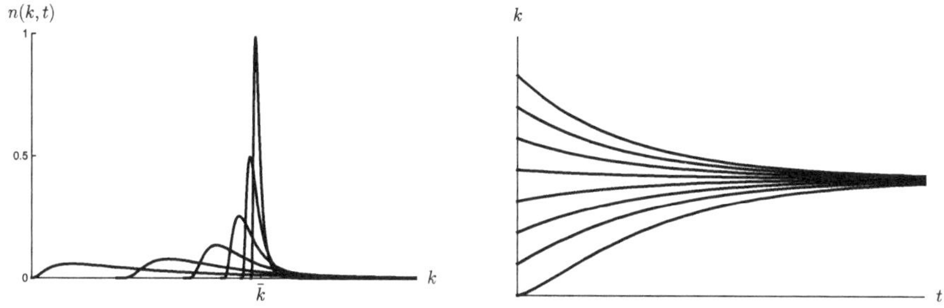

(a) Household density for increasing t (b) Projected characteristics

Fig. 13.11 Solow dynamics with initial heterogeneous capital distribution

$$n(t,k) = n_0\left(R(t,k)\right)e^{\beta(t,k)t}, \text{ for each } (t,k) \in \mathbb{R}^2_{++},$$

where

$$R(t,k) = \bar{k}\left[1 - \left(1 - \left(\frac{k}{\bar{k}}\right)^{1-\alpha}\right)e^{\delta(1-\alpha)t}\right]^{\frac{1}{1-\alpha}}$$

and

$$\beta(t,k) = \delta(1-\alpha) - \frac{\alpha}{1-\alpha}\left[\frac{\left(\frac{k}{\bar{k}}\right)^{1-\alpha}}{1 - \left(1 - \left(\frac{k}{\bar{k}}\right)^{1-\alpha}\right)e^{\delta(1-\alpha)t}}\right]\frac{1}{t}.$$

Figure 13.11a displays the density $n(k,t)$ for increasing t, and Fig. 13.11b shows the projected characteristics. As we can observe, even starting from a heterogeneous initial capital distribution, the density converges to a degenerate density concentrated at $k = \bar{k}$ (it is a Dirac-delta $\delta(k - \bar{k})$). In this case, differently from the linear case, the average rate of growth, and the variance, converge to zero. In the economic growth literature, this is a case of σ-convergence: all economies converge asymptotically to the same per capita capital and income level.

Example 13.12 (*Age-structured population dynamics*) The Malthusian model of population dynamics, studied in Example 2.5, $\dot{n} = \mu\, n$, where μ is the natural rate of growth, assumes that both the fertility and the mortality rates are age independent. Although that model may be a good approximation asymptotically, in the shorter run, it displays a large deviation from the actual population dynamics. McKendrick (1926) observed that the Malthusian model is unrealistic because the basic drivers of demography, fertility, and mortality, are age-dependent. If we introduce age-dependent mortality and fertility rates, then the dynamics of the population is governed by a first-order PDE, together with an integral equation.

Let $N(t,a)$ be the number of females of age a at time t in a population. The total population at time t is $N(t) = \int_0^{a_{max}} N(t,a)\,da$ where $N(t,a)$ is the number of females of age a at time t. Denoting $n(t,a)$ the density function for the population (of females) of age a at time t, we have, $n(a,t) \geq 0$, and $\int_0^{a_{max}} n(t,a)\,da = 1$.

The proportion of population, measured by the proportion of females, between ages a_1 and $a_2 > a_1$ at time t is $N(t, [a_1, a_2]) = \int_{a_1}^{a_2} n(t, a)da$. If there is no mortality, the instantaneous change in $N(t, [a_1, a_2])$ is $\frac{d}{dt}\int_{a_1}^{a_2} n(t, a)\, da = n(t, a_1) - n(t, a_2)$, where $n(t, a_1)$ and $n(t, a_2)$ is the flow of females entering and leaving the age-interval $[a_1, a_2]$, at time t. As $\frac{d}{dt}\int_{a_1}^{a_2} n(t, a)\, da = \int_{a_1}^{a_2} \partial_t n(t, a)\, da$ and, using the fundamental theorem of calculus, $n(t, a_1) - n(t, a_2) = -\int_{a_1}^{a_2} \partial_a n(t, a)\, da$ then

$$\frac{d}{dt}\int_{a_1}^{a_2} n(t, a)da - \big(n(t, a_1) - n(t, a_2)\big) = \int_{a_1}^{a_2} \big(\partial_t n(t, a) + \partial_a n(t, a)\big)\, da = 0.$$

If we consider mortality, then population in the interval $[a_1, a_2]$ does not remain constant. If the instantaneous rate of mortality at age a and time t is denoted by $\mu(a, t)$, the population in that interval is

$$\int_{a_1}^{a_2} \big(\partial_t n(t, a) + \partial_a n(t, a)\big)\, da = -\int_{a_1}^{a_2} \mu(t, a)\, n(t, a)\, da, \text{ for each } t \in \mathrm{T}.$$

Therefore, we have the McKendrick (1926) equation for the evolution of a cohort having age a at time t, in the presence of mortality,

$$\partial_t n(t, a) + \partial_a n(t, a) = -\mu(t, a)\, n(t, a), \text{ for } (t, a) \in [0, a_{\max}] \times \mathbb{R}_+. \quad (13.55)$$

The evolution of every cohort for an age-dependent population according to the McKendrick (1926) model is given by the solution for PDE (13.55) complemented with two side-conditions: one for the initial size of the cohort, i.e., for the size of the population with age $a = 0$ at every time t, and another for the age distribution of the whole population at the initial time $t = 0$.[14]

$$\begin{cases} \partial_t n(t, a) + \partial_a n(t, a) = \mu(t, a)\, n(t, a), & \text{for each } (t, a) \in \mathrm{T} \times \mathrm{A}, \\ n(t, 0) = b(t), & (t, a) \in \mathrm{T} \times \{a = 0\}, \\ n(0, a) = n_0(a), & (t, a) \in \{t = 0\} \times \mathrm{A}, \end{cases} \quad (13.56)$$

where $\mathrm{A} = [0, a_{\max}]$, $n_0(a)$ is the initial age distribution of the population, and the number of newborns (approximated by a real number) is

$$b(t) = \int_0^{a_{max}} \beta(t, a)\, n(t, a)\, da,$$

where $\beta(t, a)$ is the age-dependent fertility rate at time t. The McKendrick model has two noteworthy features: first, it is a two-point boundary problem with two boundary conditions, an initial (at $t = 0$) age distribution of the population and the population at age $a = 0$ over time; and, second, the boundary condition for the initial size of a

[14] See Keyfitz and Caswell (2005) and Kot (2001, Chap. 23).

cohort is non-local, depending on the part of the distribution of the population that is fertile. This last feature implies that it does not have in general a closed form solution, requiring the solution of an integral equation. Therefore, the total population size at time t, is the sum of the sizes of every cohort still alive at time t.

If we disregard the global nature of the initial size of each cohort, we can still observe that, in the Mc-Kendrick equation, there is a different type of dynamics depending on the difference between a and t: for $a < t$ the dynamics depends on the newborns, i.e., population with age $a = 0$, while for $a > t$ the dynamics is governed by the initial age distribution of the population. We can infer that the first type of behavior prevails asymptotically.

Example 13.13 (*Age-structured population dynamics: constant mortality cases*) Consider the simple case in which the mortality rate is age- and time-independent, and the initial size of a cohort is exogenously given by function $\phi(t)$,

$$\begin{cases} \partial_t n(t,a) + \partial_a n(t,a) = -\mu\, n(t,a), & \text{for each } (t,a) \in \mathrm{T} \times \mathrm{A}, \\ n(t,0) = \phi(t), & (t,a) \in \mathrm{T} \times \{a = 0\}, \\ n(0,a) = n_0(a), & (t,a) \in \{t = 0\} \times \mathrm{A}. \end{cases}$$

where $n_0(a)$ is the initial distribution of population, and $\phi(t)$ is the number of offspring assumed to be exogenous, i.e., independent of the distribution of population. The solution is

$$n(t,a) = \begin{cases} \phi(t-a)p(a), & \text{if } a < t \\ n_0(a-t)\frac{\pi(a)}{p(a-t)}, & \text{if } a > t. \end{cases}$$

where $p(a) = e^{-\mu a}$ is the probability of survival at age a. Therefore, $\frac{p(a)}{p(a-t)} = e^{-\mu t}$. To understand the effects of the age-dependent fertility, let us consider the following case, in which the initial distribution is free, or is consistent with the evolution of the population since the "Archimedian time", $t = -\infty$, and the only fertile age is $a = \alpha > 0$

$$\begin{cases} \partial_t n(t,a) + \partial_a n(t,a) = -\mu\, n(t,a), & \text{for each } (t,a) \in \mathbb{R} \times \mathbb{R}_+ \\ n(t,0) = \beta\, n(t,\alpha), & (t,a) \in \mathbb{R} \times \{a = 0\}. \end{cases}$$

The number of newborns at time t is $n(t,0) = \int_0^{a_{max}} \beta\, \delta(a-\alpha)\, n(t,a)\, da = \beta\, n(t,\alpha)$, where $\beta > 0$ is the fertility rate. The solution to the PDE is $n(t,a) = f(a-t)\, e^{-\mu t}$. Substituting in the constraint yields a linear algebraic functional equation $f(-t) = \beta\, f(\alpha - t)$. The solution to this equation is $f(x) = h(x)\, \beta^{-\frac{x}{\alpha}}$, where $h(x)$ is a periodic function with period α. Therefore, the solution to this simplified Mckendrick model is

$$n(t,a) = h(a-t)\, \beta^{\frac{t-a}{\alpha}}\, e^{-\mu t}, \text{ for each } (t,a) \in \mathbb{R} \times R_+.$$

Figure 13.12 provides an illustration to this solution. Two features are important: first, after birth, the size of each cohort evolves along the projected characteristic

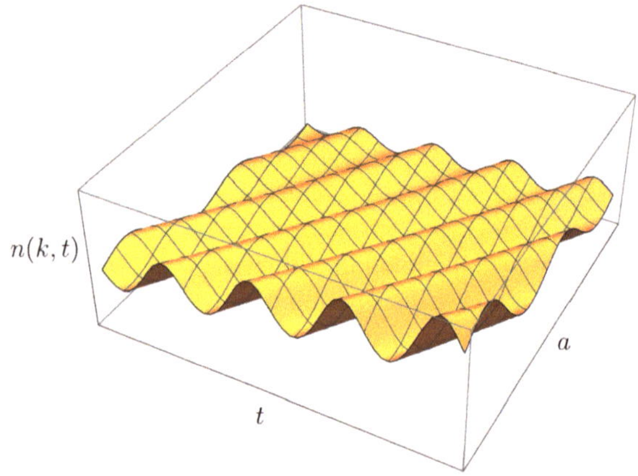

Fig. 13.12 Age-dependent dynamics for Dirac-delta fertility, and $h(x) = \exp\{\gamma \sin\left(\frac{\pi(t-a)}{\alpha}\right)\}$, where γ is a positive constant

$a = t - t_0$, where t_0 is the birth year, and the solution displays an "echo effect" with period equal to α.

Example 13.14 (*A cohort's budget constraint*) The budget constraint studied in Example 2.8 is realistic at an aggregate level. However, it does not incorporate a fundamental feature of the household behavior at a micro level: the existence of a life-cycle behavior, and its implication in the existence of a age dependence of savings. The incorporation of life-cycle features into an household's budget constraint transforms it into a first-order PDE.

Let $w(t, a)$ be the net financial net wealth of an agent belonging to a cohort having age a at time t. The budget constraint of a representative member of the a−cohort is

$$\partial_t w(t, a) + \partial_a w(t, a) = s(t, a) + r\, w(t, a), \text{ for each } (t, a) \in \mathbb{R}^2_+, \tag{13.57}$$

where $s(t, a)$ is the savings at age a at time t, and r is the interest rate. Let us assume that the initial stock of wealth is free, then $w : \mathbb{R}^2_+ \to \mathbb{R}$, that the wealth at birth for any cohort is zero, then $w(t, 0) = 0$, that there are no bequests between cohorts, and that $w(0, a) = \phi(a)$ is the age distribution of wealth at time $t = 0$. The general solution to the age distribution of net wealth over time is (13.57) is

$$w(t, a) = \left(\int_0^a s(z, z - a + t)e^{-rz}dz + \phi(t - a)\right) e^{ra}, \text{ for each } (t, a) \in \mathbb{R}^2_+.$$

Exercise 13.4 Prove this.

If savings of cohort a at time t is $s(t, a) = e^{ba(K-a)+gt} - c$, assuming a Mincerian income profile across every lifetime and a constant consumption, then the net wealth of cohort that was born at time $t - a$ is

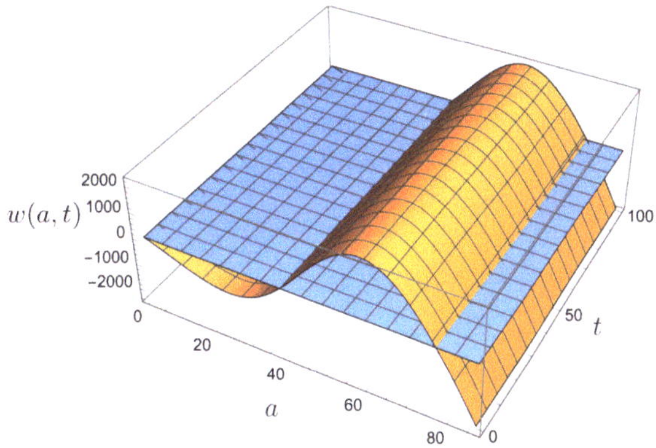

Fig. 13.13 Illustration of equation (13.58) for $K = 88$, $b = 0.0029$, $r = 0.02$, $g = 0.02$ and $c = 50$

$$w(t, a) = \frac{\sqrt{\pi}}{2\sqrt{b}}\left(\Phi\left(\frac{Kb + g - r}{2\sqrt{b}}\right) - \Phi\left(\frac{(K - 2a)b + g - r}{2\sqrt{b}}\right)\right)e^{\gamma(t,a)} - \frac{c}{r}\left(1 - e^{ra}\right) \tag{13.58}$$

where $\Phi(x) = \text{erf}(x) = (2/\sqrt{\pi})\int_0^x e^{-z^2}dz$, and

$$\gamma(t, a) \equiv \frac{K^2b^2 + ((2K + 4(t - a))g - 2(K - 2a)r)b + (g - r)^2}{4b}.$$

Figure 13.13 illustrates Eq. (13.58). It displays a life-cycle behavior of savings: a representative agent tends to be a net borrower when young and a lender at middle age, although it dissaves later in life.

Example 13.15 (*Interest rate term-structure*) Consider a financial market in which there are two types of risk-free assets: a bank account and a continuum of bonds with maturity dates, x, ranging from zero to infinity, $x \in [0, \infty)$.[15] We show that, if there is perfect foresight (i.e., there is a deterministic setting), and absence of arbitrage opportunities, then the forward rate at time t, for a bond maturing at time $t + x$, perfectly anticipates the future spot interest rate, at time $t + x$.

Let a bank account balance be $B(0)$, at time $t = 0$. If there are no deposits or withdrawals, it will be $B(t) = B(0)e^{\int_0^t r(\tau)d\tau}$ at time $t > 0$, where $r(t)$ is the spot interest rate at each time $t > 0$. Therefore, the instantaneous rate of return for a bank account is

$$\frac{dB(t)/dt}{B(t)} = r(t), \text{ for each } t \in \mathbb{R}_+.$$

The price at time t of a bond[16] that matures at time $t + x \geq t$ is

[15] See Santa-Clara and Sornette (2001).

[16] It is implicitly assumed that the bond has face value equal to one and does not pay coupons.

$$P(t,x) = e^{-\int_0^x f(t,y)dy}, \tag{13.59}$$

where $f(t,x)$ is the forward interest rate for maturity x. The spot interest rate is taken as a forward rate with maturity zero, $f(t,0) = r(t)$ for each $t \geq 0$.

As time and the duration of the bond satisfy $dx + dt = 0$, then the change in the price of the bond satisfies

$$dP(t,x) = \partial_t P(t,x)\,dt + \partial_x P(t,x)\,dx = \Big(\partial_t P(t,x) - \partial_x P(t,x)\Big)\,dt.$$

Taking derivatives in Eq. (13.59), we find $\partial_t P(t,x) = -P(t,x) \int_0^x \partial_t f(t,y)dy$, and $\partial_x P(t,x) = -P(t,x)\,f(t,x)$. Invoking the mean value theorem, $f(t,x) - f(t,0) = \int_0^x \partial_x f(t,y)\,dy$, and as $f(t,0) = r(t)$, the rate of return for a bond with maturity x is

$$\begin{aligned} \frac{dP(t,x)/dt}{P(t,x)} &= f(t,x) - \int_0^x \partial_t f(t,y)\,dy = \\ &= -\int_0^x \big(\partial_t f(t,y) - \partial_x f(t,y)\big)\,dy + r(t), \text{ for each } t \in \mathbb{R}_+. \end{aligned}$$

If there is absence of arbitrage opportunities then the instantaneous rate of return for any two investments should be equal. In particular, it implies that the rate of return for the bank account should be equal to the rate of return for a bond of any maturity $x \geq 0$,

$$\frac{dB(t)/dt}{B(t)} = \frac{dP(t,x)/dt}{P(t,x)} \iff \int_0^x \big(\partial_t f(t,y) - \partial_x f(t,y)\big)\,dy = 0, \text{ for each } x \geq 0.$$

Therefore, the forward rate, in a deterministic setting, is the solution to the following Cauchy problem

$$\begin{cases} \partial_t f(t,x) - \partial_x f(t,x) = 0, & (t,x) \in \mathbb{R}_+^2, \\ f(t,0) = r(t), & (t,x) \in \mathbb{R}_+ \times \{t=0\}. \end{cases}$$

As the PDE has the general solution $f(t,x) = h(t+x)$ (see Eq. (13.3)), then the particular solution to the Cauchy problem is

$$f(t,x) = r(t+x), \text{ for each } (t,x) \in \mathbb{R}_+^2.$$

Therefore, if there are no arbitrage opportunities, the market forward rate at time t for a bond maturing at time $t+x$ is a perfect predictor to the future spot rate at time $t+x$.

Example 13.16 (*Optimality condition for a consumer choice problem*) The previous theory has a surprising application in the derivation of indirect utility functions for static (primal) consumer problems. Consider the following textbook consumer problem of finding the demand for two goods, indexed by $i = 1, 2$, to maximize a continuous, differentiable, increasing, and concave utility function, $u(c_1, c_2)$, subject to a budget constraint setting that expenditure equal to income. To simplify, let us assume there is an interior solution, such that $(c_1, c_2) > \mathbf{0}$. The (primal) consumer problem is

$$\max_{c_1, c_2} \left\{ u(c_1, c_2) : \; E(c_1, c_2) = p_1 c_1 + p_2 c_2 = W \right\}$$

where c_j and p_j are the quantity and the price of good $j = 1, 2$, respectively, and $E(\cdot)$ and W denote the household's expenditure and income, respectively.

The first-order conditions for an (interior) optimum (c_1^*, c_2^*) are

$$p_2 \, \partial_{c_1} u(c_1^*, c_2^*) = p_1 \, \partial_{c_2} u(c_1^*, c_2^*), \tag{13.60}$$

$$E(c_1^*, c_2^*) = W. \tag{13.61}$$

Equation (13.60) is a first-order PDE with solution

$$u(c_1^*, c_2^*) = v\left(\frac{p_1 c_1^* + p_2 c_2^*}{p_1} \right)$$

and, if we use Eq. (13.61) and define $w \equiv W/p_1$, in the optimum we have $u(c_1^*, c_2^*) = v(w)$. If the utility function is strictly concave then with very weak conditions (differentiability), we have a unique interior optimum. It is clear that the budget constraint is, in real terms, a projected characteristic.

13.5 References

Important references for the mathematics of first-order PDE are: introductory Olver (2014), more advanced Evans (1998, Chap. 3) or Chechkin and Goritsky (2009). On the dynamics of first-order PDE, a useful reference close to the approach in this chapter is Sachdev and Rao (2010). Applications to economics can be found in Basov (2007), and, with an emphasis in vintage capital models, Hritonenko and Yatsenko (2013). For applications to mathematical demography, see Kot (2001, Chap. 23).

13.6 Appendix: Laplace Transforms and Inverse Laplace Transforms

Consider function $f(x)$, where $x \in \mathbb{R}_+$. The (unilateral) Laplace transform (LT) of function $f(x)$ is a functional

Table 13.1 Laplace transforms and inverse Laplace transforms

$f(x) = \mathcal{L}^{-1}[F(s)](x)$	$F(s) = \mathcal{L}[f(x)](s)$	
$f'(x)$	$s\,F(s) - f(0)$	
a	$\frac{1}{s}$	$a \in \mathbb{R}$
x	$\frac{1}{s^2}$	
e^{ax}	$\frac{1}{s-a}$	
$H(a-x)$	$\frac{e^{-as}}{s}$	if $a > 0$
$f(x-a)\,H(x-a)$	$F(s)\,e^{-as}$	if $a > 0$

$$F(s) = \mathcal{L}[f(x)](s) = \int_0^{\infty} e^{-sx} f(x)dx,$$

parameterized by $s \in \mathbb{C}$, if the integral exists. In economic applications, in general, $s \in \mathbb{R}$ is a real number. Under some conditions, the Bromwich integral (also known as Fourier-Mellin integral) allows for recovering function $f(x)$,

$$f(x) = \mathcal{L}^{-1}[F(s)](x) = \frac{1}{2\pi\, i} \lim_{z\to\infty} \int_{\gamma - z i}^{\gamma + z i} e^{s\, t}\, F(s)\, ds,$$

and we say that $f(x)$ is the inverse Laplace transform (ILT) of $F(s)$. The Laplace transform can be extended to the bilateral domain $\mathbb{R} = (-\infty, \infty)$ and to functions of more than one variable. The LT of a derivative of a function $f(x)$, $f^{'}(x) = df(x)/dx$ is

$$\begin{aligned}
\mathcal{L}[f^{'}(x)](s) &= \frac{d}{dx}\left(\int_0^{\infty} e^{-sx} f^{'}(x)dx\right) \\
&= s\int_0^{\infty} e^{-sx} f(x)dx + \int_0^{\infty} e^{-sx} f(x) \\
&= sF(s) - f(0),
\end{aligned}$$

as $\lim_{x\to\infty} e^{-s\,x} f(x) = 0$ is a condition for the existence of a LT. Laplace transforms are used to solve ODE and PDE because they transform, directly or indirectly, those functional equations into algebraic equations. The transformation of the ODE is immediate, as can be seen in the next example, while the transformation of a PDE is indirect, because applying a LT to a PDE transforms it into a parameterized ODE, as seen in Sect. 13.2.

Table 13.1 presents some Laplace transforms used in this book.

Example 13.17 (*Laplace transform method for solving a linear ODE*) Consider the ODE $f'(x) - af(x) = 0$, where $f : \mathbb{R}_+ \to \mathbb{R}$. Applying the LT to both sides, yields

$$\begin{aligned}\mathcal{L}\Big[f'(x) - a\,f(x)](s)\Big] &= \int_0^\infty e^{-s\,x}\Big(f'(x) - a\,f(x)\Big)\,dx \\ &= \int_0^\infty e^{-s\,x}\,f'(x)dx - a\int_0^\infty e^{-s\,x}\,f(x)\,dx \\ &= s\,F(s) - f(0) - a\,F(s).\end{aligned}$$

Then the original ODE in $f(x)$ is equivalent to an algebraic equation $(s - a)F(s) = f(0)$, with solution

$$F(s) = \frac{f(0)}{s - a}.$$

To obtain the solution to the ODE we apply the ILT from Table 13.1

$$f(x) = \mathcal{L}^{-1}[F(s)](x) = f(0)\mathcal{L}^{-1}\Big[\frac{1}{s - a}\Big](x).$$

But as $\mathcal{L}^{-1}[\frac{1}{s-a}](x) = e^{a\,x}$, then the general solution to the ODE is

$$f(x) = f(0)\,e^{ax}, \text{ for each } x \in \mathbb{R}_+.$$

References: Doetsch (1974) is one important modern contribution to the field, see also Schiff (1999). For applications, we can use tables of LT and ILT in Abramowitz and Stegun (1972), or Weisstein (2024b) and Weisstein (2024a) for using MATTHEMATICA®.

Chapter 14
Optimal Control of First-Order PDE

This chapter introduces calculus of variations and optimal control problems for functions of two independent variables, thus extending Chaps. 9 and 10. Although we present results for unknown functions of two independent variables, we focus particularly on the case in which time is one independent variable. This is useful for applications in which there is an intrinsic heterogeneity, over an ordered continuum $X \subseteq \mathbb{R}$, that evolves over time, and our goal is to find the optimal evolution of that heterogeneity. That is, this chapter is concerned with the optimal dynamics of distributions, that are infinite dimensional, and not to the optimal dynamics of variables as in Part II. This chapter is confined to the study of the dynamics of variables governed by a transport-like behavior, leaving the optimal dynamics for variables governed by a diffusion-like behavior to Chap. 16.

The approach followed here is introductory, informal, and heuristic. Our goal is restricted to suggesting the potential usefulness of using optimization of PDE to address the dynamics of heterogeneity. In solving those problems, the type of boundary conditions raises a paradox. Assuming finite domains and their associated boundary conditions, on one hand, simplifies the derivation of the necessary conditions for an optimum (they are called simplest problems), but, on the other hand, they may imply ill-posedness to the problem and they largely increase the complexity of the process of obtaining particular solutions, as was already seen in Sect. 13.2. In order to obtain solutions, we may have to enlarge its domain to generalized function spaces. To concentrate on the general structure of the solutions to CV and OC problems, we sidestep this complexity either by presenting only general solutions to the dynamic optimality conditions or by restraining to problems having strong or classic solutions.

In Sect. 14.1, we present the (necessary) first-order conditions for calculus of variations (CV), and derive the implications for time-dependent distributions. In Sect. 14.2, a version of the Pontryagin's maximum principle (PMP) for an infinite horizon optimal control problems of first-order PDE is (heuristically) derived and presented. Those conditions are applied to solving the problem of optimal wealth

P. B. Brito, *Economic Dynamics and Distributions*, Dynamic Modeling and Econometrics in Economics and Finance 32, https://doi.org/10.1007/978-3-031-94717-9_14

distribution for an inequality averse central planner in the presence of aggregate social mobility.

14.1 Two-Dimensional Calculus of Variations

In this section, we assume that the vector of independent variables is two dimensional and the dependent variable is one dimensional. A calculus of variations (CV) problem in this framework entails maximizing an objective functional in which the known objective function has partial derivatives of the dependent variable as arguments. In Sect. 14.1.1, we set up and present the optimality conditions for generic independent variables, and in Sect. 14.1.2, we specialize to cases in which one independent variable is time, thus addressing the optimal dynamics of distributions.

14.1.1 General Independent Variables

Let the vector of independent variables be $\mathbf{x} = (x_1, x_2)$ with domain $\mathbf{X} = \mathrm{X}_1 \times \mathrm{X}_2$, where $\mathrm{X}_i \subseteq \mathbb{R}$, for $i = 1, 2$, and let the dependent variable be a mapping $y(\mathbf{x}) : \mathbf{X} \to \mathrm{Y} \subseteq \mathbb{R}$. The **objective functional** is defined by

$$J[y] = \int_{\mathrm{X}_1} \int_{\mathrm{X}_2} F(\mathrm{D}_{\mathbf{x}} y(\mathbf{x}), y(\mathbf{x}), \mathbf{x})\, dx_2\, dx_1, \tag{14.1}$$

where $y(\mathbf{x}) = y(x_1, x_2)$, and $\mathrm{D}_{\mathbf{x}} y(\mathbf{x}) = \left(\partial_{x_1} y(x_1, x_2), \partial_{x_2} y(x_1, x_2)\right)^{\top}$ is the gradient of $y(\cdot)$.[1] The objective function $F(\cdot)$ is assumed to be a continuous and continuously differentiable function of the dependent variable, and of the gradient $\mathrm{D}_{\mathbf{x}} y(\cdot)$, such that $\mathrm{D}_{\mathrm{D}_{\mathbf{x}} y} F(\mathrm{D}_{\mathbf{x}} y(\mathbf{x}), y(\mathbf{x}), \mathbf{x}) \neq 0$ for each $\mathbf{x} \in \mathbf{X}$. We say this problem is regular.

The simplest two-dimensional CV problem. Let $\mathbf{X}$ be a bounded and compact set specified by $\mathbf{X} = \mathrm{X}_1 \times \mathrm{X}_2$, where $\mathrm{X}_i = [\underline{x}_i, \overline{x}_i]$ for $\underline{x}_i < \overline{x}_i$, for $i = 1, 2$.

The simplest two-dimensional CV problem is defined by

$$\max_{y(\cdot)} J[y] = \int_{\mathrm{X}_1} \int_{\mathrm{X}_2} F(\mathrm{D}_{\mathbf{x}} y(\mathbf{x}), y(\mathbf{x}), \mathbf{x})\, dx_2\, dx_1, \tag{14.2a}$$

subject to

$$y(\underline{x}_1, x_2) = \underline{y}_1(x_2), \;\; y(\overline{x}_1, x_2) = \overline{y}_1(x_2), \tag{14.2b}$$

$$y(x_1, \underline{x}_2) = \underline{y}_2(x_1), \;\; y(x_1, \overline{x}_2) = \overline{y}_2(x_1), \tag{14.2c}$$

[1] Throughout this chapter, we denote $\partial_{x_j} y(\mathbf{x})) = \frac{\partial y(\mathbf{x})}{\partial x_j}$, for $j = 1, 2$.

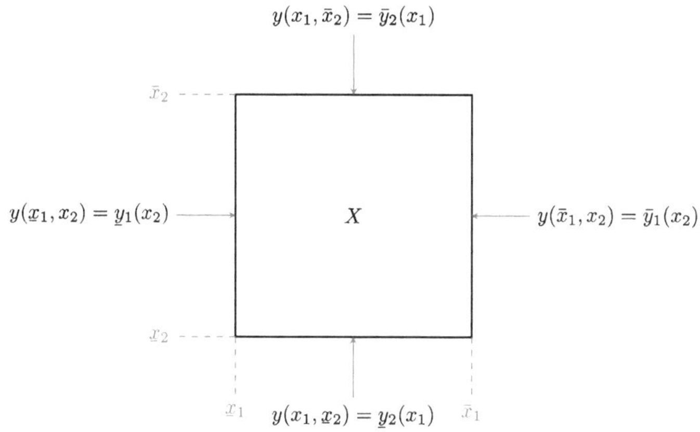

Fig. 14.1 The domain **X** and the boundary conditions for CV problem (14.2)

where $\underline{y}_i(x_i)$, and $\overline{y}_i(x_i)$, for $i = 1, 2$, are fixed (known) functions.[2] Denoting the boundary of set **X** by $\partial\mathbf{X}$, in this case, $y(\mathbf{x})$ is a fixed function for any $\mathbf{x} \in \partial\mathbf{X}$, as shown in Fig. 14.1.[3]

Proposition 14.1 (Necessary conditions for the fixed boundary values problem) *Let* $\mathbf{y}^* = (y^*(\mathbf{x}))_{\mathbf{x}\in\mathbf{X}}$ *be the solution to problem 14.2. Then it satisfies the Euler-Lagrange (EL) PDE*[4]

$$\partial_y F^*(\mathbf{x}) - \frac{\partial}{\partial x_1}\left(\partial_{\partial_{x_1} y} F^*(\mathbf{x})\right) - \frac{\partial}{\partial x_2}\left(\partial_{\partial_{x_2} y} F^*(\mathbf{x})\right) = 0, \text{ for all } \mathbf{x} \in \mathbf{X}, \tag{14.3}$$

and the constraints

$$y^*(\underline{x}_1, x_2) = \underline{y}_1(x_2),\ y^*(\overline{x}_1, x_2) = \overline{y}_1(x_2), \tag{14.4a}$$

$$y^*(x_1, \underline{x}_2) = \underline{y}_2(x_1),\ y^*(x_1, \overline{x}_2) = \overline{y}_2(x_1). \tag{14.4b}$$

Proof The objective functional at the optimum, $y^*(\mathbf{x})$, is

$$\mathsf{J}[y^*] = \int_{X_1}\int_{X_2} F(\mathrm{D}_{\mathbf{x}} y^*(\mathbf{x}), y^*(\mathbf{x}), \mathbf{x})\, dx_2\, dx_1.$$

We introduce an admissible perturbation to $y^*(\mathbf{x})$, such that $y(\mathbf{x}) = y^*(\mathbf{x}) + \varepsilon\,\eta(\mathbf{x})$, for a constant ϵ, and $\eta(\mathbf{x}) : \mathbf{X} \to \mathrm{Y}$ is a differentiable function. The perturbation is admissible if $\eta(\tilde{\mathbf{x}}) = 0$ for $\tilde{\mathbf{x}} \in \partial\mathbf{X}$, that is, if $\eta(\underline{x}_1, x_2) = \eta(\overline{x}_1, x_2) = \eta(x_1, \overline{x}_2) =$

---

[2] We use the notation $x_{-i} = \{x_{j\neq i} : j = 1, \ldots, n\}$. In the two-dimensional case, we have $x_{-1} = x_2$ and $x_{-2} = x_1$.

[3] Some requirements should be satisfied for the existence of compatibility between constraints.

[4] The following notation is used $F^*(\mathbf{x}) = F(\mathrm{D}_{\mathbf{x}} y^*(\mathbf{x}), y^*(\mathbf{x}), \mathbf{x})$, $\partial_y F^*(\mathbf{x}) = \frac{\partial}{\partial y} F(\mathrm{D}_{\mathbf{x}} y^*(\mathbf{x}), y^*(\mathbf{x}), \mathbf{x})$, and $\partial_{\partial_i y} F^*(\mathbf{x}) = \frac{\partial}{\partial_{x_i} y} F(\mathrm{D}_{\mathbf{x}} y^*(\mathbf{x}), y^*(\mathbf{x}), \mathbf{x})$ for $i = 1, 2$,

$\eta(x_1, \underline{x}_2) = 0$. A first order Taylor expansion of the value functional, evaluated at y^*, yields $\mathsf{J}[y^* + \epsilon\,\eta] = \mathsf{J}[y^*] + \mathrm{D}_{\eta(\cdot)}\mathsf{J}[y^*]\epsilon + O(\epsilon^2)$. At the optimum, we should have $\mathsf{J}[y^*] \geq \mathsf{J}[y^* + \epsilon\,\eta]$, for any $\eta(\cdot)$. This condition is satisfied only if $\mathrm{D}_{\eta(\cdot)}\mathsf{J}[y^*] = 0$.

The Gâteaux differential (see Eq. (8.32)) is

$$
\begin{aligned}
\mathrm{D}_{\eta(\cdot)}\mathsf{J}[y^*] = \int_{\mathrm{X}_1}\int_{\mathrm{X}_2} \Big(&\partial_y F^*(\mathbf{x})\,\eta(\mathbf{x}) + \\
&+ \partial_{\partial_{x_1} y} F^*(\mathbf{x})\,\partial_{x_1}\eta(\mathbf{x}) + \partial_{\partial_{x_2} y} F^*(\mathbf{x})\,\partial_{x_2}\eta(\mathbf{x})\Big)\,dx_2\,dx_1 .
\end{aligned}
\tag{14.5}
$$

Integration by parts yields

$$
\begin{aligned}
\int_{\mathrm{X}_1}\int_{\mathrm{X}_2} \partial_{\partial_{x_i} y} F^*(\mathbf{x})\,\partial_{x_i}\,\eta(\mathbf{x})\,dx_2\,dx_1 = &-\int_{\mathrm{X}_1}\int_{\mathrm{X}_2} \partial_{x_i}\Big(\partial_{\partial_{x_i} y} F^*(\mathbf{x})\Big)\,\eta(\mathbf{x})\,dx_2\,dx_1 + \\
&+ \int_{\underline{x}_{-i}}^{\bar{x}_{-i}} \partial_{\partial_{x_i} y} F^*(\mathbf{x})\,\eta(\mathbf{x})\,dx_{-i}\Big|_{x_i=\underline{x}_i}^{x_i=\bar{x}_i}, \quad \text{for } i = 1, 2,
\end{aligned}
$$

where $x_{-1} = x_2$, and $x_{-2} = x_1$. Then

$$
\begin{aligned}
\mathrm{D}_{\eta(\cdot)}\mathsf{J}[y^*] = \int_{\mathrm{X}_1}\int_{\mathrm{X}_2} \Big(&\partial_y F^*(\mathbf{x}) - \frac{\partial}{\partial x_1}\Big(\partial_{\partial_{x_1} y} F^*(\mathbf{x})\Big) - \\
&- \frac{\partial}{\partial x_2}\Big(\partial_{\partial_{x_2} y} F^*(\mathbf{x})\Big)\Big)\,\eta(\mathbf{x})\,dx_2,\ dx_1 \\
&+ \int_{\underline{x}_2}^{\bar{x}_2} \partial_{\partial_{x_1} y} F^*(\mathbf{x})\,\eta(\mathbf{x})\,dx_2\Big|_{x_1=\underline{x}_1}^{x_1=\bar{x}_1} + \\
&+ \int_{\underline{x}_1}^{\bar{x}_1} \partial_{\partial_{x_2} y} F^*(\mathbf{x})\,\eta(\mathbf{x})\,dx_1\Big|_{x_2=\underline{x}_2}^{x_2=\bar{x}_2}
\end{aligned}
\tag{14.6}
$$

and

$$
\begin{aligned}
\mathrm{D}_{\eta(\cdot)}\mathsf{J}[y^*] = \int_{\mathrm{X}_1}\int_{\mathrm{X}_2} \Big(&\partial_y F^*(\mathbf{x}) - \frac{\partial}{\partial x_1}\Big(\partial_{\partial_{x_1} y} F^*(\mathbf{x})\Big) - \\
&- \frac{\partial}{\partial x_2}\Big(\partial_{\partial_{x_2} y} F^*(\mathbf{x})\Big)\Big)\,\eta(\mathbf{x})\,dx_2\,dx_1,
\end{aligned}
$$

for admissible perturbations. Then $\mathrm{D}_{\eta(\cdot)}\mathsf{J}[y^*] = 0$ if the EL PDE (14.3) holds. □

The PDE (14.3) is a second-order PDE, as the next example shows.

Example 14.1 Consider the problem of maximizing the objective functional

$$
\mathsf{J}[y] = \int_{\underline{x}_1}^{\overline{x}_1}\int_{\underline{x}_2}^{\overline{x}_2} \ln\big(\partial_{x_1} y(x_1, x_2) - \partial_{x_2} y(x_1, x_2)\big)\, e^{-(\alpha_1 x_1 + \alpha_2 x_2)} dx_1 dx_2
$$

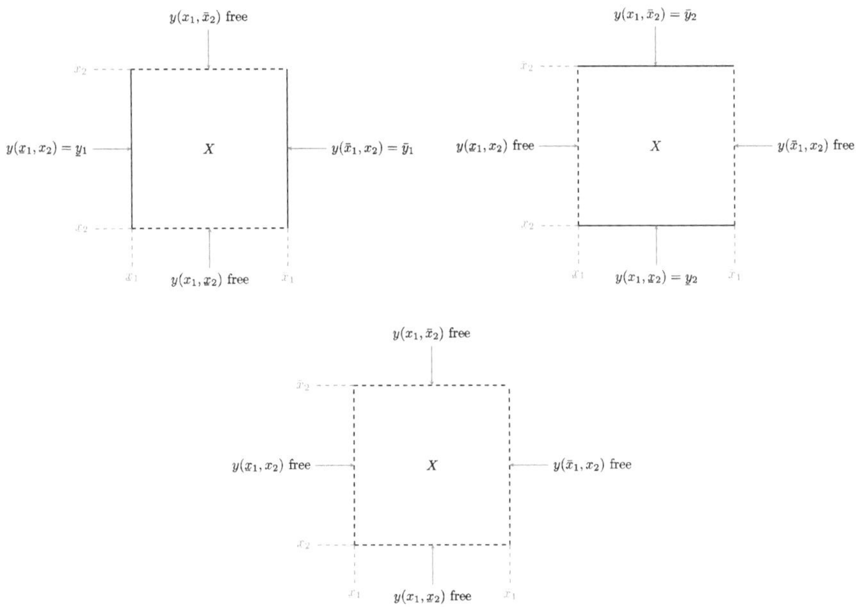

Fig. 14.2 The domain **X** and the boundary conditions for CV problems with constraints (14.7), (14.8), and (14.9). The solid line corresponds to fixed values and the dashed lines to free values for $y(\cdot)$

where $\alpha_1 > 0$, and $\alpha_2 > 0$. The EL PDE is

$$\partial^2_{x_1, x_1} y(\mathbf{x}) + 2\, \partial^2_{x_1 x_2} y(\mathbf{x}) + \partial^2_{x_2 x_2} y(\mathbf{x}) - (\alpha_1 + \alpha_2)\, y(\mathbf{x}) = 0$$

The general solution to the EL equation is

$$y(\mathbf{x}) = f_1(x_2 - x_1) + e^{-x_2\,(\alpha_1+\alpha_2)}\, f_2(x_2 - x_1)$$

where $f_1(\cdot)$ and $f_2(\cdot)$ arc arbitrary functions.

Free boundary values CV problems. Next, we consider the following problems: maximize the objective functional (14.2a) subject to one of the alternative constraints, illustrated in Fig. 14.2:

1. free boundary values for $x_2 \in \{\underline{x}_2, \overline{x}_2\}$

$$y(\underline{x}_1, x_2) = \underline{y}_1,\;\; y(\overline{x}_1, x_2) = \overline{y}_1, \;\text{ and } y(x_1, \underline{x}_2), \text{ and } y(x_1, \overline{x}_2) \text{ free;} \tag{14.7}$$

2. free boundary values for $x_1 \in \{\underline{x}_1, \overline{x}_1\}$

$$y(x_1, \underline{x}_2) = \underline{y}_2,\;\; y(x_1, \overline{x}_2) = \overline{y}_2, \;\text{ and } y(\underline{x}_1, x_2)\; y(\overline{x}_1, x_2) \text{ free;} \tag{14.8}$$

3. free boundary values for $x_1 \in \{\underline{x}_1, \overline{x}_1\}$ and $x_2 \in \{\underline{x}_2, \overline{x}_2\}$

$$y(x_1, \underline{x}_2),\; y(x_1, \overline{x}_2),\; y(\underline{x}_1, x_2), \text{ and } y(\overline{x}_1, x_2) \text{ free.} \tag{14.9}$$

Proposition 14.2 (Necessary conditions for CV problems with boundary free values for the dependent variable) *Let* $\mathbf{y}^* = (y^*(\mathbf{x}))_{\mathbf{x}\in\mathbf{X}}$ *be the solution to the problem of maximizing the objective functional* (14.2a) *subject to one of the alternative constraints* (14.7), (14.8), *or* (14.9). *Then,* $\mathbf{y}^*$ *satisfies the Euler-Lagrange (EL) PDE* (14.3), *and one of the following boundary conditions:*

1. *if constraint* (14.7) *is assumed,*

$$\partial_{\partial_{x_2} y} F^*(x_1, \underline{x}_2) = \partial_{\partial_{x_2} y} F^*(x_1, \overline{x}_2) = 0, \text{ for each } x_1 \in [\underline{x}_1, \overline{x}_1]; \tag{14.10}$$

2. *if constraint* (14.8) *is assumed,*

$$\partial_{\partial_{x_1} y} F^*(\underline{x}_1, x_2) = \partial_{\partial_{x_1} y} F^*(\overline{x}_1, x_2) = 0, \text{ for each } x_2 \in [\underline{x}_2, \overline{x}_2]; \tag{14.11}$$

3. *if constraint* (14.9) *is assumed,*

$$\begin{aligned} \partial_{\partial_{x_2} y} F^*(x_1, \underline{x}_2) &= \partial_{\partial_{x_2} y} F^*(x_1, \overline{x}_2) = 0, \text{ for each } x_1 \in [\underline{x}_1, \overline{x}_1], \\ \partial_{\partial_{x_1} y} F^*(\underline{x}_1, x_2) &= \partial_{\partial_{x_1} y} F^*(\overline{x}_1, x_2) = 0, \text{ for each } x_2 \in [\underline{x}_2, \overline{x}_2]. \end{aligned} \tag{14.12}$$

Proof Let $y^*(\mathbf{x})$ be the solution to problem of maximizing (14.1) subject to (14.7). Following the same steps of the proof of Proposition 14.1, the necessary conditions hold if the integral derivative (14.6) is equal to zero, which is the case if the Euler equation (14.3) holds and

$$\int_{\underline{x}_2}^{\overline{x}_2} \partial_{\partial_{x_1} y} F^*(\overline{x}_1, x_2)\, \eta(\overline{x}_1, x_2)\, dx_2 = \int_{\underline{x}_2}^{\overline{x}_2} \partial_{\partial_{x_1} y} F^*(\underline{x}_1, x_2)\, \eta(\underline{x}_1, x_2)\, dx_2 = 0,$$

$$\int_{\underline{x}_1}^{\overline{x}_1} \partial_{\partial_{x_2} y} F^*(x_1, \overline{x}_2)\, \eta(x_1, \overline{x}_2)\, dx_1 = \int_{\underline{x}_1}^{\overline{x}_1} \partial_{\partial_{x_2} y} F^*(x_1, \underline{x}_2)\, \eta(x_1, \underline{x}_2)\, dx_1 = 0.$$

If constraint (14.7) is imposed, then the two first integral are equal to zero because $\eta(\overline{x}_1, x_2) = \eta(\underline{x}_1, x_2) = 0$ for any $x_2 \in \mathrm{X}_2]$. As $\eta(x_1, \overline{x}_2)$ and $\eta(x_1, \underline{x}_2)$ are arbitrary, and function $F(\cdot)$ is continuously differentiable in the two derivative terms, then the two last integrals are equal to zero if condition (14.10) holds. The proof that the boundary conditions are (14.11) and (14.12) for the cases in which the boundary conditions are (14.8) and (14.9), respectively, is obtained in an analogous way. □

14.1.2 Dynamic Problems

For applications, particularly in macroeconomics and growth theory, one of the independent variables is time, and the other real independent variable refers to any other source of heterogeneity of the variation type. We denote the independent variable vector as (t, x) and let $t \in \mathrm{T} \subseteq \mathbb{R}_+$ and $x \in \mathrm{X} \subseteq \mathbb{R}$.

The simplest time-dependent distributional CV problem. Let the sets of the independent variables be bounded, in particular, $\mathrm{T} = [0, T]$ and $\mathrm{X} = [0, X]$, where $T > 0$ and $X > 0$, and the dependent variable be a mapping $y : \mathrm{T} \times \mathrm{X} \to \mathrm{Y} \subseteq \mathbb{R}$. The simplest time-dependent distributional CV problem is defined by:

$$\max_{y(\cdot)} \mathrm{J}[y] = \int_0^T \int_0^X F\big(\partial_t y(t,x), \partial_x y(t,x), y(t,x), t, x\big)\, dx\, dt, \tag{14.13a}$$

subject to

$$y(0, x) = \phi_0(x), \text{ and } y(T, x) = \phi_T(x), \text{for } x \in \mathrm{X}, \tag{14.13b}$$

$$y(t, 0) = \varphi_0(t), \text{ and } y(t, X) = \varphi_X(t) \text{for } t \in \mathrm{T}. \tag{14.13c}$$

Let $y^*(t, x)$ by the solution to the CV problem (14.13). Then, a direct application of Proposition 14.1 yields the necessary conditions, consisting in the EL PDE

$$\partial_y F^*(t,x) = \frac{\partial}{\partial t}\left(\partial_{\partial_t y} F^*(t,x)\right) + \frac{\partial}{\partial x}\left(\partial_{\partial_x y} F^*(t,x)\right), \ (t,x) \in \mathrm{T} \times \mathrm{X}, \tag{14.14}$$

and the boundary conditions

$$\begin{cases} y^*(0,x) = \phi_0(x), & \text{for } (t,x) \in \{t=0\} \times \mathrm{X}, \\ y^*(T,x) = \phi_T(x), & \text{for } (t,x) \in \{t=T\} \times \mathrm{X}, \\ y^*(t,0) = \varphi_0(t), & \text{for } (t,x) \in \mathrm{T} \times \{x=0\}, \\ y^*(t,X) = \varphi_X(t), & \text{for } (t,x) \in \mathrm{T} \times \{x=X\}, \end{cases} \tag{14.15}$$

Finite time free boundary distributional CV problem. If we assume instead a CV distributional problem in which the constraints refer to the initial time and the lower bound of the set X, as $y(0, x) = \phi_0(x)$, and $y(t, 0) = \varphi_0(t)$, then the necessary conditions include the EL PDE (14.14) together with the following constraints

$$\begin{cases} y^*(0,x) = \phi_0(x), & \text{for } (t,x) \in \{t=0\} \times \mathrm{X}, \\ \partial_{\partial_t y} F^*(T,x) = 0, & \text{for } (t,x) \in \{t=T\} \times \mathrm{X}, \\ y^*(t,0) = \varphi_0(t), & \text{for } (t,x) \in \mathrm{T} \times \{x=0\}, \\ \partial_{\partial_x y} F^*(t,X) = 0, & \text{for } (t,x) \in \mathrm{T} \times \{x=X\}. \end{cases} \tag{14.16}$$

Infinite horizon distributional CV problem. We can extend the previous case to the following infinite horizon distributional CV problem, in which we can have both time and x-related weighting functions, $g(\cdot)$ and $n(\cdot)$ respectively,[5]

$$\max_{y(\cdot)} \mathsf{J}[y] = \int_0^\infty \int_0^\infty f(\partial_t y(t,x), \partial_x y(t,x), y(t,x))\, n(x)\, g(t)\, dx\, dt, \tag{14.17a}$$

subject to

$$y(0,x) = \phi_0(x), \text{ for } x \in \mathrm{X}, \tag{14.17b}$$

$$y(t,0) = \varphi_0(t), \text{ for } t \in \mathrm{T}. \tag{14.17c}$$

The necessary conditions for an optimum $\mathbf{y}^* = \left(y^*(t,x)\right)_{(t,x)\in \mathrm{T}\times\mathrm{X}}$, are the EL PDE

$$f_y^*(t,x) = \frac{g'(t)}{g(t)} f^*_{\partial_t y}(t,x) + \frac{n'(x)}{n(x)} f^*_{\partial_x y}(t,x) + \tag{14.18}$$

$$+ f^*_{\partial_t yy}(t,x)\partial_t y^*(t,x) + f^*_{\partial_x yy}(t,x)\, \partial_x y^*(t,x) + \tag{14.19}$$

$$+ f^*_{\partial_t y \partial_t y}(t,x)\, \partial_{tt} y^*(t,x) + f^*_{\partial_t y \partial_x y}(t,x) \partial_{tx} y^*(t,x) + \tag{14.20}$$

$$+ f^*_{\partial_x y \partial_t y}(t,x)\, \partial_{xt} y^*(t,x) + f^*_{\partial_x y \partial_x y}(t,x) \partial_{xx} y^*(t,x) \tag{14.21}$$

and the (formally identical) boundary conditions (14.16).

Example 14.2 (*A age-dependent resource depletion problem*) As a simple example, consider an extension of the resource depletion Eq. 10.3 to an age-dependent population. Denoting by $w(t,a)$ the size of the endowment allocated to the members of the population with age $a \in \mathrm{A} = [0, A]$, at time $t \in \mathrm{T} = [0, \infty)$, this size is driven by the PDE $\partial_t w(t,a) + \partial_a w(t,a) = -c(t,a)$, where $c(t,a)$ is the consumption of the members with age a at time t. We assume that the utility functional involve both time discounting and the change in the size of every cohort by mortality,

$$\mathsf{U}[c] = \int_0^\infty \int_0^A \ln\left(c(t,a)\right) e^{-\mu a}\, e^{-\rho t}\, da\, dt,$$

where $\rho > 0$ is the social rate of time preference, and μ is the instantaneous mortality rate.

Substituting the PDE constraint, we have a distribution CV problem for the distribution of the resource between members of the age-dependent population over time

[5] If the objective functional is bounded then we can change the order of integration by invoking Fubini's theorem.

$$\max_{c(\cdot)} \mathsf{U}[c] = \int_0^\infty \int_0^A \ln\left(c(t,a)\right) e^{-\mu a}\, e^{-\rho t}\, da\, dt, \tag{14.22a}$$

subject to

$$\int_0^A w(0,a)\, da = \phi, \tag{14.22b}$$

$$\lim_{t\to\infty} w(t,a) = 0, \ \text{ for each } a \in \mathrm{A}, \tag{14.22c}$$

where ϕ is the initial size of the resource. The EL equation is the second-order PDE

$$\begin{aligned} &\partial_{tt} w(t,a) + 2\,\partial_{at} w(t,a) + \partial_{aa} w(t,a) + \\ &+ (\rho+\mu)\left(\partial_t w(t,a) + \partial_a w(t,a)\right) = 0, \ \text{ for each } (t,a) \in \mathrm{A}\times\mathrm{T}. \end{aligned} \tag{14.23}$$

This equation has the general solution of the form

$$w(t,a) = f_1(a-t) + f_2(t-a) e^{-(\rho t + \mu a)t},$$

where $f_1(.)$ and $f_2(.)$ are arbitrary functions defined over the domain $\mathrm{A}\times\mathrm{T}$. Setting $f_1(a-t) = 0$ and $f_2(t-a) = \phi\left(\frac{\mu}{1-e^{-\mu A}}\right)$ yields the solution

$$w^*(t,a) = \phi\left(\frac{\mu}{1-e^{-\mu A}}\right) e^{-(\rho t + \mu a)}, \ \text{ for each } (t,a) \in \mathrm{T}\times\mathrm{A},$$

that satisfies the EL equation (14.23), and both constraints (14.22b) and (14.22c). Observe that, as the size of each cohort reduces over time by mortality by a factor $e^{-\mu a}$, this solution entails an equal distribution per member of the population, at every point in time.

14.2 Optimal Control: Pontriyagin's Principle

Let the independent variables be time and another independent variable $(t,x) \in \mathrm{T}\times\mathrm{X}$, where $\mathrm{T} \subseteq \mathbb{R}_+$ and $\mathrm{X} \subseteq \mathbb{R}$, and let mappings $y : \mathrm{T}\times\mathrm{X} \to \mathrm{Y} \subseteq \mathbb{R}$ and, for simplicity, $u : \mathrm{T}\times\mathrm{X} \to \mathrm{U} \subseteq \mathbb{R}$ denote the state and control variables, respectively. We consider the optimal control problem governed by a first-order PDE,

$$\max_{u(\cdot)} \int_{\mathrm{T}} \int_{\mathrm{X}} F(y(t,x), u(t,x), t, x)\, dx\, dt, \tag{14.24a}$$

subject to

$$\partial_t y(t,x) + \partial_x y(t,x) = G(y(t,x), u(t,x), t, x), \ \ (t,x) \in \mathrm{T}\times\mathrm{X}, \tag{14.24b}$$

$$y(0,x) = y_0(x), \ \ (t,x) \in \{t=0\}\times\mathrm{X}, \tag{14.24c}$$

where $y_0(\cdot)$ is fixed. We can include or not terminal conditions referring to $t = T$, (or $t \to \infty$), and/or boundary conditions referring to $x \in \partial \mathrm{X}$.

Proposition 14.3 (PMP for a first-order PDE optimal control problem) *Assume there is a solution* $(\mathbf{y}^*, \mathbf{u}^*) = (y^*(t,x), u^*(t,x))_{(t,x)\in \mathrm{T}\times\mathrm{X}}$ *to problem* (14.24). *Then there is a function* $\lambda : \mathrm{T}\times\mathrm{X} \to \mathbb{R}$*, and an Hamiltonian function*

$$H(t,x) \equiv H(y,u,\lambda,t,x) \equiv F(y,u,t,x) + \lambda\, G(y,u,t,x)$$

such that $(\mathbf{y}^*, \mathbf{u}^*, \boldsymbol{\lambda})$ *satisfy the (algebraic) optimality condition, the adjoint PDE equation, and the terminal and boundary conditions*

$$\partial_u H^*(t,x) = 0, \text{ for each } (t,x) \in \mathrm{T}\times\mathrm{X}, \tag{14.25a}$$
$$\partial_t \lambda(t,x) + \partial_x \lambda(t,x) + \partial_y H^*(t,x) = 0, \text{ for each } (t,x) \in \mathrm{T}\times\mathrm{X}, \tag{14.25b}$$
$$y^*(T,x) \text{ fixed or } \lambda(T,x) = 0, \text{ for } (t,x) \in \{t = T\}\times\mathrm{X}, \tag{14.25c}$$
$$y^*(t,x) \text{ fixed or } \lambda(t,x) = 0, \text{ for } (t,x) \in \mathrm{T}\times\partial\mathrm{X}. \tag{14.25d}$$

and the constraints (14.24b) *and* (14.24c).

Proof At the optimum, the value functional is $\mathsf{J}[y^*, u^*] = \int_{\mathrm{T}}\int_{\mathrm{X}} F^*(t,x)\,dx\,dt$, where $F^*(t,x) = F(y^*(t,x), u^*(t,x), t, x)$. Introducing the co-state variable $\lambda : \mathrm{T}\times\mathrm{X} \to \mathbb{R}$, and using integration by parts yields

$$\begin{aligned}
\mathsf{J}[y^*, u^*] &= \int_{\mathrm{T}}\int_{\mathrm{X}} \Big(F^*(t,x) + \lambda(t,x)\left(G^*(t,x) - \partial_t y^*(t,x) - \partial_x y^*(t,x)\right)\Big)\,dx\,dt \\
&= \int_{\mathrm{T}}\int_{\mathrm{X}} \Big(H^*(t,x) + (\partial_t \lambda(t,x) + \partial_x \lambda(t,x))\, y^*(t,x)\Big)\,dx\,dt \\
&\quad - \int_{\mathrm{T}}\int_{\mathrm{X}} \lambda(t,x)\, y^*(t,x)\,dx - \int_{\mathrm{T}}\int_{\mathrm{X}} \lambda(t,x)\, y^*(t,x)\,dt,
\end{aligned}$$

where $G^*(t,x) = G(y^*(t,x), u^*(t,x), t, x)$ and $H^*(t,x) = F^*(t,x) + \lambda(t,x)$ $G^*(t,x)$. Introducing a perturbation $\boldsymbol{\eta} : \mathrm{T}\times\mathrm{X} \to \mathrm{Y}\times\mathrm{U}$, such that $(y(t,x), u(t,x))$ $= (y^*(t,x), u^*(t,x)) + \epsilon\,(\eta_y(t,x), \eta_u(t,x))$, yields the Gâteaux differential for the value functional

$$\begin{aligned}
\mathrm{D}_{\boldsymbol{\eta}(\cdot)}\mathsf{J}[y^*, u^*] &= \int_{\mathrm{T}}\int_{\mathrm{X}} \partial_u H^*(t,x)\,\eta_u(t,x)\,dx\,dt + \\
&\quad + \int_{\mathrm{T}}\int_{\mathrm{X}} \left(\partial_y H^*(t,x) + \partial_t \lambda(t,x) + \partial_x \lambda(t,x)\right)\eta_y(t,x)\,dx\,dt \\
&\quad - \int_{\mathrm{T}}\int_{\mathrm{X}} \lambda(t,x)\,\eta_y(t,x)\,dx - \int_{\mathrm{T}}\int_{\mathrm{X}} \lambda(t,x)\,\eta_y(t,x)\,dt.
\end{aligned}$$

As $(y^*(\cdot), u^*(\cdot))$ is the optimum only if $\mathrm{D}_{\boldsymbol{\eta}(\cdot)}\mathsf{J}[y^*, u^*] = 0$, we find conditions (14.25a)–(14.25d). □

Differently from the calculus of variations approach, a necessary condition for an optimum is that the solution satisfies the **system of first-order PDE**. If $\partial^2_{u\,u} H^*(t,x) \neq 0$, for all (t,x),[6] then we can apply the inverse function theorem to condition (14.25a) to obtain $u^*(t,x) = U(y(t,x), \lambda(t,x), t, x)$. Substituting in the PDE (14.24b) and (14.25b), yields a MHDS PDE system

$$\partial_t y(t,x) + \partial_x y(t,x) = M_1(y(t,x), \lambda(t,x), t, x), \tag{14.26a}$$

$$\partial_t \lambda(t,x) + \partial_x \lambda(t,x) = M_2(y(t,x), \lambda(t,x), t, x), \tag{14.26b}$$

where $M_1(y(t,x), \lambda(t,x), t, x) \equiv G(y(t,x), U(y(t,x), \lambda(t,x), t, x), t, x)$, and

$$M_2(y(t,x), \lambda(t,x), t, x) \equiv -\Big(\partial_y F(y(t,x), U(t,x,y(t,x), \lambda(t,x))), t, x) +$$
$$\lambda(t.x)\, \partial_y G(y(t,x), U(t,x,y(t,x), \lambda(t,x)), t, x)\Big),$$

evaluated at the optimum. This is an infinite-dimensional extension of the MHDS for optimal control problems of ODE. As the objective of the planner is to find the optimal dynamics of a distribution, $y(t,x)$, over time, given an initial distribution $y_0(x)$, the shadow price, represented by $\lambda(t,x)$ is a dynamic valuation distribution, inducing incentives for changing the state variable both over time and across the heterogeneity support. Obtaining (or characterizing) the general solution to first-order PDE systems (14.26a)–(14.26b) is beyond the scope of this book.[7]

Exercise 14.1 Setup and solve Example 14.2 using the PMP presented in Proposition 14.3.

Optimal transportation problem. In Example 13.8, we presented the transportation equation, in which the transportation technology was exogenous. Next, we assume that the technology is endogenous and is optimally controlled such that an objective functional can be maximized. This leads to a slightly different example of an optimal control of a first-order PDE, a simple optimal transportation problem specified as follows:

$$\max_{u(\cdot)} \mathrm{J}[y,u] = \int_{\mathrm{T}} \int_{\mathrm{X}} F(y(t,x), u(t,x), t, x)\, dx\, dt, \tag{14.27a}$$

subject to

$$\partial_t y(t,x) + \partial_x \Big(g\big(u(t,x), t, x)\, y(t,x)\Big), \quad \text{for each } (t,x) \in \mathrm{T} \times \mathrm{X}, \tag{14.27b}$$

$$y(0,x) = y_0(x), \quad \text{for each } (t,x) \in \{t=0\} \times \mathrm{X}, \tag{14.27c}$$

$$y(t,x) = 0, \quad \text{for each } (t,x) \in \mathrm{T} \times \partial\mathrm{X}, \tag{14.27d}$$

---

[6] Or $\det \mathrm{D}^2_{uu}(H^*) \neq 0$ if there is more than one control variable.

[7] References on multi-dimensional first-order PDE include Evans (1998, Chap. 11) and Benzoni-Gavage and Serre (2007).

where $y_0(x)$ is a known bounded function.

Proposition 14.4 (PMP for the optimal transportation problem) *Let* $(\mathbf{y}^*, \mathbf{u}^*) = (y^*(t,x), u^*(t,x))_{(t,x)\in \mathrm{T}\times \mathrm{X}}$ *be the solution to the optimal control problem* (14.27). *Then there exists a continuously differentiable function* $\lambda : \mathrm{T}\times \mathrm{X} \to \mathbb{R}$ *such that the following conditions are satisfied: the optimality condition*[8]

$$F_u^*(t,x) + g_u^*(t,x)\, y^*(t,x)\, \partial_x \lambda(t,x) = 0, \text{ for each } (t,x) \in \mathrm{T}\times \mathrm{X}; \tag{14.28}$$

the adjoint equation,

$$F_y^*(t,x) + \partial_t \lambda(t,x) + g^*(t,x)\, \partial_x \lambda(t,x) = 0, \text{ for each } (t,x) \in \mathrm{T}\times \mathrm{X}; \tag{14.29}$$

one of the following terminal conditions,

$$\lambda(T,x) = 0, \text{ if } \mathrm{T} = [0,T], \text{ for each } x \in \mathrm{X}, \tag{14.30a}$$

$$\lim_{t\to\infty} \lambda(t,x) = 0, \text{ if } \mathrm{T} = \mathbb{R}_+, \text{ for each } x \in \mathrm{X}; \tag{14.30b}$$

and the admissibility conditions,

$$\partial_t y^*(t,x) + \partial_x \Big(g^*(t,x)\, y^*(t,x) \Big), \text{ for each } (t,x) \in \mathrm{T}\times \mathrm{X}, \tag{14.31a}$$

$$y^*(0,x) = y_0(x), \text{ for each } (t,x) \in \{t=0\}\times \mathrm{X}, \tag{14.31b}$$

$$y^*(t,x) = 0, \text{ for each } (t,x) \in \mathrm{T}\times \partial\mathrm{X}. \tag{14.31c}$$

Proof The value functional is $\mathsf{J}[y^*, u^*] = \int_{\mathrm{T}}\int_{\mathrm{X}} F^*(t,x)\, dx\, dt$. Then, introducing the co-state variable $\lambda(\cdot)$, we have

$$\begin{aligned} \mathsf{J}[y^*, u^*] &= \int_{\mathrm{T}}\int_{\mathrm{X}} \Big(F^*(t,x) - \lambda(t,x) \Big(\partial_t y^*(t,x) + \partial_x \big(g^*(t,x)\, y^*(t,x) \big) \Big) \Big)\, dx\, dt \\ &= \int_{\mathrm{T}}\int_{\mathrm{X}} \Big(F^*(t,x) + \Big(\partial_t \lambda(t,x) + \partial_x \lambda(t,x)\, g^*(t,x) \Big)\, y^*(t,x) \Big)\, dx\, dt \\ &\quad - \int_{\mathrm{X}} \lambda(t,x)\, y^*(t,x)\, dx \Big|_{t\in\partial T} - \int_{\mathrm{T}} \lambda(t,x)\, g^*(t,x)\, y^*(t,x)\, dt \Big|_{x\in\partial X}. \end{aligned}$$

Introducing admissible perturbations to the control and state variables $u^*(t,x) \to u^*(t,x) + \epsilon\, \eta_u(t,x)$, and $y^*(t,x) \to y^*(t,x) + \epsilon\, \eta_y(t,x)$, for $\eta(0,x) = 0$, for every $x \in \mathrm{X}$, and $\eta(t,x) = 0$ for $(t,x) \in \mathrm{T}\times\partial\mathrm{X}$, we have the Gâteaux differential

[8] We use notation $F^*(t,x) = F(y^*(t,x), u^*(t,x), t, x)$, $g^*(t,x) = g(u^*(t,x), t, x)$, and an analogous notation for their derivatives, as regards both y and u.

$$
\begin{aligned}
D_{\eta(\cdot)} J[y^*, u^*] = & \int_T \int_X \Big(F_u^*(t,x) + \partial_x \lambda(t,x)\, g^*(t,x)\, y^*(t,x) \Big)\, \eta_u(t,x)\, dx\, dt \\
& + \int_T \int_X \Big(F_y^*(t,x) + \partial_t \lambda(t,x) + \partial_x \lambda(t,x)\, g^*(t,x) \Big)\, \eta_y(t,x)\, dx\, dt \\
& - \int_X \lambda(t,x)\, \eta_y(t,x)\, dx \Big|_{t \in \partial T} \\
& - \int_T \lambda(t,x) \Big(g_u^*(t,x)\, y^*(t,x)\, \eta_u(t,x) + g^*(t,x)\, \eta_y(t,x) \Big)\, dt \Big|_{x \in \partial X}
\end{aligned}
$$

Given the constraints, (14.27c) and (14.27d), and the associated admissible perturbation values $\eta_y(t, x) = 0$, for $x \in \partial X$, then $D_{\eta(\cdot)} J[y^*, u^*] = 0$ if conditions (14.28), (14.29), and (14.28) are satisfied. □

14.3 Application to Economics

Example 14.3 (*Social welfare with aggregate mobility*) In Example 13.9, we showed one application of the transport equation to model the dynamics of capital distribution driven by aggregate social mobility. The basic mechanism for the distribution dynamics was related to the effect of savings in propagating net wealth, across the continuum of heterogeneous households, over time. In this example, we assume there is a social planner that optimally allocates consumption, and therefore savings, in order to maximize an intertemporal KAS social welfare functional. The state variable is the distribution of households over the net wealth continuum, $k \in K = \mathbb{R}_+$, for each time $t \in T = \mathbb{R}_+$, mapping $n : T \times K \to [0, 1]$, such that $\int_K n(t, k)\, dk = 1$, and the control variable is consumption $c : T \times K \to C \subseteq \mathbb{R}_+$. We assume the social welfare functional (8.60) with the social welfare function (8.61), displaying inequality aversion,[9] subject to the constraint (13.51):

$$
\max_{c(\cdot)} \int_0^\infty \int_0^\infty \left(\frac{c(t,k)^{1-\xi} - 1}{1-\xi} \right) n(t,k)\, e^{-\rho t}\, dk\, dt, \tag{14.32a}
$$

subject to

$$
\partial_t n(t,k) + \partial_k \big((A\,k - c(t,k))\, n(t,k) \big) = 0, \;\; (t,k) \in T \times K, \tag{14.32b}
$$

$$
n(0,k) = n_0(k), \;\; k \in K, \tag{14.32c}
$$

$$
\lim_{k \to \infty} n(t,k) = n(t,0) = 0, \;\; t \in T, \tag{14.32d}
$$

where $\xi > 0$ represents inequality aversion, and the initial density satisfies the normalization constraint $\int_K n_0(k)\, dk = 1$, for each $t \in T$. Defining the current value co-state variable $q(t, k) = e^{\rho t} \lambda(t, k)$, the Proposition 14.4 allows us to find the

---

[9] We sidestep the problem of the difference between inequality aversion related to household consumption or to household utility, by assuming it relates to aversion to inequality in consumption.

optimality conditions

$$c(t,k)^{-\xi} = \partial_k q(t,k) \tag{14.33a}$$

$$\frac{c(t,k)^{1-\xi}-1}{1-\xi} + \partial_t q(t,k) - \rho\, q(t,k) + \big(A\,k - c(t,k)\big)\,\partial_k q(t,k) = 0 \tag{14.33b}$$

$$\lim_{t\to\infty} e^{-\rho t} q(t,k) = 0. \tag{14.33c}$$

Substituting $c(\cdot)$, from Eq. (14.33a), into Eq. (14.33b) yields the non-linear first-order PDE

$$\partial_t q(t,k) - \rho\, q(t,k) + A\,k\,\partial_k q(t,k) - \frac{1}{1-\xi} + \frac{\xi}{1-\xi}\big(\partial_k q(t,k)\big)^{\frac{\xi-1}{\xi}} = 0.$$

Integrating for t

$$q(t,k) = e^{\rho t}\left(q(0,k) - \int_0^t e^{-\rho s}\Big(A\,k\,\partial_k q(s,k) - \frac{1}{1-\xi} + \frac{\xi}{1-\xi}\big(\partial_k q(s,k)\big)^{\frac{\xi-1}{\xi}}\Big)\,ds\right)$$

taking the limit $t \to \infty$ and using the transversality constraint (14.33c) yields

$$q(0,k) = \int_0^\infty e^{-\rho t}\Big(A\,k\,\partial_k q(t,k) - \frac{1}{1-\xi} + \frac{\xi}{1-\xi}\big(\partial_k q(t,k)\big)^{\frac{\xi-1}{\xi}}\Big)\,ds, \text{ for } k \in \mathrm{K}.$$

As the right-hand side should be time independent, let $q(t,k) = v(k)$, for each $t \in [0,\infty)$. Then we obtain a HJB-like equation

$$\rho\, v(k) + A\,k\,v'(k) - \frac{1}{1-\xi} + \frac{\xi}{1-\xi}\Big(v'(k)\Big)^{\frac{\xi-1}{\xi}} = 0.$$

Using the method of undetermined coefficients, we find the solution

$$v(k) = \frac{1}{1-\xi}\left(\frac{\xi}{\rho + (\xi-1)\,A}\Big(\frac{\rho+(\xi-1)\,A}{\xi}\,k\Big)^{1-\xi} - \frac{1}{\rho}\right).$$

Therefore, the optimal consumption for a household with net wealth k at time t is

$$c^*(t,k) = v'(k)^{-\frac{1}{\xi}} = \Big(\frac{\rho+(\xi-1)\,A}{\xi}\Big)\,k, \text{ for each } (t,k) \in \mathrm{T}\times\mathrm{K}.$$

Substituting in the constraint (14.32b), yields the linear first-order PDE

$$\partial_t n(t,k) + \partial_k\Big(\gamma^*\,k\,n(t,k)\Big) = 0$$

where $\gamma^* = \frac{A-\rho}{\xi}$. Solving this equation together with the initial distribution (14.32c) allows us to find the optimal density

$$n^*(t,k) = e^{-\gamma^* t}\, n_0\Big(k\, e^{-\gamma^* t}\Big). \tag{14.34}$$

You can find an illustration and a characterization of this solution in Example 13.10 for $n_0(k) \sim LN(\mu, \sigma^2)$. The solution displays endogenous growth, aggregate social mobility, but pure social immobility, translated by the fact that the usual inequality measures (the Gini, Theil and quantile ratios) are constant over time.

14.4 References

Calculus of variations for first-order PDE are dealt in Gel'fand and Fomin (1963, Sect. 5, Chap. 7).

Chapter 15
Scalar Parabolic Partial Differential Equations

Parabolic partial differential equations (PDE) are functional equations defined by a function depending on an unknown function of at least two independent variables, and its partial derivatives of at least second order. In this chapter, we assume that the independent variables are time, t, and another variable, x, the unknown function is denoted by $y(t, x)$, and the PDE takes the general form

$$F(\partial_{xx} y(t,x), \partial_t y(t,x), \partial_x y(t,x), y(t,x), t, x) = 0, \text{ for each } (t,x) \in \mathrm{T} \times \mathrm{X}, \tag{15.1}$$

for $y : \mathrm{T} \times \mathrm{X} \to \mathbb{R}$, where $\mathrm{T} \subseteq \mathbb{R}_+$ and $\mathrm{X} \subseteq \mathbb{R}$, and $\partial_t y(t,x) \equiv \frac{\partial y(t,x)}{\partial t}$, $\partial_x y(t,x) = \frac{\partial y(t,x)}{\partial x}$, and $\partial_{xx} y(t,x) = \frac{\partial^2 y(t,x)}{\partial x^2}$. Equation (15.1) represents the dynamics of a distribution across the support X, driven not only by a transport term ($\partial_x y$), and by a growth or decay factor (y), as in first-order PDE, but also by the diffusion term ($\partial_{xx} y$). Sometimes, independent variables (t and x) appear in function $F(.)$ if there are some time or support specific components. The diffusion property is sometimes related to the close connection between parabolic PDE and probability distributions associated with stochastic differential equations (SDE).

In economics and finance applications, it is important to distinguish between **forward** (FPDE) and **backward** (BPDE) parabolic PDE. While the first are complemented with an initial distribution and generate a flow of distributions forward in time, the latter are complemented with a terminal distribution and its solution generates a flow of distributions consistent with that terminal constraint. While for BPDE, the terminal distribution is fixed, for FPDE, the distribution at time $t = 0$ is fixed. In mathematical finance, most applications, such as option pricing models, are PDE of the backward type. In economics, there is recent interest in FPDE related to the topical importance of heterogeneous agent macrodynamic models, distribution issues, and spatial dynamic models. Optimal control of PDE and

P. B. Brito, *Economic Dynamics and Distributions*, Dynamic Modeling and Econometrics in Economics and Finance 32, https://doi.org/10.1007/978-3-031-94717-9_15

mean-field games usually lead to forward-backward partial differential equations (FBPDE).[1]

The body of theory and applications of parabolic PDE in several fields of science and mathematics is immense. In this chapter, we only present some introductory results on linear scalar parabolic PDE having explicit solutions, and some applications related to the distributions generated by the solution of some SDE, and by simple models of heterogeneous household wealth dynamics.

The rest of this chapter presents some definitions and facts concerning scalar parabolic PDE in Sect. 15.1. The solution to forward and backward linear parabolic PDE, as well as some applications, can be found in Sects. 15.2 and 15.3 respectively. At last, in Sect. 15.4, we provide a very brief introduction to linear stability analysis for semi-linear parabolic PDE. In Appendices 15.6, 15.7, and 15.7.2, we gather some useful facts on Fourier transforms and Gaussian integrals.

15.1 A General Overview on Scalar Parabolic PDE

Let $y(t, x)$ be a mapping $y : \mathrm{T} \times \mathrm{X} \subseteq \mathbb{R} \times \mathbb{R}_+ \rightarrow \mathrm{Y} \subseteq \mathbb{R}$ and assume it is almost everywhere continuous and (at least) differentiable to the first order as regards t, and to the second order as regards x. Consider the parabolic PDE (15.1). There are several types of parabolic PDE.

Definition 15.1 (*Non-linear scalar parabolic PDE*) A scalar parabolic PDE is non-linear if, in functional equation (15.1), function $F(\cdot)$ is non-linear in any of the derivatives of function $y(\cdot)$.

Definition 15.2 (*Quasi-linear scalar parabolic PDE*) A scalar parabolic PDE is quasi-linear if it of the form,

$$\partial_t y = a(y, t, x)\, \partial_{xx} y + b(y, t, x)\, \partial_x y + c(y, t, x), \quad \text{for each } (t, x) \in \mathrm{T} \times \mathrm{X}. \tag{15.2}$$

This equation is linear in the derivatives of y, and their coefficients are functions of y.

Definition 15.3 (*Semi-linear scalar parabolic PDE*) A scalar parabolic PDE is semi-linear if it of the form,

$$\partial_t y = a(t, x)\, \partial_{xx} y + b(t, x)\, \partial_x y + c(y, t, x), \quad \text{for each } (t, x) \in \mathrm{T} \times \mathrm{X}. \tag{15.3}$$

This equation is linear in the derivatives, their coefficients are independent from y, but the term independent of the derivatives is a non-linear function of y.

---

[1] Planar parabolic PDE, can be of forward, backward or both forward-backward (FBPDE) type. The last case can be considered a generalization of the saddle-path dynamics for ODE. But these equations, in their general form are beyond the object of this book.

Definition 15.4 (*Linear scalar parabolic PDE*) A scalar parabolic PDE is linear if it of the form,

$$\partial_t y = a(t,x)\,\partial_{xx} y + b(t,x)\,\partial_x y + c(t,x)\,y + d(t,x), \text{ for each } (t,x) \in \mathrm{T} \times \mathrm{X}. \tag{15.4}$$

This equation is linear in both the derivatives, and in function y. As the coefficients depend on (t, x) we say the equation is **non-autonomous**. If all the coefficients are constant, as in

$$\partial_t y = a\,\partial_{xx} y + b\,\partial_x y + c\,y + d, \text{ for each } (t,x) \in \mathrm{T} \times \mathrm{X}. \tag{15.5}$$

we say the PDE is **autonomous**. If coefficient $d(t, x) = 0$, for each (t, x), we say the PDE is **homogeneous**. If there are (t, x) such that $d(t, x) \neq 0$ we say that the PDE is **non-homogeneous**.

Equation (15.4) is sometimes called a diffusion equation with transport, and growth or decay. This is because the time behavior of y depends on three terms: a diffusion term, $a\,\partial_{xx} y$, a transport term, $b\,\partial_x y$, a growth/decay term $c\,y$ and a position term d. If $a > 0$, and there is an initial side-condition, referring to the minimum of T, the equation is a called **forward** PDE (FPDE). However, If $a < 0$, and there is a terminal side-condition, referring to the maximum of T, the equation is a called **backward** PDE (BPDE). While in the first case, the diffusion operator works forward in time, in the second case it works backward in time. The coefficient b involves a transportation across the support set X similar to the first-order PDE, in the forward direction if $b < 0$ or in the backward direction if $b > 0$. The coefficient c produces a change in the whole distribution $y(x, .)$ over time in a way similar to an ODE, and, therefore, there is stability if $c < 0$ or instability if $c > 0$.

In the case of a parabolic PDE, the stability or instability properties are related to the whole distribution: we have **stability in a distributional sense** if there is a solution $y(t, x) = \overline{y}(x)$ such that

$$\lim_{t\to\infty} y(t,x) = \overline{y}(x)$$

where $\overline{y}(x)$ is a stationary distribution, that can be non-degenerate or degenerate (in the sense of converging to a Dirac-delta like distribution). There are several types of **instability in a distributional sense**. In particular, we can have blow-up if $y(t, x)$ becomes unbounded at a finite t, or a non-stationary unbounded distribution if it only becomes unbounded in infinite time (as in economic growth models).

An important element regarding the existence and characterization of the solution of PDE, and the availability of closed form solutions, concerns the **specification of the support** space X. The most common applications assume that the set X can be a line or a circle (a ring). In the first case, we can distinguish between three main cases: the unbounded or infinite case $\mathrm{X} = (-\infty, \infty)$, the semi-bounded of semi-infinite case $\mathrm{X} = [\underline{x}, \infty)$ or $\mathrm{X} = (-\infty, \overline{x}]$, or the bounded case $\mathrm{X} = (\underline{x}, \overline{x})$, where both $\underline{x}$ and $\overline{x}$ are finite numbers.

A **problem involving a parabolic PDE** is specified by a PDE supplement with a side-condition taking the form of a distribution referred to a point in time (an initial distribution for a FPDE or terminal distribution for a BPDE), and possibly conditions involving known values for the values of $y(t, x)$ at the boundaries of X (called boundary conditions), i.e., values for $y(x, \cdot)$ for points $x \in \partial X$, where ∂X denotes the boundary of X.

A **boundary-value problem** is defined by a PDE supplemented with conditions on their boundaries. The conditions on the boundaries can take several forms. In particular: we say we have a **Cauchy problem** if the PDE is supplemented with a condition on the level of y at the boundaries of space X, $y(t, x) = y(t, \underline{x})$ and/or $y(t, x) = y(t, \overline{x})$ for $x \in \{\underline{x}, \overline{x}\} \in \partial X$; or we say we have a **Neumann problem** if the PDE is supplemented with a condition on the spatial derivative of y at the boundaries: $\partial_x y(t, x) = \partial_x y(t, \underline{x})$ and/or $\partial_x y(t, x) = \partial_x y(t, \overline{x})$, for $x \in \{\underline{x}, \overline{x}\} \in \partial X$. However, boundary conditions may not be specified, which is sometimes the case when $X = \mathbb{R}$, and there are other types of boundary side-conditions.

A problem is said to be **well-posed** if there is a solution to the PDE that satisfies both the initial (or terminal) conditions, referring to time, and the boundary conditions, and it is continuous at all those points. In this case we say we have a **classic solution**. If a problem is not well-posed, it is **ill-posed**. In this case, there are no solutions, or we can have generalized or weak solutions. A necessary condition for a problem involving a FPDE to be well posed is that it is supplemented with an initial condition in time, and a necessary condition for a problem BPDE to be well-posed is that it involves a terminal time condition.

In this introduction to parabolic PDE, we will concentrate on equations and problems having explicit solutions. Explicit, or classic solutions, only exist for some parabolic PDE, defined on particular sets of independent variables. There are several methods for solving PDE.[2] Next, we will use mainly Fourier transforms allowing for the transformation of parabolic PDE into parameterized ODE.[3]

15.2 Forward Linear PDE

In this section, we solve several linear autonomous PDE when set X is infinite and semi-infinite in Sects. 15.2.1 and 15.2.2, respectively, and some common non-autonomous parabolic PDE in Sect. 15.2.3.

---

[2] Complete lists are available, for instance, in Polyanin and Zaitsev (2004) or Polyanin and Nazaikinskii (2016).

[3] See the Appendix 15.6 for definitions and applications.

15.2.1 Autonomous PDE in the Infinite Domain

This subsection presents solutions to linear autonomous scalar parabolic PDE, as in Eq. (15.5), in which $\mathrm{X} = \mathbb{R} = (-\infty, \infty)$, that is, equations defined in the infinite "spatial" domain. We also consider forward equations having as side conditions an arbitrary function or a given initial condition of the Cauchy type.

15.2.1.1 The Simplest Equation

The simplest forward linear PDE is

$$\partial_t y(t, x) = a\, \partial_{xx} y(t, x), \text{ for each } (t, x) \in \mathbb{R}_+ \times \mathrm{X}, \tag{15.6}$$

where a is a real positive number.

Proposition 15.1 (Solution to the simplest parabolic PDE) *Let $y(0, x)$ be an arbitrary $L^1(\mathbb{R})$ function, i.e., satisfying $\int_{-\infty}^{\infty} |y(0, x)| dx < \infty$. Then the solution to PDE (15.6) is*

$$y(t, x) = \begin{cases} y(0, x), & (t, x) \in \{t = 0\} \times \mathrm{X}, \\ \dfrac{1}{\sqrt{4\pi a t}} \displaystyle\int_{-\infty}^{\infty} y(0, \xi) e^{-\frac{(x-\xi)^2}{4at}} d\xi, & (t, x) \in \mathbb{R}_{++} \times \mathbb{R}. \end{cases} \tag{15.7}$$

Proof The proof proceeds in three steps. First step: we apply the Fourier operator to PDE (15.6) and transform it into a characteristic ODE. Using Table 15.3, the Fourier transform of the PDE yields

$$\begin{aligned} \mathcal{F}\big[\partial_t y(t, x) - a\, \partial_{xx} y(t, x)\big](\omega) &= \int_{-\infty}^{\infty} \Big(\partial_t y(t, x) - a\, \partial_{xx} y(t, x)\Big)\, e^{-2\pi \iota \omega x}\, dx \\ &= \partial_t Y(t, \omega) + a\, (2\pi\, \omega)^2\, Y(t, \omega), \end{aligned}$$

where $Y(t, \omega) = \mathcal{F}[y(t, x)](\omega)$, for $\omega \in \mathbb{R}$. Calling $\lambda(\omega) \equiv -a\,(2\pi\omega)^2$ an eigenfunction, the PDE is equivalent to the characteristic (spectral) ODE

$$\partial_t Y(t, \omega) = \lambda(\omega)\, Y(t, \omega), \text{ for each } (t, \omega) \in \mathrm{T} \times \mathbb{R},$$

which is a linear ODE with time as the independent variable, and the frequency ω as a parameter. Second step: we solve the characteristic ODE, to obtain

$$Y(t, \omega) = Y(0, \omega)\, G(t, \omega), \text{ for each } (t, \omega) \in \mathrm{T} \times \mathbb{R}$$

where $Y(0, \omega)$ is an arbitrary function of ω, and function

$$G(\omega, t) \equiv \begin{cases} 1, & t = 0, \\ e^{\lambda(\omega)\, t}, & t > 0, \end{cases}$$

is called the **Gaussian kernel**.Third step: applying the inverse Fourier transform operator, yields

$$y(t, x) = \mathcal{F}^{-1}[Y(t, \omega)](x) = \mathcal{F}^{-1}[Y(0, \omega)\, G(t, \omega)] = y(0, x) * g(t, x)$$

where $*$ is the convolution operator (see Eq. (15.50)) where $y(0, x) = \mathcal{F}^{-1}[Y(0, \omega)](x)$, and[4]

$$g(t, x) = \mathcal{F}^{-1}[e^{\lambda(\omega) t}](x) = \begin{cases} \delta(x), & t = 0, \\ \dfrac{e^{-\frac{x^2}{4at}}}{\sqrt{4\pi\, a\, t}}, & t > 0, \end{cases}$$

where $\delta(.)$ is the Dirac's delta generalized function. Then we obtain the general solution (15.7) if $y(0, x) \in L^1(\mathbb{R})$ because $\int_{-\infty}^{\infty} y(0, \xi)\delta(x - \xi)d\xi = y(0, x)$. □

Two observations can be made concerning the solution to this PDE, in Eq. (15.7): first, applying the Fourier transform, we change from a distribution in the original coordinate x to a distribution in frequencies ω. Then the transformed PDE becomes a linear ODE where the coefficient is the eigenfunction

$$\lambda(\omega) = -a\, (2\pi\omega)^2, \ \text{for}\, \omega \in \mathbb{R},$$

that is a real and non-positive function of any $\omega \in \mathbb{R}$ such that: $\lambda(0) = 0$ and $\lambda(\omega) < 0$ for $\omega \neq 0$ and, $\lim_{\omega \to \pm\infty} \lambda(\omega) = -\infty$. This means that $\lim_{\omega \to \pm\infty} Y(t, \omega) = 0$, for any t, if $Y(0, \omega)$ is bounded; second, associated to the previous property, the solution involves a Gaussian perturbation, with a mean zero and a time varying variance equal to $2\, a\, t$, to an initial distribution $y(0, x)$. This implies that asymptotically $\lim_{t \to \infty} y(t, x) = 0$ for each $x \in \mathbb{R}$: the solution displays rarefaction.

We have a **Cauchy problem**, if the initial function is specified, as in

$$\begin{cases} \partial_t y = a\, \partial_{xx} y, & (t, x) \in (0, \infty) \times \mathrm{X}, \\ y(0, x) = y_0(x) & (t, x) \in \{t = 0\} \times \mathrm{X}, \end{cases} \tag{15.8}$$

where $a > 0$ and $y_0(x)$ is a **known function** (and not a known number as in an ODE).

The problem is **well-posed** if $y_0(x)$ is bounded, which means that the solution to the problem exists and is continuous in time. Substituting in the general solution to the PDE, function (15.7), yields the particular solution to the Cauchy problem (15.8)

[4] See Table 15.2 in the Appendix.

$$y(t,x) = \begin{cases} y_0(x), & (t,x) \in \{t=0\} \times \mathrm{X}, \\ \displaystyle\int_{-\infty}^{\infty} \frac{y_0(\xi)}{\sqrt{4\pi\,a\,t}} e^{-\frac{(x-\xi)^2}{4at}} d\xi, & (t,x) \in \mathbb{R}_{++} \times \mathrm{X}. \end{cases} \tag{15.9}$$

Next, we illustrate the solution to several examples, having different initial distributions.

Example 15.1 (*Dirac's delta initial distribution*) Let $a = 1$ and assume that the initial distribution is concentrated at point $x = x_0$. Then $y_0(x) = \delta(x - x_0)$, satisfying $\int_{-\infty}^{\infty} \delta(x - x_0)\,dx = 1$. From Eq. (15.7), the solution is

$$y(t,x) = \frac{1}{\sqrt{4\,\pi\,t}} e^{-\dfrac{(x-x_0)^2}{4\,t}}, \quad \text{for } (t,x) \in \mathrm{T} \times \mathrm{X}.$$

The solution follows again a time-varying density, $y(t,x) \sim N(x_0, 2\,t)$, with mean x_0 and variance $2\,t$. Therefore, it has three noteworthy properties: first, it has a constant mean equal to x_0, second it has an increasing variance decaying over time to a spatially homogeneous distribution $\lim_{t\to\infty} y(t,x) = 0$ for every $x \in \mathbb{R}$, and it satisfies a conservation law at every point in time

$$\int_{-\infty}^{\infty} y(t,x)\,dx = 1, \text{ for every } t \geq 0.$$

Example 15.2 (*An initial Gaussian distribution*) Assume instead that the initial distribution is a standard Gaussian function $y_0(x) \sim N(0,1) = \frac{e^{-\frac{(x)^2}{2}}}{\sqrt{2\pi}}$.[5] From Eq. (15.7), the solution is[6]

$$\begin{aligned} y(t,x) &= \int_{-\infty}^{\infty} \frac{1}{\sqrt{2\pi}\sqrt{4\,\pi\,a\,t}} \exp\left\{-\frac{1}{2}\left(\xi^2 + \frac{(x-\xi)^2}{2at}\right)\right\} d\xi \\ &= \frac{1}{\sqrt{2\,\pi}\,\sqrt{4\,\pi\,a\,t}} \int_{-\infty}^{\infty} \frac{\sqrt{2\,a\,t}}{\sqrt{1+2\,a\,t}} \exp\left\{-\frac{z^2}{2} - \frac{x^2}{2\,(1+2\,a\,t)}\right\} dz \\ &= \frac{1}{\sqrt{2\,\pi\,(1+2\,a\,t)}} e^{-\dfrac{x^2}{2\,(1+2\,a\,t)}}, \quad \text{for each } (t,x) \in \mathrm{T} \times \mathbb{R}, \end{aligned}$$

---

[5] See Appendix Sect. 15.7.2.

[6] As $a > 0$ and $t > 0$, we introduce the variable, $z = \frac{(1+2\,a\,t)\,\xi - x}{\sqrt{(1+2\,a\,t)(2\,a\,t)}}$, integrate by substitution, and use the property of the Gaussian integral $\int_{-\infty}^{\infty} \exp\{-\frac{z^2}{2}\}\,dz = \sqrt{2\pi}$, see Appendix 15.7.2. Recall the formula for integration by substitution of variables: setting $z = Z(\xi)$ and $\xi \in (a,b)$ then

$$\int_{Z(a)}^{Z(b)} f(z)dz = \int_a^b f(Z(\xi))\,Z'(\xi)\,d\xi.$$

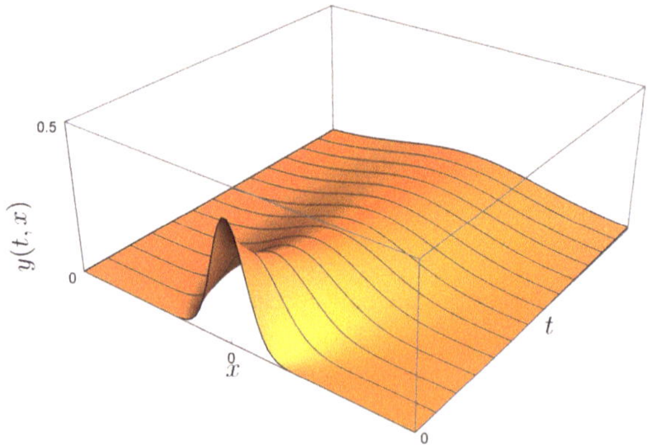

Fig. 15.1 Solution to forward PDE (15.6) with $a = 1$ and $y_0(x) \sim N(0, 1)$

which is Gaussian $N(0, 1 + 2\,a\,t)$ distribution with a variance increasing linearly over time. Figure 15.1 illustrates the behavior of this solution, showing that it has the same properties as the previous example: it satisfies a conservation law, decays over time, and converges to a flat distribution.

Example 15.3 (*Piecewise-constant initial condition*) Let the initial distribution be a discontinuous function defined on the infinite domain

$$y_0(x) = \begin{cases} \phi_0, & \text{if } x \in [\underline{x}, \overline{x}], \\ 0 & \text{if } x \notin [\underline{x}, \overline{x}], \end{cases} \tag{15.10}$$

where $\underline{x} < \overline{x}$ are both finite and ϕ_0 is a constant.[7] Applying the solution (15.9) yields

$$y(t, x) = \phi_0 \left[\Phi\left(\frac{x - \underline{x}}{\sqrt{2at}} \right) - \Phi\left(\frac{x - \overline{x}}{\sqrt{2at}} \right) \right], \tag{15.11}$$

where $\Phi(z)$ is the Gaussian cumulative distribution function (see Eq. (15.54)). Observe that as $\int_{-\infty}^{\infty} e^{-\frac{z^2}{2}}\, dz = \sqrt{2\pi}$ then $\Phi(y) \in (0, 1)$. The solution is illustrated in Fig. 15.2.

To prove this result, we take the general solution (15.7), and apply the initial condition to obtain $y(t, x) = \frac{\phi_0}{\sqrt{4\pi\,a\,t}} \int_{\underline{x}}^{\overline{x}} e^{-\frac{(x-\xi)^2}{4at}}\, d\xi$. To simplify the expression, we make the transformation of variables $z = Z(\xi) \equiv (x - \xi)/\sqrt{2at}$. As $\overline{z} = Z(\overline{x})$, $\underline{z} = Z(\underline{x})$, and $dz = -\frac{1}{\sqrt{2at}}\, d\xi$, then

[7] This problem should not be confused to a problem defined on a set X with a finite width, as $X = [\underline{x}, \overline{x}]$. In this case we should use Fourier series instead of Fourier transforms.

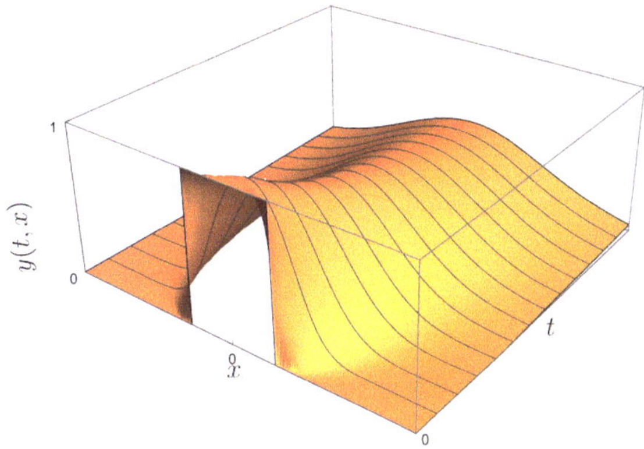

Fig. 15.2 Solution to the forward PDE (15.6) for the piecewise-constant initial condition (15.10) with $\phi_0 = 1$

$$\frac{1}{\sqrt{4\pi at}}\int_{\underline{x}}^{\overline{x}} e^{-\frac{(x-\xi)^2}{4at}}\,d\xi = -\frac{\sqrt{2at}}{\sqrt{4\pi at}}\int_{\frac{x-\underline{x}}{\sqrt{2at}}}^{\frac{x-\overline{x}}{\sqrt{2at}}} e^{-\frac{z^2}{2}}\,dz =$$

$$= \Phi\left(\frac{x-\underline{x}}{\sqrt{2at}}\right) - \Phi\left(\frac{x-\overline{x}}{\sqrt{2at}}\right).$$

Example 15.4 (*The heat equation*) The historically first, most popular, and simplest, linear parabolic PDE is the heat equation. The first formulation of the heat equation is attributed to Jean-Baptiste Joseph Fourier in a presentation to the *Institut de France* and in Fourier (1822), both in 1807.

Denoting the temperature at location $x \in \mathrm{X} = \mathbb{R}$ at time t by $y(t, x)$, the dynamics of the heat distribution across set X is modelled by the linear forward parabolic PDE

$$\partial_t y(t,x) - \partial_{xx} y(t,x) = 0, \text{ for each } (t,x) \in \mathrm{T}\times\mathrm{X}. \tag{15.12}$$

As this model is foundational for the history of parabolic PDE, it is useful to understand the ideas that lead to the derivation of the heat equation. The basic principle is the following: the change in time of the temperature at some point in space is driven by the differences in the temperature between neighboring locations. Consider a homogeneous rod with infinite width, $\mathrm{X} = \mathbb{R}$, and let $y(t, x)$ be the temperature at location $x \in \mathrm{X}$ at time $t \geq 0$. Consider a small segment of the rod $[x, x + \Delta x]$, such as $\Delta x > 0$. The average temperature in the segment at time t is $\int_x^{x+\Delta x} y(t,z)\,dz$, and the instantaneous change of the average temperature in the segment is

$$\frac{d}{dt}\Big(\int_x^{x+\Delta x} y(t,z)\,dz\Big) = \int_x^{x+\Delta x} \partial_t y(t,z)\,dz.$$

It is assumed that this change in temperature is generated by the heat flow through the segment, that cross the two boundaries of the segment, x and $x + \Delta x$. If there is a hotter spot located outside the segment, for instance in a leftward region, and because the heat flows from hot to colder regions, then temperature in the segment $(x, x + \Delta x)$ is lower than in the leftward region, implying $\partial_x y(t, x) < 0$, and it is higher than in the rightward region, implying $\partial_x y(t, x + \Delta x) < 0$, and the gradient in the leftward boundary is higher in absolute terms than in the rightward boundary, $\partial_x y(t, x + \Delta x) - \partial_x y(t, x) > 0$. Invoking the mean-value theorem, the heat flow across the segment is

$$\partial_x y(t, x + \Delta x) - \partial_x y(t, x) = \int_x^{x+\Delta x} \partial_{xx} y(t, z)\, dz > 0.$$

If we **assume** that the instantaneous change in the average temperature of the segment is equal to the heat flow through the segment, then

$$\int_x^{x+\Delta x} \partial_t y(t, z)\, dz = \int_x^{x+\Delta x} \partial_{xx} y(t, z)\, dz.$$

This is equivalent to

$$\int_x^{x+\Delta x} \big(\partial_t y(t, z) - \partial_{xx} y(t, z)\big)\, dz = 0,$$

which holds if and only if Eq. (15.12) is satisfied.

15.2.1.2 The Autonomous Homogeneous Equation

The general forward homogeneous linear scalar parabolic PDE in the infinite domain, $\mathrm{X} = \mathbb{R}$, is

$$\partial_t y(t, x) = a\, \partial_{xx} y(t, x) + b\, \partial_x y(t, x) + c\, y(t, x), \;\; (t, x) \in \mathrm{T} \times \mathrm{X}, \tag{15.13}$$

where $a > 0$, and b and c are arbitrary real numbers.

Proposition 15.2 (Solution to the linear autonomous homogenous parabolic PDE) *Let $y(0, x)$ be an arbitrary $L^1(\mathbb{R})$ function. Then the solution to PDE* (15.13) *is*

$$y(t, x) = \int_{-\infty}^{\infty} \frac{y(0, \xi)}{\sqrt{4\pi a t}}\, e^{\gamma(t, x-\xi)}\, d\xi, \;\; \textit{for each } (t, x) \in \mathrm{T} \times \mathrm{X}, \tag{15.14}$$

where

$$\gamma(t, x) \equiv -\frac{x^2 + 2\, b\, x\, t + (b^2 - 4ac)\, t^2}{4\, a\, t}. \tag{15.15}$$

Proof There are, at least, two alternative methods for solving PDE (15.13): either we can use the same method as in the proof of Proposition 15.1, or we can transform the PDE (15.13) into the simpler PDE (15.6), use solution (15.7), and transform back to the original coordinates.

First method: applying the Fourier operator to PDE (15.13)

$$\mathcal{F}\big[\partial_t y(t,x) - a\,\partial_{xx} y(t,x) - b\,\partial_x y(t,x) - c\,y(t,x)\big](\omega) = 0, \text{ for each } (t,\omega),$$

yields the characteristic ODE

$$\partial_t Y(t,\omega) = \lambda(\omega)\,Y(t,\omega), \text{ for each } (t,\omega),$$

where the eigenfunction $\lambda(\omega)$ is a complex-valued function

$$\lambda(\omega) \equiv -\Big(a(2\pi\omega)^2 - c - b\,2\pi\,\omega\,i\Big), \text{ for } \omega \in \mathbb{R}.$$

The solution to the characteristic ODE is $Y(t,\omega) = Y(0,\omega)\,G(t,\omega)$, where $Y(0,\omega) = \mathcal{F}[y(0,x)](\omega)$, and $G(t,\omega) = \exp\{\text{-}\lambda(\omega)t\}$, for $t > 0$. Applying the inverse Fourier transform operator yields $y(t,x) = y(0,x) * g(t,x)$ where $y(0,x) = \mathcal{F}^{-1}[Y(0,\omega)](x)$ and (see the Appendix 15.6 Table 15.2)

$$g(t,x) = \mathcal{F}^{-1}\left[e^{\text{-}\lambda(\omega)t}\right] = \frac{1}{\sqrt{4\pi a t}} e^{\gamma(t,x)}, \tag{15.16}$$

because $a\,t > 0$, where $\gamma(t,x)$ is given in Eq. (15.15).

Method 2: defining a transformed function $v(t,z) = e^{-ct}y(t,x)$ where $z = x + b\,t$, Eq. (15.13) is equivalent to the simpler linear PDE $\partial_t v(t,z) = a\,\partial_{zz} v(t,z)$, that has solution

$$v(t,z) = \int_{-\infty}^{\infty} v(0,s) \frac{e^{-\frac{(z-s)^2}{2at}}}{\sqrt{4\pi\,a\,t}}\,ds.$$

Transforming back to the original independent variables yields function (15.14). □

The solution to the Cauchy problem, in which $y(0,x) = y_0(x)$, for $x \in \mathrm{X}$, where $y_0(\cdot)$ is a specified function, is

$$y(t,x) = \int_{-\infty}^{\infty} \frac{y_0(\xi)}{\sqrt{4\pi a t}}\, e^{-\dfrac{(x-\xi)^2 + 2\,b\,(x-\xi)\,t + (b^2 - 4ac)\,t^2}{4\,a\,t}}\, d\xi. \tag{15.17}$$

This solution contains three terms: a forward diffusion term, $a > 0$, a transport term, if $b \neq 0$, and a growth or decay term if $c > 0$ or $c < 0$. We can define the functional

$$\mathsf{y}(t) = \int_{\mathrm{X}} y(t,x)dx$$

aggregating the spatial dimension of the solution for every point in time. The solution has one of the following properties: either $\mathsf{y}(t) = e^{c\,t}$, for any $t > 0$, or $\lim_{t\to\infty} \mathsf{y}(t) = \lim_{t\to\infty} e^{c\,t}$. Therefore, considering the dynamics of the distribution functional, we can have three types of solutions, depending on the sign of c: (1) if $c = 0$, the solution is **conservative**, that is $\mathsf{y}(t) = \bar{y}$, for each $t > 0$, where $\bar{y}$ is a constant; (2) if $c < 0$, the solution is **dissipative**, that is $\lim_{t\to\infty} \mathsf{y}(t) = 0$; or (3) if $c > 0$ the solution is **unbounded**that is $\lim_{t\to\infty} \mathsf{y}(t) = \infty$.

The dynamics in a distributional sense, measured by the dynamics of the first and second centered moments, defined by the functionals

$$\mathsf{m}(t) = \int_{\mathrm{X}} x\, y(t,x) dx, \text{ and } \mathsf{v}(t) = \int_{\mathrm{X}} (x - \mathsf{m}(t))^2\, y(t,x) dx,$$

can be better understood by assuming a specific initial $L^1(\mathrm{X})$ function $y_0(x)$.

Example 15.5 (*Initial Gaussian distribution*) Let $y_0(x) = \frac{e^{-x^2}}{\sqrt{\pi}}$, then solution (15.14) becomes

$$y(t,x) = \frac{e^{c\,t - \frac{(x+b\,t)^2}{1+4\,a\,t}}}{\sqrt{\pi\,(1+4\,a\,t)}}, \quad \text{for each } (t,x) \in \mathrm{T} \times \mathrm{X}. \tag{15.18}$$

Then, $\mathsf{y}(t) = e^{c\,t}$, $\mathsf{m}(t) = -b\,t\,e^{c\,t}$ and $\mathsf{v}(t) = (1 + 2\,a\,t)\,e^{c\,t}$. The effects of the three parameters on the dynamics of the functional y and of the statistics m and v are clear: while a has an effect on the variance, and b on the location, c determines the asymptotic distributional stability properties. Figure 15.3 illustrates the behavior of $y(t,x)$, according to solution (15.18), for all the sign combinations of parameters b and c : while parameter b transports the mean to ∞ if $b < 0$ (see Fig. 15.3a, b, and c) or to $-\infty$ if $b > 0$ (see Fig. 15.3g, h, and i), parameter c forces the solution to converge exponentially to zero if $c < 0$ (see Fig. 15.3a, d, and g), or to ∞ (see Fig. 15.3c, f, and i) for each $x \in \mathrm{X}$. If $c = 0$, the mean increases (decreases) linearly over time if $b < 0$ (if $b > 0$) and the variance also grows linearly over time (see Fig. 15.3b, e, and h).

Example 15.6 (*Probability distribution associated to a Brownian motion*) The conditional probability density for a Brownian motion, $p(t,y) = \mathbb{P}[Y(t) = y|Y(0) = y_0]$, where $p : \mathrm{T} \times \mathbb{R} \to (0,1)$ (see Sect. 18.1.1.1) is the solution to the Cauchy problem, specified by the Fokker-Planck equation and a Dirac-delta generalized function,

$$\begin{cases} \partial_t p(t,y) = \frac{\sigma^2}{2} \partial_{yy} p(t,y) - \mu\, \partial_y p(t,y), & (t,y) \in \mathrm{T} \times \mathbb{R}, \\ p(0,y) = \delta(y - y_0) & (t,y) \in \{t = 0\} \times \mathbb{R}. \end{cases} \tag{15.19}$$

This is a particular case of Eq. (15.13) with $a = \frac{\sigma^2}{2}$, $b = -\mu < 0$, and $c = 0$. Substituting those coefficients in Eq. (15.14) yields the solution to problem (15.19),

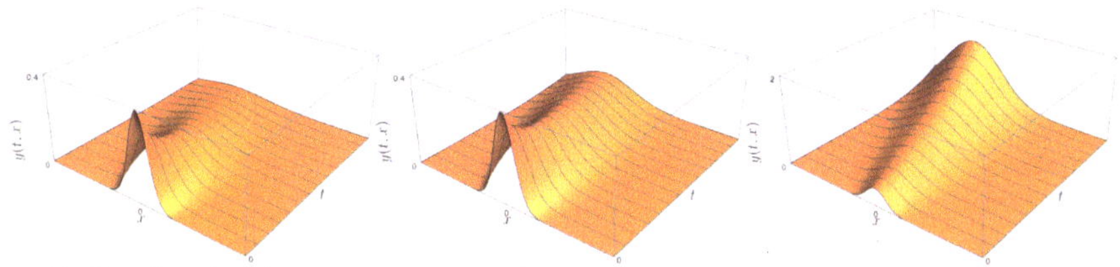

(a) Case $b < 0$ and $c < 0$ (b) Case $b < 0$ and $c = 0$ (c) Case $b < 0$ and $c > 0$

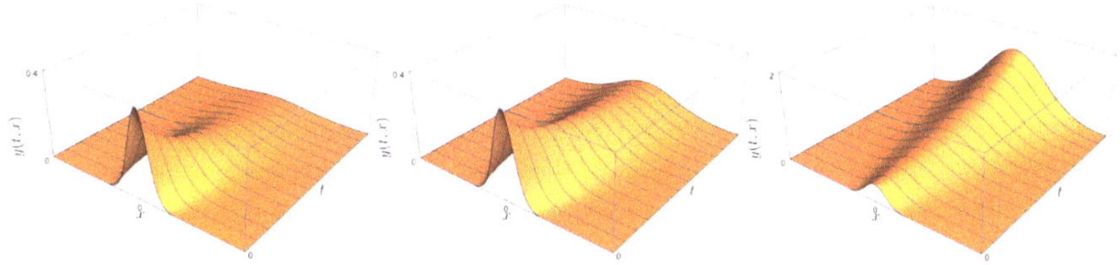

(d) Case $b = 0$ and $c < 0$ (e) Case $b = 0$ and $c = 0$ (f) Case $b = 0$ and $c > 0$

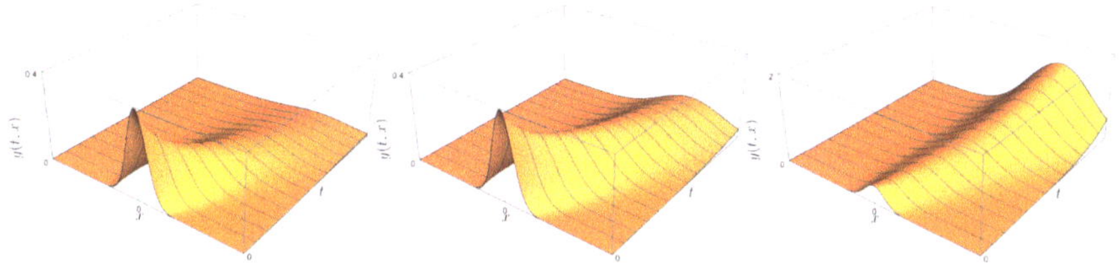

(g) Case $b > 0$ and $c < 0$ (h) Case $b > 0$ and $c = 0$ (i) Case $b > 0$ and $c > 0$

Fig. 15.3 Solutions to the forward PDE (15.13) for $a = 1$ and $y_0(x) \sim N(0, 1)$

Fig. 15.4 Solution to PDE (15.19) for $y_0 = 5$, $\mu = 1$ and $\sigma = 0.5$

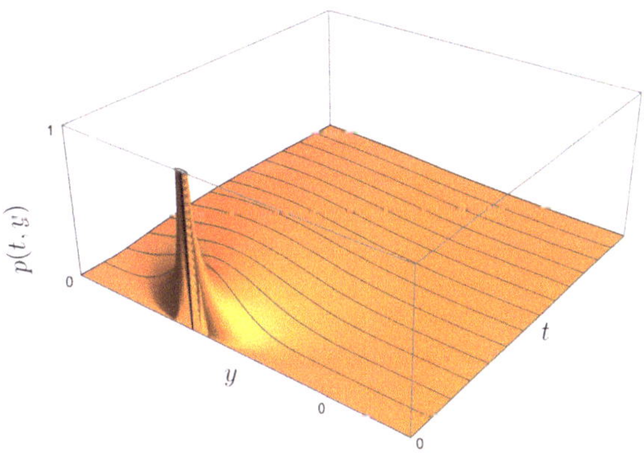

$$p(t, y) = \frac{e^{-\left(\frac{y-y_0-\mu t}{2\sigma^2 t}\right)^2}}{\sqrt{2\pi\sigma^2 t}}, \quad \text{for} \quad (t, y) \in \mathrm{T} \times \mathbb{R}. \tag{15.20}$$

Then $p(t, y) \sim N(y_0 + \mu t, \sigma^2 t)$ illustrated in Fig. 15.4.

Example 15.7 (*A distributional Solow AK model*) In Brito (2006), we prove that in an economy in which the capital stock is heterogeneously distributed across space, and can move freely, the macroeconomic equilibrium condition can be represented by a parabolic PDE. Let the set of locations be specified by $\mathrm{X} = \mathbb{R}$, and denote by $K(t, x)$ the capital stock at location $x \in \mathrm{X}$ at time $t \geq 0$. Let there be a single good that can be produced in every location x and can be used in consumption and investment. The good's market equilibrium condition in every location x at time t requires an equality between internal and external investments, and savings, $I(t, x) + T(t, x) = S(t, x)$, where $T(\cdot)$ represents the net capital flows. If $T(t, x) > 0$ ($T(t, x) < 0$) location x is a capital exporter (importer).

Consider a region of length $\Delta x \geq 0$ and assume that the net capital flow is proportional to the symmetric to the gradient of the capital stock as regards the neighboring regions, $T(t, x)\Delta x = \tau^2\ (\partial_x K(t, x) - \partial_x K(t, x + \Delta x))$, where τ^2 represents the barrier to regional capital flows. Intuitively, capital flows from capital rich to capital poor locations. Therefore, the net capital flow for every location x is determined by letting $\Delta x \to 0$, that is $T(t, x) = -\tau^2\, \partial_{xx} K(t, x)$.

Let the production technology be linear and homogeneous over space, such that income is $Y(t, K) = AK$, for $A > 0$. If the savings behavior is add-hoc, as in the Solow (1956) model, and is homogeneous across space, then savings in location x is $S(t, x) = s\, A\, K(t, x)$, where $0 < s < 1$. At last, assuming that there is no capital depreciation, then $I(t, x) = \partial_t K(t, x)$, the equilibrium capital accumulation at location $x \in \mathrm{X}$ is represented by the following parabolic PDE

$$\partial_t K(t, x) = \partial_{xx} K(t, x) + s\, A\, K(t, x), \text{ for each } (t, x) \in \mathrm{T} \times \mathrm{X}. \tag{15.21}$$

As there is homogeneity across space of the technology and savings behavior, there is a common positive growth rate $\gamma = s\, A$. This allows us to define the detrended capital stock $k(t, x) = K(t, x)\, e^{-\gamma t}$, that is governed by the PDE $\partial_t k(t, x) = \tau^2\, \partial_{xx} k(t, x)$.

Assuming an initial spatial distribution of the capital stock $K(x, 0) = k_0(x)$, from Eq. (15.13), the solution for this spatial AK model is

$$K(t, x) = e^{\gamma t} \int_{-\infty}^{\infty} \frac{k_0(\xi)}{\sqrt{4\pi\, \tau^2\, t}} e^{-\frac{(x-\xi)^2}{4\tau^2 t}}\, d\xi, \text{ for each } (t, x) \in \mathrm{T} \times \mathrm{X},$$

and the solution is similar to the case depicted in Fig. 15.3, for the case in which $b = 0$ and $c > 0$. As the weight of location x is negligible, we have asymptotically $\lim_{t\to\infty} K(t, x) = 0$ for every bounded initial distribution $k_0(x)$. However, if we consider a region aggregating locations, and, in particular, the whole economy we have $\lim_{t\to\infty} \mathsf{k}(t) = \lim_{t\to\infty} \int_{\mathrm{X}} K(t, x) dx = \lim_{t\to\infty} e^{\gamma t} \bar{k}$, if $\int_{\mathrm{X}} k_0(x) dx = \bar{k}$, that is the distribution tends to a spatially-homogenous balanced growth path (BGP) growing at the rate $\gamma = s\, A$. The main conclusions are: (1) if there is constant returns to scale and positive savings, there is long run growth; (2) if there are homogenous technologies and preferences the asymptotic distribution will become spatially homogeneous. This model is characterized by the existence of both so-called β- and σ-convergence.

15.2.1.3 Non-homogeneous Autonomous Linear Equations

At last, the solution to the autonomous linear PDE (15.5), where $a > 0$, and b, c, and d are real numbers, can be obtained.

Proposition 15.3 (Solution to the linear non-autonomous PDE) *The solution to Eq.* (15.5) *is*

$$y(t,x) = y^h(t,x) + \frac{d}{c}\,(e^{c\,t} - 1),\ \textit{for each}\ (t,x) \in \mathbb{R}_+ \times \mathbb{R} \tag{15.22}$$

where $y^h(t,x)$ is the solution to the homogenous equation (15.14)*, and $y(0,x)$ is an arbitrary $L^1(\mathrm{X})$ function.*[8]

Proof In this case, the Fourier transform method simplifies the proof. From Table 15.1 the Fourier transform of the left-hand side of equation (15.5) is

$$\mathcal{F}\big[\partial_t y - a\,\partial_{xx} y - b\,\partial_x y - c\,u + d\big](\omega) = \partial_t Y(t,\omega) - \lambda(\omega)\,Y(t,\omega) - d\,\delta(\omega)$$

where $\delta(\cdot)$ is Dirac's delta distribution, and we have again the eigenfunction $\lambda(\omega) \equiv -a(2\pi\,\omega)^2 + 2\pi\,b\,\omega\,i + c$. The characteristic ODE, $\partial_t Y(t,\omega) = \lambda(\omega)\,Y(t,\omega) + d\,\delta(\omega)$, has the explicit solution

$$Y(t,\omega) = Y(0,\omega)\,G(t,\omega) + \int_0^t d\,\delta(\omega)\,G(t-s,\omega)\,ds,\ \text{ for } t \in \mathrm{T}$$

where $G(t,\omega) = e^{-\lambda(\omega)\,t}$. Applying the IFT operator yields $y(t,x) = y(0,x) * g(t,x) + \int_0^t d * g(t-s,x)\,ds$, where $y(0,x) * g(t,x)$ is the solution to the homogeneous equation, in (15.14), and, because $a > 0$, $\int_0^t d * g(t-s,x)\,ds = \frac{d}{c}\,(e^{c\,t} - 1)$. □

Example 15.8 Consider forward PDE (15.5), where all the coefficients are non-zero, and $y_0(x) = \frac{e^{-\frac{x^2}{2}}}{\sqrt{2\pi}}$. we obtain the solution to the initial value problem,

$$y(t,x) = \frac{e^{c\,t - \frac{(x+b\,t)^2}{2\,(1+2\,a\,t)}}}{\sqrt{2\,\pi\,(1+2a\,t)}} - \frac{d}{c}\left(1 - e^{c\,t}\right).$$

For $d > 0$ and $b = 0$, Fig. 15.5a and b illustrates a case in which $c > 0$, and $c < 0$, respectively. In the first case, the solution is distributionaly unbounded, and in the second converges to a homogeneous distribution $\lim_{t\to\infty} y(t,x) = -\dfrac{d}{c}$ for any $x \in \mathrm{X}$. In both cases, the asymptotic state is driven by the dynamics of the deterministic skeleton $\dot{y} = d + cy$.

---

[8] Note that if $c = 0$ the l'Hôpital rule yields $\lim_{c\to 0} \frac{d}{c}\,(e^{c\,t} - 1) = d\,t$.

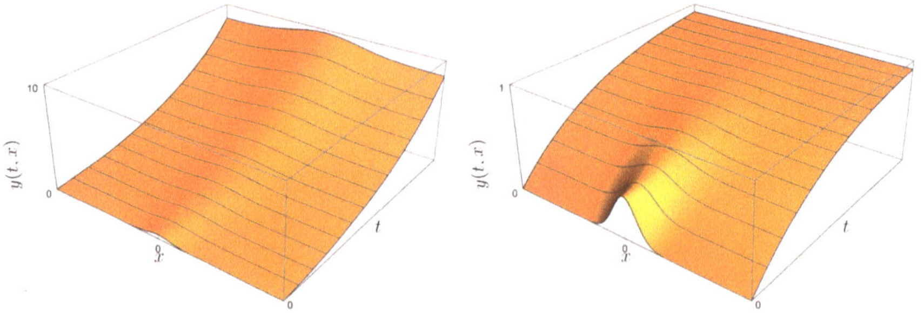

(a) For $b = 0$, $c > 0$ and $d > 0$ (b) For $b = 0$, $c = -0.5$ and $d > 0$

Fig. 15.5 Solution to PDE (15.5) for $a > 0$, $b = 0$, and $y_0(x) \sim N(0, 1)$

15.2.2 The Forward Linear Equation in the Semi-infinite Domain

In this subsection, we address scalar parabolic PDE in which the domain of the independent variable is called semi-infinite, that is $\mathrm{X} = \mathbb{R}_+$. This support space appears in economic applications in which the independent variable cannot take negative values, for instance when x refers to a price or a physical stock variable. This case should not be confused with the case in which $\mathrm{X} = \mathbb{R}$ but the initial condition takes zero values for subsets of the support space. Consider the simplest FPDE, where again $a > 0$,

$$\partial_t y(t, x) = a\, \partial_{xx} y(t, x), \ (t, x) \in \mathbb{R}^2_+. \tag{15.23}$$

Proposition 15.4 (Solution to the simplest PDE in the semi-infinite domain) *Let $y(0, x) \in L^1(\mathbb{R}_+)$. Then the solution to Eq.* (15.23) *depends on the type of function $y(0, x)$:*

1. *if it is an odd function we have*

$$y(t, x) = \int_0^\infty \frac{y(0, \xi)}{\sqrt{4 \pi a t}} \left(e^{-\frac{(x-\xi)^2}{4at}} - e^{-\frac{(x+\xi)^2}{4at}} \right) d\xi, \ (t, x) \in \mathrm{T} \times \mathrm{X}; \tag{15.24}$$

2. *if it is an even function we have*

$$y(t, x) = \int_0^\infty \frac{y(0, \xi)}{\sqrt{4 \pi a t}} \left(e^{-\frac{(x-\xi)^2}{4at}} + e^{-\frac{(x+\xi)^2}{4at}} \right) d\xi, \ (t, x) \in \mathrm{T} \times \mathrm{X}. \tag{15.25}$$

Proof If $\mathrm{X} = \mathbb{R}$ Eq. (15.7) provides the solution to the PDE. The **method of images** allows us to obtain the solution on the half space $\mathrm{X} = \mathbb{R}_+$ from the solution over the space $\mathbb{R}$. In order to apply it, we need to distinguish odd from even functions. The function $f(x)$, where $x \in \mathbb{R}$, is said to be an odd function if $f(x) = -f(-x)$, and if it is said to be an even function if $f(x) = f(-x)$. Let $y(0, x)$, with domain $x \in \mathbb{R}_+$, be an odd function. Then we define the extended function

$$\tilde{y}(0,x) = \begin{cases} y(0,x), & \text{if } x \geq 0 \\ -y(0,-x) & \text{if } x < 0. \end{cases}$$

Substituting in Eq. (15.7), for $t > 0$, yields

$$\begin{aligned} y(t,x) &= \frac{1}{\sqrt{4\pi a t}} \int_{-\infty}^{\infty} \tilde{y}(0,\xi) e^{-\frac{(x-\xi)^2}{4at}} d\xi = \\ &= \frac{1}{\sqrt{4\pi a t}} \left(\int_{-\infty}^{0} \tilde{y}(0,\xi) e^{-\frac{(x-\xi)^2}{4at}} d\xi + \int_{0}^{\infty} \tilde{y}(0,\xi) e^{-\frac{(x-\xi)^2}{4at}} d\xi \right) = \\ &= \frac{1}{\sqrt{4\pi a t}} \left(-\int_{0}^{\infty} y(0,\xi) e^{-\frac{(x+\xi)^2}{4at}} d\xi + \int_{0}^{\infty} y(0,\xi) e^{-\frac{(x-\xi)^2}{4at}} d\xi \right) \end{aligned}$$

where the last step involves integration by substitution: i.e., if we define $u = -x$ for $x \in [0,\infty)$ then $\int_{-\infty}^{0} f(u)du = -\int_{\infty}^{0} f(-x)dx = \int_{0}^{\infty} f(-x)dx$. Then the solution (15.24) for an arbitrary odd function is obtained.

Let $y(0,x)$, with domain $x \in \mathbb{R}_+$, be an even function. Then we define the extended function

$$\tilde{y}(0,x) = \begin{cases} y(0,x), & \text{if } x \geq 0 \\ y(0,-x) & \text{if } x < 0. \end{cases}$$

Then the solution (15.25) for an arbitrary even function is obtained, in an analogous way. □

Example 15.9 Consider the Cauchy problem specified by the simplest PDE (15.23), and by the initial distribution that is a truncated Gaussian distribution $y_0(x) = \frac{1}{\Phi(0)} \frac{e^{\frac{-x^2}{2}}}{\sqrt{2\pi}}$, where $\Phi(x) = \frac{1}{2}\left(1 + \text{erf}\left(\frac{x}{\sqrt{2}}\right)\right)$, for $x \in \mathrm{X} = \mathbb{R}_+$. Observe that $\int_0^\infty y_0(x)dx = 1$. As this is an even function, substituting in solution (15.25) yields the solution to the Cauchy problem,

$$y(t,x) = 2 \frac{e^{-\frac{x^2}{2(1+2at)}}}{\sqrt{2\pi\,(1+2at)}}, \quad \text{for } (t,x) \in \mathrm{T} \times \mathrm{X}$$

that is illustrated in Fig. 15.6. This solution is conservative satisfying $\int_0^\infty y(t,x)\,dx = 1$ for every $t > 0$.

15.2.3 Non-autonomous Equations

The general linear scalar parabolic PDE is the non-autonomous equation (15.4), in which the coefficients are functions of the independent variables (t,x). This structure covers a large number of cases. It has been proved that the homogeneous version of

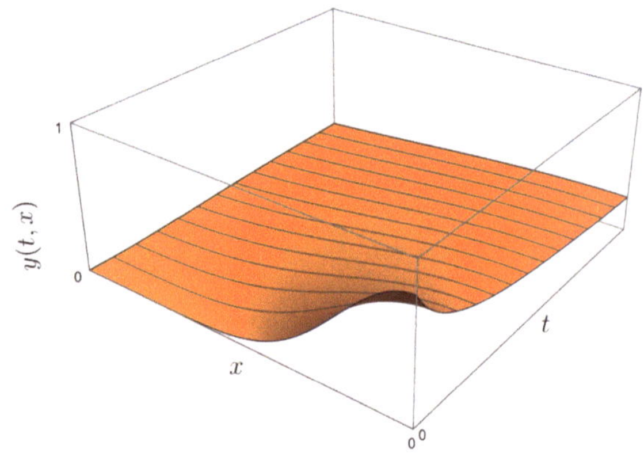

Fig. 15.6 Solution to PDE (15.23), for a truncated standard Gaussian distribution

equation (15.4), such that $d(t, x) = 0$, can be transformed to a PDE of the form[9]

$$\partial_t y = \partial_{xx} y + \tilde{c}(t, x) y$$

A general method of solution consists in finding a mapping $y(t, x) \to Y(\tau, z)$ such that the non-autonomous PDE (15.4), over $y(t, x)$, can be transformed into an autonomous PDE, over $Y(\tau, z)$. This can be done by finding functions $T(\cdot)$ and $X(\cdot)$ such that $t = T(\tau, z)$ and $x = X(\tau, z)$.

Given its potentially important application in modelling the dynamics of heterogeneity in economics, in particular related to Fokker-Planck-Kolmogorov (FPK) PDE, we concentrate next on some particular non-autonomous cases in which the coefficients are time-independent but "spatially"-dependent.

15.2.3.1 Heterogeneous Transport

Consider the parabolic FPDE in which the "transport" term is time-independent but is a linear function of the other independent variable x,

$$\partial_t y = a\, \partial_{xx} y + (b_0 + b_1\, x)\, \partial_x y + c\, y \text{ for each } (t, x) \in \mathrm{T} \times \mathrm{X}, \tag{15.26}$$

in which $a > 0$ and the initial distribution $y(0, x)$ is an arbitrary bounded function.

---

[9] See Gazizov and Ibragimov (1998, p. 384) citing Lie (1881).

Proposition 15.5 *Consider Eq.* (15.26) *for* $\mathrm{X} = \mathbb{R}$ *assuming that* $a > 0$. *The solution to the Cauchy problem, constrained by an initial bounded function* $y(0, x)$, *is*

$$y(t,x) = \int_{\mathrm{X}} \frac{\tilde{y}(0,\xi)}{\sqrt{\frac{2\pi a}{b_1}\left(1-e^{-2b_1 t}\right)}} \times$$

$$\times \exp\left\{(c-b_1)\,t - \frac{\left(x-\xi+\frac{b_0}{b_1}\left(1-e^{-b_1 t}\right)\right)^2}{\frac{2a}{b_1}\left(1-e^{-2b_1 t}\right)}\right\} d\xi, \qquad (15.27)$$

where $\tilde{y}(0,x) = \mathcal{F}^{-1}[\mathcal{F}[y(0,x)](\omega e^{-b_1 t})](x)$.

Proof Defining $y(t,x) = e^{c\,t}\,z(t,x)$, allows us to transform the PDE (15.26) into $\partial_t z = a\,\partial_{xx} y + (b_0 + b_1\,x)\,\partial_x z$. Applying Fourier transforms yields the transformed first-order PDE

$$\partial_t Z(t,\omega) + b_1\,\omega\,\partial_\omega Z(t,\omega) = \ell(\omega)\,Z(t,\omega), \text{ for } (t,\omega) \in \mathrm{T}\times\mathbb{R}$$

where $Z(t,\omega) = \mathcal{F}[z(t,x)]$, $\ell(\omega) = -a\,(2\pi\,\omega)^2 + b_0\,2\pi\,i\,\omega - b_1$ and $Z(0,\omega) = \mathcal{F}[y(0,x)](\omega)$. Using the method of characteristics, and defining $\omega = w(t)$, we obtain the characteristic ODE $\dot{w} = b_1\,w(t)$, and the consistency ODE $\dot{\hat{Z}} = l(t)\,\hat{Z}(t)$ where $\hat{Z}(t) = Z(t, w(t))$, and $l(t) = \ell(w(t))$. Solving those ODE yields $\omega = w(t) = w(0)\,e^{b_1 t}$ and $\hat{Z}(t) = \hat{Z}(0)\,e^{\int_0^t l(s)\,ds}$, where

$$\int_0^t l(s)\,ds = \frac{a}{2\,b_1}\left(1-e^{2\,b_1 t}\right)(2\pi\,w(0))^2 - \frac{b_0}{b_1}\left(1-e^{b_1 t}\right)2\pi i\,w(0) - b_1\,t.$$

Therefore, $Z(t,\omega) = Z\left(0, \omega\,e^{-b_1 t}\right)G(t,\omega)$ where $G(t,\omega) = e^{-\tilde{a}(2\pi\,\omega)+\tilde{b}\,(2\pi\,i\,\omega)+\tilde{c}}$, and $\tilde{a} = \frac{a}{2\,b_1}\left(1-e^{-2\,b_1 t}\right)$, $\tilde{b} = \frac{b_0}{b_1}\left(1-e^{-b_1 t}\right)$, and $\tilde{c} = -b_1\,t$. Applying the inverse Fourier operator to $Z(t,\omega)$, and transforming back to the original variable, we obtain the general solution to the Cauchy problem (15.27). □

Example 15.10 (*The probability distribution of the Ornstein-Uhlenbeck process*) The probability distribution of the Ornstein-Uhlenbeck process,[10] see (18.7), is represented by the solution to Fokker-Planck-Kolmogorov equation, given the initial value of the stochastic process $Y(0) = y_0$. The probability distribution for the process $p(t,y) = \mathbb{P}[Y(t) = y | Y(0) = y_0]$, mapping $p : \mathrm{T}\times\mathrm{Y} \to (0,1)$, where $\mathrm{Y} = \mathbb{R}$, is the solution to the forward PDE:

$$\begin{cases} \partial_t p(t,y) = -\partial_y\left((\mu_0 + \mu_1\,y)\,p(t,y)\right) + \dfrac{\sigma^2}{2}\partial_{yy} p(t,y), & (t,x) \in \mathbb{R}_{++}\times\mathrm{Y}, \\ p(0,y) = \delta(y-y_0), \quad (t,y) \in \{t=0\}\times\mathrm{Y}. \end{cases} \qquad (15.28)$$

[10] First presented in Uhlenbeck and Ornstein (1930).

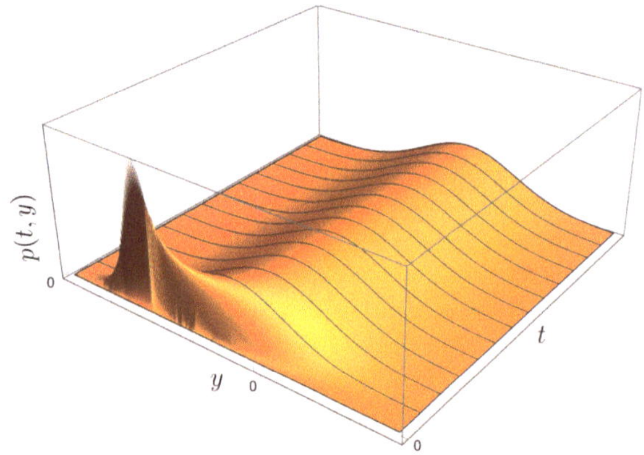

Fig. 15.7 Solution to PDE (15.28) for $x_0 = 5$, $\mu_0 = 1$, $\mu_1 = -1$ and $\sigma = 0.5$

This is a special case of the forward PDE (15.26). Evaluating the solution (15.27) with the specific initial condition in problem (15.28), yields the solution

$$p(t, y) = \frac{\exp\left\{-\frac{\left(y+\frac{\mu_0}{\mu_1}-\left(\frac{\mu_0}{\mu_1}+y_0\right)e^{\mu_1 t}\right)^2}{\frac{\sigma^2}{\mu_1}\left(e^{2\mu_1 t}-1\right)}\right\}}{\sqrt{\pi\,\frac{\sigma^2}{\mu_1}\left(e^{2\mu_1 t}-1\right)}}, \text{ for each } (t, y) \in \mathrm{T} \times \mathbb{R}. \tag{15.29}$$

This is a Gaussian density function with time-varying mean and variance,

$$p(t, y) \sim N\left(\frac{\mu_0}{\mu_1} - \left(\frac{\mu_0}{\mu_1} + y_0\right)e^{\mu_1 t}, \frac{\sigma^2}{2\,\mu_1}\left(e^{2\mu_1 t} - 1\right)\right).$$

If $\mu_1 < 0$ then the asymptotic probability is ergodic: for any initial value y_0 it tends asymptotically to a normal distribution with finite time-dependent parameters, $\lim_{t\to\infty} p(t, y) \sim N\left(-\frac{\mu_0}{\mu_1}, -\frac{\sigma^2}{2\mu_1}\right)$. Figure 15.7 illustrates the dynamics of the probability distribution.

15.2.3.2 Heterogeneous Diffusion and Transport

Next, we consider the parabolic PDE in which both the diffusion and the transport terms are time-independent but depend on the other independent variable, defined on the semi-infinite domain, $x \in \mathrm{X} = \mathbb{R}_+$,

$$\partial_t y = (ax)^2\,\partial_{xx} y + b\,x\,\partial_x y + c\,y \text{ for each } (t, x) \in \mathrm{T} \times \mathrm{X}. \tag{15.30}$$

Again, we assume that it is a forward equation in which $a > 0$ and the initial distribution $y(0, x)$ is an arbitrary bounded function.

Proposition 15.6 *Consider Eq.* (15.26) *for* $\mathrm{X} = \mathbb{R}_+$ *assuming that* $a > 0$*. The solution to the Cauchy problem, constrained by an initial bounded function* $y(0, x)$*, is*

$$y(t, x) = \int_{\mathrm{X}} \frac{y(0, \xi)}{\xi \sqrt{4 \pi a^2 t}} \exp\left\{ - \frac{(\ln(x/\xi) + (b - a^2) t)^2}{4 a^2 t} + c\, t \right\} d\xi, \quad (15.31)$$

for each $(t, x) \in \mathrm{T} \times \mathrm{X}$.

Proof Introducing the transformation $y(t, x) = u(t, Z(x))$, where $z = Z(x) = \ln(x)$ and $x = X(z) = e^z$, yields the linear PDE $\partial_t u(t, z) = a^2\, \partial_{zz} u(t, z) + (b - a^2)\, \partial_z u(t, z) + c\, u(t, z)$ over $u : \mathrm{T} \times \mathbb{R} \to \mathbb{R}$. Then, we can invoke Proposition 15.2 to find the solution to this PDE, and after transforming back to the original independent variable x, we find the general solution (15.31). □

Example 15.11 (*The probability distribution for the geometric Brownian motion*) The probability distribution for the geometric Brownian motion process (see equation (18.4)) is the mapping $p : \mathrm{T} \times \mathrm{Y} \to (0, 1)$, where $\mathrm{Y} = \mathbb{R}_+$, that solves the parabolic FPDE

$$\begin{cases} \partial_t p(t, y) = -\partial_y\big(\mu\, y\, p(t, y)\big) + \partial_{yy}\Big(\frac{\sigma^2}{2}\, y^2\, p(t, y)\Big), & (t, y) \in \mathrm{T} \times \mathbb{R}_+, \\ p(0, y) = \delta(y - y_0), & (t, y) \in \{t = 0\} \times \mathbb{R}_+, \end{cases} \quad (15.32)$$

where $\sigma > 0$ and $y_0 > 0$. The solution to this problem is the probability $p(t, y)$ of the realization $y \in \mathrm{Y} = \mathbb{R}_{++}$, at time t

$$p(t, y) = \frac{1}{y \sqrt{2 \pi \sigma^2 t}} e^{-\frac{\left(\ln(y/y_0) - \left(\mu - \frac{1}{2}\sigma^2\right) t\right)^2}{2\sigma^2 t}}, \quad (t, y) \in \mathbb{R}^2_+. \quad (15.33)$$

To prove this result, we expand the FPK equation to obtain

$$\partial_t p(t, y) = \frac{\sigma^2}{2}\, y^2\, \partial_{yy} p(t, y) + (2\sigma^2 - \mu)\, y\, \partial_y p(t, y) + (\sigma^2 - \mu)\, p(t, y),$$

which is an equation of the form (15.26), with $a = \frac{\sigma}{\sqrt{2}}$, $b = 2\,\sigma^2 - \mu$ and $c = \sigma^2 - \mu$. Substituting in the solution (15.31) yields the probability density function (15.33). Figure 15.8 illustrates the dynamics of this probability density function over time.

Example 15.12 (*Social mobility*) In this example, we extend Example 13.10 by incorporating pure social mobility. Starting with Shorrocks (1976), Shorrocks (1978b), and Shorrocks (1978a), the dynamics of household income distribution, referred as social mobility, is modelled by a Markov chain process. In this example we extend this approach to continuous time.

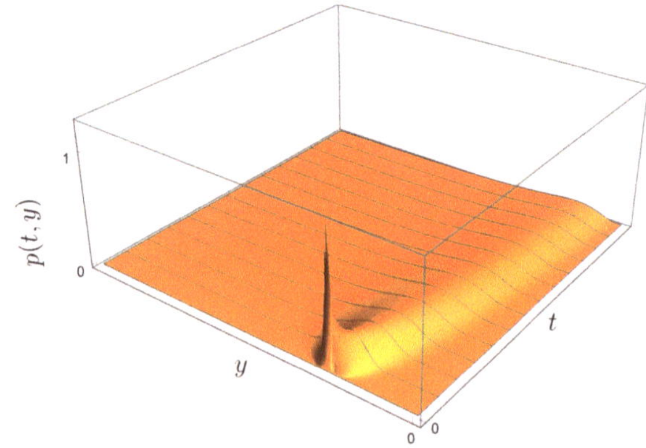

Fig. 15.8 Solution to PDE (15.32) for $x_0 = 20$, $\mu_0 = 0.005$, and $\sigma = 0.2$

Consider an economy having a continuum of households with heterogeneous income, where $\mathrm{Y} \subseteq \mathbb{R}_+$ is the income space. Let household i, have income level y^i at time $t = 0$, and denote it by, $y^i(0) = y^i \in \mathrm{Y}$, and let its income at time $t > 0$, $y^i(t)$, be governed by the linear stochastic differential equation (SDE)

$$dy^i(t) = \mu_1^i\, y^i(t)\, dt + \sigma_1^i\, y^i(t)\, dW(t), \quad \text{for each } t > 0,$$

where μ_1^i and σ_1^i are known constants, and $(W(t))_{t \in \mathrm{T}}$ is a Wiener process.[11] Given this process, the probability that the household income would take value $y \in \mathrm{Y}$, at time $t > 0$, when it is y^i at time $t = 0$, denoted by $f^i(t, y) = \mathbb{P}[Y^i(t) = y | y^i(0) = y^i]$, is the solution to the Fokker-Planck-Kolmogorov (FPK) equation,

$$\partial_t f^i(t, y) = -\partial_y \Big(\mu_1^i\, y\, f^i(t, y) \Big) + \frac{1}{2}\, \partial_{yy}^2 \Big(\sigma_1^{i2}\, y^2\, f^i(t, y) \Big),$$

that is a parabolic FPDE over $f^i(t, y)$.

Let us order all households according to their income and let $Y(t)$ be the distribution of income among all households. If we order households in this way, we find that the income of household i at each time t, $y^i(t)$, can be assigned to a quantile of the whole distribution $Y(t)$, $y^q(t) = \Big\{ y : F(t, y) = q \Big\}$, for $0 < q < 1$, where $F(t, y) = \int_0^y f(t, s) ds$.

We say there is **aggregate social mobility** (in a statistical sense) if every household i do not change their income quantile but the mean income of all households change over time

$$y^i(t) = y^q(t), \;\; y^i(t') = y^q(t'), \;\; \text{and } \mathbb{E}[Y(t)] \neq \mathbb{E}[Y(t')], \;\; \text{for } t \neq t' \in \mathrm{T}.$$

[11] See Chaps. 17 and 18.

We say there is **pure social mobility** if household's i income can change it's income quantile position but the average income for all households remains constant over time,

$$y^i(t) = y^q(t),\ \ y^i(t') \in (y^q(t'))_{q\in(0,1)},\ \text{and } \mathbb{E}[Y(t)] = \mathbb{E}[Y(t')],\ \text{ for } t \neq t' \in \mathrm{T}.$$

There is **aggregate and pure social mobility** if households can change their quantile position, and the average income of all households also changes,

$$y^i(t) = y^q(t),\ \ y^i(t') \in (y^q(t'))_{q\in(0,1)},\ \text{and } \mathbb{E}[Y(t)] \neq \mathbb{E}[Y(t')],\ \text{ for } t \neq t' \in \mathrm{T}.$$

In the first case, the whole distribution moves but the income of each quantile moves proportionally meaning that there is social immobility, in a statistical sense. In the two other cases, there is social mobility, in the sense that there are households that move up and there are households that move down the income ladder, with or without growth in the mean income for the whole distribution.

Assuming there is heterogeneity in household income distribution, but the parameters characterizing the stochastic income dynamics, μ_1 and σ_1 , are common to all households, the probability density function for the income distribution of households is the solution to the parabolic FPDE,

$$\partial_t f(t,y) = -\partial_y\Big(\mu\, y\, f(t,y)\Big) + \partial^2_{yy}\Big(\frac{\sigma^2}{2}\, y^2\, f(t,y)\Big),\ (t,y)\in \mathrm{T}\times\mathrm{Y}, \tag{15.34a}$$

$$f(0,y) = f_0(y),\ (t,y)\in\{t=0\}\times\mathrm{Y}, \tag{15.34b}$$

where the initial distribution of households per income levels, $f_0(y)$, satisfying $\int_{\mathrm{Y}} f_0(y)\,dy = 1$, is satisfied. This is a particular case of forward PDE (15.26) with solution in Eq. (15.27).

If $f_0(y)$ is assumed to be a lognormal distribution with parameters μ_0 and σ_0, we obtain, after simplification, an explicit solution

$$f(t,y) = \frac{1}{y\sqrt{2\pi\,(\sigma_0^2+\sigma_1^2\, t)}}\, e^{\gamma(t,y)},\ \text{for each } (t,y)\in\mathbb{R}^2_+, \tag{15.35}$$

where

$$\gamma(t,y) \equiv -\frac{\Big(\ln y - \big(\mu_0 - \frac{1}{2}\sigma_0^2 + (\mu - \frac{1}{2}\sigma_1^2)\, t\big)\Big)^2}{2\,(\sigma_0^2+\sigma_1^2 t)}.$$

Therefore, the income distribution follows a time-varying log-normal density, $f(t,y) \sim LN(\mu_0 + \mu_1\, t, \sigma_0^2 + \sigma_1^2\, t)$, satisfying the following properties: first, as a density function it satisfies $0 < f(t,y) < 1$, for each $y \in \mathrm{Y}$, and $\int_0^\infty f(t,y)\,dy = 1$ for each $t \in [0,\infty)$; second, as the mean and the Gini index[12]

[12] See Example 13.10 for the formulas that were used to obtain these statistics.

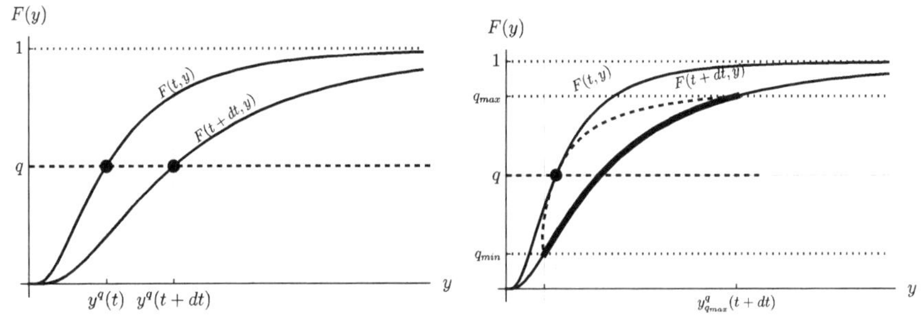

(a) Case: $\mu_1 > 0$ and $\sigma_1 = 0$. (b) Case: $\mu_1 > 0$ and $\sigma_1 > 0$.

Fig. 15.9 Aggregate mobility without (**a**), or with (**b**) pure mobility. Panel (**b**) shows the 95% region of confidence for $t > 0$, for a household starting from quantile q at $t = 0$

$$\mathsf{M}[Y(t)] = \int_0^\infty y\, f(t, y)\, dy = e^{\mu_0 + \mu_1 t},$$

$$\mathsf{G}[Y(t)] = \text{erf}\left(\frac{\sqrt{\sigma_0^2 + \sigma_1^2 t}}{2}\right).$$

then the mean income is changing over time if $\mu_1 \neq 0$, and the Gini index always increases over time if $\sigma_1 > 0$. Therefore, the following cases are possible if $\mu_1 = 0$ and $\sigma_1 = 0$ there is neither aggregate nor pure social mobility; if $\mu_1 \neq 0$ and $\sigma_1 = 0$ there is aggregate mobility and pure social immobility; if $\mu_1 = 0$ and $\sigma_1 > 0$ there is pure social mobility but not aggregate mobility, and if $\mu_1 \neq 0$ and $\sigma_1 > 0$, there is both aggregate and pure social mobility. Furthermore, the aggregate income increases if $\mu_1 > 0$, and decreases if $\mu_1 < 0$.

Figure 15.9 displays the distribution functions, $F(t, y)$ and $F(t + dt, y)$, for $dt > 0$, and for all $y \in \mathrm{Y}$, and the evolution of the relative position of a representative household starting from the quantile q at time t. Figure 15.9a illustrates a case in which there is aggregate income growth and pure social immobility, and Fig. 15.9b illustrates a case in which there is both aggregate income growth and pure social mobility. In the last case, we observe that with 95% probability, a household starting from the quantile q at time $t = 0$, can find itself in the quantile interval $q_{max} < q < q_{min}$ at time $t > 0$. The Gini coefficient remains constant over time in the case depicted in Fig. 15.9a and grows over time in the case depicted in Fig. 15.9b, which means that this model reproduces the observed statistics for several years for several countries.

15.3 Linear Backward PDE

Backward parabolic PDE (BPDE) appear as adjoint equations in distributional optimal control problems, as probability distributions of stochastic processes (backward Kolmogorov equations), as HJB equations of stochastic optimal control problems, and in mathematical finance. Next, we present a brief introduction to linear backward PDE, with relevance to economics.

15.3.1 *The Backward Heat Equation*

Let $\mathrm{T} = [0, T]$ where $T > 0$ is a constant, and $\mathrm{X} = \mathbb{R}$. The simplest parabolic BPDE equation for $y : \mathrm{T} \times \mathrm{X} \to \mathrm{Y} = \mathbb{R}$ is

$$\partial_t y + a\, \partial_{xx} y = 0, \quad \text{for each } (t, x) \in \mathrm{T} \times \mathrm{X}, \tag{15.36}$$

where $a > 0$.

Proposition 15.7 (Solution to the backward heat equation) *The general solution to the linear scalar BPDE* (15.36) *is function* $y = y(t, x) \in \mathbb{R}$

$$y(t, x) = \int_{-\infty}^{\infty} \frac{y(T, \xi)}{\sqrt{4\pi a(T-t)}}\, e^{-\frac{(x-\xi)^2}{4a(T-t)}}\, d\xi, \quad \textit{for each } (t, x) \in \mathrm{T} \times \mathrm{X}. \tag{15.37}$$

Proof Let $\tau = T - t$ and $v(\tau, x) = y(t(\tau), x)$ where $t(\tau) = T - \tau$. As $\partial_\tau v(\tau, x) = -\partial_t y(t(\tau), x)$, and $\partial_{xx} v(\tau, x) = \partial_{xx} y(t(\tau), x)$, then the backward PDE (15.36) in (τ, x) is transformed into the forward heat equation $\partial_\tau v(\tau, x) = a\, \partial_{xx} v(\tau, x)$ in (τ, x). Then, from Eq. (15.7), $v(\tau, x) = \int_{-\infty}^{\infty} v(0, \xi)\, (4\pi a\tau)^{-1/2}\, e^{-\frac{(x-\xi)^2}{4a\tau}}\, d\xi$, for $\tau \in (0, T)$. Transforming back to $y(t, x)$, and observing that $t = T$ if $\tau = 0$, yields solution (15.37). □

A Cauchy backward problem is well posed if together with the PDE there is a known terminal function $y_T(x)$, such as

$$\partial_t y = -a\, \partial_{xx} y, \ (t, x) \in \mathrm{T} \times \mathrm{X}, \tag{15.38a}$$

$$y(T, x) = y_T(x), \ (t, x) \in \{t = T\} \times \mathrm{X}, \tag{15.38b}$$

for $a > 0$. It has the solution

$$y(t, x) = \frac{1}{\sqrt{4\pi a(T-t)}} \int_{-\infty}^{\infty} y_T(\xi) e^{-\frac{(x-\xi)^2}{4a(T-t)}} d\xi, \ (t, x) \in (0, T) \times \mathrm{X}. \tag{15.39}$$

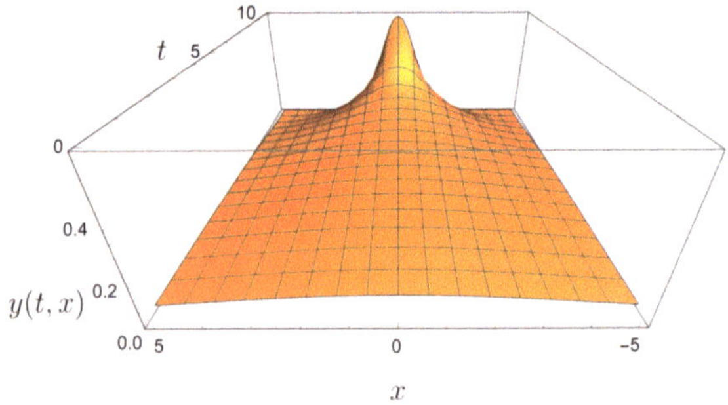

Fig. 15.10 Solutions for backward heat equation with $a = 1$ and $y_T(x) = \frac{e^{-x^2}}{\sqrt{\pi}}$

The particular solution for time $t = 0$ is a function of the terminal time, T, and of the fixed terminal function, $y_T(x)$,

$$y(0, x) = \frac{1}{\sqrt{4\pi\, a\, T}} \int_{-\infty}^{\infty} y_T(\xi) e^{-\frac{(x-\xi)^2}{4aT}} d\xi.$$

Example 15.13 Assuming a terminal distribution $y_T(x) = \frac{e^{-x^2}}{\sqrt{\pi}}$, solution (15.37) simplifies to

$$y(t, x) = \frac{e^{-\frac{x^2}{1+4a(T-t)}}}{\sqrt{\pi\,(1 + 4\,a\,(T-t))}}, \quad \text{for } t \in (0, T].$$

Figure 15.10 illustrates this solution. It can be compared to the analog forward case in Fig. 15.1.

15.3.2 General Backward Linear PDE

The autonomous homogeneous backward linear scalar parabolic PDE (BPDE), for $y = y(t, x) \in \mathbb{R}$, where $(t, x) \in \mathrm{T} \times \mathrm{X} = [0, T] \times \mathbb{R}$, is

$$\partial_t y = -a\, \partial_{xx} y + b\, \partial_x y + c\, y, \tag{15.40}$$

where $a > 0$, and $(b, c) \in \mathbb{R}^2$. Using a similar approach as in the proof of proposition 15.7, we can find its solution to be

$$y(t, x) = e^{-c\,(T-t)} \int_{-\infty}^{\infty} \frac{y(T, \xi)\, e^{-\frac{(x-\xi-b\,(T-t))^2}{4\,a^2\,(T-t)}}}{\sqrt{4\,\pi\, a^2\,(T-t)}} d\xi, \quad (t, x) \in \mathrm{T} \times \mathrm{X}. \tag{15.41}$$

An important non-autonomous backward linear PDE, defined over $\mathrm{X} = \mathbb{R}_+$, is

$$\partial_t y = -(a\, x)^2\, \partial_{xx} y + b\, x\, \partial_x y + c\, y, \tag{15.42}$$

has solution

$$y(t, x) = e^{-c\,(T-t)} \int_0^\infty \frac{y(T, \xi)\, e^{-\frac{(\ln(x/\xi) - (b+a^2)\,(T-t))^2}{4\,a^2\,(T-t)}}}{\sqrt{4\,\pi\, a^2\,(T-t)}}\, d\xi,\ \ (t, x) \in \mathrm{T} \times \mathrm{X}. \tag{15.43}$$

These two solutions, in Eq. (15.40) and (15.42), can be compared with the analog forward PDE, in Eq. (15.22) and (15.31).

Example 15.14 (*The option pricing model*) In Example 17.5, we present the Black and Scholes (1973) European option's pricing model. The value of an option is a stochastic process, $(V(t))_{t \in \mathrm{T}}$, depending on the process for the price of the underlying asset, $(S(t))_{t \in \mathrm{T}}$, on the expiration date $T > 0$ and on the exercise price, $K > 0$, such that the realization at time $V(t) = v(t, s)$, is a function of time $t \in \mathrm{T} = [0, T]$ and of the realization for the underlying price $S(t) = s \in \mathbb{R}_+$, given by the solution to the parabolic BPDE

$$\partial_t v(t, s) = -\frac{\sigma^2}{2}\, s^2\, \partial_{ss} v(t, s) - r\, s\, \partial_s v(t, s) + r\, v(t, s),\ \ t < T, \tag{15.44a}$$

$$v(T, s) = \max\{s - k, 0\},\ \ t = T, \tag{15.44b}$$

where $r > 0$ is the interest rate for the risk free asset, and σ^2 is the volatility of the underlying risky asset.[13]

Proposition 15.8 (Black-Scholes formula) *The price of the European call option, for each realization of the asset price process at time t, $S(t) = s$, is*

$$v(t, s) = s\,\Phi(d_+) - k\, e^{-r(T-t)}\, \Phi(d_-),\ \ t \in [0, T], \tag{15.45}$$

where $\Phi(d) = \mathbb{P}(x \leq d)$ is Gaussian cumulative density function (CDF) where

$$d_\pm = \frac{\ln\left(\frac{s}{k}\right) + (T - t)\left(r \pm \frac{\sigma^2}{2}\right)}{\sigma \sqrt{T - t}}. \tag{15.46}$$

Proof BPDE (15.44a) is a special case of BPDE (15.42) in which $a = \frac{\sigma}{\sqrt{2}}$, $b = -r$ and $c = r$. Therefore, its general solution is

$$v(t, s) = \frac{e^{-r\,(T-t)}}{\sqrt{2\,\pi\,\sigma^2\,(T-t)}} \int_0^\infty \frac{v(T, \xi)\, e^{-\frac{(\ln(s/\xi)\ \ (r - \sigma^2/2)\,(T-t))^2}{2\sigma^2\,(T-t)}}}{\xi}\ \ (t, x) \in \mathrm{T} \times \mathbb{R}_+.$$

[13] The two values $v(t, 0) = 0$, and $\lim_{s \to \infty} v(t, s) = s$ are sometimes introduced as conditions. We can see that they are particular solutions to the BPDE (15.44a)–(15.44b) and are not side-conditions.

Substituting the terminal condition (15.44b), the particular solution is

$$
\begin{aligned}
v(t,s) =& \frac{e^{-r\,(T-t)}}{\sqrt{2\,\pi\,\sigma^2\,(T-t)}} \int_k^\infty \frac{\xi-k}{\xi}\, e^{-\frac{(\ln(^s/_\xi)-(r-^{\sigma^2}/_2)\,(T-t))^2}{2\,\sigma^2\,(T-t)}}\, d\xi \\
=& \frac{e^{-r\,(T-t)}}{\sqrt{2\,\pi\,\sigma^2\,(T-t)}} \left(\int_k^\infty e^{-\frac{(\ln(^s/_\xi)-(r-^{\sigma^2}/_2)\,(T-t))^2}{2\,\sigma^2\,(T-t)}}\, d\xi \right. \\
& \left. - k \int_k^\infty \frac{e^{-\frac{(\ln(^s/_\xi)-(r-^{\sigma^2}/_2)\,(T-t))^2}{2\,\sigma^2\,(T-t)}}}{\xi}\, d\xi \right).
\end{aligned}
$$

Evaluating the two integrals, and using the properties of the Gaussian integral, from the Appendix 15.7.2, yields the Black-Scholes formula (15.45). □

15.4 Linear Stability Analysis for Semi-Linear Parabolic PDE

The existence, well-posedness, and properties of solutions to a parabolic PDE depend on several features: on the type of the set X, on the type of boundary conditions, and on the functional form of the equation. In order to concentrate on the properties associated to the functional form, let us assume that $\mathrm{X} = \mathbb{R}$ and there are no boundary constraints. For ODE, the Grobman-Hartmann theorem (see Chap. 3 Theorem 5.1) provides conditions for the characterization of local dynamics. There is not a Grobman-Hartmann theorem for general parabolic PDE.[14] However, we can conduct a local stability analysis by using our previous results concerning linear parabolic PDE.[15]

Consider the forward semi-linear scalar parabolic PDE

$$
\partial_t y = a\,\partial_{xx} y + b\,\partial_x y + f(y), \quad \text{for } y = y(t,x) \in \mathrm{Y} \subseteq \mathbb{R}, \tag{15.47}
$$

where $a > 0$, and $\mathrm{X} \subseteq \mathbb{R}$. We call **uniform steady state** to a value of the dependent variable $\bar{y} \in \mathrm{Y}$ that is independent of time and is homogeneous across X. Thus, the **uniform steady state set** is formally defined by $\bar{\mathrm{Y}} \equiv \{y \in \mathrm{Y} : f(y) = 0\}$, such that $y(t,x) = \bar{y} \in \bar{\mathrm{Y}}$ for each $(t,x) \in \mathrm{T} \times \mathrm{X}$. From now on we assume that $\bar{\mathrm{Y}}$ is non-empty.

A uniform steady state is **stable** if $f'(\bar{y}) < 0$, and it is unstable if $f'(\bar{y}) > 0$. If $f'(\bar{y}) = 0$ it can correspond to a bifurcation state (or to a linear PDE in which $f(y) = \text{const}$).

---

[14] See Galaktionov and Vázquez (2004, p. 25).

[15] We draw upon the spectral linearization approach by e.g., van Saarloos (2003), Cross and Greenside (2009), and Kuehn (2019, Chaps. 8, 9, 10).

Example 15.15 (*Non-linear scalar parabolic PDE*) In the literature,[16] there are several semi-linear PDE, most of which assuming that $b = 0$. The Fisher equation, where $f(y) = y\,(1 - y)$, has two uniform states in which one is unstable ($\bar{y}_1 = 0$) and one is stable ($\bar{y}_2 = 1$). The Newel-Whitheead or amplitude equation, where $f(y) = y\,(1 - y^2)$, has two stable uniform states ($\bar{y}_1 = -1$ and $\bar{y}_3 = 1$) and one unstable uniform state ($\bar{y}_2 = 0$). The Zeldovich equation, where $f(y) = y^2\,(1 - y)$, has two stable uniform states ($\bar{y}_1 = -1$ and $\bar{y}_3 = 1$) and one bifurcation point ($\bar{y}_2 = 0$). At last, the Naguno equation, where $f(y) = y\,(1 - y)\,(y - \alpha)$, for $0 < \alpha < 1$, has two stable uniform states ($\bar{y}_1 = 0$ and $\bar{y}_3 = 1$) and one unstable uniform state ($\bar{y}_2 = \alpha$). While the Fisher equation is said to be monostable, the other PDE are said to be bistable.

To consider the dynamics of the uniform states, let us introduce an arbitrary perturbation $u(t, x)$ in a small neighborhood of an uniform state, $\bar{y}$, such that $y(t, x) = \bar{y} + u(t, x)$. A first-order approximation of semi-linear PDE, Eq. (15.47) is

$$\partial_t u = a\,\partial_{xx} u + b\,\partial_x u + \bar{c}\,u, \text{ for } u = u(t, x) \in \mathrm{Y} \subseteq \mathbb{R},\ \mathrm{X} = \mathbb{R}, \tag{15.48}$$

where $\bar{c} = f'(\bar{y})$, which is a particular case of PDE (15.13). Taking Fourier transforms, we have $U(t, \omega) = \mathcal{F}[u(t, x)](\omega)$, as a function of the Fourier mode ω, and the characteristic ODE, displaying its dynamics, has the solution

$$U(t, \omega) = U(0, \omega)\,e^{\lambda(\omega)\,t},$$

where $\lambda(\omega) = -a(2\,\pi)^2\,\omega^2 + \bar{c} + b\,2\,\pi\,i\,\omega$ is the eigenfunction, a complex-valued function of ω, that can be interpreted as the rate of change over time of $U(\cdot, \omega)$ for each frequency ω. Taking the inverse transform,

$$u(t, x) = \int_{-\infty}^{\infty} U(0, \omega)\,e^{2\pi\,x\,i\,\omega}\,e^{\lambda(\omega)\,t}\,d\omega, \tag{15.49}$$

we can show that the dynamics of the perturbation, for any x, is driven by the rates of growth associated to all the Fourier modes. Denoting the real part and the complex part of the eigenfunction by $\gamma(\omega) = \mathrm{Re}[\lambda(\omega)] = -a(2\,\pi)^2\,\omega^2 + \bar{c}$ and $\sigma(\omega) = \mathrm{Im}[\lambda(\omega)] = b\,2\,\pi\,\omega$, we see that: first, for Fourier mode $\omega = \omega'$ the transformed perturbation $U(t, \omega')$ decays (grows) over time if $\gamma(\omega') < 0$ ($\gamma(\omega') > 0$), and is constant if $\gamma(\omega') = 0$; and, second, the real part is a function of parameters a and $\bar{c}$, and the complex part is a function of parameter b.

Function $\gamma(\omega)$ is concave and has the following properties: (i) it has a maximum at the critical mode $\omega^* = 0$, where $\gamma(\omega^*) = 0$; (ii) if $\bar{c} \le 0$ then $\gamma(\omega) \le 0$ for all $\omega \in \mathbb{R}$; (iii) and, if $\bar{c} > 0$ then $\gamma(\omega) > 0$ if $\omega \in \Omega_u = (\omega_-, \omega_+)$, or $\gamma(\omega) \le 0$ if $\omega \in \mathbb{R}/\,\Omega_u$ where $\omega_{\pm} = \pm\frac{1}{2\pi}\sqrt{\frac{\bar{c}}{a}}$.

[16] See e.g., Gilding and Kersner (2004, Chap. 1).

Therefore, there is a bifurcation value for the parameter $\bar{c}$, $\bar{c}^* = 0$, such that: if $\bar{c} < 0$ the steady state $\bar{y}$ is **uniformly stable**, because $\lim_{t\to\infty} u(t,x) = 0$ for each $x \in \mathrm{X}$, in the sense that a local perturbation from it will converge back to the uniform steady state over time; and if $\bar{c} > 0$ the steady state $\bar{y}$ is **uniformly unstable** because $\lim_{t\to\infty} u(t,x) = \infty$ for every $x \in \mathrm{X}$, in the sense that a small deviation from it will not converge back to the uniform state but rather to a uniform balanced growth path $\lim_{t\to\infty} u(t,x) \propto \lim_{t\to\infty} e^{\bar{c}t}$. Figure 15.3 provides an illustration of the three types of behavior for the case in which there is only one steady state at $\bar{y} = 0$.

If there is **more than one uniform steady state**, that is if $\dim(\bar{Y}) > 1$, and we consider the monostable PDE case, deviations from an unstable uniform steady state will tend to converge to another stable uniform state if it exists. In the absence of explicit (global) solutions to the semi-linear PDE over $y(\cdot)$, we can have a further characterization of the propagation of the distribution across space X in the neighborhood of an unstable uniform state.

Let $\bar{c} > 0$ and assume that the initial perturbation is a Dirac-delta, with a mass concentrated at $x = 0$, $u_0(x) = \delta(x)$. The evolving distribution takes the form of a front propagation and has two flanks, a right flank for $x > 0$ and a left flank for $x < 0$. Let us consider a linear frame in the $\mathrm{T} \times \mathrm{X}$ space $x = \xi + \nu t$, where ν, a real number, is called the linear speed of propagation.[17] Along every line in this frame, the perturbed solution, represented by Eq. (15.49), can be written as

$$u(t, \xi, \nu) = \int_{-\infty}^{\infty} \tilde{U}(0, \omega, \xi)\, e^{\rho(\omega,\nu)\, t}\, d\omega$$

where $\tilde{U}(0, \omega, \xi) = U(0, \omega) e^{2\pi\, i\, \xi\, \omega}$ and

$$\rho(\omega, \nu) = 2\pi\, i\, \nu\, \omega + \lambda(\omega) = \bar{c} + 2\pi\, i\, \Big((\nu + b)\, \omega + a\, (2\pi)^2\, \omega^2\, i\Big),$$

is a complex-valued convex function of ω that determines the propagation of the solution across set X. The **linear spreading speed**, ν^*, is a speed that is a minimum of $\rho(\omega, \cdot)$ such that there is a Fourier mode ω^* satisfying $\rho(\omega^*, \nu^*) = 0$, i.e., $(\omega^*, \nu^*) = \{(\omega, \nu) : \partial_\omega \rho(\omega, \nu) = 0,\ \rho(\omega, \nu) = 0\}$. Thus we find $\omega^* = \pm \frac{1}{2\pi}\sqrt{\frac{\bar{c}}{a}}$ and $\nu^* = -b \pm 2\sqrt{a\,\bar{c}}$. A wave propagating with speed ν^* is called a **pulled front**. This means that, if $\nu^*_+ > 0$ ($\nu^*_+ < 0$), the right flank of $u(t, x)$ is pulled to the right (left) into the unstable uniform state by a front with speed $|\nu^*|$.

The non-linear component of the solution to PDE (15.47) generates another type of front propagation, called **pushed front** that will have a propagation speed greater than ν^*.

[17] Observe that, for a fixed ξ this is similar to a characteristic line for a first-order linear autonomous PDE as in Fig. 13.1.

The linear spreading speed ν^* provides an intuition on the effects of the transport, diffusion, and growth components in the propagation of the distribution of $y(t, x)$ over time. If the transport component b is positive and very large, it can dominate the movement of the distribution across set X.

15.5 References

There is an overwhelming literature on parabolic partial differential equations. We have particularly followed introductory textbooks as Pinsky (2003, Chap. 5), Olver (2014), Salsa (2016), and the more advanced textbook Evans (2010, Chap. 3). A very complete reference is still Ladyženskaja et al. (1968). There is a recent interest in applications of parabolic PDE to economic theory, not restricted to mathematical finance. For an advanced survey see Achdou et al. (2014).

15.6 Appendix: Fourier Transforms and Inverse Fourier Transforms

15.6.1 Fourier Transforms

Consider a function $f(x)$, mapping $f : \mathbb{R} \to \mathbb{R}$, such that f is a $L^1(\mathbb{R})$ function, satisfying $\int_{-\infty}^{\infty} |f(x)|dx < \infty$. We define a pair of generalized functions, the **Fourier transform** of $f(x)$ (FT),[18] by applying the Fourier operator $\mathcal{F}(\cdot)$,

$$F(\omega) = \mathcal{F}[f(x)] \equiv \int_{-\infty}^{\infty} f(x)\, e^{-2\pi i \omega x}\, dx,$$

where $i^2 = -1$, and the **inverse Fourier transform** (IFT) of $F(\omega)$, obtained by applying the inverse Fourier operator $\mathcal{F}^{-1}(\cdot)$,

$$f(x) = \mathcal{F}^{-1}[F(\omega)] \equiv \int_{-\infty}^{\infty} F(\omega) e^{2\pi i \omega x}\, d\omega.$$

A **convolution** of two functions $f(\cdot)$ and $g(\cdot)$ is the integral

$$f(x) * g(x) = \int_{-\infty}^{\infty} f(y)\, g(x - y)\, dy. \tag{15.50}$$

---

[18] See Kammler (2000).

Table 15.1 Fourier and inverse Fourier transforms: properties

	$f(x)$ for $x \in \mathbb{R}$	$F(\omega)$ for $\omega \in \mathbb{R}$	Obs.
1	a	$a\,\delta(\omega)$	$a \in \mathbb{C}$, a complex number, $\delta(\cdot)$ Dirac's delta
2	$a\,\delta(x)$	a	$a \in \mathbb{C}$, a complex number
3	$a\,f(x)$	$a\,F(\omega)$	$a \in \mathbb{C}$, a complex number
4	$a\,f(x) + b\,g(x)$	$a\,F(\omega) + b\,G(\omega)$	$a, b \in \mathbb{C}$, complex numbers, $g \in L^1(\mathbb{R})$
5	$f(x)\,g(x)$	$F(\omega) * G(\omega)$	$F(\omega) * G(\omega)$ is a convolution
6	$f(x) * g(x)$	$F(\omega)\,G(\omega)$	$f(x) * g(x)$ is a convolution
7	x	$-\dfrac{\delta'(\omega)}{2\pi i}$	$\delta'(\cdot)$ is the first derivative of $\delta(\cdot)$
8	x^2	$\dfrac{\delta''(\omega)}{(2\pi)^2}$	$\delta''(\cdot)$ is the second derivative of $\delta(\cdot)$
9	$x\,f(x)$	$-\dfrac{F'(\omega)}{2\pi i}$	
10	$f'(x)$	$2\pi\,i\,\omega\,F(\omega)$	
11	$x\,f'(x)$	$-(F(\omega) + \omega\,F'(\omega))$	if $\omega \in \mathbb{R}$
12	$f''(x)$	$-(2\pi\,\omega)^2\,F(\omega)$	
13	$x\,f''(x)$	$\dfrac{2\pi\,\omega\,(2\,F(\omega) + \omega\,F'(\omega))}{i}$	
14	$x^2\,f''(x)$	$-\omega^2\,F''(\omega)$	
15	$f(t, x)$	$F(t, \omega)$	t is a real number

Table 15.2 Fourier and inverse Fourier transforms of some functions

$f(x)$ for $x \in \mathbb{R}$	$F(\omega)$ for $\omega \in \mathbb{R}$	Obs.
$\delta(x - a)$	$e^{-2\pi\,i\,\omega\,a}$	
$\frac{1}{\sqrt{4\pi a}} e^{-\frac{x^2}{4a}}$	$e^{-a(2\pi\omega)^2}$	$a > 0$
$\frac{1}{\sqrt{4\pi a}} e^{-\frac{(x+b)^2}{4a}}$	$e^{-a(2\pi\omega)^2+b(2\pi i\omega)}$	$a > 0,\ b \in \mathbb{R}$
$\frac{1}{\sqrt{4\pi a}} e^{c-\frac{x^2}{4a}}$	$e^{-a(2\pi\omega)^2+c}$	$a > 0,\ c \in \mathbb{R}$
$\frac{1}{\sqrt{4\pi a}} e^{c-\frac{(x+b)^2}{4a}}$	$e^{-a(2\pi\omega)^2+b(2\pi i\omega)+c}$	$a > 0,\ (b, c) \in \mathbb{R}^2$

Table 15.1 presents some properties of FTs, and Table 15.2 presents the FTs for some functions that appear in the main text.[19]

[19] There are different definitions of Fourier transforms, we use the definition by, e.g., Kammler (2000). See also https://dlmf.nist.gov/1.14

The proofs of some of the entries in Table 15.1 are:

1. the FT of $a \in \mathbb{C}$[20] is $F(a) = \mathcal{F}[a] = a \int_{-\infty}^{\infty} e^{-2\pi i \omega x}\, dx = a \int_{-\infty}^{\infty} \delta(\omega - x)\, dx = a\, \delta(\omega)$, and the IFT of $a\, \delta(\omega)$ is $\mathcal{F}^{-1}[a\delta(\omega)] = \int_{-\infty}^{\infty} a\, \delta(\omega)\, e^{2\pi i \omega x}\, d\omega = a$;
2. the FT of $a \in \mathbb{C}$ multiplied by a Dirac's delta generalized function is $F(a\, \delta(x)) = \mathcal{F}[a\, \delta(x)] = a \int_{-\infty}^{\infty} \delta(x)\, e^{-2\pi i \omega x}\, dx = a$, and the IFT of a constant is $\mathcal{F}^{-1}[a] = \int_{-\infty}^{\infty} a\, e^{2\pi i \omega x}\, ds = a\, \delta(x)$;
3. the FT and IFT preserve multiplication by a complex number $a \in \mathbb{C}$: $\mathcal{F}[a\, f(x)] = \mathcal{F}[a\, f(x)] = \int_{-\infty}^{\infty} a\, f(x)\, e^{-2\pi i \omega x}\, dx = a \int_{-\infty}^{\infty} f(x)\, e^{-2\pi i \omega x}\, dx = a\, F(\omega)$, and $\mathcal{F}^{-1}[a\, F(\omega)] \equiv \int_{-\infty}^{\infty} a\, F(\omega) e^{2\pi i \omega x} d\omega = a \int_{-\infty}^{\infty} F(\omega) e^{2\pi i \omega x} d\omega = a\, f(x)$,
4. the FT and IFT preserve linearity: $\mathcal{F}[a\, f(x) + b\, g(x)] = a\, F(\omega) + b\, G(\omega)$, and, $\mathcal{F}^{-1}[a\, F(\omega) + b\, G(\omega)] = a\, f(x) + b\, g(x)$,
5. the FT does not preserve multiplication of two functions, but there is a relationship between convolution and multiplication. The FT of a product of two functions is a convolution.

$$
\begin{aligned}
\mathcal{F}[f(x)g(x)](\omega) &= \int_{-\infty}^{\infty} f(x)g(x)e^{-2\pi i \omega x} dx = \\
&= \int_{-\infty}^{\infty} \Big[\int_{-\infty}^{\infty} F(z)e^{2\pi i z x} dz\Big] g(x)e^{-2\pi i \omega x} dx = \\
&= \int_{-\infty}^{\infty} F(z)\Big[\int_{-\infty}^{\infty} e^{-2\pi i (\omega - z) x} dx\Big] dz = \\
&= \int_{-\infty}^{\infty} F(z)\, G(\omega - z)\, dz = F(\omega) * G(\omega)
\end{aligned}
$$

The IFT of a convolution is a product: $\mathcal{F}^{-1}[F(\omega) * G(\omega)](x) = f(x)g(x)$;

6. the FT of a convolution is a product of Fourier transforms:

$$
\begin{aligned}
\mathcal{F}[f(x) * g(x)](\omega) &= \int_{-\infty}^{\infty} \Big[\int_{-\infty}^{\infty} f(z)\, g(x - z) dz\Big] e^{-2\pi i \omega x} dx \\
&= \int_{-\infty}^{\infty} \int_{-\infty}^{\infty} f(z)\, g(x - z)\, e^{-2\pi i \omega x}\, dz\, dx \\
&= \int_{-\infty}^{\infty} \int_{-\infty}^{\infty} f(z)\, g(u)\, e^{-2\pi i \omega (u+z)}\, dz\, du \\
&= \int_{-\infty}^{\infty} f(z)\, e^{-2\pi i \omega z}\, dz \int_{-\infty}^{\infty} g(u)\, e^{-2\pi i \omega u}\, du = F(\omega)\, G(\omega).
\end{aligned}
$$

The IFT of a product is a convolution: $\mathcal{F}^{-1}[F(\omega)\, G(\omega)](x) = f(x) * g(x)$;

---

[20] Some useful facts concerning the Dirac's delta generalized function: $\int_{-\infty}^{\infty} e^{2\pi i \omega x} \delta(\omega) d\omega = 1$, and $\int_{-\infty}^{\infty} e^{2\pi i \omega x}\, \delta(\omega) = 0$.

7. let $x \in \mathbb{R}$ then

$$\mathcal{F}[x](\omega) = \int_{-\infty}^{\infty} x\, e^{-2\pi i \omega x}\, dx = -\frac{1}{2\pi i} \int_{-\infty}^{\infty} -2\pi i\, x\, e^{-2\pi i \omega x}\, dx =$$
$$= -\frac{1}{2\pi i} \int_{-\infty}^{\infty} \frac{d}{d\omega}\left(e^{-2\pi i \omega x}\right) dx = \frac{1}{2\pi i} \frac{dF[1](\omega)}{d\omega} = -\frac{1}{2\pi i}\delta'(\omega),$$

and the IFT is $\mathcal{F}[-\frac{1}{2\pi i}\delta'(\omega)] = x$ as $\int_{-\infty}^{\infty} e^{2\pi i \omega x}\delta(\omega) d\omega = 1$ and $\int_{-\infty}^{\infty} e^{2\pi i \omega x}\, \delta(\omega) = 0$.

8. $\mathcal{F}[x^2] = \mathcal{F}[x\, x] = F(\omega) * F(\omega) = \dfrac{1}{(2\pi)^2}\delta''(\omega)$, because $F(\omega) * F(\omega) = -\frac{1}{(2\pi)^2} \int_{-\infty}^{\infty} \delta'(z)\, \delta'(\omega - z)\, dz = -\frac{1}{(2\pi)^2}\Big(\int_{-\infty}^{\infty} \delta(z)\delta'(\omega - z) - \int_{-\infty}^{\infty} \delta(z)\delta''(\omega - z)dz\Big) = \dfrac{1}{(2\pi)^2}\delta''(\omega)$;

9 $\mathcal{F}[x\, f(x)] = \mathcal{F}[x] * \mathcal{F}[f(x)] = \int_{-\infty}^{\infty} -\dfrac{1}{2\pi i}\, \delta'(z)\, F(\omega - z) dz = -\dfrac{1}{2\pi i} \int_{-\infty}^{\infty} \delta(z)\, F'(\omega - z) dz = -\dfrac{F'(\omega)}{2\pi i}$

10. $\mathcal{F}[f'(x)] = \int_{-\infty}^{\infty} f'(x)\, e^{-2\pi i \omega x}\, dx = \int_{-\infty}^{\infty} f(x)\, e^{-2\pi i \omega x} - \int_{-\infty}^{\infty} f(x)\, \dfrac{\partial}{\partial x}\left(e^{-2\pi i \omega x}\right) dx = 2\pi i\, \omega\, F(\omega)$

11. $\mathcal{F}[x f'(x)] = \mathcal{F}[x] * \mathcal{F}[f'(x)] = -\big(F(\omega) + \omega F'(\omega)\big)$ if $\omega \in \mathbb{R}$, because

$$\mathcal{F}[x] * \mathcal{F}[f'(x)] = -\int_{-\infty}^{\infty} \delta'(z)\, (\omega - z)\, F(\omega - z) dz$$
$$= -\omega \int_{-\infty}^{\infty} \delta'(z)\, F(\omega - z) dz + \int_{-\infty}^{\infty} \delta'(z)\, z\, F(\omega - z) dz$$
$$= -\omega\, F'(\omega) + \int_{-\infty}^{\infty} \delta(z)\, z\, F(\omega - z) - \int_{-\infty}^{\infty} \delta(z)\, F(\omega - z) dz$$
$$= -(\omega\, F'(\omega) + F(\omega));$$

12. $\mathcal{F}[f''(x)] = -\int_{-\infty}^{\infty} f'(x)\, \frac{\partial}{\partial x}\left(e^{-2\pi i \omega x}\right) dx = -(2\pi\, \omega)^2\, F(\omega)$
13. $\mathcal{F}[x\, f''(x)] = \dfrac{2\pi\, \omega}{i}\Big(2\, F(\omega) + \omega\, F'(\omega)\Big)$
14. $\mathcal{F}[x^2\, f''(x)] = -\omega^2\, F''(\omega)$
15. if $f(t, x)$, where t is a real variable, then $F(t, \omega) = \mathcal{F}[f(t, x)]$ and $f(t, x) = \mathcal{F}^{-1}[F(t, \omega)]$. Also $\partial_t F(t, \omega) = \mathcal{F}[\partial_t f(t, x)]$ and $\partial_t f(t, x) = \mathcal{F}^{-1}[\partial_t F(t, \omega)]$.

15.7 Appendix: Fourier Transforms and the Solution of Parabolic PDE

15.7.1 *Fourier Transforms and the Solution to Parabolic PDE*

There are several methods for solving linear parabolic PDE. A general method consists in transforming the PDE into a parameterized ODE. There are several possible transformations—sine, cosine, Laplace, Mellin, or Fourier transforms. The choice of the method usually depends on the domain of the "spatial" variable, X.

When the domain of the independent variable x is $\mathrm{X} = \mathbb{R} = (-\infty, \infty)$, the most direct method to find a solution is to use Fourier and inverse Fourier transforms. The method of obtaining a solution follows three steps.

In the first step, we transform function $y(t, x)$ into function $Y(t, \omega)$, where $\omega \in \Omega = \mathbb{R}$ represents frequencies, by invoking the (spatial) **Fourier transform** of the integrable function $y(t, x)$ over x, taking t as a parameter

$$\mathcal{F}[y(t, x)](\omega) \equiv \int_{-\infty}^{\infty} y(t, x)e^{-2\pi i \omega x}\, dx = Y(t, \omega), \tag{15.51}$$

and apply this transformation to the PDE. Table 15.3 can help us in this task, as in the proofs of the main text. In most cases, this allows us to transform the PDE over $y(t, x)$ into a simpler equation over $Y(t, \omega)$, called characteristic equation, where t is a variable and ω is treated as a parameter.

In the second step, we try to solve the characteristic dynamic equation, that can be an ODE or a first-order PDE, in the cases studied in this chapter. In the cases were presented, we obtained closed form solutions of type

$$Y(t, \omega) = Y(0, \omega)\, G(t, \omega)$$

where $Y(0, \omega) = \mathcal{F}[y(0, x)](\omega)$ and $G(t, \omega) = \mathcal{F}[g(t, x)](\omega)$.

In the last step, we invoke the **inverse Fourier transform**

Table 15.3 Fourier spatial transforms

	$y(t, x)$ for $x \in \mathbb{R}$	$Y(t, \omega)$ for $\omega \in \mathbb{R}$	Obs.
1	$y(t, x)$	$Y(t, \omega)$	t is a real number
2	$\partial_t y(t, x)$	$\partial_t Y(t, \omega)$	
3	$\partial_x y(t, x)$	$2\pi\, i\, \omega\, Y(t, \omega)$	
4	$\partial_{xx} y(t, x)$	$-(2\pi\, \omega)^2\, Y(t, \omega)$	

$$\mathcal{F}^{-1}[Y(t,\omega)](x) \equiv \int_{-\infty}^{\infty} Y(t,\omega)e^{2\pi i\omega x}d\omega = y(t,x), \tag{15.52}$$

to obtain the solution of the PDE, as a function of the initial dependent variable x. In principle, this is possible if we are able to find the inverse Fourier transforms, $y(0,x) = \mathcal{F}^{-1}[Y(0,\omega)](x)$ and $g(t,x) = \mathcal{F}^{-1}[G(t,\omega)](x)$, then we can obtain $y(t,x)$ as the inverse Fourier transform of $Y(t,\omega)$,

$$y(t,x) = \mathcal{F}^{-1}[Y(t,\omega)](x) = \mathcal{F}^{-1}[Y(0,\omega)\, G(t,\omega)](x) = y(0,x) * g(t,x).$$

If functions $y(0,x)$ and $g(t,x)$ are integrable functions in the domain of x then the solution to the PDE takes the form

$$y(t,x) = y(0,x) * g(t,x) \equiv \int_{-\infty}^{\infty} y(0,\xi)\, g(t,x-\xi)\, d\xi.$$

When the domain of x is not the double-infinity, we may have to adapt this method, or use the method of images.

15.7.2 The Gaussian Integral

In the book, we frequently use the (parameterized) **Gaussian function**

$$g(x) = e^{-a\,x^2}, \tag{15.53}$$

where a is a positive real number. If $a > 0$, it has the well known bell shape. Let $h : \mathbb{R} \to \mathbb{R}$ and $\left(\int_{-\infty}^{\infty} \left|h(x)\, g(x)\right|^2 dx\right)^{\frac{1}{2}} < \infty$. A **Gaussian integral** is defined by $\int_{-\infty}^{\infty} h(x)g(x)dx$, where $g(\cdot)$ is the Gaussian function. It is finite in the $L^2(\mathbb{R})$ sense. Some properties of the Gaussian integral are: (1) $\int_{-\infty}^{\infty} e^{-a\,x^2}dx = \sqrt{\frac{\pi}{a}}$; (2) $\int_{-\infty}^{\infty} x\, e^{-a\,x^2}dx = 0$; (3) $\int_{-\infty}^{\infty} |x|\, e^{-a\,x^2}dx = \frac{1}{a}$; (4) $\int_{-\infty}^{\infty} x^2\, e^{-ax^2}dx = \frac{1}{2a}\sqrt{\frac{\pi}{a}}$. The **Gaussian cumulative density** function (CDF),

$$\Phi(x) = \frac{1}{\sqrt{2\pi}} \int_{-\infty}^{x} e^{-\frac{s^2}{2}}\, ds, \text{ for } x \in \mathbb{R}. \tag{15.54}$$

involves a parameterized Gaussian function with $a = \frac{1}{2}$. Using the previous properties of the Gaussian integral, then $\Phi(\infty) = 1$.

A related function is the **Gaussian error function**

$$\operatorname{erf}(x) = \frac{2}{\sqrt{\pi}} \int_{0}^{x} e^{-z^2} dz, \tag{15.55}$$

that is related to the Gaussian CDF by the relationship $\Phi(x) = \frac{1}{2}\left(1 + \mathrm{erf}\left(\frac{x}{\sqrt{2}}\right)\right)$. Some other useful properties, relating the Gaussian integral with the Gaussian CDF are, assuming that $a > 0$ and $x_1 > 0$:

$$\int_{x_1}^{\infty} e^{-\left(\frac{a-\ln(x)}{\sqrt{2}\,b}\right)^2} dx = \sqrt{2\pi}\, b\, e^{a+\frac{b^2}{2}}\, \Phi\left(\frac{a+b^2-\ln(x_1)}{b}\right)$$

and

$$\int_{x_1}^{\infty} \frac{1}{x} e^{-\left(\frac{a-\ln(x)}{\sqrt{2}\,b}\right)^2} dx = \sqrt{2\pi}\, b\, \Phi\left(\frac{a-\ln(x_1)}{b}\right).$$

Chapter 16
Optimal Control of Parabolic Partial Differential Equations

The optimal control of PDE is sometimes called optimal control of distributions because it seeks to find the optimal path of a distribution over time. Differently from the optimal control of a first-order PDE, the dynamics of the distribution includes a diffusion component, in addition to transport and growth or decaying components. Similarly to the optimal control of ODE, and of first-order PDE, the first-order conditions include a system of forward-backward parabolic PDE, together with side conditions and/or boundary conditions for co-state variables. The requirements for the existence of solutions are usually very strong, and, therefore, ill-posedness is usually a problem to be addressed.

In this chapter, we assume there are two independent variables, time $t \in \mathrm{T} = \mathbb{R}_+$, and another independent variable $x \in \mathrm{X} \subseteq \mathbb{R}$, and two dependent variables, the control variable $u = u(t, x)$, mapping $u : \mathrm{T} \times \mathrm{X} \to \mathbb{R}$, and the state variable $y = y(t, x)$, mapping $y : \mathrm{T} \times \mathrm{X} \to \mathbb{R}$. The optimal control problems that interest us here differ as regards the state space, X, that can be infinite or semi-infinite, the objective functional, and the constraint PDE. We continue to assume that the initial distribution for the state variable, $y(0, x)$ is known. Boundary conditions allowing for well-posedness of the control problems are assumed to hold. Next, we present the necessary conditions for three different optimal control of parabolic PDE: a simple infinite horizon problem, in Sect. 16.1, an average optimal control problem, in Sect. 16.2, and the optimal control of a Fokker-Planck-Kolmogorov (FPK) equation, in Sect. 16.3.

16.1 Optimal Control Problem of a Semi-linear Parabolic PDE

In this section, a simple optimal control problem for a system governed by a parabolic PDE, featuring functions $y = y(t, x)$ and $u = u(t, x)$, in which the distribution domain is $\mathrm{X} = \mathbb{R}$, is postulated. The specific problem we consider has the following components: a semi-linear parabolic PDE having only diffusion and growth com-

P. B. Brito, *Economic Dynamics and Distributions*, Dynamic Modeling and Econometrics in Economics and Finance 32, https://doi.org/10.1007/978-3-031-94717-9_16

ponents, $\partial_t y = \partial_{xx} y + G(y, u, t, x)$, where $G(\cdot)$ is a regular (i.e., continuous and continuously differentiable up to the second order) function of (y, u); an initial condition $y(0, x) = y_0(x)$, where fixed function $y_0 : \mathrm{X} \to \mathbb{R}$ is bounded; a Neumann boundary condition $\lim_{x\to\pm\infty} \partial_x y(t, x) = 0$ for each $t \in \mathrm{T}$, requiring the state variable to be "flat" for very large absolute values of independent variable x; and the value functional $\mathsf{V}[y, u] = \int_{\mathrm{T}\times\mathrm{X}} F(y(t, x), u(t, x), t, x)\, dx\, dt$, where the objective function $F(\cdot, u, y)$ is smooth, concave, and measurable, in the sense of satisfying $\int_{\mathrm{T}\times\mathrm{X}} |F(t, x)| d(t, x) < \infty$.[1]

Therefore, our benchmark problem is to find the optimal distributions over time $(\mathbf{y}^*, \mathbf{u}^*) = \left(y^*(t, x), u^*(t, x)\right)_{(t,x)\in\mathrm{T}\times\mathrm{X}}$, that solve the problem:

$$\max_{u(\cdot)} \int_{\mathrm{T}} \int_{\mathrm{X}} F(y(t, x), u(t, x), t, x)\, dx\, dt, \tag{16.1a}$$

subject to

$$\partial_t y(t, x) = \partial_{xx} y(t, x) + G(y(t, x), u(t, x), t, x), \;\; (t, x) \in \mathrm{T} \times \mathrm{X}, \tag{16.1b}$$

$$y(0, x) = y_0(x), \;\; (t, x) \in \{t = 0\} \times \mathrm{X}, \tag{16.1c}$$

$$\lim_{x\to\pm\infty} \partial_x y(t, x) = 0, \;\; (t, x) \in \mathrm{T} \times \partial\mathrm{X}, \tag{16.1d}$$

A distributional Pontriyagin maximum principle can be established by defining the Hamiltonian function

$$H(y, u, \lambda, t, x) = F(y(t, x), u(t, x), t, x) + \lambda(t, x) G(y(t, x), u(t, x), t, x)$$

where $\lambda = \lambda(t, x)$ is the distributional co-state variable.

Proposition 16.1 (Necessary first-order conditions for Problem (16.1)) *Let* $(\mathbf{y}^*, \mathbf{u}^*)$ *be the solution to problem (16.1). Then there is a co-state variable* $\lambda : \mathrm{T} \to \mathrm{X}$ *such that admissible solutions, satisfying Eqs. (16.1b)–(16.1d), also satisfy the following optimality conditions*

$$\frac{\partial H^*(t, x)}{\partial u} = 0, \;\; \textit{for each } (t, x) \in \mathrm{T} \times \mathrm{X}, \tag{16.2a}$$

$$\partial_t \lambda(t, x) + \partial_{xx} \lambda(t, x) + \frac{\partial H^*(t, x)}{\partial y} = 0, \;\; \textit{for each } (t, x) \in \mathrm{T} \times \mathrm{X}, \tag{16.2b}$$

$$\lim_{t\to\infty} \lambda(t, x) = 0, \;\; \textit{for each}\, x \in \mathrm{X}, \tag{16.2c}$$

$$\lim_{x\to\pm\infty} \partial_x \lambda(t, x) = 0, \textit{for each } t \in \mathrm{T}, \tag{16.2d}$$

where $H^*(t, x) = H(y^*(t, x), u^*(t, x), \lambda(t, x), t, x)$.

[1] This condition requires that the function is bounded for every value of $u(.)$ and $y(.)$ and satisfies the requirements of the Fubini theorem. Therefore, we can interchange the order of integration between t and x.

Proof At the optimum, $(\mathbf{y}^*, \mathbf{u}^*)$, the value functional is

$$\mathsf{V}^* = \mathsf{V}[y^*, u^*] = \int_{\mathrm{T}} \int_{\mathrm{X}} F^*(t, x)\, dx\, dt,$$

where $F^*(t, x) = F(y^*(t, x), u^*(t, x), t, x))$. For any pair $(y(t, x), u(t, x))$, introducing the co-state function $\lambda(t, x)$, and noting that $F(t, x) = H(t, x) + \lambda(t, x)\left(\partial_{xx} y(t, x) - \partial_t y(t, x)\right)$, then the value functional can be expanded as

$$\begin{aligned}
V[y, u] &= \int_{\mathrm{T}} \int_{\mathrm{X}} F(t, x)\, dx, dt \\
&= \int_{\mathrm{T}} \int_{\mathrm{X}} \left[H(t, x) - \lambda(t, x)\left(\partial_t y(t, x) - \partial_{xx} y(t, x)\right)\right] dx\, dt \\
&= \int_{\mathrm{T}} \int_{\mathrm{X}} \left[H(t, x) + \left(\partial_t \lambda(t, x) + \partial_{xx} \lambda(t, x)\right) y^*(t, x)\right] dx\, dt + \\
&- \int_{\mathrm{T}} \int_{\mathrm{X}} \lambda(t, x)\, y(t, x)\, dx + \\
&+ \int_{\mathrm{T}} \int_{\mathrm{X}} \left(\lambda(t, x)\, \partial_x y(t, x) - \partial_x \lambda(t, x)\, y(t, x)\right)\, dt.
\end{aligned}$$

Consider admissible perturbations to the control and state variables, $\boldsymbol{\eta}(\cdot) = (\eta_u(\cdot), \eta_y(\eta))$, such that for a neighborhood of the solution pair (u^*, y^*) we have, $u(t, x) = u^*(t, x) + \varepsilon\, \eta_u(t, x)$, and $y(t, x) = y^*(t, x) + \varepsilon\, \eta_y(t, x)$, for an arbitrary constant ϵ, and satisfying $\eta_y(0, x) = 0$, for every $x \in \mathrm{X}$, and $\lim_{x \to \pm\infty} \partial_x \eta_y(t, x) = 0$, for every $t \in \mathrm{T}$. The Gâteaux differential of the value functional associated to admissible perturbations for the control and state variables, is

$$\begin{aligned}
\mathrm{D}_{\eta(\cdot)} \mathsf{V}[\mathrm{y}^*, \mathrm{u}^*] = &\int_{\mathrm{T}} \int_{\mathrm{X}} \left[\frac{\partial H^*(t, x)}{\partial u} \eta_u(t, x) + \right. \\
&\left. + \left(\partial_t \lambda(t, x) + \partial_{xx} \lambda(t, x) + \frac{\partial H^*(t, x)}{\partial y}\right) \eta_y(t, x)\right] dx\, dt - \\
&- \lim_{t \to \infty} \int_{\mathrm{X}} \lambda(t, x)\, \eta_y(t, x)\, dx - \lim_{x \pm \infty} \int_{\mathrm{T}} \partial_x \lambda(t, x)\, \eta_y(t, x)\, dt.
\end{aligned}$$

The variational principle states that (y^*, u^*) is optimal if $\mathsf{V}[y^*, u^*] \geq \mathsf{V}[y, u]$ for any admissible pair (y, u), which holds only if $\mathrm{D}_{\eta(\cdot)} \mathsf{V}[\mathrm{y}^*, \mathrm{u}^*] = 0$. This condition is satisfied if conditions in Eqs. (16.2a)–(16.2d) jointly hold. □

Equation (16.2a) is a static optimality condition,

$$\frac{\partial H^*(t, x)}{\partial u} = \frac{\partial F^*(t, x)}{\partial u} + \lambda(t, x) \frac{\partial G^*(t, x)}{\partial u} = 0, \text{ for each } (t, x) \in \mathrm{T} \times \mathrm{X},$$

representing a static trade-off between benefits and costs (in terms of the value of the state variable) of changing the control variable. If function $H(.)$ is sufficiently smooth as regards the control variable u, the implicit function theorem allows to determine the optimal control u^* as a function of the co-state and the state variables. If $\frac{\partial^2 H^*(t,x)}{\partial u^2} \neq 0$, for all $(t, x) \in \mathrm{T} \times \mathrm{X}$, then the optimal control is $u^*(t, x) = U(y(t, x), \lambda(t, x), t, x)$.

Equation (16.2b) is a generalized adjoint equation, and provides an optimal (distributional) rule for changing the state variable. As

$$\frac{\partial H^*(t,x)}{\partial y} = \frac{\partial F^*(t,x)}{\partial y} + \lambda(t,x)\frac{\partial G^*(t,x)}{\partial y}, \text{ for each } (t,x) \in \mathrm{T} \times \mathrm{X},$$

where $\frac{\partial F^*(t,x)}{\partial y} = \frac{\partial F(y^*(t,x),u^*(t,x),t,x)}{\partial y}$ and $\frac{\partial G^*(t,x)}{\partial y} = \partial G(y^*(t,x), u^*(t,x), t, x)$, that it is a parabolic BPDE for $\lambda(\cdot)$, also depending on the state variable $y(\cdot)$. Equation (16.2c) is a distributional transversality condition, that has to hold for each $x \in \mathrm{X}$, and equation (16.2d) introduces boundary conditions to the co-state variable that are dual to boundary constraints for the state variable (16.1d). The last two conditions are necessary and jointly imply that the value functional is bounded at the optimum.

Denoting $G^*(y, \lambda, t, x) = G(y^*(t, x), u^*(t, x))$ and $H^*_y(y, \lambda, t, x) = \frac{\partial}{\partial y} H(y^*, u^*(t, x), \lambda(t, x), t, x)$, the first-order conditions are represented by a distributional maximized Hamiltonian dynamic system (MHDS) as a planar parabolic FBPDE

$$\begin{cases} \partial_t y = \partial_{xx} y + G^*(y, \lambda, t, x), & \text{for } (t,x) \in \mathrm{T} \times \mathrm{X}, \\ \partial_t \lambda = -\partial_{xx}\lambda - H^*_y(y, \lambda, t, x), & \text{for } (t,x) \in \mathrm{T} \times \mathrm{X}, \end{cases}$$

that should be solved jointly with the initial, transversality and boundary conditions. This system contains two semi-linear parabolic PDE: a forward parabolic PDE for the state variable, and a backward parabolic PDE for the co-state variable.[2] It is a distributional generalization of the MHDS of the optimal control problem of ODE, as presented in equation (10.50). A general discussion on the nature of the dynamics of this distributional MHDS, in particular on the conditions for the existence of a distributional version of the saddle-path dynamics, is beyond the scope of this book. However, we will next present a simple model in which a closed form solution can be obtained, that may suggest an intuition on the qualitative properties of the distributional MHDS.

[2] Studying its dynamic and distributional properties is not only beyond the object of this book but also has not been addressed in the literature, as far as we know.

16.2 Average Optimal Control of a Semi-Linear Parabolic PDE

In some problems, the value functional is the average of discounted flows of heterogenous objective functions depending on the control and state variables, $(y(t,x), u(t,x))$, mapping $\mathrm{T} \times \mathrm{X} \rightarrow \mathbb{R}^2$. To concentrate on the dynamic features of the optimal distributions, assume that the heterogeneity domain is unbounded, $\mathrm{X} = \mathbb{R}$, and assume that the value functional is

$$\mathsf{V}[y,u] = \lim_{x\rightarrow\infty} \frac{1}{2x} \int_{-x}^{x} \int_{0}^{\infty} f(y(t,\xi), u(t,\xi))\, e^{-\rho t}\, dt\, d\xi,$$

where $\rho > 0$, and $F(.)$ is continuous, differentiable and concave in (y, u). This is the value functional for social welfare problems having a Millian utilitarian criterium, displaying an absence of inequality aversion.[3]

Next, we consider a problem in which the constraints include an autonomous semi-linear parabolic PDE, initial and terminal time constrains, and average Dirichlet boundary constraints. The average optimal control problem we consider is

$$\max_{u} \lim_{x\rightarrow\infty} \frac{1}{2x} \int_{-x}^{x} \int_{0}^{\infty} f(y(t,\xi), u(t,\xi)) e^{-\rho t}\, dt\, d\xi, \tag{16.3a}$$

subject to

$$\partial_t y = \sigma^2 \partial_{xx} y + g(y,u), \text{ for each } (t,x) \in \mathrm{T} \times \mathbb{R}, \tag{16.3b}$$

$$y(x,0) = y_0(x), \text{ for each } x \in \mathbb{R}, \tag{16.3c}$$

$$\lim_{t\rightarrow\infty} y(t,x) \geq 0, \text{ for each } x \in \mathbb{R}, \tag{16.3d}$$

$$\lim_{x\rightarrow+\infty} \frac{y(t,x)}{x} = 0, \text{ for each } t \in \mathrm{T}. \tag{16.3c}$$

The (current-value) Hamiltonian function is $h(y,u,q) = f(y,u) + q\, g(y,u)$, where q is the current value co-state variable. The necessary first-order conditions, according to the Pontryagin's maximum principle, are the following.

Proposition 16.2 (Necessary first-order conditions for average optimal control problem (16.3)) *Let* $(\mathbf{y}^*, \mathbf{u}^*) = \big(y^*(t,x), u^*(t,x)\big)_{(t,x)\in \mathrm{T}\times\mathrm{X}}$ *be the solution to problem (16.3). Then there is a co-state variable* $q : T \times X \rightarrow \mathbb{R}$ *such that admissible solutions, satisfying Eqs. (16.3b)–(16.3e), also satisfy the following optimality conditions*

[3] See Sect. 8.1.5.8 for social welfare functionals displaying inequality aversion.

$$\frac{\partial h^*(t,x)}{\partial u} = 0, \tag{16.4a}$$

$$\partial_t q(t,x) + \sigma^2 \partial_{xx} q(t,x) = \rho\, q(t,x) - \frac{\partial h^*(t,x)}{\partial y}, \tag{16.4b}$$

$$\lim_{t\to\infty} \lim_{x\to\infty} e^{-\rho t} \frac{1}{2x} \int_{-x}^{x} q(t,\xi)\, y(t,\xi)\, d\xi = 0, \tag{16.4c}$$

$$\lim_{x\to\pm\infty} e^{-\rho t} \frac{q(t,x)}{x} = 0, \tag{16.4d}$$

where $h^*(t,x) = h(y^*(t,x), u^*(t,x), q(t,x), t, x)$.

Proof Let $(\mathbf{y}^*, \mathbf{u}^*)$ be known. The value function, at the optimum, is

$$\mathsf{V}^* = \mathsf{V}[y^*, u^*] = \lim_{x\to\infty} \frac{1}{2x} \int_{-x}^{x} \int_0^{\infty} f^*(t,\xi)\, e^{-\rho t}\, dt\, d\xi.$$

For any path $(\mathbf{y}, \mathbf{u}) = \big(y(t,x), u(t,x)\big)_{(t,x)\in \mathrm{T}\times\mathrm{X}}$, there is a co-state variable $\lambda(t,x) = e^{-\rho t} q(t,x)$ such that the value functional can be expanded as

$$\begin{aligned}
\mathsf{V}[y,u] &= \lim_{x\to\infty} \frac{1}{2x} \int_{-x}^{x} \int_0^{\infty} f(y(t,\xi), u(t,\xi))\, e^{-\rho t}\, d\xi\, dt = \\
&= \lim_{x\to\infty} \frac{1}{2x} \int_{-x}^{x} \int_0^{\infty} e^{-\rho t} \Big[f(y(t,\xi), u(t,\xi)) + \\
&+ q(t,\xi) \Big(g(y(t,\xi), u(t,\xi)) + \sigma^2 \partial_{\xi\xi} y(t,\xi) - \partial_t y(t,\xi) \Big) \Big] dt\, d\xi \\
&= \lim_{x\to\infty} \frac{1}{2x} \int_{-x}^{x} \int_0^{\infty} e^{-\rho t} \Big[h(t,\xi) + \\
&+ \Big(\partial_t q(t,\xi) - \rho\, q(t,\xi) + \sigma^2 \partial_{\xi\xi} q(t,\xi) \Big) y^*(t,\xi) \Big] dt\, d\xi \\
&- \lim_{x\to\infty} \frac{1}{2x} \int_{t=0}^{\infty} \int_{-x}^{x} e^{-\rho t} q(t,\xi)\, y(t,\xi)\, d\xi \\
&+ \lim_{x\to\infty} \frac{1}{2x} \int_{t=0}^{\infty} \int_{-x}^{x} e^{-\rho t} \Big(q(t,\xi)\, \partial_\xi y(t,\xi) - \partial_\xi q(t,\xi)\, y(t,\xi) \Big) dt.
\end{aligned}$$

As we have a terminal constraint (16.3d) we define the Lagrangean functional,

$$\mathsf{L}[y,u] = \mathsf{V}[y,u] + \lim_{t\to\infty} \lim_{x\to\infty} \frac{1}{2x} \int_{-x}^{x} \mu(t,\xi)\, y(t,\xi) d\xi,$$

satisfying $\lim_{t\to\infty} \lim_{x\to\infty} \frac{1}{2x} \int_{-x}^{x} \mu(t,\xi)\, y(t,\xi) d\xi = 0$, where $\mu(\cdot)$ is a KKT multiplier.

For any admissible perturbation, $\boldsymbol{\eta} = (\eta_u\ \eta_y)$, to the control and state variables in the neighborhood of the optimum, $u(t,x) = u^*(t,x) + \varepsilon\, \eta_u(t,x)$, and $y(t,x) = y^*(t,x) + \varepsilon\, \eta_y(t,x)$, for an arbitrary constant ε, satisfying $\eta_y(0,x) = 0$, for every

$x \in \mathrm{X}$, and $\lim_{x\to\pm\infty} \eta_y(t,x) = 0$, for every $t \in \mathrm{T}$, the Gâteaux differential of the Lagrange functional is[4]

$$
\begin{aligned}
\mathrm{D}_{\boldsymbol{\eta}(\cdot)}\mathsf{L}^* = & \lim_{x\to\infty} \frac{1}{2x} \int_{-x}^{x} \int_0^\infty e^{-\rho t} \left[\frac{\partial H^*(t,\xi)}{\partial u} \eta_u(t,\xi) + \right. \\
& + \left. \left(\partial_t q(t,\xi) - \rho\, q(t,\xi) + \partial_{\xi\xi} q(t,\xi) + \frac{\partial H^*(t,\xi)}{\partial y} \right) \eta_y(t,\xi) \right] dt\, d\xi \\
& - \lim_{t\to\infty} \lim_{x\to\infty} \frac{1}{2x} \int_{-x}^{x} \left(e^{-\rho t} q(t,\xi) - \mu(t,\xi) \right) \eta_y(t,\xi) d\xi \\
& + \lim_{x\to\infty} \frac{1}{2x} \int_{-x}^{x} q(0,\xi)\, \eta_y(0,\xi) d\xi + \\
& + \lim_{x\to\infty} \frac{1}{2x} \int_{-x}^{x} \int_0^\infty e^{-\rho t} \left(q(t,\xi)\, \partial_\xi \eta_y(t,\xi) - \partial_\xi q(t,\xi)\, \eta_y(t,\xi) \right) dt.
\end{aligned}
$$

The variational principle states a necessary condition for an optimum is $\mathrm{D}_{\boldsymbol{\eta}(\cdot)}\mathrm{L}[\mathrm{y}^*, \mathrm{u}^*] = 0$. The first integral in equal to zero if first-order conditions (16.4a) and (16.4a) are satisfied; the second integral is equal to zero if $\mu(t,x) = e^{-\rho t} q(t,x)$, that together with the KKT condition yields a generalization of the transversality condition (16.4c); the third integral is equal to zero for admissible perturbations; and the last integral is also equal to zero for admissible perturbations and if the dual boundary conditions for the co-state variable is satisfied. □

Example 16.1 The distributional AK **model** In this example, we consider the problem of the optimal allocation of capital among a continuum of economies, located along an infinite domain $\mathrm{X} = \mathbb{R}$, in which the capital stock for every location is constrained by a PDE analogous to the one presented in Example 15.7, but in which consumption is endogenous. Denoting by $K(t,x)$ and $C(t,x)$ the stock of capital and consumption at location $x \in \mathrm{X}$ at time $t \in \mathrm{T} = \mathbb{R}_+$, and assuming an uniform linear technology of production, where A is the net total factor productivity, the capital accumulation constraint for location x is

$$
\partial_t K(t,x) = \sigma^2\, \partial_{xx} K(t,x) + AK(t,x) - C(t,x), \quad \text{for each} (t,x) \in \mathrm{T} \times \mathrm{X}. \tag{16.5}
$$

The optimal allocation of consumption over time and space, and therefore the allocation of savings and capital, is done by a Millian social planer having the social welfare functional

$$
\mathsf{V}[K,C] = \lim_{x\to\infty} \frac{1}{2x} \int_{-x}^{x} \int_0^\infty \frac{C(t,\xi)^{1-\theta} - 1}{1-\theta} e^{-\rho t}\, dt\, d\xi, \tag{16.6}
$$

where $\theta > 0$ is the uniform inverse of the EIS for the representative household in every location. In addition to the PDE constraint (16.5) the following initial, terminal,

[4] Again, we are assuming that the conditions for validity of the Fubini's theorem are satisfied.

and boundary constrains

$$K(x,0) = k_0(x), \text{ given for each } x \in \mathrm{X}, \tag{16.7a}$$
$$\lim_{t\to\infty} K(t,x) \geq 0, \text{ for each } x \in \mathrm{X}, \tag{16.7b}$$
$$\lim_{x\to\mp\infty} \frac{K(t,x)}{x} = 0, \text{ for each } t \in \mathrm{T}, \tag{16.7c}$$

should also be satisfied. From Propositon 16.2, the optimal allocation $(K^*, C^*)_{(t,x)\in \mathrm{T}\times\mathrm{X}}$ can be obtained from the solution to the distributional MHDS

$$\partial_t K(t,x) = \sigma^2 \partial_{xx} K(t,x) + AK(t,x) - C(t,x), \tag{16.8a}$$
$$\partial_t C(t,x) = -\sigma^2 \left(\partial_{xx} C(t,x) - \frac{1+\theta}{C(t,x)} \left(\partial_x C(t,x)\right)^2\right) + \gamma\, C(t,x), \tag{16.8b}$$
$$K(0,x) = k_0(x),\ x \in \mathrm{X}, \tag{16.8c}$$
$$\lim_{t\to\infty}\lim_{x\to\infty} \frac{1}{2x}\int_{-x}^{x} e^{-\rho t} K(t,\xi)\, C(t,\xi)^{-\theta} d\xi = 0,\ x \in \mathrm{X}, \tag{16.8d}$$
$$\lim_{x\to\pm\infty} \left(e^{\rho t} C(t,x)^{\theta} x\right)^{-1} = 0,\ t \in \mathrm{T}. \tag{16.8e}$$

The endogenous growth rate

$$\gamma \equiv \frac{A-\rho}{\theta}, \tag{16.9}$$

is the same as the endogenous growth rate of the benchmark (homogeneous) AK model, and is assumed to be positive. The PDE system (16.8a)–(16.8b) is recursive, which allows us to find an explicit solution

$$K(t,x) = e^{\gamma t} k(t,x),\ C(t,x) = e^{\gamma t} c(t,x),\ \text{for each } (t,x) \in \mathrm{T}\times\mathrm{X}, \tag{16.10}$$

that take the form of a **distributional BGP** where

$$k(t,x) = \int_{-\infty}^{\infty} \frac{k_0(\xi)}{2\sigma\sqrt{\pi\theta t}}\, e^{-\left(\frac{x-\xi}{2\sigma}\right)^2 \frac{1}{\theta t}} d\xi,$$

and

$$c(t,x) = \int_{-\infty}^{\infty} \frac{k_0(\xi)\, \phi(t, x-\xi)}{2\sigma\sqrt{\pi\theta t}}\, e^{-\left(\frac{x-\xi}{2\sigma}\right)^2 \frac{1}{\theta t}} d\xi.$$

where $\phi(t,z) \equiv \left[A - \gamma + (\theta-1)\left(\frac{1}{2\theta t} - \left(\frac{z}{2\sigma\theta t}\right)^2\right)\right]$.

A necessary condition for the existence of a solution to the centralized problem is that $A > \gamma$. This model displays convergence to a time-unbounded balanced growth

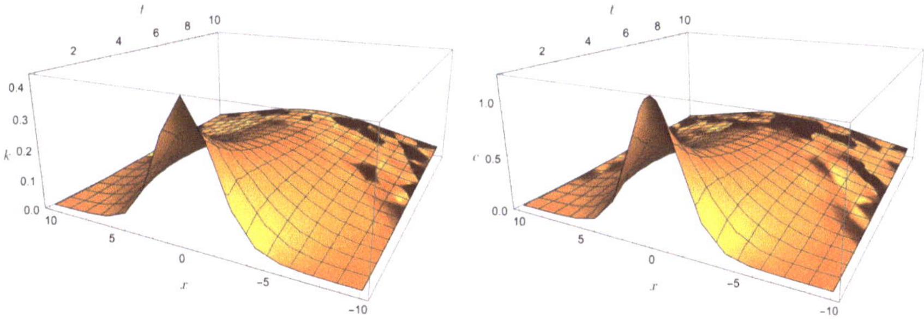

(a) Solution for $k(t, x)$ (b) Solution for $c(t, x)$

Fig. 16.1 Detrended solutions to the spatial AK model

path similar to the homogeneous-agent AK model, with a uniform capital distribution among locations. Figure 16.1 illustrates the optimal detrended solutions for the distribution of capital and consumption over time for an initial heterogeneous distribution of the capital stock $k_0(x) = e^{-|x|}$. We observe that the initial distribution of consumption is also non-homogeneous, and is higher in the central locations. However, it still allows for positive savings in those locations that will not only increase the local capital accumulation but also will diffuse over space, allowing for a general increase in capital accumulation across all economies. Although the benchmark spatially homogeneous AK model does not display transition dynamics (see Example 10.5), we have just shown that, if there is spatial heterogeneity plus spatial mobility of capital, there will be transition dynamics resulting from the process of spatial reallocation of capital.

16.3 Optimal Control of Fokker-Planck-Kolmogorov Equations

In this section we consider optimal control problems constrained by a time-varying probability function obtained from a Fokker-Planck-Kolmogorov (FPK) equation, associated to a stochastic differential equation. More generally, this problem can be seen as an optimal control problem of a probability density function modeling an heterogeneous population, in which the heterogeneity is captured by variable $x \in \mathrm{X}$, evolving over time, and in which the drift and volatility components are endogenously controlled. It is also an extension of problem (14.27) for the case in which there is a diffusion component.

The problem that interest us here has several specific features: first, the state variable, $y(t, x)$, enters in the objective functional as a weighting variable; second, the diffusion and the volatility components depend on the control variable, $u(t, x)$; third, the initial condition satisfies a conservation law $\int_{\mathrm{X}} y_0(x)\, dx = 1$; at last, the state variable takes value zero at the boundaries of set X, at every point in time.

Both the FPK constraint and the initial and the boundary conditions provide necessary conditions for the satisfaction of the following conservation property (with or without a normalization)

$$\int_{\mathrm{X}} y(t,x)\,dx = 1, \text{ for each } t \in \mathrm{T},$$

for the time-varying density $y(\cdot)$ that can be optimally controlled. Although first-order PDE satisfy a conservation law for every $t > 0$ if the initial condition, for $t = 0$, does satisfy it, this property does not hold in general for parabolic PDE. This is particularly the case when the set X is not the real line, in which case a normalization needs to be introduced.[5]

Let $T = [\underline{t}, \bar{t}]$ and $X = [\underline{x}, \bar{x}]$, and the state variable be the mapping $y : T \times X \to \mathbb{R}$ and the control variable be the mapping $u : T \times X \to \mathbb{R}$. The **optimal control problem of a FPK equation** is to find $(\mathbf{y}^*, \mathbf{u}^*) = \big(u^*(t,x), y^*(t,x)\big)_{(t,x)\in \mathrm{T}\times\mathrm{X}}$ that solve the problem

$$\max_{u(\cdot)} \int_{\mathrm{T}}\int_{\mathrm{X}} f(u(t,x),t,x)\,y(t,x)\,dx\,dt, \tag{16.11a}$$

subject to

$$\partial_t y = -\partial_x\Big(g(u,x)\,y\Big) + \partial_{xx}\Big(\sigma(u,x)\,y\Big), \text{ a.e } (t,x) \in \mathrm{T}\times\mathrm{X}, \tag{16.11b}$$

$$y(\underline{t},x) = y_0(x), \text{ s.t. } \int_{\mathrm{X}} y_0(x)dx = 1, \text{ for } (t,x) \in \{t = \underline{t}\} \times X, \tag{16.11c}$$

$$y(t,x) = 0, \text{ for } (t,x) \in T \times \{x = \underline{x}, x = \bar{x}\}. \tag{16.11d}$$

Let functions $f(\cdot)$, $g(\cdot)$ and $\sigma(\cdot)$ be at least continuous and continuously differentiable. The problem simplifies if the volatility function, $\sigma(\cdot)$, is linear in (u, x).

Proposition 16.3 (Optimal control of the FPK equation) *Let (u^*, y^*) be a solution to problem (16.11a)–(16.11d). Then there is a co-state variable $\lambda : T \times X \to \mathbb{R}$ such that admissible solutions, satisfying Eqs. (16.11b)–(16.11d), also satisfy the following: optimality and adjoint conditions*

$$\frac{\partial f^*(t,x)}{\partial u} + \frac{\partial g^*(t,x)}{\partial u}\,\partial_x\lambda(t,x) + \frac{\partial \sigma^*(t,x)}{\partial u}\,\partial_{xx}\lambda(t,x) = 0, \tag{16.12a}$$

$$f^*(t,x) + \partial_t\lambda(t,x) + g^*(t,x)\,\partial_x\lambda(t,x) + \sigma^*(t,x)\,\partial_{xx}\lambda(t,x) = 0, \tag{16.12b}$$

for $(t,x) \in \mathrm{T}\times\mathrm{X}$, a transversality condition

$$\lambda(\bar{t},x) = 0, \ (t,x) \in \{t = \bar{t}\} \times \mathrm{X}. \tag{16.13}$$

If $\sigma(\cdot)$ is non-linear in (u, x), the adjoint boundary conditions

$$\lambda(t,\bar{x}) = \lambda(t,\underline{x}) = 0, \ (t,x) \in \mathrm{T} \times \{\mathrm{x} = \underline{\mathrm{x}}, \mathrm{x} = \bar{\mathrm{x}}\}. \tag{16.14}$$

are also necessary.

[5] Therefore, the version of optimal control we present may not satisfy a conservation law.

Proof Consider the solution to the problem, denoted by (u^*, y^*). At the optimum the value functional is $\mathsf{V}[y^*, u^*] = \int_{\mathrm{T}} \int_{\mathrm{X}} f^*(t,x)\, y^*(t,x)\, dx\, dt$, where $f^*(t,x) = f\big(u^*(t,x), t, x\big)$.

For any other control-state pair (u, y), and writing $f(t,x) = f(u(t,x), t, x)$, $g(t,x) = g(u(t,x), t, x)$, and $\sigma(t,x) = \sigma(u(t,x), t, x)$, there is a co-state variable, λ, such that the value functional satisfies

$$
\begin{aligned}
\mathsf{V}[y,u] &= \int_{\mathrm{T}} \int_{\mathrm{X}} f(t,x)\, y(t,x)\, dx\, dt = \\
&= \int_{\mathrm{T}} \int_{\mathrm{X}} f(t,x)\, y(t,x) - \lambda(t,x) \Big(\partial_t y(t,x) + \partial_x \big(g(t,x)\, y(t,x) \big) - \\
&\quad - \partial_{xx} \big(\sigma(t,x)\, y(t,x) \big) \Big)\, dx\, dt \\
&= \int_{\mathrm{T}} \int_{\mathrm{X}} \Big[f(t,x) + \partial_t \lambda(t,x) + g(t,x)\, \partial_x \lambda(t,x) + \\
&\quad + \sigma(t,x)\, \partial_{xx} \lambda(t,x) \Big]\, y(t,x)\, dx\, dt \\
&\quad - \int_{\mathrm{T}} \int_{\mathrm{X}} \lambda(t,x)\, y(t,x) dx - \int_{\mathrm{T}} \int_{\mathrm{X}} \Big[\lambda(t,x)\, g(t,x)\, y(t,x) - \\
&\quad - \lambda(t,x)\, \partial_x \big(\sigma(t,x)\, y(t,x) \big) + \partial_x \lambda(t,x)\, \sigma(t,x)\, y(t,x) \Big]\, dt.
\end{aligned}
$$

In particular, consider the perturbations to the control and state variables, $u(t,x) = u^*(t,x) + \varepsilon\, \eta_u(t,x)$ and $y(t,x) = y^*(t,x) + \varepsilon\, \eta_y(t,x)$, for each $(t,x) \in \mathrm{T} \times \mathrm{X}$, satisfying the admissibility constraints $\eta_y(\underline{t}, x) = 0$, for any $x \in \mathrm{X}$, and $\eta_y(t,x) = 0$ for $x \in \partial X = \{\underline{x}, \overline{x}\}$ (note that $\eta_y(\overline{t}, x)$ is arbitrary for $x \in \mathrm{Int}(X)$). Denoting $g^*(t,x) = g\big(u^*(t,x), t, x\big)$ and $\sigma^*(t,x) = f\big(u^*(t,x), t, x\big)$, the Gâteaux differential of the value functional, evaluated at the optimum, is

$$
\begin{aligned}
&\mathrm{D}_{\eta(\cdot)} \mathsf{V}[\mathrm{y}^*, \mathrm{u}^*] = \\
&\int_{\mathrm{T}} \int_{\mathrm{X}} \left[\frac{\partial f^*(t,x)}{\partial u} + \frac{\partial g^*(t,x)}{\partial u}\, \partial_x \lambda(t,x) + \frac{\partial \sigma^*(t,x)}{\partial u}\, \partial_{xx} \lambda(t,x) \right] y^*(t,x) \eta_u(t,x) + \\
&+ \Big[f^*(t,x) + \partial_t \lambda(t,x) + g^*(t,x)\, \partial_x \lambda(t,x) + \sigma^*(t,x)\, \partial_{xx} \lambda(t,x) \Big] \eta_y(t,x)\, dx\, dt \\
&- \int_{\mathrm{T}} \int_{\mathrm{X}} \lambda(t,x)\, \eta_y(t,x) dx \\
&- \int_{\mathrm{T}} \int_{\mathrm{X}} \Big\{ \lambda(t,x) \Big[\Big(\frac{\partial g^*(t,x)}{\partial u}\, y^*(t,x) - \frac{\partial}{\partial u} \Big(\partial_x \big(g^*(t,x)\, y^*(t,x) \big) \Big) \Big) \eta_u(t,x) + \\
&+ \Big(g^*(t,x) - \frac{\partial}{\partial y} \Big(\partial_x \big(g^*(t,x)\, y^*(t,x) \big) \Big) \Big) \eta_y(t,x) \Big] \\
&+ \partial_x \lambda(t,x) \Big(\frac{\partial \sigma^*(t,x)}{\partial u}\, y^*(t,x)\, \eta_u(t,x) + \sigma^*(t,x) \eta_y(t,x) \Big) \Big\}\, dt.
\end{aligned}
$$

For admissible perturbations we have

$$
\begin{aligned}
\mathrm{D}_{\eta(\cdot)}\mathrm{V}[\mathrm{y}^*, \mathrm{u}^*] = & \int_{\mathrm{T}}\int_{\mathrm{X}} \left[\frac{\partial f^*(t,x)}{\partial u} + \right. \\
& + \left.\frac{\partial g^*(t,x)}{\partial u}\,\partial_x\lambda(t,x) + \frac{\partial \sigma^*(t,x)}{\partial u}\,\partial_{xx}\lambda(t,x)\right] y^*(t,x)\eta_u(t,x) + \\
& + \Big[f^*(t,x) + \partial_t\lambda(t,x) + g^*(t,x)\,\partial_x\lambda(t,x) + \\
& + \sigma^*(t,x)\,\partial_{xx}\lambda(t,x)\Big]\eta_y(t,x)\,dx\,dt - \int_{\mathrm{X}} \lambda(\bar{t},x)\,\eta_y(\bar{t},x)dx + \\
& + \int_{\mathrm{T}}\int_{\mathrm{X}} \lambda(t,x)\,\frac{\partial}{\partial u}\Big(\partial_x\big(\sigma^*(t,x)\,y^*(t,x)\big)\Big)\,\eta_u(t,x)\,dt,
\end{aligned}
$$

where the last integral is equal to zero if the volatility function $\sigma(\cdot)$ is linear in (u, x). A necessary condition for the optimum is that $\mathrm{D}_{\eta(\cdot)}\mathrm{V}[\mathrm{y}^*, \mathrm{u}^*] = 0$. This will be the case if and only if conditions (16.12a) to (16.13) are satisfied if $\sigma(\cdot)$ is linear in (u, x). If $\sigma(\cdot)$ is not linear in (u, x) the adjoint boundary conditions (16.14) are also necessary. □

Example 16.2 (*Optimal social mobility*) In this example the problem for maximizing social welfare presented in Example 14.3 is extended by incorporating uncertainty in the dynamics of income for each household following the approach in Example 15.12. This extension modifies the previous two examples. It introduces a diffusion component to the constraint of the optimal social welfare problem dealt in Example 14.3, and provides an economic justification for the diffusion components of the social dynamic function of Example 15.12. By doing this we offer an example in which the social planner optimally manages both aggregate and pure social mobility with a goal of generating both optimal growth and income distribution.[6]

Assume an economy with heterogeneous households, in which the net wealth of a household with capital $k(0)$ at time $t = 0$, is driven by the stochastic differential equation $dk(t) = (Ak(t) - c(t))\,dt + \sigma k(t)dW(t)$, where $(W(t))_{t>0}$ is a Wiener process, and let $n(t, k)$ be the density of households with capital $k(t)$ at time t, satisfying $0 < n(t, k) < 1$, for each $(t, k) \in \mathbb{R}^2_+$, and $\int_0^\infty n(t, k)dt = 1$, for every $t \in [0, \infty)$.

It can be proved that $n(t, k)$ represents both the proportion of the population with net-wealth k at time t, and the probability that an household starting in the quantile $k_q = k$, at time $t = 0$ will remain in the same quantile over time. Then we can invoke the FPK equation (see Sect. 17.3.3) to show that the dynamics of the wealth distribution is driven by the PDE

[6] This is an abridge version of Brito (2024). In this paper, we present a more general problem in which the social planner also controls volatility, thus introducing a further instrument related to providing optimal social insurance.

$$\partial_t n(t,k) + \partial_k\Big(\big(Ak - c\big)\, n(t,k)\Big) - \frac{1}{2}\partial_{kk}\Big((\sigma k)^2\, n(t,k)\Big) = 0,\ \ (t,k) \in \mathbb{R}^2_+, \tag{16.15}$$

subject to the initial net-wealth distribution $n(0,k) = n_0(k)$, and the boundary constraints $n(t,0) = \lim_{k\to\infty} n(t,k) = 0$.

The social welfare problem is to maximize the Kolm-Atkinson-Sen social welfare functional[7]

$$\int_0^\infty \int_0^\infty \frac{c(t,k)^{1-\xi} - 1}{1-\xi}\, n(t,k) e^{-\rho t}\, dk\, dt,\ \ \rho > 0,$$

where ξ parameterizes the degree of inequality aversion, by choosing the paths $(c(t,k), n(t,k))_{(t,k)\in\mathbb{R}^2_+}$ subject to the constraint (16.15), given the initial distribution $n(0,k) = n_0(k)$, and the boundary conditions $\lim_{k\to 0} n(t,k) = \lim_{k\to\infty} n(t,k) = 0$. This is a special case of optimal control problem (16.11). Applying the generalized PMP in Proposition 16.3, the necessary first-order conditions lead to the forward-backward parabolic PDE-system, and the transversality condition

$$\partial_t q + A\,k\,\partial_k q + \frac{\xi\left(\partial_k q\right)^{\frac{\xi-1}{\xi}} - 1}{1-\xi} + \frac{\sigma^2}{2} k^2 \partial_{kk} q - \rho q(t,k) - 1 = 0, \tag{16.16a}$$

$$\partial_t n + \partial_k\Big(\big(A\,k - (\partial_k q)^{-\frac{1}{\xi}}\big) n\Big) - \frac{\sigma^2}{2}\partial_{kk}\big(k^2 n\big) = 0, \tag{16.16b}$$

$$\lim_{t\to\infty} e^{-\rho t} q(t,k) = 0, \tag{16.16c}$$

where $q = q(t,k) = e^{\rho t}\lambda(t,k)$ is the current distributional co-state variable.[8]

As the PDE system is recursive, we can integrate the Euler equation (16.16a) over time, and imposing the transversality condition (16.16c), we find that $q(t,k)$ is time-independent. Setting $q(t,k) = \bar{q}(k)$, the Euler equation can be reduced to a second-order ODE, similar to an HJB equation, that has an explicit solution. This allows us to find the optimal consumption as a linear function of k. Substituting the optimal consumption policy in the FPK equation, (16.16b), yields a linear forward parabolic PDE for the optimal net-wealth distribution

$$\partial_t n^*(t,k) + \partial_k\left(\gamma\, k\, n^*(t,k)\right) - \frac{\sigma^2}{2}\partial_{kk}\left(k^2\, n^*(t,k)\right) = 0, \tag{16.17}$$

where

$$\gamma \equiv \frac{A - \rho + \xi\,(\xi - 1)\frac{\sigma^2}{2}}{\xi}. \tag{16.18}$$

---

[7] See Sect. 8.1.5.8 for discussion on social welfare functionals.

[8] As the analog to the constraint $\sigma(u,x)$ is linear, the boundary conditions (16.14) on the co-state variable, are not necessary.

The optimal distribution of net wealth, that maximizes intertemporal social welfare is

$$n^*(t,k) = \int_0^\infty \frac{n_0(\xi)}{\xi\sqrt{2\pi\sigma^2 t}} \exp\left\{-\frac{\left(\ln\left(\frac{k}{\xi}\right) - \left(\gamma - \frac{\sigma^2}{2}\right)t\right)^2}{2\sigma^2 t}\right\} d\xi,\ (t,k) \in \mathbb{R}^2_+. \tag{16.19}$$

If the initial distribution is a log-normal density with parameters μ_0 and σ_0^2, then the density for capital distribution follows a time varying lognormal density function $n^*(t,k) \sim LN\left(\mu_0 + \left(\gamma - \frac{\sigma^2}{2}\right)t, \sigma_0^2 + \sigma^2 t\right)$. Therefore, the average capital stock and the Gini index are, respectively,

$$\mathsf{M}[K(t)] = e^{\mu_0 + \gamma t}, \text{ and } \mathsf{G}[K(t)] = \text{erf}\left(\frac{\sigma_0^2 + \sigma^2 t}{2}\right).$$

Transforming from the net wealth distribution to income distribution, by observing that $f(t,y) = n(t,k/A)$, comparing with the solution to the "ad-hoc" FPK equation (15.34), we observe that they have the same functional form. However, the current example provides a theory on the fundamentals of the dynamics of the distribution. Assuming an efficient distribution of income by an KAS inequality averse social planner that wants to maximizing intertemporal welfare, we can have both economic growth, if $\gamma > 0$, and an increase in inequality, measured by the Gini index. Furthermore, in an absence of a social insurance intervention, that would make σ endogenous to policy, an increase in social mobility induces both an increase in the average growth rate and in inequality, and a higher inequality aversion will only have an effect on the growth rate of the economy, not in the Gini index.

16.4 References

Butkovskiy (1969), Lions (1971), Derzko et al. (1984) or Neittaanmäki and Tiba (1994) present results on optimal control problems of PDE, also called optimal distributed control problems, with varying levels of generality. We draw mainly upon the last two references. Other references include Fattorini (1999) and Tröltzsch (2010), and, with applications in economics and related fields, Carlson et al. (1996, Chap. 9). The spatial AK model appears in Brito (2004) but has its roots in Isard and Liossatos (1979). Several papers deal rigorously with its mathematical properties, and extend it into several directions: Brock and Xepapadeas (2008), Boucekkine et al. (2009), Fabbri (2016), Boucekkine et al. (2013, 2019),, and Brito (2022). Augeraud-Véron et al. (2019) survey applications to resource economics. Optimal control problems of FPK equations are dealt in the mean field optimal control literature, see Gomes and Saúde (2014) and Bensoussan et al. (2013). Nuño and Moll (2018) is one rare example of an application to macroeconomics.

Part IV
Stochastic Differential Equations

Chapter 17
Introduction to Stochastic Calculus and Stochastic Differential Equations

An ordinary differential equation (ODE), having time as the independent variable, $\dot{y} = f(y)$, where $y = y(t)$, can be extended by introducing a random perturbation, $\epsilon(t)$, linearly as in $\dot{Y} = f(Y) + \epsilon(t)$, or non-linearly, as in $\dot{Y} = f(Y, \epsilon(t))$, where $Y = Y(t, \epsilon(t))$. To the last two equations, we call generically **random differential equations** (RDE). While a scalar ODE contains a deterministic mapping $y : T \to Y \subseteq \mathbb{R}$, a RDE contains a random mapping $Y : T \times \Omega \to Y \subseteq \mathbb{R}$, where $(\Omega, \mathbb{P})$ is a probability space. In this case, a distinction is introduced between a random variable, $Y(t) = (y(t, \omega))_{\omega\in\Omega}$, where $y(t, \omega) \in Y$ is a realization conditional on the state of nature $\omega \in \Omega$, and an ex-post realization, $y(t)$.

Part I addresses the determination and characterized of solutions to ODE. In general, if function $f(.)$ is continuous and differentiable a solution $y(t)$ always exists, it is unique, and it is a continuous and differentiable function of time. In addition, it addressed the existence of steady states, their stability properties, and the existence of bifurcations. For RDE, solutions can be conceived as a (very large) family of functions, or realizations, parameterized by the outcomes of a random experiment. This is why we use $Y(t)$ instead of $y(t)$. Indeed, for a particular realization of the random variable, such as $\epsilon(t) = \epsilon_t$, where ϵ_t is a number, it has local properties similar to an ODE, v.g, local stability properties. However, the introduction of noise implies that solutions of a RDE display some of the following new features: non-continuity, non-differentiability (in the classic sense), and non-convergence (except in particular cases) to a deterministic steady state, even if the deterministic part (or skeleton) has a stable fixed point.

Stochastic dynamics refer generically to the behavior over time of solutions to RDE. They include the characterization of their sample path properties over time, and of their moment or statistic properties. **Sample path dynamics** refer to the behavior over time of solutions to an RDE, for a particular or several realizations of the stochastic process, according to the specification of the RDE. Furthermore, the existence of a probability measure, associated to the stochastic component of those equations, allow for characterizing the **statistical or distributional dynamics**

P. B. Brito, *Economic Dynamics and Distributions*, Dynamic Modeling and Econometrics in Economics and Finance 32, https://doi.org/10.1007/978-3-031-94717-9_17

properties of the solutions. Although we are unaware of a general bifurcation theory for RDE, we can characterize their distributional stability properties, by drawing on their relationship with partial differential equations (PDE).

A possible taxonomy of stochastic dynamics considers the types and effects of random perturbations relative to a related deterministic (linear or non-linear) underlying skeleton model. We can consider the following four cases: (1) high noise can generate large deviations; (2) high noise can generate small deviations; (3) low noise can generate small deviations; or (4) low noise can generate high deviations.

In continuous time environments, applied to economics and finance, **stochastic differential equations** (SDE) are the main types of RDE.[1] Two main types of SDE are considered: stochastic differential equations with jumps for modeling rare events with a local high impact, or stochastic differential equations of the diffusive type (SDE), for modeling frequent events having a local small impact. In **stochastic differential equations with jumps** , the random component is a **Poisson process**, $\left(Q(t)\right)_{t\in T}$, such that $dQ(t) = 1$ with probability $\lambda\, dt$ and $dQ(t) = 0$ with probability $(1-\lambda)\, dt$, and take the form

$$dY(t) = \mu(Y(t), t)dt + v(Y(t^-), t^-)\, dQ(t), \tag{17.1}$$

where $Y(t^-) = \lim_{s\uparrow t} Y(s)$, and $\mu(.)$ and $v(.)$ are continuous and differentiable known functions. In **stochastic differential equations of the diffusive type** uncertainty is introduced via a **Wiener process**, $(W(t))_{t\in T}$, and take the form

$$dY(t) = \mu(Y(t), t)dt + \sigma(Y(t), t)dW(t), \tag{17.2}$$

where $\mu(.)$ and $\sigma(.)$ are known continuous and differentiable functions.

The reason for specifying those equations as a differential, in both cases, is related to the fact that $Y(t)$ is not differentiable in the classic sense. Before solving any of those equations a specific stochastic calculus has to be developed. This means that the integral versions of those equations are well defined. For SDEJ we have

$$Y(t) = Y(0) + \int_0^t f(Y(s), s)ds + \int_0^t v(Y(s^-), s^-)\, dQ(s), \tag{17.3}$$

where the first integral in the right-hand side is a Riemann integral, but the second is a **Poisson integral**, and for SDE we have

$$Y(t) = Y(0) + \int_0^t f(Y(s), s)ds + \int_0^t \sigma(Y(s), s)\, dW(s), \tag{17.4}$$

where the first integral in the right-hand side is a Riemann integral, but the second is an **Itô integral**.

[1] For a clear discussion, see Merton (1982).

We can also have stochastic differential equations having both jump and diffusion components,[2]

$$dY(t) = \mu(Y(t), t)dt + \sigma(Y(t), t)dW(t) + v(Y(t^-), t^-)\, dQ(t).$$

In this book, we restrain to Itô's SDE, (17.2) or (17.4), because they are the most frequently used in economics and finance. We restrain even further to linear SDE, by presenting a very brief (pragmatic) introduction to Itô's stochastic calculus, and by emphasizing their relationship to ODE and PDE. In Sect. 17.1, we define and describe the properties of the Wiener process, in Sect. 17.2, we define and present the properties of Itô's processes, and, in Sect. 17.3, we present their application to solving and characterizing the solutions of SDE.

17.1 The Wiener Process

17.1.1 Stochastic Processes: A Brief Description

Let time, t, be an independent variable belonging to the set $T \subseteq \mathbb{R}_+$, and let $(\Omega, \mathcal{F}, \mathbb{P}, \mathbb{F})$ denote a filtered probability space, where the sample space is Ω, the set of all events (i.e., the set of all subsets of Ω) is $\mathcal{F}$, and the probability measure over events, $\mathbb{P}$, is a mapping $\mathbb{P} : \Omega \to (0, 1)$ such that $\mathbb{P}(\Omega) = 1$, and the filtration is the sequence of event sets $\mathbb{F} = \big(\mathcal{F}(t)\big)_{t \in T}$. Informally, a filtration specifies the way in which the flow information allows for determining the flow of probability distributions $(\mathbb{P}(t))_{t \in T}$, over time.

A one-dimensional **stochastic process**[3]can be understood as a flow of random variables $\big(X(t)\big)_{t \in T}$, where $X(t) = \big(x(t, \omega)\big)_{\omega \in \Omega}$, is a random variable at time t, where $x(t, \omega)$ is the realization conditional on the state of nature $\omega \in \Omega$, such that $x(t, \omega) : T \times \Omega \to X \subseteq \mathbb{R}$, where X is the image set of all possible realizations. Intuitively, $X(t) : T \to X$ represents the **possible realizations** of the process at time t, that is **before** nature (or a pseudo random number generator of a computer) makes a draw at time t. The **realization** of a stochastic process at time t, denoted by $X(t) = x(t) \in X$, is a particular realization that is observed **after** nature (or the computer) makes a draw. An **history at time** T is a particular path of realizations $\big(x(t)\big)_{t \in [0,T]}$ of a stochastic process.

A stochastic process is called **non-anticipating** if the probability assigned to a particular event is determined from past events, meaning that we can only ascertain the probability of a future event on the base of past information. A stochastic process is called an **anticipating process** if we condition the probability of present events on the occurrence of future events. A stochastic process is **adapted to a filtration** if

---

[2] For applications of jump-diffusion processes to economics, see Sennewald and Wälde (2006).

[3] The seminal contribution to the analytical approach to stochastic processes is Kolmogorov (1931).

it has a probability distribution drawn from a particular filtration, that is, depending on the flow of information implicit in the filtration.

We only consider non-anticipating processes, or **processes adapted** to a non-anticipating filtration. This implies that a particular stochastic process is specified by a particular flow of information. This is the case of Markov processes and of martingales:

Definition 17.1 (*Markov process*) The stochastic process $\big(X(t)\big)_{t\in T}$ is called a Markov process if, for any $0 < s < t$, and for $A \subset X$, it satisfies

$$\mathbb{P}[X(t) \in A|\big(X(r)\big)_{r\in[0,s]} = \big(x(r)\big)_{r\in[0,s]}] = \mathbb{P}[X(t) \in A|X(s) = x(s)], \quad (17.5)$$

where $\big(x(r)\big)_{r\in[0,s]}$ and $x(s)$ denote a particular history until time s and a particular realization at time s, respectively.

Definition 17.2 (*Martingales*) The stochastic process $\big(X(t)\big)_{t\in T}$ with bounded mathematical expectation, $\mathbb{E}[|X(t)|] < \infty$, is called a **martingale** if it satisfies, for any history until time t, $\mathbf{x}_t \in \mathbf{X}_t = \big(X(s)\big)_{s\in[0,t)}$, the martingale property

$$\mathbb{E}[X(t)|\mathbf{X}_t = \mathbf{x}_t] = x(s), \text{ for any } s \le t, \quad (17.6)$$

and any realization $X(s) = x(s)$. If $\mathbb{E}[X(t)|\mathbf{X}_t = \mathbf{x}_t] \ge x(s)$ the process is called a **submartingale**, and if $\mathbb{E}[X(t)|\mathbf{X}_t = \mathbf{x}_t] \le x(s)$ the process is called a **supermartingale**. In particular, a martingale satisfies $\mathbb{E}[X(t)|\big(X(s)\big)_{s\in[0,t]}] = x_0$ for any $t \ge 0$.

Informally, a Markov process is a memory-less process, in the sense that the probability of a future event is independent from the past history (only the last realization matters), and a martingale is a stationary process in expected value terms. A Markov process satisfies the martingale property if $\mathbb{E}[X(t)|X(s) = x(s)] = x(s)$, for any $s \le t$.

A particular type of continuous time process satisfying both the Markov and the martingale properties, in Eqs. (17.5) and (17.6), is the Wiener process.

17.1.2 Wiener Process: Definition

There are several ways to define and characterize a **Wiener process**, also called **standard Brownian process**.[4] The Wiener process is defined by the particular type of information on the flow of events and the associated probability distribution.

Definition 17.3 A **Wiener process**, denoted by $(W(t))_{t\ge 0}$, is a stochastic process for which $W : T \to \mathbb{R}$ has the following properties:

[4] First introduced by Bachelier (1900) and proved to exist by Norbert Wiener around 1920.

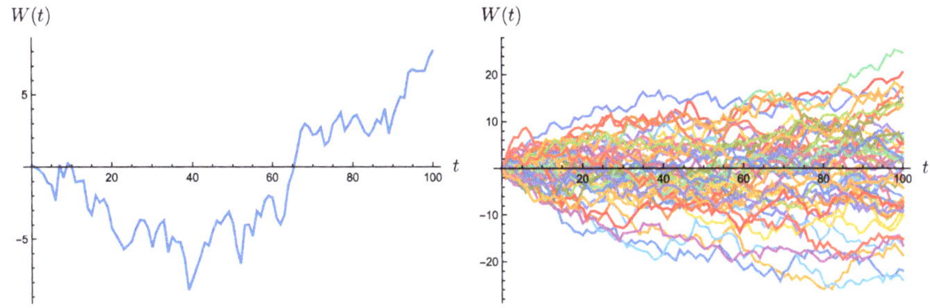

(a) One sample path (b) 100 sample paths

Fig. 17.1 Sample paths for the Wiener process

1. the initial value is equal to 0 with probability one: $\mathbb{P}[W(0) = 0] = 1$ (also written as $W(0) = 0$ a.s.)[5];
2. it has a continuous version, in the sense that each sample path is a continuous function of time with probability one (i.e., discontinuous jumps can occur with probability zero);
3. the path increments are independent and have a Gaussian probability distribution with zero mean and variance equal to the time increment: $dW(t) = W(t + dt) - W(t) \sim N(0, dt)$, for $dt \geq 0$.

Consider a positive time increment $dt > 0$ and let $\mathbb{P}[W(t + dt) = w' | W(t) = w]$ be the probability that the realization of a Wiener process at time $t + dt$ would be w' if its realization at time t is w. If we can write $\mathbb{P}[W(t + dt) \in dw' | W(t) = w] = dP_t(w', w)\, dw'$, we call $dP_t(w', w)$ a **transition probability**. The **transition probability of a Wiener process** is the Gaussian probability density function

$$dP_t(w', w) = \frac{1}{\sqrt{2\pi dt}} e^{-\frac{(dw')^2}{2dt}}.$$

Therefore, a Wiener process is a Markov process having a Gaussian conditional transition probability.

Figure 17.1 depicts one sample path, in Fig. 17.1a, and 100 sample paths, in Fig. 17.1b, for a Wiener process. They illustrate the stochastic properties of the Wiener process: first, every sample path is strongly jagged, second, the distribution displays increasing dispersion over time, although it tends to be located most of the time close to $W = 0$. Furthermore, although sample paths are continuous, they are not smooth enough to allow differentiability (i.e., at every point in time, slope limits do not look like being unique).

We can provide informal proofs of those **sample path** and **statistical** properties of the Wiener process.

---

[5] Almost surely.

Proposition 17.1 (Sample path of the Wiener process: are non-differentiable) *The Wiener process is not first-order-differentiable in time, in the classic sense.*

Proof Let $t \in (0, \infty)$ and $dt > 0$. Defining $|dW(t)| = |W(t + dt) - W(t)|$ then $\mathbb{E}\left[\left|\frac{dW(t)}{dt}\right|\right] = \frac{1}{dt}\mathbb{E}[|W(t + dt) - W(t)|]$, and from the definition of a Wiener process,

$$\mathbb{E}[|dW(t)|] = \int_{-\infty}^{\infty} \frac{|dw|}{\sqrt{2\pi dt}} e^{-\dfrac{dw^2}{2dt}} d(dw) = \frac{\sqrt{2dt}}{\sqrt{\pi}} \int_{-\infty}^{\infty} \frac{|dw|}{\sqrt{2dt}} e^{-\dfrac{dw^2}{2dt}} \frac{d(dw)}{\sqrt{2dt}}$$
$$= \frac{\sqrt{2dt}}{\sqrt{\pi}} \int_{-\infty}^{\infty} |y|\, e^{-y^2}\, dy = \sqrt{\frac{2dt}{\pi}}.$$

Then

$$\mathbb{E}\left[\left|\frac{dW(t)}{dt}\right|\right] = \frac{d}{dt}\left(\sqrt{\frac{2dt}{\pi}}\right) = \sqrt{\frac{2}{\pi dt}} = o\left(\frac{1}{\sqrt{dt}}\right).$$

As the time derivative is of order $1/\sqrt{dt}$ then $(W(t))$ is not first-order differentiable in time, because $\lim_{dt \to 0} \mathbb{E}\left[\left|\frac{dW(t)}{dt}\right|\right] = \infty$ □

Although the derivative $\frac{dW(t)}{dt}$ is not well defined,[6] the variation $dW(t)$ is well defined and has the following statistic properties:

Proposition 17.2 (Statistic properties of the variation of a Wiener process) *Assume that the time variation is positive $dt > 0$. The time variation of Wiener process has the following statistic properties:*

(1) it is a martingale, or it is ***stationary*** *in expected value terms: $\mathbb{E}[dW(t)] = 0$, for each $t \in T$;*
(2) the mathematical expectation of the square variation of the Wiener process and its variance are both equal to the time increment: $\mathbb{V}[dW(t)] = \mathbb{E}[(dW(t))^2] = dt$, for each $t \in T$.

Proof From the transition probability of a Wiener process, and from the properties of the Gaussian integral see Sect. 15.7.2, then the expected value of a variation is

$$\mathbb{E}[dW(t)] = \int_{-\infty}^{\infty} \frac{dw}{\sqrt{2\pi dt}} e^{-\frac{(dw)^2}{2dt}} d(dw) = \sqrt{\frac{2dt}{\pi}} \int_{-\infty}^{\infty} xe^{-x^2} dx = 0;$$

the expected value of the square variation is

$$\mathbb{E}[(dW(t))^2] = \int_{-\infty}^{\infty} \frac{dw^2}{\sqrt{2\pi dt}} e^{-\frac{(dw)^2}{2dt}} d(dw) = \frac{2dt}{\sqrt{\pi}} \int_{-\infty}^{\infty} x^2 e^{-x^2} dx = dt;$$

[6] This is the reason why we need a particular calculus to deal with SDE that are functions of Wiener processes.

and $\mathbb{V}[dW(t)] = \mathbb{E}\left[dW(t)^2\right] - \mathbb{E}[dW(t)]^2 = dt.$ □

By the fundamental theorem of calculus, the process specified in the integral form

$$W(t) = \int_0^t dW(t),\ a.s.,$$

is well defined, and has the following statistic properties:

Proposition 17.3 (Statistic properties of the Wiener process) *Assume that the time variation is positive $dt > 0$. The Wiener process has the following statistic properties:*

(1) the expected value is time-independent and equal to zero: $\mathbb{E}[W(t)] = 0$, for each $t \in T$;
(2) the mathematical expectation of the square of the process is equal to its variance and they are both equal to t: $\mathbb{V}[W(t)] = \mathbb{E}[W(t)^2] = t$, for each $t \in T$;
(3) the covariance of the process between times s and t is $Cov[W(s), W(t)] = s$, for any $s > t \in T$;
(4) the correlation coefficient between t and s is $Corr[W(s), W(t)] = \sqrt{\frac{s}{t}}$, for any $s > t \in T$.

Proof (1) As $W(t) = W(0) + \int_0^t dW(t)$ and $\mathbb{P}[W(0) = 0] = 1$, then $\mathbb{E}[W(t)] = \mathbb{E}\left[\int_0^t dW(t)\right] = \int_0^t \mathbb{E}[dW(s)] = 0$, for every $t \in T$; (2) as $W(t)^2 = \int_0^t dW(t)^2$ then $\mathbb{E}[W(t)^2] = \mathbb{E}\left[\int_0^t dW(t)^2\right] = \int_0^t \mathbb{E}[dW(s)^2] = \int_0^t ds = t$, for every $t \in T$; (3) as the increments of the Wiener process are i.i.d and Markovian

$$\begin{aligned}\mathrm{Cov}[W(s), W(t)] &= \mathrm{Cov}(W(s), W(s) - (W(s) - W(t))) = \\ &= \mathrm{Cov}(W(s), W(s)) - \mathrm{Cov}(W(s), W(s) - W(t)) = \\ &= \mathbb{V}(W(s)) - \mathrm{Cov}(W(s), dW(t))) = s.\end{aligned}$$

□

As $\mathrm{Corr}[W(s), W(t)] = \frac{\mathrm{Cov}[W(s), W(t)]}{\sigma_{W(s)}\sigma_{W(t)}}$ where $\sigma_{W(t)}$ is the standard deviation, then, for a Wiener process = $\mathrm{Corr}[W(s), W(t)]\frac{s}{\sqrt{s\,t} = \sqrt{\frac{s}{t}}}$.

17.2 Itô Processes

The last summand in SDE (17.4) is the integral involving the Wiener process $\int_0^t \sigma(Y(s))dW(s)$. As the Wiener process is non-differentiable, in the Riemann's sense, we need to generalize the concept of differentiability in order to solve stochastic differential equations. Next we present Itô (1951) interpretation of the integral.

Definition 17.4 (*Itô integral and Itô process*) Let $f(t)$ be a bounded function of time and $\left(W(t)\right)_{t\geq 0}$ by a Wiener process. We call **Itô integral** to the functional

$$I(t) = \int_0^t f(s)\, dW(s). \tag{17.7}$$

The flow of Itô's integrals, $(I(s))_{s \in (0,t)}$, is called **Itô's process**.

This definition can be extended to functions of type $f(t, \omega)$, where ω is adapted to the Wiener process. If the function is square-integrable, in the sense $\mathbb{E}\Big[\int_0^t f(t)^2 dt\Big] < \infty$, a more general definition of an Itô integral is

$$I(t, w) = \int_0^t f(s, w) dW(s), \tag{17.8}$$

where w is a realization of a non-anticipating Wiener process, i.e., $W(s) = w$ for $s \leq t$.

An Itô process shares some properties with the classic (Riemannian) integral, but other properties are a consequence of its dependence on the Wiener process.

Proposition 17.4 (Properties of the Itô integral) *Let $(W(t))$ be a Wiener process, then:*

(1) the Itô integral is not first-order differentiable, even if $f(.)$ is differentiable;
(2) the Itô integral is additive as regards the integrand: $\int_0^t (f_1(s) + f_2(s))dW(s) = \int_0^t f_1(s)dW(s) + \int_0^t f_2(s)dW(s)$, for each $t \in T$;
(3) the Itô integral is additive as regards the time limit: $\int_0^T f(s)dW(s) = \int_0^t f(s)dW(s) + \int_t^T f(s)dW(s)$, for any $0 < t < T \in T$;
(4) the expected value of an Itô integral is equal to zero, $\mathbb{E}[I(t)] = 0$, for each $t \in T$;
(5) the variance of the Itô integral is $\mathbb{V}[I(t)] = \mathbb{E}[I(t)^2] = \int_0^t f(s)^2 ds$, for each $t \in T$.

Proof (4) $\mathbb{E}[I(t)] = \mathbb{E}\Big[\int_0^t f(s)\, dW(s)\Big] = \int_0^t f(s)\mathbb{E}[dW(s)] = 0$; (5) for $\mathbb{V}[I(t)]$ see the proof in Example 17.1. □

Therefore: an Itô process is a martingale, as it is stationary in expected value terms, from (4), and its variance increases over time, from (5).

17.2.1 The Itô Integral and Stochastic Calculus

The Itô integral in integral form, in Eq. (17.7), is equivalent to the **Itô differential**

$$dI(t) = f(t)\, dW(t) \tag{17.9}$$

where $dW(t)$ is a variation of the Wiener process.

Lemma 17.1 **(Itô's formula)** *Let $X(t)$ be an Itô integral, and assume that $g(t, X)$ is a $C^{1,2}(T \times X)$ function. Then the integral $Y(t) = g(t, X(t))$ satisfies, in its differential form, the* ***Itô's formula***

$$dY(t) = g_t(t, X(t))\,dt + g_x(t, X(t))\,dX(t) + \frac{1}{2}g_{xx}(t, X(t))(dX(t))^2. \quad (17.10)$$

where, for any realization x, we denote $g_t(t,x) \equiv \frac{\partial g(t,x)}{\partial t}$, $g_x(t,x) \equiv \frac{\partial g(t,x)}{\partial x}$, and $g_{xx}(t,x) \equiv \frac{\partial^2 g(t,x)}{\partial x^2}$.

In using the Itô's formula, the following ***Itô's rules*** *apply:*

$$(dt)^2 = dt,\ dW(t) = 0,\ \text{and } (dW(t))^2 = dt,\ \text{for each } t \in T.$$

Proof See Itô (1951) □

From Itô differential to SDE. Let $dX(t) = f(t)\,dW(t)$ and $Y(t) = g(t, X(t))$, where $g(\cdot) \in C^{1,2}(\mathbb{R}^2)$, and assume that $g^{-1}(\cdot, x)$ exists, then the stochastic process $(Y(t))_{t\in T}$ is governed by a SDE in the Itô sense. Applying Itô's formula (17.10) we have

$$dY(t) = \left(g_t(t, X(t)) + \frac{1}{2}g_{xx}(t, X(t))\,f^2(t)\right)dt + g_x(t, X(t))\,f(t)\,dW(t).$$

If function $g(\cdot, x)$ is invertible, then we can obtain $x = h(t, y) = g^{-1}(t, y)$, and substituting in $dY(t)$ yields the SDE

$$\begin{aligned} dY(t) =&\left(g_t(t, h(t, Y(t))) + \frac{1}{2}g_{xx}(t, (t, h(t, Y(t))))\,f^2(t)\right)dt+ \\ &+ g_x(t, (t, h(t, Y(t))))\,f(t)\,dW(t),\ \text{for each } t \in T. \end{aligned}$$

From now on, we call stochastic differential equation (SDE) to a stochastic differential equation in the Itô's sense.

Exercise 17.1 (*Itô stochastic calculus*) Let $dX(t) = dW(t)$ and let $Y(t) = g(X(t))$. Prove that:

(1) if g is linear, such as $g(x) = a\,x + b$, then $dY(t) = a\,dW(t)$
(2) if g is a power function $g(x) = x^a$, for $a \neq 0$, then $dY(t) = aY(t)^{\frac{a-2}{a}}\left(\frac{a-1}{2}dt + Y(t)\,dW(t)\right)$;
(3) if g is an exponential function $g(x) = e^{\lambda x}$, for $\lambda \neq 0$, then $dY(t) = \frac{\lambda^2}{2}Y(t)\,dt + \lambda Y(t)\,dW(t)$;
(4) if g is a logarithmic function $g(x) = \ln(x)$, then $dY(t) = \frac{1}{2}e^{-2Y(t)}dt + e^{-Y(t)}\,dW(t)$.

Example 17.1 We can use the previous formulas to show that $\mathbb{E}[I(t)^2] = \int_0^t f(s)^2\,ds$. To prove this, start by considering the Itô integral $dI(t) = f(t)\,dW(t)$, where $Y(t) = I(t)^2$. Applying the Itô's formula, we have

$$d\big(I(t)^2\big) = 2\,I(t)\,dI(t) + (dI(t))^2 = f(t)^2\,dt + 2\,f(t)\,I(t)\,dW(t).$$

Integrating and using the properties of the Wiener process yields

$$I(t)^2 = I(0)^2 + \int_0^t dI(s)^2 = \int_0^t dI(s)^2 = \int_0^t f(s)^2\,ds + 2\int_0^t f(s)\,I(s)\,dW(s).$$

Taking the expected value, we finally obtain $\mathbb{E}[I(t)^2] = \int_0^t f(s)^2\,ds$ because $\mathbb{E}\Big[\int_0^t f(s)\, I(s)\,dW(s)\Big] = 0.$

17.2.2 Itô's Formula for Multi-Dimensional Processes

Consider the multi-dimensional stochastic process, such that $\mathbf{X}(t) = (X_1(t), \cdots, X_n(t))^\top$, where $n > 1$, in which every component is driven by an Itô's SDE $dX_i(t)$. The Itô formula for a function of a multi-dimensional process, $Y(t) = g(t, \mathbf{X}(t))$, extends formula (17.10) for one-dimensional processes:

$$\begin{aligned} dY(t) = g_t(t, \mathbf{X}(t))dt + D_x g(t, \mathbf{X}(t))^\top d\mathbf{X}(t) + \\ + \frac{1}{2}(\mathbf{X}(t))^\top D_{xx} g(t, \mathbf{X}(t)) d\mathbf{X}(t), \end{aligned} \tag{17.11}$$

where $D_x g(\cdot, \mathbf{x})$ is the Jacobian, and $D_{xx} g(t, \mathbf{x})$ is its Hessean of $g(\cdot, \mathbf{x})$

$$D_x g(t, \mathbf{x}) = \begin{pmatrix} g_{x_1}(t, \mathbf{x}) \\ \cdots \\ g_{x_n}(t, \mathbf{x}) \end{pmatrix}, \quad D_{xx} g(t, \mathbf{x}) = \begin{pmatrix} g_{x_1 x_1}(t, \mathbf{x}) & \dots & g_{x_1 x_n}(t, \mathbf{x}) \\ \vdots & \ddots & \vdots \\ g_{x_n x_1}(t, \mathbf{x}) & \dots & g_{x_n x_n}(t, \mathbf{x}), \end{pmatrix}$$

for any realization of $\mathbf{X}(t) = \mathbf{x}$. For each Wiener process, $W_i(t)$, we use the rules, $dW_i(t)dt = 0$ and $dW_i(t)dW_i(t) = dt$, for every i, and, if there are at least two independent Wiener processes, we have $dW_i(t)dW_j(t) = 0$, for each $t \in T$, and for any $i \neq j$.

There are several useful applications of formula (17.11). In particular:

Exercise 17.2 (*Product rule*) Let $Y(t) = X_1(t)X_2(t)$. Prove that

$$dY(t) = X_2(t)dX_1(t) + X_1(t)dX_2(t) + dX_1(t)dX_2(t), \quad \text{for each } t \in T \tag{17.12}$$

and show it is equivalent to

$$\frac{dY(t)}{Y(t)} = \frac{dX_1(t)}{X_1(t)} + \frac{dX_2(t)}{X_2(t)} + \frac{dX_1(t)}{X_1(t)}\frac{dX_2(t)}{X_2(t)}, \quad \text{for each } t \in T,$$

where the presence of the last term distinguishes the product rule of Itô's stochastic calculus from product rule of elementary calculus.

Exercise 17.3 (*Quocient rule*) Let $Y(t) = X_1(t)/X_2(t)$. Prove that

$$dY(t) = \frac{X_1(t)}{X_2(t)}\left(\frac{dX_1(t)}{X_1(t)} - \frac{dX_2(t)}{X_2(t)} + \frac{dX_2(t)^2}{X_2(t)^2} - \frac{dX_1(t)\,dX_2(t)}{X_1(t)\,X_2(t)}\right), \tag{17.13}$$

and show it is equivalent to

$$\frac{dY(t)}{Y(t)} = \left(1 - \frac{dX_2(t)}{X_2(t)}\right)\left(\frac{dX_1(t)}{X_1(t)} - \frac{dX_2(t)}{X_2(t)}\right), \text{ for each } t \in T.$$

17.3 Diffusion Processes

In this section, we consider a pervasive SDE (in the Itô interpretation), of the form (17.2), which is also called **diffusion equation**. The autonomous diffusion SDE is, in its differential form

$$dX(t) = \mu(X(t))\,dt + \sigma(X(t))\,dW(t), \text{ for each } t \in T, \tag{17.14}$$

where $\mu(\cdot)$ and $\sigma(\cdot)$ are called the **drift** and the **diffusion** coefficients, or, in its integral form,

$$X(t) = X(0) + \int_0^t \mu(X(s))ds + \int_0^t \sigma(X(s))dW(s), \text{ for each } t \in T, \tag{17.15}$$

where the solution $(X(t))_{t\in T}$ is called a **diffusion process**. Next, we deal with the one-dimensional diffusion, $X : T \to X \subseteq \mathbb{R}$. There are several results that allow us to characterize the properties of the diffusion process. In the next chapter we will apply them to obtain explicit solutions to linear SDE.

17.3.1 *Functions of a Diffusion Process*

Before determining the statistics for the process $\left(X(t)\right)_{t\in T}$, it is useful to apply the Itô's formula to a function of a diffusion.

Proposition 17.5 (Functions of a diffusion process) *Consider the diffusion process $(X(t))_{t\in T}$ generated by the SDE* (17.14). *The process $(Y(t))_{t\in T}$ such that $Y(t) = f(X(t))$, where $f(\cdot)$ is a point-wise mapping $f : X \to \mathbb{R}$ which is at least $C^2(X)$, and such that $x = f^{-1}(y) = g(y)$, where $g(\cdot)$ is continuous, is generated by the diffusion SDE*

$$dY(t) = \mu_y(Y(t))dt + \sigma_y(Y(t))dW(t), \text{ for each } t \in T, \tag{17.16}$$

where the drift and diffusion components are

$$\begin{aligned}\mu_y(y) &= \frac{\partial f(g(y))}{\partial x}\,\mu(g(y)) + \frac{1}{2}\frac{\partial^2 f(g(y))}{\partial x^2}(\sigma(g(y)))^2,\\ \sigma_y(y) &= \frac{\partial f(g(y))}{\partial x}\,\sigma(g(y)).\end{aligned}$$

Proof To prove this, we use Itô's formula to find $dY(t) = d(f(X(t))$,

$$\begin{aligned}dY(t) &= \frac{\partial f(X(t))}{\partial x}dX(t) + \frac{1}{2}\frac{\partial^2 f(X(t))}{\partial x^2}\,(dX(t))^2\\ &= \frac{\partial f(X(t))}{\partial x}\,(\mu(X(t))dt + \sigma(X(t))dW(t)) + \frac{1}{2}\frac{\partial^2 f(X(t))}{\partial x^2}(\sigma(X(t)))^2 dt =\\ &= \left(\frac{\partial f(X(t))}{\partial x}\mu(X(t)) + \frac{1}{2}\frac{\partial^2 f(X(t))}{\partial x^2}(\sigma(X(t)))^2\right)dt +\\ &\quad + \frac{\partial f(X(t))}{\partial x}\sigma(X(t))dW(t).\end{aligned}$$

If the function $f(.)$ is invertible then we substitute, for every realization, $x = f^{-1}(y) = g(y)$ into the last equation. □

From SDE (17.16), we can obtain sample paths of process $(Y(t))$ by using

$$Y(t) = Y(0) + \int_0^t \mu_y(Y(s))ds + \int_0^t \sigma_y(Y(s))dW(s),$$

that can be obtained in explicit form in some cases.

17.3.2 Moment Equations and Their Dynamics

In this subsection, we derive the first and the second moments for the diffusion process $(X(t))_{t\in\mathbb{R}_+}$ generated by the one-dimensional diffusion equation in integral form in Eq. (17.15).

Proposition 17.6 (Moments of a diffusion process) *Consider the diffusion in integral form* (17.15)*, and assume that* $X(0) = x_0$ *is deterministic. Then*

(1) the first moment of the diffusion process is

$$\mathbb{E}[X(t)] = x_0 + \int_0^t \mathbb{E}\left[\mu(X(s))\right]ds; \tag{17.17}$$

(2) the second moment of the diffusion process is

$$\mathbb{E}[X(t)^2] = x_0^2 + \int_0^t \left(2\,\mathbb{E}[X(s)\mu(X(s))] + \mathbb{E}[\sigma(X(s))^2]\right) ds; \qquad (17.18)$$

(3) and the variance of the diffusion process is

$$\mathbb{V}[X(t)] = x_0^2 + \int_0^t \left(2\mathbb{E}[X(s)\mu(X(s))] + \mathbb{E}[\sigma(X(s))^2]\right) ds - \\ - \left(x_0 + \int_0^t \mathbb{E}\left[\mu(X(s))\right] ds\right)^2. \qquad (17.19)$$

Proof As $\sigma(X(t))$ is a non-anticipating random variable, from the properties of the Wiener process, then

$$\mathbb{E}[X(t)] = x_0 + \mathbb{E}\left[\int_0^t \mu(X(s))ds\right] + \mathbb{E}\left[\int_0^t \sigma(X(s))dW(s)\right] \\ = x_0 + \int_0^t \mathbb{E}\left[\mu(X(s))\right] ds.$$

As, from Itô's formula, the square of a diffusion is, in differential form,

$$dX(t)^2 = \left(2X(t)\mu(X(t)) + \sigma(X(t))^2\right) dt + 2X(t)\sigma(X(t))dW(t)),$$

and, in integral form,

$$X(t)^2 = x_0^2 + \int_0^t \left(2X(s)\mu(X(s)) + \sigma(X(s))^2\right) ds + \int_0^t 2X(s)\sigma(X(s))\,dW(s)).$$

then

$$\mathbb{E}[X(t)^2] = x_0^2 + \mathbb{E}\left[\int_0^t (2X(s)\mu(X(s)) + \sigma(X(s))^2)ds\right] + \\ + \mathbb{E}\left[\int_0^t 2X(s)\sigma(X(s))\,dW(s))\right] = \\ = x_0^2 + \int_0^t \left(2\mathbb{E}[X(s)\mu(X(s))] + \mathbb{E}[\sigma(X(s))^2]\right) ds$$

The variance is obtained from $\mathbb{V}[X(t)] = \mathbb{E}[X(t)^2] - (\mathbb{E}[X(t)])^2$. □

Taking time derivatives, we obtain differential equivalents of the first moment

$$\frac{d\mathbb{E}[X(t)]}{dt} = \mathbb{E}[\mu(X(t))],$$

of the second moment

$$\frac{d\mathbb{E}[X(t)^2]}{dt} = 2\mathbb{E}[X(t)\mu(X(t))] + \mathbb{E}[\sigma(X(t))^2],$$

and of the variance

$$\frac{d\mathbb{V}[X(t)]}{dt} = 2\mathbb{E}[X(t)\mu(X(t))] + \mathbb{E}[\sigma(X(t))^2] - \mathbb{E}[\mu(X(t))]^2.$$

Depending on the functional forms of the drift and diffusion coefficients, $\mu(\cdot)$ and $\sigma(\cdot)$, the moments of solutions to SDE can occasionally be represented by ordinary differential equations (ODE). If this is the case, we can use the knowledge of these equations to characterize their dynamics, and, in particular, their asymptotic properties. This is the case if the coefficients are linear.

Example 17.2 (*Linear diffusion equation*)Let

$$dX(t) = -\gamma\, X(t)\, dt + \sigma\, dW(t), \text{ for each } t \in T,$$

where $X(0) = x_0$ is fixed, and γ and σ are two constants. Denote $\mathsf{m}_j(t) = \mathbb{E}[X(t)^j]$, the jth moment, and $\mathsf{v}(t) = \mathbb{V}[X(t)]$ the variance of $X(t)$. The first moment satisfies the ODE $\frac{d\mathsf{m}(t)}{dt} = -\gamma\, \mathsf{m}(t)$, then

$$\mathsf{m}(t) = \mathbb{E}[X(t)] = x_0 e^{-\gamma t},$$

because $\mathsf{m}(0) = x_0$. The second moment satisfies the ODE $\frac{d\mathsf{m}_2(t)}{dt} = \sigma^2 - 2\,\gamma\, \mathsf{m}_2(t)$, then

$$\mathsf{m}_2(t) = \mathbb{E}[X(t)^2] = \frac{\sigma^2}{2\gamma} + \left(x_0^2 - \frac{\sigma^2}{2\gamma}\right) e^{-2\gamma t};$$

because $\mathsf{m}_2(0) = x_0^2$. The variance satisfies the ODE $\frac{d\mathsf{v}(t)}{dt} = \sigma^2 - 2\,\gamma\, \mathsf{v}(t)$, then

$$\mathsf{v}(t) = \mathbb{V}[X(t)] = \frac{\sigma^2}{2\gamma}\left(1 - e^{-2\gamma t}\right),$$

because $\mathsf{v}(0) = 0$. Therefore, the dynamic properties of the moments of the process $(X(t))_{t \in T}$ depend on the sign and level of γ:

1. if $\gamma > 0$ then $\lim_{t\to\infty} \mathbb{E}[X(t)] = 0$ and $\lim_{t\to\infty} \mathbb{V}[X(t)] = \lim_{t\to\infty} \mathbb{E}[X(t)^2] = \frac{\sigma^2}{2\gamma}$, and the process tends asymptotically to a Gaussian distribution $N\left(0, \frac{\sigma^2}{2\gamma}\right)$;

2. if $\gamma = 0$ then $\mathbb{E}[X(t)] = x_0$, for all $t \in T$, and the second moments are linear functions of time, $\mathbb{E}[X(t)^2] = x_0^2 + \sigma^2 t$ and $\mathbb{V}[X(t)] = \sigma^2 t$, and, therefore $\lim_{t\to\infty} \mathbb{E}[X(t)^2] = \lim_{t\to\infty} \mathbb{V}[X(t)] = \infty$ and the process has constant mean and unbounded variance[7];
3. if $\gamma < 0$ then the moments are exponential functions of time and, as $\lim_{t\to\infty} \mathbb{E}[X(t)] = \lim_{t\to\infty} \mathbb{V}[X(t)] = \lim_{t\to\infty} \mathbb{E}[X(t)^2] = \infty$, then both the mean and the variance of the process are unbounded in the distributional sense.

17.3.3 Distribution Dynamics: The Kolmogorov Equations

Kolmogorov (1933), a seminal contribution to the theory of Markov processes, contains two PDE, the Kolmogorov backward (KB) equation (see Kolmogorov, 1933, pg. 125) and the Kolmogorov forward equation (see Kolmogorov, 1933, pg. 133), also known by Fokker-Planck-Kolmogorov (FPK) equation[8], that provide analytical tools for studying the dynamics of the probability distributions of diffusion processes.

Let $\mathbb{P}[X(t) \in A|X(s) = x_s]$ be the conditional probability that the realizations of the process $(X(t))_{t\in T}$ belong at time t to subset $A \subseteq X$, conditional on the realization $X(s) = x \in X$ at time $s < t$. We can take this probability as a parameterized function, with range (0,1) by setting

$$P(t, A, s, x) = \mathbb{P}[X(t) \in A|X(s) = x].$$

By the Chapman-Kolmogorov equation, there is a transition probability $P(t, dy, s, x)$ such that

$$P(t, A, s, x) = \int_X P(t, A, \tau, y)\, P(\tau, dy, s, x).$$

If we can write $P(\tau, dy, s, x) = p(\tau, y, s, x)\, dy$, we call $p(t, y, s, x)$ the **transition probability density** function.

Let $f(\cdot)$ be a bounded function and denote the expectation of $f(X(t))$, for a given $t > s$, by

$$q(s, x) = \mathbb{E}[f(X(t))|X(s) = x] = \int_X f(y)\, p(t, y, s, x)\, dy. \tag{17.20}$$

The KB equation provides a PDE for the conditional mathematical expectation of a function of the stochastic process $q(s, x)$, and the FPK equation provides a PDE for

[7] Observe that, by l'Hôpital's rule, $\lim_{\gamma\to 0} \frac{\sigma^2(1-e^{-2\gamma t})}{2\gamma} = \lim_{\gamma\to 0} \frac{\sigma^2\, 2t\, e^{-2\gamma t}}{2} = \gamma^2 t$.

[8] The same Kolmogorov recognized that a similar equation have already been suggested by A. Fokker and M. Planck, and a version of those equations already appeared in Bachelier (1900).

the transition probability $p(t, y, s, x)$, for a diffusion process. Next we present those two PDE in a way allowing for solving them by the methods presented in Part III.

17.3.4 Forward Density Dynamics: The Fokker-Planck-Kolmogorov Equation

We define the **transition probability** as a mapping $p : T \times X \to (0, 1)$,

$$p(t, x) = \mathbb{P}[X(t) = x | X(0) = x_0], \text{ for each } (t, x) \in T \times X,$$

where the initial realization $x_0 \in X$ is fully observed (deterministic).

Definition 17.5 (*Adjoint operator of a diffusion*) Let $X(t) = x$ be the realization of the diffusion process (17.14) at time $t > 0$, and let $p(t, x)$ be its transition probability. Then

$$\mathcal{G}^*[p](t, x) = -\frac{\partial(\mu(x)\, p(t, x))}{\partial x} + \frac{1}{2}\frac{\partial^2(\sigma(x)^2\, p(t, x))}{\partial x^2}$$

is called **adjoint operator of the diffusion**.

The following PDE, called **Fokker-Planck-Kolmogorov equation** (FPK)

$$\partial_t p(t, x) = -\partial_x\big(\mu(x)\, p(t, x)\big) + \frac{1}{2}\partial_{xx}\big(\sigma(x)^2 p(t, x)\big), \tag{17.21}$$

states that the instantaneous change in transitional probability, for a diffusion process, is equal to the adjoint operator of a diffusion, that is $\partial_t p(t, x) = \mathcal{G}^*[p](t, x)$. To have an unconditional probability, we postulate the probability at time $t = 0$ to be Dirac-delta distribution centred at x_0, $p(0, x) = \delta(x - x_0)$. Furthermore, we assume that $X = \mathbb{R}$, that $\lim_{x\to\pm\infty} p(t, x) = 0$, and that the normalization condition $\int_{-\infty}^{\infty} p(t, x)\, dx = 1$ is satisfied, for each $t \geq 0$.

Proposition 17.7 (Forward density dynamics) *Let the stochastic process* $(X(t))_{t\geq 0}$ *be generated by the SDE* (17.14)*, and let the initial state be* $X(0) = x_0$. *Then the unconditional probability distribution of* $X(t)$ *at time* $t > 0$*, is the solution to the following forward PDE problem*

$$\begin{cases} \partial_t p(t, x) = \mathcal{G}^*[p](t, x), & \text{for each } (t, x) \in T \times X, \\ p(0, x) = \delta(x - x_0), & \text{for each } (t, x) \in \{t = 0\} \times X, \end{cases} \tag{17.22}$$

where $\mathcal{G}^*[\cdot]$ *is the adjoint operator of the diffusion* (17.14).

Proof (Heuristic proof)[9] We introduce a test function $f(\cdot)$. Let $t \in T = [0, T]$ and $x \in X = \mathbb{R}$, and consider an arbitrary stationary and bounded function $f(t, x)$

[9] See (Kolmogorov, 1933, Sect. 14) for the original proof.

such that $f(0, X(0)) = f(T, X(T)) = 0$, for $X(0) = x_0$, and for any realization of $X(T) \in X$. In addition, let $\lim_{x\to\pm\infty} f(t, x) = 0$, for any $t \in T$.

Therefore, $f(t, X(t)) = \int_0^t df(s, X(s))$, and

$$\mathbb{E}[f(T, X(T))] = \mathbb{E}\Big[\int_0^T df(t, X(t))\Big] = 0.$$

From Itô's Lemma, for any realization of the diffusion process generated by equation (17.14) at time t, $X(t) = x$, we have

$$df(t,x) = \Big[\partial_t f(t,x) + \mu(x)\partial_x f(t,x) + \frac{1}{2}\sigma^2(x)\partial_{xx} f(t,x)\Big] dt + \\ + \Big(\sigma(x)\partial_x f(t,x)\Big) dW(t),$$

and the variation of f, from $t = 0$ to $t = T$, is

$$\int_0^T df(t,x) = \int_0^T \Big[\partial_t f(t,x) + \mu(x)\partial_x f(t,x) + \frac{1}{2}\sigma^2(x)\partial_{xx} f(t,x)\Big] dt + \\ \int_0^T \Big(\sigma(x)\partial_x f(t,x)\Big) dW(t).$$

Assume there is an unconditional probability distribution $p(t, x)$. The unconditional expected value of the variation of f from $t = 0$ to $t = T$ is

$$\begin{aligned}\mathbb{E}\Big[\int_0^T df((t, X(t)))\Big] &= \mathbb{E}\Big[\int_0^T \Big(\partial_t f((t, X(t))) + \mu(X(t))\partial_x f((t, X(t))) + \frac{1}{2}\sigma^2(X(t))\partial_{xx} f((t, X(t)))\Big) dt\Big] + \\ &+ \mathbb{E}\Big[\int_0^T \Big(\sigma(X(t))\partial_x f((t, X(t)))\Big) dW(t)\Big] \\ &= \mathbb{E}\Big[\int_0^T \Big(\partial_t f(t, X(t)) + \mu(X(t))\partial_x f((t, X(t))) + \frac{1}{2}\sigma^2(X(t))\partial_{xx} f(t,x)\Big) dt\Big] \\ &= \int_X \int_0^T \Big(\partial_t f(t,x) + \mu(x)\partial_x f(t,x) + \frac{1}{2}\sigma^2(x)\partial_{xx} f(t,x)\Big) p(t,x)\, dt\, dx \\ &= \int_X \int_0^T \Big(-\partial_t p(t,x) - \partial_x\big(\mu(x)\, p(t,x)\big) + \\ &+ \frac{1}{2}\partial_{xx}\big(\sigma^2(x) p(t,x)\big)\Big) f(t,x)\, dt\, dx,\end{aligned}$$

from the assumptions on $f(\cdot)$, and integrating by parts. Therefore, $\mathbb{E}\Big[\int_0^T df(t)\Big] = 0$ if and only if Eq. (17.21) is satisfied. □

Equation (17.21) provides an analytical method for finding the moments of a diffusion process, that is an alternative to the method presented in Sect. 17.3.2. If $p(t, x)$ is a solution to the FPK equation, then we can find the mathematical expectation and the variance, for each $t \in T$, by calculating

$$\mathbb{E}[X(t)] = \int_{-\infty}^{\infty} x\, p(t,x)\, dx, \text{ and } \mathbb{V}[X(t)] = \int_{-\infty}^{\infty} \left(x - \mathbb{E}[X(t)]\right)^2 p(t,x)\, dx.$$

Definition 17.6 (*Ergodic process*) A process $(X(t))_{t\in T}$ is said to be **ergodic** if the probability density function is asymptotically time independent, as

$$\lim_{t\to\infty} p(t,x) = p^*(x),$$

where p^* is unique and bounded, for any initial probability density function $p(0,x)$.

Then, if the process $(X(t))_{t\in T}$ is ergodic, the mean and variance are asymptotically constant

$$\lim_{t\to\infty} \mathbb{E}[X(t)] = \int_{-\infty}^{\infty} x\, p(t,x)\, dx = \mu_X^*,$$

and

$$\lim_{t\to\infty} \mathbb{V}[X(t)] = \int_{-\infty}^{\infty} \left(x - \mathbb{E}[X(t)]\right)^2 p(t,x)\, dx = \sigma_X^{*2} > 0.$$

The intuition is: for an ergodic process small or large random perturbations do not generate large long-run variations on the sample paths, with high probability.

Example 17.3 (*Moments for a linear diffusion process*) Let $dX(t) = \mu\, dt + \sigma dW(t)$, where μ and $\sigma > 0$ are constants, and let $X(0) = x_0$. As we have $\mu(x) = \mu$ and $\sigma(x) = \sigma$ the adjoint operator is $\mathcal{G}^*[p](t,x) = -\mu\partial_x p(t,x) + \frac{\sigma^2}{2}\partial_{xx} p(t,x)$, and $p(t,x)$ is the solution to the FPK equation

$$\begin{cases} \partial_t p(t,x) = -\mu\, \partial_x p(t,x) + \frac{\sigma^2}{2}\partial_{xx} p(t,x), & (t,x) \in T \times X, \\ p(0,x) = \delta(x - x_0), & t = 0. \end{cases}$$

We saw in Example (15.6), in Eq. (15.20), that the solution to this problem is

$$p(t,x) = \frac{1}{\sqrt{2\pi\sigma^2 t}} e^{-\frac{(x-x_0-\mu t)^2}{2\sigma^2 t}} \text{ for each } (t,x) \in T \times X. \tag{17.23}$$

Thus, the mathematical expectation is

$$\mathbb{E}[X(t)] = \int_{-\infty}^{\infty} x\, p(t,x)\, dx = x_0 + \mu\, t,$$

and the variance is

$$\mathbb{V}[X(t)] = \int_{-\infty}^{\infty} (x - x_0 - \mu\, t)^2\, p(t,x)\, dx = \sigma^2\, t.$$

The process is not ergodic, for any value of $\mu \in \mathbb{R}$.

Exercise 17.4 Consider the diffusion equation in Example 17.2, $dX(t) = -\gamma\, X(t)\, dt + \sigma dW(t)$. Assuming that $X(0) = x_0 \in X$, find the unconditional probability density function $p(t, x)$ by using the FPK equation. Find the moments and prove that this process is ergodic.

17.3.5 Backward Kolmogorov Equation

Assume again we have a diffusion process generated by SDE (17.14). First, we need to introduce the following definition:

Definition 17.7 (*Infinitesimal generator*) Let $(X(t))_{t\in T}$ be a Markovian process, let $X(t) = x \in X$ be a realization of the stochastic process at time $t \geq 0$, and let $f(x) \in C^2(X)$ be a smooth function. The **infinitesimal generator** of $f(X(t))$ is the operator

$$\begin{aligned}\mathcal{G}[f](t,x) &= \frac{d\mathbb{E}[f(X(t))|X(t)=x]}{dt} = \\ &= \lim_{\Delta t\to 0}\frac{\mathbb{E}[f(X(t+\Delta t))|X(t)=x] - f(x)}{\Delta t}\end{aligned}$$

The generator is defined for every time, $t \geq t^+$, and is conditional on the realization of the process at time t, $X(t) = x$. The **generator of a function** $f(X)$ **of the diffusion**, $dX(t) = \mu(X(t))dt + \sigma(X(t))\, dW(t)$, conditional on $X(t) = x$, is

$$\mathcal{G}[f](t,x) = \mu(x)\frac{\partial f(x)}{\partial x} + \frac{1}{2}\sigma(x)^2\frac{\partial^2 f(x)}{\partial x^2},\quad (t,x) \in T \times X. \tag{17.24}$$

We can prove this by just using Itô's formula. The generator of differentiable function of a diffusion is a **directional derivative of** $f\cdot$ averaged over the paths generated by the diffusion, starting from the realization $X(t) = x$.

Recalling the definition of the expected value of a function of a stochastic process, in Eq. (17.20), if the stochastic process is a diffusion its dynamics is driven by a diffusion generator:

Proposition 17.8 (Backward Kolmogorov equation) *Let* $T = [0, T]$, $X = \mathbb{R}$, *and let function* $f(x) \in C^2(X)$ *be bounded. Define*

$$q(t,x) \equiv \mathbb{E}[f(X(\tau)|X(t)=x),\ \textit{for}\ \tau > t \in T,\ \textit{and}\ x \in X,$$

and let $(X(t))_{t\in T}$ *be a diffusion process where* $\mu(x)$ *and* $\sigma(x)$ *are* $C^2(X)$ *functions. Then* $q(t, x)$ *is the solution to the backward PDE*

$$\begin{cases}\partial_t q(t,x) = -\mathcal{G}[p](t,x), & \textit{for each}\ (t,x) \in (0,T] \times X,\\ q(T,x) = f(x), & \textit{for each}\ (t,x) \in \{t = T\} \times X,\end{cases} \tag{17.25}$$

where $\mathcal{G}[\cdot]$ *is the infinitesimal generator of a diffusion in Eq.* (17.24).

Proof See Kolmogorov (1933, Sect. 13) or Pavliotis (2014, p. 45). □

The PDE over $q(t, x)$

$$\partial_t q(t,x) = -\mu(x)\partial_x q(t,x) - \frac{1}{2}\sigma(x)^2\,\partial_{xx}q(t,x), \text{ for } t \in (0,T], \tag{17.26}$$

is called **Kolmogorov backward** (KB) equation for a diffusion process. To find a particular solution to the BK equation it should be solved together with a terminal condition.

Example 17.4 Let $dX(t) = \sigma dW(t)$ and find the flow of expected values for this process, for each $t \in [0, T]$, such that $X(T) = x_T$, is a constant. The BK equation is $\partial_t q(t, x) = -\frac{\sigma^2}{2}\partial_{xx}q(t, x)$. Therefore, $q(t, x)$ is the solution to the parabolic BPDE

$$\begin{cases} \partial_t q(t,x) = -\frac{\sigma^2}{2}\partial_{xx}q(t,x), & \text{for } 0 \le t < T, \\ q(T,x) = \delta(x - x_T), & t = T. \end{cases} \tag{17.27}$$

The solution is

$$q(t,x) = \frac{e^{-\frac{-(x-x_T)^2}{2\sigma^2(T-t)}}}{\sigma\sqrt{2\pi(T-t)}}, \text{ for } 0 < t < T,$$

and is illustrated in Fig. 17.2a. Let us now assume that we have a standard Gaussian terminal distribution $q_T(x) = e^{-x^2}/\sqrt{\pi}$, then $q(t, x)$ is the solution to the problem,

$$\begin{cases} q_t(t,x) = -\frac{\sigma^2}{2}q_{xx}(t,x), & 0 < t < T, \\ q(T,x) = e^{-x^2}/\sqrt{\pi}, & t = T, \end{cases} \tag{17.28}$$

which is analytically

$$q(t,x) = \frac{e^{-\frac{-x^2}{1+2\sigma^2(T-t)}}}{\sigma\sqrt{1+2\pi(T-t)}}, \text{ for } 0 < t < T,$$

and geometrically illustrated in Fig. 17.2b.

17.3.6 The Feynman-Kac Formulas

The Feynman-Kac formula allows us to determine the mathematical expectation, at time $0 < t < T$, of the present value of a known terminal distribution, at time T, conditional on a particular realization of a diffusion process $(X(t))_{t\in[0,T]}$. The discount factor, $r(X(t))$, is a known function.

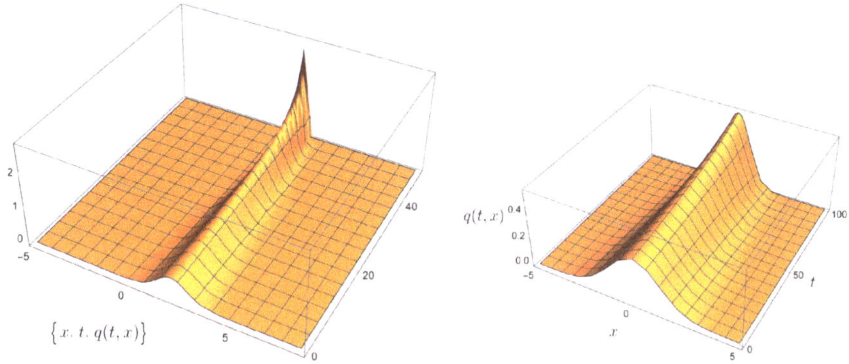

(a) Terminal condition $X(t) = 0$ (b) Terminal condition $X(t) \sim N(0, 0.1)$

Fig. 17.2 Solution to the backward Kolmogorov equation (17.27) for $\sigma = 0.1$

Let $t \in [0, T]$, for a finite T and $x \in X = \mathbb{R}$. Assume $(X(t))_{t \in T}$ is a diffusion process, generated by the SDE (17.14), and let $v(t, x)$ be the expected value, at time $t \in T$, for a realization $X(t) = x \in X$. The expected value solves the BK equation, together with a terminal condition,

$$\begin{cases} \partial_t v(t,x) = -\mathcal{G}[v](t,x) + r(x)\, v(t,x), \text{ for each } (t,x) \in T \times X, & (17.29a) \\ v(T, X(T)) = f(T, X(T)), \text{ for } t = T, \text{ and } X(T) \subset X, & (17.29b) \end{cases}$$

where $f(\cdot)$ is known, and

$$\mathcal{G}[v](t,x) = \mu(x)\, \partial_x v(t,x) + \frac{1}{2}\sigma(x)^2 \partial_{xx} v(t,x)$$

is the infinitesimal generator of $v(.)$. The terminal condition can be considered a set-valued function, where $X(T)$ is a subset of realizations of the diffusion process. If the terminal condition is a known function of a specific realization of the underlying process, $X(T) = x_T \in X$, we can write it as $v(T, x) = f(T, x_T)$. Observe that the PDE has a growth/decay component when compared to the backward Kolmogorov PDE in Eq. (17.25).

Proposition 17.9 (Feynman-Kac formula for a diffusion) *The solution to the PDE problem* (17.29) *satisfies the* ***Feynman-Kac*** *formula:*

$$v(t,x) = \mathbb{E}\left[f(T, X(T))\, e^{-\int_t^T r(X(s))ds} | X(t) = x \right]. \tag{17.30}$$

Proof Let us write the discounted probability distribution $V(t) = V(t, X(t)) = v(t, X(t))H(t)$, where $H(t) \equiv e^{-Z(t)} = e^{-\int_0^t r(X(\tau))d\tau}$ and $v(t, X(t))$ is driven by the PDE (17.29a). To obtain $V(T)$, we follow three steps: first, as $dZ(t) = r(X(t))dt$

then $dH(t) = -H(t)\,dZ(t) = -H(t)r(X(t))dt$; second, from Itô's formula, and substituting Eq. (17.29a), yields

$$dv(t, X(t)) = r(X(t))v(t, X(t))dt + v_x(t, X(t))\sigma(X(t))dW(t);$$

third, writing $v(t) = v(t, X(t))$ and $r(t) = r(X(t))$ from Itô's multiplication rule,

$$dV(t) = H(t)dv(t) + v(t)dH(t) + dv(t)dH(t) = H(t)v_x(t)\sigma(t)dW(t);$$

at last, integrating forward starting from t, we obtain

$$V(T) = V(t) + \int_t^T dV(s) = V(t) + \int_t^T e^{-\int_t^s r(X(\tau))d\tau} v_x(s, X(s))\sigma(X(s))dW(s).$$

Therefore, the expected value, conditional on $X(t) = x$, is $\mathbb{E}\,[V(T)|X(t) = x] = \mathbb{E}[V(t)|X(t) = x]$. As $\mathbb{E}[V(t)|X(t) = x] = v(t, x)e^{-\int_0^t r(s)ds}$, and $V(T, X(T)) = e^{-\int_0^T r(X(s))ds}\, v(T, X(T))$, then the Feynman-Kac formula (17.30) is obtained. □

Then $v(t, x)$ is equal to mathematical expectation of the present value of a terminal value $v(T, X(T))$, discount at rate $r(X(t))$, conditional on having $X(t) = x$, if $r\cdot$ is a discount rate that is independent of the underlying process $(X(t))$. If the discount factor is constant, $r(x) = r$, then the Feynman-Kac formula simplifies to

$$v(t, x) = e^{-r(T-t)}\,\mathbb{E}\,[f(T, X(T))|X(t) = x]\,, \text{ for each } (t, x) \in T \times X. \quad (17.31)$$

For the parabolic BPDE problem,

$$\begin{cases} \partial_t v(t,x) = -\mathcal{G}[v](t,x) + r(x)\,v(t,x) - b(t,x),\ (t,x) \in T \times X, & (17.32a) \\ v(T,x) = f(T,x),\ (t,x) \in \{t = T\} \times X, & (17.32b) \end{cases}$$

the Feynman-Kac formula is

$$v(t, x) = \mathbb{E}\Big[\int_t^T e^{-\int_t^\tau r(X(s))ds} b(\tau, X(\tau))\,d\tau + \\ + f(T, X(T))\,e^{-\int_t^T r(X(s))ds}\Big|X(t) = x\Big]. \quad (17.33)$$

17.3.7 Applications to Finance

Example 17.5 (*Pricing European options*) Black and Scholes (1973) is a seminal contribution to the pricing of European options, and to mathematical finance. As we will show, the Black-Scholes equation is a backward Kolmogorov equation, with a discount factor, for the value of an option, that satisfies the Feynman-Kac formula.

The financial market is characterized by the existence of two assets, a risk free and a risky asset. The price of the risk-free asset follows the deterministic process $db(t) = rb(t)dt$, where r is the interest rate, assumed to be constant. The price of the risky asset follows the diffusion process $dS(t) = \mu S(t)dt + \sigma S(t)dW(t)$, where the rate of return, μ, and volatility, σ, are constant. At the time of the contract, $t = 0$, the prices of both assets, $b(0) = b_0$ and $S(0) = s_0$, are observed.

An European call option is a contract offering the option (but not the obligation) to buy, at the expiration time $T > 0$, the risky asset at a price s_T. A purchaser of the option would only be interested to exercise the option if the price of the risky asset at time T, $S(T)$, is greater than the exercise price s_T. If $s_T < S(T)$ then the purchaser would not exercise the option.

Denote by $V(t, S(t))$ the value of the option on the risky asset at time t, for $0 \le t \le T$, where $V(0, S(0)$ is the price of the contract, and the value of the option at time of the exercise T is $V(T, S(T)) = \max\{S(T) - s_T, 0\}$. The main question in the option pricing problem is: what would be the price of the option at the moment of the contract, i.e., at time $t = 0$, given the fact that the observed price of the underlying risky asset is s_0 ? That is, what should be the value $V(0, s_0)$?

In order to answer this question, we need to introduce conditions allowing for the derivation the option's valuation, by taking as parameters the previously introduced market data. Assuming the financial markets are characterized by the absence of arbitrage opportunities and by market completeness, then it is possible to build a replicating portfolio, combining investments in both the risk-free and risky assets, such that its value, $R(t)$, is equal to the value of the option, that is $R(t) = V(t)$ for each $t \in T$. The value of the replicating portfolio is

$$R(t) = (1 - \theta(t))B(t) + \theta(t)S(t), \text{ for each } t \in [0, T],$$

where θ is the proportion of the risky asset in the replicating portfolio. Using Itô's formula, we obtain the process for $R(t)$,

$$\begin{aligned} dR(t) &= (1 - \theta(t))\, dB(t) + \theta(t)\, dS(t) \\ &= \big(r\, R(t) + (\mu - r)\, S(t)\big)\, dt + \theta(t)\, \sigma\, S(t)\, dW(t). \end{aligned}$$

Given the process for the price of the risky asset, the value of the option according to Itô's formula is

$$\begin{aligned} dV(t) &= \partial_t V(t, S(t))\, dt + \partial_s V(t, S(t))\, dS(t) + \frac{1}{2} \partial_{ss} V(t, S(t))\, (dS(t))^2 = \\ &= \left(\partial_t V(t, S(t)) + \mu\, S(t)\, \partial_s V(t, S(t)) + \frac{\sigma^2}{2}\, S(t)^2\, \partial_{ss} V(t, S(t))\right) dt + \\ &+ \sigma\, S(t) \partial_s V(t, S(t)) dW(t). \end{aligned}$$

Then $dV(t) = dR(t)$, if and only if $\theta(t) = \partial_s V(t, S(t))$ and $rR(t) + (\mu - r)S(t) = V_t(t, S) + \mu\, S(t)\partial_s V(t, S(t)) + \frac{\sigma^2}{2} S(t)^2 \partial_{ss} V(t, S(t))$ for each t. Setting

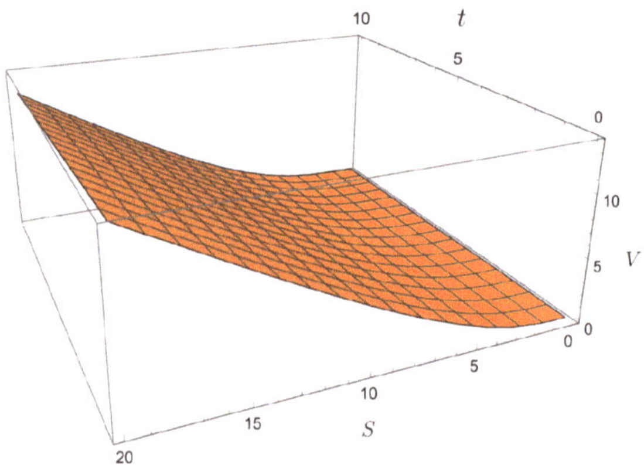

Fig. 17.3 Solution for the Black and Scholes model, for $r = 0.02$, $\sigma = 0.2$, $T = 20$, and $s_T = 10$

$R(t) = V(t, S(t))$, we obtain the Black and Scholes (1973) equation, a semi-linear parabolic BPDE.

The value of the option, for any time $t \in [0, T]$, is the solution to the following option valuation problem:

$$\partial_t V(t, S) = -\frac{\sigma^2}{2} S^2 \partial_{ss} V(t, S) - r\, S\, \partial_s V(t, S) + r\, V(t, S), \;\; (t, S) \in [0, T] \times \mathbb{R}_+,$$
$$V(T, S) = \max\{S - s_T, 0\}, (t, S) \in \{t = T\} \times \mathbb{R}_+.$$

In Proposition 15.8, we showed that value of the option for any $t \in [0, T]$, for any realization of the underlying asset price, $S(t) = s > 0$, is

$$V(t, s) = s\Phi(d_+(t)) - s_T\, e^{-r(T-t)}\Phi(d_-(t)), \;\; t \in [0, T],$$

where $\Phi(\cdot)$ is the Gaussian distribution function, and $d_{\mp}(t) = \frac{\ln\left(\frac{S(t)}{s_T}\right)+(T-t)\left(r\mp\frac{\sigma^2}{2}\right)}{\sigma\sqrt{T-t}}$. Figure 17.3 illustrates the option value as a function of (t, s). Therefore, the price of the option for a given observed price $S(0) = s_0$ is

$$V(0, s_0) = s_0\, \Phi(d_+(0)) - s_T\, e^{-r\,T}\, \Phi(d_-(0)).$$

where s_0 is observed, s_T and T are specified in the option contract, and r and σ are estimated or conjectured from actual market data.

Example 17.6 (*Pricing credit risk*) The Merton (1974) model (see also Leland 1994) is a seminal paper on the pricing of default bonds. It was the first model on the so-called structural approach to modeling credit risk.[10] In essence, this model considers the issuance of a bond with an expiring date T whose indenture entitles absolute

---

[10] This model is the inspiration of credit risk models used by rating agencies, according to Duffie and Singleton (2003).

priority on the value of the firm at the expiry date. This means that either the creditor takes possession of the firm, if the value of the firm is smaller than the face value of the bond, or it recovers the face value of the credit, in the opposite case. Therefore, we can interpret the position of the equity owner as holding an European call option over the value of the firm with strike price equal to the face value of the debt, and, by symmetry, we can interpret the position of the creditor as having an European put option.

Assume that the value of the firm, V, follows a linear diffusion process $dV(t) = \mu V(t)dt + \sigma V(t)dW(t)$. If $f(t, V)$ denotes the value of equity, as a function of time and the value of the firm, Merton (1974) proves that it is governed by the BK equation

$$\partial_t f(t, V) = -\frac{\sigma^2}{2} V^2 \partial_{vv} f(t, V) - r V \partial_v f(t, V) + r f(t, V), \ (t, V) \in [0, T] \times \mathbb{R}_+,$$
$$f(T, V) = \max\{V - b, 0\}, (t, V) \in \{t = T\} \times \mathbb{R}_+,$$

where r is the risk-free interest rate and b is the credit extended by bondholders.

Exercise 17.5 Find the price of the default bond at the time of the issuance, $t = 0$.

17.4 References

Very useful and pragmatic textbooks on SDE are Gardiner (2009), and Särkkä and Solin (2019). For a more complete study of Brownian motions, see Schilling (2021). More advanced textbooks on SDE include Karatzas and Shreve (1991), and Øksendal (2003). For textbooks emphasizing an analytical approach, see Pavliotis (2014), and emphasizing the relationship between dynamic systems theory and SDE, see Cai and Zhu (2017). For references containing explicit solutions to SDE, see Kloeden and Platen (1992), and Gardiner (2009). On the Fokker-Planck-Kolmogorov equations, see Risken (1996), and Frank (2005).

Chapter 18
Linear Scalar Stochastic Differential Equations

Linear scalar stochastic differential equations (SDE) are equations of the form

$$dY(t) = \Big(\mu_0(t) + \mu_1(t)\, Y(t)\Big)\, dt + \Big(\sigma_0(t) + \sigma_1(t)\, Y(t)\Big)\, dW(t), \quad \text{for } t \in \mathrm{T}, \tag{18.1}$$

where $\mu_0(\cdot)$, $\mu_1(\cdot)$, $\sigma_0(\cdot)$, and $\sigma_1(\cdot)$ are known functions, and $\big(W(t)\big)_{t\in\mathbb{R}_+}$ is a standard one-dimensional Wiener process, and therefore, a non-anticipating stochastic process. The solution to equation (18.1), if it exists, is a process $(Y(t))_{t\in\mathrm{T}}$ adapted to the filtration generated by the Wiener process. For linear forward stochastic differential equations (FSDE), in which $Y(0) = y_0$ is fixed, the existence of solutions is usually not problematic. This is not the case for backward stochastic differential equations (BSDE).[1] There is one important difference between deterministic and stochastic differential equations: while time can be reversed for deterministic equations, it cannot be reversed for stochastic equations.

In Sect. 18.1, we solve, characterize, and include applications to economics of linear FSDE. Although BSDE is beyond the object of this book, we briefly discuss in Sect. 18.2 a method to address their solution, and apply it to solve linear equations, and to deal with the "sustainability" issue.

18.1 Forward SDE

Section 18.1.1 focuses on the most usual autonomous linear FSDE, their explicit solutions, and their sample path and statistical properties. Section 18.1.2 compares the dynamic properties of linear FSDE with those of linear scalar ODE, dealt with in Chap. 2. Section 18.1.3 briefly addresses non-autonomous equations, and Sect. 18.1.4 contains applications to economics.

---

[1] And also for forward-backward SDE as can be seen in Ma and Yong (2007).

P. B. Brito, *Economic Dynamics and Distributions*, Dynamic Modeling and Econometrics in Economics and Finance 32, https://doi.org/10.1007/978-3-031-94717-9_18

18.1.1 Autonomous Equations

In this section, we consider the **autonomous linear** FSDE,

$$dY(t) = \Big(\mu_0 + \mu_1\, Y(t)\Big)\, dt + \Big(\sigma_0 + \sigma_1\, Y(t)\Big)\, dW(t),\ t \in \mathrm{T}, \tag{18.2}$$

where $\mathrm{T} = \mathbb{R}_+$, and the coefficients are fixed reals numbers. We assume a known initial realization $Y(0) = y_0$. The most well known cases are the Brownian motion, the geometric brownian motion, and the Ornstein-Uhlenbeck equation.

We can compare them with the analogous (deterministic) autonomous ODE, $dy(t) = \big(\mu_0 + \mu_1\, y(t)\big)\, dt$, sometimes called its skeleton. In Chap. 2 we showed that this skeleton has an explicit solution taking one of the following forms: if $\mu_1 \neq 0$ the solution is $y(t) = -\frac{\mu_0}{\mu_1} + \big(y(0) + \frac{\mu_0}{\mu_1}\big)\, e^{\mu_1 t}$, or, if $\mu_1 = 0$ the solution is $y(t) = y(0) + \mu_0\, t$. We also saw that: (1) if $\mu_1 < 0$ the solution is asymptotically stable, and $\lim_{t\to\infty} y(t) = -\frac{\mu_0}{\mu_1}$; (2) if $\mu_1 > 0$, or if $\mu_1 = 0$ and $\mu_0 \neq 0$, the solution is unstable; (3) if $\mu_0 = \mu_1 = 0$ the solution is stationary, that is, time independent.

18.1.1.1 Brownian Motion

The Brownian motion is the usual name for a process $\big(Y(t)\big)_{t\in\mathrm{T}}$, that is generated by solution to the Itô SDE

$$dY(t) = \mu\, dt + \sigma\, dW(t),\ \text{ for each } t \in \mathrm{T}, \tag{18.3}$$

where $\sigma > 0$, and any realization $Y(t) = y$, for each $t \in \mathrm{T}$, has the domain $\mathrm{Y} = \mathbb{R}$. This is a special case of equation (18.2) with $\mu_1 = \sigma_1 = 0$ and $\mu_0 = \mu$ and $\sigma_0 = \sigma$.

Proposition 18.1 (Solution to the Brownian motion) *The solution to Eq.* (18.3), *for* $Y(0) = y_0 \in \mathrm{Y}$ *fixed, is*

$$Y(t) = y_0 + \mu\, t + \sigma\, W(t),\ a.s.\ for\ t \in \mathbb{R}_+.$$

Proof Writing equation (18.3) in the integral form yields

$$Y(t) = Y(0) + \int_0^t dY(s) = y_0 + \int_0^t \mu\, ds + \int_0^t \sigma dW(s) = y_0 + \mu\, t + \sigma\, W(t)$$

because, form the properties of the Wiener process, $W(0) = 0$ a.s. □

For a case in which $\mu < 0$, Fig. 18.1a and b illustrates one sample path, and 100 sample paths for the Brownian motion. Although we have $\mu < 0$ we can observe that not all sample paths decrease in the beginning of the sampling interval. The probability density function, conditional to $Y(0) = y_0$, $p(t, y) = \mathbb{P}\,[Y(t) = y | Y(0) = y_0]$ was already derived in Chap. 13, see Eq. (15.20),

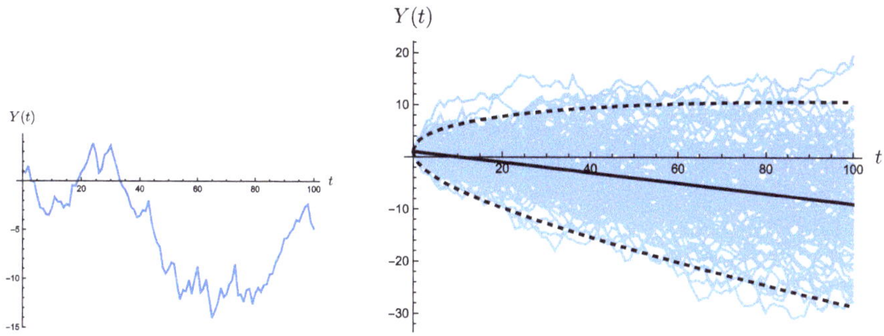

(a) One sample path

(b) 100 sample paths, confidence region, and mean

Fig. 18.1 Sample paths for the Brownian process for $\mu = -0.5$ and $\sigma = 1$

$$p(t, y) = \frac{1}{\sigma\sqrt{2\pi t}} e^{-\frac{(y-y_0-\mu t)^2}{2\sigma^2 t}}, \quad (t, y) \in \mathrm{T} \times \mathbb{R}.$$

Then, we can find the conditional cumulative distribution function,

$$\begin{aligned} F(t, y) = \mathbb{P}[Y(t) \leq y | Y(0) = y_0] &= \int_{-\infty}^{y} p(t, z) dz \\ &= \frac{1}{2}\left(1 - \operatorname{erf}\left(\frac{y_0 + \mu t - y}{\sigma\sqrt{2t}}\right)\right), \end{aligned}$$

and the conditional mean and variance,

$$\begin{aligned} \mathbb{E}[Y(t)] &= \int_{-\infty}^{\infty} y\, p(t, y) dy = y_0 + \mu t, \ \text{for } t \in \mathrm{T}, \\ \mathbb{V}[Y(t)] &= \int_{-\infty}^{\infty} \left(y - \mathbb{E}[Y(t)]\right)^2 p(t, y) dy = \sigma^2 t, \ \text{for } t \in \mathrm{T}. \end{aligned}$$

The mean is equal to the solution to the deterministic skeleton, $dy(t) = \mu\, dt$, and therefore evolves linearly over time, except in the case in which $\mu = 0$. The variance always increases linearly over time. Therefore, the Brownian motion is not ergodic because the first- and second-centered-moments are non-stationary and unbounded

$$\lim_{t\to\infty} \mathbb{E}[Y(t)] = \begin{cases} y_0 & \text{if } \mu_0 = 0 \\ \pm\infty & \text{if } \mu_0 \neq 0 \end{cases} \text{ and } \lim_{t\to\infty} \mathbb{V}[Y(t)] = +\infty.$$

Figure 18.1b also shows that the 95% confidence region is bounded by the two time-varying limits $\underline{y}(t) = \{y : F(t, y) = \alpha\}$ and $\overline{y}(t) = \{y : F(t, y) = 1 - \alpha\}$, for $\alpha = 0.025$. For a general degree of confidence 2α they can be explicitly obtained from

$$\underline{y}(t) = y_0 + \mu\, t - \underline{\beta}\sqrt{t}\sigma, \text{ and } \overline{y}(t) = y_0 + \mu\, t - \overline{\beta}\sqrt{t}\sigma$$

where $\underline{\beta} \equiv \mathrm{erf}^{-1}(1 - 2\,\alpha)\sqrt{2} \approx 2$ and $\overline{\beta} \equiv \mathrm{erf}^{-1}(2\,\alpha - 1)\sqrt{2} \approx -2$. Both limits of the confidence region are also unbounded over time, even in the case in which $\mu = 0$.

18.1.1.2 Geometric Brownian Motion

The geometric Brownian motion is the usual name of a process $\big(Y(t)\big)_{t\in\mathbb{R}_+}$ that is generated by the solution to the Itô SDE

$$dY(t) = \mu\, Y(t)dt + \sigma\, Y(t)dW(t),\ t \in \mathrm{T}, \tag{18.4}$$

where it is assumed that $\sigma > 0$. This is a special case of equation (18.2) with $\mu_0 = \sigma_0 = 0$ and $\mu_1 = \mu$ and $\sigma_1 = \sigma$. Observe that, in this case, it is required that the domain for every realization is the set of non-negative numbers, thus $\mathrm{Y} = \mathbb{R}_+$.

Proposition 18.2 (Solution to the geometric Brownian motion) *The explicit solution to Eq.* (18.4), *when* $Y(0) = y_0 > 0$, *is*

$$Y(t) = y_0\, e^{\left(\mu - \frac{\sigma^2}{2}\right) t + \sigma\, W(t)},\ \textit{a.s. for } t \in \mathrm{T}. \tag{18.5}$$

Proof Using Itô's formula, to the transformation $\ln Y(t)$, yields the Brownian motion $d \ln Y(t) = \big(\mu - \frac{\sigma^2}{2}\big)dt + \sigma dW(t)$. Integrating, and invoking the properties of the Wiener process, we have $\ln Y(t) = \ln y(0) + \big(\mu - \frac{\sigma^2}{2}\big)t + \sigma\, W(t)$, with probability one. As $Y(t) = e^{\ln Y(t)}$, then the solution (18.5) is obtained. □

The conditional probability density function $p(t, y)$ is obtained by solving the FPK equation

$$\begin{cases} \partial_t\, p(t, y) = -\frac{\partial}{\partial x}\left(\mu\, y\, p(t, y)\right) + \frac{\sigma^2}{2}\frac{\partial^2}{\partial x^2}\left(x^2\, p(t, y)\right), & \text{for } t > 0, \\ p(0, y_0) = \delta(y - y_0), \text{ where } y_0 > 0, & \text{for } t = 0, \end{cases}$$

that has solution (see Example 15.11)

$$p(t, y) = \frac{1}{y\sqrt{2\pi\,\sigma^2\, t}} e^{-\frac{\left(\ln(y/y0) - \left(\mu - \frac{1}{2}\sigma^2\right)t\right)^2}{2\sigma^2 t}}, \text{ for } (t, y) \in \mathbb{R}^2_+. \tag{18.6}$$

Then we can find, the conditional cumulative distribution function,

$$F(t, y) = \int_0^y p(t, z)dz = \frac{1}{2}\left(1 + \mathrm{erf}\left(\frac{-\left(\mu - \frac{\sigma^2}{2}\right)t + \ln\left(\frac{y}{y_0}\right)}{\sqrt{2\sigma^2 t}}\right)\right),$$

and the conditional mean and variance

$$\mathbb{E}[Y(t)] = y_0\, e^{\mu t} \text{ and } \mathbb{V}[Y(t)] y_0^2\, e^{2\mu t}\left(e^{\sigma^2 t} - 1\right), \text{ for } t \in \mathrm{T}.$$

Again, we find that the formula for the average coincides with the solution to the skeleton, $dy(t) = \mu\, dt$, $y(t) = y_0\, e^{\mu t}$. Therefore, it tends exponentially to a steady state $y = 0$ if $\mu < 0$, and diverges exponentially from it if $\mu > 0$. In the stochastic environment, the realization $Y = y = 0$ is said to be an **absorbing state** in the first case and a **repelling state** in the second. In the first case, although the model is stochastic, all the trajectories converge to a (measure zero) realization, in which the variance is zero asymptotically, and all the mass of the distribution will be concentrated at point $0 \in \mathrm{Y}$. In the second case, as the variance tends to infinity, there is an asymptotic rarefaction of the distribution. More formally, we conclude that: (1) if $\mu < 0$, for any $\sigma \neq 0$, then $\lim_{t\to\infty} \mathbb{E}[Y(t)] = \lim_{t\to\infty} \mathbb{V}[Y(t)] = 0$; or, if $\mu > 0$, for any $\sigma \neq 0$, then $\lim_{t\to\infty} \mathbb{E}[Y(t)] = \pm\infty$ and $\lim_{t\to\infty} \mathbb{V}[Y(t)] = \infty$.

Figure 18.2 presents sample paths, means, and 95% confidence regions, with boundaries given by

$$\underline{y}(t) = y_0\, e^{\left(\mu - \frac{\sigma^2}{2}\right) t + \overline{\beta}\sqrt{t}\sigma}, \text{ and } \overline{y}(t) = y_0\, e^{\left(\mu - \frac{\sigma^2}{2}\right) t + \underline{\beta}\sqrt{t}\sigma}.$$

Figure 18.2a and b illustrates a case in which $\mu < 0$ and all sample paths converge to $\lim_{t\to\infty} Y(t) = 0$. Figure 18.2c and d illustrates a case in which $\mu > 0$, and most sample paths they tend to diverge over time.

18.1.1.3 Ornstein-Uhlenback Process

The mean-reverting, or Ornstein-Uhlenbeck process,[2] $\left(Y(t)\right)_{t\in\mathbb{R}_+}$ is generated by solution to the Itô linear SDE

$$dY(t) = \theta\left(\mu - Y(t)\right) dt + \sigma\, dW(t), \text{ for } t \in \mathrm{T}, \tag{18.7}$$

where it is assumed that $\theta > 0$, and the space of realizations is $\mathrm{Y} = \mathbb{R}$. This is a special case of equation (18.2) with $\mu_0 = \theta\,\mu$, $\mu_1 = -\theta$, $\sigma_0 = \sigma$ and $\sigma_1 = 0$.

Proposition 18.3 (Solution to the Ornstein-Uhlenbeck process) *The solution to Eq.* (18.7) *for $\theta > 0$, for $Y(0) = y_0 \in \mathrm{Y}$ fixed, is*

$$Y(t) = \mu + e^{-\theta t}\left(y_0 - \mu + \sigma \int_0^t e^{\theta s} dW(s)\right), \text{ a.s for } t \in \mathrm{T}, \tag{18.8}$$

Proof Let $Z(t) = Y(t)\, e^{\theta t}$. By Itô's formula, $d\left(Y(t)\, e^{\theta t}\right) = e^{\theta t}\left(\theta\mu dt + \sigma dW(t)\right)$. Integrating over time we have

[2] Named after Uhlenbeck and Ornstein (1930).

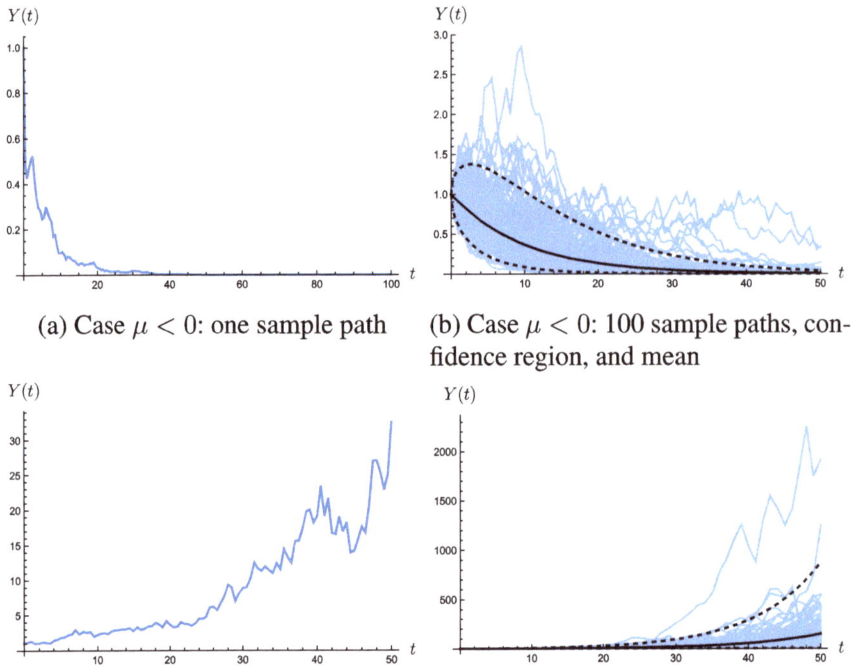

(a) Case $\mu < 0$: one sample path

(b) Case $\mu < 0$: 100 sample paths, confidence region, and mean

(c) Case $\mu > 0$: One sample path

(d) Case $\mu > 0$: 100 sample paths, confidence region, and mean

Fig. 18.2 Sample paths for the geometric Brownian motion for different signs of μ

$$Y(t)\, e^{\theta t} = y_0 + \theta\, \mu \int_0^t e^{\theta s} ds + \sigma \int_0^t e^{\theta s} dW(s).$$

Multiplying by $e^{-\theta t}$, and integrating, yields (18.8). □

The conditional probability density function is obtained by solving the FPK equation associated to the SDE (18.7)

$$\begin{cases} \partial_t p(t, y) = -\frac{\partial}{\partial x}\Big(\theta\, (\mu - y)\, p(t, y)\Big) + \frac{1}{2}\frac{\partial^2}{\partial x^2}\Big(\sigma p(t, y)\Big) \\ p(0, y) = \delta(y - y_0). \end{cases}$$

In Example 15.10, we showed that the solution to this FPK equation is

$$p(t, y) = \frac{1}{\sqrt{2\pi\, \frac{\sigma^2}{\theta}\left(1 - e^{-2\theta t}\right)}}\, e^{-\frac{\left(y - \mu - (y_0 - \mu)\, e^{-\theta t}\right)^2}{2\frac{\sigma^2}{\theta}\left(1 - e^{-2\theta t}\right)}}, \quad (t, x) \in \mathrm{T} \times \mathrm{X}.$$

Then the CDF is

$$F(t, y) = \int_{-\infty}^{y} p(t, z)dz$$
$$= \frac{1}{2}\left[1 - \text{erf}\left(\left(y_0 - \mu - (y - \mu)\, e^{\theta\, t}\right)\sqrt{\frac{\theta}{2\,\sigma^2\,(e^{2\,\theta\, t} - 1)}}\right)\right],$$

as $\theta > 0$. The conditional expected value and variance are

$$\mathbb{E}[Y(t)] = \mu + (y_0 - \mu)e^{-\theta t}, \text{ and } \mathbb{V}[Y(t)] = \frac{\sigma^2}{2\theta}\left(1 - e^{-2\theta t}\right).$$

The skeleton $dy(t) = \theta\,(\mu - y(t))\,dt$ has the solution $y(t) = \mu + (y_0 - \mu)\, e^{-\theta\, t}$ which is asymptotically stable as $\theta > 0$, and converges to the steady state $y = \mu$, independently from the initial value y_0. This process is ergodic, as both the mean and the variance tend asymptotically to finite constants, that are independent of the initial state, y_0.

$$\lim_{t\to\infty} \mathbb{E}[Y(t)] = \mu \;\; \lim_{t\to\infty} \mathbb{V}[Y(t)] = \frac{\sigma^2}{2\theta}.$$

Thus, the Ornstein-Uhlenbeck process is asymptotically Gaussian with time-independent parameters, $\lim_{t\to\infty} Y(t) \sim N\left(\mu, \frac{\sigma^2}{2\theta}\right)$. Therefore, the OU process does not has a (degenerate) absorbing state, but has instead a **asymptotic non-degenerate distribution**, whose mass is concentrated around the asymptotic mean, as can be observed in Fig. 18.3. This can be proved by the fact that the limits to the 95% confidence interval[3] converge to $\lim_{t\to\infty} \underline{y}(t) = \mu - \frac{\underline{\beta}\,\sigma}{\sqrt{\theta}}$ and $\lim_{t\to\infty} \overline{y}(t) = \mu - \frac{\overline{\beta}\,\sigma}{\sqrt{\theta}}$.

18.1.2 Stochastic Dynamic Properties of the Linear Autonomous SDE

The general linear scalar Itô-SDE (18.2), subject to the initial condition $Y(0) = y_0$, has the explicit solution

---

[3] They are

$$\underline{y}(t) = \mu + \frac{e^{-\theta\, t}}{\theta}\left(\theta\,(y_0 - \mu) - \underline{\beta}\,\sigma\sqrt{\theta\,(e^{2\,\theta\, t} - 1)}\right),$$
$$\overline{y}(t) = \mu + \frac{e^{-\theta\, t}}{\theta}\left(\theta\,(y_0 - \mu) - \overline{\beta}\,\sigma\sqrt{\theta\,(e^{2\,\theta\, t} - 1)}\right).$$

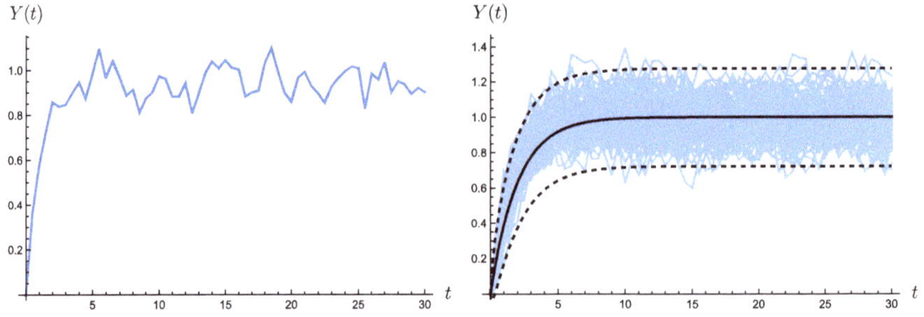

(a) One sample path

(b) 100 sample paths, confidence region, and mean

Fig. 18.3 Sample paths for Ornstein-Uhlenbeck process for $\theta > 0$ and $\mu = 1$

$$Y(t) = \Phi(t)\Big(y_0 + (\mu_0 - \sigma_0\,\sigma_1)\int_0^t \Phi(s)^{-1}\,ds + \sigma_0 \int_0^t \Phi(s)^{-1}\,dW(s)\Big), \text{ a.s., } t \in \mathbb{R}_+, \tag{18.9}$$

where $\Phi(t)$ is the solution to the geometric Brownian motion, $d\Phi(t) = \mu_1\,\Phi(t)dt + \sigma_1\,\Phi(t)dW(t)$, where $\Phi(0) = 1$.

Exercise 18.1 Prove the last result. Hint: conjecture that $Y(t) = \Phi(t)\,Z(t)$, where $\Phi(t)$ follows a geometric Brownian motion. Use the Itô formula to derive $dY(t)$. Match with Eq. (18.2) to find a linear SDE for the process $dZ(t)$. Solve this equation with the initial condition $Z(0) = y_0$. Use again the transformation $Y(t) = \Phi(t)\,Z(t)$ to find Eq. (18.9).

It can be proved that the conditional moments are,

$$\mathsf{m}(t) = \mathbb{E}[Y(t)] = \bar{\mathsf{m}} + (y_0 - \bar{\mathsf{m}})\,e^{\mu_1 t}, \text{ for } t \in \mathrm{T}, \tag{18.10}$$

and

$$\begin{aligned}\mathsf{v}(t) =& \mathbb{V}[Y(t)] = \bar{\mathsf{v}}\left(1 - e^{\gamma t}\right) + e^{\mu_1 t}\Big\{(\bar{\mathsf{m}} - y_0)^2\left(1 - e^{\mu_1 t}\right) + \\ &+ \left(y_0^2 - \bar{\mathsf{m}}^2 - \frac{2\,(\mu_0 + \sigma_0\,\sigma_1)\,(\bar{\mathsf{m}} - y_0)}{\gamma - \mu_1}\right)\left(e^{(\gamma-\mu_1)t} - 1\right)\Big\}\end{aligned} \tag{18.11}$$

where

$$\bar{\mathsf{m}} = -\frac{\mu_0}{\mu_1},\ \bar{\mathsf{v}} = -\frac{(\mu_0\,\sigma_1 - \mu_1\,\sigma_0)^2}{\mu_1^2\,\gamma}, \text{ and } \gamma = 2\,\mu_1 + \sigma_1^2.$$

Exercise 18.2 Prove the last result. Hint: take the time derivative of the expected value of the solution (18.9), to obtain the ODE $d\mathsf{m}(t) = (\mu_1(\mathsf{m} - \bar{\mathsf{m}}(t))\,dt$, and solve it. To obtain the variance, recall that $\mathsf{v}(t) = \mathsf{m}^2(t) - \mathsf{m}(t)^2$, where $\mathsf{m}^2(t)$ is found by applying the Itô's formula, to find the ODE of type $d\mathsf{m}^2(t) = \Big(\psi_0 + \psi_1\,e^{\mu_1 t} +$

$\gamma\, \mathsf{m}^2(t)\Big)\, dt$, where ψ_0 and ψ_1 are constants; solve the ODE and substitute back in $\mathsf{v}(t)$.

Those two moments allow for a general characterization of the **dynamic statistic properties** of the general linear autonomous Itô-SDE, Eq. (18.2). The parameter μ_1 acts as a bifurcating parameter:

1. if $\mu_1 > 0$ then $\lim_{t\to\infty} \mathsf{m}(t) = \infty$ and $\lim_{t\to\infty} \mathsf{v}(t) = \infty$, for any value of the other parameters;
2. if $\mu_1 = 0$ then $\mathsf{m}(t) = y_0 + \mu_0\, t$, and as

$$\mathsf{v}(t) = \frac{(\sigma_0 + \sigma_1\, y_0)^2 + 2\mu_0\, (y_0 + \mu_0 + \sigma_0\, \sigma_1)}{\sigma_1^2}\left(e^{\sigma_1^2\, t} - 1\right) - \\ -2\, \mu_0\, (y_0 + \mu_0 + \sigma_0\, \sigma_1)\, t - \mu_0^2\, t^2,$$

   the variance tends to be asymptotically unbounded;
3. if $\mu_1 < 0$ then the expected value converges to a finite level, $\lim_{t\to\infty} \mathsf{m}(t) = \bar{\mathsf{m}}$, independently from y_0, but the dynamics of the variance depends on the other parameters:
   a. if $\gamma = 2\, \mu_1 + \sigma_1^2 < 0$ and $\mu_1\, \sigma_0 \neq \mu_1 \sigma_1$ then $\lim_{t\to\infty}(\mathsf{m}(t), \mathsf{v}(t)) = (\bar{\mathsf{m}}, \bar{\mathsf{v}})$ the process is ergodic and tends asymptotically to a non-degenerate distribution;
   b. if $\gamma < 0$ and $\mu_1\, \sigma_0 = \mu_1 \sigma_1$ then $\lim_{t\to\infty}(\mathsf{m}(t), \mathsf{v}(t)) = (\bar{\mathsf{m}}, 0)$ the process tends to an absorbing state $\bar{\mathsf{m}}$ which is a deterministic steady state;
   c. if $\gamma = 0$ then $\lim_{t\to\infty} \mathsf{v}(t) = y_0^2 - \bar{\mathsf{m}}^2 - \frac{2\, (\mu_0 + \sigma_0\, \sigma_1)\, (\bar{\mathsf{m}} - y_0)}{\mu_1 + \sigma_1^2}$, that should be non-negative;
   d. if $\gamma > 0$ the variance is unbounded $\lim_{t\to\infty} \mathsf{v}(t) \propto \lim_{t\to\infty} e^{\gamma t} = \infty$. However, two cases are possible as regards the mean: if $\mu_1 < 0 < \gamma$ then $\lim_{t\to\infty} \mathsf{m}(t) = \bar{\mathsf{m}}$, however, if $\mu_1 > 0$ then $\lim_{t\to\infty} \mathsf{m}(t) \propto \lim_{t\to\infty} e^{\mu_1 t} = \infty$ as well.

18.1.3 Non-Autonomous Equation

The general non-autonomous equation (18.1) solution is

$$Y(t) = \Phi(t)\left(y_0 + \int_0^t \Phi(s)^{-1}(\mu_0(s) - \sigma_0(s)\sigma_1(s))ds + \int_0^t \Phi(s)^{-1}\sigma_0(s)dW(s)\right)$$

where $\Phi(t)$ is the solution to

$$d\Phi(t) = \mu_1(t)\Phi(t)dt + \sigma_1(t)\, \Phi(t)dW(t), \;\; \text{for } t \in [0, \infty)$$

and $\Phi(0) = 1$. The existence of explicit solutions depends on the functional form of the coefficients.

18.1.4 Applications to Economics

Example 18.1 (*Economic growth: a stochastic AK model*) In this example, we present a stochastic version of the Solow (1956) model when the technology of production displays constant marginal productivity $Y = A\,K$,[4] where $A > 0$ is the productivity parameter, and Y and K denote output and the stock of capital.[5] The equilibrium condition in the output market, at time t, is $S(t) = I(t)$, where I and S denote gross investment and savings, respectively. Both the gross investment and savings are specified in variation terms, between times t and $t + dt$. The gross investment variation is $dI(t) = dK(t) + \delta\, K(t)dt$, where $dK(t)$ is the change in net investment and $\delta \in (0, 1)$ is the (constant) depreciation rate. The variation of savings is stochastic and follows a diffusion process,

$$dS(t) = s\,Y(t)\,dt + \psi\,Y(t)dW(t), \text{ for } t \in \mathbb{R}_+,$$

where $\psi > 0$, reflecting stochastic variations in consumption expenditure, around a savings rate s. As $dY(t) = A\,dK(t)$, we obtain a linear forward SDE for capital accumulation

$$dK(t) = (s\,A - \delta)\,K(t)\,dt + \psi\,AK(t)\,dW(t), \text{ for each } t \in \mathbb{R}_+, \tag{18.12}$$

where $K(0) = k_0$ is known. This stochastic linear capital accumulation equation can be compared to its deterministic analog in Example 2.7. As Eq. (18.12) is a geometric Brownian motion, its solution is

$$K(t) = k_0 \exp\left\{\left(s\,A - \delta - \frac{1}{2}\,\psi^2\,A^2\right)t + \psi\,A\,W(t)\right\} \text{for each } t \in \mathbb{R}_+.$$

As

$$\mathbb{E}[K(t)] = k_0\,e^{(s\,A-\delta)\,t}, \text{ and } \mathbb{V}[K(t)] = k_0^2\,e^{2\,(s\,A-\delta)\,t}\left(e^{\psi^2\,A^2\,t} - 1\right),$$

the capital accumulation displays the following properties: first, if $s\,A > \delta$ both the average stock of capital (and therefore the average output) is non-stationary and converges in the mean to a balanced growth path $\lim_{t\to\infty}\mathbb{E}[K(t)] \propto \lim_{t\to\infty} e^{(s\,A-\delta)t}$ and the variance is asymptotically unbounded; second, if $s\,A = \delta$ the average stock of capital is stationary, satisfying $\mathbb{E}[K(t)] = k_0$, but the variance grows over time; or

[4] As in Frankel (1962),

[5] It is assumed that population is constant and is normalized to $L = 1$. All variables are in real per capita terms.

if $s\,A < \delta$ the average stock of capital is non-stationary but converges asymptotically to zero, $\lim_{t\to\infty}\mathbb{E}[K(t)] = 0$, but the variance grows over time, or converges to zero if $0 < \psi^2 A^2 < 2(\delta - s\,A)$.

Example 18.2 (*The Solow stochastic growth model*) Several papers, starting with Merton (1975) and Bourguignon (1974)[6] study the stochastic Solow (1956) model with constant returns to scale but decreasing marginal returns. Assume that population follows a geometric Brownian motion,

$$dL(t) = \mu L dt + \sigma L dW(t), \quad \text{for } t \in \mathrm{T} = \mathbb{R}_+,$$

where μ and σ are both positive. The equilibrium equation for the product market is $dK(t) = sF(K(t), L(t))dt$ where $F(.)$ has the neoclassical properties (increasing, concave, homogeneous of degree one, and Inada). The savings rate satisfies $0 < s < 1$. We define the capital intensity as usual $k \equiv K/L$. Then $F(K, L) = Lf(k)$. and $dk(t) = sL(t)f(k(t))dt$. Writing $k = \kappa(K/L)$, the Itô's formula yields[7]

$$dk(t) = \left(sf(k) - (\mu - \sigma^2)k\right)\, dt - k\sigma dW(t).$$

For a Cobb-Douglas function, and in the absence of depreciation of capital, we have a diffusion process for the capital intensity, which is a non-linear scalar FSDE,

$$dk(t) = \left(s\,A\,k(t)^\alpha - (\mu - \sigma^2)k(t)\right) dt - \sigma k(t) dW(t), \tag{18.13}$$

where $0 < \alpha < 1$, and $A > 0$. This equation is a stochastic analog to the deterministic capital accumulation equation in Example 3.3.

Defining $Z(t) = k(t)^{1-\alpha}$ we obtain a linear autonomous FSDE, a special case of equation (18.2), where $\mu_0 = (1-\alpha)\,s\,A$, $\mu_1 = -(1-\alpha)\left(\mu - \left(1 - \frac{\alpha}{2}\right)\sigma^2\right)$, $\sigma_0 = 0$, and $\sigma_1 = (1 \quad \alpha)\sigma$. Applying the formulas from Eqs. (18.10) and (18.11), we obtain

$$\bar{\mathsf{m}}_z = \frac{s\,A}{\mu - \left(1 - \frac{\alpha}{2}\right)\sigma^2}, \text{ and } \bar{\mathsf{v}}_z = \mathsf{m}_z^2\,\frac{(1-\alpha)\,\sigma^2}{2\left(\mu - \left(\frac{3}{2} - \alpha\right)\sigma^2\right)}.$$

---

[6] See Malliaris and Brock (1982, Chap. 3).

[7] Observe that

$$\begin{aligned}
dk &= \kappa_K dK + \kappa_L dL + \frac{.1}{2}\left(\kappa_{KK}(dK)^2 + 2\kappa_{KL} dK dL + \kappa_{LL}(dL)^2\right)\\
&= sf(k)dt - k(\mu dt + \sigma dW) + \frac{1}{2}\left(- sf(k)dt(\mu dt + \sigma dW) + 2k(\mu dt + \sigma dW)^2\right)\\
&= \left(sf(k) - (\mu - \sigma^2)k\right)\, dt - k\sigma dW(t).
\end{aligned}$$

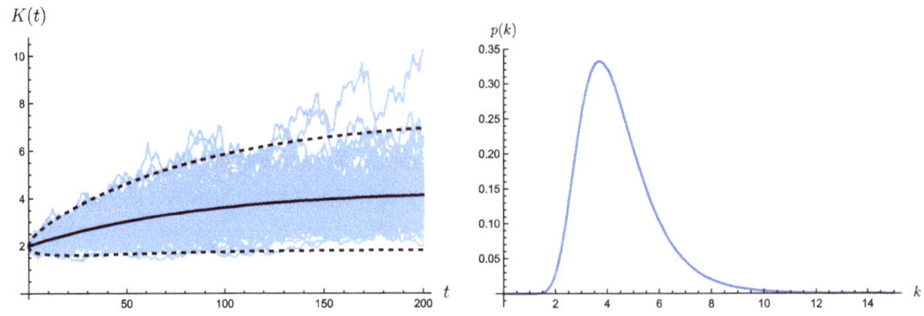

(a) 100 sample paths and approximations of $\mathsf{m}_k(t)$ and of 95% confidence region

(b) Asymptotic probability distribution

Fig. 18.4 Stochastic Solow model with $s = 0.1, \alpha = 0.3, \mu = 0.01, \sigma = 0.1$

If $\mu > \left(\frac{3}{2} - \alpha\right)\sigma^2$, then $\gamma > 0$, $\bar{\mathsf{v}}_z > 0$, and $\mu > \left(1 - \frac{\alpha}{2}\right)\sigma^2$, which implies $\mu_1 < 0$ and $\bar{\mathsf{m}}_z > 0$. In this case, the process for $Z(t)$ is ergodic and $\lim_{t\to\infty}(\mathsf{m}_z(t), \mathsf{v}_z(t)) = (\bar{\mathsf{m}}_z, \bar{\mathsf{v}}_z)$.

Figure 18.4a illustrates 100 sample paths for $k(t)$, and approximations of the mean, and of the limits of the 95% confidence interval. We approximate the expected value of capital by $\mathsf{m}_k = \mathbb{E}[k(t)] = (\mathsf{m}_z(t))^{\frac{1}{1-\alpha}}$, and the limits to the confidence region by $\left(\mathsf{m}_z(t) \pm \sqrt{\mathsf{v}_z(t)}\right)^{\frac{1}{1-\alpha}}$.

We are not aware of a closed form solution to the FPK equation both for the non-linear SDE in k, (18.13), or for the associated linear SDE in the transformed variable $Z(t)$. However, Merton (1975) (see also Malliaris and Brock (1982, p. 146)) finds the asymptotic distribution, if $2\,\mu > \sigma^2$ and $A = 1$ to be,

$$p(k) = \frac{\bar{p}}{\sigma^2 k^2} \exp\left(2 \int^k \frac{sf(\xi) - (\mu - \sigma^2)\xi}{\sigma^2 \xi^2} d\xi\right),$$

where $\bar{p}$ is a normalizing constant such that $\int_0^\infty p(k)dk = 1$. For the Cobb-Douglas case it is[8]

$$p(k) = \bar{p}\, k^{-2\mu/\sigma^2} \exp\left(\frac{2s}{(1-\alpha)\sigma^2} k^{-(1-\alpha)}\right).$$

The asymptotic distribution is illustrated in Fig. 18.4b.

Exercise 18.3 Extend the last case by assuming capital depreciation.

Example 18.3 (*Stochastic budget constraint dynamics*) A deterministic financial constraint is represented by the linear forward ODE (see Eq. (2.33) in Example 2.8)

[8] For this case $m^{-1} = \frac{1}{1-\alpha}\left(\frac{(1-\alpha)\sigma^2}{2s}\right)^{\zeta} \Gamma(\zeta)$, where $\zeta \equiv \frac{2\mu - \sigma^2}{(1-\alpha)\sigma^2}$ and $\Gamma(\cdot)$ is the Gamma distribution.

$\dot{a} = r(t)\,a(t) + s(t)$, where $a(t)$ is the net wealth at time t, $r(\cdot)$ is the rate of return, and $s(\cdot)$ is the difference between non-financial income and expenditure, or primary surplus. With the adequate denomination of its components, this equation can take the form of a household budget constraint, of the government budget constraint, or of the current account balance in an open economy. The equation can refer to nominal or to deflated variables, in which case the more frequent deflators are a price index, income or the aggregate product. When referring to deflated variables, the net rate of return can take positive or negative values.

There are several extensions to a stochastic environment in which the process for the net wealth, $(A(t))_{t\in\mathrm{T}}$, is adapted to a Wiener process, $(W(t))_{t\in\mathrm{T}}$. To obtain forward sample paths and statistic dynamics properties, starting from an initial fixed net wealth $A(0) = a_0$, the specific FSDE depends on the assumptions regarding the evolution of the primary surplus and the rate of return. We consider next just a few.

If the variation of the primary surplus is stochastic and is driven by a Brownian motion, as $dS(t) = \mu_s(t)\,dt + \sigma_s(t)\,dW(t)$, in which $\mu_s(t)$ and $\sigma_s(t)$ are deterministic or predictable,[9] and the rate of return is deterministic, the stochastic budget constraint is the FSDE

$$\begin{cases} dA(t) = (r\,A(t) + \mu_s(t))\,dt + \sigma_s(t)\,dW(t), \text{ for } t > 0, & (18.14a)\\ A(0) = a_0 \text{ fixed}, \text{ for } t = 0. & (18.14b)\end{cases}$$

The solution to this FSDE is

$$A(t) = a_0\,e^{r\,t} + \int_0^t \mu_s(z)e^{-r\,z}\,dz + \int_0^t \sigma_s(z)e^{-r\,z}\,dW(z).$$

If $\mu_s(t) = \mu_s$ and $\sigma_s(t) = \sigma_s$ are constants the net wealth can follow a Ornstein-Uhlenbeck process if $r < 0$. However, if $r > 0$ the process is not non-stationary.

If both the primary surplus and the (net) rate of return are stochastic, satisfying $dR(t) = \mu_r(t)\,dt + \sigma_r(t)\,dW(t)$, then we have the FSDE instead

$$\begin{cases} dA(t) = (\mu_s(t)A(t) + \mu_s)\,dt + (\sigma_r(t)\,A(t) + \sigma_s(t))\,dW(t), & (18.15a)\\ A(0) = a_0 \text{ fixed}. & (18.15b)\end{cases}$$

The solution to this FSDE is now

$$A(t) = \Gamma(t)\left(a_0 + \int_0^t \Gamma(z)^{-1}(\mu_s(z) - \sigma_s(z)\,\sigma_r(z))\,dz + \int_0^t \Gamma(z)^{-1}\sigma_s(z)\,dW(z)\right),$$

where the capitalization and discount factors, $\Gamma(t)$ and $\Gamma(t)^{-1}$, are now stochastic and are the solution to the geometric Brownian motion, for $\Gamma(0) = 1$,

$$d\Gamma(t) = \Gamma(t)\left(\mu_r(t)\,dt + \sigma_r\,dW(t)\right).$$

[9] They are functions of past values of the Wiener process $(W(s))_{s\leq t}$.

If the coefficients μ_s, μ_r, σ_s, and σ_r are constant, then the solution to the stochastic budget constraint is

$$A(t) = \Gamma(t)\left(a_0 + \int_0^t \Gamma(z)^{-1}(\mu_s - \sigma_s\,\sigma_r)\,dz + \int_0^t \Gamma(z)^{-1}\sigma_s\,dW(z)\right),$$

where $\Gamma(t) = \exp\{\left(\mu_r - \frac{\sigma_r^2}{2}\right)t + \sigma_r\,W(t)\}$. The types of stochastic dynamics we can expect are the same as the ones discussed in Sect. 18.1.2. The long-run statistic properties depend crucially on the sign of μ_r. If $\mu_r > 0$ then both the mean and the variance of the net wealth will be asymptotically unbounded. A necessary condition for the net wealth to be an ergodic process, converging to a distribution with finite moments, or to converge to a deterministic absorbing steady state, is that $\mu_s < 0$.

18.2 Backward SDE: A Brief Introduction

"One of the main differences between a SDE and a ODE is that one cannot reverse the time".[10] Solving backward SDE requires a background knowledge on stochastic processes and stochastic calculus that is beyond the scope of this book. Next, we provide just a brief introduction.

Consider the ODE $\dot{y} = 0$, where $y = y(t)$ is a mapping $y : \mathrm{T} = [0, T] \rightarrow \mathrm{Y} \subseteq \mathbb{R}$. We can have the same particular solution $y(t) = \upsilon$ for each $t \in [0, T]$, for a forward ODE problem, defined by $\dot{y} = 0$ and $y(0) = \upsilon$, or for a backward ODE defined by $\dot{y} = 0$ and $y(T) = \upsilon$.[11] Consider next the simple backward stochastic differential equation (BSDE) defined over the process $(Y(t))_{t\in\mathrm{T}}$, adapted to a filtration associated to the Wiener process,

$$\begin{cases} dY(t) = 0, \ \text{for each } t \in [0, T), & (18.16a) \\ y(T) = \Upsilon, \ \text{for } t = T, & (18.16b) \end{cases}$$

in which Υ is $\mathcal{F}_T$ measurable square-integrable random variable. If $\Upsilon = \upsilon$ is a constant then the solution to the BSDE is $Y(t) = \upsilon$, for each $t \in \mathrm{T}$. However, if Υ is a random variable then $Y(t) = \Upsilon$ would not be a solution, for each $t < T$, because it is not $\mathcal{F}_t$-adapted.

A way to solve this problem involves invoking the **martingale representation theorem** (MRT).[12] As $(Y(t))$ is square integrable, we introduce an unique by $\mathcal{F}_t$-adaptable process $Z(t)$, allowing us to represent the solution by

[10] Ma and Yong (2007, p. 1).

[11] We borrow this example from Ma and Yong (2007, pp. 1–2). A similar discussion is available in Zhang (2017, Chap. 4).

[12] See Øksendal (2003, p. 53). Informally, it states that for every square-integrable martingale $(M(t))_{t\in\mathrm{T}}$ there exists a unique $\mathcal{F}_t$-adapted stochastic process $(f(t))_{t\in\mathrm{T}}$, satisfying $\mathbb{E}[\int_0^t f(t)^2 ds] < \infty$, such that $M(t) = \mathbb{E}[M_0] + \int_0^t f(t)dW(t)$.

$$Y(t) = y(0) + \int_0^t Z(s)\,dW(s), \text{ for each } t \in [0, T]. \tag{18.17}$$

From now on, we call BSDE, in differential form, to

$$\begin{cases} dY(t) = Z(t)\,dW(t), \text{ for each } t \in [0, T), & (18.18a) \\ y(T) = \Upsilon, \text{ for } t = T. & (18.18b) \end{cases}$$

From Eq. (18.17), and condition (18.18b), we have

$$Y(T) = y(0) + \int_0^T Z(s)\,dW(s) = \Upsilon,$$

solving for $Y(0)$ and substituting in (18.17) yields an integral form of the BSDE (18.18)

$$Y(t) = \Upsilon - \int_t^T Z(s)\,dW(s), \text{ for each } t \in [0, T]. \tag{18.19}$$

The solution to this equation is the pair of $\mathcal{F}_t$-adapted processes $((Y(t), Z(t))_{t \in \mathrm{T}}$ that solve it. The solution satisfies

$$Y(t) = \mathbb{E}_t[\Upsilon] = \mathbb{E}[\Upsilon|\mathcal{F}_t] \text{ for each } t \in [0, T]. \tag{18.20}$$

As an example, if $\Upsilon = \upsilon$ is a constant, then the solution is $Y(t) = \upsilon$ and $Z(t) = 0$, for each $t \in \mathrm{T}$.

The SDE in which $dY(t) = \mu_0\,dt$, for each $t \in [0, T)$, satisfying the terminal condition $y(T) = \Upsilon$ for $t = T$, can be represented by processes adapted to a non-anticipating filtration over a Wiener process that is, by the BSDE, in differential form,

$$\begin{cases} dY(t) = \mu_0\,dt + Z(t)\,dW(t), \text{ for each } t \in [0, T), & (18.21a) \\ y(T) = \Upsilon, \text{ for } t = T, & (18.21b) \end{cases}$$

or in integral form

$$Y(t) = \Upsilon - \int_t^T \mu_0 ds - \int_t^T Z(s)\,dW(s), \text{ for each } t \in [0, T]. \tag{18.22}$$

It has the explicit solution,

$$Y(t) = \mathbb{E}_t[\Upsilon - \int_t^T \mu_0\,ds] = \mathbb{E}_t[\Upsilon] - \mu_0\,(T - t) \text{ for each } t \in [0, T]. \tag{18.23}$$

The slightly more general linear SDE in which $dY(t) = (\mu_0 + \mu_1\,Y(t)\,dt$, for each $t \in [0, T)$, satisfying the terminal condition$y(T) = \Upsilon$ for $t = T$, can be represented by the BSDE, in differential form,

$$\begin{cases} dY(t) = (\mu_0 + \mu_1 Y(t))\, dt + Z(t)\, dW(t), \text{ for each } t \in [0, T), & (18.24a) \\ y(T) = \Upsilon, \text{ for } t = T, & (18.24b) \end{cases}$$

or in the integral form

$$Y(t) = \Upsilon - \int_t^T (\mu_0 + \mu_1 Y(s))\, ds - \int_t^T Z(s)\, dW(s), \; t \in [0, T]. \tag{18.25}$$

This BSDE has the explicit solution

$$Y(t) = e^{\mu_1 t}\, \mathbb{E}_t\Big[e^{-\mu_1 T}\, \Upsilon - \int_t^T \mu_0\, e^{-\mu_1 s}\, ds\Big], \text{ for each } t \in [0, T]. \tag{18.26}$$

To prove this, we consider the deterministic adjoint process $d\Gamma(t) = -\mu_1\, \Gamma(t)\, dt$, with $\Gamma(0) = 1$. Then, defining $\hat{Y}(t) = \Gamma(t)\, Y(t)$, then from Itô's formula

$$\begin{aligned} d\hat{Y}(t) &= \Gamma\, dY(t) + Y(t)\, d\Gamma(t) + d\Gamma(t)\, dY(t) = \Gamma(t)\, (\mu_0\, dt + Z(t)\, dW(t)) = \\ &= \hat{\mu}_0\, dt + \hat{Z}(t)\, dW(t), \end{aligned}$$

where $\hat{Z}(t) = \Gamma(t)\, Z(t)$, $\hat{\Upsilon}(t) = \Gamma(t)\, \Upsilon$, and $\hat{\mu}_1(t) = \Gamma(t)\, \mu_1$. As this equation has the same form of Eq. (18.21), then the solution is $\hat{Y}(t) = \mathbb{E}_t\Big[\hat{\Upsilon}(t) - \int_t^T \hat{\mu}_0\, ds\Big]$, that is equivalent to (18.26).

The previous BSBE contained a deterministic multiplicative factor. To have a linear BSDE with a stochastic multiplicative factor, we can consider the BSDE in differential form

$$\begin{cases} dY(t) = (\mu_0 + \mu_1 Y(t) + \sigma_1 Z(t))\, dt + Z(t)\, dW(t), \; t \in [0, T) & (18.27a) \\ y(T) = \Upsilon, \quad t = T, & (18.27b) \end{cases}$$

where we assume that $\sigma_1 > 0$, or in the integral form

$$Y(t) = \Upsilon - \int_t^T (\mu_0 + \mu_1 Y(s) + \sigma_1 Z(s))\, ds - \int_t^T Z(s)\, dW(s), \; t \in [0, T]. \tag{18.28}$$

This BSDE has the explicit solution[13]

$$Y(t) = \Gamma(t)^{-1}\, \mathbb{E}_t\Big[\Gamma(T)\, \Upsilon - \int_t^T \mu_0\, \Gamma(s)\, ds\Big], \text{ for } t \in [0, T], \tag{18.29}$$

where the multiplicative exponential factor is stochastic

[13] See Karoui et al. (1997, p. 22) or Zhang (2017, p. 81).

$$\Gamma(t) = \exp\Big\{ - \Big((\mu_1 - \frac{\sigma_1^2}{2}) t + \sigma_1 W(t) \Big) \Big\}, \ t \in [0, T].$$

To prove this, we assume the adjoint process is

$$d\Gamma(t) = -(\mu_1 \, dt + \sigma_1 \, dW(t)) \, \Gamma(t),$$

with $\Gamma(0) = 1$, and follow the same approach as in derivation of the solution (18.29) to BSDE (18.28).

Example 18.4 (*Government debt sustainability in a stochastic environment*) is the case for several popular buzzwords, "sustainability" has an ambiguous (not to say void) meaning before a formal definition and a measurement formula derived from it, are specified.

In a deterministic environment, in which an asset dynamics is governed by a budget constraint, the usual definition is: there is sustainability if the asymptotic present value of the asset is equal to zero. This is the case if the asset level converges to a constant. We say the net asset position is sustainable if the actual level of the net asset at time $t = 0$, v.g. $a(0)$, is the particular solution of a backward ODE, represented by a budget constraint and by the terminal sustainability condition. This was proved in Example 2.9, where we used the time reversal property of ODE.

In a stochastic environment, there are (at least) two difficulties, a conceptual difficulty and a mathematical difficulty. First, even if we define "sustainability" in analogy with the deterministic case, as the asymptotic present value of the level of the asset, we should specify if it refers to a deterministic state or to the moments of a random terminal state. The second difficulty is introduced by the "non-reversibility of time", as discussed previously, if the stochastic budget dynamics is specified by an Itô SDE as in Example 18.3.

Taking the stochastic budget constraint (18.15a), in Example 18.3, and assuming that $\sigma_s = 0$, the BSDE is a particular case of the Eq. (18.30), in differential form,

$$\begin{cases} dA(t) = (\mu_s + \mu_r \, A(t) + \sigma_r \, Z(t)) \, dt + Z(t) \, dW(t), \ t \in [0, T) & (18.30a) \\ A(T) = \Upsilon, \quad t = T, & (18.30b) \end{cases}$$

where Υ is $\mathcal{F}_T$-adapted, or of (18.31) in the integral form

$$A(t) = \Upsilon - \int_t^T (\mu_s + \mu_r \, A(s) + \sigma_r \, Z(s)) \, ds - \int_t^T Z(s) \, dW(s), \ t \in [0, T]. \tag{18.31}$$

The solution to this BSDE is

$$A(t) = \Gamma(t)^{-1} \, \mathbb{E}_t \Big[\Gamma(T) \, \Upsilon - \int_t^T \mu_s \, \Gamma(s) \, ds \Big] \, t \in [0, T], \tag{18.32}$$

where the multiplicative exponential factor is stochastic

$$\Gamma(t) = \exp\Big\{ - \Big((\mu_r - \frac{\sigma_r^2}{2}) t + \sigma_r W(t) \Big) \Big\}, \; t \in [0, T].$$

If we say that the net asset position is sustainable if it satisfies

$$\lim_{T \to \infty} \mathbb{E}_0 \Big[\Gamma(T) A(T) \Big] = 0,$$

then the average sustainable level of the asset, at time $t = 0$, is

$$A(0) = -\mathbb{E}_0 \Big[\int_0^{\infty} \mu_s \Gamma(s) \, ds \Big], \tag{18.33}$$

provided the integral exists.

18.3 References

The following textbooks present not only the mathematical foundations and tools for continuous-time SDEs but also many applications to macroeconomics, microeconomics, and finance: Malliaris and Brock (1982), Dixit and Pindyck (1994), Cvitanić and Zapatero (2004), Stokey (2009), and Dumas and Luciano (2017). For textbooks on SBDE, see Ma and Yong (2007), Pardoux and Răşcanu (2014), or Zhang (2017).

Chapter 19
Stochastic Optimal Control

In this chapter, we define the **stochastic optimal control problems** (SOCP) as an optimal control problem constrained by an Itô forward stochastic differential equation (FSDE) subject to additional side constraints, such as an initial and terminal condition on the state variable. We address the simplest problems, by presenting heuristic proofs of the optimality conditions, and by finding the solutions to them.

There are three approaches to solving the SOCP: (1) the stochastic principle of dynamic programming (SDPP); (2) the stochastic Pontriyagin maximum principle (SPMP); and (3) the convex duality method.[1] The first method is the most well-known[2] and leads to the solution of a parabolic PDE, for finite horizon problems, or a second-order ODE, for infinite horizon problems, and it is presented in Sect. 19.1. The second method is less well known and leads directly to a system of forward-backward stochastic differential equations (FBSDE),[3] and is presented in Sect. 19.2. In Sect. 19.3, we provide a stochastic, and heuristic, version of the envelope theorem linking the SDPP and SPMP solutions to a SOCP.

19.1 Stochastic Dynamic Programming

Necessary conditions for finite horizon and for the discounted infinite horizon problems are next presented in Sects. 19.1.1 and 19.1.2, respectively.

---

[1] This third method is used in association with the Malliavin calculus (see Pham (2009)). It is not presented in the following text.

[2] See Fleming and Rishel (1975) or Malliaris and Brock (1982) for applications in economics and finance.

[3] See Yong and Zhou (1999).

P. B. Brito, *Economic Dynamics and Distributions*, Dynamic Modeling and Econometrics in Economics and Finance 32, https://doi.org/10.1007/978-3-031-94717-9_19

19.1.1 Finite Horizon Problems

Consider the filtered probability space $(\Omega, \mathcal{F}, \mathbb{P}, \mathbb{F})$, endowed with a non-anticipating filtration $\mathbb{F} = \left(\mathcal{F}_t\right)_{t\in T}$, for $T \subseteq \mathbb{R}_+$, where each σ-algebra $\mathcal{F}_t$ encodes the information generated by past realizations of a Wiener process $\left(W(s)\right)_{s\in(0,t)}$. The $\mathcal{F}_t$-adapted processes for the state and control variable are denoted by $(Y(t))_{t\in T}$ and $(U(t))_{t\in T}$ respectively, where $Y(t) = (y(t,\omega))_{\omega\in\mathcal{F}_t}$ and $U(t) = (u(t,\omega))_{\omega\in\mathcal{F}_t}$ for each $t \in T$. Furthermore, every realization of the state variable, $y(t,\omega)$, belongs to the state set $Y \subseteq \mathbb{R}$, and every realization of the control variable, $u(t,\omega)$, belongs to the control set $U \subseteq \mathbb{R}$, respectively. We address only simple problems in which there are no additional constraints on the state and control sets, Y and U.

Definition 19.1 (*Finite horizon stochastic optimal control problem*) Let $T = [0, T]$, for a finite and known $T > 0$, be the time domain. We call finite horizon stochastic optimal control problem (SOCP) to the following optimal control problem of a forward SDE, in the Itô sense:

$$\max_{(U(t))_{t\in T}} \mathsf{V}_0[Y,U] = \mathbb{E}_0\left[\int_0^T f(Y(t),U(t),t)\,dt\right], \tag{19.1a}$$

subject to

$$dY(t) = g(Y(t),U(t),t)\,dt + \sigma(Y(t),U(t),t)\,dW(t),\ \ t \in T, \tag{19.1b}$$

$$Y(0) = y_0,\ a.s. \tag{19.1c}$$

It is postulated that the objective, the drift, and the volatility functions, $f(.)$, $g(.)$ and $\sigma(.)$, respectively, are known. Function $g(.)$ is assumed to be of class H[4] and functions $f(.)$ and $\sigma(.)$ are of class N.[5] Informally, this means that, for any realizations of the state and control stochastic processes over time, they are bounded in expected value terms.

One important difference as regards deterministic optimal control problems, dealt with Chaps. 10, 11, and 12, is that while in the OCP case the control variable, together with the transversality condition, can be conceived a backward looking variable, in the SOCP case the control variable should be a $\mathcal{F}_t$-adapted process. Therefore, some type of consistency between the forward propagation of information and the backward nature of the optimum conditions should be guaranteed.

The stochastic dynamic programming principle (SDPP) is the analog to the deterministic dynamic programming principle, presented in Chap. 11, for the optimal control of ODE. It provides a local necessary condition for optimality.

Proposition 19.1 (SDPP for the finite horizon problem) *Let $(Y^*(t), U^*(t))_{t\in[0,T]}$ be the processes that solve the SOCP problem (19.1). Then, every realization of*

[4] Satisfying $\mathbb{E}\left[\int_T |g(Y(t),U(t),t)|\,dt\right] < \infty$.

[5] Satisfying $\mathbb{E}\left[\int_T \sigma(Y(t),U(t),t)^2\,dt\right] < \infty$, and $\mathbb{E}\left[\int_T f(Y(t),U(t),t)^2\,dt\right] < \infty$.

the state and control variables, $(Y^*(t), U^*(t)) = (y^*, u^*) \in Y \times U$, for each $t \in T$, satisfies the ***Hamilton-Jacobi-Bellman*** *partial differential equation*

$$- \partial_t V(t, y) = \max_u \Big\{ f(y, u, t) + g(y, u, t)\, \partial_y V(t, y) + \\ + \frac{1}{2} \sigma(y, u, t)^2\, \partial_{yy} V(t, y) \Big\}, \text{ for each } (t, y) \in T \times Y. \quad (19.2)$$

Proof (Heuristic) Assume we know the solution processes for the SOCP problem $(Y^*(t), U^*(t))_{t \in T}$. If we consider the solution starting from any time $t \in (0, T)$, given the initial realization of the process, $Y^*(t) = y$, then the objective functional at the optimum, $\mathsf{V}_t[Y^*, U^*]$, is a function of the pair (t, y). We call it value function, and denote it by $V(t, y)$. Thus

$$V(t, y) = \mathsf{V}_t[Y^*, U^*] = \max_{\left(U(s)\right)_{s=t}^{T}} \mathbb{E}_t\left[\int_t^T f(Y(s), U(s), s)\, ds \right].$$

From the principle of dynamic programming and the law of iterated expectations

$$\begin{aligned} V(t, y) = & \max_{\left(U(s)\right)_{s=t}^{t+dt}} \mathbb{E}_t\Bigg[\int_t^{t+dt} f(Y(s), U(s), s)\, ds + \\ & + \max_{\left(U(s)\right)_{s=t+dt}^{T}} \mathbb{E}_{t+dt}\Big[\int_{t+dt}^{T} f(Y(s), U(s), s)\, ds \Big]\Bigg] \\ = & \max_{\left(U(s)\right)_{s=t}^{t+dt}} \mathbb{E}_t\Bigg[\int_t^{t+dt} f(Y(s), U(s), s)\, ds + V(t + dt, y + dY(t)) \Bigg]. \end{aligned}$$

For a small interval dt, we have

$$V(t, y) = \max_{U(t)} \mathbb{E}_t\Big[f\big(Y(t), U(t)\big)\, dt + V(t + dt, y + dY(t)) \Big].$$

Let $dV(t, Y(t)) = V(t + dt, y + dY(t)) - V(t, y)$. If V is continuously differentiable of (at least) the second order as regards y, then from Itô's formula we obtain

$$dV(t, Y(t)) = \partial_t V(t, Y(t))\, dt + \partial_y V(t, Y(t))\, dY(t) + \frac{1}{2}\, \partial_{yy} V(t, Y(t))\, (dY(t))^2,$$

where $dY(t)$ is given by Eq. (19.1b), and $(dY(t))^2 = (\sigma(Y(t), U(t), t))^2 dt$. Then, at the optimum

$$\max_{U(t)} \mathbb{E}_t\Big[f(Y(t), U(t), t)\, dt + \partial_t V(t, Y(t))\, dt + \partial_y V(t, Y(t))\, g(Y(t), U(t), t)\, dt \\ + \frac{1}{2} \partial_{yy} V(t, Y(t))\, (\sigma(Y(t), U(t), t))^2\, dt + \partial_y V(t, Y(t))\, \sigma(Y(t), U(t), t)\, dW(t) \Big] = 0.$$

As $\mathbb{E}_t[dW(t)] = 0$, for each $t \in T$, any realization of the process $(Y^*(t), U^*(t)) = (y^*, u^*)$, is optimum only if it satisfies the HJB equation (19.2).

The existence of a solution thus requires that a $C^{1,2}(T, Y)$ function $V(t, y)$, solving the HJB equation, exists.

19.1.2 Discounted Infinite Horizon Problem

In macroeconomic and growth models, the autonomous discounted infinite horizon problem is the benchmark:

Definition 19.2 (*Discounted autonomous infinite horizon SOCP*) Let $T = \mathbb{R}_+ = [0, \infty)$ be the time domain. We call infinite horizon discounted stochastic optimal control problem (SOCP) to the following optimal control problem of a forward SDE, in Itô's sense:

$$\max_{(U(t))_{t\in T}} \mathsf{V}_0[Y, U] = \mathbb{E}_0\left[\int_0^\infty f(Y(t), U(t))\, e^{-\rho t} dt\right], \tag{19.3a}$$

subject to

$$dY(t) = g(Y(t), U(t))\, dt + \sigma(Y(t), U(t))\, dW(t), \;\; t \in T, \tag{19.3b}$$

$$Y(0) = y_0, \;\; a.s., \tag{19.3c}$$

where $\rho > 0$, and functions $f(.)$, $g(.)$ and $\sigma(.)$ are bounded measurable functions as previously assumed.

Proposition 19.2 (SDPP for the discounted infinite horizon problem) *Let $(Y^*(t), U^*(t))_{t\in[0,\infty)}$ be the processes that solve the SOCP problem (19.3). Then, every realization of the state and control variables, $(Y^*(t), U^*(t)) = (y^*, u^*) \in Y \times U$, for each $t \in T$, satisfies the* **Hamilton-Jacobi-Bellman** *ordinary differential equation*

$$\rho\, v(y) = \max_u \left(f(y, u) + g(y, u)v'(y) + \frac{1}{2}\sigma(y, u)^2\, v''(y) \right). \tag{19.4}$$

Proof Write $V(t, y) = e^{-\rho t}\, v(y)$. Therefore $\partial_t V(t, y) = -\rho\, e^{-\rho t}\, v(y)$, $\partial_y V(t, y) = e^{-\rho t}\, v'(y)$, and $\partial_{yy} V(t, y) = e^{-\rho t}\, v''(y)$. Substituting in HJB equation (19.2) yields the new HJB equation (19.4).

In this case, the HJB equation is an implicit second-order ODE, and not an implicit parabolic PDE as in Eq. (19.2).

Differently from the deterministic case, less is known about the dynamics of the solutions to SOCP. In particular, how can we characterize the stochastic analog to the dynamics of the MHDS of deterministic optimal control problems (see Sect. 10.2.3)?

Example 19.1 illustrates a case of **stochastic MHDS dynamics** for which we can determine explicit solutions. In that example, the optimal control, obtained by solving the HJB equation (19.4), is a linear function of the state variable, say $U^*(t) = \beta\, Y(t)$, where β is a constant depending on the deep parameters of the problem. Substituting in the constraint (19.3b) we have a FSDE

$$dY^*(t) = G(Y^*)\, dt + S(Y^*(t))\, dW(t),\ t \in T,$$
$$Y^*(0) = y_0,\ a.s.$$

Let us assume that we can find a monotonic transformation $h(\cdot)$ such that the transformed variable $Z(t) = h(Y^*(t))$ is driven by a linear FSDE that can be solved explicitly (or whose FPK equation can be solved explicitly). Thus, we can obtain explicit solutions to the mean and variance, say $\mathsf{m}_z(t)$, and $\mathsf{v}_z(t)$, for every point in time.

In Sect. 18.1.2, we showed that there are several types of statistic dynamic properties for linear a FSDE: (1) it can converge over time to an asymptotic distribution with finite moments; (2) it can have an asymptotic absorbing state; (3) it can converge to a rarefaction distribution; or (4) it can be unbounded over time, with the moments converging to an exponential function with positive growth rates.

In an absence of an explicit transversality condition[6] let us assume that the moments of the solution to the SOCP problem converge to finite asymptotic constants, $\lim_{t\to\infty} \mathsf{m}_z(t) = \bar{\mathsf{m}}_z > 0$ and $\lim_{t\to\infty} \mathsf{v}_z(t) = \bar{\mathsf{v}}_z > 0$. Making the inverse transformations, we can obtain asymptotic approximations to the mean of both the state and control variables,

$$\lim_{t\to\infty} \mathsf{m}_y(t) = \bar{\mathsf{m}}_y = h^{-1}(\bar{\mathsf{m}}_z),\ \text{and}\ \lim_{t\to\infty} \mathsf{m}_u(t) = \bar{\mathsf{m}}_u = \beta\Big(h^{-1}(\bar{\mathsf{m}}_z)\Big).$$

We can also obtain 95% asymptotic confidence regions for those variables,

$$\mathbb{P}[\underline{y} < \lim_{t\to\infty} Y^*(t) < \overline{y}] = \mathbb{P}[\underline{u} < \lim_{t\to\infty} U^*(t) < \overline{u}] = 0.95,$$

where $\underline{y} = h^{-1}\Big(\bar{\mathsf{m}}_z - 2\sqrt{\bar{\mathsf{v}}_z}\Big)$, $\overline{y} = h^{-1}\Big(\bar{\mathsf{m}}_z + 2\sqrt{\bar{\mathsf{v}}_z}\Big)$,$\underline{u} = \beta\, \underline{y}$, and $\overline{u} = \beta\, \overline{y}$.

Therefore, this version of the **stochastic optimal dynamics** has the following properties: (1) at time $t = 0$, given an observed level for the state variable, the solution is $(y_0, u^*(0))$, where the level of the control is deterministic $u^*(0) = \beta\, y_0$; (2) over time, the solution is a stochastic process $(Y^*(t), U^*(t))$ contingent of the realizations of the Wiener process, meaning that there is an infinite number of possible realizations. In our example, for any realization path for the state and control vari-

---

[6] That would allow, for instance, for showing that a solution of the fourth type can exist. However, if there is a solution to the time independent HJB equation (19.4), then the discounted value function $V(t, y)$ is bounded when time tends to infinity, which suggests that an optimal solution to a SOCP can be unbounded and have unbounded moments asymptotically. This was the case for some deterministic optimal control problems, and it is the case for the next examples.

ables, $(y^*(t), u^*(t))_{t\in T}$, **there is a monotonic relationship between the control and the state variable** $u^*(t) = \beta\, y^*(t)$, for each $t \in T$; (3) asymptotically, most realizations will converge, with 95% probability, to their asymptotic regions of confidence.

Example 19.1 (*The benchmark household consumption-portfolio problem*) The continuous time benchmark household consumption-portfolio problem, or the Merton (1971) model,[7] is a standard micro model for the simultaneous determination of the strategies of consumption and portfolio investment for the representative household, in a finance economy characterized by imperfect information. This model can be considered a micro-finance model, or as a component of a DSGE macro-financial model.[8] As the financial income of a creditor, or the financial cost of a debtor, is stochastic then the household consumption also becomes stochastic. In addition to the intertemporal allocation of its resources to finance consumption, that is present in the deterministic version, the household has to weight both the intertemporal dimension and its dependence on the states of nature. In the presence of both impatience and risk aversion, there is both smoothing out of consumption over time but also across states of nature. That is, there is also a precautionary, or insurance, dimension involved in the choice of both consumption and portfolio strategies.

This example extends the deterministic version presented in Example 10.4 to a stochastic setting, in which there is both a risk-free and a risky asset, where the price of the risk-free asset, $B(t)$, is generated by a deterministic process $dB(t) = r\, B(t)dt$, where r is the risk-free interest rate, and the price of the risky asset is generated by the diffusion process $dS(t) = \mu\, S(t)\, dt + \sigma\, S(t)\, dW(t)$, where the constants μ and σ denote the rate of return and the volatility for the risky asset. Denoting a non-financial (deterministic) income by y, consumption expenditures by $C(t)$, and the net financial wealth by $A(t)$, the budget constraint of the household is the FSDE

$$dA(t) = \Big[\Big(r\,(1 - w(t)) + \mu\, w(t)\Big)\, A(t) + y - C(t)\Big]\, dt + \sigma\, w(t)\, A(t)\, dW(t), \quad \text{for } t \geq 0, \tag{19.5}$$

where $w(t)$ is the share of risky asset in the net wealth.[9]

Assuming a dynastic (or a uncertain infinite life-time) household, the consumer-portfolio problem is

$$\max_{C(\cdot), w(\cdot)} \mathbb{E}_0\Big[\int_0^\infty u(C(t))\, e^{-\rho t}\, dt\Big], \tag{19.6}$$

---

[7] Other references include. Merton (1990), Duffie (1996), Cvitanić and Zapatero (2004), or Chang (2004, p. 191).

[8] As e.g., in Basak and Cuoco (1998), and Brunnermeier and Sannikov (2014).

[9] We assume that there are no financial frictions, and the household is free to participate in both financial markets. Then the household can have long and/or short positions in both assets. Therefore, $w \in \mathbb{R}$ and it is not restricted to the $(0, 1)$ interval, as would be the case if the household could not participate freely in the risky financial markets.

subject to the instantaneous budget constraint (19.5), given $A(0) = a_0$, and assuming that the utility function is increasing and concave.[10]

This is a SOCP with infinite horizon, and two control variables, C and w. According to the SDPP (see Proposition 19.1), any optimal realizations of the net wealth, consumption, and portfolio composition at time t, $(A^*(t), C^*(t), w^*(t)) = (a^*, c^*, w^*)$, satisfy the Hamilton-Jacobi-Bellman equation (19.4)

$$\rho\, v(a) = \max_{c,w} \left\{ u(c) + v'(a) \left((r + (\mu - r)\, w)\, a + y - c \right) + \frac{\sigma^2}{2} w^2 a^2 v''(a) \right\}.$$

The optimal policies for consumption and portfolio composition are

$$u'(c^*) = v'(a), \tag{19.7}$$

$$w^* = W(a) = \frac{(\mu - r)}{\varrho_v(a)\, \sigma^2}, \tag{19.8}$$

where $\frac{(\mu-r)}{\sigma}$ is the Sharpe index and $\varrho_v(a) \equiv -\frac{a\, v''(a)}{v'(a)}$ is elasticity of the value function. While the optimum portfolio composition is explicitly given by Eq. (19.8), the optimal consumption can only be implicitly derived from Eq. (19.7), assuming the value function $v(a)$ exists and can be found. If the utility function is globally concave (as in the Inada case) then, from the implicit function theorem, the consumption policy function can be written as $c^* = C(a) \equiv (u')^{-1}\big(v'(a)\big)$. Substituting the policy functions, the maximized HJB equation is the implicit ODE, for function $v(a)$,

$$\rho\, v(a) = u\big(C(a)\big) + v'(a)\big(y + r\, a - C(a)\big) + \frac{1}{2} \left(\frac{\mu - r}{\sigma} \right)^2 \frac{(v'(a))^2}{v''(a)}. \tag{19.9}$$

In some cases, this HJB equation can be solved explicitly, in particular when the utility function displays constant relative risk aversion and the constraint is a linear SDE.

Example 19.2 (*The CRRA case*) Let the utility function display constant relative risk aversion (CRRA)

$$u(c) = \frac{c^{1-\eta} - 1}{1 - \eta}, \quad \text{for } \eta > 0,$$

where $\eta = \frac{1}{\varrho} > 0$ is the inverse of the coefficient of relative risk aversion. We assume that $\rho + (\eta - 1) \left(r + \frac{1}{2\eta} \left(\frac{\mu - r}{\sigma} \right)^2 \right) > 0$. Let the household's total net wealth be defined as the sum of the human and the net financial wealth, $N(t) = \frac{y}{r} + A(t)$. Then the optimal share of the risky asset on net wealth is proportional to the Sharpe

[10] On the implicit assumptions regarding this utility functional, refer to Sect. 8.1.5.6.

index $\frac{w^*(t)\,A(t)}{N(t)} = \frac{\varrho}{\sigma}\left(\frac{(\mu - r)}{\sigma}\right)$, and the optimal consumption[11] is also proportional to total net wealth

$$C(t) = \beta\, N(t), \text{ where } \beta \equiv \frac{1}{\eta}\left(\rho + (\eta - 1)\left(r + \frac{1}{2\eta}\left(\frac{\mu - r}{\sigma}\right)^2\right)\right), \tag{19.10}$$

where $\beta > 0$ if our previous assumption is satisfied. As, in the deterministic analog, with only risk-free asset, optimal consumption would be $c^* = \frac{\rho + r(\eta - 1)}{\eta}\, n$ then last term in Eq. (19.10) is associated to the extra financial income provided by investments in the risky asset, if the household is a net creditor (or a financial cost if it is a debtor).

The household cannot completely eliminate uncertainty in consumption (at least if asset markets are not complete) if $\mu \neq r$. To see this, substitute the optimal consumption policy, from Eq. (19.10), into the budget constraint (19.5). Then the optimal net wealth process, $(N(t))_{t \in T}$, follows a FSDE

$$dN(t) = \left(\mu^*\, dt + \sigma^*\, dW(t)\right) N(t), \text{ for each } t \in T,$$

where

$$\mu^* = \frac{1}{\eta}\left(r - \rho + \frac{1 + \eta}{2\,\eta}\left(\frac{\mu - r}{\sigma}\right)^2\right), \text{ and } \sigma^* = \frac{1}{\eta}\left(\frac{\mu - r}{\sigma}\right),$$

implying that the optimal household consumption also follows a FSDE

$$dC(t) = \left(\mu^* dt + \sigma^*\, dW(t)\right) C(t), \text{ for each } t \in T.$$

From the solution to the geometric Brownian motion, in Eq. (18.5), we find the solution to optimal (contingent) consumption,

$$C(t) = c(0) \exp\left\{\frac{1}{\eta}\left(r - \rho + \frac{1}{2}\left(\frac{\mu - r}{\sigma}\right)^2\right) t + \frac{1}{\eta}\left(\frac{\mu - r}{\sigma}\right) W(t)\right\}, \text{ for each } t \in T,$$

conditional on the realization of the Wiener process, and on the optimal consumption level at time $t = 0$, $c(0) = \beta\, n_0$, for $n_0 = a_0 + \frac{y}{r}$, that is deterministic. The expected value for consumption at time t is

$$\mathbb{E}_0[C(t)] = c(0)\, e^{\mu^* t}, \text{ for each } t \in T.$$

As $\mu^* = r + \eta\,\sigma^{*2} - \beta$ then three types of solutions are possible: (1) if $\beta < r + \eta\,\sigma^{*2}$ then $\lim_{t \to \infty} \mathbb{E}_0[C(t)] = \infty$; (2) if $\beta = r + \eta\,\sigma^{*2}$ then $\mathbb{E}_0[C(t)] = c(0)$ for each $t \in T$; or (3) if $\beta > r + \eta\,\sigma^{*2}$ then $\lim_{t \to \infty} \mathbb{E}_0[C(t)] = 0$. It is easy to check that, if $\beta > 0$, the value function, $\mathsf{V}_0[A^*]$, is bounded in all the previous cases.

Exercise 19.1 (Solving the HJB equation (19.9)) We can solve HJB equation (19.9) by using the method of undetermined coefficients. Conjecture a solution

---

[11] See Exercise 19.1.

of type[12] $\alpha v(a) \propto n(a)^{1-\eta}$, where α is an arbitrary constant. This function satisfies $(1-\eta)\,(v(a)-\alpha) = n(a)\,v'(a)$ and $n(a)\,v''(a) = \eta\,v'(a)$. Then, the optimality condition (19.7) becomes $(c^*)^{-\eta} = v'(a)$, and (19.8) becomes $w^* = \frac{(\mu-r)}{\sigma^2}\frac{n(a)}{\eta\,a}$. Substituting in the HJB equation (19.9), we obtain

$$\rho\left(\alpha + \frac{1}{1-\eta}n(a)v'(a)\right) = \frac{1}{\eta-1} + v'(a)\left[\frac{\eta}{1-\eta}c^* + \left(r + \frac{1}{\eta}\left(\frac{\mu-r}{\sigma}\right)^2\right)n(a)\right]$$

If we set $\alpha\,\rho\,(\eta-1) = 0$, and factor, we obtain the algebraic equation

$$\rho\,n(a) = \eta\,c^* + (1-\eta)\left(r + \frac{1}{\eta}\left(\frac{\mu-r}{\sigma}\right)^2\right)n(a)$$

that can be solved for c^*.

Example 19.3 (*The stochastic Ramsey model*) The seminal papers Brock and Mirman (1972), and Merton (1975) extended the Ramsey (1928) model of optimal accumulation of capital to an uncertainty environment.

Let K be the stock physical capital and L be the stock of labor, that is assumed to be equal to the population size (there is no unemployment, diseases, etc). Both stocks are fixed at time $t=0$, $K(0) = K_0$ and $L(0) = L_0$. Over time, they are driven by the macroeconomic equilibrium condition and by an exogeneously given geometric Brownian motion, respectively,

$$\begin{aligned} dK(t) &= \Big(A\,F(K(t), L(t)) - C(t)\Big)\,dt,\ t \geq 0, \\ dL(t) &= \mu\,L(t)\,dt + \sigma\,L(t)\,dW(t),\ t \geq 0, \end{aligned}$$

where the total factor productivity, A, and the volatility labor supply, σ, are both positive. It is assumed that the production function, $F(K, L)$, is linearly homogeneous. Defining $k(t) \equiv \frac{K(t)}{L(t)}$ and $c(t) \equiv \frac{C(t)}{L(t)}$, and using Itô's formula, yields the dynamic equation for capital intensity as a scalar FSDE[13]

$$dk(t) = \big(f(k) - c(t) - (\mu-\sigma^2)\,k(t)\big)\,dt - \sigma\,k(t)\,dW(t),\ \text{for } t \in \mathbb{R}_+ \quad (19.11)$$

given $k(0) = k_0$, where the production function in intensity terms is $f(k) = F\left(\frac{K}{L}, 1\right)$.

At last we assume the social welfare functional

$$\mathsf{V}[c] = \mathbb{E}_0\left[\int_0^\infty u(c(t))e^{-\rho t}dt\right].$$

[12] See Chang (2004, pp. 191–4) for a justification of this conjecture.

[13] We now denote the stochastic process for the capital intensity by $(k(t))_{t\in T}$. From now on $k(t)$ refers to the level of the process at time t, a random variable, and k refers to any realization of the process.

The optimal allocation problem is the SOCP, of maximizing the social welfare functional subject to the macroeconomic constraint (19.11), given $k(0) = k_0$. Using the SDPP, the HJB equation (19.4) takes the form

$$\rho\, v(k) = \max_c \left\{ u(c) + v'(k) \left(f(k) - c - (\mu - \sigma^2)\, k \right) + \frac{\sigma^2}{2}\, k^2\, v''(k) \right\}.$$

The optimality condition is similar to the household problem, in Eq. (19.7), $u'(c) = v'(k)$. If $u(\cdot)$ is a smooth function, then maximized HJB is the implicit second-order ODE

$$\rho\, v(k) = u(C(k)) + v'(k) \left(f(k) - C(k) - (\mu - \sigma^2)\, k \right) + \frac{\sigma^2}{2}\, k^2\, v''(k).$$

The benchmark version of this model in the literature assumes a CRRA utility function $u(c) = \frac{c^{1-\eta}-1}{1-\eta}$, where $\eta > 0$,[14] and a Cobb-Douglas production function, $f(k) = k^{\alpha}$, where $0 < \alpha < 1$. In this case, the optimal policy function is $c^* = C(k) = v'(k)^{-\frac{1}{\eta}}$, and the maximized HJB equation is

$$\rho\, v(k) = \frac{1}{1-\eta} \left(\eta\, v'(k)^{\frac{\eta-1}{\eta}} - 1 \right) + v'(k)\, \left(k^{\alpha} - (\mu - \sigma^2)\, k \right) + \frac{\sigma^2}{2} k^2\, v''(k). \tag{19.12}$$

This implicit ODE does not seem to have a closed form solution.

Exercise 19.2 (Particular case with $\eta = \alpha$) To compare the previous model with the stochastic Solow model, studied in Example 18.2, and with the deterministic RCK model, studied in Chap. 11, we assume for pedagogical reasons the (unrealistic) case in which $\eta = \alpha$. This implies the maximized HJB (19.12) takes the form

$$\rho\, v(k) = \frac{\alpha}{1-\alpha}\, v'(k)^{\frac{\alpha-1}{\alpha}} - \frac{1}{1-\alpha} + v'(k)\, \left(k^{\alpha} - (\mu - \sigma^2)\, k \right) + \frac{\sigma^2}{2} k^2\, v''(k).$$

If we conjecture a solution of the form $v(k) = b_0 + b_1\, k^{1-\alpha}$, we find it solves the maximized HJB equation if (b_0, b_1) jointly solve the two equations $\rho\,(1-\alpha)\, b_0 = (1-\alpha)^2\, A\, b_1 - 1$ and $\alpha \left((1-\alpha)\, b_1 \right)^{-\frac{1}{\alpha}} = \rho + (1-\alpha) \left(\mu - \left(1 - \frac{\alpha}{2} \right) \sigma^2 \right)$.

Then the optimal consumption process is linearly related to the process for k by

$$c^*(t) = \varrho\, k(t), \;\; \text{where } \varrho \equiv \frac{\rho + (1-\alpha) \left(\mu - \left(1 - \frac{\alpha}{2} \right) \sigma^2 \right)}{\alpha},$$

[14] Refer to Chap. 8 for the several implicit assumptions in this functional regarding not only the choice over time and the states of nature by households but also the social welfare measurement. In particular, from Sects. 8.1.5.3 and 8.1.5.6, parameter η represents both the coefficient of relative risk aversion and the inverse of the elasticity of intertemporal substitution.

under the condition $\rho + (1-\alpha)\,\mu > (1-\alpha)\left(1-\frac{\alpha}{2}\right)\sigma^2 > 0$. The optimal process for the capital intensity is the non-linear FSDE

$$dk^*(t) = \left(A\,k^*(t)^\alpha - \psi\,k^*(t)\right)dt - (1-\alpha)\,\sigma\,dW(t),\ t \geq 0,$$
$$k^*(0) = k_0,$$

where $\psi \equiv \mu - \sigma^2 + \varrho = \frac{1}{\alpha}\left(\rho + \mu - \left(1 - \frac{\alpha\,(1-\alpha)}{2}\right)\sigma^2\right)$. As explained earlier, the process for consumption is linearly related to the process for capital by $dc^* = \beta\,dk^*(t)$, for each $t \in [0,\infty)$. In order to solve the SOCP, we need to find the solution to the FSDE for $k^*(t)$. This SDE has the same functional form as the stochastic Solow model (18.13). Defining, again, $Z(t) = k(t)^{1-\alpha}$ we obtain a linear FSDE,

$$dZ(t) = (1-\alpha)\left[\left(A - \frac{\rho + \mu - \left(1-\frac{\alpha}{2}\right)\sigma^2)}{\alpha}\right)Z(t)\,dt - \sigma\,Z(t)\,dW(t)\right],\ t \geq 0.$$

Observe that $Z(t) = \frac{A}{r(t)}$ where $r(t) = A\,k(t)^{\alpha-1}$, is the ratio between TFP and the marginal productivity of capital. Then from the analysis in Exercise 18.1, we can conclude that the asymptotic rate of growth of this ratio is

$$\gamma = -\frac{2\,(1-\alpha)}{\alpha}\left(\rho + \mu + \sigma^2\left(\frac{\alpha\,(1-\alpha)}{2} - \frac{2-\alpha}{2}\right)\right) < 0$$

if $\rho + \mu > \frac{2-\alpha}{2}\,\sigma^2$, which implies that

$$\lim_{t\to\infty}\left(\mathbb{E}\left[\frac{1}{r^*(t)}\right], \mathbb{V}\left[\frac{1}{r^*(t)}\right]\right) = \left(\mathsf{m}_r, \mathsf{v}_r\right),$$

where $\mathsf{m}_r = \frac{\alpha}{\rho+\mu-\frac{2-\alpha}{2}\sigma^2} > 0$ and $\mathsf{v}_r = -\frac{\left((1-\alpha)\,\sigma\,\mathsf{m}\right)^2}{\gamma}$. Therefore, we can approximate the asymptotic averages for the stock of capital and consumption by

$$\lim_{t\to\infty}\mathsf{m}_k(t) = \bar{\mathsf{m}}_k \approx \left(A\,\mathsf{m}_r\right)^{\frac{1}{1-\alpha}},\quad \lim_{t\to\infty}\mathsf{m}_c(t) = \bar{\mathsf{m}}_c \approx \varrho\,\bar{\mathsf{m}}_k,$$

and the asymptotic 95% confidence region for capital by $(\underline{k}, \overline{k})$, where

$$\underline{k} \approx \left(A\,\mathsf{m}_r - 2\sqrt{A^2\mathsf{v}_r}\right)^{\frac{1}{1-\alpha}},\ \text{and}\ \overline{k} \approx \left(A\,\mathsf{m}_r + 2\sqrt{A^2\mathsf{v}_r}\right)^{\frac{1}{1-\alpha}},$$

and $(\varrho\,\underline{k}, \varrho\,\overline{k})$ for consumption. The dynamics for both variables is similar to those illustrated in Fig. 18.4, in which each optimal path corresponds to a contingent plan for optimal capital accumulation.

19.2 The Stochastic Pontryiagin's Maximum Principle Approach

In the economics literature, we are unaware of the use of the stochastic Pontriyagin maximum principle (SPMP). In this section, we try to fill this gap, by presenting, in an informal way, the necessary optimality conditions according to this principle, and apply it to simple models. Another purpose of our presentation is to derive stochastic maximized Hamiltonian dynamic systems (SMHDS) as FBSDE, and solve them for simple cases.

An advantage of the Pontryiagin's maximum principle (PMP) approach for solving deterministic optimal control problems, discussed in Chap. 10, is related to possible characterization of their solutions by studying the maximized Hamiltonian dynamic system (MHDS), which is a backward-forward ODE. The main theoretical difficulty for solving stochastic optimal control problems is related to the fact that the control variable is determined from a backward-looking optimality condition, while the state variable should be $\mathcal{F}_t$-adapted, that is, non-anticipating and forward-looking. This difficulty is somewhat concealed when using the SDPP approach, but clearly arises when using the SPMP approach.

In the rest of this section, we present two types of problems sharing the same value functional (19.3a), but having different FSDE constraints, leading to two different versions of the SPMP, for infinite horizon problems. First, we consider problems in which the diffusion coefficient is independent from the control variable, in Sect. 19.2.1, and next, we consider problems in which the diffusion coefficient depends on the control variable, in Sect. 19.2.2.

19.2.1 Volatility Coefficient Independent of the Control Variable

Consider the constraints of the problem

$$dY(t) = g(Y(t), U(t))\,dt + \sigma(Y(t))dW(t), \ |text for\ t \in T, \tag{19.13a}$$

$$Y(0) = y_0. \tag{19.13b}$$

Proposition 19.3 (SPMP for uncontrolled volatility) *Let $(Y^*(t), U^*(t))_{t\in[0,T]}$ be the solution to the SOCP problem of maximizing the functional* (19.3a) *under the constraints (19.13). Then, there are two processes $(P_1(t), Q_1(t))_{t\in[0,T]}$, and a Hamiltonian function , that, for each realization of $(Y(t), U(t), P_1(t), Q_1(t))$ is defined by*

$$H(y, u, p_1, q_1) = f(y, u) + p_1\, g(y, u) + q_1\, \sigma(y), \tag{19.14}$$

such that the processes $(Y^*(t), U^*(t), P_1(t), Q_1(t))$, *satisfy the adjoint equation and the terminal condition*

$$dP_1(t) = \left(\rho\, P_1(t) - \frac{\partial}{\partial y} H(Y^*(t), U^*(t), P_1(t), Q_1(t))\right) dt + \\ + Q_1(t)\, dW(t), \tag{19.15a}$$

$$\lim_{t\to\infty} e^{-\rho t} P_1(t) = 0, \tag{19.15b}$$

and the optimal control satisfies for every realizations of the state and the control variables, $Y^*(t) = y^*(t)$ *and* $U^*(t) = u^*(t)$,

$$H(y^*(t), u^*(t), p_1(t), q_1(t)) = \max_u H(y^*(t), u(t), p_1(t), q_1(t)), \text{ for each } t \in T, \tag{19.16}$$

and the admissibility conditions

$$dY^*(t) = g(Y^*(t), U^*(t))\, dt + \sigma(Y^*(t))\, dW(t).\ t \in T, \tag{19.17a}$$

$$Y^*(0) = y_0. \tag{19.17b}$$

Proof See Yong and Zhou (1999, pp. 123–137).

The variables $P_1(\cdot)$ and $Q_1(\cdot)$ are called first-order adjoint variables, and the SDE (19.15a) is called first-order adjoint equation. If function $H(\cdot)$ is a smooth function of u, then a necessary condition for the maximum is

$$\frac{\partial H(y^*, u^*, p_1, q_1)}{\partial u} = 0,$$

for every optimal realization of the state co-state and control variables. By the implicit function theorem, the optimal control is a function of the state and co-state variables $u^* = U^*(y, p_1, q_1)$. Substituting in the derivative in Eq. (19.15a),

$$\frac{\partial H(y, u, p_1, q_1)}{\partial y} = \frac{\partial f(y, u)}{\partial y} + p_1 \frac{\partial g(y, u)}{\partial y} + q_1 \frac{\partial \sigma(y)}{\partial y},$$

then the stochastic maximized Hamiltonian system (SMHDS) is the FBSDE

$$dY(t) = \frac{\partial H^*(t)}{\partial p_1} dt + \frac{\partial H^*(t)}{\partial q_1} dW(t),$$

$$dP_1(t) = \left(\rho\, P_1(t) - \frac{\partial H^*(t)}{\partial y}\right) dt + \frac{\partial H^*(t)}{\partial \sigma} dW(t),$$

where, for any realization of $(Y_1(t), P_1(t))$, the Hamiltonian function is $H^*(y, p_1, q_1) = H(y, U^*(y, p_1, q_1), p_1, q_1)$, and after eliminating q_1 in the way we show in the next example, the FBSDE system can hopefully be solved for the stochastic processes

$(Y^*, P_1(t), Q_1(t))$ together with the initial and the transversality conditions, (19.13b) and (19.15b), respectively.

Example 19.4 (*The stochastic AK model*) The stochastic version of the simplest endogenous growth model, in which we assume that the volatility is independent of consumption, the control variable, is :

$$\max_{C(.)} \int_0^\infty \ln(C(t))e^{-\rho t}dt, \tag{19.18a}$$

subject to

$$dK(t) = \big(\mu K(t) - C(t)\big)\,dt + \sigma K(t)\,dW(t), \tag{19.18b}$$

$$K(0) = k_0, \ a.s. \tag{19.18c}$$

The Hamiltonian function is

$$H(c, k, p_1, q_1) = \ln(c) + p_1(\mu k - c) + q_1 \sigma k.$$

The first-order adjoint equation and the transversality condition are

$$dP_1(t) = \Big((\rho - \mu)\,P_1(t) + \sigma\, Q_1(t)\Big)dt + Q_1(t)\,dW(t), \ t \in T,$$

$$\lim_{t\to\infty} e^{-\rho t} P_1(t) = 0, \ a.s.$$

Any realization of the optimal consumption satisfies $C^*(t) = P_1^{-1}(t)$, for each t. Applying the Itô's formula, we have a FSDE that is a stochastic version of the Keynes-Ramsey rule,

$$dC^*(t) = C^*(t)\left[\left(\mu - \rho + \sigma\frac{Q_1(t)}{P_1(t)} + \left(\frac{Q_1(t)}{P_1(t)}\right)^2\right)dt - \frac{Q_1(t)}{P_1(t)}dW(t)\right].$$

We have a stochastic differential equation for $P_1(t)$ but we do not have one equation allowing for the determination of $Q_1(.)$. Based on our knowledge of the related deterministic model, we introduce a trial relationship $C^*(t) = \phi\, K(t)$ where ϕ is a constant to be determined. Applying the Itô's formula, yields another SDE, in this case a BSDE, for C,

$$dC^*(t) = \phi\,\left((\mu K(t) - C(t))\,dt + \sigma K(t) dW(t)\right).$$

Matching the deterministic and the stochastic components of the backward and forward SDE for C, the following relations should be true, for any realization of $(C(t), K(t), P_1(t), Q_1(t)) = (c, k, p_1, q_1)$,

$$c\left(\mu - \rho + \sigma\frac{q_1}{p_1} + \left(\frac{q_1}{p_1}\right)^2\right) = \phi\,(\mu k - c), \text{ and } -c\frac{q_1}{p_1} = \phi\,\sigma\, k,$$

yielding $q_1 = -\sigma\, p_1$ and $\phi = \rho$. Therefore, the optimal consumption is a linear function of the stock of capital (as in the deterministic analog), $C^*(t) = \rho K^*(t)$. The transversality condition, Eq. (19.15b), is satisfied if $\lim_{t\to\infty} \frac{e^{(-\mu+\frac{1}{2}\sigma^2)t-\sigma W(t)}}{\rho\, k_0} = 0$, that is the case if $\mu > \frac{1}{2}\sigma^2 > 0$.

Furthermore, substituting the optimal consumption in Eq. (19.18b), then both optimal capital and consumption follow two perfectly correlated geometric Brownian motions (see Sect. 18.1.1.2). For the case of capital, we have the linear FSDE

$$dK^*(t) = K^*(t)\left((\mu-\rho)dt + \sigma dW(t)\right),$$

that has a solution

$$K^*(t) = k_0 e^{(\mu-\rho-\frac{1}{2}\sigma^2)t+\sigma W(t)}, \text{ for each } t \in T,$$

and the statistics

$$\mathbb{E}[K^*(t)] = k_0 e^{(\mu-\rho)t} \text{ and } \mathbb{V}[K^*(t)] = k_0^2\, e^{2(\mu-\rho)t}\left(e^{\sigma^2 t} - 1\right).$$

The following statistic dynamics are possible: first, if $\mu > \rho$ the economy follows a stochastic balanced growth path (SBGP); second, if $\mu = \rho$, there is rarefaction with a stationary mean, $\lim_{t\to\infty} \mathbb{V}[K^*(t)] = \infty$, and $\mathbb{E}[K(t)] = k_0$, for each $t \in T$; or the economy collapses to a deterministic steady state $\lim_{t\to\infty} K(t) = 0$ if $\mu < \rho$.

19.2.2 Volatility Coefficient Depending on the Control Variable

Proposition 19.4 (SPMP for the general SOCP) *Let* $\left(Y^*(t), U^*(t)\right)_{t\in T}$ *be the solution to the SOCP problem of maximizing the functional* (19.3a) *under the constraints* (19.3a)–(19.3b). *Then, there are four processes* $\left(P_1(t), Q_1(t)\right)_{t\in T}$ *and* $\left(P_2(t), Q_2(t)\right)_{t\in T}$, *and two Hamiltonian functions that, for every realization of those processes, are defined by*

$$H(y, u, p_1, q_1) = f(y, u) + p_1\, g(y, u) + q_1\, \sigma(y, u). \tag{19.19}$$

and

$$\mathcal{H}(y, u, p_1, q_1, p_2) = G(t, y, u, p_1, p_2) + \\ +\sigma(y, u, t)\left(q_1 - p_2\, \sigma(t, y^*, u^*)\right), \tag{19.20}$$

where $G(y, u, p_1, p_2) = f(y, u) + p_1\, g(y, u) + p_2\, \frac{1}{2}\sigma(y, u)^2$, *such that the processes* $\big(Y^*(t), U^*(t), P_1(t), Q_1(t), P_2(t), Q_2(t)\big)_{t\in T}$ *satisfy the following conditions: the first-order adjoint equation and the transversality condition*

$$\begin{aligned} dP_1(t) &= \Big(\rho\, P_1(t) - \frac{\partial}{\partial y} H(Y^*(t), U^*(t), P_1(t), Q_1(t))\Big)\, dt + \\ &\quad + Q_1(t)\, dW(t), && (19.21a)\\ \lim_{t\to\infty} e^{-\rho t} P_1(t) &= 0, && (19.21b) \end{aligned}$$

the second-order adjoint equation and the transversality condition

$$\begin{aligned} dP_2(t) &= \Big\{\rho\, P_2(t) - \frac{\partial^2}{\partial y^2} H(Y^*(t), U^*(t), P_1(t), Q_1(t)) - \\ &\quad - 2\, P_2(t)\, \frac{\partial}{\partial y} g(Y^*(t), U^*(t)) - P_2(t)\Big(\frac{\partial}{\partial y}\sigma(Y^*(t), U^*(t))\Big)^2 - \\ &\quad - 2\, Q_2(t) \frac{\partial}{\partial y}\sigma(Y^*(t), U^*(t))\Big\}\, dt + Q_2(t)\, dW(t) && (19.22a)\\ \lim_{t\to\infty} e^{-\rho t} P_2(t) &= 0, && (19.22b) \end{aligned}$$

the optimality condition for every realization of the optimal control, $U^*(t) = u^*$,

$$\mathcal{H}(t, y^*, u^*, p_1, q_1, p_2) = \max_u \mathcal{H}(t, y^*, u, p_1, q_1, p_2), \qquad (19.23)$$

and the admissibility conditions

$$\begin{aligned} dY^*(t) &= g(Y^*(t), U^*(t))\, dt + \sigma(Y^*(t), U^*(t))\, dW(t),\ t \in T, && (19.24a)\\ Y^*(0) &= y_0. && (19.24b) \end{aligned}$$

Proof See Yong and Zhou (1999, pp. 123–137).

Example 19.5 (*The* Merton (1971) *model in the SPMP version*) Consider again the problem of maximizing the intertemporal utility functional (19.6) subject to the stochastic differential equation (19.5), as in the problem in Example 19.1 but simplifying it by assuming a logarithmic utility function. Therefore, the problem is

$$\max_{C,w} \mathbb{E}_0\Big[\int_0^\infty \ln(C(t)) e^{-\rho t} dt\Big],$$

subject to

$$dN(t) = \Big[\big(r + (\mu - r)w(t)\big)\, N(t) - C(t)\Big]\, dt + \sigma\, w(t)\, N(t)\, dW(t),\ t \in T,$$

$$N(0) = n_0,$$

where the state variable, N, is financial net wealth. Note that both the drift and the volatility of the net wealth dynamics depend on the share of risky assets in net wealth, w. The Hamiltonian function (19.19) takes the form

$$H(n, c, w, p_1, q_1) = \ln(c) + p_1 \Big((r + (\mu - r) w)\, n - c \Big) + q_1\, \sigma\, w\, n,$$

implying that the first-order adjoint equation is the BSDE

$$\begin{aligned} dP_1(t) &= \Big((\rho - r + (\mu - r)\, w(t))\, P_1(t) + \sigma\, w(t)\, Q_1(t) \Big)\, dt + \\ &\quad + Q_1(t)\, dW(t), \qquad (19.26a)\\ \lim_{t\to\infty} &P_1(t) e^{-\rho t} = 0. \qquad (19.26b) \end{aligned}$$

The second-order adjoint equation is the BSDE

$$\begin{aligned} dP_2(t) &= \Big\{ \rho\, P_2(t) - 2\Big(r + (\mu - r) w(t)\Big)\, P_2(t) - (\sigma\, w(t))^2\, P_2(t) - \\ &\quad - 2\,\sigma\, w(t)\, Q_2(t) \Big\}\, dt + Q_2(t)\, dW(t), \\ \lim_{t\to\infty} &e^{-\rho t}\, P_2(t) = 0. \end{aligned}$$

To find the optimal controls, we write the generalized Hamiltonian

$$\mathcal{H}(n, c, w) = G(n, c, w, p_1, p_2) + \sigma\, w\, n\, \big(q_1 - p_2\, \sigma\, w^*\, n^* \big),$$

where

$$G(n, c, w, p_1, p_2) = \ln(c) + p_1\, \big((r + (\mu - r)\, w)\, n - c \big) + p_2\, \frac{\sigma^2}{2} w^2 n^2.$$

The optimal realizations of the control variables maximize $\mathcal{H}(n, c, w)$. Therefore,

$$C^*(t) = P_1(t)^{-1}, \ \text{ for each } t \in T, \qquad (19.27)$$

and the condition

$$\begin{aligned} P_1(t)\, (\mu - r)\, N^*(t) + \sigma^2\, w^*(t)\, N^*(t)^2\, P_2(t) + \sigma\, N^*(t)\, \big(Q_1(t) - \\ - \sigma\, w^*(t)\, N^*(t)\, P_2(t) \big) = 0, \end{aligned}$$

is equivalent to

$$Q_1(t) = -P_1(t) \left(\frac{\mu - r}{\sigma} \right).$$

Substituting in Eq. (19.26a) yields

$$dP_1(t) = -P_1(t)\left(rdt + \left(\frac{\mu - r}{\sigma}\right)dW(t)\right).$$

Observe that the structure of the model simplifies our problem by implying that the second-order shadow value of volatility functions P_2 and Q_2 have no effect on the first-order shadow value functions associated with the drift component P_1 and Q_1, which simplifies the solution.

Applying the Itô's formula to consumption (19.27), and using this expression for the adjoint variable Q_2, we have again a BSDE for consumption

$$\begin{aligned} dC(t) &= -\frac{C(t)}{P_1(t)}dP_1(t) + \frac{C(t)}{P_1^2(t)}(dP_1(t))^2 = \\ &= C(t)\left\{\left(r - \rho + \left(\frac{\mu - r}{\sigma}\right)^2\right)dt + \left(\frac{\mu - r}{\sigma}\right)dW(t)\right\}. \end{aligned}$$

Again, conjecturing that consumption is a linear function of net wealth $C = \xi N$, which, if true, would allow us to obtain the optimal portfolio composition w^*. From Itô's formula we obtain again a FSDE for consumption

$$dC(t) = \xi\, dN(t) = C(t)\left(\left(r + (\mu - r)\, w(t) - \xi\right)dt + \sigma\, w(t)\, dW(t)\right)$$

To match the two SDE for consumption, we find that w^* should be time-independent, and furthermore, solving

$$r - \rho + \left(\frac{\mu - r}{\sigma}\right)^2 = r + (\mu - r)w - \xi, \text{ and } \frac{\mu - r}{\sigma} = \sigma\, w,$$

we find $\xi = \rho$ and $w = \frac{\mu - r}{\sigma^2}$. Therefore, optimal consumption is again linearly related to net wealth and the portfolio share is constant,

$$C^*(t) = \rho\, N^*(t) \tag{19.28}$$

$$w^*(t) = \frac{\mu - r}{\sigma^2}. \tag{19.29}$$

Substituting in the budget constraint, we find again that the optimal net wealth follows a geometric Brownian motion process,

$$\frac{dN^*(t)}{N^*(t)} = \left(r - \rho + \left(\frac{\mu - r}{\sigma}\right)^2\right)dt + \left(\frac{\mu - r}{\sigma}\right)dW(t), \text{ for each } t \in T.$$

The properties of the solution to this problem have already been studied in Example 19.2.

19.3 Envelope Theorem Interpretation

Consider again the discounted infinite horizon problem (19.3) of Sect. 19.1.2. The SDPP establishes that a necessary condition for the optimum is that the value function satisfies the HJB equation at the optimum, for any realization of the process $Y^*(t) = y$

$$\rho\, v(y) = f(y, u^*) + g(y, u^*) v'(y) + \frac{1}{2}\sigma(y, u^*)^2\, v''(y), \tag{19.30}$$

where, if function $v(\cdot)$ is sufficiently smooth, $u^* = U^*(y)$ is the policy function. We next show that an application of a version of the **stochastic envelope theorem** allows us to obtain a relationship between the SDPP and SPMP, which provides an interpretation of its co-state variables.

Proposition 19.5 (SPMP for the general SOCP) *Consider the optimal control problem (19.3), and the associated HJB equation at the optimum* (19.30). *Assume that $v(\cdot) \in C^4(\mathbb{R})$. Then, if $p_1 = v'(y)$, $q_1 = \sigma(y, u^*)\, v''(y)$, $p_2 = v''(y)$, $q_2 = \sigma(y, u^*)\, v'''(y)$, we obtain the adjoint equations of the stochastic Pontriyagin's maximum principle in Proposition 19.4.*

Proof First, observe that the optimum condition, using the SDPP approach is

$$f_u^*(y) + g_u^*(y)\, v'(y) + \sigma^*(y)\, \sigma_u^*(y)\, v''(y) = 0, \text{ for each } Y^* = y,$$

where $\sigma^*(y) = \sigma(y, U^*(y))$, $f_u^*(y) = f_u(y, U^*(y))$, and the same notation for the other partial derivatives. It also satisfies the condition

$$\begin{aligned}\Big(f_{uu}^*(y) + g_{uu}^*(y)\, v'(y) + \sigma_u^*(y)^2\, v''(y) + \sigma^*(y)\, \sigma_{uu}^*(y)\, v''(y)\Big)\, U^{*'}(y) +\\ + f_{uy}^*(y) + g_{uy}^*(y)\, v'(y) + g_u^*(y)\, v''(y) + \sigma_y^*(y)\, \sigma_u^*(y)\, v''(y) +\\ + \sigma_{uy}^*(y) \sigma^*(y)\, v''(y) + \sigma_u^*(y)\, \sigma^*(y)\, v'''(y) = 0, \text{ for each } Y^* = y.\end{aligned}$$

Introduce two processes $(P_1(t))$ and $(Q_1(t))$, and for each realization of the processes $(Y(t), U(t), P_1(t), Q_1(t)) = (y, u, p_1, q_1)$ define the Hamiltonian function

$$H(y, u, p_1, q_1) = f(y, u) + p_1\, g(y, u) + q_1\, \sigma(y, u).$$

At the optimum, let $P_1(t) = v'(Y(t))$ and $Q_1(t) = \sigma(Y(t), U^*(t))\, v''(Y(t))$. From the Itô's formula, we have

$$
\begin{aligned}
d P_1(t) &= \Big(g(Y(t), U^*(t)) v''(Y(t)) + \frac{1}{2}\sigma(Y(t), U^*(t))^2\, v'''(Y(t))\Big)\, dt + \\
&+ \sigma(Y(t), U^*(t)) v''(Y(t))\, dW(t) \\
&= \Big(g(Y(t), U^*(t)) v''(Y(t)) + \frac{1}{2}\sigma(Y(t), U^*(t))^2\, v'''(Y(t))\Big)\, dt + \\
&+ Q_1(t)\, dW(t).
\end{aligned} \tag{19.31}
$$

To find the drift as a function of the co-state variables, we use the envelope theorem. To show this, take a derivative of the HJB equation (19.30), for any optimal realization,

$$
\begin{aligned}
\rho\, v'(y) &= f_y^*(y) + g_y^*(y)\, v'(y) + g^*(y)\, v''(y) + \sigma^*(y)\, \sigma_y^*(y)\, v''(y) + \\
&+ \frac{1}{2}\,(\sigma^*(y))^2\, v'''(y) + \Big(f_u^*(y) + g_u^*(y)\, v'(y) + \sigma^*(y)\, \sigma_u^*(y)\, v''(y)\Big)\, U_y^*(y) \\
&= f_y^*(y) + g_y^*(y)\, v'(y) + g^*(y)\, v''(y) + \sigma^*(y)\, \sigma_y^*(y)\, v''(y) + \\
&+ \frac{1}{2}\,(\sigma^*(y))^2\, v'''(y) \\
&= H_y(y, u^*(y), v'(y), \sigma^*(y) v''(y)) + g^*(y)\, v''(y) + \frac{1}{2}\,(\sigma^*(y))^2\, v'''(y)
\end{aligned}
$$

Then $g^*(y)\, v''(y) + \frac{1}{2}\,(\sigma^*(y))^2\, v'''(y) = \rho\, v'(y) - H_y(y, u^*(y), v'(y), \sigma^*(y)\, v''(y))$, and substituting $v'(y) = p_1$ and $\sigma^*\, v''(y) = q_1$ in Eq. (19.31) yields the first-order adjoint equation (19.21a) for the SPMP.

Next, we need to deal with the second-order conditions, as a consequence of the nature of Itô's processes. At the optimum, let $P_2(t) = v''(Y(t))$ and $Q_2(t) = \sigma(Y(t), U^*(t)) v'''(Y(t))$. From the Itô's formula, we have

$$
\begin{aligned}
d P_2(t) &= \Big(g(Y(t), U^*(t)) v'''(Y(t)) + \frac{1}{2}\sigma(Y(t), U^*(t))^2\, v''''(Y(t))\Big)\, dt + \\
&+ \sigma(Y(t), U^*(t)) v'''(Y(t))\, dW(t) \\
&= \Big(g(Y(t), U^*(t)) v'''(Y(t)) + \frac{1}{2}\sigma(Y(t), U^*(t))^2\, v''''(Y(t))\Big)\, dt + \\
&+ Q_2(t)\, dW(t).
\end{aligned} \tag{19.32}
$$

Very tedious calculations of the derivative

$$
\begin{aligned}
&\frac{\partial}{\partial y}\Big[\rho\, v'(y) - \Big(f_y^*(y) + g_y^*(y)\, v'(y) + g^*(y)\, v''(y) + \sigma^*(y)\, \sigma_y^*(y)\, v''(y) + \\
&\frac{1}{2}\,(\sigma^*(y))^2\, v'''(y) + \Big(f_u^*(y) + g_u^*(y)\, v'(y) + \sigma^*(y)\, \sigma_u^*(y)\, v''(y)\Big)\, U_y^*(y)\Big] = 0,
\end{aligned}
$$

and using the envelope theorem, allows for finding

$$
\begin{aligned}
& g^*(y)\, v'''(y) + \frac{1}{2}\,(\sigma^*(y))^2\, v''''(y) = \rho\, v''(y) - \\
& \quad - \Big(\frac{\partial^2}{\partial y^2} H\big(y, U^*(y), v'(y), \sigma(y, U^*(y))\big)\, v''(y) + 2\frac{\partial}{\partial y} g(y, U^*(y))\, v''(y) + \\
& \quad + \Big(\frac{\partial}{\partial y}\sigma(y, U^*(y))\Big)^2 v''(y) + 2\,\frac{\partial}{\partial y}\sigma(y, U^*(y))\,\sigma(y, U^*(y))\, v'''(y)\Big).
\end{aligned}
$$

Substituting $v'(y) = p_1$, $\sigma^* v''(y) = q_1$, $v''(y) = p_2$, and $\sigma^* v'''(y) = q_2$ in Eq. (19.32) yields the second-order adjoint equation (19.21b) for the SPMP.

19.4 References

References on the SDPP approach for solving stochastic optimal control problems can be found in Fleming and Rishel (1975) and Seierstad (2009), and Pham (2009). The SPMP approach can be found in Bensoussan (1988), Peng (1990), and Yong and Zhou (1999, Chap. 3). Application to economics can be found in Malliaris and Brock (1982), Chang (2004) and Wälde (2012). More specialized references on DSGE financial macromodels in continuous time with frictions are surveyed in Brunnermeier and Sannikov (2016) and Brunnermeier et al. (2013). Applications to industrial economics can be found in Dixit and Pindyck (1994), Stokey (2009), to asset pricing in Björk (2004) and Cvitanić and Zapatero (2004), to credit risk Bielecki and Rutkowski (2004), and to management science in Sethi (2019). Kushner (2014) is a survey paper.

References

Abramowitz, M., & Stegun, I. A. (1972). *Handbook of mathematical functions*. New York: Dover.

Acemoglu, D. (2009). *Introduction to modern economic growth*. Princeton University Press.

Achdou, Y., Buera, F. J., Lasry, J.-M., Lions, P.-L., & Moll, B. (2014). Partial differential equation models in macroeconomics. Philosophical Transactions of the Royal Society, 372.

Alogoskoufis, G. (2019). *Dynamic macroeconomics*. MIT Press.

Amit, R. (1986). Petroleum reservoir exploitation: Switching from primary to secondary recovery. *Operations Research, 34*(4), 34.

Arrow, K. J., & Kurz, M. (1970). Public investment, the rate of return and optimal fiscal policy. The Johns Hopkins Press.

Atkinson, A. B. (1970). On the measurement of inequality. Journal of Economic Theory, 2.

Augeraud-Véron, E., & Boucekkine, R. (2019). Veliov (Vol. M). A review. Mathematical Modelling of Natural Phenomena: Distributed optimal control models in environmental economics.

Azariadis, C. (1993). Intertemporal macroeconomics. Blackwell's.

Bacaër, N. (2011). *A short history of mathematical population dynamics*. Springer.

Bachelier, L. (1900). Théorie de la Spéculation. *Annales Scientifiques de l'École Normale Supérieure, 17*, 21–86.

Banks, R. B. (1994). *Growth and diffusion phenomena*. Springer.

Barnett, S. (1990). *Matrices*. Methods and Applications: Oxford University Press.

Barnett, W. A., & He, Y. (2002). Stabilization policy as bifurcation selection: Would stabilization policy work if the economy really were unstable? *Macroeconomic Dynamics, 6*, 713–747.

Barro, R. J., & Sala-i-Martin, X. (2004). Economic growth (2nd e.). MIT Press.

Basak, S., & Cuoco, D. (1998). An equilibrium model with restricted stock market participation. *Review of Financial Studies, 11*(2), 309–341.

Basov, S. (2007). Partial Differential Equations in Economics and Finance. Nova Science Publishers.

Beardmore, R. E., Laister, R., & Peplow, A. (2004). Trajectories of a DAE near a pseudo-equilibrium. *Nonlinearity, 17*, 253–279.

Beckmann, M. J. (1968). Dynamic programming of economic decisions. Ökonometrie und Unternehmensforschung Econometrics and Operations Research? (Vol. 9, 1st ed.). Springer.

Beckmann, M. J., & Ryder, H. E. (1969). Simultaneous price and quantity adjustment in a single market. *Econometrica, 37*(3), 470–484.

Bellman, R. (1957). *Dynamic programming*. Princeton University Press.

Bendixson, I. (1901). Sur les courbes définies par des équations différentielles. *Acta Mathematica, 24*(1), 1871–2509.

P. B. Brito, *Economic Dynamics and Distributions*, Dynamic Modeling and Econometrics in Economics and Finance 32, https://doi.org/10.1007/978-3-031-94717-9

Benhabib, J., & Farmer, R. (1994). Indeterminacy and increasing returns. *Journal of Economic Theory, 63*, 19–41.

Benhabib, J., & Farmer, R. (1996). Indeterminacy and sector specific externalities. *Journal of Monetary Economics, 37*, 421–43.

Benhabib, J., Schmitt-Grohé, S., & Uribe, M. (2001). The Perils of Taylor rules. *Journal of Economic Theory, 96*, 40–69.

Bensoussan, A. (1988). Perturbation methods in optimal control. Wiley/Gauthier-Villars.

Bensoussan, A., Frehse, J., & Yam, P. (2013). Mean field games and mean field type control theory. Springer briefs in mathematics (1st ed.). Springer.

Benzoni-Gavage, S., & Serre, D. (2007). *Multi-dimensional hyperbolic partial differential equations: First-order systems and applications*. Oxford mathematical monographs: Oxford University Press.

Bielecki, T. R., & Rutkowski, M. (2004). *Credit risk: Modeling, valuation and hedging*. Springer.

Björk, T. (2004). Arbitrage theory in continuous time. Finance (2nd ed.). Oxford University Press.

Black, F., & Scholes, M. (1973). The pricing of options and corporate liabilities. *Journal of Political Economy, 81*(3), 637–659.

Blanchard, O. (2000). What do we know about macroeconomics that fisher and Wicksell did not? *The Quarterly Journal of Economics, 115*(4), 1375–1409.

Blanchard, O., & Fischer, S. (1989). *Lectures in macroeconomics*. MIT Press.

Blanchard, O., & Khan, C. M. (1980). The solution of linear difference models under rational expectations. *Econometrica, 48*(5), 1305–1311.

Blinder, A., & Solow, R. (1973). Does fiscal policy matter? *Journal of Public Economics, 2*(4), 319–337.

Blinder, A. S., & Solow, R. M. (1976). Does fiscal policy matter?: A correction. *Journal of Public Economics, 5*(1–2), 183–184.

Boucekkine, R., Camacho, C., & Fabbri, G. (2013). Spatial dynamics and convergence: The spatial AK model. *Journal of Economic Theory, 148*(6), 2719–2736.

Boucekkine, R., Camacho, C., & Zou, B. (2009). Bridging the gap between growth theory and the new economic geography: The spatial Ramsey model. *Macroeconomic Dynamics, 13*, 20–45.

Boucekkine, R., Fabbri, G., Federico, S., & Gozzi, F. (2019). Growth and agglomeration in the heterogeneous space: A generalized AK approach. *Journal of Economic Geography, 19*(6), 1287–1318.

Boucekkine, R., Saglan, C., & Vallée, T. (2004). Technology adoption under embodiment: A two-stage optimal control approach. *Macroeconomic Dynamics, 8*(2), 250–271.

Bourguignon, F. (1974). A particular class of continuous-time stochastic growth models. *Journal of Economic Theory, 9*, 141–58.

Bourguignon, F. (1979). Decomposable income inequality measures. *Econometrica, 47*(4), 901–920.

Brito, P. (1999). Local dynamics for optimal control problems of three-dimensional ODE systems. *Annals of Operations Research, 89*, 195–214.

Brito, P. (2000). Locating local bifurcations in optimal control problems of 4-dimensional ODE systems. Econometric society world congress 2000 contributed papers 0506. Econometric Society.

Brito, P. (2004). The dynamics of growth and distribution in a spatially heterogeneous world. Working Papers of the Department of Economics, ISEG-UTL. http://ideas.repec.org/p/ise/isegwp/wp142004.html.

Brito, P. (2006). *A spatial Solow mdel with unbounded growth*. In Ensaios de Homenagem a António Simões Lopes: ISEG/UTL.

Brito, P. (2017). Government debt, fiscal rules and singular growth dynamics. In B. Bökemeier & A. Greiner (Eds.), *Inequality and finance in macrodynamics* (pp. 43–74). Springer International Publishing.

Brito, P., Costa, L., & Dixon, H. (2013). Non-smooth dynamics and multiple equilibria in a Cournot-Ramsey model with endogenous markups. *Journal of Economic Dynamics and Control, 37*(11), 2287–2306.

Brito, P., Costa, L., & Dixon, H. (2017). From sunspots to Black Holes: Singular dynamics in macroeconomic models. In Nishimura, K., & all, (Eds.), Sunspots and non-linear dynamics, chapter 3. Springer.

Brito, P., & Dixon, H. (2013). Fiscal policy, entry and capital accumulation: Hump-shaped responses. *Journal of Economic Dynamics and Control, 37*(10), 2123–2155.

Brito, P., Fabião, F., & StAubyn, A. (2008). Euler, Lambert, and the Lambert W function today. The Mathematical Scientist, 33(2).

Brito, P., & Venditti, A. (2010). Local and global indeterminacy in two sector models of endogenous growth. *Journal of Mathematical Economics, 46*(5), 893–911.

Brito, P. B. (2022). The dynamics of growth and distribution in a spatially heterogeneous world. *Portuguese Economic Journal, 21*(3), 311–350.

Brito, P. B. (2024). Inequality aversion, social mobility, social insurance, and optimal income growth and distribution.

Brock, W., & Xepapadeas, A. (2008). Diffusion-induced instability and pattern formation in infinite horizon recursive optimal control. *Journal of Economic Dynamics and Control, 32*, 2745–2787.

Brock, W. A., & Malliaris, A. G. (1989). *Differential equations, stability and chaos in dynamic economics*. North-Holland.

Brock, W. A., & Mirman, L. (1972). Optimal economic growth and uncertainty: The discounted case. *Journal of Economic Theory, 4*, 479–513.

Brock, W. A., & Scheinkman, J. A. (1976). Global asymptotic stability of optimal control systems with applications to the theory of economic growth. *Journal of Economic Theory, 12*, 164–90.

Brunnermeier, M., Merkel, S., & Sannikov, Y. (2020). Lecture notes on macro, money and finance: A heterogeneous-agent continuous time approach. Princeton University.

Brunnermeier, M. K., Eisenbach, T., & Sannikov, Y. (2013). Macroeconomics with financial frictions: A survey. In Advances in economics and econometrics, tenth world congress of the econometric society. Cambridge University Press.

Brunnermeier, M. K., & Sannikov, Y. (2014). A macroeconomic model with a financial sector. *American Economic Review, 104*, 379–421.

Brunnermeier, M. K., & Sannikov, Y. (2016). Macro, money and finance: A continuous time approach. NBER Working Papers 22343, National Bureau of Economic Research Inc.

Bucci, A., & Ushchev, P. (2016). Specialization versus competition: An anatomy of increasing returns to scale. HSE Working papers WP BRP 134/EC/2016, National Research University Higher School of Economics.

Buiter, W. H. (1984). Saddlepoint problems in continuous time rational expectations models: A general method and some macroeconomic examples. *Econometrica, 52*(3), 665–680.

Bullen, P. (2003). Handbook of means and their inequalities. Number 560 in mathematics and its applications (2nd ed.). Springer.

Burgers, J. (1948). A mathematical model illustrating the theory of turbulence. In R. Von Mises & T. Von Kármán (Eds.), *Advances in applied mechanics, volume 1 of Advances in applied mechanics* (pp. 171–199). Elsevier.

Burmeister, E., & Dobell, A. R. (1970). *Mathematical Theories of Economic Growth*. Macmillan.

Burr, I. W. (1942). Cumulative frequency functions. *The Annals of Mathematical Statistics, 13*(2), 215–232.

Butkovskiy, G. (1969). *Distributed control systems*. American Elsevier.

Cai, G.-Q., & Zhu, W.-Q. (2017). *Elements of stochastic dynamics*. World Scientific.

Canada, A., Drabek, P., & Fonda, A. (2004). Handbook of differential equations. Ordinary differential equations (Vol. 1, 1st ed.). North Holland.

Cardin, P. T., da Silva, P. R., & Teixeira, M. A. (2012). Implicit differential equations with impasse singularities and singular perturbation problems. *Israel Journal of Mathematics, 189*, 307–322.

Carlson, D. A., Haurie, A. B., & Leizarowitz, A. (1996). Infinite horizon optimal control (2nd ed.). Springer.
Cass, D. (1965). Optimum growth in an aggregative model of capital accumulation. *Review of Economic Studies, 32*, 233–40.
Cass, D., & Stiglitz, J. E. (1970). The structure of investor preferences and asset returns, and separability in portfolio allocation: A contribution to the pure theory of mutual funds. *Journal of Economic Theory, 2*(2), 122–160.
Chang, F.-R. (2004). *Stochastic optimization in continuous time*. Cambridge University Press.
Chechkin, G. A., & Goritsky, A. Y. (2009). S. N. Kruzhkov's lectures on first-order quasilinear PDEs. In Emmrich, E., & Wittbold, P., editors, De Gruyter proceedings in mathematics, analytical and numerical aspects of partial differential equations (pp. 1–68). De Gruyter. traduit du russe par B. Andreianov.
Chiang, A. (1992). *Elements of dynamic optimization*. McGraw-Hill.
Christ, C. F. (1968). A simple macroeconomic model with a government budget restraint. *Journal of Political Economy, 76*, 155–170.
Chua, L. O., Shilnikov, L. P., Shilnikov, A. L., & Turaev, D. V. (2001). Methods of qualitative theory in nonlinear dynamics part 2, volume 5 of A (illustrated ed.). World Scientific Publishing Company.
Clarke, F. (2013). Functional analysis, calculus of variations and optimal control. Graduate texts in mathematics (2013 ed.). Springer.
Colombo, A., di Bernardo, M., Hogan, S., & Jeffrey, M. (2012). Bifurcations of piecewise smooth flows: Perspectives, methodologies and open problems. Physica D: Nonlinear Phenomena, 241(22), 1845–1860. Dynamics and bifurcations of nonsmooth systems.
Corless, R. M., Gonnet, G., Hare, D. E. G., Jeffrey, D. J., & Knuth, D. E. (1996). On the Lambert W Function. *Advances in Computational Mathematics, 5*, 329–59.
Cover, T. M., & Thomas, J. A. (2006). Elements of information theory (2nd ed.). Wiley-Interscience.
Cowell, F. A. (1980). On the structure of additive inequality measures. *Review of Economic Studies, 47*, 521–531.
Cross, M., & Greenside, H. (2009). Pattern formation and dynamics in nonequilibrium systems.
Cvitanić, J., & Zapatero, F. (2004). *Introduction to the economics and mathematics of financial markets*. MIT Press.
Cvitanić, J., & Zhang, J. (2013). Contract theory in continuous-time models. Springer finance (1st ed.). Springer.
Dafermos, C. M. (2000). Hyperbolic conservation laws in continuum physics, volume 325 of Grundlehren der mathematischen Wissenschaften. Springer.
Dalton, H. (1920). The measurement of the inequality of incomes. *The Economic Journal, 30*(119), 348–361.
Dana, R.-A., & Jeanblanc, M. (2003). *Financial markets in continuous time*. Springer.
Dasgupta, P., Sen, A. K., & Starrett, D. (1973). Notes on the measurement of inequality. *Journal of Economic Theory, 6*, 180–187.
de la Fuente, A. (2000). *Mathematical methods and models for economists*. Cambridge University Press.
Dechert, W. D., & Nishimura, K. (1983). A complete characterization of optimal growth paths in an aggregate model with a non-concave production function. *Journal of Economic Theory, 31*, 322–54.
Derzko, N. A., Sethi, S. P., & Thompson, G. L. (1984). Necessary and sufficient conditions for optimal control of quasilinear partial differential equations systems. *Journal of Optimization Theory and Applications, 43*(1), 89–101.
Dhingra, S., & Morrow, J. (2019). Monopolistic competition and optimum product diversity under firm heterogeneity. *Journal of Political Economy, 127*(1), 196–232.
di Bernardo, M., Budd, C. J., Champneys, A. R., & Kowalczyk, P. (2008a). Piecewise-smooth dynamical systems. Theory and applications, volume 163 of Applied mathematical sciences. Springer.

di Bernardo, M., Budd, C. J., Champneys, A. R., Kowalczyk, P., Nordmark, A. B., Tost, G. O., & Piiroinen, P. T. (2008). Bifurcations in nonsmooth dynamical systems. *SIAM Review, 50*(4), 629–701.
Diamond, P. A. (1998). Optimal income taxation: An example with a U-shaped pattern of optimal marginal tax rates. *The American Economic Review, 88*(1), 83–95.
Dixit, A. K., & Pindyck, R. S. (1994). *Investment under Uncertainty*. Princeton University Press.
Dixit, A. K., & Stiglitz, J. E. (1977). Monopolistic competition and optimum product diversity. *American Economic Review, 67*(3), 297–308.
Doetsch, G. (1974). *Introduction to the theory and applications of the laplace transform*. Springer.
Dornbusch, R. (1976). Expectations and exchange rate dynamics. *Journal of Political Economy, 84*, 1161–76.
Drabek, P., & Milota, J. (2013). Methods of nonlinear analysis: applications to differential equations. Birkhauser advanced texts (1st ed.). Birkhäuser.
Duffie, D. (1996). Dynamic asset pricing theory (2nd ed.). Princeton University Press.
Duffie, D., & Singleton, K. J. (2003). Credit risk, pricing, measurement and management. Princeton series in finance. Princeton University Press.
Dumas, B., & Luciano, E. (2017). *The economics of continuous-time finance*. The MIT Press.
Ekeland, I. (2010). From ramsey to rené thom: A classical problem in the calculus of variations leading to an implicit differential equation. *Discrete and Continuous Dynamical Systems, 28*(3), 1101–1119.
Epstein, L. G., & Zin, S. E. (1989). Substitution, risk aversion, and the temporal behavior of consumption and asset returns: A theoretical framework. *Econometrica, 57*, 937–969.
Evans, L. C. (1998). Partial differential equations, volume 19 of Graduate series in mathematics. American Mathematical Society.
Evans, L. C. (2010). Partial differential equations, volume 19 of Graduate series in mathematics (2nd ed.). American Mathematical Society.
Fabbri, G. (2016). Geographical structure and convergence: A note on geometry in spatial growth models. Journal of Economic Theory, 162(C), 114–136.
Fattorini, H. O. (1999). *Infinite dimensional optimization and control theory*. Cambridge
Feigin, M. I. (1970). Doubling of the oscillations period with c-bifurcations in piecewise continous systems. *PMM, 34*, 861–869.
Fenichel, N. (1979). Geometric singular perturbation theory for ordinary differential equations. *Journal of Difference Equations and Applications, 31*, 53–98.
Fields, G. (2021). Exploring the concepts of social mobility. In V. Iversen, A. Krihna, & K. Sen (Eds.), *Social mobility in developing countries: concepts, methods, and determinants, chapter 3* (pp. 54–74). Oxford University Pres.
Filippov, A. F. (1988). *Differential equations with discontinuous right-hand sides*. Kluwer.
Fleming, W. H., & Rishel, R. W. (1975). *Deterministic and stochastic optimal control*. Springer.
Forster, B. A. (1975). Optimal pollution control with a nonconstant exponential rate of decay. *Journal of Environmental Economics and Management, 2*(1), 1–6.
Fourier, M. (1822). *Théorie Analytique de la Chaleur*. Pére et Fils, Paris: Firmin Didot.
Frank, T. (2005). Nonlinear Fokker-Planck equations: Fundamentals and applications. Springer series in synergetics, Springer complexity (1st ed.). Springer.
Frankel, M. (1962). The production function in allocation and growth: A synthesis. *American Economic Review, 52*(5), 996–1022.
Fujita, M., & Thisse, J.-F. (2002). *Economics of agglomeration*. Cambridge
Gabaix, X., Lasry, J., Lions, P., & Moll, B. (2016). The dynamics of inequality. *Econometrica, 84*, 2071–2111.
Galaktionov, V. A., & Vázquez, J. L. (2004). A stability technique for evolution partial differential equations: A dynamical systems approach. Progress in nonlinear differential equations and their applications 56. Birkhäuser.
Gandolfo, G. (1997). *Economic dynamics*. Springer.

Gardiner, C. (2009). Stochastic methods. Handbook of stochastic methods for physics, chemistry, and the natural sciences. Springer series in synergetics (4th ed.). Springer.

Gazizov, R. K., & Ibragimov, N. H. (1998). Lie symmetry analysis of differential equations in finance. *Nonlinear Dynamics, 17*(4), 387–407.

Gel'fand, I. M., & Fomin, S. V. (1963). *Calculus of variations*. Dover.

Gil, P. M., Brito, P., & Afonso, Ó. (2013). Growth and firm dynamics with horizontal and vertical R&D. *Macroeconomic Dynamics, 17*(7), 1438–1466.

Gilding, B. H., & Kersner, R. (2004). *Travelling wavesin nonlinear diffusion-convection reaction*. Springer.

Goldstine, H. H. (1980). A history of the calculus of variations from the 17th through the 19th century. Studies in the history of mathematics and physical sciences (Vol. 5, 1st ed.). Springer.

Gollier, C. (2001). *The economics of risk and time*. MIT Press.

Golubitsky, M., & Schaeffer, D. G. (1985). *Singularities and groups in bifurcations theory (Vol.* I): Springer.

Gomes, D. A., & Saúde, J. (2014). Mean field games models. A brief survey. Dynamic Games and Applications, 4(2), 110–154.

Gompertz, B. (1825). On the nature of the function expreive of the law of mortality. *Philosophical Transactions, 27*, 213–221.

Grandmont, J.-M. (1988). Nonlinear difference equations bifurcations and chaos: An introduction. CEPREMAP Working Paper, No. 8811.

Grandmont, J.-M. (2008). Nonlinear difference equations, bifurcations and chaos: An introduction. *Research in Economics, 62*(3), 122–177.

Grass, D., Caulkins, J. P., Feichtinger, G., Tragler, G., & Behrens, D. A. (2008). Optimal control of nonlinear processes. With applications in drugs, corruption, and terror. Springer.

Grattan-Guinness, I. (2009). On Jevons's handling of differential equations in the context of his principle of exchange, and the early reactions. *History of Political Economy, 41*(2), 297–309.

Grobman, D. (1959). Homeomorphisms of systems of differential equations. *Dokl. Akad. Nauk SSSR, 129*, 880–881.

Guckenheimer, J., & Holmes, P. (1990). Nonlinear oscillations and bifurcations of vector fields (2nd ed.). Springer.

Hale, J., & Koçak, H. (1991). *Dynamics and bifurcations*. Springer.

Halkin, H. (1974). Necessary conditions for optimal control problems with infinite horizon. *Econometrica, 42*, 267–73.

Hartl, R. F., Sethi, S. P., & Vickson, R. G. (1995). A survey of the maximum principles for optimal control problems with state constraints. SIAM Review, 37.

Hartman, P. (1964). *Ordinary differential equations*. Wiley.

He, Y., & Barnett, W. A. (2006). Singularity bifurcations. *Journal of Macroeconomics, 28*, 5–22.

Hirsch, M., & Smale, S. (1974). *Differential equations, dynamical systems, and linear algebra*. Academic.

Honkapohja, S., & Ito, T. (1983). Stability with regime switching. *Journal of Economic Theory, 29*, 22–48.

Hritonenko, N., & Yatsenko, Y. (2013). Mathematical modeling in economics, ecology and the environment. Springer optimization and its applications (Vol. 88). Springer, US.

Hubbard, J. H. (1994). What it means to understand a differential equation. *The College Mathematics Journal, 25*(5), 372–384.

Hubbard, J. H., & West, B. H. (1991). Differential equations: A dynamical systems approach: Ordinary differential equations. Number 5 in texts in applied mathematics (1st ed.). Springer.

Inada, K.-I. (1964). Economic growth and factor substitution. *International Economic Review, 5*(3), 318–327.

Infante, E. F., & Stein, J. (1976). Does fiscal policy matter? *Journal of Monetary Economics, 2*(4), 473–500.

Intriligator, M. D. (2002). Mathematical optimization and economic theory. Classics in applied mathematics. Society for industrial mathematics (1st ed.).

Isard, W., & Liossatos, P. (1979). *Spatial dynamics and optimal space-time development*. North-Holland.

Isohätälä, J., Klimenko, N., & Milne, A. (2016). Post-crisis macrofinancial modeling: Continuous time approaches. In E. Haven, P. Molyneux, J. O. S. Wilson, S. Fedotov, & M. Duygun (Eds.), *The handbook of post crisis financial modeling* (pp. 235–282). Palgrave Macmillan UK.

Itô, K. (1951). On stochastic differential equations. *Memoirs of the American Mathematical Society, 4*, 289–302.

Jaynes, E. T. (1957a). Information theory and statistical mechanics. *Physical Review, 106*, 620–630.

Jaynes, E. T. (1957b). Information theory and statistical mechanics. *Physical Review, 108*, 171–190.

Jaynes, E. T. (2003). Probability theory: The logic of science. Cambridge University Press.

Jeffrey, M. R., & Hogan, S. J. (2011). The geometry of generic sliding bifurcations. *SIAM Review, 53*(3), 505–525.

Jeffrey, M. R., Seidman, T. I., Teixeira, M. A., & Utkin, V. I. (2022). Into higher dimensions for nonsmooth dynamical systems. *Physica D, 434*, 1–13.

Jevons, W. S. (1871 (1907)). The theory of political economy (1.ed 1871, 2nd ed. 1879, 3rd ed. 1888, 4th ed. 1911. 5th ed. 1907). Macmillan and Co., Available at https://socialsciences.mcmaster.ca/econ/ugcm/3ll3/jevons/TheoryPoliticalEconomy.pdf. <!– Wrong Number: [1]3 –>

Johnson, N. L., Kotz, S., & Balakrishnan, N., (Eds.), (1994). Continuous univariate distributions, volume 1 of Wiley series in probability and statistics (2nd ed.). Wiley-Interscience.

Judge, G. G., & Mittelhammer, R. C. (2012). *An information theoretic approach to econometrics*. Cambridge University Press.

Kamien, M. I., & Schwartz, N. L. (1991). Dynamic optimization (2nd ed.). North-Holland.

Kamihigashi, T. (2001). Necessity of transversality conditions for infinite horizon problems. *Econometrica, 69*(4), 995–1012.

Kammler, D. W. (2000). *A first course in fourier analysis*. Prentice-Hall.

Karatzas, I., & Shreve, S. (1991). Brownian motion and stochastic calculus (2nd ed.). Springer.

Karoui, N. E., Peng, S., & Quenez, M. C. (1997). Backward stochastic differential equationsin finance. *Mathematical Finance, 7*(1), 1–71.

Keeler, E., Spence, A., & Zeckhauser, R. (1972). The optimal control of pollution. *Journal of Economic Theory, 4*(1), 19–34.

Keyfitz, N., & Caswell, H. (2005). Applied mathematical demography. Statistics for biology and health (3rd ed.). Springer.

King, R. B. (1996). Beyond the quartic equation. Birkäuser.

Kleiber, C., & Kotz, S. (2003). Statistical size distributions in economics and actuarial sciences. Wiley series in probability and statistics. Wiley.

Kloeden, P. E., & Platen, E. (1992). Numerical solutions of stochastic differential equations, volume 23 of Applications of mathematics. Springer.

Kolm, S.-C. (1969). The optimal production of social justice. In J. Margolis & H. Guitton (Eds.), *Public economics* (pp. 145–200). Macmillan.

Kolmogorov, A. N. (1931). Über die analytischen methoden in der wahrscheinlichkeitsrechnung. Mathematische Annalen, 104:415–458. English version: "On analytical methods in probability theory", in A. N. Shiryayev, ed. (1992). Selected Works of A. N. Kolmogorov, volume II, Probability theory and mathematical statistics (pp. 62–108).

Kolmogorov, A. N. (1933). *Grundbegriffe der Wahrscheinlichkeittsrechnung*. Foundations of the theory of probabikity: Springer.

Koopmans, T. (1965). *On the concept of optimal economic growth*. Pontificiae Acad: Sci. In The econometric approach to development planning.

Kot, M. (2001). *Elements of Mathematical Biology*. Cambridge

Kotz, S., & Vicari, D. (2005). Survey of developments in the theory of continuous skewed distributions. Metron - International Journal of Statistics, 0(2), 225–261.

Kremer, M. (1993). The O-ring theory of economic development. *The Quarterly Journal of Economics, 108*(3), 551–575.

Krugman, P. (1980). Scale economies, product differentiation, and the pattern of trade. *The American Economic Review, 70*(5), 950–959.

Kuehn, C. (2015). Multiple time scale dynamics. Applied mathematical sciences (Vol. 191, 1st ed.). Springer.

Kuehn, C. (2019). *PDE dynamics. An Introduction*. SIAM: Philadelphia.

Kushner, H. J. (2014). A partial history of the early development of continuous-time nonlinear stochastic systems theory. *Automatica, 50*, 303–334.

Kuznetsov, Y. A. (2005). Elements of applied bifurcation theory (3rd ed.). Springer.

Kuznetsov, Y. A. (2023). Elements of applied bifurcation theory. Applied mathematical sciences (Vol. 112, 4th ed.). Springer.

Kuznetsov, Y. A., Rinaldi, S., & Gragnani, A. (2003). One-parameter bifurcations in planar Filippov systems. *International Journal of Bifurcation & Chaos in Applied Sciences & Engineering, 13*(8), 2157–2188.

Köhler, M. (1980). Pointwise maximum principle for convex optimal control problems with mixed control-phase variable inequality constraints. *Journal of Optimization Theory and Applications, 30*(2), 30.

Ladyženskaja, O. A., Solonnikov, V. A., & Ural'tseva, N. N. (1968). Linear and quasi-linear equations of parabolic type, volume 23 of Translations of mathematical monographs (reprint ed.). American Mathematical Society.

Lee, C., Famoye, F., & Alzaatreh, A. Y. (2013). Methods for generating families of univariate continuous distributions in the recent decades. *WIREs Computational Statistics, 5*(3), 219–238.

Leine, R., & Nijmeijer, H. (2004). Dynamics and bifurcations of non-smooth mechanical systems. Lecture notes in applied and computational mechanics (Vol. 18). Springer.

Leine, R. I., & van de Wouw, N. (2008). *Stability and convergence of mechanical systems with unilateral constraints*. Springer.

Leland, H. E. (1994). Corporate debt value, bond covenants, and optimal capital structure. *The Journal of Finance, 49*(4), 1213–1252.

Liberzon, D. (2012). Calculus of variations and optimal control theory: A concise introduction. Princeton UP.

Lie, S. (1881). On the integration of linear partial differential equations by means of definite integrals. Archiv for Mathematik og Naturvidenskab, V I(3), 328–368. In Gernan. English translation in N. H. Ibragimov (ed.) CRC Handbook of lie group analysis of differential equations (Vol. 2), 1995. CRC Press.

Lions, J. L. (1971). *Optimal Control of Systems Governed by Partial Differential Equations*. New-York: Springer-verlag.

Ljungqvist, L., & Sargent, T. J. (2018). Recursive macroeconomic theory (4th ed.). MIT Press.

Llibre, J., Sotomayor, J., & Zhitomirskii, M. (2002). Impasse bifurcations of constrained systems. *Fields Institute Communications, 31*, 235–256.

Lucas, R. E. (1988). On the mechanics of economic development. *Journal of Monetary Economics, 22*(1), 3–42.

Lucas, R. J. (1976). Econometric policy evaluation: A critique. *Carnegie-Rochester Conference Series on Public Policy, 1*(1), 19–46.

Ma, J., & Yong, J. (2007). Forward-backward stochastic differential equations and their applications. Lecture notes in mathematics. (corrected ed.). Springer.

Magnus, J. R., & Neudecker, H. (1988). *Matrix differential calculus with applications in statistics and econometrics*. Wiley.

Makris, M. (2001). Necessary conditions for infinite-horizon discounted two-stage optimal control problems. *Journal of Economic Dynamics and Control, 25*(12), 25.

Malliaris, A., & Brock, W. (1982). *Stochastic methods in economics and finance*. North-Holland.

Marszalek, W. (2009). Fold points and singularities in hall mhd differential-algebraic equations. *IEEE Transactions on Plasma Science, 37*(1), 254–260.

Matsuyama, K., & Ushchev, P. (2023). Love-for-Variety. CEPR Discussion Papers 18184, C.E.P.R. Discussion Papers.

McKendrick, A. G. (1926). Applications of mathematics to medical problems. *Proceedings of the Edinburgh Mathematical Society, 44*, 98–130.
Melitz, M. J. (2003). The impact of trade on intra-industry reallocations and aggregate industry productivity. *Econometrica, 71*(6), 1695–1725.
Merton, R. C. (1971). Optimum consumption and portfolio rules in a continuous time model. *Journal of Economic Theory, 3*, 373–413.
Merton, R. C. (1973). An intertemporal capital asset pricing model. *Econometrica, 41*, 867–88.
Merton, R. C. (1974). On the pricing of corporate debt: The risk structure of interest rates. *The Journal of Finance, 29*(2), 449–470.
Merton, R. C. (1975). An asymptotic theory of growth under uncertainty. Review of Economic Studies, 42(3), 375–393. Reproduced in R. C. Merton, "Continuous-Time Finance", Blackwell, pp. 579–605.
Merton, R. C. (1982). On the mathematics and economics assumptions of continuous-time models. In Essays in honor of paul cootner. Prentice-Hall. Reproduced in R. C. Merton, "Continuous-Time Finance", Blackwell, pp. 57–93.
Merton, R. C. (1990). *Continuous time finance*. Blackwell.
Miao, J. (2020). Economic dynamics in discrete time (2nd ed.). MIT Press.
Michel, P. (1982). On the transversality condition in infinite-horizon problems. *Econometrica, 50*, 975–85.
Mirrlees, J. A. (1971). An exploration in the theory of optimum income taxation. *Review of Economic Studies, 38*, 175–208.
Mortensen, D. T. (1999). Equilibrium unemployment dynamics. *International Economic Review, 40*(4), 889–914.
Moscati, I. (2023). The history and methodology of expected utility. Elements in decision theory and philosophy. Cambridge University Press.
Müller, A., & Stoyan, D. (2002). *Comparison methods for stochastic models and risks*. Wiley.
Neittaanmäki, P., & Tiba, D. (1994). *Optimal control of nonlinear parabolic systems*. Marcel Dekker.
Nuño, G., & Moll, B. (2018). Social optima in economies with heterogeneous agents. *Review of Economic Dynamics, 28*, 150–180.
Ok, E. A. (2007). *Real analysis with economic applications (tenho)*. Princeton University Press.
Øksendal, B. (2003). Stochastic differential equations (6th ed.). Springer.
Olver, P. J. (2014). Introduction to partial differential equations. Undergraduate texts in mathematics (1st ed.). Springer International Publishing.
Pardoux, E., & Răşcanu, A. (2014). Stochasic differential equations, backward sdes, partial differential equations. Number 69 in Stochastic modelling and applied probability. Springer.
Parenti, M., Ushchev, P., & Thisse, J.-F. (2017). Toward a theory of monopolistic competition. Journal of Economic Theory, 167(C):86–115.
Park, S. Y., & Bera, A. K. (2018). Information theoretic approaches to income density estimation with an application to the U.S. income data. The Journal of Economic Inequality, 16(4), 461–486.
Pavliotis, G. A. (2014). Stochastic processes and applications: Diffusion processes, the Fokker-Planck and Langevin Equations. Texts in applied mathematics (Vol. 60, 1st ed.). Springer.
Pearson, K. (1894). Contributions to the mathematical theory of evolution. ii. Skew variation in homogeneous material. *Proceedings of the Royal Society of London, 57*, 257–260.
Peng, S. (1990). A general stochastic maximum principle for optimal control problems. SIAM Journal on Control and Optimization, 28.
Perko, L. (1996). Differential equations and dynamical systems (2nd ed.). Springer.
Peschl, H. J., & Plail, M. (2009). The maximum principle of optimal control: A history of ingenious ideas and missed opportunities. *Control and Cybernetics, 38*(4A), 973–995.
Pham, H. (2009). Continuous-time stochastic control and optimization with financial applications. Stochastic modelling and applied probability (1st ed.). Springer.
Pinsky, M. A. (2003). Partial differential equations and bounday-value problems with applications (3rd ed.). Waveland.
Pissarides, C. A. (1990). *Equilibrium unemployment theory*. Basil Blackwell.

Polyanin, A. D., & Nazaikinskii, V. E. (2016). Handbook of linear partial differential equations for engineers and scientists (2nd ed.). Chapman and Hall/CRC.
Polyanin, A. D., & Zaitsev, V. F. (2004). Handbook of nonlinear partial differential equations (1st ed.). Chapman & Hall/CRC.
Pontryagin, L. S., Boltyanskii, V. G., Gamkrelidze, R. V., & Mishchenko, E. F. (1962). *The mathematical theory of optimal processes*. Interscience Publishers.
Pratt, J. W. (1964). Risk aversion in the small and in the large. *Econometrica, 32*, 122–136.
Ramsey, F. P. (1928). A mathematical theory of saving. Economic Journal, 38(Dec):543–59.
Rebelo, S. (1991). Long run policy analysis and long run growth. *Journal of Political Economy, 99*(3), 500–21.
Riaza, R. (2008). Differential-algebraic systems: Analytical aspects and circuit applications. World Scientific.
Riesz, F., & Sz.-Nagy, B. (1955). Functional analysis. Frederick Ungar Publishing Co., republished by dover (1990) edition.
Risken, H. (1996). The Fokker-Planck Equation: Methods of solutions and applications (Springer Series in Synergetics). Springer series in synergetics (2nd ed.). Springer.
Rogerson, R., Shimer, R., & Wright, R. (2005). Search-theoretic models of the labor market: A survey. *Journal of Economic Literature, 43*(4), 959–988.
Romer, P. (1986a). Cake eating, chattering and jumps: Existence results for variational problems. *Econometrica, 54*(4), 897–908.
Romer, P. (1986b). Increasing returns and long-run growth. *Journal of Political Economy, 94*(5), 1002–37.
Rotemberg, J. J. (1982). Monopolistic price adjustment and aggregate output. *Review of Economic Studies, 49*(4), 517–531.
Sachdev, P., & Rao, C. S. (2010). Large time asymptotics for solutions of nonlinear partial differential equations. Springer monographs in mathematics (1st ed.). Springer.
Salop, S. (1979). Monopolistic competition with outside goods. *Bell Journal of Economics, 10*, 141–156.
Salsa, S. (2016). Partial differential equations in action: From modelling to theory. Number 99 in Unitext. Springer International Publishing, third edition.
Samuelson, P. (1947). Foundations of economic analysis, volume 80 of Harvard economic studies (Atheneum). Harvard University Press.
Santa-Clara, P., & Sornette, D. (2001). The dynamics of the forward interest rate curve with stochastic string shocks. *Review of Financial Studies, 14*(1), 149–185.
Sargeant, R. W. H. (2000). Optimal control. *Journal of Computational and Applied Mathematics, 124*, 361–371.
Sargent, T. J. (2008). Evolution and intelligent design. *The American Economic Review, 98*(1), 3–37.
Sargent, T. J., & Wallace, N. (1973). The stability of models of money and growth with perfect foresight. *Econometrica, 41*, 1043–8.
Särkkä, S., & Solin, A. (2019). *Applied Stochastic Differential Equations*. Cambridge University Press.
Schiff, J. L. (1999). *The laplace transform*. Springer.
Schilling, R. (2021). Brownian motion. A guide to random processes and stochastic calculus (3rd ed.). De Gruyter.
Seierstad, A. (2009). *Stochastic control in discrete and continuous time*. Springer.
Seierstad, A., & Sydsæter, K. (1987). *Optimal control theory with economic applications*. North-Holland.
Sen, A. K. (1973). On economic inequality. Clarendon Press and Norton.
Sennewald, K., & Wälde, K. (2006). "itô's lemma" and the bellman equation for poisson processes: An applied view. Journal of Economics, 89.
Sethi, S. (1977). Dynamic optimal control models in advertising: a survey. *Journal of Optimization Theory and Applications, 19*, 685–725.

Sethi, S. P. (2019). Optimal control theory: Applications to management science and economics (3rd ed.). Springer.
Shannon, C. E. (1948). A mathematical theory of communication. *The Bell System Technical Journal, 27*(379–423), 623–656.
Shapiro, A. (1990). On concepts of directional differentiability. *Journal of Optimization Theory and Applications, 66*, 477–487.
Shilnikov, L. P., Shilnikov, A. L., Turaev, D. V., & Chua, L. O. (1998). Methods of qualitative theory in nonlinear dynamics. Part 1, volume 4 of A (illustrated ed.). World Scientific Publishing Company.
Shorrocks, A. (1978a). Income inequality and income mobility. *Journal of Economic Theory, 19*(2), 376–393.
Shorrocks, A. F. (1976). Income mobility and the markov assumption. *The Economic Journal, 86*(343), 566–578.
Shorrocks, A. F. (1978b). The measurement of mobility. *Econometrica, 46*(5), 1013–1024.
Shorrocks, A. F. (1980). The class of additively decomposable inequality measures. *Econometrica, 48*(3), 613–625.
Shorrocks, A. F. (1984). Inequality decomposition by population groups. *Econometrica, 52*, 1369–1385.
Siddiqi, A. H. (2018). *Functional analysis and applications*. Industrial and applied mathematics: Springer.
Simmons, G. F. (2017). Differential equations with applications and historical notes. Textbooks in mathematics (3rd ed.). CRC Press Chapman and Hall.
Simon, C. P., & Blume, L. (1994). *Mathematics for economists*. W. W: Norton & Co.
Skiba, A. K. (1978). Optimal growth with a convex-concave production function. *Econometrica, 46*, 527–39.
Smith, V. (1974). Optimal costly firm entry in general equilibrium. *Journal of Economic Theory, 9*, 397–417.
Solow, R. M. (1956). A contribution to the theory of economic growth. *Quarterly Journal of Economics, 70*(1), 65–94.
Stokey, N. L. (2009). *The economics of inaction*. Princeton
Strang, G. (1988). *Linear algebra and its applications*. Brace and Jovanovich: Hartcourt.
Strogatz, S. (2019). Infinite powers: The story of calculus - the language of universe. Atlantic Books.
Szpiro, G. (2020). Risk, choice, and uncertainty: Three centuries of economic decision-making. Columbia University Press.
Tähvonen, O., & Withagen, C. (1996). Optimality of irreversible pollution accumulation. *Journal of Economic Dynamics and Control, 20*, 1775–1795.
Takayama, A. (1994). Analytical methods in economics. Harvester Wheatsheaf.
Thom, R. (1975). Structural stability and morphogenesis. Benjamin, Reading, Massachusetts. translated by D. H. Fowler.
Tomiyama, K. (1985). Two-stage optimal control problems and optimality conditions. Journal of Economic Dynamics and Control, vol. 9 iss. 3, 9.
Tröltzsch, F. (2010). Optimal control of partial differential equations. Number 112 in Graduate studies in mathematics. American Mathematical Society.
Tsallis, C. (2009). Introduction to nonextensive statistical mechanics: Approaching a complex world (1st ed.). Springer.
Tu, P. (1994). Dynamical systems (2nd ed.). Springer.
Tu, P. N. V. (1984). Introductory optimization dynamics: Optimal control with economics and management science applications. Springer.
Turnovsky, S. (1977). *Macroeconomic analysis and stabilization policy*. Cambridge University Press.
Turnovsky, S. (1995). *Methods of macroeconomic dynamics*. MIT Press.
Uhlenbeck, G. E., & Ornstein, L. S. (1930). On the theory of the brownian motion. *Physical Review, 36*, 823–841.

Uzawa, H. (1965). Optimal technical change in an aggregative model of economic growth. *International Economic Review, 6*, 12–31.
van Brunt, B. (2010). *The calculus of variations*. Springer: Universitext.
van Marrewijk, C., & Verbeek, J. (1993). Sector-specific capital, "bang-bangâŁž investment, and the filippov solution. *Journal of Economics, 57*(2), 131–146.
van Saarloos, W. (2003). Front propagation into unstable states. *Physics Reports, 386*, 29–222.
Venkatasubramanian, V., Schättler, H., & Zaborszky, J. (1995a). Dynamics of large constrained nonlinear systems-a taxonomy theory. *Proceedings of the IEEE, 83*(11), 1530–1561.
Venkatasubramanian, V., Schättler, H., & Zaborszky, J. (1995b). Local bifurcations and feasibility regions in differential-algebraic equations. *IEEE Transactions on Automatic Control, 40*, 1992–2013.
Verhulst, P.-F. (1845). Recherches mathématiques dur la loi d'accroissement de la population. *Nouveaux Mémoires de l'Académie Royale des Sciences et Belles-Lettres de Bruxelles, 18*, 1–42.
Wälde, K. (2012). Applied intertemporal optimization. Know Thyself - Academic Publishers. www.waelde.com/KTAP.
Ważewski, T. (1947). Sur un principe topologique de l'examen de l'allure asymptotique des integrales des equations differentielles ordinaires. *Ann. Soc. Polon. Math., 20*, 279–313.
Weber, T. A. (2011). *Optimal Control Theory with Applications in Economics*. The MIT Press.
Weder, M. (2001). Indeterminacy in a small open economy ramsey growth model. *Journal of Economic Theory, 98*, 339–356.
Weisstein, E. W. (2024). Bromwich integral. From MathWorld–A Wolfram Web Resource. https://mathworld.wolfram.com/BromwichIntegral.html. Last visited on 25/11/2024.
Weisstein, E. W. (2024). Laplace transform. From MathWorld–A Wolfram Web Resource. https://mathworld.wolfram.com/LaplaceTransform.html. Last visited on 25/11/2024.
Yong, J., & Zhou, X. Y. (1999). Stochastic controls. Hamiltonian systems and HJB equations. Number 43 in Applcations of mathematics. Stochastic modelling and applied probability. Springer.
Zaitsev, V. F., & Polyanin, A. D. (2003). Handbook of exact solutions for ordinary differential equations (2nd ed.). Chapman & Hall/CRC.
Zeeman, W. C. (1977). *Catastrophe theory - Selected papers 1972–1977*. Addison-Wesley.
Zhang, J. (2017). Backward stochastic differential equations. From linear to fully nonlinear theory. Number 86 in Probability theory and stochastic modelling. Springer.
Zhelobodko, E., Kokovin, S., Parenti, M., & Thisse, J.-F. (2012). Monopolist competition: Beyond the constant elasticity of substitution. Econometrica, 80.
Zhitomirskii, M. (1993). Local normal forms for constrained systems on 2-manifolds. *Boletim da Sociedade Brasileira de Matemática, 24*, 211–232.
Zwillinger, D. (1998). Handbook of differential equations (3rd ed.). Academic.

Author Index

P. B. Brito, *Economic Dynamics and Distributions*, Dynamic Modeling and Econometrics in Economics and Finance 32, https://doi.org/10.1007/978-3-031-94717-9

Index

P. B. Brito, *Economic Dynamics and Distributions*, Dynamic Modeling and Econometrics in Economics and Finance 32, https://doi.org/10.1007/978-3-031-94717-9

D

P